W9-BQZ-015

HANDBOOK OF MATHEMATICAL ECONOMICS

VOLUME II

EHEHEHEHEHEHEHEHEHEHEHEHEHEHEHEHEHEHEHEHEHEHEHEHEHEHEHEHE

HANDBOOKS IN ECONOMICS

BOOK 1

Editors

KENNETH J. ARROW

MICHAEL D. INTRILIGATOR

ELSEVIER

AMSTERDAM • LAUSANNE • NEW YORK • OXFORD • SHANNON • SINGAPORE • TOKYO

HANDBOOK OF MATHEMATICAL ECONOMICS

VOLUME II

Edited by

KENNETH J. ARROW
Stanford University

and

MICHAEL D. INTRILIGATOR
University of California, Los Angeles

ELSEVIER
AMSTERDAM • LAUSANNE • NEW YORK • OXFORD • SHANNON • SINGAPORE • TOKYO

ELSEVIER SCIENCE PUBLISHERS B.V.
Sara Burgerhartstraat 25
P.O. Box 211, 1000 AE Amsterdam, The Netherlands

First edition: 1982
Second impression: 1984
Third impression: 1987
Fourth impression: 1991
Fifth impression: 1993
Sixth impression: 1998

Library of Congress Cataloging in - Publication Data
Main entry under title:

Handbook of mathematical economics.

(Handbooks in economics; bk. 1)
Partial contents: v. I. Historical introduction. Mathematical methods in economics -- v. 2. Mathematical approaches to microeconomic theory. Mathematical approaches to competitive equilibrium -- v. 3. Mathematical approaches to welfare economics. Mathematical approaches to economic organization and planning.
1. Economics, Mathematical-- Adresses, essays, lectures. I. Arrow, Kenneth Joseph, 1921 – II. Intriligator, Michaël D. III. Series.
HB135.H357 330'.01'51 81-2820
ISBN 0-444-86126-2 AACR2

ISBN: 0 444 86127 0

This book is printed on acid-free paper.

Printed in The Netherlands

INTRODUCTION TO THE SERIES

The aim of the *Handbooks in Economics* series is to produce Handbooks for various branches of economics, each of which is a definitive source, reference, and teaching supplement for use by professional researchers and advanced graduate students. Each Handbook provides self-contained surveys of the current state of a branch of economics in the form of chapters prepared by leading specialists on various aspects of this branch of economics. These surveys summarize not only received results but also newer developments, from recent journal articles and discussion papers. Some original material is also included, but the main goal is to provide comprehensive and accessible surveys. The Handbooks are intended to provide not only useful reference volumes for professional collections but also possible supplementary readings for advanced courses for graduate students in economics.

CONTENTS OF THE HANDBOOK

VOLUME I

VOLUME II

Part 2 – MATHEMATICAL APPROACHES TO MICROECONOMIC THEORY

Part 3 – MATHEMATICAL APPROACHES TO COMPETITIVE EQUILIBRIUM

PREFACE TO THE HANDBOOK

The field of mathematical economics

Mathematical economics includes various applications of mathematical concepts and techniques to economics, particularly economic theory. This branch of economics traces its origins back to the early nineteenth century, as noted in the historical introduction, but it has developed extremely rapidly in recent decades and is continuing to do so. Many economists have discovered that the language and tools of mathematics are extremely productive in the further development of economic theory. Simultaneously, many mathematicians have discovered that mathematical economic theory provides an important and interesting area of application of their mathematical skills and that economics has given rise to some important new mathematical problems, such as game theory.

Purpose

The *Handbook of Mathematical Economics* aims to provide a definitive source, reference, and teaching supplement for the field of mathematical economics. It surveys, as of the late 1970's, the state of the art of mathematical economics. Bearing in mind that this field is constantly developing, the Editors believe that now is an opportune time to take stock, summarizing both received results and newer developments. Thus all authors were invited to review and to appraise the current status and recent developments in their presentations. In addition to its use as a reference, the Editors hope that this Handbook will assist researchers and students working in one branch of mathematical economics to become acquainted with other branches of this field. Each of the chapters can be read independently.

Organization

The Handbook includes 29 chapters on various topics in mathematical economics, arranged into five parts: *Part 1* treats *Mathematical Methods in Economics*, including reviews of the concepts and techniques that have been most useful for the mathematical development of economic theory. *Part 2* elaborates on *Mathematical Approaches to Microeconomic Theory*, including consumer, producer, oligopoly, and duality theory. *Part 3* treats *Mathematical Approaches to Competitive Equilibrium*, including such aspects of competitive equilibrium as existence, stability, uncertainty, the computation of equilibrium prices, and the core of an economy. *Part 4* covers *Mathematical Approaches to Welfare Economics*, including social choice theory, optimal taxation, and optimal economic growth. *Part 5* treats *Mathematical Approaches to Economic Organization and Planning*, including organization design and decentralization.

Level

All of the topics presented are treated at an advanced level, suitable for use by economists and mathematicians working in the field or by advanced graduate students in both economics and mathematics.

Acknowledgements

Our principal acknowledgements are to the authors of chapters in the *Handbook of Mathematical Economics*, who not only prepared their own chapters but also provided advice on the organization and content of the Handbook and reviewed other chapters.

KENNETH J. ARROW
Stanford University

MICHAEL D. INTRILIGATOR
University of California, Los Angeles

CONTENTS OF VOLUME II

Chapter 14
Market Demand and Excess Demand Functions
WAYNE SHAFER and HUGO SONNENSCHEIN 671

Part 3 – MATHEMATICAL APPROACHES TO COMPETITIVE EQUILIBRIUM

Chapter 15
Existence of Competitive Equilibrium
GERARD DEBREU 697

Chapter 16
Stability
FRANK HAHN 745

PART 2

MATHEMATICAL APPROACHES TO MICROECONOMIC THEORY

Chapter 9

CONSUMER THEORY*

ANTON P. BARTEN

C.O.R.E. and Catholic University of Louvain, Belgium

and

VOLKER BÖHM†

Universität Mannheim, Fed. Rep. Germany

1. Introduction

The main objective of consumer theory is to determine the impact on observable demands for commodities of alternative assumptions on the objectives and on the behavioral rules of the consumer, and on the constraints which he faces when making a decision. The traditional model of the consumer takes preferences over alternative bundles to describe the objectives. Its behavioral rule consists of maximization of these preferences under a budget restriction which determines the trading possibilities. The principal results of the theory consist of the qualitative implications on observed demand of changes in the parameters which determine the decision of the consumer.

The historical development of consumer theory indicates a long tradition of interest of economists in the subject, which has undergone substantial conceptual changes over time to reach its present form. A detailed survey of its history can be found in Katzner (1970).

*We would like to thank Angus Deaton, Michael Intriligator, Andreu Mas-Colell, and Hugo Sonnenschein for valuable suggestions and criticism. Of course, we bear the sole responsibility for any remaining errors.

†Parts of this paper were written while Volker Böhm was at the Center for Operations Research and Econometrics and at the Université Libre de Bruxelles. Financial support under Ford Foundation Grant no. 775/0022 is gratefully acknowledged.

Handbook of Mathematical Economics, vol. II, edited by K.J. Arrow and M.D. Intriligator

2. Commodities and prices

Commodities can be divided into goods and services. Each commodity is completely specified by its physical characteristics, its location, and date at which it is available. In studies of behavior under uncertainty an additional specification of the characteristics of a commodity relating to the state of nature occurring may be added [see Arrow (1953) and Debreu (1959)], which leads to the description of a contingent commodity. With such an extensive definition of a commodity the theory of the consumer subsumes all more specific theories of consumer choice of location and trade, of intertemporal choice over time, and of certain aspects of the theory of decisions under uncertainty. Formally these are equivalent. The traditional theory usually assumes that there exists a finite number ℓ of commodities implying that a finite specification of physical characteristics, location, etc. suffices for the problems studied. Quantities of each commodity are measured by real numbers. A *commodity bundle*, i.e. a list of real numbers (x_h), $h=1,\dots,\ell$, indicating the quantity of each commodity, can be described therefore as an ℓ-dimensional vector $x=(x_1,\dots,x_\ell)$ and as a point in ℓ-dimensional Euclidean space R^ℓ, the *commodity space*. Under perfect divisibility of all commodities any real number is possible as a quantity for each commodity so that any point of the commodity space R^ℓ is a possible commodity bundle. The finite specification of the number of commodities excludes treatment of situations in which characteristics may vary continuously. Such situations arise in a natural way, for example, in the context of quality choice of commodities, and in the context of location theory when real distance on a surface is the appropriate criterion. As regards the time specification of commodities, each model with an infinite time horizon, whether in discrete or continuous time, requires a larger commodity space. This is also the case for models of uncertainty with sufficiently dispersed random events. These situations lead in a natural way to the study of infinite dimensional commodity spaces taken as vector spaces over the reals. Applications can be found in Gabszewicz (1968) and Bewley (1970, 1972), with an interpretation regarding time and uncertainty. Mas-Colell (1975) contains an analysis of quality differentiation. As far as the theory of the consumer is concerned many of the usual results of the traditional theory in a finite dimensional commodity space can be extended to the infinite dimensional case. This survey restricts itself to the finite dimensional case.

The *price* p_h of a commodity h, $h=1,\dots,\ell$, is a real number which is the amount paid in exchange for one unit of the commodity. With the specification of location and date the theory usually assumes the general convention that the prices of all commodities are those quoted now on the floor of the exchange for delivery at different locations and at different dates. It is clear that other conventions or constructions are possible and meaningful. A price system or a price vector $p=(p_1,\dots,p_\ell)$ can thus be represented by a point in Euclidean space R^ℓ. The *value* of a commodity bundle given a price vector p is $\sum_{h=1}^{\ell} p_h x_h = p\cdot x$.

3. Consumers

Some bundles of the commodity space are excluded as consumption possibilities for a consumer by physical or logical restrictions. The set of all consumption bundles which are possible is called the *consumption set*. This is a non-empty subset of the commodity space, denoted by X. Traditionally, inputs in consumption are described by positive quantities and outputs by negative quantities. This implies, in particular, that all labor components of a consumption bundle x are non-positive. One usually assumes that the consumption set X is closed, convex, and bounded below, the lower bound being justified by finite constraints on the amount of labor a consumer is able to perform. A consumer must choose a bundle from his consumption set in order to exist.

Given the sign convention on inputs and outputs and a price vector p, the value $p\cdot x$ of a bundle $x\in X$ indicates the *net outlay*, i.e. expenses minus receipts, associated with the bundle x. Since the consumer is considered to trade in a market, his choices are further restricted by the fact that the value of his consumption should not exceed his initial wealth (or income). This may be given to him in the form of a fixed non-negative number w. He may also own a fixed vector of initial resources ω, a vector of the commodity space R^ℓ. In the latter case his *initial wealth* given a price vector p is defined as $w=p\cdot\omega$. The set of possible consumption bundles whose value does not exceed the initial wealth of the consumer is called the *budget set* and is defined by

$$\beta(p,w)=\{x\in X \mid p\cdot x\leqq w\}. \tag{3.1}$$

The ultimate decision of a consumer to choose a bundle from the consumption set depends on his tastes and desires. These are represented by his *preference relation* $\precsim$ which is a binary relation on X. For any two bundles x and y, $x\in X$, $y\in X$, $x\succsim y$ means that x is at least as good as y. Given the preferences the consumer chooses a most preferred bundle in his budget set as his *demand*. This is defined as

$$\varphi(p,w)=\{x\in\beta(p,w) \mid x'\in\beta(p,w) \text{ implies } x\succsim x' \text{ or not } x'\succsim x\}. \tag{3.2}$$

Much of consumer theory, in particular the earlier contributions, describe consumer behavior as one of utility maximization rather than of preference maximization. Historically there has been a long debate whether the well-being of an individual can be measured by some real valued function. The relationship between these two approaches has been studied extensively after the first rigorous axiomatic introduction of the concept of a preference relation by Frisch (1926). It has been shown that the notion of the preference relation is the more basic concept of consumer theory, which is therefore taken as the starting point of any analysis of consumer behavior. The relationship to the concept of a utility function, however, as a convenient device or as an equivalent approach to describe consumer behavior, constitutes a major element of the foundations of consumer theory. The following analysis therefore is divided basically into two parts. The first part, Sections 4–9, deals with the axiomatic foundations of preference theory and utility theory and with the existence and basic continuity results of consumer demand. The second part, Sections 10–15, presents the more classical results of demand theory mainly in the context of differentiable demand functions.

4. Preferences

Among alternative commodity bundles in the consumption set, the consumer is assumed to have preferences represented by a binary relation $\succsim$ on X. For two bundles x and y in X the statement $x \succsim y$ is read as "*x is at least as good as y*". Three basic axioms are usually imposed on the preference relation which are often taken as a definition of a rational consumer.

Axiom 1 (Reflexivity)

For all $x \in X$, $x \succsim x$, i.e. any bundle is as good as itself.

Axiom 2 (Transitivity)

For any three bundles x, y, z in X such that $x \succsim y$ and $y \succsim z$ it is true that $x \succsim z$.

Axiom 3 (Completeness)

For any two bundles x and y in X, $x \succsim y$ or $y \succsim x$.

A preference relation $\succsim$ which satisfies these three axioms is a *complete preordering* on X and will be called a *preference order*. Two other relations can be derived immediately from a preference order. These are the relation of strict preference $\succ$ and the relation of indifference $\sim$.

Definition 4.1

A bundle x is said to be strictly preferred to a bundle y, i.e. $x \succ y$ if and only if $x \succsim y$ and not $y \succsim x$.

Definition 4.2

A bundle x is said to be indifferent to a bundle y, i.e. $x \sim y$ if and only if $x \succsim y$ and $y \succsim x$.

With $\succsim$ being reflexive and transitive the strict preference relation is clearly irreflexive and transitive. It will be assumed throughout that there exist at least two bundles x' and x'' such that $x' \succ x''$. The indifference relation $\sim$ defines an *equivalence relation* on X, i.e. $\sim$ is reflexive, symmetric, and transitive.

The validity of these three axioms is not questioned in most of consumer theory. They do represent assumptions, however, which are subject to empirical tests. Observable behavior in many cases will show inconsistencies, in particular with respect to transitivity and completeness. Concerning the latter, it is sometimes argued that it is too much to ask of a consumer to be able to order all possible bundles when his actual decisions will be concerned only with a certain subset of his consumption set. Empirical observations or experimental results frequently indicate intransitivities of choices. These may be due to simple errors which individuals make in real life or in experimental situations. On the other hand, transitivity may also fail for some theoretical reasons. For example, if a consumer unit is a household consisting of several individuals where each individual's preference relation satisfies the three axioms above, the preference relation of the household may be non-transitive if decisions are made by majority rule. Some recent developments in the theory of consumer demand indicate that some weaker axioms suffice to describe and derive consistent demand behavior [see, for example, Chipman et al. (1971), Sonnenschein (1971), Katzner (1971), Hildenbrand (1974), and Shafer (1974)]. Some of these we will indicate after developing the results of the traditional theory.

The possibility of defining a strict preference relation $\succ$ from the weaker one $\succsim$, and vice versa, suggests in principle an alternative approach of starting with the strict relation $\succ$ as the primitive concept and deriving the weaker one and the indifference relation. This may be a convenient approach in certain situations, and it seems to be slightly more general since the completeness axiom for the strict relation has no role in general. On the other hand, a derived indifference relation will generally not be transitive. Other properties are required to make the two approaches equivalent, which then, in turn, imply that the derived weak relation is complete. However, from an empirical point of view the weak relation $\succsim$ seems to be the more natural concept. The observed choice of a bundle y over a bundle x can only be interpreted in the sense of the weak relation and not as an indication of strict preference.

Axioms 1–3 describe order properties of a preference relation which have intuitive meaning in the context of the theory of choice. This is much less so with the topological conditions which are usually assumed as well. The most common one is given in Axiom 4 below.

Axiom 4 (Continuity)

For every $x \in X$ the sets $\{y \in X \mid y \succsim x\}$ and $\{y \in X \mid x \succsim y\}$ are closed relative to X.

The set $\{y \in X \mid y \succsim x\}$ is called the *upper contour set* and $\{y \in X \mid x \succsim y\}$ is called the *lower contour set*. Intuitively Axiom 4 requires that the consumer behaves consistently in the "small", i.e. given any sequence of bundles y^n converging to a bundle y such that for all n each y^n is at least as good as some bundle x, then y is also at least as good as x. For a preference order, i.e. for a relation satisfying Axioms 1–3, the intersection of the upper and lower contour sets for a given point x defines the *indifference class* $I(x) = \{y \in X \mid y \sim x\}$ which is a closed set under Axiom 4. For alternative bundles x these are the familiar indifference curves for the case of $X \subset R^2$. Axioms 1–4 together also imply that the upper and the lower contour sets of the derived strict preference relation $\succ$ are open, i.e. $\{y \in X \mid y \succ x\}$ and $\{y \in X \mid x \succ y\}$ are open.

Many known relations do not display the continuity property. The most commonly known of them is the lexicographic order which is in fact a strict preference relation which is transitive and complete. Its indifference classes consist of single elements.

Definition 4.3

Let $x = (x_1, \ldots, x_\ell)$ and $y = (y_1, \ldots, y_\ell)$ denote two points of R^ℓ. Then x is said to be lexicographically preferred to y, $x \,\text{Lex}\, y$, if there exists k, $1 \leqq k \leqq \ell$, such that

$$\begin{array}{ll} x_j = y_j, & \\ x_k > y_k, & j < k. \end{array}$$

It can easily be seen that the upper contour set for any point x is neither open nor closed.

The relationship between the order properties of Axioms 1–3 and the topological property of Axiom 4 has not been studied extensively. Schmeidler (1971), however, indicated that continuity of a preference relation which is transitive implies completeness. The case when transitivity is implied by continuity and completeness is examined by Sonnenschein (1965).

Theorem 4.1 [Schmeidler (1971)]

Let $\succsim$ denote a transitive binary relation on a connected topological space X. Define the associated strict preference relation $\succ$ by $x \succ y$ if and only if $x \succsim y$ and not $y \succsim x$, and assume that there exists at least one pair $\bar{x}$, $\bar{y}$ such that $\bar{x} \succ \bar{y}$. If for every $x \in X$

(i) $\{y \in X \mid y \succsim x\}$ and $\{y \in X \mid x \succsim y\}$ are closed

and

(ii) $\{y \in X \mid y \succ x\}$ and $\{y \in X \mid x \succ y\}$ are open,

then $\succsim$ is complete.

Proof

The proof makes use of the fact that the only non-empty subset of a connected topological space which is open and closed is the space itself. First it is shown that for any x and y such that $x \succ y$

$$\{z \mid z \succ y\} \cup \{z \mid x \succ z\} = X.$$

By definition one has

$$\{z \mid z \succ y\} \cup \{z \mid x \succ z\} \subset \{z \mid z \succsim y\} \cup \{z \mid x \succsim z\}.$$

The set on the left-hand side of the inclusion is open by assumption (ii), the one on the right-hand side is closed by assumption (i). Therefore equality of the two sets proves the assertion. Suppose $u \in \{z \mid z \succsim y\}$ and $u \notin \{z \mid z \succ y\}$. Then $y \succsim u$. Since $x \succ y$, one obtains $x \succ u$. Therefore $u \in \{z \mid x \succ z\}$, which proves equality of the two sets since symmetric arguments hold for all points of the set on the right-hand side.

Now assume that there are two points v and w in X which are not comparable. Since $\bar{x} \succ \bar{y}$ and $\{z \mid z \succ \bar{y}\} \cup \{z \mid \bar{x} \succ z\} = X$, one must have $v \succ \bar{y}$ or $\bar{x} \succ v$. Without loss of generality, assume $v \succ \bar{y}$. Applying the above result to v and $\bar{y}$ yields

$$\{z \mid z \succ \bar{y}\} \cup \{z \mid v \succ z\} = X.$$

Since v and w are not comparable one has to have $w \succ \bar{y}$ and $v \succ \bar{y}$. By assumption the two sets $\{z \mid v \succ z\}$ and $\{z \mid w \succ z\}$ are open and so is their intersection.

The intersection is non-empty and not equal to X by the non-comparability of v and w. It will be shown that

$$\{z \mid v \succ z\} \cap \{z \mid w \succ z\} = \{z \mid v \succsim z\} \cap \{z \mid w \succsim z\},$$

contradicting the connectedness of X. Suppose $v \succsim z$ and $w \succsim z$. $z \succsim v$ and transitivity implies $w \succsim v$, contradicting non-comparability. Similarly, $z \succsim w$ and $v \succsim z$ implies $v \succsim w$. Therefore $v \succ z$ and $w \succ z$, which proves equality of the two sets and the theorem. Q.E.D.

5. Utility functions

The problem of the representability of a preference relation by a numerical function was given a complete and final solution in a series of publications by Eilenberg (1941), Debreu (1954, 1959, 1964), Rader (1963), and Bowen (1968). Historically the concept of a utility function was taken first as a cardinal concept to measure a consumer's well-being. Pareto (1896) seemed to be the first to recognize that arbitrary increasing transformations of a given utility function would result in identical maximization behavior of a consumer. The importance and methodological consequence, however, were only recognized much later by Slutsky (1915) and especially by Wold (1943–44) who gave the first rigorous study of the representation problem.

Definition 5.1

Let X denote a set and $\succsim$ a binary relation on X. Then a function u from X into the reals R is a representation of $\succsim$, i.e. a utility function for the preference relation $\succsim$, if, for any two points x and y, $u(x) \geqq u(y)$ if and only if $x \succsim y$.

It is clear that for any utility function u and any increasing transformation $f\colon R \to R$ the function $v = f \circ u$ is also a utility function for the same preference relation $\succsim$. Some weaker forms of representability were introduced in the literature [see, for example, Aumann (1962) and Katzner (1970)]. But they have not proved useful in consumer theory.

One basic requirement of a utility function in applications to consumer theory is that the utility function be continuous. It is seen easily that Axioms 1–4 are necessary conditions for the existence of a continuous utility function. That this is true for Axioms 1–3 follows directly from the definition of a representation. To demonstrate necessity of Axiom 4 for the continuity of the function u one observes that, for any point x, the upper and lower contour sets of the preference relation coincide with the two sets $\{z \in X \mid u(z) \geqq u(x)\}$ and $\{z \in X \mid u(x) \geqq u(z)\}$ which are closed sets by the continuity of u. The basic result of utility theory is that Axioms 1–4 combined with some weak assumption on the consumption set X is also sufficient for the existence of a continuous utility function.

Theorem 5.1 (Debreu, Eilenberg, Rader)

Let X denote a topological space with a countable base of open sets (or a connected, separable topological space) and $\succsim$ a continuous preference order defined on X, i.e. a preference relation which satisfies Axioms 1–4. Then there exists a continuous utility function u.

We will give a proof for the case where X has a countable base. It divides into two parts. The first is concerned with a construction of a representation function

v given by Rader (1963). The second applies a basic result of Debreu (1964) to show continuity of an appropriate increasing transformation of the function v.

Proof

(i) *Existence.* Let $O_1, O_2, \ldots$ denote the open sets in the countable base. For any x consider $N(x)=\{n \mid x \succ z, \forall z \in O_n\}$ and define

$$v(x)=\sum_{n \in N(x)} \frac{1}{2^n}.$$

If $y \succsim x$, then $N(x) \subset N(y)$, so that $v(x) \leqq v(y)$. On the other hand, if $y \succ x$ there exists $n \in N(y)$ such that $x \in O_n$ but not $n \in N(x)$. Therefore $N(x) \subsetneqq N(y)$ and $v(y) > v(x)$. Hence v is a utility function.

(ii) *Continuity.* Let S denote an arbitrary set of the extended real line which later will be taken to be $v(X)$. S as well as its complement may consist of non-degenerate and degenerate intervals. A *gap* of S is a maximal non-degenerate interval of the complement of S which has an upper and a lower bound in S. The following theorem is due to Debreu (1964).

Theorem 5.2

If S is a subset of the extended real line $\bar{R}$ there exists an increasing function g from S to $\bar{R}$ such that all the gaps of $g(S)$ are open.

Applying the theorem one defines a new utility function $u = g \circ v$. According to the theorem all the gaps of $u(X)$ are open. For the continuity of u it suffices to show that for any $t \in \bar{R}$ the sets $u^{-1}([t, +\infty])$ and $u^{-1}([-\infty, t])$ are closed.

If $t \in u(X)$, there exists $y \in X$ such that $u(y)=t$. Then $u^{-1}([t,+\infty])=\{x \in X \mid x \succsim y\}$ and $u^{-1}([-\infty, t])=\{x \in X \mid y \succsim x\}$. Both of these sets are closed by assumption.

If $t \notin u(X)$ and t is not contained in a gap, then

(a) $t \leqq \operatorname{Inf} u(X)$, or

(b) $t \geqq \operatorname{Sup} u(X)$, or

(c) $$[t, +\infty] = \bigcap_{\substack{\alpha < t \\ \alpha \in u(X)}} [\alpha, +\infty]$$

and

$$[-\infty, t] = \bigcap_{\substack{t < \alpha \\ \alpha \in u(X)}} [-\infty, \alpha].$$

(a) implies $u^{-1}([t,+\infty])=X$ and $u^{-1}([-\infty,t])=\varnothing$,

(b) implies $u^{-1}([t,+\infty])=\varnothing$ and $u^{-1}([-\infty,t])=X$.

Both X and the empty set $\varnothing$, are closed.

In the case of (c),

$$u^{-1}([t,+\infty])=\bigcap_{\substack{\alpha<t\\ \alpha\in u(X)}} u^{-1}([\alpha,+\infty])$$

and

$$u^{-1}([-\infty,t])=\bigcap_{\substack{\alpha>t\\ \alpha\in u(X)}} u^{-1}([-\infty,\alpha])$$

are closed as intersections of closed sets.

If t belongs to an open gap, i.e. $t\in]a,b[$, where a and b belong to $u(X)$, then

$$u^{-1}([t,+\infty])=u^{-1}([b,+\infty]) \quad \text{which is closed}$$

and

$$u^{-1}([-\infty,t]=u^{-1}([-\infty,a]) \quad \text{which is closed.} \quad \text{Q.E.D.}$$

Theorem 5.1 indicates that the concepts of utility function and of the underlying preferences may be used interchangeably to determine demand, provided the preferences satisfy Axioms 1–4. In many situations, therefore, it becomes a matter of taste or of mathematical convenience to choose one or the other.

6. Properties of preferences and of utility functions

In applications additional assumptions on the preferences and/or on the utility function are frequently made. For some of them there exist almost definitional equivalence of the specific property of preferences and of the corresponding property of the utility function. Others require some demonstration. We will discuss the ones most commonly used in an order which represents roughly an increasing degree of mathematical involvement to demonstrate equivalence.

6.1. Monotonicity, non-satiation, and convexity

6.1.1. Monotonicity

Definition 6.1

A preference order on R^ℓ is called monotonic if $x\geqq y$ and $x\neq y$ implies $x\succ y$.

This property states that more of any one good is preferred, which means that all goods are desired (desirability). The associated utility function of a monotonic preference order is an increasing function on R^{ℓ}.

6.1.2. Non-satiation

Definition 6.2

A point $x \in X$ is called a satiation point for the preference order $\succsim$ if $x \succsim y$ holds true for all $y \in X$.

A satiation point is a maximal point with respect to the preference order. Most parts of consumer theory discuss situations in which such global maxima do not exist or, at least for the discussion of demand problems, where an improvement of the consumer can be achieved by a change of his chosen consumption bundle. In other words, the situations under discussion will be points of non-satiation. If at some point x an improvement can be found in any neighborhood of x, one says that the consumer is locally not satiated at x. More precisely:

Definition 6.3

A consumer is locally not satiated at $x \in X$ if for every neighborhood V of x there exists a $z \in V$ such that $z \succ x$.

This property excludes the possibility of indifference classes with non-empty interiors and it implies that the utility function is non-constant in the neighborhood of x.

6.1.3. Convexity

Definition 6.4

A preference order on $X \subset R^{\ell}$ is called convex if the set $\{y \in X \mid y \succsim x\}$ is convex for all $x \in X$.

This definition states that all upper contour sets are convex. The associated concept for a utility function is that of quasi-concavity.

Definition 6.5

A function u: $X \to R$ is called quasi-concave if $u(\lambda x + (1-\lambda)y) \geqq \min\{u(x), u(y)\}$ for all x, $y \in X$ and any λ, $0 \leqq \lambda \leqq 1$.

It is easy to see that the upper level sets of a quasi-concave function are convex. Therefore, a utility function u for a preference order $\succsim$ is quasi-concave if and only if the preference order is convex. Thus, quasi-concavity is a property which relates directly to the ordering. Therefore it is preserved under increasing transformations. Such properties of a utility function are sometimes called ordinal properties as opposed to cardinal properties which are related to some specific representation function u. Concavity, for example, is one such cardinal property. Since the class of concave functions is a subclass of the class of quasi-concave functions, one may ask which additional properties of a convex preference order are required such that there exists a concave utility function. A final answer to this question was given by Kannai (1977). A treatment of this problem would go beyond the scope of this survey.

Definition 6.6

A preference order is called strictly convex if for any two bundles x and x', $x \neq x'$, $x \succsim x'$, and for λ, $0<\lambda<1$, $\lambda x+(1-\lambda)x' \succ x'$.

Any associated utility function of a strictly convex preference relation is a strictly quasi-concave function. Strict convexity excludes all preference relations whose indifference classes have non-empty interiors.

6.2. Separability

Separable utility functions were used in classical consumer theory long before an associated property on preferences had been defined. The problem has been studied by Sono (1945), Leontief (1947), Samuelson (1947a), Houthakker (1960), Debreu (1960), Koopmans (1972), and others. Katzner (1970) presented a general characterization of essentially two notions of separability which we will also use here.

Let $N=\{N_j\}_{j=1}^k$ denote a partition of the set $\{1,\ldots,\ell\}$ and assume that the consumption set $X=S_1 \times \cdots \times S_k$. Such partitions arise in a natural way if consumption is considered over several dates, locations, etc. Loosely speaking, separability then implies that preferences for bundles in each element of the partition (i.e. at each date, location, etc.) are independent of the consumption levels outside. Let $J=\{1,\ldots,k\}$ and, for any $j \in J$ and any $x \in X$, write $x_{\hat{j}}=(x_1,\ldots,x_{j-1},x_{j+1},\ldots,x_k)$ for the vector of components different from x_j. For any fixed $x_{\hat{j}}^0$ the preference ordering $\succsim$ induces a preference ordering on S_j which is defined by $x_j \succsim_{x_{\hat{j}}^0} x_j'$ if and only if $(x_{\hat{j}}^0, x_j) \succsim (x_{\hat{j}}^0, x_j')$ for any x_j and x_j' in S_j. In general, such an induced ordering will depend on the particular choice of $x_{\hat{j}}^0$. The first notion of separability states that these orderings for a particular element j of the partition are identical for all $x_{\hat{j}}$.

Definition 6.7 (weak separability of preferences)

A preference order $\succsim$ on $X=\prod_{j\in J} S_j$ is called weakly separable if for each $j \in J$ $x_j \succsim_{x_{\hat{j}}^0} x_j'$ implies $x_j \succsim_{x_{\hat{j}}} x_j'$ for all $x_{\hat{j}} \in \prod_{i\neq j} S_i$.

The induced ordering can then be denoted by $\succsim_j$. The corresponding notion of a weakly separable utility function is given in the next definition.

Definition 6.8 (weak separability of utility functions)

A utility function u: $\prod_{j\in J} S_j \to R$ is called weakly separable if there exist continuous functions v_j: $S_j \to R$, $j \in J$ and V: $R^k \to R$ such that $u(x) = V(v_1(x_1),\dots, v_k(x_k))$.

The following theorem establishes the equivalence of the two notions of weak separability. Its proof is straightforward and can be found in Katzner (1970).

Theorem 6.1

Let $\succsim$ be a continuous preference order. Then $\succsim$ is weakly separable if and only if every continuous representation of it is weakly separable.

Historically a stronger notion of separability was used first. Early writers in utility theory thought of each commodity as having its own intrinsic utility represented by some scalar function. The overall level of utility then was simply taken as the sum of these functions. With the more general development of preference and utility theory the additive utility function has remained an interesting case used frequently in certain classes of economic problems. Its relationship to the properties of the underlying preferences is now well understood, the essential result being due to Debreu (1960).

Definition 6.9 (strong separability of utility functions)

A utility function u: $\prod_{j\in J} S_j \to R$ is called strongly separable if there exist continuous functions v_j: $S_j \to R, j\in J$ and V: $R \to R$ such that

$$u(x) = V\left(\sum_{j\in J} v_j(x_j)\right).$$

Since V is continuous and increasing one observes immediately that the function $V^{-1}\circ u$ is additive, representing the same preference relation. The problem of finding conditions on preference relations which yield strong separability for all of its representations is therefore equivalent to the one of establishing conditions under which an additive representation exists.

Let $u(x)=\sum_{j\in J} u_j(x_j)$ denote an additive utility function with respect to the partition N. Consider any non-empty proper subset $I\subset J$ and two bundles x and x' where all components j belonging to $J\setminus I$ are kept at the same level x_j^0, $j\in J\setminus I$. We can write therefore $x=(x_I, x_{J\setminus I}^0)$ and $x'=(x_I', x_{J\setminus I}^0)$. Since u is additive it is immediately apparent that the induced utility function on $\prod_{j\in I} S_j$ is independent of the particular choice of $x_{J\setminus I}^0$, which also makes the underlying induced preference order independent. This property holds true for any non-empty proper subset $I\subset J$. It also supplies the motivation for the definition of a strongly separable preference relation.

Definition 6.10 (strong separability of preferences)

A preference order $\succsim$ on $X=\prod_{j\in J}S_j$ is called strongly separable if it is weakly separable with respect to all proper partitions of all possible unions of $N_1,\dots, N_k$.

Equivalently, a preference order is strongly separable if for any proper subset I of J, the induced preference order on $\prod_{j\in I}S_j$ is independent of the particular choice of $x^0_{J\setminus I}$.

Theorem 6.2 [Debreu (1964)]

Let $\succsim$ be a continuous preference order. Then $\succsim$ is strongly separable if and only if every continuous representation is strongly separable.

6.3. *Smooth preferences and differentiable utility functions*

The previous sections on preferences and utility functions indicated the close relationship between the continuity of the utility function and the chosen concept of continuity for a preference order. It was shown that the two concepts are identical. Section 7 uses this fact to demonstrate that, under some conditions, demand behaves continuously when prices and wealth change. When continuous differentiability of the demand function is required, continuity of the preference order will no longer be sufficient. It is clear that some degree of differentiability of the utility function is necessary. Since the utility function is the derived concept the differentiability problem has to be studied with respect to the underlying preferences. The first rigorous attempt to study differentiable preference orders goes back to Antonelli (1886) and an extensive literature has developed studying the so-called integrability problem, which was surveyed completely by Hurwicz (1971). Debreu (1972) has chosen a more direct approach to characterize differentiable preference orders. His results indicate that sufficiently "smooth", i.e. differentiable preference orders, are essentially equivalent to sufficiently differentiable utility functions and that a solution to the integrability problem can be found along the same lines. [See also Debreu (1976) and Mas-Colell (1975).]

To present the approach of Debreu we will assume, for the purpose of this section as well as for all later sections which deal with differentiability problems, that the consumption set X is the interior of the positive cone of R^ℓ which will be denoted P. The preference order $\succsim$ is considered as a subset of $P\times P$, i.e. $\succsim\subset P\times P$, and it is assumed to be continuous and monotonic. To describe a smooth preference order, differentiability assumptions will be made on the graph of the indifference relation in $P\times P$.

Let C^k, $k\geqq 1$, denote the class of functions which have continuous partial derivatives up to order k, and consider two open sets X and Y in R^n. A function h

from X onto Y is a C^k-diffeomorphism if h is one-to-one and both h and h^{-1} are of class C^k. A subset M of R^n is a C^k-hypersurface if for every $z \in M$, there exists an open neighborhood U of z, a C^k-diffeomorphism of U onto an open set $V \subset R^n$, and a hyperplane $H \subset R^n$ such that $h(M \cap U) \subset V \cap H$. Up to the diffeomorphism h, the hypersurface M has locally the structure of a hyperplane. The hypersurface of interest here with respect to the preference order is the indifference surface $\tilde{I}$ defined as $\tilde{I} = \{(x, y) \in P \times P \mid x \sim y\}$. *Smooth preference orders* are those which have an indifference hypersurface of class C^2, and we will say that $\succsim$ is a C^2-preference order. The result of Debreu states that C^2-utility functions are generated by preference orders of class C^2 and vice versa.

Theorem 6.3 [Debreu (1972)]

Let $\succsim$ denote a continuous and monotonic preference order on the interior of the positive cone of R^ℓ. There exists a monotonic utility function of class C^2 with no critical point for $\succsim$ if and only if $\tilde{I}$ is a C^2-hypersurface.

7. Continuous demand

Given a price vector $p \neq 0$ and the initial wealth w the consumer chooses the best bundle in his budget set as his demand. For preference orders satisfying Axioms 1–3 any best element with respect to the preference relation is also a maximizer for any utility function representing it, and vice versa. Thus, preference maximization and utility maximization lead to the same set of demand bundles. We now study the dependence of demand on its two exogenous parameters, price and wealth.

The budget set of a consumer was defined as $\beta(p, w) = \{x \in X \mid p \cdot x \leqq w\}$. Let $S \subset R^{\ell+1}$ denote the set of price–wealth pairs for which the budget set is non-empty. Then β describes a correspondence (i.e. a set valued function) from S into R^ℓ. The two notions of continuity of correspondences used in the sequel are the usual ones of upper hemi-continuity and lower hemi-continuity [see, for example, Hildenbrand (1974), Hildenbrand and Kirman (1976), or Green and Heller, Chapter 1 in Volume I of this Handbook].

Definition 7.1

A correspondence ψ from S into T, a compact subset of R^ℓ, is upper hemi-continuous (u.h.c.) at a point $y \in S$ if, for all sequences $z^n \to z$ and $y^n \to y$ such that $z^n \in \psi(y^n)$, it follows that $z \in \psi(y)$.

With T being compact this definition says that ψ is upper hemi-continuous if it has a closed graph. Furthermore, one immediately observes that any upper hemi-continuous correspondence which is single valued is in fact a continuous function.

Definition 7.2

A correspondence ψ from S into an arbitrary subset T of R^{ℓ} is lower hemi-continuous (l.h.c.) at a point $y \in S$ if for any $z^0 \in \psi(y)$ and for any sequence $y^n \to y$ there exists a sequence $z^n \to z^0$ such that $z^n \in \psi(y^n)$ for all n.

A correspondence is continuous if it is both lower and upper hemi-continuous. With these notions of continuity the following two lemmas are easily established. [For proofs see, for example, Debreu (1959) or Hildenbrand (1974).]

Lemma 7.1

The budget set correspondence β: $S \to X$ has a closed graph and is lower hemi-continuous at every point (p, w) for which $w > \min\{p \cdot x | x \in X\}$ holds.

The condition $w > \min\{p \cdot x | x \in X\}$ is usually referred to as the *minimal wealth condition*.

It was argued above that preference maximization and utility maximization lead to the same set of demand bundles if the preference relation is reflexive, transitive, and complete. Therefore, if u: $X \to R$ is a utility function the *demand* of a consumer can be defined as

$$\varphi(p, w) = \{x \in \beta(p, w) | u(x) \geqq u(x'), x' \in \beta(p, w)\}, \tag{7.1}$$

which is equivalent to definition (3.2). With a continuous utility function the demand set $\varphi(p, w)$ will be non-empty if the budget set is compact. In this case an application of a fundamental theorem by Berge (1966) yields the following lemma on the continuity of the demand correspondence.

Lemma 7.2

For any continuous utility function u: $X \to R$ the demand correspondence φ: $S \to X$ is non-empty valued and upper hemi-continuous at each $(p, w) \in S$, where $\beta(p, w)$ is compact and $w < \min\{p \cdot x | x \in X\}$.

From the definition of the budget correspondence and the demand correspondence it follows immediately that $\varphi(\lambda p, \lambda w) = \varphi(p, w)$ for any $\lambda > 0$ and for any price–wealth pair (p, w). This property states in particular that demand is homogeneous of degree zero in prices and wealth. For convex preference orders the demand correspondence will be convex-valued, a property which plays a crucial role in existence proofs of competitive equilibrium. If the preference order is strictly convex then the demand correspondence is single-valued, i.e. one obtains a demand function. Upper hemi-continuity then implies the usual continuity. To summarize this section, we indicate in the following lemma the weakest assumptions of traditional demand theory which will generate a continuous demand function.

Lemma 7.3

Let $\succsim$ denote a strictly convex and continuous preference order. Then the demand correspondence φ: $S \to X$ is a continuous function at every $(p,w) \in S$ for which $\beta(p,w)$ is compact and $w > \min\{p \cdot x \mid x \in X\}$ holds. Moreover, for all $\lambda > 0$, $\varphi(\lambda p, \lambda w) = \varphi(p,w)$, i.e. φ is homogeneous of degree zero in prices and wealth.

For the remainder of this survey, following the traditional notational convention, the letter f will be used to denote a demand function.

8. Demand without transitivity

Empirical studies of demand behavior have frequently indicated that consumers do not behave in a transitive manner. This fact has been taken sometimes as evidence against the general assumption that preference maximization subject to a budget constraint is the appropriate framework within which demand theory should be analyzed. Sonnenschein (1971) indicates, however, that the axiom of transitivity is unnecessary to prove existence and continuity of demand. A related situation without transitivity is studied by Katzner (1971) where preferences are defined locally and thus "local" results for demand functions are obtained.

Recall that without transitivity a utility function representing the preference relation cannot be defined. Let $\succsim$ denote a preference relation which is complete but not necessarily transitive. Then the definition of demand as in Definition 3.2 can be given in the following form:

Definition 8.1

The demand correspondence φ: $S \to X$ is defined as $\varphi(p,w) = \{x \in \beta(p,w) \mid x \succsim x'$ for all $x' \in \beta(p,w)\}$.

Theorem 8.1 (Sonnenschein)

Let $\varphi(p,w) \neq \varnothing$ for all $(p,w) \in S$ and assume that β is continuous at $(p^0, w^0) \in S$. If the preference relation is continuous, then the demand correspondence φ is u.h.c. at (p^0, w^0).

The assumption that $\varphi(p,w) \neq \varnothing$ for all $(p,w) \in S$ is implied by some modified convexity assumption on the strict preference relation, as indicated by the next theorem.

Theorem 8.2 (Sonnenschein)

Let $\succsim$ denote a continuous preference relation on X such that the set $\{x' \in X | x' \succ x\}$ is convex for all $x \in X$. Then $\varphi(p, w) \neq \varnothing$ whenever $\beta(p, w) \neq \varnothing$.

Therefore, the two results by Sonnenschein indicate that continuous demand functions will be obtained if the transitivity is replaced by the convexity of preferences.

A further result in the theory of the non-transitive consumer was given by Shafer (1974). This approach formulates the behavior of a consumer as one of maximizing a continuous numerical function subject to a budget constraint. This function, whose existence and continuity does not depend on transitivity, can be considered as an alternative approach to represent a preference relation.

Let $X = R^\ell_+$ and let $\succsim$ denote a preference relation on X. Since any binary relation uniquely defines a subset of $R^\ell_+ \times R^\ell_+$, and vice versa, one writes $\succsim \subset R^\ell_+ \times R^\ell_+$, such that $(x, y) \in \succsim$ if and only if $x \succsim y$. Define:

$$\succsim(x) = \{y | (y, x) \in \succsim\},$$
$$\succsim^{-1}(x) = \{y | (x, y) \in \succsim\},$$
$$\succ(x) = \{y | (y, x) \in \succsim \quad \text{and} \quad (x, y) \notin \succsim\}.$$

With this notation the usual properties of completeness, continuity, and strict convexity which will be needed below are easily redefined.

Definition 8.2

The relation $\succsim$ is complete if and only if $(x, y) \in \succsim$ or $(y, x) \in \succsim$ for any x and y.

Definition 8.3

The relation $\succsim$ is continuous if for all x the sets $\succsim(x)$ and $\succsim^{-1}(x)$ are closed.

Definition 8.4

The relation $\succsim$ is strictly convex if for $(x, z) \in \succsim(y)$ and $0 < \alpha < 1$ it follows that $\alpha x + (1 - \alpha) z \in \succ(y)$.

We can now state Shafer's representation result.

Theorem 8.3 (Shafer (1974))

Let $\succsim \subset R^\ell_+ \times R^\ell_+$ denote a continuous, complete, and strictly convex preference relation. Then there exists a continuous function k: $R^\ell_+ \times R^\ell_+ \to R$ satisfying

(i) $k(x, y) > 0$ if and only if $x \in \succ(y)$,
(ii) $k(x, y) < 0$ if and only if $y \in \succ(x)$,
(iii) $k(x, y) = 0$ if and only if $x \in \succsim(y)$ and $y \in \succsim(x)$,
(iv) $k(x, y) = -k(y, x)$.

The assumptions of the theorem are the traditional ones except that the transitivity axiom is excluded. If the latter were assumed then a utility function u exists and the function k can be defined as $k(x, y)=u(x)-u(y)$.

As before, let $\beta(p,w)$ denote the budget set of the consumer. Then the demand of the consumer consists of all points in the budget set which maximize k. More precisely, the demand is defined as

$$\varphi(p,w)=\{x\in\beta(p,w)|k(x,y)\geqq 0,\quad \text{all } y\in\beta(p,w)\}$$

or, equivalently,

$$\varphi(p,w)=\{x\in\beta(p,w)|(x,y)\in\succsim,\quad \text{all } y\in\beta(p,w)\}.$$

The strict convexity assumption guarantees that there exists a unique maximal element. The following theorem makes the maximization argument precise and states the result of the existence of a continuous demand function by Sonnenschein in an alternative form, which may be used to derive demand functions for a non-transitive consumer explicitly.

Theorem 8.4 (Shafer)

Under the assumptions of Theorem 8.2 and for each strictly positive price vector p and positive wealth w, the demand $x=f(p,w)=\{x\in\beta(p,w)|k(x,y)\geqq 0$, all $y\in\beta(p,w)\}$, exists and the function f is continuous at (p,w).

9. Demand under separability

Separability of the preference order and the utility function, whether weak or strong, has important consequences for the demand functions. Using the notation and definitions of section 6.2, under separability the utility function can be written as

$$u(x)=V(v_1(x_1),\dots,v_k(x_k)), \tag{9.1}$$

where the x_j, $j=1,\dots,k$, are vectors of quantities of commodities in S_j and $X=S_1\times\cdots\times S_k$. The $v_j(x_j)$ are utility functions defined on S_j. We will use the vector p_j for the prices of the commodities in N_j.

Definition 9.1

For any $w_j\in R^{\ell}_+$ define a sub-budget set

$$\beta^j(p_j,w_j)\equiv\{x_j\in S_j|\,p_j'x_j\leqq w_j\}. \tag{9.2}$$

We are now in a position to introduce the concept of conditional demand functions $f_j^j(p_j, w_j)$ describing the x_j which maximize $v_j(x_j)$ over the sub-budget set.

Definition 9.2

Conditional demand functions are given as

$$f_j^j(p_j, w_j) \equiv \{x_j \in \beta^j(p_j, w_j) | v_j(x_j) > v_j(x_j^0), x_j^0 \in \beta^j(p_j, w_j)\}. \tag{9.3}$$

These conditional demand functions share all the properties of the usual demand functions, except that their domain and range is limited to p_j, w_j, and S_j. Given $v_j(x_j)$, p_j, and w_j, then demand x_j is known. However, w_j is not given exogenously, but as part of the overall optimization problem. Let $f_j(p, w)$ be the j-subvector of the demand function $f(p, w)$. Then, w_j is given by

$$w_j^*(p, w) = p_j' f_j(p, w). \tag{9.4}$$

Note that in general the full price vector p is needed to determine w_j^*.[1] When using the w_j^* generated by $w_j(p, w)$ in the conditional demand functions one would expect to obtain the same demand vector as the one given by $f_j(p, w)$. Indeed, one has

Theorem 9.1

Under separability of the utility function

$$f_j^j(p_j, w_j^*(p, w)) = f_j(p, w), \quad \text{for all } j. \tag{9.5}$$

Proof

Consider a particular (p^0, w^0). Let $x_j^* = f_j^j(p_j^0, w_j^*(p^0, w^0))$ for some j and $x^0 = f(p^0, w^0)$. Clearly, $x_j^0 \in \beta^j(p_j^0, w_j^*(p^0, w^0))$. Assume $x_j^* \neq x_j^0$. Then $v_j(x_j^*) > v_j(x_j^0)$ and

$$\begin{aligned} &V(v_1(x_1^0), \dots, v_j(x_j^*), \dots, v_k(x_k^0)) \\ &\quad > V(v_1(x_1^0), \dots, v_j(x_j^0), \dots, v_k(x_k^0)) = u(x^0) \end{aligned} \tag{9.6}$$

because V is monotone increasing in $v_j(x_j)$. Since $(x_{\hat{j}}^0, x_j^*)$ is an element of the

[1]Gorman (1959) has investigated under what condition one can replace the full ℓ-vector p in (9.4) by a k-vector of price indexes of the type $P_j(p_j)$. These conditions are restrictive.

budget set $\beta(p,w)$, inequality cannot hold. Equality, however, means $v_j(x_j^*)=v_j(x_j^0)$ and $x_j^*=x_j^0$ since x_j is the unique vector maximizing $v_j(x_j)$ over $x_j\in\beta_j^j(p_j^0\cdot w_j^*(p^0,w^0))$. Therefore (9.5) holds for (p^0,w^0). Since (p^0,w^0) is arbitrary, it holds for all admissible (p,w) and the theorem is proved.

The interest of Theorem 9.1 is twofold. First, it shows that the other prices affect the demand for x_j only by way of the scalar function $w_j^*(p,w)$, implying a considerable restriction on the scope of the impact of p_j. Second, if one can observe the w_j empirically one can concentrate on the conditional demand functions for which only the p_j are needed. An example of the latter is to consider demand behavior for a certain period, say a year. Under the (usually implicit) assumption of separability over different time periods one only needs to know total expenditure for that period (w_j) and the corresponding price vector (p_j). In this context (9.4) may then be considered as the consumption function, relating total consumer expenditure to total wealth and prices for all periods.

10. Expenditure functions and indirect utility functions

An alternative approach in demand analysis, using the notion of an expenditure function, was suggested by Samuelson (1947a). It was only fully developed after 1957 [see, for example, Karlin et al. (1959) and McKenzie (1957)]. It later became known as the duality approach in demand analysis [see also Diewert (1974) and Diewert, Chapter 12 in Volume II of this Handbook]. In certain cases it provides a more direct analysis of the price sensitivity of demand and enables a shorter and more transparent exposition of some classical properties of demand functions. Without going too much into the details of this approach we will describe its basic features and results for a slightly more restrictive situation than the general case above. These restrictions will be used in all later sections.

From now on it will be assumed that the consumption set X is equal to the positive orthant R_+^ℓ and that all prices and wealth are positive. This implies that the budget set is compact and that the minimum wealth condition is satisfied. Therefore for any continuous utility function the demand correspondence φ is upper hemi-continuous. Furthermore, the assumption of local non-satiation will be made on preferences or on the utility function, respectively. This implies that the consumer spends all his wealth when maximizing preferences.

Given an attainable utility level $v=u(x), x\in X$, the *expenditure function* is the minimum amount necessary to be spent to obtain a utility level at least as high as v at given prices p. Hence, the expenditure function E: $R_{++}^\ell\times R\to R$ is defined as

$$E(p,v)=\min\{p\cdot x\,|\,u(x)\geqq v\}. \tag{10.1}$$

The following properties of the expenditure function are easily established.

Lemma 10.1

If the continuous utility function satisfies local non-satiation then the expenditure function is:

(i) strictly increasing and continuous in v for any price vector p, and
(ii) non-decreasing, positive linear homogeneous, and concave in prices at each utility level v.

Let $y=E(p,v)$ denote the minimum level of expenditures. Since E is continuous and strictly increasing in v, its inverse $v=g(p,y)$ with respect to v expresses the utility level as a function of expenditures and prices which is called the *indirect utility function*. It is easy to see that

$$g(p,y)=\max\{u(x)|\,p\cdot x=y\}. \tag{10.2}$$

Due to the properties of the expenditure function the indirect utility function is

(i) strictly increasing in y for each price vector p, and
(ii) non-increasing in prices and homogeneous of degree zero in income and prices.

From the definitions of E and g the following relations hold identically:

$$v\equiv g(p,E(p,v)) \quad \text{and} \quad y\equiv E(p,g(p,y)). \tag{10.3}$$

Given a price vector p and a utility level v the expenditure minimum $E(p,v)$ will be attained on some subset of the plane defined by $E(p,v)$ and p. If preferences are strictly convex there will be a unique point $x\in X$ minimizing expenditures and we denote the function of the minimizers by $x=h(p,v)$. By definition one has therefore

$$E(p,v)=p\cdot h(p,v). \tag{10.4}$$

The function h is the so-called *Hicksian "income-compensated" demand function*. h is continuous in both of its arguments and homogeneous of degree zero in prices.

Considering the original problem of the maximization of utility subject to the budget constraint $p\cdot x\leqq w$, our assumptions of local non-satiation and strict convexity imply that one obtains a continuous function of maximizers $f(p,w)$. This function is the so-called *Marshallian "market" demand function* which satisfies the property

$$p\cdot f(p,w)=w. \tag{10.5}$$

From these definitions one obtains a second pair of identities which describe the fundamental relationship between the Hicksian and the Marshallian demand functions:

$$\begin{aligned} f(p,w)&=h(p,g(p,w)), \quad \text{for all } (p,w),\\ h(p,v)&=f(p,E(p,v)), \quad \text{for all } (p,v). \end{aligned} \tag{10.6}$$

One important property of the Hicksian demand function can be obtained immediately. For a fixed utility level v, consider two price vectors, p and p', and the associated demand vectors, $x=h(p,v)$ and $x'=h(p',v)$. Using the property that x and x' are expenditure minimizers, one obtains

$$(p-p')(x-x')\leqq 0. \tag{10.7}$$

For the change $\Delta p_k=p_k-p'_k$ of the price of a single commodity k with all other prices held constant, i.e. $\Delta p_h=0$, $h\neq k$, (10.7) implies

$$\Delta p_k \Delta x_k \leqq 0. \tag{10.8}$$

In other words, a separate price increase of one particular commodity will never result in a larger demand for this commodity. The Hicksian demand function of any commodity is therefore never upward sloping in its own price. This property is commonly known as the *negativity (non-positivity) of the own substitution effect*. A detailed discussion in the differentiable context can be found in Section 13.

11. Properties of differentiable utility functions

The following sections treat utility and demand in the context of differentiability which is the truly classical approach to the theory of consumer demand [see, for example, Slutsky (1915), Hicks (1939), and Samuelson (1947a)].

Let u: $X\to R$ be a C^2 utility function with no critical point representing a complete and continuous preference order of class C^2 on X characterized by monotonicity and strict convexity. Then this function is (i) continuous, (ii) increasing, i.e. $u(x)>u(y)$ for $x\geqslant y$ and $x\neq y$, and (iii) strictly quasi-concave, i.e. $u(\alpha x+(1-\alpha)y)>u(y)$ for $\alpha\in(0,1)$ and $u(x)\geqq u(y)$. It is twice continuously differentiable with respect to x, i.e. all its second-order partial derivatives exist and are continuous functions of x. It will be assumed that all first-order derivatives, namely $\partial u/\partial x_i=u_i$, $i=1,\dots,\ell$, are positive. They are called *marginal utilities*. In the following the symbol u_x will be used to denote the ℓ-vector of marginal utilities. Since the second-order derivatives are continuous functions of their arguments, applying Young's theorem yields the *symmetry property*

$$u_{ij} = \frac{\partial^2 u}{\partial x_i \partial x_j} = \frac{\partial^2 u}{\partial x_j \partial x_i} = u_{ji}.$$

Let U_{xx} be the $\ell \times \ell$ Hessian matrix of the utility function, i.e. the matrix of the second partial derivatives of u with typical element u_{ij}. The symmetry property means that U_{xx} is a symmetric matrix, i.e. $U_{xx} = U'_{xx}$ (where the prime denotes transposition).

The property of strict quasi-concavity of the utility function implies some further restrictions on the first- and second-order partial derivatives of the utility function.

Theorem 11.1

Under strict quasi-concavity of the utility function

$$z'U_{xx}z \leqq 0 \quad \text{for every element of } \{z \in R^\ell \,|\, u'_x z = 0\},$$

where the derivatives are evaluated at the same but arbitrarily selected bundles $x \in X$.

Proof

Take an arbitrary bundle $x \in X$. Select another bundle $y \neq x$ such that $u(y) = u(x)$. For scalar real α consider the bundle $m = \alpha y + (1-\alpha)x = \alpha z + x$ with $z = y - x$. For fixed x and y and varying α, define the function $f(\alpha) = u(m)$, which is a differentiable and strictly quasi-concave function, such that $f(\alpha) > f(0) = f(1)$, $\alpha \in (0,1)$ and $f'(0) > 0$, $f'(1) < 0$. Thus, there exists a value $\hat{\alpha} \in (0,1)$ such that $f'(\hat{\alpha}) = 0$ corresponding to a maximum of $f(\alpha)$ implying $f''(\hat{\alpha}) \leqq 0$. This maximum is unique since f is *strictly* quasi-concave. Now

$$f'(\hat{\alpha}) = \sum_i \frac{\partial u}{\partial m_i} \frac{\mathrm{d}m_i}{\mathrm{d}\alpha} = u'_x z = 0,$$

$$f''(\hat{\alpha}) = \sum_i \sum_j \frac{\mathrm{d}m_i}{\mathrm{d}\alpha} \frac{\partial^2 u}{\partial m_i \partial m_j} \frac{\mathrm{d}m_j}{\mathrm{d}\alpha} = z'U_{xx}z \leqq 0,$$

with the derivatives evaluated at $\hat{w} = \hat{\alpha} z + x$. Since y can be chosen arbitrarily close to x, the property holds also when the derivatives are evaluated at bundles in a small neighborhood of x and by virtue of continuity at x itself.

The property of strict quasi-concavity of the utility function is not strong enough to obtain everywhere differentiable demand functions. As a regularity condition the weak inequality of the theorem above is changed into a strong one. The result is known as strong quasi-concavity.

Definition 11.1

A strictly quasi-concave utility function is said to be strongly quasi-concave if

$$z'U_{xx}z<0 \quad \text{for every element of } \{z\in R^{\ell} \mid u'_x z=0, z\neq 0\}.$$

This additional condition is equivalent to non-singularity of the so-called *bordered Hessian matrix*

$$H=\begin{bmatrix} U_{xx} & u_x \\ u'_x & 0 \end{bmatrix}. \tag{11.1}$$

Theorem 11.2

The bordered Hessian matrix H associated with a strictly quasi-concave monotone increasing utility function is non-singular if and only if the utility function is strongly quasi-concave.

Proof[2]

(i) *Sufficiency.* Assume that H is singular. Then there exist an ℓ-vector z and a scalar r such that simultaneously

$$U_{xx}z+u_x r=0; \qquad u'_x z=0; \qquad (z',r)\neq 0. \tag{11.2}$$

The case of $z=0$, $r\neq 0$ can be ruled out because then $u_x r=0$, which implies $u_x=0$, contradicting the monotonicity of the utility function. The case of $z\neq 0$ can also be ruled out, since premultiplying (11.2) by z' would result in $z'U_{xx}z=0$, $u'_x z=0$, $z\neq 0$, contradicting the property of strong quasi-concavity. Consequently, no non-zero vector (z',r) exists such that $(z',r)H=0$ and thus H is non-singular.

(ii) *Necessity.* We will proceed in three steps. First it will be shown that if H is non-singular there exist real numbers $\alpha<\alpha^*$ such that the matrix $A(\alpha)=U_{xx}+\alpha u_x u'_x$ is non-singular, where the typical element of $A(\alpha)$ is $u_{ij}+\alpha u_i u_j$. Secondly, we will show that under strict quasi-concavity there exist real numbers $\beta\leqslant\beta^*$ such that $A(\beta)$ is a negative semi-definite matrix. The third step will combine the first two steps.

Step 1. Non-singularity means that for all ℓ-vectors c_1 such that $u'_x c_1=0$, $A(\alpha)c_1\neq 0$, for all α. Consider next all vectors c_2 such that $u'_x c_2\neq 0$, and

[2]See also Debreu (1952), Dhrymes (1967), and Barten et al. (1969).

normalize c_2 such that $u_x'c_2=1$. $A(\hat{\alpha})c_2=0$ means $\hat{\alpha}=-c_2'U_{xx}c_2$. Let $\alpha^*=\min_{c_2}\{-c_2'U_{xx}c_2 \mid c_2'u_x=1\}$. For $\alpha<\alpha^*$, $A(\alpha)c_2\neq 0$ and $A(\alpha)$ is non-singular.

Step 2. If $A(\beta)$ is negative semi-definite, i.e. $c'A(\beta)c\leqslant 0$, then clearly for all β, $z'U_{xx}z\leqslant 0$ for all z such that $u_x'z=0$. Furthermore, if $z'U_{xx}z\leqslant 0$ for all z such that $u_x'z=0$, then also $z'A(\beta)z\leqslant 0$ for all β. Consider next all vectors c such that $u_x'c\neq 0$ and normalize these such that $u_x'c=1$. $c'A(\beta)c\leqslant 0$ requires $\beta\leqslant -c'U_{xx}c$. Let $\beta^*=\min_c\{-c'U_{xx}c \mid c'u_x=1\}$. Therefore, $A(\beta)$ is negative semi-definite if $\beta\leqslant\beta^*$.

Step 3. There exist real numbers $\gamma<\beta^*=\alpha^*$ such that $A(\gamma)$ is both non-singular and negative semi-definite. A negative semi-definite matrix which is non-singular is a negative definite matrix. Therefore, $z'A(\gamma)z=z'U_{xx}z<0$ for all $u_x'z=0$, $z\neq 0$, i.e. u is strongly quasi-concave. Q.E.D.

What has been said about the property of the derivatives of u is also true for any differentiable increasing transformation of u. This is evident in the case of the positive sign of the marginal utilities and the consequences of strict quasi-concavity which are based directly on properties of the preference ordering, i.e. on monotonicity and convexity, respectively. Still, it is useful to write down explicitly the consequences of such transformations for the derivatives. Let F be a twice continuously differentiable increasing transformation F: $R\to R$, i.e. $F'>0$ and F'' continuous. Define $v(x)=F(u(x))$. The following relations hold between the first- and second-order partial derivatives of $v(x)$ and $u(x)$:

$$\partial v/\partial x_i = F'\partial u/\partial x_i \quad \text{or} \quad v_x=F'u_x,$$

$$\partial^2 v/\partial x_i\partial x_j = F'\partial^2 u/\partial x_i\partial x_j + F''(\partial u/\partial x_i)(\partial u/\partial x_j),$$

or

$$V_{xx}=F'U_{xx}+F''u_xu_x'.$$

Since F' is positive, v_x has the same sign as u_x. The elements of V_{xx} do not necessarily have the same sign as those of U_{xx}. However, $z'U_{xx}z<0$ for every element of $\{z\in R^\ell \mid u_x'z=0\}$ implies $z'V_{xx}z<0$ for every element of $\{z\in R^\ell \mid v_x'z=0\}$. Indeed, $v_x'z=F'u_x'z=0$ and thus $z'V_{xx}z=F'z'U_{xx}z+F''(z'u_x)^2=F'z'U_{xx}z<0$. Note that this is not a new result, but simply another demonstration of the fact that strict and strong quasi-concavity reflect properties of the preference ordering.

Since the marginal utilities $\partial u/\partial x_i$ are not invariant under monotone increasing transformations, one sometimes reasons in terms of ratios of a pair of marginal utilities, for instance $(\partial u/\partial x_i)/(\partial u/\partial x_j)$ which is clearly invariant. Keeping the level of utility constant and varying x_i and x_j alone, one has, locally,

$$(\partial u/\partial x_i)\mathrm{d}x_i^* + (\partial u/\partial x_j)\mathrm{d}x_j^* = 0 \tag{11.3}$$

or

$$R_{ij} = \frac{\partial u/\partial x_i}{\partial u/\partial x_j} = -\frac{\mathrm{d}x_j^*}{\mathrm{d}x_i^*}, \tag{11.4}$$

where R_{ij} is the *marginal rate of substitution* (MRS) between commodities i and j. It represents the amount of commodity j to be sacrificed in exchange for an increase in commodity i, keeping utility constant.

R_{ij} is assumed to be a decreasing function of x_i, i.e. at the same level of utility less of x_j has to be sacrificed to keep utility constant when x_i is large than when x_i is small. This assumption of *diminishing marginal rate of substitution* for any pair (i, j) follows from strong quasi-concavity of the utility function.

Decreasing MRS means that

$$\frac{\partial R_{ij}}{\partial x_i} - R_{ij}\frac{\partial R_{ij}}{\partial x_j} < 0, \tag{11.5}$$

which yields

$$\frac{1}{u_j^3}\left(u_{ii}u_j^2 - 2u_iu_ju_{ij} + u_{jj}u_i^2\right) < 0.$$

The term in parentheses is equal to $z'U_{xx}z$ for $z_k = 0$, $k \neq i, j$ and $z_i = -u_j$ and $z_j = u_i$. Since $u_j > 0$ and $u_x'z = 0$, strong quasi-concavity implies the negativity of (11.5). The reverse implication holds under some additional assumptions which are discussed by Arrow and Enthoven (1961).

Traditionally the concept of MRS has been used in connection with weak and strong separability. Before turning to this issue it is useful to investigate the consequences of the differentiability of $u(x)$ in the case of (weak) separability. Given that under separability

$$u(x) = V(v_1(x_1), \ldots, v_k(x_k)), \tag{11.6}$$

differentiability implies that for $i \in N_j$

$$\frac{\partial u}{\partial x_i} = \frac{\partial V}{\partial v_j}\frac{\partial v_j}{\partial x_i} \tag{11.7}$$

exists and hence $\partial V/\partial v_j$ and $\partial v_j/\partial x_i$ exist. Since $v_j(x_j)$ has all the properties of a utility function, hence $\partial v_j/\partial x_i>0$; also $\partial V/\partial v_j$ is positive because $\partial u/\partial x_i>0$. Next for $i, k\in N_J$

$$\frac{\partial^2 u}{\partial x_i \partial x_k}=\frac{\partial V}{\partial v_j}\frac{\partial^2 v_j}{\partial x_i \partial x_k}+\frac{\partial^2 V}{\partial v_j^2}\frac{\partial v_j}{\partial x_i}\frac{\partial v_j}{\partial x_k}, \tag{11.8}$$

and for $i\in N_j$, $k\in N_g$, $j\neq g$

$$\frac{\partial^2 u}{\partial x_i \partial x_k}=\frac{\partial^2 V}{\partial v_j \partial v_g}\frac{\partial v_j}{\partial x_i}\frac{\partial v_g}{\partial x_k}. \tag{11.9}$$

Therefore, the existence and symmetry of U_{xx} implies the existence and symmetry of the Hessian matrix V_{vv}.

In the case of strong separability $\partial V/\partial v_j=V'$, i.e. equal for all j. Then

$$\frac{\partial u(x)}{\partial x_i}=V'\frac{\partial v_j(x_j)}{\partial x_i}, \tag{11.10}$$

while the Hessian matrix V_{vv} has all elements equal.

Separability and properties of the MRS are related to each other by the following two theorems.

Theorem 11.3

The MRS between two commodities i and k within the same element N_j of the partition is independent of consumption levels outside of N_j if and only if the utility function is weakly separable.

This means that for all $i\in N_j$ the $\partial u/\partial x_i$ consist of the product of a common factor $\alpha_j(x)$ and a specific factor $\beta_{ji}(x_j)$ which is a function of x_j only, i.e.

$$\partial u(x)/\partial x_i=\alpha_j(x)\beta_{ji}(x_j).$$

This corresponds to (11.7) for $\alpha_j(x)=\partial V/\partial v_j$ and $\beta_{ji}(x_j)=\partial v_j/\partial x_i$.

Theorem 11.4

The MRS between a commodity $i\in N_j$ and another commodity $f\in N_g$, $g\neq j$, in different elements of the partition, can be written as the ratio of two functions $\beta_{ji}(x_j)$ and $\beta_{gf}(x_g)$, respectively, if and only if the utility function is strongly separable.

In the literature, for example Goldman and Uzawa (1964), strong separability is identified with independence of the MRS between i and f, $i \in N_j$, $f \in N_g$, $g \neq j$, of the consumption levels of commodities in other groups. This definition requires a partition into at least three groups, which is unsatisfactory since strong separability can exist also for two groups. The present theorem covers also the case of a partition into two groups and is inspired by Samuelson (1947a). Note that in general weak separability into two groups does not imply strong separability.

12. Differentiable demand

In Lemma 7.3 conditions are given for the existence of continuous demand functions $f(p, w)$, which are moreover homogeneous of degree zero in prices and wealth. This section will focus on the consequences of the assumption of differentiability of the utility function for the demand functions. In particular, the differentiability of the demand functions will be studied.

We will confine ourselves to the case where the consumption set X is the open positive cone P of R^ℓ. To obtain demand bundles in P we will further assume that preferences are monotonic, of class C^2, and that the closures of the indifference hypersurfaces are contained in P. Then, under positive prices and positive wealth the demand function is well defined and it maps into the positive cone of R^ℓ. Moreover, the consumer will spend all his wealth maximizing preferences. Therefore his choice can be considered as being limited to all bundles $x \in P$ which satisfy $p'x = w$.

If $u(x)$ is twice continuously differentiable then demand $x = f(p, w)$, as defined in (3.2) or in (7.1), can be determined as the solution of a classical maximization problem: maximize $u(x)$ with respect to x subject to $p'x = w$. One forms the Lagrangean

$$L(x, \lambda, p, w) = u(x) - \lambda(p'x - w), \tag{12.1}$$

where λ is a Lagrange multiplier. First-order conditions for a stationary value of $u(x)$ are

$$\partial L/\partial x = u_x - \lambda p = 0, \tag{12.2a}$$

$$\partial L/\partial \lambda = w - p'x = 0. \tag{12.2b}$$

As in the previous section it will be assumed that $u_i > 0$, $i = 1, \ldots, n$. Then, with u_x and p both strictly positive, the first condition implies a positive value of λ. A necessary second-order condition for a relative maximum is that

$$z'L_{xx}z \leqslant 0 \quad \text{for every } z \in R^{\ell} \text{ such that } p'z=0, \tag{12.3}$$

where $L_{xx}=\partial^2 L/\partial x\,\partial x'$, evaluated at a solution of (12.2). Under strict quasi-concavity of the utility function (see Theorem 11.1) this condition is satisfied since $L_{xx}=U_{xx}$ and $p'z=0$ implies $u_x'z=0$ in view of (12.2a) and positive λ.

System (12.2) is a system of $\ell+1$ equations in $2(\ell+1)$ variables: the ℓ-vectors x and p and the scalars λ and w. For our purpose p and w are taken as given, and x and λ are the "unknown" variables. Lemma 7.3 guarantees the existence of a unique solution for $x=f(p,w)$. Then there also exists a unique solution for λ, namely $\theta(p,w)=u_x^{0\prime}f(p,w)/w$, with u_x^0 evaluated at $x=f(p,w)$.

It can easily be verified that the solution of (12.2) for x is invariant under monotone increasing transformations of $u(x)$ but the one for λ is not. For such a transformation F the first-order conditions (12.2) are changed into $F'u_x-\lambda^*p=0$, $w-p'x=0$. The first one becomes (12.2a) again after division by $F'>0$, with $\lambda=\lambda^*/F'$. The solution for x is thus invariant, while $\lambda^*=F'\lambda$ is the solution for the Lagrange multiplier for the transformed problem.

We now turn to the matter of differentiability of $f(p,w)$ and $\theta(p,w)$. To prepare the ground we start by writing down the differential form of system (12.2) at (x^0, λ^0, p, w) with $x^0=f(p,w)$ and $\lambda^0=\theta(p,w)$:

$$U_{xx}^0\,\mathrm{d}x-\lambda^0\mathrm{d}p-p\,\mathrm{d}\lambda=0, \tag{12.4a}$$

$$\mathrm{d}w-p'\,\mathrm{d}x-x^{0\prime}\mathrm{d}p=0, \tag{12.4b}$$

where U_{xx}^0 is U_{xx} evaluated at x^0. After some rearrangement of terms one obtains the *Fundamental Matrix Equation of Consumer Demand*:

$$\begin{bmatrix} U_{xx}^0 & p \\ p' & 0 \end{bmatrix}\begin{bmatrix} \mathrm{d}x \\ -\mathrm{d}\lambda \end{bmatrix}=\begin{bmatrix} \lambda^0 I & 0 \\ -x^{0\prime} & 1 \end{bmatrix}\begin{bmatrix} \mathrm{d}p \\ \mathrm{d}w \end{bmatrix}, \tag{12.5}$$

where I is the $\ell\times\ell$ identity matrix. One may write formally

$$\mathrm{d}x=X_p\,\mathrm{d}p+x_w\,\mathrm{d}w; \qquad \mathrm{d}\lambda=\lambda'_p\,\mathrm{d}p+\lambda_w\,\mathrm{d}w \tag{12.6}$$

or

$$\begin{bmatrix} \mathrm{d}x \\ -\mathrm{d}\lambda \end{bmatrix}=\begin{bmatrix} X_p & x_w \\ -\lambda'_p & -\lambda_w \end{bmatrix}\begin{bmatrix} \mathrm{d}p \\ \mathrm{d}w \end{bmatrix}. \tag{12.7}$$

It is clear that $f(p,w)$ and $\theta(p,w)$ are continuously differentiable if and only if the matrix on the right-hand side of (12.7) is the unique matrix of derivatives. Combining (12.5) and (12.7) one obtains

$$\begin{bmatrix} U_{xx}^0 & p \\ p' & 0 \end{bmatrix}\begin{bmatrix} X_p & x_w \\ -\lambda_p & -\lambda_w \end{bmatrix} = \begin{bmatrix} \lambda^0 I & 0 \\ -x^{0\prime} & 1 \end{bmatrix} \tag{12.8}$$

for arbitrary $(\mathrm{d}p, \mathrm{d}w)$. We can now state the result on the differentiability of demand, first given by Katzner (1968), who also gives an informative counterexample.

Theorem 12.1 (Differentiability of Demand)

The system of demand functions $f(p,w)$ is continuously differentiable with respect to (p,w) if and only if the matrix

$$\begin{bmatrix} U_{xx}^0 & p \\ p' & 0 \end{bmatrix} \tag{12.9}$$

is non-singular at $x^0 = f(p,w)$.

Proof

(i) *Sufficiency.* Applying the Implicit Function Theorem to the system (12.2) requires the non-singularity of the Jacobian matrix with respect to (x, λ), which is equal to the matrix (12.9).

(ii) *Necessity.* If $f(p,w)$ is differentiable, the chain rule implies that $\theta(p,w)$ is differentiable and that the matrix on the right-hand side of (12.7) is the matrix of derivatives. The $(\ell+\ell)\times(\ell+1)$ matrix on the right-hand side of (12.8) is of full rank. Thus, the two $(\ell+1)\times(\ell+1)$ matrices in the product on the left-hand side of (12.8) are of full rank. Therefore (12.9) is non-singular.

Lemma 12.2

The matrix (12.9) is non-singular if and only if the matrix

$$H = \begin{bmatrix} U_{xx} & u_x \\ u_x' & 0 \end{bmatrix}$$

is non-singular.

We have now established that differentiability of the demand functions derived from strictly quasi-concave monotone increasing differentiable utility functions is equivalent to non-singularity of H. This latter condition is implied by strong quasi-concavity of these utility functions (Theorem 11.2), which, in turn, is equivalent to the assumption of diminishing marginal rate of substitution. As far as these two latter properties exclude singular points, the four properties mentioned here are equivalent and interchangeable.

Thus far we have studied the relationship between differentiability of the (direct) utility function and the Marshallian demand functions $f(p,w)$. The next theorems establish differentiability properties of the indirect utility function (10.2), the expenditure function (10.1), and the Hicksian demand function $h(p,v)$ in (10.4).

Theorem 12.3 (Differentiability of the Indirect Utility Function)

Assume that the direct utility function $u(x)$ is twice continuously differentiable and that the demand function $f(p,w)$ is continuously differentiable. Then, the indirect utility function

$$g(p,w)=\max\{u(x)|\, p'x=w\}$$

is twice continuously differentiable with respect to (p,w).

Proof

From the differentiability properties of u and f and from the identity

$$g(p,w)=u(f(p,w)) \tag{12.10}$$

and from $p'f(p,w)=w$ one obtains

$$\frac{\partial g}{\partial p'}=u'_x\frac{\partial f}{\partial p'}=\lambda p'\frac{\partial f}{\partial p'}=-\theta(p,w)f(p,w)', \tag{12.11}$$

since

$$p'\frac{\partial f}{\partial p'}=-f(p,w)'.$$

Similarly, differentiation with respect to w yields

$$\frac{\partial g}{\partial w}=u'_x\frac{\partial f}{\partial w}=\lambda p'\frac{\partial f}{\partial w}=\theta(p,w), \tag{12.12}$$

since

$$p'\frac{\partial f}{\partial w}=1.$$

Since $\theta(p,w)$ and $f(p,w)$ are both continuously differentiable with respect to (p,w) the indirect utility function is twice continuously differentiable.

Two comments can be made. Note that λ, the Lagrange multiplier associated with the budget constraint, is the derivative of the indirect utility function with respect to wealth: the *marginal utility of wealth* (*income*, *money*), i.e.

$$\lambda = \frac{\partial g}{\partial w} = \frac{\partial u(f(p,w))}{\partial w}.$$

Next, one has a result due to Roy (1942):

Corollary 12.4 (Roy's Identity)

$$f(p,w) = -\frac{1}{\partial g/\partial w}\partial g/\partial p. \tag{12.13}$$

Proof

Combine (12.11) and (12.12).

According to (10.3) the expenditure function $E(p,v)$ is the solution of $v = g(p,w)$ with respect to w.

Theorem 12.5 (Differentiability of the Expenditure Function)

The expenditure function is continuously differentiable if and only if the indirect utility function is continuously differentiable.

Proof

$g(p,w)$ is the inverse of $E(p,v)$ with respect to v, i.e.

$$E(p,v) = g^{-1}(p,v), \tag{12.14}$$

$$g(p,w) = E^{-1}(p,w). \tag{12.15}$$

Differentiation of $E(p, E^{-1}(p,w)) = w$ then yields

$$\frac{\partial E}{\partial p}\cdot\frac{1}{\partial E/\partial v} = -\frac{\partial E^{-1}}{\partial p} = -\frac{\partial g}{\partial p} \tag{12.16}$$

and

$$\frac{1}{\partial E/\partial v} = \frac{\partial E^{-1}}{\partial w} = \frac{\partial g}{\partial w}. \tag{12.17}$$

Since $\partial E/\partial v>0$, differentiability of E implies differentiability of g. Conversely, differentiation of $g(p, g^{-1}(p,v))=v$ yields

$$\frac{\partial g}{\partial p}\cdot\frac{1}{\partial g/\partial w}=-\frac{\partial g^{-1}}{\partial p}=-\frac{\partial E}{\partial p} \tag{12.18}$$

and

$$\frac{1}{\partial g/\partial w}=\frac{\partial g^{-1}}{\partial v}=\frac{\partial E}{\partial v}. \tag{12.19}$$

Since $\partial g/\partial w>0$, differentiability of $g(p,w)$ implies differentiability of $E(p,v)$. Q.E.D.

Theorem 12.6 (Differentiability of the Hicksian Demand Function)

If the expenditure function $E(p,v)$ is differentiable then

$$\frac{\partial E}{\partial p}(p,v)=h(p,v). \tag{12.20}$$

Consequently, if $E(p,v)$ is twice continuously differentiable then

$$\frac{\partial(\partial E/\partial p)}{\partial p'}=\frac{\partial h}{\partial p'} \tag{12.21}$$

which implies that $\partial h/\partial p'$ is a symmetric matrix.

Proof

One only has to establish (12.20).

Consider an arbitrary commodity k and let t_k denote the ℓ-vector with $t>0$ in the kth position and zero elsewhere. Then

$$\begin{aligned}\frac{E(p+t_k,v)-E(p,v)}{t}&=\frac{p'[h(p+t_k,v)-h(p,v)]}{t}+h_k(p+t_k,v)\\&\geqq h_k(p+t_k,v).\end{aligned}$$

On the other hand,

$$\begin{aligned}\frac{E(p-t_k,v)-E(p,v)}{-t}&=\frac{p'[h(p-t_k,v)-h(p,v)]}{-t}+h_k(p-t_k,v)\\&\leqq h_k(p-t_k,v).\end{aligned}$$

Taking the limit for $t\to 0$ yields

$$\frac{\partial E}{\partial p_k}(p,v)=h_k(p,v). \quad \text{Q.E.D.}$$

13. Properties of first-order derivatives of demand functions

Since $u(x)$ is twice continuously differentiable, the Hessian matrix U_{xx} is symmetric. Then also matrix (12.9) is symmetric, and the same is true for its inverse. Let Z be an $\ell\times\ell$ matrix, z be an ℓ-vector, and ζ a scalar defined by

$$\begin{bmatrix} Z & z \\ z' & \zeta \end{bmatrix}=\begin{bmatrix} U^0_{xx} & p \\ p' & 0 \end{bmatrix}^{-1}. \tag{13.1}$$

Note that $Z=Z'$, i.e. matrix Z is symmetric. Premultiply both sides of (12.8) by this inverse to obtain

$$\begin{bmatrix} X_p & x_w \\ -\lambda'_p & -\lambda_w \end{bmatrix}=\begin{bmatrix} \lambda Z-zx' & z \\ \lambda z'+\zeta x' & \zeta \end{bmatrix}, \tag{13.2}$$

where zx' denotes the matrix $(z_i x_j)$. Clearly $z=x_w$ and $\zeta=-\lambda_w$. Furthermore, writing $K=\lambda Z$ one has

$$X_p=K-x_w x', \tag{13.3a}$$

with $x_w x'$ denoting the matrix $[(\partial f_i/\partial w)x_j]$ and $x=f(p,w)$. In scalar form this equation, sometimes named the *Slutsky equation*, can be written as

$$\frac{\partial f_i(p,w)}{\partial p_j}=k_{ij}(p,w)-\frac{\partial f_i(p,w)}{\partial w}f_j(p,w). \tag{13.3b}$$

The matrix $K=(k_{ij}(p,w))$ is known as the *Slutsky matrix*. Its properties are of considerable interest.

Before proceeding to derive a set of properties for x_w, X_p, and K, it is useful to give the following implications of (13.1) and (13.2):

$$U^0_{xx}Z+px'_w=I, \tag{13.4a}$$

$$U^0_{xx}x_w=\lambda_w p, \tag{13.4b}$$

$$p'Z=0, \tag{13.4c}$$

$$p'x_w=1. \tag{13.4d}$$

Theorem 13.1

Let $f(p,w)$ denote the set of ℓ differentiable demand functions resulting from maximizing the utility function $u(x)$ subject to $p'x=w$. Let $\partial f/\partial w = x_w$, $\partial f/\partial p' = X_p$, and $K = X_p + x_w x'$. Then one has the following properties:

(i) Adding up $\quad p'x_w = 1$ (Engel aggregation),
$p'X_p = -x'$ (Cournot aggregation),
$p'K=0$.
(ii) Homogeneity $\quad X_p p + x_w w = 0$ or $Kp=0$.
(iii) Symmetry $\quad K=K'$.
(iv) Negativity $\quad y'Ky<0$, for $\{y\in R^\ell \mid y\neq \alpha p\}$.
(v) Rank $\quad r(K)=\ell-1$.

Proof

(i) *Adding-up.* Follows from (13.4d), $K=\lambda Z$, (13.4c), and (13.3).
(ii) *Homogeneity.* Apply Euler's theorem to $f(p,w)$ which is homogeneous of degree zero in prices and wealth. Use (13.3) and $p'x=w$.
(iii) *Symmetry.* $K=\lambda Z$ is symmetric since Z is symmetric.
(iv) *Negativity.* Since $K=\lambda Z$ and $\lambda>0$, the negativity condition for K is equivalent to the negativity condition for Z. Premultiplication of (13.4a) by Z gives $ZU^0_{xx}Z=Z$. Consider the quadratic form $y'ZU^0_{xx}Zy=y'Zy$ for some ℓ-vector $y\neq 0$. Let $q=Zy$. Note that $p'q=0$ or for $u_x=\lambda p$, $u'_x q=0$. The property of strong quasi-concavity (Definition 11.1) thus implies

$$y'Zy=q'U^0_{xx}q<0, \quad \text{for } q\neq 0.$$

The case of $q=0$ occurs when $y=\alpha p$. Hence Z and K are negative semi-definite matrices. Moreover, all diagonal elements of K are negative, since $e_i'Ke_i=K_{ii}<0$, where $e_i\neq\alpha p$ denotes the ith unit vector in R^ℓ.
(v) *Rank.* Since $Kp=0$, matrix K cannot have full rank, i.e. $r(K)\leqslant \ell-1$. It follows from (13.4a) that $U_{xx}Z=I-px'_w$, a matrix of rank $\ell-1$. Thus, $r(Z)=r(K)\geqslant \ell-1$. Therefore $r(K)=\ell-1$.

The Slutsky matrix K deserves some further discussion. It is invariant under monotone increasing transformations of the utility function. This follows from the fact that $f(p,w)$ is invariant and so are its derivatives X_p and x_w. Consequently, $K=X_p+x_w x'$ is invariant, too.

Next it can be shown that K is the matrix of price derivatives of the Hicksian demand function. Differentiation of (10.6) yields

$$\frac{\partial h}{\partial p'} = \frac{\partial f}{\partial p'} + \frac{\partial f}{\partial w}\cdot\frac{\partial E}{\partial p'}.$$

Then (12.20) and (13.3) imply

$$\frac{\partial h}{\partial p'}=X_p+x_w x'=K.$$

Therefore, for small price changes $\mathrm{d}p$, $K\mathrm{d}p$ describes the changes in the composition of the demand bundle in response to price changes, if utility is kept constant. Since Hicks (1934) and Allen (1934) this is known as the *substitution effect*.

The remaining part of the price effect, namely $-x_w x'\mathrm{d}p$, is known as the *income effect*. Its effect is similar to that of a change in w (wealth, income). This effect can be neutralized by an appropriate change in w equal to $x'\mathrm{d}p$. Obviously, the level of utility is then kept constant. For this reason the substitution effect is also called the *income compensated price effect*.

Finally, using (12.20) again, it is easily seen that

$$K=\frac{\partial^2 E}{\partial p\, \partial p'}, \tag{13.6}$$

i.e. K is the Hessian matrix of the expenditure function when v is kept constant. According to Lemma 10.1 E is concave in prices. Then, its Hessian matrix K is negative semi-definite. This result may suffice to indicate that the properties of the matrix K, derived in the context of Marshallian demand functions, can also be obtained in the context of expenditure functions and Hicksian demand functions.

Concluding this section, it should be pointed out that only very weak properties of a consumer's market demand function have been established. In particular, without any further assumptions on preferences no general statement can be made about the direction of change of the demand of a particular commodity when its price changes. From Theorem 13.1 one only has

$$\frac{\partial x_i}{\partial p_i}+x_i\frac{\partial x_i}{\partial w}<0. \tag{13.7}$$

If $\partial x_i/\partial w$ is positive (the case of a *superior good*) the own-price effect is obviously negative. If commodity i is an *inferior good*, i.e. when $\partial x_i/\partial w$ is negative, then the own-price effect may be positive. Commodities which display a positive own-price effect are known as *Giffen goods* – see Marshall (1895) and Stigler (1947). It all depends on the sign of the derivative with respect to w. On the basis of the adding-up condition one knows that at least one $\partial x_i/\partial w$ has to be positive. Therefore, all theory has to offer is that for at least one commodity the own-price effect is negative.

14. Separability and the Slutsky matrix

Separability of the utility function implies restrictions on the Slutsky matrix K because of the structured impact of p_j on demand for commodities in other elements of the partition than the one to which j belongs. Let N_j and N_g be two elements of the partition $N_1,\ldots,N_k$, and let K be partitioned into k^2 blocks of the type K_{jg}. From (13.3) one has

$$K_{jg}=\frac{\partial x_j}{\partial p'_g}+\frac{\partial x_j}{\partial w}x'_g, \tag{14.1}$$

where, as before, $x_j=f_j(p,w)$ is the vector of Marshallian demand functions for the commodities in N_j. Symmetry of K yields

$$K_{jg}=K'_{gj}. \tag{14.2}$$

For any $j\in J=\{1,\ldots,k\}$, let $x_j^j=f_j^j(p_j,w_j)$ as defined above in (9.3).

Theorem 14.1

Under weak separability of the utility function

$$K_{jg}=\psi_{jg}(p,w)\frac{\partial x_j^j}{\partial w_j}\frac{\partial x_g^{g\prime}}{\partial w_g} \tag{14.3}$$

where $\psi_{jg}(p,w)=\psi_{gj}(p,w)$.

Proof

Differentiate both sides of (9.5) with respect to w to obtain

$$\frac{\partial x_j^j}{\partial w_j}\frac{\partial w_j^*}{\partial w}=\frac{\partial x_j}{\partial w}. \tag{14.4}$$

Next, differentiate both sides of (9.5) with respect to the vector p'_g. Using (14.1) and (14.4) yields

$$\frac{\partial x_j^j}{\partial p'_g}=\frac{\partial x_j^j}{\partial w_j}\frac{\partial w_j^*}{\partial p'_g}=K_{jg}-\frac{\partial x_j^j}{\partial w_j}\frac{\partial w_j^*}{\partial w}x'_g.$$

Therefore,

$$K_{jg}=\frac{\partial x_j^j}{\partial w_j}\left[\frac{\partial w_j^*}{\partial p'_g}+\frac{\partial w_j^*}{\partial w}x'_g\right]=\frac{\partial x_j^j}{\partial w_j}z'_{jg}, \tag{14.5}$$

where z'_{jg} is the row vector between square brackets. One also has

$$K_{gj} = \frac{\partial x_g^g}{\partial w_g} z'_{gj}.$$

Since $K_{gj} = K'_{jg}$, it follows that

$$z_{jg}(p,w) = \psi_{jg}(p,w) \frac{\partial x_g^g}{\partial w_g}$$

which, when inserted into (14.5), gives (14.3). The equality of $\psi_{jg}(p,w)$ and $\psi_{gj}(p,w)$ follows directly from the symmetry of the matrix K.

One sometimes finds in the literature the derivation of a property similar to (14.3), namely:

Corollary 14.2

Under weak separability of the utility function and for $\partial w_j^*/\partial w \neq 0$, $\partial w_g^*/\partial w \neq 0$,

$$K_{jg} = \sigma_{jg}(p,w) \frac{\partial x_j}{\partial w} \frac{\partial x'_g}{\partial w}, \qquad j \neq g, \tag{14.6}$$

where

$$\sigma_{jg}(p,w) = \sigma_{gj}(p,w) = \psi_{jg}(p,w) \Big/ \left(\frac{\partial w_j^*}{\partial w} \frac{\partial w_g^*}{\partial w} \right).$$

Proof

Use (14.4) in (14.3).

The additional conditions of this corollary are of an empirical and not of a theoretical nature. If the conditions are not both true, $\sigma_{jg}(p,w)$ is not defined. Theorem 14.1 remains valid, however.

Before turning to the structure of the diagonal blocks of K, i.e. the K_{jj}, it is useful to note that also for the case of conditional demand equations one has a Slutsky matrix, K^j, say, which is formally analogous to the matrix K of the full set of the unconditional demand function. This K^j, defined by

$$K^j = \frac{\partial x_j^j}{\partial p_j} + \frac{\partial x_j}{\partial w_j} x_j^{j\prime}, \tag{14.7}$$

has all the properties given earlier for K in Theorem 13.1.

The first matrix on the right-hand side of (14.7) does not represent the complete effect of a change in p_j on x_j, because w_j is taken to be fixed. For the full impact, one also has to take into account the effect which p_j has on x_j via a change in $w_j^*(p, w)$. From (9.4) it follows that

$$\frac{\partial w_j^*}{\partial p_j'} = x_j + p_j' \frac{\partial x_j}{\partial p_j'}. \tag{14.8}$$

Thus, for the complete effect one has, using (14.7), (14.8), and (9.5),

$$\frac{\partial x_j^j}{\partial p_j'} = K^j + \frac{\partial x_j^j}{\partial w_j} p_j' \frac{\partial x_j}{\partial p_j'}. \tag{14.9}$$

In view of (9.5) and (14.4) this has to be equal to

$$\frac{\partial x_j}{\partial p_j'} = K_{jj} - \frac{\partial x_j^j}{\partial w_j} \frac{\partial w_j^*}{\partial w} x_j'. \tag{14.10}$$

This prepares the ground for

Theorem 14.3

Under weak separability of the utility function

$$K_{jj} = K^j + \psi_{jj}(p, w) \frac{\partial x_j^j}{\partial w_j} \frac{\partial x_j^{j'}}{\partial w_j}, \tag{14.11}$$

where $\psi_{jj}(p, w)$ is a scalar function.

Proof

Equating the right-hand sides of (14.9) and (14.10) gives

$$K_{jj} = K^j + \frac{\partial x_j^j}{\partial w_j} \left[p_j' \frac{\partial x_j}{\partial p_j'} + \frac{\partial w_j^*}{\partial w} x_j' \right].$$

Symmetry of K_{jj} implies that the term in square brackets is proportional to $\partial x_j^j / \partial w_j$, with the scalar factor of proportionality being a function of (p, w).

Corollary 14.4

$$\psi_{jj}(p, w) = - \sum_{g \neq j} \psi_{jg}(p, w). \tag{14.12}$$

Proof

First note that for all g

$$\frac{\partial x_g^{g'}}{\partial w_g} p_g = 1$$

as a consequence of the adding-up condition for the conditional demand functions. Furthermore, $K^j p_j = 0$. It then follows from (14.3) and (14.11) that for all j and g

$$p_j' K_{jg} p_g = \psi_{jg}(p,w).$$

Since $Kp = 0$,

$$\sum_{g=1}^{k} K_{jg} p_g = 0$$

and thus

$$\sum_{g=1}^{k} \psi_{jg}(p,w) = 0.$$

Therefore, (14.12) follows immediately.

Result (14.11) deserves some further discussion. One can also write it as

$$K^j - K_{jj} - \psi_{jj}(p,w) \frac{\partial x_j^j}{\partial w_j} \frac{\partial x_j^{j'}}{\partial w_j}, \tag{14.13}$$

where now K_{jj} is a negative definite matrix. The Slutsky matrix of the conditional demand equations for x_j can be decomposed into a negative definite matrix and into the outer vector product of the derivatives with respect to w_j multiplied by a scalar. A similar decomposition was derived by Houthakker (1960) for the Slutsky matrix K of the full unconditional system, namely

$$K = \theta(p,w) U_{xx}^{-1} + \sigma(p,w) \frac{\partial x}{\partial w} \frac{\partial x'}{\partial w} \tag{14.14}$$

with

$$\sigma(p,w) = -1 \Big/ \frac{\partial \ln \theta(p,w)}{\partial w}.$$

The first matrix on the right-hand side of (14.14) is called the *specific substitution matrix* and the second the *general substitution matrix*. Unlike (14.13), decomposition (14.14) is not invariant under monotone increasing transformations of the utility function, because neither $\sigma(p,w)$ nor $\theta(p,w)U_{xx}^{-1}$ are invariant. If one considers a "full unconditional system" as a conditional component of a still wider system, decomposition (14.13) is the appropriate one, where one may employ the same interpretation of the two components as in the case of (14.14). Both decompositions do not seem particularly restrictive. Indeed, further restrictions on K_{jj} in (14.13) or on $\theta(p,w)U_{xx}^{-1}$ in (14.14) are needed to give the decomposition operational significance. Note that if any symmetric pair of non-diagonal elements of U_{xx}^{-1} in (14.14) are set equal to zero, the arbitrariness (and cardinal nature) of the decomposition is removed.

As one expects intuitively, *strong* separability is more restrictive than weak separability. Indeed, one has the following theorem.

Theorem 14.5

If the utility function is strongly separable with respect to a partition $N_1,\ldots,N_k$ and if the Hessian matrix of its form $u(x)=\sum_{j\in J}v_j(x_j)$ is non-singular, then, for any pair j, $g\in J$,

$$K_{jg}=\sigma(p,w)\frac{\partial x_j}{\partial w}\frac{\partial x_g'}{\partial w},\qquad j\neq g, \tag{14.15}$$

i.e. the $\sigma_{jg}(p,w)$ of Corollary 14.2 is not specific for the pair j, g.

Proof

The Hessian matrix U_{xx} is block diagonal and so is its inverse. For this matrix (14.14) applies. The j, g-th off-diagonal block is given by (14.15).

Some remarks are in order. First, since K and $\partial f/\partial w$ are ordinal concepts, $\sigma(p,w)$ is invariant under monotone increasing transformations of the utility function. Second, the condition of a non-singular block diagonal U_{xx} is perhaps special, because $r(U_{xx})=\ell-1$ is also acceptable, as can be seen from (13.4a). However, a singular block diagonal Hessian matrix has rather unrealistic implications, for example, that for all except one element of the partition, $w_j^*(p,w)$, the amount to be spent on the partition does not depend on w. This will not be investigated further. Third, under strong separability the *marginal budget shares* $\partial w_j^*/\partial w$ are either all positive or all but one negative, the positive one being larger than unity. This property will be derived for the case of strong separability in elementary commodities.

15. Additive utility and demand

Consider the case that the utility function is separable in elementary commodities. This is clearly a limiting situation. It is known as *additive utility* and has very restrictive implications. In spite of this, it is not only of historical interest but also frequently used in empirical applications. Therefore, some attention will be paid to the consequences of additive utility for demand, although we will limit ourselves to the case of non-singular Hessian matrices.

Strong separability in elementary commodities clearly means that for all off-diagonal elements of the Slutsky matrix, Theorem 14.5 applies and thus, for any pair $i, j \in \{1,\dots,\ell\}$, $i \neq j$

$$k_{ij} = \sigma(p,w)\frac{\partial x_i}{\partial w}\frac{x_j}{\partial w}. \tag{15.1}$$

There is no specific substitution effect for any pair of goods. From (14.14) one can derive that

$$k_{ii} = \theta(p,w)/u_{ii} + \sigma(p,w)\left(\frac{\partial x_i}{\partial w}\right)^2. \tag{15.2}$$

Since $\sum_j k_{ij} p_j = 0$, it follows that

$$k_{ii} = -\sigma(p,w)\frac{1}{p_i}\frac{\partial x_i}{\partial w}\sum_{j\neq i} b_j(p,w)$$

$$= -\sigma(p,w)\frac{1}{p_i^2} b_i(p,w)[1 - b_i(p,w)], \tag{15.3}$$

where $b_i(p,w) = p_i \partial x_i/\partial w$ is the marginal propensity to spend on commodity i (the marginal budget share for i). Note that

$$\sum_i b_i(p,w) = 1 \tag{15.4}$$

and that combining (15.1) and (15.2) results in

$$b_i(p,w) = \zeta p_i^2/u_{ii}, \tag{15.5}$$

where $\zeta = -\theta/\sigma$.

The own-price derivative of demand is thus seen to be

$$\frac{\partial x_i}{\partial p_i} = k_{ii} - \frac{\partial x_i}{\partial w} x_i$$

$$= \frac{1}{p_i^2} b_i[-\sigma(1-b_i) - p_i x_i]. \tag{15.6}$$

In terms of *price elasticities* $\varepsilon_{ii} = \partial \ln x_i / \partial \ln p_i$ and *wealth (or income) elasticities* $\eta_i = \partial \ln x_i / \partial \ln w$

$$\varepsilon_{ii} = \eta_i[-\sigma^*(1-b_i) - a_i], \tag{15.7}$$

where $\sigma^* = \sigma / w$ and $a_i = p_i x_i / w$, the *average budget share*.

It is argued by Deaton (1974) that, for large ℓ, a_i and b_i will be very small on the average. Then one has *Pigou's Law* $\varepsilon_{ii} \sim -\sigma^* \eta_i$, i.e. the own-price elasticities are virtually proportional to the wealth (or income) elasticities. This is empirically a rather restrictive implication.

Another consequence is given in

Theorem 15.1

If the utility function is strongly separable in elementary commodities then the marginal budget shares $b_i(p, w)$ are either (i) all positive and smaller than one, or (ii) all but one negative, the only positive one being larger than unity.

Proof

According to the negativity condition, $k_{ii} < 0$. Then also $p_i^2 k_{ii} < 0$ or $-\sigma(p, w) b_i(p, w)[1 - b_i(p, w)] < 0$. Consider first the case $\sigma(p, w) > 0$. Then, $b_i(p, w)[1 - b_i(p, w)] > 0$ or $b_i(p, w) > b_i(p, w)^2$. Hence, $0 < b_i(p, w) < 1$. Next consider the case $\sigma(p, w) < 0$. Then $b_i(p, w)[1 - b_i(p, w)] < 0$. This means that either $b_i(p, w) < 0$ or $b_i(p, w) > 1$. In view of (15.4) not all $b_i(p, w)$ can be negative; nor can all be larger than one. Thus, some $b_i(p, w)$ are negative while others are positive. Since θ is positive, ζ in (15.5) is also positive. Therefore, the sign of u_{ii} is equal to the sign of $b_i(p, w)$. For the type of utility function under discussion, the strong quasi-concavity condition specializes to $\sum_j z_j^2 u_{jj} < 0$, $\sum_j z_j u_j = 0$. This implies that at most one of the u_{jj} can be positive, and all others have to be negative. Thus, one and not more than one $b_i(p, w)$ is positive and hence larger than unity. All others are negative.

For practical purposes one can take the b_i to be all positive. The other alternative is clearly pathological. The theorem is of importance because it is one of the few statements of a qualitative nature that can be made in demand analysis. Under additive utility inferior commodities ($b_i < 0$) are ruled out as exceptions. Actually, inferiority of a commodity occurs usually only when close substitutes, of a superior nature, are available. Strong separability of the utility function does not allow for the existence of such close substitutes. Another consequence of a qualitative nature is given by

Corollary 15.2

If the utility function is strongly separable in the elementary commodities, the cross-substitutions k_{ij}, $i \neq j$, are either (i) all positive or (ii) negative when i and j are inferior commodities and positive when i or j is the only superior commodity.

Proof

Use (15.1) and the proof of Theorem 15.1.

A discussion of the *Linear Expenditure System* (*L.E.S.*) may serve to illustrate some of the issues mentioned in this section. Assume that the preference ordering can be adequately represented by the *Stone–Geary utility function*

$$u = \sum_i \beta_i \log(x_i - \gamma_i) \tag{15.8}$$

which was first proposed in Geary (1949) and used in Stone (1954). In this utility function β_i and γ_i are constants – at least they do not depend on the quantities of commodities. The existence of a real-valued utility function requires that $\gamma_i \leqq x_i$. To be differentiable with respect to the x_i, $\gamma_i = x_i$ has to be ruled out. For the utility function to be increasing the β_i have to be positive. Without loss of generality one may assume that $\sum_i \beta_i = 1$ and thus $\beta_i \leqq 1$.

Utility is maximized under the budget condition $\sum_i p_i x_i = w$ if and only if the x_i satisfy

$$\frac{\beta_i}{x_i - \gamma_i} = \lambda p_i, \tag{15.9}$$

where $\lambda = \theta(p, w)$ is a positive factor of proportionality. Solving (15.9) for x_i and eliminating λ to take the budget condition into account, one finds

$$x_i = \gamma_i + \frac{\beta_i}{p_i}\left(w - \sum_j p_j \gamma_j\right) \tag{15.10}$$

as the typical demand equation. It is usually presented as

$$p_i x_i = p_i \gamma_i + \beta_i \left(w - \sum_j p_j \gamma_j\right), \tag{15.11}$$

where the left-hand side is the expenditure on commodity i. This expenditure depends linearly on the prices and w. This explains its name – linear expenditure function – which is attributed to Klein and Rubin (1947). Note that it is not linear in the β's and γ's.

As is clear from (15.11) the marginal budget share, $b_i(p,w)=\beta_i$, i.e. is constant and positive. Apply (13.3) to obtain

$$k_{ii}=-\sigma(p,w)\frac{(1-\beta_i)\beta_i}{p_i^2}, \tag{15.12a}$$

and for $i\neq j$

$$k_{ij}=\sigma(p,w)\frac{\beta_i\beta_j}{p_i p_j}, \tag{15.12b}$$

where

$$\sigma(p,w)=\left(w-\sum_j p_j\gamma_j\right)>0.$$

Since $\beta_i/p_i=\partial x_i/\partial w$, implication (15.1) is verified by (15.12b), i.e. there is no specificity in the relationship among the demands for various goods. The own-price elasticity (15.7) can be expressed as

$$\begin{aligned}\varepsilon_{ii}&=-(1-\beta_i)\beta_i\frac{\sigma}{p_i\gamma_i+\beta_i\sigma}-\beta_i\\&=-\beta_i\frac{p_i\gamma_i+\sigma}{p_i\gamma_i+\beta_i\sigma}\end{aligned} \tag{15.13}$$

indicating that the own-price elasticity is, in absolute value, larger than β_i. For sufficiently small γ_i, $\varepsilon_{ii}\sim-1$. The implied income elasticity is given by

$$\eta_i=\beta_i\frac{w}{p_i\gamma_i+\beta_i\sigma}. \tag{15.14}$$

Consequently,

$$\varepsilon_{ii}=-\eta_i\frac{p_i\gamma_i+\sigma}{w}.$$

For sufficiently small γ_i, Pigou's Law is clearly satisfied.

An interesting interpretation of (15.11) is due to Samuelson (1947b). Let γ_i be the a priori decided amount of commodity i to be acquired anyway. It is the committed quantity of i. The committed quantities require $\sum_j p_j\gamma_j$ of the budget. What remains, i.e. $w-\sum_j p_j\gamma_j$, is *supernumerary income* to be allocated according to the constant fractions β_i. This interpretation suggests that the γ_i are positive, which is theoretically not a necessary condition.

In applying the L.E.S. or any other demand system to actual data on an aggregated level one should realize that the theory outlined here is a comparative static one referring to the individual consumer deciding on quantities of elementary commodities. This theory may suggest certain properties and interpretations of demand systems, extremely useful for their empirical application. The additional requirements for the empirical validity of these systems are of considerable importance and fall outside the scope of this chapter. Consequently, a further discussion of demand systems, which are first of all tools of empirical analysis, is not undertaken here.[4]

References

Allen, R. G. D. (1934), "A reconsideration of the theory of value, II", Economica, New Series, 1:196–219.

Antonelli, G. B. (1886), Sulla Teoria Matematica della Economia Politica. Pisa. Translated as: "On the mathematical theory of political economy", in. Chipman et al. (eds.) (1971), 333–364.

Arrow, K. J. (1953), "Le rôle des valeurs boursières pour la répartition la meilleure des risques", Econométrie, Colloques Internationaux du Centre National de la Recherche Scientifique. Paris. Vol. 11:41–47. English translation: "The role of securities in the optimal allocation of risk-bearing", Review of Economic Studies, 31 (1964): 91–96.

Arrow, K. J., S. Karlin and P. Suppes (eds.) (1960), Mathematical methods in the social sciences, 1959, Proceedings of the First Stanford Symposium. Stanford: Stanford University Press.

Arrow, K. J. and A. C. Enthoven (1961), "Quasi-concave programming", Econometrica, 29:779–800.

Aumann, R. J. (1962), "Utility theory without the completeness axiom", Econometrica, 30:445–462.

Barten, A. P. (1977), "The systems of consumer demand functions approach: A review", Econometrica, 45:23–51.

Barten, A. P., T. Kloek and F. B. Lempers (1969), "A note on a class of utility and production functions yielding everywhere differentiable demand functions", The Review of Economic Studies, 36:109–111.

Berge, C. (1966), Espaces topologiques. Fonctions multivoques. Paris: Dunod. English translation: Topological spaces. Edinburgh: Oliver and Boyd (1973).

Bewley, T. F. (1972), "Existence of equilibria in economics with infinitely many commodities", Journal of Economic Theory, 4:514–540.

Bowen, R. (1968), "A new proof of a theorem in utility theory", International Economic Review, 9:374.

Brown, A. and A. Deaton (1972), "Surveys in applied economics: Models of consumer behaviour", Economic Journal, 82:1145–1236.

[4]See Brown and Deaton (1972), Phlips (1974), Powell (1974), Theil (1975, 1976), Barten (1977), and Deaton (forthcoming).

Chipman, J. S., L. Hurwicz, M. K. Richter and H. F. Sonnenschein (eds.) (1971), Preferences, utility, and demand. A Minnesota Symposium. New York: Harcourt Brace Jovanovich.
Deaton, A. S. (1974), "A reconsideration of the empirical implications of additive preferences", Economic Journal, 84:338–348.
Deaton, A. S. (forthcoming), "Demand analysis", in: Z. Griliches and M. D. Intriligator (eds.) Handbook of Econometrics. Amsterdam: North-Holland Publishing Co.
Debreu, G. (1952), "Definite and semidefinite quadratic forms", Econometrica, 20:295–300.
Debreu, G. (1954), "Representation of a preference ordering by a numerical function", in: R. M. Thrall et al. (eds.) Decision Processes. New York: John Wiley & Sons, Inc., 159–165.
Debreu, G. (1959), Theory of value. An axiomatic analysis of economic equilibrium. New York: John Wiley & Sons, Inc.
Debreu, G. (1960), "Topological methods in cardinal utility theory", in: K. J. Arrow et al. (1960), 12–26.
Debreu, G. (1964), "Continuity properties of Paretian utility," International Economic Review, 5:285–293.
Debreu, G. (1972), "Smooth preferences", Econometrica, 40:603–615.
Debreu, G. (1976), "Smooth preferences. A corrigendum", Econometrica, 44:831–832.
Dhrymes, P. J. (1967), "On a class of utility and production functions yielding everywhere differentiable demand functions", Review of Economic Studies, 34:399–408.
Diewert, W. E. (1974), "Applications of duality theory", in: M. D. Intriligator and D. A. Kendrick (eds.) Frontiers of quantitive economics, Vol. II. Amsterdam: North-Holland Publishing Co., 106–199.
Eilenberg, S. (1941), "Ordered topological spaces", American Journal of Mathematics, LXIII:39–45.
Frisch, R. (1926), "Sur un problème d'économie pure", Norsk Matematisk Forenings Skrifter, Serie 1, No. 16:1–40.
Gabszewicz, J. (1968), Coeurs et allocations concurrentielles dans des économies d'échange avec un continu de biens. Thesis, Université Catholique de Louvain, Belgium.
Geary, R. C. (1949), "A note on 'A constant utility index of the cost of living'", Review of Economic Studies, 18:65–66.
Goldman, S. M. and H. Uzawa (1964), "A note on separability in demand analysis", Econometrica, 32:387–398.
Gorman, W. M. (1959), "Separable utility and aggregation", Econometrica, 27:469–481.
Hicks, J. R. (1934), "A reconsideration of the theory of value, I", Economica, New Series, 1:52–75.
Hicks, J. R. (1939), Value and capital. Oxford: Clarendon Press.
Hildenbrand, W. (1974), Core and equilibria of a large economy. Princeton: Princeton University Press.
Hildenbrand, W. and A. P. Kirman (1976), Introduction to Equilibrium Analysis. Amsterdam: North-Holland Publishing Co.
Houthakker, H. S. (1960), "Additive preferences", Econometrica, 33:797–801.
Hurwicz, L. and H. Uzawa (1971), "On the integrability of demand functions," in: Chipman et al. (eds.) (1971), 114–148.
Hurwicz, L. (1971), "On the problem of integrability of demand functions", in: J. S. Chipman et al. (eds.) (1971), 174–214.
Kannai, Y. (1977), "Concavifiability and constructions of concave utility functions", Journal of Mathematical Economics, 4:1–56.
Karlin, S. (1959), Mathematical methods and theory in games, programming and economics, Vol. I. Reading: Addison-Wesley Publishing Co.
Katzner, D. W. (1968), "A note on the differentiability of consumer demand functions", Econometrica, 36:415–418.
Katzner, D. W. (1970), Static demand theory. London: Macmillan.
Katzner, D. W. (1971), "Demand and exchange analysis in the absence of integrability conditions", in: J. S. Chipman et al. (eds.) (1971), 254–270.
Klein, L. R. and H. Rubin (1947), "A constant utility index of the cost of living", Review of Economic Studies, 15:84–87.

Koopmans, T. (1972), "Representation of preference orderings with independent components of consumption", in: C. B. McGuire and R. Radner (eds.) Decision and organization. Amsterdam: North-Holland Publishing Co., 57–78.

Leontief, W. (1947), "Introduction to a theory of the internal structure of function relationship", Econometrica, 15:361–373. Reprinted in: K. J. Arrow (ed.) (1971), Selected readings in economic theory from Econometrica. Cambridge, Mass.: MIT Press, 179–191.

Marshall, A. (1895), Principles of economics. London: Macmillan.

Mas-Colell, A. (1975), "A model of equilibrium with differentiated commodities", Journal of Mathematical Economics, 2:263–295.

McKenzie, L. (1957), "Demand functions without a utility index", Review of Economic Studies, 25:185–189.

Pareto, V. (1896), Cours d'économie politique. Lausanne: Rouge.

Phlips, L. (1974), Applied consumption analysis. Amsterdam: North-Holland Publishing Co.

Powell, A. (1974), Empirical analysis of demand systems. Lexington: Heath Lexington.

Rader, T. (1963), "The existence of a utility function to represent preferences", Review of Economic Studies, 30:229–232.

Roy, R. (1942), De l'utilité, contribution à la théorie des choix. Paris: Hermann.

Samuelson, P. A. (1947a), Foundations of economic analysis. Cambridge, Mass.: Harvard University Press.

Samuelson, P. A. (1947b), "Some implications of linearity", Review of Economic Studies, 15:88–90.

Schmeidler, D. (1971), "A condition for the completeness of partial preference relations", Econometrica, 39:403–404.

Shafer, W. J. (1974), "The nontransitive consumer", Econometrica, 42:913–919.

Slutsky, E. (1915), "Sulla teoria del bilancio del consumatore", Giornale degli Economistri e Rivista di Statistica, 51:1–26. English translation: "On the theory of the budget of the consumer", Chapter 2 in: G. J. Stigler and K. E. Boulding (eds.) (1953), Readings in Price Theory. London: George Allen and Unwin, 27–56.

Sonnenschein, H. (1965), "The relationship between transitive preference and the structure of the choice space", Econometrica, 33:624–634.

Sonnenschein, H. (1971), "Demand theory without transitive preferences, with applications to the theory of competitive equilibrium", in: J. S. Chipman et al. (eds.) (1971), 215–223.

Sono, M. (1945), "The effect of price changes on the demand and supply of separable goods", Kokumni Keizai Zasski, 74:1–51 [in Japanese]. English translation: International Economic Review, 2 (1960):239–271.

Stigler, G. J. (1947), "Notes on the history of the Giffen paradox", Journal of Political Economy, 55:152–156.

Stone, J. R. N. (1954), "Linear expenditure system and demand analysis: An application to the pattern of British demand", Economic Journal, 64:511–527.

Theil, H. (1975), Theory and measurement of consumer demand, Vol. I. Amsterdam: North-Holland Publishing Co.

Theil, H. (1976), Theory and measurement of consumer demand, Vol. II. Amsterdam: North-Holland Publishing Co.

Wold, H. O. A. (1943), "A synthesis of pure demand analysis", Skandinavisk Aktuarietidskrift, 26:85–118, 220–263.

Wold, H. O. A. (1944), "A synthesis of pure demand analysis", Skandinavisk Aktuarietidskrift, 27:69–120.

Wold, H. O. A. and L. Juréen (1952), Demand analysis. New York: John Wiley & Sons, Inc.

Wold, H. O. A. (1952), "Ordinal preferences or cardinal utility", Econometrica, 20:661–664.

Chapter 10

PRODUCERS THEORY

M. ISHAQ NADIRI*

New York University

0. Introduction

Producers theory is concerned with the behavior of firms in hiring and combining productive inputs to supply commodities at appropriate prices. Two sets of issues are involved in this process: one is the technical constraints which limit the range of feasible productive processes, while the other is the institutional context, such as the characteristics of the market where commodities and inputs are purchased and sold. In recent years these two elements of producers theory have been refined extensively, extended in new directions, and applied to a variety of new subjects. Also, the underlying assumptions of the theory have been questioned, giving rise to a set of unresolved issues.

It is impossible to discuss all the important theoretical developments concerning technical constraints facing the firm and its decisions under different market environments. We shall deliberately avoid discussion of market structures; both space limitations and the fact that these issues are discussed in other chapters of this Handbook suggest our choice. In our discussion we adopt the assumption, albeit highly unrealistic, of "perfect competition" in both commodity and input markets. The main focus of discussion will be the description and analysis of technical constraints facing the firm, but some attention will also be given to dynamic input demand, output and price responses when regulatory constraints are present, and price and output decisions when the firm produces several products.

The chapter is organized into four sections. Section 1 describes the set of axiomatic approaches to production technology and the neoclassical theory of the multi-product and multi-input firm. This section is an attempt to set forth the general framework of analysis that underlies neoclassical producer decision theory. Section 2 specifies the properties of the production function, its various forms, and applications of the duality principles. Section 3 provides a brief

*I am indebted to Professors William Baumol, Rolf Färe, Dietrich Fischer, Ingmar Prucha, and Andrew Schotter for their helpful comments. I should like to thank Ms. Holly Ochoa and Ms. Irene Yew for their help in preparing the manuscript.

Handbook of Mathematical Economics, vol. II, edited by K.J. Arrow and M.D. Intriligator

discussion of the Cambridge–Cambridge controversy insofar as it pertains to the existence and usefulness of production functions. Section 4 deals with three special cases of the theory of the firm in which new research has appeared, specifically dynamic input disequilibrium models, response to regulatory constraints, and optimal price and output decisions in multi-product firms.

1. Production technology

1.1. Properties of the production technology set and production function

Production technology describes the technical constraints which limit the range of productive processes for an individual firm. A production technology consists of certain alternative methods of transforming materials and services to produce goods and services. The distinct goods and services used as inputs to the technology are called factors of production. To formalize the characteristics of production technology in the steady state,[1] let $y=(y_1, y_2,\dots, y_n)$ be the combination of n non-negative outputs produced using the services of various combination of m non-negative inputs, $x=(x_1, x_2,\dots, x_m)$. The production possibility set L is the set of all input vector x producing the output combination y and has the following regularity properties:

(i) L is a closed set, i.e. if $v^n \in L$, $n=1,2,\dots$, $\lim_{n\to\infty} v^n = v^0$ implies $v^0 \in t$;
(ii) L is a convex set, i.e. if $v' \in t$, $v'' \in L$, $0 \leqslant \lambda \leqslant 1$ implies $\lambda v' + (1-\lambda)v'' \in L$;
(iii) if $(y; x) \in L$, $x' \geqslant x$, then $(y; x') \in L$;
(iv) if $(y; x) \in L$, $0_n \leqslant y' \leqslant y$, then $(y'; x) \in L$, and
(v) for every finite input vector $x \geqslant 0_m$, the output set y is bounded from above.

Property (i) is a weak mathematical regularity condition that cannot be contradicted by empirical data; Property (ii) states that the production technology is subject to diminishing marginal rates of transmission of outputs for inputs, increasing marginal rates of substitution of outputs for outputs and diminishing marginal rates of substitution of inputs for inputs; properties (iii) and (iv) are the free disposal assumptions; property (iv) suggests that a bounded vector of inputs can produce only a bounded vector of outputs. It is possible that properties (ii) to (iv) be contradicted by a finite body of data while property (i) must be satisfied.

Not all technologically feasible transformations are of interest a priori; some may require greater inputs to yield technically efficient production subset E which is called a transformation function. The set efficient input–output combinations can be described symmetrically as a set of $(y; x)$ which satisfy the equation $E'(y; x)=0$, where E' is the transformation function. The efficient set for one

[1]For a formal development of the theory, see Shephard (1970), Diewert (1973), and Fuss and McFadden, Eds. (1980). The statements in this section are based on Diewert's work which can be consulted for proofs and further formal development of the relevant theory.

output, y_1, can be defined by $y_1 = E'$ $y_1 = E'(y_2, \ldots, y_n, x)$, i.e. the maximum production of y_1 given the vectors of inputs and other outputs. The unsymmetric transformation function, $y_1 = E(y_2, \ldots, y_n, x)$, may not be well defined for all non-negative vectors of other outputs and of inputs. For if the components of other outputs, $\hat{y} \equiv (y_2, y_3, \ldots, y_n)$, are chosen to be large while the components of x remain small, then it may be impossible to produce any non-negative amount of y_1. In this case, $E(\hat{y}, x) = -\infty$. The unsymmetric transformation function, E, is defined as

$$E(\hat{y}, x) \equiv \begin{cases} \max_{y_1}\{y_1 : (y_1, y, x) \in L\}, & \text{if there exist a } y_1 \text{ such that} \\ & \text{there exist } (y_1, \hat{y}; x) \in T, \\ -\infty, & \text{otherwise,} \end{cases} \tag{1.1}$$

for all $y \geqslant 0_{n-1}$, $x \geqslant 0_m$.

It follows that if L satisfies properties (i)–(v), the transformation function defined as above will satisfy the conditions:

> E is a continuous from above function; it is a concave function; it is non-decreasing in components for every $x > 0_m$, there exists a number α large enough such that if any components of vector $\hat{y}$ is greater than α, then $E(\hat{y}, x) = -\infty$.

From the general assumptions about the production possibility set which in turn determine the properties of the transformation functions, it is possible to generate different functional forms for the transformation and production functions to be used in empirical applications. Some of the specific functional forms for single and multiple outputs are described in Section 3.

1.2. Neoclassical theory of the firm

Given these characteristics of the technology set and the production function, we shall consider the equilibrium conditions for firm decisions with respect to production of commodities and employment of inputs. In particular, we consider two specific decisions: determination of the optimal output plan of the firm and behavior of its profit function – postponing discussion of some issues in the theory of the firm to Section 3.

1.2.1. Optimal output plan of the firm

Consider a firm that produces n products and employs m inputs; its objective is to maximize *profits* given as

$$\Pi = \sum_{i=1}^{m+n} p_i y_i = \sum_{i=1}^{n} p_i y_i + \sum_{i=n+1}^{m} p_i y_i, \tag{1.2}$$

where y_i $(i=1,\dots,n)$ are the outputs and p_i $(i=n+1,\dots,m)$ are output prices; $y_{n+i}=-x_i$ $(i=1,\dots,m)$ are the inputs, and p_{n+i} $(i=1,\dots,m)$ are the input prices. Profit, Π, is maximized subject to the production function

$$f(y_1, y_2,\dots,y_n, y_{n+1},\dots,y_{m+n})\leqslant 0. \tag{1.3}$$

$y_1, y_2,\dots,y_{m+n}$ are often called *net outputs*: they have positive signs for outputs and negative signs for inputs.

Assuming that the production function $f(\cdot)$ is: (a) twice differentiable, i.e. $\partial f/\partial y_i = f_i$ and $\partial^2 f/\partial y_i \partial y_j = f_{ij}$, $(i, j=1,\dots,m+n)$, exist; (b) increasing in the *net* outputs, i.e. the derivatives f_i are always positive; and (c) convex (subject to the condition $f(\cdot)=0$, the function is *strictly* convex); the optimal production plan of the firm can be stated using the familiar Lagrangian function:

$$\begin{aligned} L(y_1, y_2,\dots,y_{m+n};\lambda) &= \Pi + \lambda f(y_1, y_2,\dots, y_{m+n}) \\ &= \sum_{i=1}^{m+n} p_i y_i + \lambda f(y_1, y_2,\dots,y_{m+n}), \end{aligned} \tag{1.4}$$

where λ is a Lagrange multiplier associated with the constraint $f(\cdot)\leqslant 0$. By the Kuhn–Tucker theorem of concave programming, an optimal production plan $\{\bar{y}_i\}$ corresponds to a saddlepoint of the Lagrangian

$$L(y_1,\dots, y_{m+n}, \bar{\lambda})=L(\bar{y}_1,\dots, \bar{y}_{m+n}, \bar{\lambda}) \tag{1.5}$$

for non-negative $\{y_i\}$, $(i=1,\dots,n)$, and non-positive $\{y_i, \lambda\}$, $(i=n+1,\dots,m+n)$. Necessary and sufficient conditions for a saddlepoint are:

$$\begin{aligned} &\text{(i)} && \frac{\partial L}{\partial y_i}\leqslant 0 \qquad (i=1,\dots, n), \\ &\text{(ii)} && \frac{\partial L}{\partial y_i}\geqslant 0 \qquad (i=n+1,\dots, m+n), \\ &\text{(iii)} && \sum_{i=1}^{m+n} \frac{\partial L}{\partial y_i}\cdot \bar{y}_i = 0, \\ &\text{(iv)} && \frac{\partial L}{\partial \lambda}\leqslant 0, \\ & && \text{and} \\ &\text{(v)} && \frac{\partial L}{\partial \lambda}\cdot\bar{\lambda}=0, \end{aligned} \tag{1.6}$$

where the derivatives $\{\partial L/\partial y_i, \partial L/\partial \lambda\}$ are evaluated at $\{\bar{y}_i, \lambda\}$.[2] These conditions give a complete characterization of the optimal production plan. We may summarize this characterization in the following proposition:

Under the assumptions mentioned, an optimal production plan $\bar{y}_1, \bar{y}_2, \dots, \bar{y}_{m+n}$ satisfies:

$$\begin{aligned} &p_i + \bar{\lambda} f_i = 0 \qquad (i=1,\dots,n), \\ &p_i + \bar{\lambda} f_i = 0 \qquad (i=n+1,\dots,m+n), \\ &\sum_{i=1}^{m+n} (p_i + \bar{\lambda} f_i)\bar{y}_i = 0. \end{aligned} \tag{1.7}$$

Furthermore,

$$f(\bar{y}_1, \bar{y}_2, \dots, \bar{y}_{m+n}) = 0,$$
$$f \cdot \bar{\lambda} = 0.$$

Since, by assumption, f is increasing, positive prices for all net outputs imply that $\bar{\lambda} < 0$ so that $f = 0$. If the production function is strictly convex, then these conditions are

$$\begin{aligned} &\frac{\partial L}{\partial \lambda} = f(\bar{y}_1, \dots, \bar{y}_{m+n}) = 0, \\ &\frac{\partial L}{\partial y_i} = p_i + \bar{\lambda} f_i = 0 \qquad (i=1,\dots,m+n), \end{aligned} \tag{1.8}$$

where the derivatives $\{f_i\}$ are evaluated at $\{\bar{y}_i, \lambda\}$. When such a solution exists, it is unique by the assumption of strict convexity of the function $f(y_1, y_2, \dots, y_{m+n}) = 0$.

The $m+n$ first-order conditions can be interpreted as equality between the marginal profitability of each net output and its revenue or cost. The Lagrange multiplier is the change in profit made by the firm with respect to a change in its production plan. Taking the ratio of any pair of these equalities, we obtain:

$$\frac{p_i}{p_j} = \frac{\partial f/\partial y_i}{\partial f/\partial y_j} = -\frac{\partial y_j}{\partial y_i} \qquad (i, j=1,\dots,m+n). \tag{1.9}$$

[2]From these conditions it follows that $\partial L/\partial y_i < 0$ only if $\bar{y}_i = 0$ $(i=1,\dots,n)$, and $\partial L/\partial y_i > 0$ only if $\bar{y}_i = 0$ $(i=n+1,\dots,m+n)$. Furthermore, if $\bar{y}_i > 0$, then $\partial L/\partial y_i = 0$ $(i=1,\dots,n)$, and if $\bar{y}_i < 0$, then $\partial L/\partial y_i = 0$ $(i=n+1,\dots,m+n)$. Finally, $\partial L/\partial \lambda < 0$ only if $\bar{\lambda} = 0$, and if $\bar{\lambda} < 0$, $\partial L/\partial \lambda = 0$. For further discussion of firm production decisions, see Malinvaud (1972).

For pairs of inputs and outputs, these equalities state that the marginal rate of transformation between the corresponding commodities is equal to the ratio of the prices of these commodities. For one input, we obtain the familiar condition that its marginal value product marginal is equal to its cost. An *expansion path* which is the locus of input combinations at different levels of output, where the marginal rate of technical substitution (MRTS) equals the input price ratio, can also be easily traced. It follows that the profit-maximizing output and input levels, $\bar{y}_i$ $(i=1,\dots,m+n)$, and the Lagrange multiplier, $\bar{\lambda}$, are functions of the prices, p_i $(i=1,\dots,m+n)$. That is,

$$\begin{aligned} &\bar{\lambda}=\bar{\lambda}(p_1,\dots,p_{m+n}) \\ &\bar{y}_i=s^i(p_1,\dots,p_{m+n}) \end{aligned} \qquad (i=1,\dots,m+n). \tag{1.10}$$

$\bar{\lambda}(\cdot)$ is homogeneous of degree one, while $s^i(\cdot)$ is homogeneous of degree zero. $\bar{y}_i$ are the *net supply functions*. For outputs, the equations $\bar{y}_i$ are the usual supply functions; for inputs, they are the negative of the demand functions. Thus, net supply functions exist provided that the marginal profitability conditions and the production function are satisfied, and provided that the production function is convex.

1.2.2. *The profit function*

The *profit function* gives the value of the objective function $\Pi=\sum_{i=1}^{m+n}p_i y_i$, subject to the equation $f(y_1,\dots,y_{m+n})=0$ and the condition that outputs and inputs are non-negative.

Substituting the net supply functions (1.10) into the objective function, we obtain

$$\Pi(p_1,\dots,p_{m+n})=\sum_{i=1}^{m+n}p_i s^i(p_1,\dots,p_{m+n}). \tag{1.11}$$

Differentiating the profit function with respect to p_i,

$$\frac{\partial \Pi}{\partial p_i}=s^i+\sum_{j=1}^{m+n}p_j\frac{\partial s^j}{\partial p_i} \tag{1.12}$$

but

$$f(s^1,\dots,s^{m+n})=0$$

so that

$$-\sum_{j=1}^{m+n}\frac{\partial f}{\partial y_j}\frac{\partial s^j}{\partial p_i}=\frac{1}{\lambda}\sum_{j=1}^{m+n}p_j\frac{\partial s^j}{\partial p_i}. \tag{1.13}$$

We therefore conclude that

$$\frac{\partial \Pi}{\partial p_i}=s^i \qquad (i=1,\dots,m+n). \tag{1.14}$$

The net supply functions are homogeneous of degree zero, so the profit function is homogeneous of degree one. Taking the second derivative of Π with respect to p_i,

$$\frac{\partial^2 \Pi}{\partial p_i \partial p_j}=\frac{\partial s^i}{\partial p_j}=\frac{\partial s^j}{\partial p_i}=\frac{\partial^2 \Pi}{\partial p_j \partial p_i}, \tag{1.15}$$

implying that the cross-effects of prices on net supply are symmetric.

The profit function can be characterized as being twice differentiable, increasing with increases in the price of output and decreasing when the prices of inputs increase; an optimal solution exists with output and input levels strictly positive for all positive price levels. The profit function is convex and, given the value of any argument of the function, it is strictly convex.

Using the conditions for optimal output and the profit function, it is possible to derive the law of supply and demand for the firm, which by definition is equal to the supply of an output and demand (with a change of sign) for an input. Specific examples will be provided in Section 3, where specific forms of the cost and profit functions are discussed. Suffice it to state here the following four properties of the net supply function:

(i) The net supply function is homogeneous of degree zero with respect to $p_1, p_2, \dots, p_{n+m}$ and for any multiplication of these prices by a constant.
(ii) The "substitution effect" of i and j is symmetric, i.e. $\partial s^i/\partial q_j=\partial s^j/\partial q_i$, where $\partial s^i/\partial q_j$ and $\partial s^j/\partial q_i$ are partial derivatives.
(iii) The net supply of good i cannot diminish when its price increases.
(iv) When the quantity of one of the inputs or outputs is fixed, then net supply function is homogeneous of degree zero in all prices, and in the special case in which all outputs are fixed, the net supply functions are negative demand functions for each of the inputs.

1.3. Scale, factor substitutions, and technical change: Some definitions

Production functions incorporate several economic effects, such as scale and substitution effects, the effects of technological change, and distributive effects. These effects are, in general, expressed in terms of the production function and its first and second derivatives. Table 1.1 provides a brief and non-exhaustive list of these effects [Fuss and McFadden (1980)].

The economic effects summarized in Table 1.1 characterize the usual comparative static properties of a production function at a given point. Given the production function $y=f(x,t)$, where x is a vector of inputs and t the index of technological change, it is possible to deduce expressions for returns to scale, shares of factors of production, price elasticity, and elasticity of substitution, as well as various indices of disembodied technical change. Other effects, such as

Table 1.1
A partial list of economic effects related to the production function.

	Economic effect	Formula	No. of distinct effects
1.	Output level	$y=f(x,t)$	1
2.	Returns to scale	$\mu=\left(\sum_{i=1}^{n} x_i f_i\right)\Big/ f$	1
3.	Distributive share	$s_i=x_i f_i \Big/ \sum_{i=1}^{n} x_i f_i$	$n-1$
4.	Own "price" elasticity	$\varepsilon_i=x_i f_{ii}/f_i$	n
5.	Elasticity of substitution	$\sigma_{ij}=\dfrac{\dfrac{f_{ii}}{f_i^2}+\dfrac{f_{ij}}{f_i f_j}-\dfrac{f_{jj}}{f_j^2}}{\dfrac{1}{x_i f_i}+\dfrac{1}{x_j f_j}}$	$\dfrac{n(n-1)}{2}$
6.	Disembodied technological change:		
	(a) Rate of technical change	$T=f_t/f$	1
	(b) Acceleration of technical change	$\dot{T}=(f_{tt}/f)-(f_t/f)^2$	1
	(c) Rate of change of marginal products	$\dot{m}_i/m_i=f_{it}/f_i$	n

Source: Adopted from Fuss and McFadden (Eds.) (1980, p. 231).

indices of embodied technical change, can also be derived. By imposing specific restrictions across these effects, different functional forms of the production function are obtained. Of this array of economic effects, those associated with returns to scale, degree of substitution among inputs, and the type and nature of technological change require brief consideration.

1.3.1. Returns to scale or scale function

Let $f=f(\lambda x_1,\dots,\lambda x_n)$, where λ is a positive constant. The returns to scale at a point $(x_1,\dots,x_n)$ is measured by the *scale function*, which is also referred to as the *function coefficient*, *elasticity of output*, and *passus coefficient* in the literature. It is defined as:

$$\mu(x)=\lim_{\lambda\to 1}\frac{\lambda}{f(\lambda x)}\frac{\partial f(\lambda x)}{\partial\lambda}=\lim_{\lambda\to 1}\frac{\partial\ln f(\lambda x)}{\partial\ln\lambda}$$

or

$$\mu(\lambda)=\sum_{i=1}^{n} x_i f_i/f. \tag{1.16}$$

Returns to scale are then decreasing, constant, or increasing accordingly as $\mu(x_1,\dots,x_n)$ is less than, equal to, or greater than unity. Note that the following relationships also hold:

(i) *Returns to scale*. The *marginal product* of input x_i is the first partial derivative of the production function with respect to x_i, $i=1,\dots,n$: $\partial f(x)/\partial x_i=MP_i(x)\geqslant 0$. $MP(x)$ is a row vector of marginal products of x_i, i.e. $MP(x)=(MP_1(x),\dots,MP_n(x))$. The returns to scale can be defined as

$$\mu(x)=\frac{MP(x)}{f(x)}\cdot x;\qquad x=(x_1,\dots,x_n). \tag{1.17}$$

(ii) *Returns to jth-input*:

$$\mu_j(x)=\frac{x_j}{f(x)}\frac{\partial f_j(x)}{\partial x_j}=f_j(x)\cdot\frac{1}{f(x)/x_j}$$

$$=\frac{MP_j(x)}{AP_j(x)};\qquad j=1,2,\dots,n, \tag{1.18}$$

where AP_j is the *average product* of factor j, defined as $f(x)/x_j$.

(iii) *Elasticity of average product*:

$$\frac{\partial(f/x_j)}{\partial x_j}\frac{x_j}{(f/x_j)}=\frac{f_j}{f/x_j}-1=\mu_j-1,$$

which suggest that there exist increasing or decreasing average returns at a given point according to whether μ_j is greater or is less than unity, respectively.

(iv) The returns to scale at any point,

$$\mu(x)=\sum_{j=1}^{n}\mu_j(x);\qquad j=1,2,\ldots,n, \tag{1.19}$$

is the sum of all elasticities of output with respect to all the inputs. Also, the returns to scale may be constant for certain types of production processes and vary for others.

Returns to scale for multi-product production functions can be stated as follows. Suppose $F(y_1,\ldots,y_m,x_1,\ldots,x_n)=F(\gamma y_1^0,\ldots,\gamma y_m^0,\lambda x_1^0,\ldots,\lambda x_n^0)$, where γ and λ are constants, y's are outputs and x's are inputs. If the multi-product production function is separable such that $h(y_1,\ldots,y_m)-f(x_1,\ldots,x_n)=0$, then the scale function of $F=h-f$ can be defined as:

$$\mu^F=\frac{\mathrm{d}\gamma/\gamma}{\mathrm{d}\lambda/\lambda}=\frac{\text{proportional relative change in all outputs}}{\text{proportional relative change in all inputs}}$$

or

$$\mu^F=\mu^f/\mu^{-h}, \tag{1.20}$$

where μ^f and μ^h are the scale functions of f and h, respectively.[3] It is difficult to

[3]If F is once differentiable, an operational definition of μ^F can be obtained by using the total differential of F,

$$\mathrm{d}F=\sum_{k=1}^{m}\frac{\partial F}{\partial y_k}\mathrm{d}y_k+\sum_{i=1}^{n}\frac{\partial F}{\partial x_i}\mathrm{d}x_i=0,$$

whence

$$\sum_{k=1}^{m}\frac{\partial F}{\partial y_k}y_k\left(\frac{\mathrm{d}y_k}{y_k}\right)=-\sum_{i=1}^{n}\frac{\partial F}{\partial x_i}x_i\left(\frac{\mathrm{d}x_i}{x_i}\right).$$

give economic meaning to μ^F if the number of outputs exceed one. In cases of multi-output production functions, the scale function μ^F, hence the measurement of returns to scale, depends on the output mix.[4] Unlike the simple output functions, return to scale cannot be indexed in a multi-product technology except under special conditions, e.g. in the trivial circumstances in which f and h are both homogeneous.

1.3.2. *Elasticity of substitution*

Substitution possibilities characterize the production function by alternative combinations of inputs which generate the same level of output. A local measure of substitution between two inputs, say x_1 and x_2, when all other inputs and the level of output are held constant, is the *elasticity of substitution*, σ, defined at a point as:

$$\sigma = \frac{\mathrm{d}\log(x_2/x_1)}{\mathrm{d}\log(f_1/f_2)} \tag{1.21a}$$

Because by assumption $\mathrm{d}y_k/y_k = \mathrm{d}\gamma/\gamma$ for each k and $\mathrm{d}x_i/x_i = \mathrm{d}\lambda/\lambda$, it follows that

$$\mu^F = \frac{-\sum_1^n \left(\frac{\partial F}{\partial x_i}\right) x_i}{\sum_1^m \left(\frac{\partial F}{\partial y_k}\right) y_k}.$$

Divide the numerator of the right-hand side of this expression by f and the denominator by $h(=-f)$. Then, $\mu^F = \mu^f/\mu^{-h}$.

[4]For example, consider the multi-product function

$$y_1 y_2^{\delta} - a x_1^{\alpha_1} x_2^{\alpha_2} x_3^{\alpha_3} = 0,$$

where $h(y_1, y_2) = y_1 y_2^{\delta}$ and $f(x_1, x_2, x_3) = a x_1^{\alpha_1} x_2^{\alpha_2} x_3^{\alpha_3}$. The scale function of this production function is

$$\mu^F = (\alpha_1 + \alpha_2 + \alpha_3)/(1+\delta), \qquad \delta \neq -1.$$

That is, if each input is doubled, and each output thereby increased proportionately, the change in each output is not by the factor $2^{\alpha_1+\alpha_2+\alpha_3}$, but by the factor $2^{(\alpha_1+\alpha_2+\alpha_3)/(1+\delta)}$. Thus, if $\delta < 0$, some technological interdependence between the two outputs leads to an increase in both outputs larger than what is implied by the input portion of the relationship. If $\delta > 0$, the reverse is true. But unless one output is in some way an input in the production of the other, it is difficult to rationalize the output interactions implied by $\delta \neq 0$. See Klein (1953).

or

$$\sigma=\frac{-f_1f_2(x_1f_1+x_2f_2)}{x_1x_2(f_{11}f_2^2-2f_{12}f_1f_2+f_{22}f_1^2)}. \tag{1.21b}$$

In other words, for a given y^0, σ is the proportional change in the factor ratio x_2/x_1 resulting from a proportional change in the marginal rate of substitution of x_1 for x_2. If factor prices do not vary with quantities demanded and the firm minimizes costs, there is a simpler interpretation of σ, i.e. as the percentage change in ratio of inputs divided by the percentage change in the marginal products or the percentage change in the relative prices of inputs.

The curvature of the isoquant is determined by the magnitude of the elasticity of substitution. The *isoquant* representing the set of inputs generating the level of output is defined as:

$$\{x\in D|\,f(x)=y^0\} \tag{1.22}$$

provided that $f(x)$ is strictly increasing along rays – i.e. $f(\lambda\cdot x)>f(x)$ if $\lambda>1$ – and continuous. Along an isoquant, by definition,

$$\sum_{j=1}^{n}\frac{\partial f(x)}{\partial x_j}\mathrm{d}x_j=0. \tag{1.23}$$

Since

$$\mathrm{d}x=(\mathrm{d}x_1,\mathrm{d}x_2,\dots,\mathrm{d}x_n);\qquad MP(x)\mathrm{d}x=0$$

which, in the case of two factors, x_1 and x_2 (assuming all other inputs are held constant), can be written as

$$MP_1(x)\mathrm{d}x_1+MP_2(x)\mathrm{d}x_2=0$$

The slope along the isoquant

$$\left.\frac{\mathrm{d}x_2}{\mathrm{d}x_1}\right|_{\text{isoquant}}=-\frac{MP_2(x)}{MP_1(x)}=MRS_{12}, \tag{1.24}$$

where MRS_{12} is the marginal rate of substitution of one unit of x_1 for a number of units of x_2 holding the level of output constant.

The form of a production function necessarily imposes restrictions on factor substitutability. For example, for the fixed proportion production function, $\sigma=0$;

i.e. it is impossible to substitute one factor for another to produce any given output. For the linear production function, $y=\alpha_1 x_1+\alpha_2 x_2$, $\alpha_i>0$, $\sigma=\infty$, so substitution among inputs is perfect. A value of σ in the interval $(0,+\infty)$ indicates a production technology allowing for some but not perfect substitutability among factors of production.

For general n-factor production functions, it is difficult to state a unique definition of the elasticity of substitution. In such cases only the concept of partial *elasticities of substitution* among the inputs can be defined. Several definitions have been proposed, of which the two best known are as follows:

(1) *The Direct Elasticity of Substitution* (*DES*) between factors i and j is a generalization of the two-factor elasticity of substitution (1.31b), and has a similar interpretation [McFadden (1963)]. It is

$$\sigma_{ij}^{\mathrm{D}}=\frac{f_i f_j(x_i f_i+x_j f_j)}{x_i x_j(f_{ii} f_j^2-2f_{ij} f_i f_j+f_{jj} f_i^2)}, \tag{1.25}$$

$$0<\sigma_{ij}^{\mathrm{D}}<\infty.$$

(2) *The Allen Partial Elasticity of Substitution* (*AES*) is a measure of change in the firm's demand for factor j given a change in the price of factor i, all other factor prices remaining constant; that is [Allen (1938)],

$$\sigma_{ij}^{\mathrm{A}}=\frac{\sum_{k=1}^{n} x_k f_k}{x_i x_j}\cdot\frac{F_{ij}}{F}; \qquad -\infty<\sigma_{ij}^{\mathrm{A}}<\infty, \tag{1.26}$$

where

$$F=\begin{vmatrix} 0 & f_1 & \cdots & f_n \\ f_1 & f_{11} & \cdots & f_{1n} \\ \vdots & \vdots & & \vdots \\ f_n & f_{1n} & \cdots & f_{nn} \end{vmatrix}$$

and F_{ij} is the co-factor of f_{ij} in F. Stability conditions require that $F_{nn}/F<0$. Since F is symmetric, the matrix of partial elasticities is also. Some of these partial elasticities can be negative, but some must be positive. When these partial elasticities are weighted, the positive elasticities together must more than counterbalance the negative ones if stability conditions are met.

Unlike the DES approach, AES possesses no straightforward interpretation. It can be shown, however, that $\sigma_{ij}^{A}=\eta_{ij}/(p_i x_i/C)$, where η_{ij} is the partial elasticity demand for on factor j with respect to i (price of input x_i); $p_i x_i$ is expenditure on factor i; and C is total cost. σ_{ij}^{D} and σ_{ij}^{A} are equal to the two-factor elasticity σ, when $\eta=2$, but in general they are not identical. The definition of DES and AES can also be used to measure the direct and partial elasticities of transformation between different products.

1.3.3. Technical progress

Technical progress deals with the process and consequences of shifts in the production function due to the adoption of new techniques. When new techniques are adopted, they either have a neutral effect on the production process or change the input–output relationships. Neutrality of technical progress can be measured by its effect on *certain* economic variables such as capital–output, output–labor, and capital–labor ratios, which should remain invariant under technical change.[5]

Several definitions of technical progress have been proposed, such as (a) product-augmenting, (b) labor or capital-augmenting, (c) input-decreasing and factor-augmenting, amongst others [Beckmann, Sato, and Schupack (1972)]. However, the most familiar definitions are the Hicks, Harrod, and Solow forms of technical progress. In Hicksian definition, changes in relative shares of the inputs is used as a measure of technical bias. The Harrod definition measures the bias along a constant capital–output ratio, and Solow's definition measures the bias along a constant labor–output ratio. Symbolically,

$$\left(\frac{\partial(f_K K)/(f_L L)}{\partial t}\right)_{\substack{K/L\\ \text{constant}}} \gtreqless 0, \quad \text{Hicks}\begin{cases}\text{labor-saving}\\ \text{neutral}\\ \text{capital-saving}\end{cases}$$

$$\left(\frac{\partial(f_K K)/f_L L)}{\partial t}\right)_{\substack{K/Y\\ \text{constant}}} \gtreqless 0, \quad \text{Harrod}\begin{cases}\text{labor-saving}\\ \text{neutral}\\ \text{capital-saving}\end{cases}$$

$$\left(\frac{\partial(f_K K)/(f_L L)}{\partial t}\right)_{\substack{L/Y\\ \text{constant}}} \gtreqless 0, \quad \text{Solow}\begin{cases}\text{labor-saving}\\ \text{neutral}\\ \text{capital-saving}\end{cases}$$

[5]By defining the relationships between these variables, say relative prices and capital–labor ratio, or wage rate and labor–output ratios, we would define different types of technical progress and also obtain the implicit production function relation which would permit such technical change to occur. See Beckmann, Sato, and Schupack (1972) for further development of these points.

If technical change is "embodied" in capital and labor, the bias in technical change will depend upon the elasticity of substitution, σ, and the differential rates of growth of labor and capital embodiment. "Embodiment" means that, because of technological advance, the new inputs are more efficient than the old inputs. Suppose a two-factor production function is written as

$$y=f(\alpha_1 L, \alpha_2 K), \tag{1.27}$$

where α_1 and α_2 are the coefficients of factor augmentation. Since the direction of technical change depends upon the ratio α_1/α_2, technical change is Hicks-neutral if the ratio is constant, Harrod-neutral (labor-augmenting) if α_2 is constant, and Solow-neutral (capital-augmenting) if α_1 is constant. The *bias* in technical change, B, can be defined as

$$B=\left[\frac{\mathrm{d}\alpha_1}{\alpha_1}-\frac{\mathrm{d}\alpha_2}{\alpha_2}\right]\left(1-\frac{1}{\sigma}\right). \tag{1.28}$$

The embodiment effect should be clearly distinguished from the augmentation effect and quality correction of the inputs. The augmentation effect means that the productivity increase of an input due to technical advances is expressed as equivalent to a specific increase in its quantity. Embodiment of technical change in capital, for example, could perfectly well produce purely labor-augmenting (but nonetheless capital embodied) technical change. Nor should all quality improvement in an input be considered equivalent to the embodiment effect. The latter refers only to quality improvement associated with vintage of capital or cohort of labor. For example, productivity increases due to sex and race characteristics (at a point in time) are *not* part of the embodiment effect, while improvements due to age and education *are* part of it.

Technological change can be *endogenous*, i.e. the firm can induce innovations by undertaking new research activities based on their payoffs in terms of new outputs or new processes. Sustained changes in relative prices and income may also lead to technological innovations. Most induced technical change occurs through the types of research activities undertaken. The productivity of different research activities, the scale of the research effort undertaken, and the rise in real income and changes in relative input prices may alter the magnitudes and direction of inventions leading to changes in demand for different factors of production.

2. Properties of the production function and duality

There are several properties of production technology which are postulated to characterize the structure of the production process. Among the most relevant of these are the concepts of homogeneity, additivity, and separability. Another important concept is the duality between production functions, normalized profit functions, or cost functions. Under the general conditions stated in Section 1, the underlying production technology can be identified employing the cost or profit functions. The duality principles are discussed here only briefly. A formal discussion of them can be found in Chapter 12 by W. E. Diewert. First, we shall discuss the homogeneity, additivity, and separability properties of the production technology.

2.1. *Homogeneity, additivity, and separability*

2.1.1. *Homogeneity and homotheticity of production functions*

A *homogeneous production function of degree k* is defined as:

$$f(\lambda x_1, \lambda x_2, \dots, \lambda x_n) = \lambda^k f(x_1, x_2, \dots, x_n); \qquad \lambda > 0. \tag{2.1}$$

A differentiable, linear homogeneous production function has the following properties:

(i) $y = \sum_i^n x_i f_i$, where, by Euler's Theorem, $f_i = \partial f / \partial x_i$.
(ii) The first derivative of a function homogeneous of degree one is homogeneous of degree zero.
(iii) $f_{ii} = (x_j^2 / x_i^2) f_{jj}$, where $f_{ii} = \partial^2 f / \partial x_i^2$ and $f_{jj} = \partial^2 f / \partial x_j^2$.
(iv) $x_i f_{ii} + x_j f_{ij} = 0$ and $x_j f_{jj} + x_i f_{ij} = 0$, where $f_{ij} = \partial^2 f / \partial x_i \partial x_j$.

From properties (iii) and (iv), it follows that:

(v) $f_{ii} - f_{jj} = f_{ij} = 0$, when one of the three is zero and there is a positive level of inputs.
(vi) $x_i^2 f_i^2 + 2 x_i x_j f_i f_j + x_j^2 f_j^2 = y^2$.

A monotonically increasing transformation of a homogeneous production function yields a *homothetic production function*. The family of homothetic production functions are characterized by straight line expansion paths through the origin, i.e. the contours of the production surface are radial blow-ups of one another. A homothetic production function is defined as

$$y = g(f(x_1, x_2, \dots, x_n)), \tag{2.2}$$

where f is a homogeneous function of some arbitrary degree ($f > 0$); f and g are twice differentiable functions and f can be interpreted as the input index of the factors of production.

2.1.2. *Additivity and homogeneity of production functions*[6]

The production possibility frontier is homogeneous and commodity-wise additive if and only if the frontier can be represented in the form:

$$\begin{aligned}&F^1(\lambda y_1)+F^2(\lambda y_2)+\cdots+F^n(\lambda y_n)\\&=F^1(y_1)+F^2(y_2)+\cdots+F^n(y_n)\\&=0,\end{aligned}\tag{2.3}$$

for any $\lambda>0$, y_i $(i=1,2,\dots,n)$ represent net output of the ith commodity, some of which are inputs to the production process. This condition can be satisfied if and only if

(a) the functions $[F^i]$ are homogeneous of the same degree, or
(b) the functions $[F^i]$ are logarithmic.

If each net output y_i $(i=1,2,\dots,n)$ consists of m mutually exclusive and exhaustive commodity groups, for example, $F(y_1,y_2,\dots,y_n)=F^1(y_1,y_2,\dots,y_n)+F^2(y_{n_1+1},\dots,y_{n_1+n_2})+\cdots+F^m(y_{n_1+n_2+\cdots+n_{m-1}+1},\dots,y_{n_1+n_2+\cdots+n_m})=0$, then the production frontier is called *group-wise additive*. The conditions (a) and (b) are still applicable when the frontier is group-wise additive. Some examples of additive homogeneous production possibility frontiers are the linear homogeneous and two-level constant elasticity of substitution production functions described briefly in Section 3.

2.1.3. *Functional separability of production functions*

If production functions of several arguments can be separated into sub-functions, then efficiency in production can be achieved by sequential optimization. Functional separability plays an important role in aggregating heterogeneous inputs and outputs, deriving value-added functions, and estimating production functions; it also opens up the possibility of consistent multi-stage estimation, which may be the only feasible procedure when large numbers of inputs and outputs are involved in the production activities of highly complex organizations.

Suppose the production function $y=f(x)=f(x_1,\dots,x_n)$ is twice differentiable and strictly quasi-concave, with n inputs denoted $N=(1,\dots,n)$, and is partitioned into r mutually exclusive and exhaustive subsets $(N_1,\dots,N_r)$, a partition called R. The $f(x)$ is *weakly separable* with respect to partition R when the marginal rate of substitution (MRS) between any two inputs x_i and x_j from any subset N_s, $s=1,\dots,r$, is independent of the quantities outside N_s. The weak separability condition is stated as [Leontief (1947), Green (1964), and Berndt and Christensen

[6]For definitions of these concepts, see Christensen, Jorgenson, and Lau (1973).

(1973)]:

$$\frac{\partial(f_i/f_j)}{\partial x_k}=0, \quad \text{for all } i, j\in N_s \text{ and } k\notin N_s. \tag{2.4}$$

Strong separability, on the other hand, exists when the MRS between any two inputs inside N_s and N_t does not depend on the quantities of inputs outside N_s and N_t, i.e.

$$\frac{\partial(f_i/f_j)}{\partial x_k}=0, \quad \text{for all } i\in N_s, j\in N_t, k\notin N_s\cup N_t \tag{2.5}$$

or, alternatively, $f_j f_{ik}-f_i f_{jk}=0$. *Weak separability* with respect to partition R is necessary and sufficient for the production function $f(x)$ to take the form $f(x^1, x^2,\dots, x^r)$, where x^s is a function of elements of N_s only. With *strong separability* it is necessary and sufficient for $f(x)$ to have the form $f(x^1+x^2+\cdots+x^r)$, where x^s is a function of elements of N_s only.

A quasi-concave homothetic production function $f(x)$ is weakly separable with respect to the partition R *at a point* if and only if the Allen elasticities of substitution among pairs of inputs are equal, i.e. $AES_{ik}=AES_{jk}$ at that point for all $i\neq j\in N_r$, $k\notin N_r$. The function is strongly separable if and only if $AES_{ik}=AES_{jk}$ for all $i\in N_s, j\in N_t, k\notin N_r\cup N_t$. If $n=R$, then all AES, $i\neq k$, are equal.[7]

These results can be extended to cost and profit functions, as well as to the case of non-homothetic production functions.[8] For example, if $f(x)$ is homothetic and separable, then the dual cost function $C(y, p)=H(y)\cdot G(p)$, where p is the vector of input prices, is weakly separable. That is to say, $C_jC_{ik}-C_iC_{jk}=0$ holds as well as $f_j f_{ik}-f_i f_{jk}=0$.

These separability theorems, if they hold, provide the conditions for existence of a consistent aggregate price index p^s and consistent quantity index X^s on the elements of N_s. Most theoretical and empirical formulations of production functions in the literature implicitly assume that separability conditions do prevail.

2.2. *Duality of production technology and cost and profit functions*

An alternative approach to production theory is to use cost and profit functions in order to specify the underlying production technology. The theory of duality ensures consistency between the cost or profit functions and production technol-

[7]See Fuss and McFadden, Eds. (1980, ch. II.1, pp. 244–248), for derivation of these concepts.

[8]For application of separability theorems to production functions with linear and non-linear parameters, see Fuss and McFadden, Eds. (1980).

ogy. The advantages of using cost and profit functions to derive the characteristics of the underlying technology are that they are computationally simple and, by utilizing economic observables, they permit testing of wider classes of hypotheses. To establish the duality between cost or profit functions and the production technology, several mathematical concepts and propositions are employed. A few of these are described below without making an attempt to provide the necessary proofs.[9]

2.2.1 *Some definitions and properties of the cost function*

(1) An *input-regular production possibility set* is defined if the set of produceable outputs y^* is non-empty and for each $y \in y^*$ the input requirement L is non-empty and closed.
(2) A *conventional production possibility set* is defined if there is free disposal of inputs [property (b) in Section 1.1], and the production sets are convex from below [property (h) in Section 1.1].

If the firm faces competitive markets with strictly positive input prices, $p = (p_1, \ldots, p_n)$, and an input-regular production possibility set, its *cost function* can be defined as

$$C(y, p) = \min\{p \cdot x \mid x \in L(y)\}. \tag{2.6}$$

The cost function $C(y, p)$ denotes the minimum total cost for input vectors to attain at least the output vector y. The function is non-negative since x and p are non-negative, and non-decreasing in input prices for a fixed output. The function is positively linear homogeneous in p and is also concave in input prices. The concavity property leads to the result that the cost function is continuous in input prices, i.e. for $p > 0$ and for all $y \geqslant 0$.

Duality theory considers the proposition that if the cost function (2.6) exists, there exists an input-regular production possibility set such that (2.6) is the minimum cost function. Application of the duality theorems yields the following results:

(i) Each input-conventional cost function determines an implicit production possibility set which is input-conventional (i.e. is input-regular and satisfies the free disposability and convexity conditions).
(ii) There is a one-to-one relationship between input-conventional production possibility sets and input-conventional cost structures.[10]

[9] Interested readers should consult the relevant chapters of Shephard (1970), and Fuss and McFadden, Eds. (1980). See also Chapter 12 of this *Handbook* by Diewert.

[10] This result is often referred to as the Shephard–Uzawa duality theorem. See Shephard (1970) and Uzawa (1964).

2.2.2. Distance and cost functions

A *distance function* is a form of transformation function often used to characterize production possibility sets. Using distance functions permits the establishment of a full and formal mathematical duality between transformation and cost functions, since both are drawn from the same class of functions and have the same properties. This in turn permits the establishment of the proposition that if property A on the transformation function implies property B on the cost function, then, by duality, property A on a cost function implies property B on a transformation function.

Consider the distance function

$$F(y,x)=\max\{\lambda>0|(1/\lambda)x\in L(y)\}, \tag{2.7}$$

which can be shown to be well defined. For $y=0$, the vector 0 may be in $L(0)$, in which case $F(y,x)$ is defined to take the value $+\infty$. By using this definition and properties of the distance function, Shephard has established the relation between input-conventional production possibilities and input-conventional distance functions [Shephard (1953, 1970); see also McFadden (1980, pp. 43–52)]. Also, by using the formal duality of cost and distance functions, he has shown that the production possibility set has property A if and only if the cost function has property B, and vice versa. Some of the duality properties between the transformation and cost functions are shown in Table 2.1 [McFadden (1980, pp. 92–112)].

Thus, by using the formal duality relationship between cost and transformation functions, it is possible to use cost data to estimate the underlying production possibility set. Duality theory can also be used to establish the properties of the "restricted profit function", which permits optimization by the firm over any set of variable inputs and outputs. Some examples of the duality principles and their applications are provided below.

2.3. Applications of the duality principle: Profit and cost functions

It follows from the duality principles that cost and profit functions can be used to characterize equivalent structural properties of the production process. Since the profit function analysis is a generalization of the cost function approach, the discussion below pertains to the profit function; however, an example of cost analysis relating variable and long-run cost function is also discussed briefly.

Assume that

$$y=F(x_1,\dots,x_m;Z_1,\dots,Z_n) \tag{2.8}$$

Table 2.1
Property A holds for an input-conventional function if and only if property B holds on its input-conventional cost function.[a]

	Property A on the transformation function $F(y, x)$	Property B on the cost function $C(y, p)$
1	Non-increasing in y	Non-decreasing in y
2	Uniformly decreasing in y	Uniformly increasing in y
3[b]	Strongly upper semi-continuous in (y, x)	Strongly lower semi-continuous in (y, p)
4[c]	Strongly lower semi-continuous in (y, x)	Strongly upper semi-continuous in (y, p)
5[d]	Strongly continuous in (y, x)	Strongly continuous in (y, p)
6[e]	Strictly quasi-concave from below in x	Continuously differentiable in positive p
7[f]	Continuously differentiable in positive x	Strictly quasi-concave from below in p
8[g]	Twice continuously differentiable and strictly differentiably quasi-concave from below in x	Twice continuously differentiable and strictly differentiably quasi-concave from below in p

[a]By the formal duality of cost and transformation functions, the implications of this table continue to hold when properties A and B are reversed, i.e. A holds for the cost function and B holds for the transformation function.

[b]This property is equivalent to the condition that the production possibility set be a closed set.

[c]Input requirement sets $x(y)$ form a *strongly lower semi-continuous correspondence* if two properties hold:

(i) if $y^i \in y^*, y^i \to y^U \notin y^*$ and A is any bounded set, then for sufficiently large i, $x(y^i)$ does not meet A;
(ii) if $y^0 \in y^*$, $x^0 \in x(y^0)$ and $y^i \in y^*$, $y^1 \to y^0$, then there exist $x^i \in x(y^i)$ such that $x^i \to x^0$

[d]A function is *strongly continuous* if it is strongly upper and strongly lower semi-continuous. This property is equivalent to a requirement that the input requirement sets $x(y)$ define a strongly continuous correspondence.

[e]This property guarantees that isoquants are "rotund" with no flat segments.

[f]This property guarantees that isoquants have no "kinks".

[g]An input-conventional transformation function with these properties is termed "neoclassical". This result, then, provides a formal duality theorem for neoclassical distance functions and neoclassical cost functions.

Note: For further discussion and formal proof of the duality principles, see Fuss and McFadden, Eds. (1980), Diewert (1974), and Chapter 12 of this Handbook by Diewert.

is the production function of the firm where the x's and Z's are the variable and fixed inputs, respectively. If the production function F is (1) a finite, non-negative real value function; (2) continuous; (3) continuously and twice differentiable; (4) monotonic in $x=(x_1 \ldots x_n)$ and $Z=(Z_1 \ldots Z_m)$; (5) concave and locally strongly concave; and (6) bounded, then the normalized profit function

$$\phi = pF(x, Z) - \sum_{i=1}^{n} p_i x_i \tag{2.9}$$

also has these properties. [See Lau (1980) for formal proof.] That is, if the assumptions about the production function hold, there exists a one-to-one correspondence between members of the normalized profit and the production function. Eq. (2.9) can be written in terms of normalized profit, $\phi^*(\Pi/P)$, or unit output price, abbreviated as "UOP". That is

$$\phi = F(x^*, Z) - Px^* = G(P, Z),$$

where x^* is the optional level of variable inputs and $P = p_i/p$ is the vector normalized price of inputs. The dual transformation relationships between the normalized profit function $G(P, Z)$ and the production function $F(X, Z)$ are summarized in Table 2.2.

The set of relationships

$$\frac{\partial G}{\partial P} = -x \tag{2.10a}$$

and

$$\frac{\partial G}{\partial Z} = \frac{\partial F}{\partial Z} \tag{2.10b}$$

are called the *Hotelling Lemma* [Hotelling (1932)], which suggest that the demand for x_i can be obtained by differentiating the profit function with respect to the price of the variable inputs and the derivatives of the production and profit

Table 2.2
Dual transformation relationships between $F(\cdot)$ and $G(\cdot)$.

	Production $F(x, Z)$	Normalized profit function $G(P, Z)$
(1)	$F(x, Z) - G(P, Z) = P'x$	
(2)	$\frac{\partial F}{\partial x} = P$	$\frac{\partial G}{\partial P} = -x$
(3)	$x = -\frac{\partial G}{\partial P}$	$P = \frac{\partial F}{\partial x}$
(4)	$\frac{\partial F}{\partial Z} = \frac{\partial G}{\partial Z}$	$\frac{\partial G}{\partial Z} = \frac{\partial F}{\partial Z}$
(5)	$F = G - P\left(\frac{\partial F}{\partial P}\right)$	$G = F - x\left(\frac{\partial F}{\partial x}\right)$
(6)	Z	Z

Source: Lau (1980, p. 140).

functions with respect to the fixed factors are equivalent. Several properties of the profit functions are of interest:

(i) If the production function, F, is homogeneous of degree k (provided $k<1$), then the profit function is homogeneous of degree $-k/(1-k)$ and the profit-maximizing cost of variable inputs is $C^*=P[k/(1-k)]\cdot G$.
(ii) The derived demand function is homogeneous of degree $-1/(1-k)$ in P if the production function is homogeneous of degree k in x.
(iii) A production function is homothetic in variable inputs, x, if and only if the normalized profit function is homothetic in P.
(iv) If the production function is given by $y=AF(x, Z)$, the normalized profit function is given as $\phi^*=AG(P/A, Z)$.
(v) The estimates of production elasticities can be derived, given the estimate of the normalized profit elasticities and vice versa. The following relationships hold:

$$\frac{\partial \ln G}{\partial \ln p_i}=-\frac{1}{(1-\varepsilon)}\frac{\partial \ln F}{\partial \ln x_i}, \quad \text{where } \varepsilon=\sum \frac{\partial \ln F}{\partial \ln x_i}, \tag{2.11a}$$

$$\frac{\partial \ln F}{\partial \ln x_i}=-\frac{1}{(1-\eta)}\frac{\partial \ln G}{\partial \ln p_i}, \quad \text{where } \eta=\sum \frac{\partial \ln G}{\partial \ln p_i}. \tag{2.11b}$$

From (2.11a) and (2.11b) it follows that $1/\varepsilon + 1/\eta=1$.

(vi) A production function is additively separable with respect to the commodity categories if and only if the normalized profit function is additively separable with respect to the corresponding price categories.
(vii) A production function and its normalized profit function are strongly separable with respect to the commodity categories and the corresponding price categories, respectively, only if they are homothetic. Both are weakly separable only if they are both homothetically separable.

In general, closed form solutions for the normalized profit function are often difficult to obtain for even simple technologies. The problem is resolved if the notion of the normalized *restricted* profit function is used. Normalized profit functions with fixed inputs are called *normalized restricted profit functions*. Some of the properties, such as homogeneity and homotheticity discussed above, are applicable to the restricted profit function but with some important modifications. Some of these modifications are:

(1) The restricted profit function $F(x, Z)$ is *almost homogeneous* of degree k_1 and k_2 in X and Z, respectively, if

$$F(\lambda x, \lambda^{k_2}Z)=\lambda^{k_1}F(X, Z); \qquad \lambda>0, \tag{2.12}$$

i.e. if the variable input is increased by some proportion (λ) and the fixed factors are increased by some power of λ, then output Y will increase by another power of that proportion [see Lau (1976)]. When $k_1 = k_2 = 1$, constant returns prevail.

(2) A production function is homogeneous of degree k in all inputs if the normalized restricted profit function is almost homogeneous of degree $-1/(1-k)$ and $-k/(1-k)$ if $k \neq 1$.

(3) If the profit function is almost homogeneous of degrees 1 and $1/k$, the underlying production is homogeneous of degree k in Z.

(4) If the production function is homogeneous of degree k, then the derived demand functions are almost homogeneous of degrees $-1/(1-k)$ and $-k/(1-k)$.

Additional results on the structure of normalized profit such as homotheticity and separability can be deduced and the results can be extended to the multi-output case [Lau (1980)]. It is sufficient to say that the normalized restricted profit function is a flexible method for analytical and empirical analysis of the structure of the production technology. For example, Lau (1976) has shown that the normalized (restricted) profit function can be used to analyze the relative efficiency between two firms. He extended the function to include adaptive and expectational lagged adjustment of inputs and took account of certain types of uncertainty about prices [Lau (1980)].

An interesting application, shown below, is the derivation of the structure of the short-run variable cost function and the resulting specifications of the long-run cost. Consider the variable cost function:

$$C_v = F(P, y, \bar{Z}), \tag{2.13}$$

and assume that the level of output y is given, there is only one fixed input $\bar{Z}$ and a vector of N variable inputs x with price vector $P = (p_1, p_2 \dots p_n)$. The variable cost function must be monotonically increasing and concave in factor prices, P. The corresponding short-run cost function is

$$C_s = F(P, y; \bar{Z}) + P_{\bar{Z}} \bar{Z}, \tag{2.14}$$

where $P_{\bar{Z}}$ is the price of $\bar{Z}$. The optimal use of the fixed factor is defined implicitly by the envelope condition

$$-\frac{\partial F(\cdot)}{\partial Z} = P_{\bar{Z}} \tag{2.15}$$

provided that the variable cost function is decreasing and convex in $\bar{Z}$. Denoting the optimal level of fixed input implied by (2.15) as $\bar{Z}=\psi(P, Y, P_{\bar{Z}})$, the long-run cost function is

$$C=H(P, y, P_{\bar{Z}}). \tag{2.16}$$

Differentiating (2.13) and (2.14) with respect to Y and rearranging yields the relationship between the cost elasticity along the variable and the short-run cost function. This relationship is expressed as

$$\eta^{s}=(1+d)^{-1}\eta^{v}, \tag{2.17}$$

where η^{s} and η^{v} are the cost elasticities of the short-run and variable cost functions and where $d=P_{\bar{Z}}\bar{Z}/C_{v}$ is the ratio of fixed to variable costs.

In contrast, the short- and long-run cost elasticities can be linked only if the envelope condition holds. In that case, differentiating (2.16) and using (2.15) yields the obvious result that the short- and long-run cost elasticities are identical: $\eta=\eta^{s}$. If the envelope condition does not hold, there is no unique relationship between η^{s} and η.

The AES along the various cost functions are also related. The variable cost AES, σ_{ij}^{v}, is defined as

$$\sigma_{ij}^{v}=\frac{F(\cdot)\partial^{2}F/\partial p_{i}\partial p_{j}}{\partial F/\partial p_{i}\partial F/\partial p_{j}}, \quad \text{for } i, j\neq k, \tag{2.18}$$

where subscripts denote inputs. Note that $\sigma_{ij}^{v}=\sigma_{ji}^{v}$. It follows immediately from (2.14) and (2.18) that $\sigma_{ij}^{v}=(1+d)\sigma_{ij}^{v}$. The AES along the long-run cost function is

$$\sigma_{ij}^{v}=\frac{H(\cdot)\partial^{2}H/\partial p_{i}\partial p_{j}}{\partial H/\partial p_{i}\partial H/\partial p_{j}}, \quad \text{for all } i, j. \tag{2.19}$$

The σ_{ij} and σ_{ij}^{v} are related by

$$\sigma_{ij}=(1+d)\left[\sigma_{ij}^{v}+S_{i}^{-1}\frac{\partial\ln x_{j}}{\partial\ln k}\frac{\partial\ln k}{\partial\ln p_{i}}\right], \qquad i, j\neq k, \tag{2.20}$$

where $S_{i}=P_{i}X_{i}/C_{v}$ is the variable cost share of input i. Eq. (2.20) links the variable cost and long-run AES for the variable factors [For further discussion see Schankerman and Nadiri (1980).]. It is easily verified that the required condition $\sigma_{ij}=\sigma_{ji}$ is met. The long-run AES involving the fixed factor is retrieved from the relation $\sum_{j}\alpha_{j}\sigma_{ij}=0$, where $\alpha_{j}=p_{j}x_{j}/C$ is the long-run cost share of input j. That is,

$$\sigma_{ik}=\alpha_k^{-1}\sum_{j\neq k}\alpha_j\sigma_{ij}.$$

Note that the relationships between the variable and long-run AES require that the envelope condition holds.

One advantage of the variable cost approach (and the restricted profit function), is that it enables calculation of the net rate of return to the fixed input. The cost minimizing use of the fixed factor is described by the envelope condition, so that analytically the envelope condition holds if and only if the net rate of return to the fixed factor is equal to its opportunity cost of funds (and the net rate of return to other assets). If the envelope condition is rejected, it is possible to compute the rate of return to the fixed factor by estimating the subset of equations excluding the envelope equation (2.14) [see Schankerman and Nadiri (1980)].

3. Functional forms of the production function: Capital stock aggregation

Considerable effort has been expended toward specifying general forms of the production function. These general forms are amenable to econometric estimation, consistent with the properties of the technological production set described in Section 1.1, and contain most of the economic effects described in Table 1.1. The criteria for choosing among various functional forms, though largely dependent upon the purpose of the particular analysis, can be stated as: (i) consistency with the theoretical properties of the production technology discussed in Section 1 – that is, the production function should be well-behaved, displaying consistency with standard hypotheses, such as positive marginal products or convexity; (ii) economy in number of parameters to be estimated ("parsimony in parameters"), and ease of interpretation of results; (iii) computational ease; and (iv) stability of the function over time or across observations.[11] A sample of specific functional forms is discussed in the first part of this section.

Important questions have been raised as to whether a neoclassical aggregate production function can exist without resort to strong and unrealistic a priori assumptions. The central issue is whether, in the presence of reswitching of techniques at different relative input prices and capital reversal, it would be possible to aggregate capital stocks embodying different techniques of production. The controversy on this issue and some attempts to resolve it are briefly discussed in the second part of this section.

[11]See Fuss, McFadden, and Mundlak (1980) and Griliches and Ringstad (1971) for excellent discussions of specifications of functional forms.

3.1. Forms of the production function

For empirical analysis it is necessary to specify the forms of the production function more precisely. The focus of recent efforts has been to remove restrictive features such as additivity, separability, constant elasticity of substitution, etc. which characterize traditional production functions. The properties of a few functional forms that have been used in econometric estimations are briefly discussed.

3.1.1. Linearly homogeneous production functions

The best known members of this class of functions are:

(i) The *fixed proportion (Leontief) production function*:

$$y=\min\left(\frac{1}{\alpha}x_1,\frac{1}{\beta}x_2\right),\quad \text{with } \sigma=0;\ \alpha,\beta>0; \tag{3.1}$$

(ii) The *Cobb–Douglas production function*:

$$y=Ax_1^{\alpha}x_2^{\beta},\quad \text{with } \sigma=1 \text{ and } \mu=\alpha+\beta\gtreqless 1 \tag{3.2}$$

and

(iii) The *CES production function*:

$$y=\gamma\left[\delta x_1^{-\rho}+(1-\delta)x_2^{-\rho}\right]^{-v/\rho},\quad \text{with } \sigma=\frac{1}{1+\rho} \text{ and } \mu=v, \tag{3.3}$$

where x_1 and x_2 are the inputs, α and β are constant elasticities of the inputs, A and γ are efficiency parameters, $\delta(0<\delta<1)$ is the input intensity, $\infty\geqslant\rho\geqslant-1$ is the substitution parameter, and v is the returns to scale parameter of the CES production function [Arrow, Chenery, Minhas, and Solow (1961)]. Several properties of the class of linearly homogeneous production functions, using the CES production function as an example, should be noted.

(a) The CES production function, like the Cobb–Douglas production function, is *strongly separable*.
(b) The marginal products from the CES production function are

$$\frac{\partial y}{\partial x_1}=\delta\gamma^{-\rho}\left(\frac{y}{x_1}\right)^{1+\rho}\quad\text{and}\quad\frac{\partial y}{\partial x_2}=(1-\delta)\gamma^{-\rho}\left(\frac{y}{x_2}\right)^{1+\rho}, \tag{3.4}$$

which are positive but decrease monotonically for all non-zero values of the inputs.

(c) The marginal rate of technical substitution of x_1 for x_2 is

$$MRS=\left(\frac{\delta}{1-\delta}\right)\left(\frac{x_2}{x_1}\right)^{1/\sigma}. \tag{3.5}$$

For any given values of MRS and σ, the larger the input intensity parameter δ the smaller will be the input ratio x_2/x_1. That is, as the production process is more x_1 intensive, the larger the value of the distribution parameter δ.

(d) The elasticity of substitution for CES is constant but not necessarily equal to unity. That is, $\sigma=1/(1+\rho)$. When $\sigma=1$, the CES production function reduces to the Cobb–Douglas function; and when $\sigma=0$, it reduces to the fixed coefficient Leontief function.

An important extension of the CES production function is the "nested" or two-level CES production function. This function is called a *"two-level" CES function* because it can be specified in two stages, with a CES function employed in each stage [Sato (1967)]. The function is strongly separable with respect to the highest level partition and then strongly separable within each subaggregate. In the first stage, factors are grouped together in clusters that are thought to have common or fairly close elasticities of substitution. This procedure can be applied to the multi-stage production process by representing the production possibility frontier by means of components of power functions and logarithmic functions. As an example, consider the following.

Let the first level input indices, z_s, be defined as functions of N inputs $x_1,\dots,x_N$, brought together in S subsets of $N_1,\dots,N_S$ inputs, respectively.

$$z_s=\left[\sum_{i\in N_s}\beta_i^s(x_i^s)^{-\rho_s}\right]^{-1/\rho_s},\qquad \beta_i^s>0,\ -1<\rho_s=\frac{1-\sigma_s}{\sigma_s}<\infty \qquad (s=1,\dots,S). \tag{3.6}$$

The second level production function then can be specified as:

$$y=\sum_{s=1}^{S}\left[\alpha_s z_s^{-\rho}\right]^{-1/\rho},\qquad \alpha_s>0,\ -1<\rho=\frac{1-\sigma}{\sigma}<\infty. \tag{3.7}$$

The direct partial elasticity of substitution (DES) for this function is:

$$\sigma_{ij}^{\mathrm{D}}=\begin{cases}\sigma_{\mathrm{s}}, & \text{if } i,\ j\in N_s,\ i\neq j\\ & \text{a harmonic mean of } \sigma,\ \sigma_{\mathrm{s}}, \text{ and}\\ \sigma_{\mathrm{r}}, & \text{if } i\in N_r,\ j\in N_s,\ r\neq s,\end{cases} \tag{3.8}$$

given by

$$\sigma_{ij}^{D}=\frac{a+b+c}{\frac{a}{\sigma_r}+\frac{b}{\sigma_s}+\frac{c}{\sigma}},$$

where

$$a=\frac{1}{\theta_i^r}\frac{1}{\theta^r},\qquad b=\frac{1}{\theta_j^s}\frac{1}{\theta^s}\quad\text{and}\quad c=\frac{1}{\theta^r}\frac{1}{\theta^s}.$$

θ^s is the relative expenditure share of the sth class of factors, and θ_j^s is the relative share of the jth element of the sth class in total expenditures.

The Allen partial elasticities of substitution are given by:

$$\sigma_{ij}^{A}=\begin{cases}\sigma+\frac{1}{\sigma_s}(\sigma_s-\sigma), & \text{if } i,j\in N_s, i\neq j,\\ \sigma, & \text{if } i\in N_s, j\in N_r, r\neq s.\end{cases} \tag{3.9}$$

Thus, σ_{ij}^{D} are constant for factors *within* a subset; σ_{ij}^{A} are constant for factors *between* subsets. Furthermore, each of these is related to the relevant substitution parameter by the same formula that applies to an n-factor CES function, namely

$$\sigma_s=\frac{1}{1-\rho_s};\qquad \sigma=\frac{1}{1-\rho}.$$

Inspection of the σ_{ij}^{D} formulae reveals that they are necessarily positive, as is σ_{ij}^{A} for inter-class elasticities. However, the Allen intra-class elasticities can be negative (only if $\sigma>\sigma_s$). Typically, it is expected that the inter-class elasticities will be less than the intra-class elasticities, i.e. $\sigma_s>\sigma$, when one subdivides factors into increasingly more homogeneous subsets. Ordinarily, therefore, σ_{ij}^{A} will also be positive.

3.1.2. *Variable elasticity of substitution functions*

There are good theoretical reasons, substantiated by empirical evidence, to assume that the elasticity of substitution is often not constant but variable. The variable elasticity production functions (VES) are often generalizations of the CES production function. In contrast to a CES production function, the VES production function requires that σ be the same along the expansion path but vary along an isoquant. Several such functions have been constructed and estimated. Consider the following examples:

(i) *Liu–Hildenbrand function:*[12]

$$y=\gamma\left[(1-\delta)K^{-\rho}+K^{-m\rho}L^{-(1-m)\rho}\right]^{-1/\rho}, \tag{3.10}$$

where m is a constant. If $m=0$, (3.10) reduces to the familiar CES production function which includes the Leontief and Cobb–Douglas functions as special cases. The elasticity of substitution for this function is

$$\sigma(m)=\frac{1}{1+\rho-m\rho/s_k}, \tag{3.11}$$

where s_k is the share of capital. (3.11) implies that y/L depends on both the wage rate, w, and the capital–labor ratio, K/L. Since m and s_k are positive, the relationship between the variable elasticity of substitution, $\sigma(m)$, and constant elasticity of substitution, σ, depends on the magnitude of ρ; if $\rho \gtreqless 0$ ($\sigma \lesseqgtr 1$), then $\sigma(m) \gtreqless \sigma$. Thus, when $\sigma<1$, $\sigma(m)$ is larger than σ and tends toward unity, and if $\sigma>1$, $\sigma(m)$ is smaller than σ and tends toward unity, again suggesting an equilibrating process.

(ii) *Transcendental production function* [Revankar (1971)]:

$$y=\gamma e^{(a_1(K/L)+a_2L)}K^{(1-\beta)}L^{\beta}, \tag{3.12}$$

where e is a constant. This reduces to the Cobb–Douglas function if $a_1=a_2=0$, but it does not generalize to contain the linear production function $y=\alpha K+\beta L$ as a special case. The elasticity of substitution depends upon the ratio of factors of production (K/L) and is given by the expression:

$$\sigma=\frac{(1-\beta+a_1K)(\beta+a_2L)}{(1-\beta)(\beta+a_2L)+\beta(1-\beta+a_1K)^2}. \tag{3.13}$$

(iii) *Constant marginal shares (CMS) function* [Bruno (1969)]:

$$y=\gamma K^{\alpha}L^{\beta}-mL, \tag{3.14}$$

where m is a constant. This function is explicitly a generalization of the Cobb–

[12]See Liu and Hildenbrand, (1965), Nerlove (1967), and Diwan (1970) for further discussion.

Douglas function (if $m=0$), and contains the linear function as a special case if $\alpha=1$, $(\alpha+\beta)=1$, and m is negative. Its elasticity of substitution is the simple expression

$$\sigma=1-\left(\frac{m\alpha}{\beta}\right)\frac{L}{y}, \tag{3.15}$$

which implies that average productivity Y/L rises as $\sigma\to 1$ and that the value of σ does depend on the level of output Y.

(iv) *Revankar's variable elasticity function.* The elasticity of substitution for the VES function is variable and may depend on the factor input ratio. Consider, for example, the function

$$y=\gamma K(1-\delta\rho)\{L+(\rho-1)K\}^{\alpha\delta\rho};\qquad \gamma>0,\ \alpha>0,\ 0<\delta<1,\ 0<\delta\rho<1. \tag{3.16}$$

The elasticity of substitution for this function is defined as

$$\sigma(K,L)=1+\frac{\rho-1}{1-\delta\rho}-1+\beta\frac{K}{L}. \tag{3.17}$$

For the production function (3.16), the marginal products are positive, and it includes Cobb–Douglas, linear, and fixed coefficient production functions as special cases. However, it cannot be generalized to include CES production functions as a special case. Also, note that σ varies linearly with K/L and stays below or above unity for the relevant range of K/L. This property is a weakness of the VES function.

A more general case of the class of VES production functions which avoids the last property can be formulated as [Kadiyala (1972)]:

$$y=f(x_1,x_2)=E\left[\omega_{11}x_1^{2\rho_1}+2\omega_{12}x_1^{\rho_1}x_2^{\rho_2}+\omega_{22}x_2^{2\rho_2}\right]^{\frac{1}{2\rho}}, \tag{3.18}$$

where E is the efficiency term; ρ, ρ_1, and ρ_2 are the substitution parameters; and ω_{11}, ω_{12}, and ω_{22} are weights. Assuming $\omega_{11}+2\omega_{12}+\omega_{22}=1$ and $\rho_1+\rho_2=2\rho$, (3.18) reduces to the CES production function if $\omega_{12}=0$, to the (VES) function developed by Lu and Fletcher (1968) if $\omega_{22}=0$, and to the Sato–Hoffman VES production function if $\omega_{11}=0$ [Sato and Hoffman (1968)]. With the assumption that $\rho_1=\rho_2=\rho$, (3.18) reduces to

$$y=G(x_1,x_2)=E\left[\omega_{11}x_1^{2\rho}+2\omega_{12}x_1^{\rho}x_2^{\rho}+\omega_{22}x_2^{2\rho}\right]^{\frac{1}{2\rho}}, \tag{3.19}$$

which has an elasticity of substitution

$$\sigma(x)=\frac{1}{1-\rho+R}, \tag{3.20}$$

where

$$R=\frac{-\rho(\omega_{11}\omega_{22}-\omega_{12}^{2})}{(\omega_{11}x^{-\rho}+\omega_{12})(\omega_{12}+\omega_{22}x^{-\rho})}$$

and

$$x=\frac{x_2}{x_1}.$$

By choosing appropriate values for ρ, (3.19) reduces to Cobb–Douglas ($\rho=0$), fixed proportion coefficient ($\rho=+\infty$), and linear ($\rho=\frac{1}{2}$ and $\omega_{12}=0$) production functions. Note that in this formulation the marginal products are non-negative and the elasticity of substitution is not a monotonic function of factor combination. Also, this function can be generalized to many input cases and to cases where the production function is homogeneous of a degree greater than one.

3.1.3. *Homothetic production functions*

Every homogeneous production function is homothetic, but the homothetic family contains non-homogeneous functions as well, i.e. production functions having variable returns to scale.[13] As an example, consider the general production function

$$y=g(f(K,L)), \tag{3.21}$$

where $f(K,L)$ is a homogeneous function of arbitrary degree. For such a function: (i) returns to scale diminish in a pre-assigned fashion with y; (ii) the associated average cost curve exhibits decreasing cost at low levels of output and increasing costs at high levels of output; and (iii) function (3.21) can be generalized to n inputs.

(a) Marginal products of (3.21) are positive:

$$\frac{\partial y}{\partial L}=\frac{\partial g}{\partial f}\frac{\partial f}{\partial L}>0 \quad\text{and}\quad \frac{\partial y}{\partial K}=\frac{\partial g}{\partial f}\frac{\partial f}{\partial K}>0.$$

[13] For the definition of homothetic production function, see Section 2.1.1, and also Zellner and Revankar (1969).

(b) The elasticity of substitution associated with $Y=g(f)$ is the same as that associated with the function f.[14]

(c) If f is a neoclassical production function homogeneous of degree a_f, and the function $y=g(f)$ has a *pre-assigned* returns to scale function $\mu(y)$, the following differential equation holds:

$$\frac{\mathrm{d}g}{\mathrm{d}f}=\frac{y}{f}\frac{\mu(y)}{\mathrm{d}f}. \tag{3.22}$$

By Euler's theorem,

$$\frac{\partial g}{\partial f}\left(L\frac{\partial f}{\partial L}+k\frac{\partial f}{\partial K}\right)=y\cdot\mu(y), \tag{3.23}$$

and since f is homogeneous of degree α_f, then

$$\left(L\frac{\partial f}{\partial L}+K\frac{\partial f}{\partial K}\right)=\alpha_f f.$$

Substituting this expression in (3.23) establishes (3.22). Thus, by specifying f, the elasticity of substitution (variable or constant) can be obtained as long as f is homogeneous. The factor shares can be deduced easily; e.g. labor's share is defined as $S_{L/g}=\mu(y)/\partial f\cdot L/f\cdot\partial f/\partial L$, which suggests that, among other things, it depends on level of output, y, factor combination K/L, and σ, which can be either constant or variable.[15]

A more general case of the homothetic production function is the class of ray-homothetic functions where the returns to scale vary with *both* input mix and level of output for any a priori given elasticity of substitution.[16] The production

[14] Let $1/\sigma=(x/z)(\partial z/\partial x)$, where $x=K/L$ and the MRS for (3.21) can be defined as

$$z=\frac{\partial g}{\partial L}\bigg/\frac{\partial g}{\partial K}=\frac{\partial g}{\partial f}\cdot\frac{\partial f}{\partial L}\bigg/\frac{\partial g}{\partial f}\cdot\frac{\partial f}{\partial K}=\frac{\partial f}{\partial L}\bigg/\frac{\partial f}{\partial K}.$$

[15] Consider some examples of the forms of $\mu(y)$: (i) $\mu(y)=\alpha_f(1-y/k)$ with $0\leqslant y<k$, where k is the limiting value for output; (ii) $\mu(y)=\alpha_f(\gamma+\beta/f)$; (iii) $\mu(y)=\alpha_f=h[(a-y)/(n+y)]$, $\alpha f>h\geqslant 0$. Note in case (iii) that returns to scale is (α_f+h) when $y=0$ and falls monotonically to reach a limiting value of (α_f-h) as y increases. Finally, (iv) $\mu(y)=\alpha/(1+\theta y)$. The corresponding production functions to be estimated for each of these cases are: (i) $y=k/(1+\mathrm{e}^{-\log f})$; (ii) $y=Cf^{\gamma}\mathrm{e}^{\beta/f}$, where C is the constant of integration, γ and β are constants; (iii) $y=[(1+\rho)Y]^{(2\rho/(1-\rho))}=k'f^{1+\rho}$, where a and k' are constants and $\rho=h/\alpha_f(0\leqslant\rho<1)$ – if $\rho=0$ then $h=0$, which implies constant returns to scale; and (iv) $y\mathrm{e}^{\theta y}=\gamma K^{\alpha(1-\delta)}L^{\alpha\delta}$. For further discussion, see Zellner and Ravankar (1969).

[16] For further discussion see Fare, Jansson, Knox, Lovell (1978).

function ϕ: $R^n_+ \to R_+$ with properties stated in Section 1 is ray-homothetic if

$$\begin{aligned}&\phi(\lambda\cdot x)=F\{\lambda^{H(x/|x|)}G(x)\} \text{ provided that } \lambda>0,\\ &\qquad F\colon R_+\to R^n_+, \text{ and } H\colon \{x/|x|\,|\,x\geqslant 0\}\to R_+,\\ &\qquad \text{where } G(x)=F^{-1}(\phi(x)).\end{aligned} \tag{3.23}$$

This class of functions includes the ray-homogeneous and homothetic functions as special cases.[17] ϕ is ray-homogeneous if (3.23) takes the form

$$\phi(\lambda\cdot x)=\lambda^{H(x/|x|)}\phi(x); \qquad \lambda>0. \tag{3.24}$$

It is homothetic if $H(x/|x|)=\alpha$, where α is a constant, i.e.

$$\phi(\lambda\cdot x)=F(\lambda^{\alpha}\cdot G(x)); \qquad \lambda>0. \tag{3.25}$$

The function (3.25) will be homogeneous if the conditions of $H(x/|x|)=\alpha$ and F as an identity function prevails, i.e.

$$\phi(\lambda\cdot x)=\lambda^{\alpha}\cdot\phi(x). \tag{3.26}$$

One property of these functions which is of interest is the *scale elasticity* μ, defined as:

$$\mu=\lim_{\lambda\to 1}\left(\frac{\partial\phi(\lambda\cdot x)}{\partial\lambda}\cdot\frac{\lambda}{\phi(x)}\right).$$

This expression takes the specific forms:

(i) $\mu_1=\mu_1(x/|x|,\phi(x))$, if ϕ is ray-homothetic,
(ii) $\mu_2=\mu_2(x/|x|)$, if ϕ is ray-homogeneous,
(iii) $\mu_3=\mu_3(\phi(x))$, if ϕ is homothetic,
(iv) $\mu_4=\mu_4=\alpha$, if ϕ is homogeneous.

The optimal output for ray-homothetic and homothetic functions is when $\mu_1=1$ or $\mu_3=1$; it is 0, indeterminate, or ∞ for ray-homogeneous or homogeneous functions.

[17]See Eichhorn (1970) for the former type, and Shephard (1953, 1970) for the latter functional forms.

An example of ray-homothetic production function provided by Fare, Jansson, and Lovell is:

$$y_{\mathrm{e}}^{\theta y}=AK^{\alpha+\gamma(K/L+\delta L/K)^{-1}}L^{\beta+\gamma\{K/L+\delta L/K\}^{-1}}, \tag{3.27}$$

with θ, $\gamma \in R$, $A, \alpha, \beta, \delta \in R_t$, and $\{\alpha+\gamma(K/L+\delta L/K)\}^{-1}>0$. $\{\beta+\gamma(K/L+\delta L/K)\}^{-1}>0$ for all K/L. If $\gamma=0$, then (3.27) collapses to the homothetic Cobb–Douglas function suggested by Zellner and Revankar (1969). If $\sigma=0$, (3.27) is ray-homogeneous, and if $\sigma=\gamma=0$, then it is a homogeneous Cobb–Douglas function. The returns to scale measure for (3.27) is defined as:

$$\mu(x/|x|, \phi(x))=\frac{\alpha+\beta}{1+\theta y}+\frac{2\gamma}{(1+\theta y)(K/L+\delta L/K)}. \tag{3.28}$$

At a technically optimal output, $\mu(x/|x|, \phi(x))=1$ and, therefore,

$$y^{*}=\frac{\alpha+\beta-1}{\theta}+\frac{2\gamma}{\theta(K/L+\delta L/K)}. \tag{3.29}$$

For the homothetic functions, returns to scale is a (monotonically decreasing) function of output only, and, therefore, technically optimal output is the same constant for all observations. For the ray-homogeneous functions, returns to scale is a (monotonically decreasing) function of input mix only, and technically optimal output is infinite for all observations. However, these implausible outcomes do not hold for the ray-homothetic function; the returns to scale vary inversely with both output and input mix, permitting technically optimal output to vary among observations, as one would expect.

3.1.4 Non-Homothetic Production Functions

These functions are characterized by *isoclines*, which are defined as the loci of points with constant marginal rates of technical substitution, but varying returns to scale and optimal input ratios. They have flexible functional forms for they do not impose the restrictive constraints a priori, such as homotheticity, constancy of the elasticity of substitution, additivity, etc. and are easily adaptable to include multiple products and multiple inputs. An example of this class of functions is the recently developed transcendental logarithmic production functions [Christensen, Jorgenson, and Lau (1973)], which are often referred to as the "translog" production frontier. The production possibility frontier, F, is represented by

$$F(y_1, y_2, y_3, \ldots, y_n, A)=0, \tag{3.30}$$

where y_1 $(i=1,2,\ldots,n)$ refer to net outputs (and inputs), and A is an index of the rate of technological progress. The translog functions are often approximated in terms of logarithmic functions, such as

$$\begin{aligned}\ln(F+1)=\alpha_0+\sum_{i=1}^{n}\alpha_i\ln y_i+\gamma_0\ln A\\ +\frac{1}{2}\sum_{i=1}^{n}\sum_{j=1}^{n}\alpha_{ij}\ln y_i\ln y_j\\ +\sum_{i=1}^{n}\gamma_i\ln A\ln y_i.\end{aligned}\tag{3.31}$$

The translog function does not assume separability; rather, separability conditions should be tested. Also the translog function F is strictly quasi-concave (strictly convex isoquants) if the corresponding bordered Hessian matrix of the first and second partial derivatives is negative definite. This is often evaluated empirically at each data point for any estimated translog function. The marginal conditions for the outputs and inputs are given by

$$\psi_i=\alpha_i+\sum_{j=1}\alpha_{ij}\ln y_j+\gamma_i\ln A,\qquad i,j=(1,\ldots,n).\tag{3.32}$$

The marginal rates of technical substitution between outputs, between inputs, and between inputs and outputs are

$$\frac{p_i y_i}{p_j y_j}=\frac{\psi_i}{\psi_j};\qquad i\neq j,\ i,j=(1,\ldots,n).$$

To obtain special functional forms, restrictions have to be imposed on (3.31). To illustrate these restrictions and flexibility of the translog functions for empirical analysis, consider the following one-output translog cost function. Similar specific functional forms can be deduced using the production function (3.31) directly. However, the cost function is often easier to estimate and the duality principle provides a description of the underlying production technology.

$$\begin{aligned}\ln C=\alpha_0+\alpha_y\ln y+\tfrac{1}{2}\gamma_{yy}(\ln y)^2+\sum_{i=1}\alpha_i\ln p_i\\ +\tfrac{1}{2}\sum_{ij}\gamma_{ij}\ln p_i\ln p_j\\ +\theta_y\ln y\ln T+\sum\theta_{it}\ln p_i\ln T+B_T\ln T\\ +\tfrac{1}{2}B_{TT}(\ln T)^2,\end{aligned}\tag{3.33}$$

where C is total cost, y is the level of output, p_i the prices of the inputs, and T is the index of time as a proxy for A, the index of technological change. The derived demand for an input, x_i, is obtained by partially differentiating the cost function with respect to the factor price of the input. This follows from duality theory and is often referred to as *Shephard's Lemma* [Christensen, Jorgenson, and Lau (1973)],

$$\frac{\partial C(\cdot)}{\partial p_i}=x_i. \tag{3.34}$$

Using Shephard's Lemma and partially differentiating (3.33):

$$\frac{\partial \ln C}{\partial \ln p_i}=m_i=\alpha_i+\tfrac{1}{2}\sum \gamma_{ij}\ln p_j+\gamma_{iy}\ln y+\theta_{iT}\ln T, \tag{3.35}$$

where $m_i=p_ix_i/C$ is the share of costs accounted by input k.[18] Duality theory and the translog approximation require the following restrictions. First, the cost function must be linearly homogeneous in factor price. This implies

$$\sum \alpha_i=1;\qquad \sum_j \gamma_{ij}=0;\qquad \sum_i \gamma_{iy}=0.$$

Second, the symmetry constraint requires that $\sum_i\gamma_{ij}-\sum_j\gamma_{ji}=0$. Third, technical change can be both neutral (β_T and β_{TT}) and biased $\theta_{iT}\neq 0$. If biased technical change is present the cost shares are affected and technical change, everything else being equal, is the ith factor saving (using) if $\theta_{iT}<0$ ($\theta_{iT}>0$). The cost function is homothetic if

$$\gamma_{iy}=\gamma_{yy}=0.$$

It is homogeneous if $\partial \ln C/\partial \ln y=k$, where k is a constant; the test of homogeneity is nested within the test for homotheticity.

The *Allen–Uzawa elasticity of substitution* ($AUES$) is defined as

$$\sigma_{ij}=\frac{\gamma_{ij}+m_i^2-m_i}{m_i^2},\quad \text{for } i=j \tag{3.36a}$$

and

$$\sigma_{ij}=\frac{\gamma_{ij}+m_im_j}{m_im_j},\quad \text{for } i\neq j, \tag{3.36b}$$

[18] Note that the coefficients in the share equations are subsets of those in the cost function. Also, since $\sum m_i=1$ only three of the four share equations are linearly independent.

where m_i are the cost shares of inputs in total cost of producing the output. Note that AUES are not constant – they vary with the cost shares. The own elasticity of factor demand, ε_{ij}, is given by

$$\varepsilon_{ij}=m_j\sigma_{ij}, \quad \text{for } i\neq j. \tag{3.37}$$

The special case of unitary elasticity of substitution clearly holds if

$$\gamma_{ij}=0, \quad \text{for all } i\neq j.$$

The scale elasticity for (3.33) is defined as

$$\mu=\left[\alpha_y+\alpha_{yy}\ln y+\sum_i \gamma_{iq}\ln p_i\right]^{-1}, \tag{3.38}$$

which varies with the level of output and factor prices unless the cost (production) function is homogeneous in output. The translog cost function (3.33) can be extended straightforwardly to multiple output cases, and, by imposing group-wise additivity among the inputs and outputs, homogeneity, symmetry, and normalization restrictions, the translog production or cost function can be specialized and tested empirically.

3.2. *Aggregate capital stock: The Cambridge–Cambridge controversy*

The neoclassical properties of production functions, such as differentiability, positive marginal products, constant returns, etc., as discussed earlier, rest upon the existence of an aggregate capital stock.[19] The usual assumptions in generating an aggregate capital good, J, called "Jelly", are: a linear factor price frontier, positive marginal products, and capital–output and capital–labor ratios as declining functions of the rate of profit (or interest rate). The elasticity of the factor price frontier should be equal to the ratio of wage-share to profit-share. Finally, no reswitching and capital reversals are permitted. *Reswitching* is said to occur when some method of production is preferable to another available method at a high rate of interest, is inferior to some other method at an intermediate interest rate, and again becomes preferable at a still lower interest rate. *Capital reversals* occur when the value of capital moves in the same direction when alternative rates of interest are considered.

[19]Similar conceptual problems of aggregation arise in cases of outputs and inputs other than capital stock. Though we concentrate on capital stock, the issues discussed here are pertinent to the aggregation of outputs of different quality and labor of different skills.

Given these assumptions, heterogeneous capital of different vintages can be aggregated into a general good J as

$$J(t)=\int_{-\infty}^{t} I(v)^{(\beta/(1-\alpha))}e^{(\lambda+\beta\delta)v/(1-\alpha)}\mathrm{d}v, \tag{3.39}$$

where $I(v)$ is investment good of vintage v and α and β are the shares of capital and labor; $(\alpha+\beta)\neq 1$. λ is the rate of embodied technical change, which can be neutral or non-neutral depending upon σ, the elasticity of substitution between capital, $J(t)$, and labor, $L(t)$. It would be capital-saving if $\sigma<1$, labor-saving if $\sigma>1$, and neutral of $\sigma=1$. δ is a constant rate of depreciation for all machines, irrespective of their vintage.

Some further assumptions are necessary before we can aggregate capital goods with different vintages and construct J: that goods of the same vintage are identical, that those of the latest vintage are more productive than those of the preceding vintage by a constant factor, that the marginal rate of substitution between capital goods of different vintages is independent of other inputs, that the marginal product of labor in all uses – i.e. over all vintages of capital – is equal, and, finally, that the production function is what is called by Fisher the "capital-generalized constant returns" function.[20]

The neoclassical production function can then be written as

$$y=F(L,J)=LF\left(1,\frac{J}{L}\right)=Lf\left(\frac{J}{L}\right) \tag{3.40}$$

which is differentiable, has positive marginal products, and is subject to constant returns, with properties:

(i)

$$w=\frac{\partial y}{\partial L}=f\left(\frac{J}{L}\right)-\frac{J}{L}\cdot f'\left(\frac{J}{L}\right) \tag{3.41}$$

and

$$r=\frac{\partial y}{\partial J}=f'\left(\frac{J}{L}\right).$$

[20]Fisher (1969) developed a class of production functions which allows capital aggregation when each firm's production function can be made to have constant returns after an appropriate (generally non-linear) stretching of the capital axis. The "capital-generalized constant returns" production functions are of the form: $g^{v}(K,L)=G^{v}(F^{v}(K),L)$, $(v=1,2,\dots,n)$, where G are homogeneous of degree one in their arguments and the F^{v} are monotonic.

(ii) The real wage rate w is an increasing function and the rate of interest (profit) r is a decreasing function of the capital–labor ratio, i.e.

$$\frac{\mathrm{d}w}{\mathrm{d}(J/L)}=-\frac{J}{L}\cdot f''\left(\frac{J}{L}\right)>0, \tag{3.42a}$$

$$\frac{\mathrm{d}r}{\mathrm{d}(J/L)}=f''\left(\frac{J}{L}\right)<0. \tag{3.42b}$$

(iii) the elasticity of factor–price relation (r,w) is equal to the ratio of relative shares:

$$-\frac{\mathrm{d}w}{\mathrm{d}r}\cdot\frac{r}{w}=\frac{Jr}{Lw}. \tag{3.43}$$

If these conditions hold, the presence of heterogeneous capital and a variety of techniques of production will not affect the conclusions of neoclassical production theory and production functions do exist.

The essential question is, however, whether these assumptions will hold in general. The Cambridge (U.K.) School has challenged most of these neoclassical assumptions, particularly those of no reswitching and no capital reversibility, thereby questioning the notion of aggregation of heterogeneous capital stock and, in turn, the existence of the aggregate production function.[21]

3.2.1. *Reswitching and capital reversibility*

The reswitching or double-switching phenomenon arises from three conditions: the staggered application of inputs to productive processes, different gestation periods for alternative technical processes, and the fact that the output of a

[21]The Cambridge School also challenges the neoclassical formulation of saving behavior. In the neoclassical life-cycle hypothesis of saving, demographic factors, such as age, number of children, and income, are often suggested as determinants of saving behavior. In the Cambridge School's view, institutional factors, such as corporations' decisions to retain substantial amounts of their earnings, and class behavior – that is, the saving propensities of the different classes of income-receivers in the community – determine saving behavior. Specification of saving behavior plays an important role in the Cambridge theory of the determination of interest rate in long-run equilibrium. The rate of interest, r, is equal to the rate of growth, n, divided by $s\pi$, the share of profit in national income, i.e. $r=n/s\pi$. Since a profit-maximizing firm, in the long-run equilibrium where all relative prices are constant, will equate the own rate of return of every capital to the rate of interest (profit), therefore $r=MPK_{ii}$, where MPK_{ii} is the marginal productivity of every capital good in terms of itself. Thus, once the rate of profit is determined by $n/s\pi$ (the ratio of rate of growth and share of profit), then it in turn determines the own marginal productivity of capital. Thus, it is not necessary to know anything about the production function and factor supplies to determine the own marginal productivity of each capital good and the capital–labor ratio. For further discussion of these sets of issues, see Pasinetti (1962), Hicks (1965), and Morishima (1966).

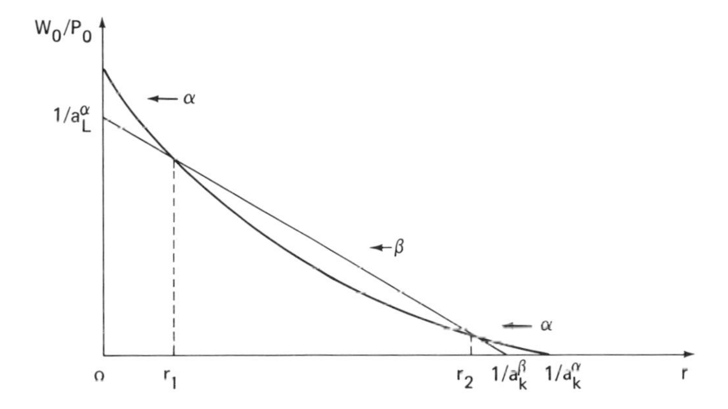

Figure 3.1. Reswitching.

process sometimes enters as an input to other processes. The rationale for reswitching is rather simple: multiple solutions to internal rate-of-return relationships arise because changes in interest rates alter the comparative cost of inputs applied at different dates in identical technical processes.[22] Consequently, because of different intertemporal usage patterns of inputs, different processes can be equally profitable at several different borrowing rates, with their relative profitability alternating in between these "cross-over points", i.e. points at which the interest rates equal the present values of their profits.

It is easily seen that reswitching prevents the unambiguous ordering of techniques in terms of capital intensity and profit rate. The simplest way to show this is to focus on two techniques yielding the two wage curves (see Figure 3.1). For $0<r<r_1$, technique α is adopted; for $r_1<r<r_2$, technique β is the more profitable; and for $r>r_2$, technique α comes back or reswitches. Since $1/a_{10}(\alpha)>1/a_{10}(\beta)$ and $1/a_{11}(\alpha)>1/a_{11}(\beta)$ (a_{10} and a_{11} are the labor and capital input requirements), as the profit rate rises monotonically from $r=0$, the firm adopts either technique α or β depending on the rate of profit. With reswitching it is not possible to say unambiguously which technique is more capital intensive [Samuelson (1966)]. Capital reversal results when the wage–price frontier is concave from below.

[22] Generally, if one of the techniques requires expenditure n periods in the future, the cross-over equation (giving interest rates at which the two techniques yield the same present value) will be of the nth degree. There would then be the possibility of n switches of techniques, i.e. n levels of interest rates at which the relative ranking of the two techniques could change. For further discussion see Robinson (1956), and Bruno, Burmeister, and Skeshinski (1966).

$$\alpha = \begin{array}{c} \\ \text{Labor} \\ \\ \text{Capital} \\ \text{type } a \end{array} \overset{\text{Capital Type } \alpha}{\begin{bmatrix} a_{\ell 0} & a_{\ell 1} \\ & \\ a_{10} & a_{11} \end{bmatrix}}; \quad \beta = \begin{array}{c} \\ \text{Labor} \\ \\ \text{Capital} \\ \text{type } b \end{array} \overset{\text{Capital Type } \beta}{\begin{bmatrix} b_{\ell 0} & b_{\ell 1} \\ & \\ b_{10} & b_{11} \end{bmatrix}}; \quad \gamma = ; \delta = ; \ldots$$

Figure 3.2. *n* Techniques.

To illustrate the issues, consider an economy with two sectors, 1 and 2, in which two commodities, y_0 and y_1, are produced in fixed proportions.[23] There are many techniques and each technique is associated with a particular specification of the capital good. For n heterogeneous capital goods (heterogeneous either in the physical sense or in the sense of different lengths of time required in producing particular capital), the technology of the economy can be described by a book of "blueprints," which is simply a set of technique matrices.

Let capital be infinitely durable.[24] In a competitive equilibrium, the price equations are:

$$p_0 = a_0 W_0 + r a_{10} p_1(\alpha), \tag{3.44}$$

$$p_1(\alpha) = a_1 W_0 + r a_{11} p_1(\alpha), \tag{3.45}$$

where $a_{1j} = L_j / y_j$ and $a_{1j} = K_{1j} / y_j$, $(j = 0, 1)$, p_0 and $p_1(\alpha)$ are the prices of y_0 and $y_1(\alpha)$, respectively. $y_1(\alpha)$ is the output of capital goods of type α; r is the rate of profit; W_0 is the nominal wage rate; $k_{1j}(\alpha)$ is the amount of capital good type α used in producing one unit of good j, $(j = 0, 1)$, and L_j equals the amount of labor employed in producing one unit of good j. For a given technique matrix α, from (3.44) and (3.45) the following wage–price relation and relative prices relations are obtained:

$$w_0 = \frac{W_0}{p_0} = \frac{1 - a_{\ell 1} r}{(1 - a_{11} r) a_{\ell 0} + a_{10} a_{\ell 1} r}, \tag{3.46}$$

$$\frac{p_1(\alpha)}{p_0} = \frac{a_{\ell 1}}{(1 - a_{11} r) a_{\ell 0} + a_{10} a_{\ell 1} r}. \tag{3.47}$$

[23] This example is taken from Brown (1980).

[24] Bruno, Burmeister and Sheshinski (1966) show that there is no essential difference between the circulating capital and fixed capital models as far as the important capital-theoretic issues are concerned.

If $\ell \equiv (a_{10}/a_{\ell 0})/(a_{11}/a_{\ell 1}) > 1$, Sector 1, and if $\ell < 1$, Sector 2 is, respectively, the more capital intensive.

The properties of the wage–price and relative price-relations, depicted by (3.46) and (3.47), are critical in the controversy on reswitching and capital reversing. The factor–price ratio $(w - r)$ can be concave or convex to the origin. In special cases where price relation (3.47) is linear, i.e. $\ell = 1$, it suggests that the capital–labor ratio in the two sectors is the same. No reswitching can take place in this case, and the value of capital is also independent of the value of r, i.e. capital reversing does not occur either. As soon as the capital–labor ratios differ between sectors, i.e. $\ell > 1$, then the factor–price frontier will not be a straight line. The frontier will be concave to the origin if the factor proportion is higher in Sector 1 than in Sector 2. If the opposite factor ratios hold, it would be convex to the origin. However, in general, the factor–price frontier may be convex for certain stretches, concave for other stretches, and become convex again, depending on the number of commodities produced and differences in capital intensity of the techniques used in producing different commodities.

The upshot of the discussion is that, in the presence of reswitching and capital reversals, the construction of an aggregate production function is not logically possible. Consequently, the results deduced from the literature on aggregate production functions rest on very precarious ground, at best. It is not necessary for both capital reversing and reswitching phenomena to be present to reach this conclusion; the reswitching phenomenon is enough to show the impossibility of constructing a rigorously justifiable measure of aggregate capital stock. Several methods have been suggested to escape this dilemma; only two of them are briefly mentioned below.[25]

3.2.2 *Pseudo-production functions*

Joan Robinson (1956) proposed to measure different elements of the capital stock in terms of labor time required to produce them, compounded over their gestation periods at various given rates of interest. The set of equipment that will be in actual use in a *given equilibrium* situation will be those with the highest rate of profit (interest) at a given wage rate.[26]

[25]See Brown (1980) for other aggregation approaches, especially his general equilibrium specification of the problem.

[26]Employing the notion of equilibrium is tantamount to abstraction from uncertainty and to fulfillment of all expectations, i.e. all issues regarding disequilibrium dynamics, expectation formation, and the adjustment process cannot be analyzed in this context. Accordingly, the construction of production functions is basically a stationary-state concept and should not be used to depict the motion of the productive process between equilibrium states.

Assume perfect competition, known wage and profit rates, and constant returns to scale in production.[27] Then, in equilibrium,

$$K = wLg(1+r)^t = \frac{y - wL_c}{r}, \tag{3.48}$$

where K is capital measured in terms of consumption goods, w is real wage rate, r is the rate of profit, Lg is labor requirement to produce a unit of capital with the investment gestation period of t periods, y is output of consumption goods produced using L_c of labor and a unit of equipment. Capital measured in terms of labor time is

$$K_L = \frac{K}{w} = Lg(1+r)^t. \tag{3.49}$$

For each unit of equipment,

$$y = wL_c + rwLg(1+r)^t \tag{3.50}$$

which can be rewritten as

$$w = \frac{y}{L_c + rLg(1+r)^t}. \tag{3.51}$$

This expression depicts a costing and valuation of each item of equipment and leads to the Robinson "Pseudo-Production Function", indicated in Figure 3.3, relating output per head to capital per head.[28] Note that: (i) points $a, b, \ldots, f$ represent stationary equilibrium positions and can be compared with one another, since both capital and output are measured in the same unit, but movements up and along the function cannot be compared; and (ii) the "pseudo-production function" cannot be differentiated to obtain the magnitude of factor payment.

From the production theory point of view, this procedure has the drawback that the same physical capital can have different values in two different equilibrium situations because it is associated with a different set of wage–profit rates.

[27]See Robinson (1956) and Harcourt (1972) for further discussion of this method.

[28]The zig-zags in the graph indicate points where one stationary state crosses to another, while the long-run stationary equilibrium consistent with a given w–r combination is indicated by the dotted lines.

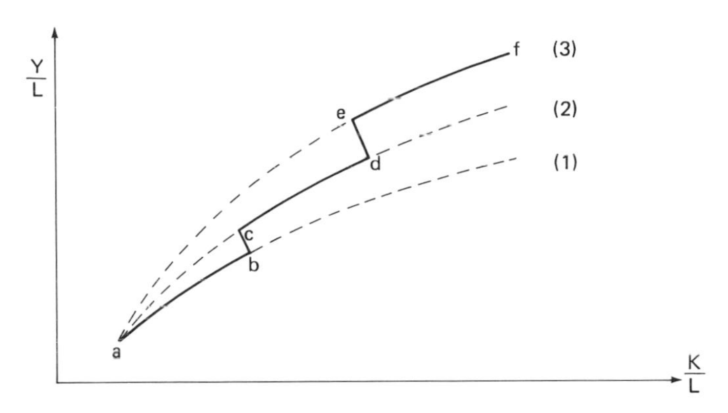

Figure 3.3. Pseudo-Production Function.

3.2.3. Chain index technique

Champernowne attempted to construct a unit to measure capital stock using his "chain-index" technique. If such a unit can be measured, then the production function can be constructed, tracing a concave relationship between output per head of a constant labor force and capital per head, allowing the marginal productivity theory to be preserved. [See Champernowne (1953) for further discussion.]

Consider an example: let the base of the index be the real value of γ equipment at r_w and call it $K(\gamma)$. Suppose that the ratio of capital costs,

$$\left[\frac{p_1(\beta)}{p_0}\sum K_{ij}(\beta)\Big/\frac{p_1(\gamma)}{p_0}K_{ij}(\gamma)\right]$$

of the β to the γ technique at r_2 is 3:1, and that the ratio of the capital costs of the α to β technique at r_1 is 6:5. Then a chain index of these three heterogeneous capital goods would be $K(\gamma)$. $(1:3:3\times 6/5)$. Thus, as the interest rate falls, the quantity of capital rises. Champernowne is then able to arrange all the alternative techniques of production in a "chain" for some "predetermined" rates of profit (chosen at equal-profitable points). Different stocks of capital are thus ordered in an unambiguous manner. The conventional production function, in which output is expressed as a unique relationship between labor and capital (here representing quantities of different capitals), can be traced out by parametric variations of the profit rates, and the familiar marginal productivity analysis can be undertaken.

Several assumptions underlie the use of the chain-index method to construct the production function. Briefly they are [Bliss (1975)]: (i) a finite number of stationary states; (ii) stationary-state equilibrium for all non-negative values of rate of interest, at least up to a maximum rate; (iii) a unique continuous functional relationship between prices and an admissible interest rate; (iv) no stationary state is repeated more than once; (v) no overlaps between two stationary states, except that *tie points* may exist between them at a given interest rate; and (vi) the set of rate of interest values is a closed connected set, i.e. any particular stationary state will appear over only one range of values of rate of interest. If these assumptions hold, then the interest rate will serve to arrange stationary states in a linear sequence. Note that given the construct of the chain-index, it is necessary to know the wage and interest rates and the associated price of output. Also, reswitching and capital reversing situations are explicitly ruled out by the above assumptions; in the presence of these phenomena the chain-index method breaks down.

The issues raised by the Cambridge School are important and the controversy is still not over. These criticisms have clearly pointed out logical difficulties in the neoclassical approach as well as the possibility of drawing misleading inferences if the neoclassical parables are accepted without reservation. The main issue that needs to be resolved is how important and frequent are the possibilities of reswitching and capital reversibility. Very little, if any, empirical work is available to answer this important question. Moreover, use of the labor input measure of capital stock, as in the work of Professor Robinson, does not solve the problem if we were to take account of human capital; the literature on human capital clearly suggests that labor input is highly heterogeneous and very often incorporates skills and attributes which make the labor input also a quasi-intermediate factor of production. The fundamental issues raised by the Cambridge School apply, if strictly interpreted, to both capital and labor aggregation. However, at present and until the critical theoretical issues are resolved and satisfactory methods of aggregation are found, the question is whether these criticisms justify abandonment of most of the received macro and microeconomic theory. The answer at present is probably no.

4. Some extensions of the theory of firm

The neoclassical model described in Section 1 is a general but static model of optimal output and input decisions. There are many interesting, special versions of this model that, due to space limitations, cannot be discussed here. Three recent developments in the areas of dynamic input decision, the effects of regulation on firm resource allocation, and output and cost analysis of multi-product firms are briefly discussed below.

4.1. Disequilibrium models of input demand

In these models an attempt is made to drop the assumption of equilibrium which underlies the model of the firm described in Sections 1 and 2. Instead, they focus on the process of moving from one equilibrium state to another. A simple disequilibrium model of factor demand is sketched below. [See Treadway (1969) for further analysis.]

Suppose the firm maximizes its present value

$$V=\int_0^{\infty}\{py-WL-rK-G\dot{K}\}e^{-\rho t}dt \tag{4.1}$$

subject to the production function $f(y, L, \dot{K}, K)=0$ and the initial condition $K(0)=K_0$. P is price of output, y is level of output, W is nominal wage rate, r is user cost of capital, G is cost of investment, K is capital, L is labor, and $\dot{K}$ is net investment. $\dot{K}$ is introduced in production on the assumption that firms produce essentially two types of outputs: y, to sell, and $\dot{K}$, the internally accumulated capital which will be used in future production. K is assumed to be neither perfectly fixed nor perfectly variable. Suppose, in addition, that the production function is characterized by the relation

$$y+C(\dot{K})-f(K,L)=0, \tag{4.2}$$

where C and f are continuous, and the marginal products f_L and f_K are positive and diminishing. The optimal behavior of the firm is characterized by maximizing the integrand

$$\int_0^{\infty}\{pf(K,L)-WL-rK-G\dot{K}-pC(\dot{K})\}e^{-\rho t}dt. \tag{4.3}$$

The necessary conditions are

(a) $pf_L=W$;

(b) $\ddot{K}=\dfrac{\rho\{G+pC'(\dot{K})\}+r-pf_K}{pC''(\dot{K})}$;

(c) $\lim\limits_{t\to\infty}\{G+pC'(\dot{K})\}e^{-\rho t}=0$;

(d) $C''(\dot{K})\geq 0$. (4.4)

Condition (a) suggests that the value of the marginal product of a perfectly variable factor equals its price; condition (b) can be rearranged and integrated to yield

$$(\mathrm{b}') \quad G+pC'(\dot{K})=\int_t^{\infty}\{pf_K-r\}\mathrm{e}^{-\rho(\tau-t)}\mathrm{d}\tau. \tag{4.5}$$

The right-hand side of (b′) is the discounted sum of future net values of the marginal product of quasi-fixed resource K, and the left-hand side is the contemporaneous marginal cost of investment, which consists of two parts: the purchase price of investment and the marginal value of real product foregone as a consequence of expansion at the rate $\dot{K}$. Condition (c) requires that the discounted marginal cost of investment approaches zero as $t\to\infty$, while condition (d) requires the adjustment cost be increasing with $\dot{K}$.

The characteristics of the production function $f(K,L)$ play an important role in determining the firm's dynamic behavior. The pattern of optimal investment and the conditions under which firms enter into or exit from the industry largely depend on the degree of returns to scale of the firm's production function. Consider the case of decreasing returns to scale:

$$H(K,L)=f_{KK}f_{LL}-f_{KL}^2>0. \tag{4.6}$$

Setting $\ddot{K}$ in (b) to zero and differentiating totally with respect to K:

$$rCp''(\dot{K})\left.\frac{\partial\dot{K}}{\partial K}\right|_{K=0}=\frac{PH(K,w)}{f_{LL}(K,w)}<0, \tag{4.7}$$

where $w=W/P$. This expression indicates that there is a saddle-point path which converges to a unique scale K^*, defined by

$$Pf_L(K^*,L^*)=W \quad \text{and} \quad pf_K(K^*,L^*)=R+r\{G+pC'(0)\}. \tag{4.8}$$

Now suppose the firm faces a U-shaped long-run average cost curve, i.e. variable returns to scale. Consider a linear approximation of (b) in a neighborhood of K^*,

$$\ddot{K}=r\dot{K}-\frac{H(K^*,w)}{C^*(0)f_{LL}(K,w)}(K-K^*). \tag{4.9}$$

If $H(K,w)$ is monotonically increasing, there exists a scale $K(w)$ at which:

$$\left(\frac{\psi}{2}\right)^2 = \frac{H\{\ddot{K}(w),w\}}{C''(0)f_{LL}\{K(w),w\}},$$
$$\left(\frac{\psi}{2}\right)^2 > \frac{H\{\cdot\}}{C''(0)f_{LL}\{\cdot\}}, \quad \text{if } K^*(w)>0. \tag{4.10}$$

No entry by new firms will take place and existing firms characterized by $K<K^*$ will leave the industry; all new capacity expansion is undertaken by the existing firms in the industry.

Based on a modified version of this dynamic model of the firm, the behavior of employment, capital accumulation, and rates of utilization in U.S. manufacturing industries has been estimated [Nadiri and Rosen (1973)]. The results indicate that (a) both labor and capital are quasi-fixed inputs; inputs can be differentiated by degree of their adjustment costs; (b) there are strong spill-over effects among the adjustment paths of the inputs and their rates of utilization; (c) the long-run solution for the optimal level of inputs and their rates of utilization are obtained when all short-term adjustments are completed; (d) in the short run, some factors of production may overshoot their long-run equilibrium values, permitting adjustment of more slowly adjusting inputs; and (e) most of the familiar employment and investment functions are special cases of the model.[29]

4.2. *Behavior of the firm under regulatory constraints*

Another extension of basic neoclassical theory has been in the area of the optimal response of firms to regulation. The basic issues are: (i) Does a regulated firm misallocate resources by using labor–capital proportions that do not minimize costs? (ii) What are the consequences of changes in the allowable rate of return on the profits and resource utilization of the regulated firm? (iii) Under what circumstances does the regulated firm favor labor-intensive techniques? Several other issues also arise which will be noted briefly.[30]

Suppose the firm's objective is to

$$\begin{aligned} &\underset{L,K}{\text{maximize}} && \pi(L,K)=R(L,K)-WL-rcK \\ &\text{subject to} && \frac{R(L,K)-WL}{cK}\leq s, \qquad s>r, \end{aligned} \tag{4.11}$$

[29] For other empirical studies, see Schramm (1970), Coen and Hickman (1970), and Brechling (1975).

[30] For a comprehensive discussion of these issues see Bailey (1973), Averch and Johnson (1962), and Baumol and Klevorick (1970); see also Cowing (1980).

where $R(L,K)=P(y)f(L,K)$ is the revenue function, $\pi(L,K)=R(L,K)-C(L,K)$ is the profit function; $C(L,K)=WL+rcK$ is the total cost, s is the allowable rate of return on investment, c is the acquisition cost per physical unit of capital, and r is the cost of capital. The solution of the constrained maximization problem can be obtained from the Lagrangian:

$$\psi(L,K,\lambda)=(1-\lambda)\{R(L,K)-WL-rcK\}+\lambda\{(s-r)cK\}. \tag{4.12}$$

The necessary conditions are:

$$\begin{aligned}
&\text{(i)} \quad \frac{\partial\psi}{\partial L}=(1-\lambda)[R_L-W]=0,\\
&\text{(ii)} \quad \frac{\partial\psi}{\partial K}=(1-\lambda)R_K-(1-\lambda)\gamma c-\lambda(s-r)c,\\
&\text{(iii)} \quad \frac{\partial\psi}{\partial \lambda}=R(L,K)-WL-scK=0,\\
&\qquad L>0, \qquad K>0, \qquad \lambda\geqslant 0.
\end{aligned} \tag{4.13}$$

From these conditions it follows that:

$$R_K=rc-\frac{\lambda}{1-\lambda}(s-r)c \tag{4.14}$$

and

$$\frac{R_K}{R_L}=\frac{f_K}{f_L}=\frac{\gamma c}{W}-\frac{\lambda}{1-\lambda}\frac{(s-r)c}{W}. \tag{4.15}$$

For the unregulated firm, efficient operation requires tangency of an isoquant and an isocline, i.e.

$$-\frac{\mathrm{d}L}{\mathrm{d}K}=\frac{f_K}{f_L}=\frac{rc}{W} \tag{4.16}$$

Contrasting (4.15) and (4.16) indicates that the regulated firm uses relatively too much capital and too little labor in its production process. Several other propositions follow:

(i) By differentiating totally eq. (iii) in (4.13), it can be shown that $\mathrm{d}K/\mathrm{d}s<0$; a rise in the rate of return ceiling actually decreases firms' demand for capital.

(ii) As the allowable rate of return, s, is set close to the cost of capital, r, the level of constrained profit declines.[31]

(iii) Employment increases (decreases) as the rate of return decreases (increases), provided that labor and capital are complements (substitutes), i.e.

$$\frac{\partial L}{\partial s}=\frac{R_{LK}}{R_{LL}}\frac{\mathrm{d}K}{\mathrm{d}s}\gtreqless 0$$

depending on whether $R_{LK}\gtreqless 0$.

(iv) Output generally increases as the allowable rate of return is set closer to the cost of capital, i.e.

$$\frac{\partial y}{\partial s}=f_L\frac{\mathrm{d}L}{\mathrm{d}s}+f_K\frac{\mathrm{d}K}{\mathrm{d}s}$$
$$=\left[\frac{R'(f_K f_{LL}-f_L f_{LK})}{R' f_{LL}+R'' f_L^2}\right]\left[\frac{cK}{R_K-sc}\right]<0. \tag{4.17}$$

(v) When the cost of capital increases, optimal levels of output, labor, and capital of the regulated firm remain unchanged but its profit falls.

(vi) If the acquisition cost of capital, c, is increased, the regulated firms' use of capital, production of output, and profit are reduced, i.e. using conditions (ii) and (iii), it follows that

$$\frac{\partial K}{\partial c}=\frac{sK}{R_K-sc}<0,$$
$$\frac{\partial y}{\partial c}=R'(f_K f_{LL}-f_L f_{KL})\frac{-sK}{R_{LL}(R_K-sc)}<0, \tag{4.18}$$

and

$$\frac{\partial \pi}{\partial c}=\frac{(s-\gamma)KR_K}{R_K-sc}<0.$$

[31] Differentiating $\pi=(s-r)cK$ with respect to K and s, and substituting for $\mathrm{d}K/\mathrm{d}s=cK/(R_K-sc)$, and after rearrangements we get:

$$\frac{\mathrm{d}\pi}{\mathrm{d}s}=(s-r)c\frac{\mathrm{d}K}{\mathrm{d}s}+cK=\frac{[(s-r)c+(R_K-sc)]cK}{R_K-sc}$$
$$=\frac{(R_K-sc)cK}{R_K-sc}>0.$$

(vii) Finally, if there is an increase in the wage rate, W, capital usage, output, and profit are reduced; i.e.

$$\frac{\partial K}{\partial W}=\frac{L}{R_K - sc}<0,$$

$$\frac{\partial \pi}{\partial W}=(s-r)c\frac{\mathrm{d}K}{\mathrm{d}W}<0, \tag{4.19}$$

and

$$\frac{\partial y}{\partial W}=\frac{R'(f_K f_{LL}-f_L f_{LK})L+f_L(R_L - sc)}{R_{LL}(R_K - sc)}<0.$$

Thus, imposition of an allowable rate of return on a firm facing a downward sloping demand curve does lead to a different behavior pattern with respect to output and resource utilization decisions than in the case of the unregulated firm.

4.3. *Economies of scale and scope in the multi-product firm*

In deriving the net supply functions in Section 1, it was assumed that diminishing returns prevailed. What happens, however, when this condition does not hold, i.e. when constant returns or increasing returns to scale prevail? Constant returns to scale prevails if the production function, f, is homogeneous of degree one. The assumption that all firms should exhibit at least constant returns to scale rests on the view that, in thc long run, an expanding firm can at least replicate its activities if no more economical options are possible. The derivation of net supply functions under constant returns to scale proceeds along the same lines as described in Section 1.2, but the further assumption of a zero unit profit level must be imposed. The procedure is to replace the function $f(y_1,\dots,y_{n+m})=0$ by $H(a_{12},\dots,q_1,a_{1,m+n})=0$, where $a_{ij}=y_i/y_1$ are the technical coefficients of the production process. The function H is twice differentiable, increasing in a's, convex or strictly convex (if $H=0$), and the unit profit level is zero, i.e. the value of all products is precisely equal to the value of the inputs. The unit profit function, $\bar{\pi}=p_1+\sum_{i=2}^{m+n}p_i\bar{a}_{1i}(p_2,\dots,p_{m+n})$, is homogeneous of degree one in prices. It can also be stated that

$$p_1+\sum_{i=2}^{m+n}p_i a_{1i}(p_2,\dots,p_{n+m})=f(p_1,p_2,\dots,p_{m+n}), \tag{4.20}$$

i.e. the technical coefficients are functions of prices and $p(p_1,p_2,\dots,p_{n+m})$ is denoted as the *price frontier* representing the set of prices for which an equilibrium exists.

Though the necessary conditions for equilibrium are the same for the cases of constant and diminishing returns to scale, they break down when increasing returns or declining long-run average costs prevail. In such cases, monopolistic organization of an industry may offer cost advantages over production by a multiplicity of firms. An interesting and important question is what are the conditions necessary and sufficient for a firm to be a natural monopoly and for the monopoly to be sustainable against entry. To illustrate these conditions, consider a multi-product firm with given technology and facing a given set of input prices. To show the conditions for sustainability of a firm's natural monopoly, some definitions are now in order [Baumol (1977)].

1. *Strict and global subadditivity of costs.* A cost function $C(y)$ is *strictly and globally subadditive* in the set of commodities $N = 1,\dots,n$, if for any m output vectors $y^1,\dots,y^m$ of the goods in N we have

$$C(y^1+\cdots+y^m)<C(y^1)+\cdots+C(y^m). \tag{4.21}$$

This definition means that it is always cheaper to have a single firm produce whatever combinations of output is supplied to the market. The criterion for a natural monopoly is that costs are strictly subadditive.

2. *Sustainability.* A *sustainable vector* is a stationary equilibrium set of product quantities and prices which does not attract rivals into the industry [Baumol, Bailey, and Willig (1977)]. Even if rivals are attracted, the monopoly may be able to protect itself from entry by changing its prices. But, by definition, only a sustainable vector can prevent entry and yet remain stationary.

Cost subadditivity and sustainability are not equivalent concepts. This can be illustrated for the single output case by Figure 4.1. For outputs less than y_1, average cost is declining. In this case the natural monopoly will exist and be sustainable. However, if quantity y_2 is demanded, the monopoly will charge price p_2. A rival firm could then anticipate positive excess profits by producing quantity y_1 at a price slightly less than p_2 but higher than p_1. Although output y_2 is not sustainable, a natural monopoly does exist at this point. The total cost of one firm producing y_1 and a second firm producing y_2-y_1 is greater than one firm producing the entire output y_2. This is due to the high cost of producing the first few units. Subsidization or restriction on entry will be necessary if a private firm is to produce the socially optimum output y_2. Conditions necessary for sustainability are (i) the products are weak gross substitutes; (ii) the cost function exhibits strictly decreasing ray average costs; and (iii) the cost function is also transray convex.

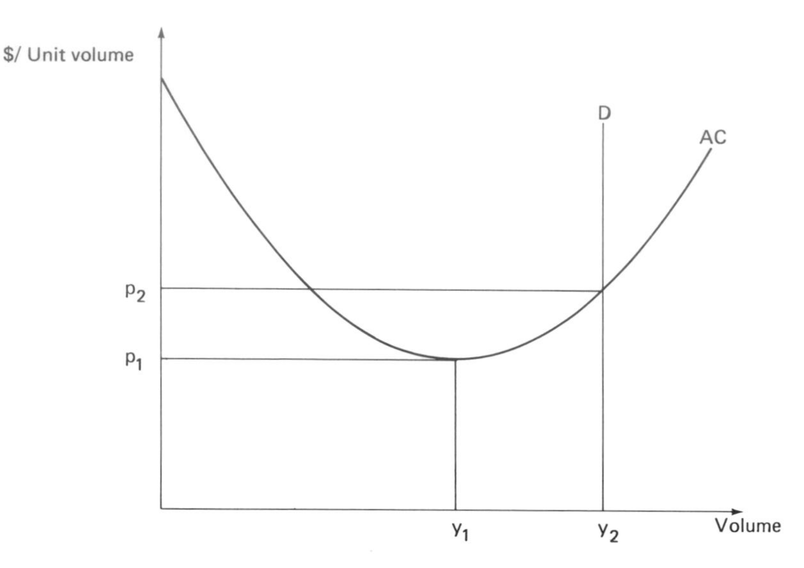

Figure 4.1. Cost subadditivity and sustainability.

Cost subadditivity is not a very practical or interpretable concept. Fortunately, it is possible to derive more intuitive conditions which imply subadditivity. However, a few further definitions are required for this discussion.

3. *Strict economies of scale* in the production of outputs are present if for any initial input–output vector $(x_1,\dots,x_r, y_1,\dots, y_n)$ and for any $\omega>1$, there is a feasible input–output vector $(\omega x_1,\dots,\omega x_r, v_1 y_1,\dots, v_n y_n)$, where all $v_i \geqslant \omega + \delta$, $\delta>0$. Scale economies occur if a proportionate ω-fold increase in all inputs implies an increase by more than the factor ω in outputs. Depending on demand growth, the firm may not wish to expand outputs proportionately. Also, if the production function is not homothetic, the firm will not want to expand inputs proportionately. Thus, economies of scale are defined entirely in terms of hypothetical input and output changes.

Declining average costs is another concept closely linked with scale economies. This concept is difficult to extend to the multi-output case since outputs do not expand proportionately and it is difficult to assign joint costs to a single output. Instead, the concept of declining ray average cost will be used. Here, output quantities are varying proportionately but input quantities follow the least cost expansion path which may not involve proportionate changes in inputs.

4. *Ray average costs* are strictly declining when $C(vy_1,\dots, vy_n)1/v< C(\omega y_1,\dots,\omega y_n)/\omega$ for $v>\omega$, where v and ω are measures of the scale of output along the ray through $y=(y_1,\dots, y_n)$.

5. *Ray concavity*. A cost function $C(y)$ is *strictly output-ray concave* (marginal costs of output bundles everywhere decreasing) if $C(kv_1 y+(1-k)v_2 y)> kC(v_1 y)+(1-k)C(v_2 y)$ for any $v_1, v_2 \geqslant 0$, $v_1 \neq v_2$, where $0<k<1$.

Economies of scale, declining ray average costs, and ray concavity are all caused by input indivisibilities or shortages of one particular factor. Each of these three concepts has been defined given that output proportions remain constant. Hence, all are ray-specific properties. In the multi-output case there must also be some way of describing the behavior of costs as output proportions vary. One such property is transray convexity.

6. *Transray convexity*. A cost function $C(y)$ is *transray convex* through $y^*=(y_1^*,\dots,y_n^*)$ if there exists any set of positive constants $\omega_1,\dots,\omega_n$, such that for every two output vectors $y^a=(y_{1a},\dots,y_{na})$, $y^b=(y_{1b},\dots,y_{nb})$ lying in the same hyperplane $\sum \omega_i y_k$ through y^*, i.e.

$$\sum_i \omega_i y_{ia} = \sum_i \omega_i y_{ib} = \sum \omega_i y_i^*$$

for which we have $C(ky^a+(1-k)y^b \leqslant kC(y^a)+(1-k)C(y^b)$ for any k, where $0<k<1$. Transray convexity exists if there are complementaries in the production of several outputs. These complementarities are in turn caused by common production processes.

Figure 4.2 illustrates a cost function with transray convexity and declining ray average costs. Total costs are shown on the vertical axis. A particular output ray is shown by a vector drawn from the origin in output space. For example, either the y_1 or y_2 axis are particular rays. Average cost is shown by the slope of a line joining the origin with any point on the cost function. Of course, all points on the minimum cost surface assume that inputs are allocated in an optimal manner. In Figure 4.2 it can be seen that average cost is declining as outputs expand along any particular ray. This indicates declining ray average cost. Transray convexity is indicated by the shape of the shaded hyperplane passing through the point (y_c, y_d). The upper surface of the hyperplane is dish-shaped and lies below a straight line joining the end points.

Two other terms often appearing in the literature are economies of scope and economies of specialization. Panzar and Willig (1977) have given the following definition of economies of scope.

7. *Economies of Scope*. A production technology exhibits *economies of scope* if $c(y^1+y^2,\omega)\leqslant C(y^1\omega)+C(y^2,\omega)$, whenever $y_j^1\cdot y_j^2=0$, $j=1,2,\dots,n$. Technically this definition can be described as strict orthogonal cost subadditivity and is similar to the criterion for natural monopoly. The difference is that the output vectors y^1 and y^2 are restricted to be orthogonal in this case. Thus, economies of scope occur if one firm can produce an output combination cheaper than two or more, each specializing in only one output. The criterion for natural monopoly is

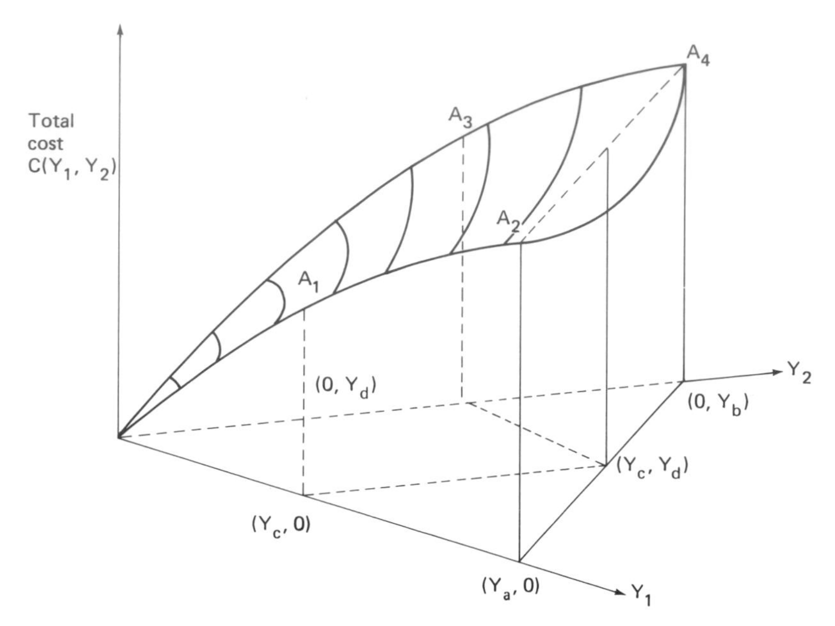

Figure 4.2. Transray convexity and declining ray average costs.

more general in that the smaller firms can also produce a mix of outputs. Diseconomies of scope are the opposite to economies of scope. Several firms specializing in one product can produce an output mix cheaper than one diversified firm.

Using these definitions and concepts, we can now determine the conditions necessary for the existence and sustainability of natural monopoly.

4.3.1. Existence

Strict subadditivity, the criterion of natural monopoly, is implied by the properties of strictly declining ray average cost (input indivisibilities) and transray convexity (joint costs) of the cost function of the firm.

Theorem

Ray average costs that are strictly declining and (non-strict) transray convexity along any one hyperplane $\sum \omega_i y_i = \omega$, $\omega_L > 0$ through an output vector y are sufficient to guarantee strict output-specific subadditivity for output y [Baumol (1977)].

4.3.2. *Sustainability*

Given that the cost properties of strictly decreasing ray average cost and transray convexity hold and the outputs are grossly weak substitutes, the use of Ramsey optimal pricing is sufficient to guarantee sustainability [Baumol, Bailey, and Willig (1977)]. Ramsey prices are derived by seeking Pareto optimal prices and outputs subject to the constraint that the firm earns profits at least equal to the maximal economic profit permitted by barriers to entry. Consider a firm producing two outputs, y_1 and y_2, and which has the subadditive cost function

$$C(y_1, y_2) = F + C_1 y_1 + C_2 y_2, \qquad (y_1, y_2) \neq (0,0)$$

For simplicity, assume that the demand for y_1 and y_2 are independent, i.e. $y_1 = y_1(p_1)$ and $y_2 = y^2(p_2)$; $\mathrm{d}y_i/\mathrm{d}p_i = y_i' < 0$; $i = 1,2$. The maximization problem can be stated as

$$\begin{aligned} L = {} & \int_{p_1}^{\infty} y_1(p)\,\mathrm{d}p + \int_{p_2}^{\infty} y_2(p)\,\mathrm{d}p \\ & + (1+\lambda)\{(p_1 - C_1)y_1(p_1) + (p_2 - C_2)y_2(p_2) - C_0\}. \end{aligned}$$

Assuming a regular interior maximum, the Ramsey optimal price pair is derived from the first-order conditions:

$$\begin{aligned} & [(p_1 - C_1)y_1' + y_1](1+\lambda) - y_1 = 0, \\ & [(p_2 - C_2)y_2' + y_2](1+\lambda) - y_2 = 0, \\ & (p_1 - C_1)y_1 + (p_2 - C_2)y_2 - C_0 = 0. \end{aligned}$$

From these equations, the well-known Ramsey inverse elasticity rule follows, i.e.

$$\frac{p_i - C_i}{p_i} = \frac{\lambda}{(1+\lambda)e_i}, \qquad i = 1,2,$$

where e_i is the price elasticity of demand for product i.

Ramsey prices are not the only set of prices which yield sustainability. However, to determine another set requires knowledge of the entire cost function at output levels which may be far removed from current or historical experience. If the cost function obeys the two properties mentioned above, the firm may use Ramsey pricing and be assured that no threat to entry exists.

5. Conclusions

It is clear that substantial effort has been made and considerable advancement achieved in the systematic specification of the properties of the technological opportunity set facing the firm and in designing more flexible functional forms of the production function. The development of new forms of production functions has opened up considerable opportunities to study different facets of the production process and to estimate empirically the structure of the production process. The development of the duality principle and its empirical applications has been a major contribution. The extension of the theory of the firm to take account of economics of scale and scope in a multi-product natural monopoly, the effect of regulation on firms' input and output decisions, and the explicit introduction of adjustment costs associated with these decisions, are new avenues of productive research.

However, considerable work remains to be done to refine the specifications of the technological set and production functions, to collect better data for firms, and to find suitable statistical techniques. Of yet greater significance is the issue of capital aggregation, which remains an important challenge to be faced both in theoretical and empirical research.

References

Allen, R. G. D. (1938), Mathematical analysis for economists. London: Macmillan.

Arrow, K. J., H. B. Chenery, B. S. Minhas, and R. M. Solow (1961), "Capital–labor substitution and economic efficiency", Review of Economics and Statistics, 43:225–235.

Averch, H. and L. Johnson (1962), "Behavior of the firm under regulatory constraints", American Economic Review, 52:1053–1069.

Bailey, E. (1973), Economic theory of regulatory constraints. Lexington: Lexington Books.

Baumol, W. J. (1977), "On the proper cost tests for natural monopoly in a multi-product industry", American Economic Review, 67:809–822.

Baumol, W. J. and A. Klevorick (1970), "Input choices and rate of return regulation: An overview of the discussion", Bell Journal of Economics and Management Sciences, 1:162–190.

Baumol, W. J., E. E. Bailey, and R. D. Willig (1977), "Weak Invisible Hand Theories on the Sustainability of Multi-product Natural Monopoly", American Economic Review, 67.

Beckmann, M. I., R. Sato, and M. Schupack (1972), "Alternative approaches to the estimation of production functions and of technical change", International Economic Review, 13:33–52.

Berndt, E. R. and L. R. Christensen (1973), "The internal structure of functional relationships: Separability substitution and aggregation", Review of Economic Studies, 40:403–410.

Bliss, C. J. (1975), Capital theory and the distribution of income. Amsterdam: North-Holland.

Brechling, F. (1975), Investment and employment decisions. Manchester: Manchester University Press.

Brown, M. (1980), "The measurement of capital aggregates: A post-reswitching problem", in: D. Asher, ed., The measurement of capital. Chicago: University of Chicago Press, 377–432.

Bruno, M. (1969), "Estimation of factor contribution to growth under structural disequilibrium", International Economic Review, 9:49–62.

Bruno, M., E. Burmeister, and E. Sheshinski (1966), "The nature and implications of the reswitching of techniques", Quarterly Journal of Economics, 80:526–553.
Champernowne, D. G. (1953), "The production function and the theory of capital: A comment", Review of Economic Studies, 21:112–135.
Christensen, L. R., D. W. Jorgenson, and L. J. Lau (1973), "Transcendental logarithmic production functions", Review of Economics and Statistics, 55:28–45.
Coen, W. and B. Hickman (1970), "Constrained Joint Estimation of Factor Demand and Production Demand", Review of Economics and Statistics, 52:287–300.
Cowing, T. G. (1980), "The effectiveness of rate of return regulation: An empirical test using profit functions", in: Fuss and McFadden (eds.) (1980) pp. 215–246.
Diwan, R. (1970), "About the growth path of firms", American Economic Review, 60(1):30–34.
Diewert, W. E. (1973), "Functional forms for profit and transformation functions", Journal of Economic Theory, 6:284–316.
Diewert, W. E. (1974), "Applications of duality theory," in: M. D. Intriligator and D. A. Kendrick (eds.), Frontiers of Quantitative Economics, Vol. II. Amsterdam: North-Holland Publishing Co.
Eichhorn, W. (1970), Theorie der Homogenen Produktions Funktion, Lecture Notes in Operation Research and Mathematical Systems, Vol. 22. Berlin, Heidelberg, New York: Springer-Verlag.
Fare, R., L. Jansson, C. A. Lovell, and Knox (1978), "On ray-homothetic production functions", in: B. Carlsson, G. Elliason, and M. I. Nadiri, eds., The importance of technology and the permanence of structure in industrial growth, Vol. 2. IUI: Stockholm.
Fisher, F. (1969), "The existence of aggregate production functions", Econometrica, 37:553–577.
Fuss, M. and D. McFadden (eds.) (1980), Production economics: A dual approach to theory and application. Amsterdam: North-Holland.
Fuss, M. and Mundlak (1980), "A survey of functional forms in economic analysis of production", in: Fuss and McFadden, eds., (1980), pp. 219–286.
Green, H. A. J. (1964), Aggregation in economic analysis: An introductory survey, Princeton: Princeton University Press.
Griliches, Z. and V. Ringstad (1971), Economics of scale and the form of the production function. Amsterdam: North-Holland.
Harcourt, G. C. (1972), Some Cambridge controversies in the theory of capital. Cambridge: Cambridge University Press.
Hicks, J. R. (1965), Capital and growth. New York: Oxford University Press.
Hotelling, H. (1932), "Edgeworth's taxation paradox and the nature of demand and supply functions", Journal of Political Economy, 40:517–616.
Kadiyala, K. R. (1972), "Production functions and elasticity of substitution", Southern Economic Journal, 38:281–284.
Klein, L. R. (1953), A textbook of econometrics. Evanston: Row, Petersen.
Lau, L. J. (1976), "A characterization of the normalized restricted profit function", Journal of Economic Theory, 12:131–163.
Lau, L. J. (1980), "Application of profit functions", in: Fuss and McFadden, eds. (1980).
Leontief, W. W. (1947), "Introduction to a theory of the internal structure of functional relationships", Econometrica, 15:361–373.
Liu, T. C. and G. H. Hildenbrand (1965), Manufacturing Production Functions in the United States, 1957. Ithaca: Cornell University School of Industrial Relations.
Lu, Y. and L. B. Fletcher (1968), "A generalization of the CES production function", Review of Economics and Statistics, 50:449–453.
Malinvaud, E. (1972), Lectures on microeconomic theory. Amsterdam: North-Holland.
McFadden, D. (1963), "Constant elasticity of substitution production functions", Review of Economic Studies, 30:73–77.
McFadden, D. (1980), "Cost, revenue, and profit functions", in Fuss and McFadden, eds. (1980).
Morishima, M. (1966), "Refutation of the non-switching theorem", Quarterly Journal of Economics, 80:520–525.
Nadiri, M. I. and S. Rosen (1973), A disequilibrium model of demand for factors of production. New York: National Bureau of Economic Research.

Nerlove, M. (1967), "Recent empirical studies of the CES and related production functions", in: M. Brown, ed., The theory and empirical analysis of production. New York: National Bureau of Economic Research.

Pasinetti, L. L. (1962), "Rate of profit and income distribution in relation to the rate of economic growth", Review of Economic Studies, 29:267–279.

Panzar, J. C. and R. D. Willig (1977), "Economies of scale in multi-output production", Quarterly Journal of Economics, 91:481–493.

Revankar, N. S. (1971), "A class of variable elasticity of substitution production functions", Econometrica, 39:61–72.

Robinson, J. (1956), The accumulation of capital. London: Macmillan.

Samuelson, P. A. (1966), "Paradoxes in capital theory, a summing up", Quarterly Journal of Economics, 80:568–583.

Sato, R. (1967), "A two-level constant elasticity of substitution production function", Review of Economic Studies, 34:201–218.

Sato, R. and R. F. Hoffman (1968), "Production functions with variable elasticity of factor substitution: Some analysis and testing", Review of Economics and Statistics, 50:453–460.

Schankerman, M. A. and M. I. Nadiri (1980), "Variable and long-run cost functions and the rate of return to R & D: An application to Bell system", mimeo. New York University, May 1980.

Schramm, R. (1970), "The influence of relative prices, production conditions, and adjustment costs on investment behavior", Review of Economic Studies, 37:361–376.

Shephard, R. W. (1953), Cost and production functions. Princeton: Princeton University Press.

Shephard, R. W. (1970), Theory of cost and production functions. Princeton: Princeton University Press.

Treadway, A. B. (1969), "On rational enterpreneurial behavior and the demand for investment", Review of Economic Studies, 38:227–239.

Uzawa, H. (1964), "Duality principles in the theory of cost and production", International Economic Review, 5:216–219.

Zellner, A. and N. Revankar (1969), "Generalized production functions", Review of Economic Studies, 36:241–250.

Chapter 11

OLIGOPOLY THEORY

JAMES FRIEDMAN*

University of Rochester

1. Overview of oligopoly theory and summary of the chapter

The term *oligopoly theory* usually refers to the partial equilibrium study of markets in which the demand side is competitive, while the supply side is neither monopolized nor competitive. Under monopoly, the sole supplying firm faces a simple, straightforward profit maximization problem. In this respect, monopoly and competition are similar. Under oligopoly, however, a firm does not stand alone nor is it so small that its interconnections with the other firms in the market are negligible. It is these interconnections which form the main matter of interest in the study of oligopoly theory. The firms' customers enter the model in a mechanical way and their behavior is completely specified by the demand functions facing the firms. Nothing that one single buyer can do has any observable effect on any firm or other buyer. The actions of any one firm, however, have quite noticeable consequences for each of the other firms. The reader may recognize that the situation from the vantage point of the firms is essentially that of an n-person game in the sense of von Neumann and Morgenstern (1954). n is the number of firms in the market, and, apart from Section 8, it is assumed throughout the chapter that the buyers are unable to form coalitions (i.e. cooperating groups).

The first treatment of oligopoly is Cournot's and dates back to 1838 [see Cournot (1927)]. His model and virtually everything to follow for over a century is essentially single period. It is true that part of Cournot (1927) and of some later writers [e.g. Edgeworth (1925), Bowley (1924), Stackelberg (1934), and Sweezy (1939)] show some concern with multi-period phenomena; however, these efforts do not really get off the ground. Section 2 is concerned exclusively with single-period models. The two models which are mainly discussed are Cournot's and a model based on Chamberlin (1956). In Cournot's model it is assumed that the

*I wish to thank Reinhard Selten for his helpful comments; however, all blame for shortcomings rests with me.

Handbook of Mathematical Economics, vol. II, edited by K.J. Arrow and M.D. Intriligator

products of the firms are perfect substitutes (i.e. the industry produces a homogeneous good) and the firms each decide output levels with price being determined "in the market." The Chamberlinian model allows for imperfect substitutability between the outputs of the firms, and the firms are each assumed to name prices for their products.

In addition to their interest for the study of single period oligopoly, the two models developed in Section 2 also provide some of the basic structure for the models of multi-period oligopoly which are the subject of Section 4. The very first dynamic elements in oligopoly are in Cournot (1927). These are his *reaction curves*. Later developments include Bowley's *conjectural variation*, Stackelberg's *leader–follower* model, and Sweezy's *kinky demand curve* model. As may be seen in Section 3, the time structure of none of these models is explicit; however, none of them makes sense when interpreted as strictly single-period models. Thus, for the best understanding of them, it is necessary to augment them by making the time relations explicit and by making an appropriate reformulation of the firms' objective functions.

In Section 4 various models are examined in which the time structure is explicit and the firms seek to maximize discounted profit streams. This work is all relatively recent, having developed over the past 15 years. Several of the models involve reaction functions and quite clearly derive their inspiration from the works surveyed in Section 3 – particularly from Cournot's reaction curves and their subsequent elaboration by Bowley and Stackelberg.

The first category of these models may be termed *lagged response*. Here it is assumed that a firm never has the opportunity to find out what others have chosen to do in the current time period prior to making its own decision. Each firm's behavior is given by a *reaction function*, which is a function that sets the current period action of a firm as a function of the previous period actions of all firms. With these models, one line of investigation has been pursued by Cyert and deGroot (1970) and another by Friedman (1977a, ch. 5). In both approaches, indeed in all models surveyed in Section 4, the time structure is crystal clear. That is to say, it is explicitly stated who makes which decisions when, what flows of income are received at which times, how the firms value income, etc. The second type of model is characterized by either one of two conditions. First, the firms can respond instantaneously to any change made by one of them. It is as if a firm cannot change its price as of the start of period t without announcing its intention beforehand and without the other firms being allowed to change their prices, also as of the start of period t, in response. The second way of characterizing these models is to have the same sort of time arrangements as, say, Friedman, along with an adjustment cost for altering price. Both versions work out in essentially the same fashion. The last of the multi-period models derives from game theory. Though it could be interpreted as a reaction function model, that is not the approach taken. What is most of interest in this model is that *non-cooperative* equilibria may exist which yield profits to the firms which could not conceivably be improved upon by collusion. For the moment, *non-cooperative* may be thought of as synonymous with non-collusive.

A characteristic of all the models in Sections 3 and 4 is that they are non-cooperative. The reason for this lies in the way oligopoly has developed, rather than in the intrinsic interest of cooperative versus non-cooperative theory. While there is a sizable body of non-cooperative oligopoly theory which is clearly distinct from, though it may draw upon, non-cooperative game theory, the same cannot be said for cooperative oligopoly theory. What little insight rigorous economic theory can bring to cooperative oligopoly is obtained from cooperative game theory, to which the reader is strongly encouraged to turn.

The remaining sections of the chapter are relatively short. Section 5 deals with some of the connections between oligopoly and game theory. Section 6 contains a brief discussion of models with entry and exit of firms. Section 7 deals with a topic which is, strictly speaking, outside the scope of oligopoly, but which is often discussed together with it, namely bi-lateral monopoly. The element in common between the two is that they are both essentially non-zero sum games of strategy. At the beginning of the chapter it was noted that oligopoly is predominately a partial equilibrium study. It is natural to wish to analyze oligopoly within a model of general equilibrium. Only very recently has such work been undertaken. Some of it is briefly sketched in Section 8. The reader of the present chapter may find Chapter 7 of this Handbook, on game theory and its applications, by Martin Shubik, to be of interest.

2. The early history – single-period models

The first consideration of oligopoly appears to have been by Cournot. His treatment of monopoly and of competitive markets, while much more precise and technically complete than anything previous to his work, essentially embodies the understanding which was "in the air" at the time he wrote. By contrast, there is nothing about oligopoly to be found earlier. The organization of his book suggests the way he might have been led to his discovery. The first type of market examined is the monopolistic. He proceeds from it to competitive markets, and, clearly, to an organized mind, there is a stop in between – the market with more than one, but not a large number, of firms. This market is examined in Section 2.1. Sections 2.2–2.4 may be thought of as comments on Cournot. Section 2.2 deals with the famous critique of Cournot's oligopoly model made by the French mathematician Joseph Bertrand (1883) and elaborated by F. Y. Edgeworth (1925).

Cournot's model of quantity choosing producers of a homogeneous good may be criticized on the ground that oligopolists choose prices, as Bertrand did, and it may also be claimed that in virtually all instances oligopolists produce differentiated products. This view is taken by Chamberlin (1956). A model in the Chamberlinian tradition is examined in Section 2.3. Section 2.4 contains a brief discussion of cooperative or collusive equilibria, along with some discussion of the circumstances under which cooperative and non-cooperative behavior might be expected.

2.1. *Cournot's homogeneous products quantity model and the Cournot equilibrium*

Imagine a market in which there are n firms producing and selling a homogeneous good. The output, which is assumed equal to sales, of the ith firm is q_i. Total sales in the market are $Q=\Sigma_{i=1}^{n}q_i$. Letting p denote market price, the *inverse demand function* in this market is $p=f(Q)$. f is assumed twice continuously differentiable and, as illustrated in Figure 2.1, it is downward sloping and cuts both axes. Each firm has a *cost function*, $C_i(q_i)$, which is twice continuously differentiable, non-negative, convex, and has a positive first derivative. That is, fixed costs are non-negative while marginal costs are both positive and non-decreasing as the level of output rises. The *profit function* for the ith firm may be written, with q denoting the vector of outputs $(q_1,\dots,q_n)$:

$$\pi_i(q)=q_i f(Q)-C_i(q_i), \qquad i=1,\dots,n. \tag{2.1}$$

This modcl is expressly single period in the sense that each firm is assumed to choose an output level exactly once, that all firms make their choices simultaneously, and that profits are given by eq. (2.1). Thus, in this model it is not possible for firms to make their decisions as a reaction to earlier decisions of others or to use the past history of the market as a guide to what they expect rivals to do in the present. There is no past history or any previous decisions to react to. The single-period nature of the model may appear objectionable on the ground that the essential relations between oligopolists cannot be captured; however, the model may be defended. First, the model does yield some fundamental insights which carry over into multi-period models. More may be learned from multi-period models, and not every result obtained in single-period models is found when the transfer is made to many periods. Yet some important results do carry over. Second, a multi-period model requires a single-period model as a building block.

One additional assumption to be made is that the profit of the ith firm is a concave function of its own output. That is, that $\partial^2\pi_i/\partial q_i^2<0$. When a *Cournot*

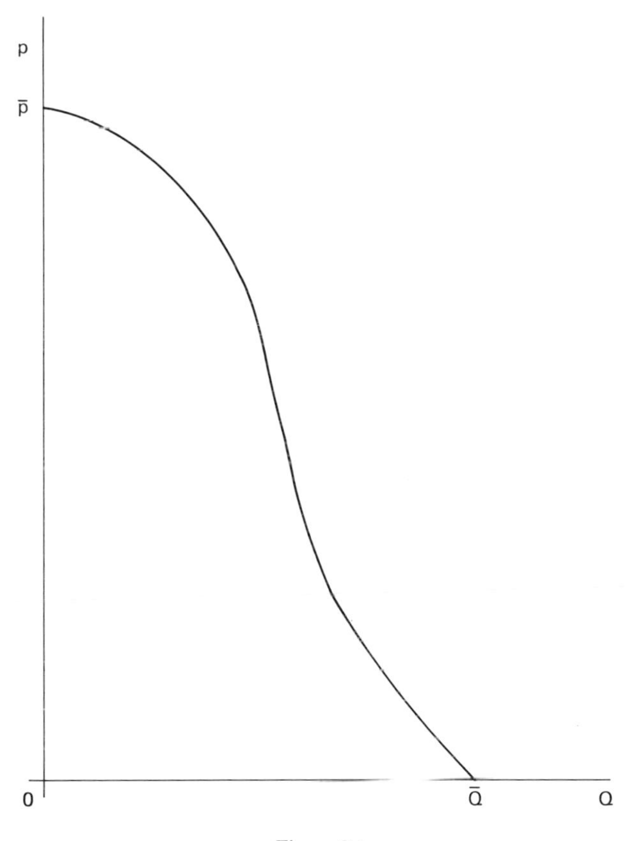

Figure 2.1.

equilibrium occurs at positive output levels for all firms and at a positive market price the following conditions of profit maximization are satisfied:

$$\frac{\partial \pi_i}{\partial q_i}=f(Q)+q_i f'(Q)-C_i'(q_i)=0, \qquad i=1,\dots,n. \tag{2.2}$$

The characteristic feature of the Cournot equilibrium is that no single firm by itself can deviate from its equilibrium output level and increases its profit when all the other firms choose their equilibrium output levels. A more general definition of the Cournot equilibrium may be based on the preceding statement and requires no reliance on first derivatives being zero. Such a statement is: q^c is a *Cournot equilibrium* if $q_i^c \geqslant 0$ $(i=1,\dots,n)$ and no single firm can increase its profits by choosing $q_i \neq q_i^c$ given that $q_j=q_j^c (j \neq i)$.

Following a formal statement of the assumptions, some theorems are given. Then there is additional discussion of the nature and reasonableness of the Cournot equilibrium.

Assumption 1

The inverse demand function for the market, $f(Q)$, is defined and continuous for all $Q \geqslant 0$. There is $\bar{Q} > 0$ such that $f(Q) = 0$ for $Q \geqslant \bar{Q}$ and $f(Q) > 0$ for $Q < \bar{Q}$. Furthermore, $f(0) = \bar{p} < \infty$, and for $0 < Q < \bar{Q}$, f has a continuous second derivative and negative first derivative.

Assumption 2

The cost function of the ith firm, $C_i(q_i)$, is defined and continuous for all non-negative output levels, $C_i(0) \geqslant 0$. C_i has a continuous second derivative for $0 < q_i$ and $C_i'(q_i) > 0$, $i = 1, \ldots, n$.

Assumption 3

For all $q_i > 0$ and $Q < \bar{Q}$, $f' + q_i f'' < 0$ and $f' - C_i'' < 0$, $i = 1, \ldots, n$.

A convenient alternative way of denoting a vector, q, is $q = (q_i, \bar{q}_i)$, where $\bar{q}_i$ denotes all coordinates of q except q_i. q^c is a Cournot equilibrium output vector if $q^c \geqslant 0$ and $\pi_i(q^c) \geqslant \pi_i(q_i, \bar{q}_i^c)$ for all $q_i \geqslant 0$ and for $i = 1, \ldots, n$. Some important theorems on single-period oligopoly are:[1]

Theorem 1

A Cournot oligopoly satisfying Assumptions 1–3 has a unique Cournot equilibrium.

Theorem 2

If $q^c \gg 0$, then there exists $q^* \geqslant 0$ such that $\pi_i(q^*) \gg \pi_i(q^c)$, $i = 1, \ldots, n$.[2]

Theorem 1 is self-explanatory. Theorem 2 states that when the Cournot equilibrium is "interior", then it does not yield Pareto optimal profits. Pareto optimality is used here in a special sense: within a partial equilibrium oligopoly model, an attainable vector of profits is said to be *Pareto optimal* if there is no other attainable vector of profits which yields more profit to some firms and no less to any firm. This is a standard definition of Pareto optimality except for the restriction of attention to outcomes to the firms alone. Figure 2.2 shows a set of attainable profit outcomes for a two-firm model with the Cournot equilibrium

[1]These theorems may be found in Friedman (1977a, ch. 2, §§ 2, 3 and ch. 7, § 6), Selten (1970, § 9.4), and Szidarovszky and Yakowitz (1977).

[2]Vector inequalities are denoted as follows: $x \geqslant y$ means $x_i \geqslant y_i$ for $i = 1, \ldots, n$. $x > y$ means $x \geqslant y$ and $x \neq y$. $x \gg y$ means $x_i > y_i$ for $i = 1, \ldots, n$.

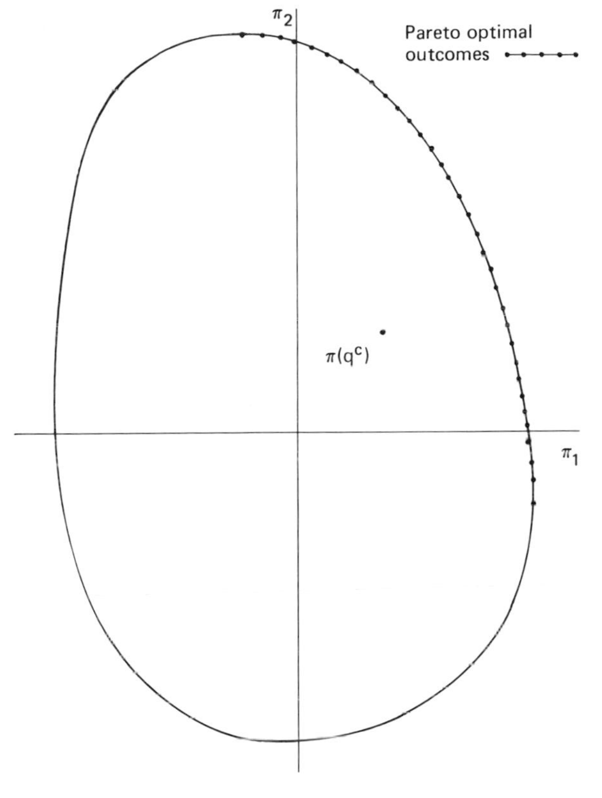

Figure 2.2.

interior to the set. It is easy to see what causes the result to be true. The conditions of the theorem ensure that eq. (2.2) is satisfied for all i at the Cournot equilibrium. Meanwhile,

$$\frac{\partial \pi_i}{\partial q_j} = q_i^c f'(Q^c) < 0, \quad \text{for all } i, j, i \neq j.$$

Thus, it is clearly possible to decrease all the q_i simultaneously by a small amount ε for which all firms' profits must rise. In the language of game theory, if the payoff of each player is a continuously differentiable function of all players' strategies, the non-cooperative equilibrium strategies (the Cournot output levels) are interior to their strategy sets, and the Jacobian of the system of payoff functions, evaluated at the non-cooperative equilibrium, is non-singular, then the equilibrium payoffs are not Pareto optimal. That the profits associated with the

Cournot equilibrium need not be Pareto optimal, indeed *cannot* be Pareto optimal in a large class of models, is a reason why some regard the Cournot equilibrium as uninteresting. This issue is pursued in Sections 2.4 and 4.3.

A very appealing proof of Theorem 1, originated by Selten (1970, Section 9.4) and also found in Szidarovszky and Yakowitz (1977), may be briefly outlined: Think of eq. (2.2) as a relationship between the two variables q_i and Q. Construct a function $q_i = \phi_i(Q)$ as follows: for each value of Q in the interval $[0, \bar{Q}]$, choose $\phi_i(Q)$ to be that value of q_i which satisfies eq. (2.2). If no such value exists, then choose 0. Thus, $\phi_i(Q)$ is the profit maximizing value of q_i for the ith firm, given that total industry output is Q. ϕ_i is a monotone, non-increasing function. Furthermore, $\phi_i(\bar{Q}) = 0$ and $\phi_i(0) \geqslant 0$. Let $\phi(Q) = \sum_{i=1}^{n} \phi_i(Q)$. Except when $\phi(Q) = 0$, it is a strictly decreasing function. It appears as illustrated in Figure 2.3. It must have a unique fixed point – that is, there must be a value of Q, call it Q^c, such that $Q^c = \phi(Q^c)$, and a fixed point of ϕ is a Cournot equilibrium with $q_i^c = \phi_i(Q^c)$. That a Cournot equilibrium is necessarily a fixed point of ϕ, and ϕ has exactly one fixed point, ensures the uniqueness of the equilibrium. Selten uses a generalized version of the technique just outlined and applies it to a broader class of

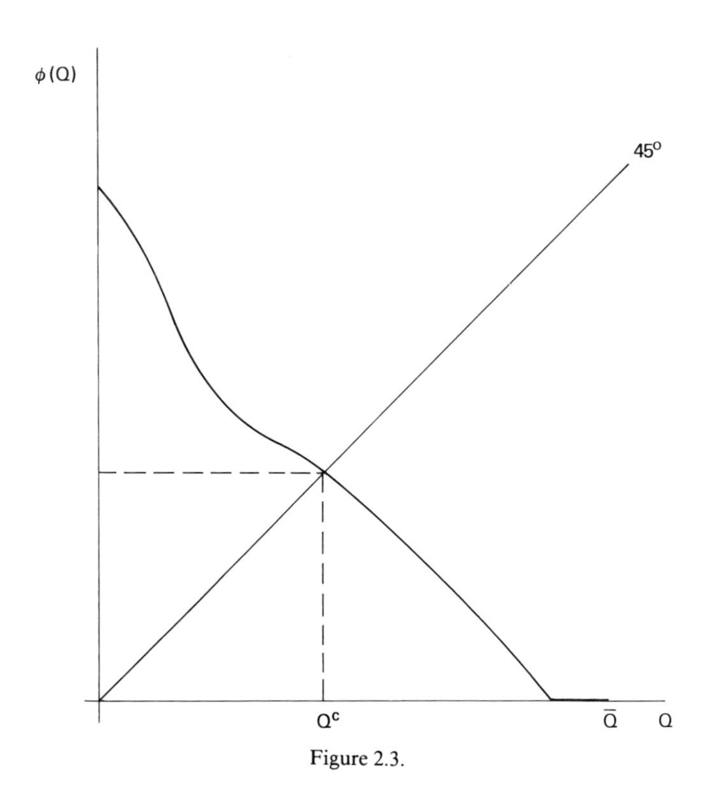

Figure 2.3.

models, which he terms *strategically agreeable games*. On this, see particularly Selten (1970, chs. 9, 10).

To appreciate more fully the nature of Cournot's equilibrium, it is instructive to think of it in relation to the possibilities for collusion which are assumed present in the model. The firms are assumed to be in one of two situations: Either they cannot communicate with each other or they can communicate but cannot make binding agreements with one another. These two alternatives are discussed in turn. If the firms cannot talk to one another or communicate in any other way, it is nearly self-evident that the only possible sort of equilibrium is that of Cournot. Assume the contrary for a moment, and say that all firms notice that the Cournot equilibrium is not Pareto optimal. Each firm realizes that if all firms choose their outputs appropriately then each would have more profit than it receives at the Cournot equilibrium. Let q^* be such an output vector. Because q^* is not a Cournot equilibrium output vector there is at least one firm, j, for which

$$\pi_j(q^*) < \max_{q_j} \pi_j(q_j, \bar{q}_j^*). \tag{2.3}$$

If firm j thought that the others, out of some sense of "public good", would choose $\bar{q}_j^*$, it would clearly not choose q_j^* because it can do better. On the other hand, even if firm i ($\neq j$) cannnot do better than choose q_i^* given that the others choose $\bar{q}_i^*$, it must realize that the others cannot be relied upon to choose q_i^* because firm j, as noted already, has an incentive to choose differently. So, spontaneous cooperation leading to an outcome which is Pareto optimal, but which is not a Cournot equilibrium, appears out of the question.

Now consider a slightly revised model. The firms can communicate, but they cannot make binding agreements. By *binding agreements* is meant that if they agree on a joint course of action, then each individual firm has no alternative except to do what it agreed to do. If binding agreements are ruled out, the managers of the firms may get together, solemnly agree on some q^*, but then each manager goes back to his office, completely free to choose any output level he sees fit. By contrast, under a binding agreement, if the firms decide on some q^*, they have entered into an unbreakable contract. Firm i, for example, is no longer able to choose any $q_i \in [0, \bar{Q}]$. It can only choose $q_i = q_i^*$. In its essentials, the situation is no different with communication (and no binding agreements) than with neither. To pursue the example of the preceding paragraph, if the firms agree on q^*, then firm j has no incentive to honor its agreement by choosing q_j^* and no other firm would expect it to choose q_j^*.

There is one exception to the foregoing remarks concerning the inability of firms to make agreements and carry then out when there is no outside enforcement mechanism. This occurs when there are multiple non-cooperative equilibria and the firms agree on one of them. Then the self-enforcing nature of the equilibrium comes into play. No firm has any incentive to do other than it said it would do, given the actions to which the others have agreed. Indeed, where there are many non-cooperative equilibria, and firms cannot communicate with one another, there is no reason to suppose that the firms would choose a non-cooperative equilibrium. Doing so requires that they all have the same equilibrium in mind.

2.2. *Bertrand's critique of Cournot's contribution*

In writing a review of Cournot's *Researches* and a book of Walras', Bertrand (1883) devotes half a page to taking Cournot to task for his chapter on oligopoly. He cites two faults: first, the firms would certainly collude and make larger profits than the Cournot equilibrium affords and, secondly, even if they were not to collude, they would choose prices rather than output levels. The first point may be rejected out of hand. This rejection is not because collusion is impossible to believe or unworthy of study. Rather, both cooperative models and non-cooperative models need to be fully developed precisely because in some circumstances binding agreements are not possible while in others they are. Cournot elected to take up the case of no binding agreements.

The second point has more substance. It is reasonable to suppose that firms are price, rather than quantity, choosers. In an oligopolistic market it is very difficult to conceive the manner in which prices are determined if they are not set by the firms. Bertand, using the assumption that the firms' outputs are homogeneous, points out that consumers would buy from the firm(s) offering the lowest price. Using the zero-cost mineral spring example, he points out that there cannot be an equilibrium at any price above zero. Should firm 1 charge a positive price, firm 2 would do well to just undercut him. But given the price of firm 2, firm 1 would be in a better position by undercutting firm 2. And so forth. Only with both prices at zero is there no way that either firm, by itself, could alter its price and increase its profit. The difficulty with Bertrand's price-choosing firms is that the relationship between sales and the price of a single firm does not seem to square with observation of the empirical world. That is to say, it does not appear to be the case that all sales go to the firm charging the lowest price. Nor does the instability which becomes possible if the firms have upper limits on the amounts they produce, discussed by Edgeworth (1925), appear to exist in the world. A reconciliation of Cournot and Bertrand is provided by Chamberlin.

2.3. *Differentiated products with prices as the firms' decision variables – Chamberlin's modification*

Chamberlin (1956) made prominent the notion of *differentiated products*. The intuitive idea is easy to grasp and even the modeling in a partial equilibrium setting offers no problem. The meaning of differentiated products in the context of a general equilibrium model of the economy does pose some difficulty. This difficulty is, for the most part, glossed over in the present chapter. At the heart of differentiated products models are the assumptions that no two firms produce identical products and that firms can be grouped according to the type of product they make. Within one group the products of two firms are very close substitutes, while between two groups the products of two firms are either complements or relatively weak substitutes. Various examples come to mind: two brands of gasoline or of aspirin might well fit the conditions outlined. In principle, however, it could very difficult to find an appropriate boundary between groups. Furthermore, the products of a given pair of firms could be extremely close substitutes under some conditions (i.e. for certain ranges of prices and certain distributions of income) and complements under other conditions. It then becomes impossible definitely to group the two products together or apart. Objections to the basic notion of Chamberlin's monopolistic competition may be found in Triffin (1940).

Although Chamberlin devotes space to oligopoly, his principal concern is with *monopolistic competition*, which is the analogue to ordinary competition for an economy in which no two firms make exactly the same good. Under monopolistic competition, as with competition, the actions of one firm have no noticeable effect on other firms in the same group (or industry). And, in any one group there are many firms, as under competition. Concern in the present chapter is with oligopoly, so the differentiated products assumption is borrowed and grafted onto a model of oligopoly. Throughout the remainder of the chapter the nature of the firms' products is absolutely fixed.[3] The consequence of assuming product differentiation is that each firm has its own demand function under which the demand faced by a firm is a continuous function of all the prices in the market. Thus, the Bertrand discontinuity is gone. There is no special significance to one firm's charging the same price as a rival, and nothing remarkable happens to the demand of a firm as its price goes from above to below that of a rival. Other things equal, its sales merely increase in an orderly fashion.

The model developed below provides an appealing price analogue to the Cournot model. It also is the basis of some of the models studied in Sections 3 and 4. The price of the ith firm is denoted p_i and the vector of prices, p. The demand function of the ith firm is $q_i = F_i(p)$. F_i is twice continuously differentiable with the first derivative of F_i with respect to p_i negative. The derivatives of F_i with respect to $p_j (j \neq i)$ are all positive, and the "own" derivative dominates. That is to say, a firm's sales fall when it raises its price, rise when some other firm raises its price, and, if all firms raise prices by an equal amount, then each firm suffers a decline of sales. The foregoing applies when the price vector at which derivatives are examined is one for which firms have positive sales. When, for

[3] Clearly the choices of which products to produce and which versions of them are very important to actual firms; however, the theoretical literature has not gotten far in analyzing such decisions.

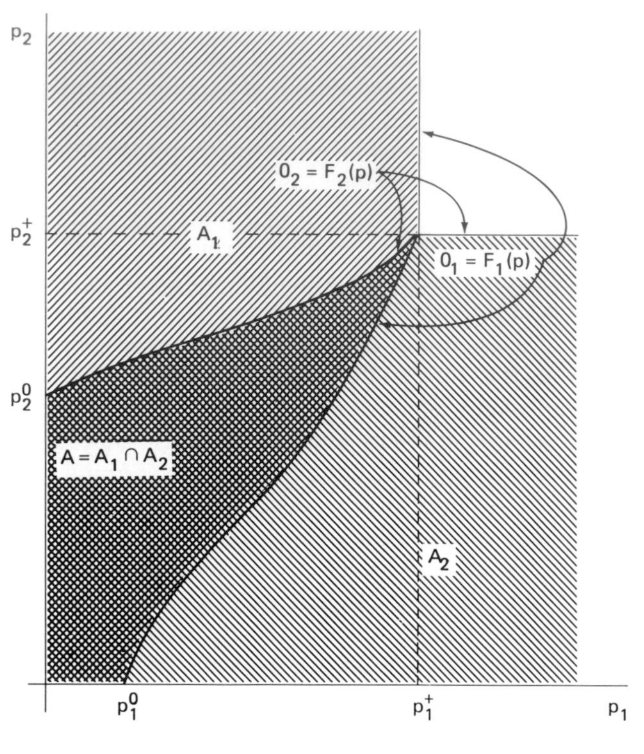

Figure 2.4.

example, $F_i(p)=0$, then price increases by firm i or decreases by any other firm cannot affect its demand. Finally, it is assumed that there are prices so high that the firm sells nothing. These conditions are illustrated in Figure 2.4. The region in which the demand of the ith firm is strictly positive is

$$\mathring{A}_i = \{ p \mid p \geqslant 0, F_i(p) > 0 \}. \tag{2.4}$$

The closure of $\mathring{A}_i$ is denoted A_i, and the set of points in A_i at which F_i is zero is denoted a_i.[4] A formal statement of the conditions governing demand is:

Assumption 4

$F_i(p)$ is defined, continuous, and bounded for all $p \geqslant 0$. $F_i(p)$ is twice continuously differentiable for all $p \gg 0$, except for $p \in \cup_j a_j$. If $p \in a_j$ for $j = i_1, \ldots, i_k$ and $p \notin a_j$ for $j = i_{k+1}, \ldots, i_n$, then continuous second partial derivatives with respect to

[4] The *closure* of a set, W, is the smallest closed set that contains W.

$p_{i_{k+1}},\ldots,p_{i_n}$ exist at p. All derivatives are bounded, and, if $F_i(p)=0$ with $p \notin \mathring{A}$, all derivatives of $F_j(j=1,\ldots,n)$ with respect to p_i are zero. For $p \in \mathring{A}_i \cap \mathring{A}_j (j \neq i)$, $F_i^j(p)>0$, and, for $p \in \mathring{A}_i \cap \mathring{A}_{i_1} \cap \ldots \cap \mathring{A}_{i_k}$, $F_i^i(p)+\Sigma_{j=1}^k F_i^{i_j}(p)<0$. A is bounded.[5]

Now the profit function of the firm may be written. It is

$$\pi_i(p)=p_iF_i(p)-C_i(F_i(p)), \qquad i=1,\ldots,n. \tag{2.5}$$

A characteristic of the demand system given by Assumption 4 is that the set A contains a unique *maximal element*, p^+. That is, there is $p^+ \in A$, $F_i(p^+)=0$ (for all i), and for any $p \in A$, $p \neq p^+$, $F_i(p)>0$ for at least one firm i.

An equilibrium in the spirit of Cournot may be defined. Although strictly speaking Cournot's equilibrium applies to his quantity model, the basic idea transfers in such an obvious way to the present model that it seems quite reasonable to call the counterpart defined below by the same name. p^c is a Cournot equilibrium price vector if $p^c \geqslant 0$ and if

$$\pi_i(p^c) \geqslant \pi_i(p_i, \bar{p}_i^c), \quad \text{for all } p_i \geqslant 0 \text{ and } i=1,\ldots,n. \tag{2.6}$$

Again, the fundamental property is that when all the other firms, $j \neq i$, choose p_j^c, the ith firm can do no better than to choose p_i^c. And this holds for all i.

There is a subset of A_i which is of special interest. It is

$$A_i^* = \{p \mid p \in A_i \text{ and } p_i \geqslant C_i'(F_i(p)), i=1,\ldots,n\}. \tag{2.7}$$

A_i^* is the set of price vectors in A_i for which the price of the ith firm is at least as large as its marginal cost. It is immediate that for any $p \notin A_i^*$ the total revenue of the firm is less than its total variable cost if its sales are also positive. Thus, at a Cournot equilibrium, for each firm either $p^c \in A_i^*$ or $p_i^c = p_i^+$. The interior of A_i^* is denoted $\mathring{A}_i^*$, $A^* = \cup_{i=1}^n A_i^*$ and $\mathring{A}^*$ is the interior of A^*. Before stating the remaining axioms and the basic theorems on the single-period price model, one more special price is needed. That is, the lowest price at which the ith firm can possibly have zero sales. This price p_i^0, satisfies the condition $(p_i^0, 0) \in a_i$. Additional assumptions are:

Assumption 5

For all $p \in \mathring{A}_i^*$, $\partial^2\pi_i/\partial p_i^2 < 0$.

[5]Superscripts are used to denote partial derivatives. For example, $F_i^j = \partial F_i/\partial p_j$ and $F_i^{jk} = \partial^2 F_i/\partial p_j \partial p_k$.

Assumption 6

For all $p \in \mathring{A}^*$,

$$\frac{\partial^2 \pi_i}{\partial p_i^2} + \sum_{j \neq i} \left| \frac{\partial^2 \pi_i}{\partial p_i \partial p_j} \right| < 0.$$

Assumption 7

$$p_i^0 > C_i'(F_i(p_i^0, 0)) = C_i'(0).$$

Assumption 5 asserts the concavity of the profit function of the ith firm with respect to p_i, the firm's own price, in the region where the firm's price is not less than its marginal cost. This assumption is used to prove existence of the Cournot equilibrium. Its role is sketched below following the statement of the existence theorem. Assumption 6, which implies Assumption 5, is used to ensure that the Cournot equilibrium price vector is the fixed point of a contraction mapping, which ensures the uniqueness of the equilibrium. Assumption 7 ensures that, no matter what prices are chosen by the other firms, the most profitable price for the ith firm is one at which its sales are strictly positive. This is in the form of a condition under which the lowest possible price at which its demand is zero is a price in excess of the firm's marginal cost at zero output. Thus, under Assumption 7 the firm can always equate marginal cost with marginal revenue at a positive output level.

Theorem 3

Under Assumptions 2, 4, and 5 a single-period oligopoly model has a Cournot equilibrium.

Theorem 4

Under Assumptions 2, 4, 5, and 7 price and output are strictly positive for all firms at any Cournot equilibrium.

Theorem 5

Under Assumptions 2, 4, 5, and 6 the Cournot equilibrium is unique. If, in addition, Assumption 7 holds, then all prices and output levels at equilibrium are strictly positive.

Theorem 6

Let p^c be a Cournot equilibrium for a market satisfying Assumptions 2 and 4 and for which $p^c \ll p^+$. Then the profits attained at the equilibrium, $\pi(p^c)$, are not Pareto optimal.

A sketch of the proof of Theorem 3 is instructive because the method of proof is widely used to show existence of Cournot-type (specifically, Nash non-cooperative) equilibria in a wide variety of models. For any single firm, say firm i, there is an optimal p_i to choose given that the other firms choose $\bar{p}_i$. As $\bar{p}_i$ is imagined to vary, so does the optimal p_i which firm i would wish to use. Denote this relationship

$$p_i = r_i(\bar{p}_i). \tag{2.8}$$

r_i is often called the *best reply function* of firm i, though there is no "replying" going on.[6] It is merely the case that if the others actually choose $\bar{p}_i$, then firm i can do no better than choose $r_i(\bar{p}_i)$. Thus, by the definition of r_i, it is clear that if there is some price vector p^* such that $p_i^* = r_i(\bar{p}_i^*), i=1,\dots,n$, then p^* is necessarily a Cournot equilibrium. No single firm can increase its profits by choosing $p_i \neq p_i^*$ when the others choose $\bar{p}_i^*$. Conversely, if p^* is a Cournot equilibrium, then $p_i^* = r_i(\bar{p}_i^*), i=1,\dots,n$. It is possible to look at all the r_i together and to consider them jointly as a function taking one price vector in R_+^n into another. Then

$$r(p) = (r_1(\bar{p}_1),\dots,r_n(\bar{p}_n)). \tag{2.9}$$

The foregoing comments should make clear that the set of fixed points of the function r [i.e. price vectors p' which map into themselves, satisfying $p' = r(p')$] and the set of Cournot equilibria are identical. The method of proof from this point on is to show that r has a fixed point. Depending on the exact nature of the model, one or another fixed point theorem is invoked; for example, whether r is a function, as here, or a correspondence; whether its argument, p, is an element of a finite dimensional space, etc.

With Theorem 3 established, Theorem 5 on uniqueness of equilibrium follows easily. The additional assumption ensures that the best reply function, r, is a contraction. This, in turn, implies that r has *at most* one fixed point.

Among the three models – Cournot, Bertrand's version of Cournot, and the differentiated products price model – the latter is the most satisfactory. It presumes firms in an oligopoly to be price choosers, and the profits of a firm vary continuously with variations in prices. If one must choose between the models of Cournot and Bertrand as providing the best simple vehicle for exhibiting the nature of oligopoly, the Cournot model is definitely superior precisely because it lacks discontinuities of profits with respect to firms' decision variables. It could be argued that the differentiated products model is better than either because it is based on more satisfactory assumptions while not being particularly more complex.

[6] In some models, the best reply is not unique, which makes r_i a best reply correspondence; however, the assumptions of the present chapter allow consideration to be limited to best reply functions.

2.4. *On cooperative versus non-cooperative equilibria*

Surely certain situations in the world allow for binding agreements and others do not. For example, within American industry and the industry of many other nations, binding agreements are mostly illegal. There are exceptions. In industries which are regulated by government commissions, the commissions have the power to force certain aspects of the firms' behavior. There is the possibility that the firms may sometimes wish to make a binding agreement that is within the scope of what the regulating agency may enact and enforce. Though labor – management negotiations do not fall within oligopoly, they share its game-like character. Labor unions and firms customarily make binding agreements. The various circumstances allowing binding agreements are quite common, though far from universal. Labor – management negotiations are appropriately modeled as a bi-lateral monopoly, which is briefly discussed in Section 7.

Returning to situations in which binding agreements cannot be made, is there no scope for agreements at all? In general, there may be some scope; however, one point is clear: if an agreement cannot be enforced by some outside mechanism (such as the courts), then, if the agreement is to be honored, it must be self-enforcing. An agreement is *self-enforcing* if the firms agree on a non-cooperative equilibrium, i.e. if they agree upon an equilibrium of the general sort of that of Cournot, one in which no single firm can change its behavior and increase its profits. If there is only one such equilibrium, then agreements are superfluous; however, if there are more than one, agreements may be the key to the attainment of any particular equilibrium. When decisions are made independently by the firms, in the sense that they cannot communicate before making them, there is no assurance that all firms will single out the same equilibrium; however, if they can communicate and they come to agreement on a particular Cournot equilibrium, it is in the interest of each to honor its word. If the others do as they said, one firm cannot increase its profit by doing other than it said. That is, the agreement is self-enforcing.

Within the framework of single-period models, it is not easy to discuss threats fully; however, a few useful points may be made.

The basic one is that a threat is not credible unless it is clear that the firm making the threat will carry out its threat if put in a situation in which it has said it would use it. There are two ways in which a threat might be made credible: one is that the maker of the threat has been able to commit itself so that it is forced to carry out its threat in the circumstances which call for it. The firm may have actually taken steps so that there is no other option, whether it likes it or not. The second sort of credible threat occurs when the threatened action is clearly the action in the best interest of the firm making the threat when it is in the circumstances in which it has said it will carry out the threat. Say, for example, there is a three-firm Cournot market with a unique Cournot equilibrium for which the outputs are $q^c = (5,7,3)$. Assume also that $q_1 = 10$ causes lower profit to firm 1 than $q_1 = 0$, no matter what choices the others make. Then firm 1 cannot successfully threaten the others by saying that if they do not agree to choose $(5,6,2)$, it will choose $q_1 = 10$. If they do not agree to what it wants, it will not be in this firm's best interest to choose 10. It can only hurt itself.

Threats do become more interesting in multi-period models, but even here the basic principles remain the same. A threat is *credible* only if, when the circumstances calling for the threat to be carried out actually arise, the threatener will have no other behavior open to it, or, if carrying out the threat gives it at least as high a (discounted) profit as any other behavior it is free to choose.

A brief word may be said about *tacit collusion*. Intuitively, that term means to most people the choosing of the sort of behavior which one would expect of colluding firms when they are not in a position to communicate. Clearly, tacit collusion which leads to something other than a non-cooperative equilibrium is hard to believe. Beyond that, it remains an open question whether tacit collusion is really possible. Given that humans in the world can talk to one another even if they cannot make contracts, it is not at all obvious that there is any scope for tacit collusion.

3. Stability and reaction functions – first steps toward multi-period models

In thinking about oligopolistic behavior, one is naturally led to consider how one firm might alter its behavior in the light of a change in some rival's policy or how one firm might try to assess the effect of a change in its own behavior on the choices of the other firms. Cournot engaged in some of this. He and others made an attempt at understanding dynamic behavior without really leaving the single-period setting. While the results of such efforts are likely to cause some confusion, they also are very suggestive of the directions in which further work ought to be carried out. The confusing element is that they obscure the value, clarity, and lessons to be learned from the single-period models. It was to avoid these confusions that the preceding section dealt strictly with single-period models. In the remainder of this section, Section 3.1 takes up Cournot's reaction curves and Bowley's variation on them. Then, in Section 3.2, the well-known models of Sweezy and Stackelberg are discussed.

3.1. *Cournot's reaction curves and Bowley's conjectural variation*

For two firms and by means of a graph, Cournot discussed the stability of his equilibrium. From eq. (2.2), assuming the implicit function theorem may be applied, it is possible to write

$$q_i = w_i(\bar{q}_i), \qquad i=1,\dots,n. \tag{3.1}$$

These are, of course, the best reply functions of the firms. In the $n=2$ example which Cournot graphs, he finds that the mapping from R^2_+ into itself defined by the mapping $w=(w_1, w_2)$ is a contraction; hence, the model is stable if $q_1' = w_1(q_2)$ is thought of as a reaction or response to q_2 and $q_2' = w_2(q_1)$ is thought of as a response to q_1. In this way a path of output pairs may be traced out which lead to q^c. Whether Cournot's intent was to establish stability or to describe the workings of a multi-period model, he made, at most, only a rudimentary beginning.[7] If there is adjustment taking place over time, it is first necessary to specify explicitly the time structure of the model. Who makes which decisions when? What payoffs are received by whom and at which point in time? These questions must be faced. Where the firms are assumed to be operating in a market over many time periods, it is more reasonable to assume they wish to maximize a (discounted) stream of profits rather than that they wish to maximize myopically. Conceivably, myopic maximization could cause the maximization of a profit stream; however, that ought to be proved instead of assumed. In any event, Cournot does not fill out these details. What he does do is to write things which might lead others to raise the questions indicated above and to do something about them.

Bowley (1924) in what is little more than a passing reference to Cournot rewrites eq. (2.2) in the following way:

$$\frac{\partial \pi_i}{\partial q_i} = f(Q) + q_i f'(Q) - C_i'(q_i) + q_i f'(Q)\frac{dq_j}{dq_i}, \qquad i \neq j, i, j = 1,2. \tag{3.2}$$

The last term in eq. (3.2) is the *conjectural variation*. The part $q_i f'(Q)$ is, of course, the partial derivative of the profit of i with respect to the output of j; and the derivative dq_j/dq_i is the change in q_j which firm i assumes is associated with a change in q_i. That is to say, firm i imagines or assumes that firm j picks q_j as a function of q_i. Say, that is, $q_j = v_j(q_i)$. dq_j/dq_i is v_j'. Bowley does not specify a time structure either; however, the valid point which he is making is that the policy which a firm follows is affected by the policies which it believes its rivals are using. It is what the firm thinks the others might do which is crucial to its own choices, and not necessarily what they actually do.

Bowley is a little bit right, but not completely so. He leaves one with the impression that how firms behave and how they are perceived to behave by their rivals can be very different. Whether that is correct depends on the nature of the

[7]The reader interested in the large recent literature on the stability of single period oligopoly models should look at Okuguchi (1976) where it is exposited and many references are given.

model and in some important cases there is no room for large persistent divergence between the actual and the assumed behavior of rivals. Such differences can exist and persist in equilibrium only when firms received information which does not contradict their beliefs. This may be illustrated with a specific multi-period version of the Cournot model, using the best reply functions as reaction functions. For purposes of this chapter, a reaction function for a firm is a function which gives its period t decision (price or output, as the case may be) as a function of the period $t-1$ decisions of the firms in the market. Thus,

$$q_{it}=w_i(\bar{q}_{i,t-1}), \qquad i=1,\dots,n, \tag{3.3}$$

defines the *Cournot reaction function*. Two conditions are required for $w_i(\bar{q}_{i,t-1})$ to give optimal behavior for firm i. One is that the firm seeks to maximize current period profits with respect to its current period decision. The other is that firm i expects q_{jt} to be the same as $q_{j,t-1}$ (for $j\neq i$). The former condition is a matter of the firm's preferences and may be allowed, for the present, to stand. The latter condition may conflict with the facts coming to the attention of the firm. Assume that $q_{t-1}\neq q^c$, and that, before making its decision for a given time period, the firm is told the true choices of the other in the preceding period. Furthermore, the firm recalls all past choices. Then, over time, it must see that the others do not generally repeat their previous period choices. Thus, the assumption that $v_j'=0$ cannot stand if it is not true – given the information assumptions of this example. The point made here holds in any instance in which the actual behavior of a firm and the behavior assumed of it by others differ in ways which time must bring to light. Both the Cournot reaction functions and the Bowley conjectural variation have long remained staples in the oligopoly diet. Though for many writers they have been a little hard to digest, they have remained partly for such insights as they bring and, until recently, partly due to the paucity of material to teach.

3.2. *The explicit behavioral hypotheses of Sweezy and Stackelberg*

Figure 3.1 shows a version of Chamberlin's famous *DD'* and *dd'* curves. Each represents the demand facing a single firm as a function of the firm's own price; however, they differ concerning the assumptions made about the prices of other firms in the market. For the *dd'* curve, it is assumed that all other prices are constant; hence, the slope of the *dd'* curve is merely F_i^i for some $\bar{p}_i$. The *DD'* curve is drawn on the assumption that all other prices change at precisely the same rate that p_i changes. Where the two curves cross they correspond to the same values of $\bar{p}_i$. Sweezy (1939) suggests an oligopoly equilibrium which is characterized by somewhat special assumptions on the part of each firm concerning the way it expects other firms to behave. To see this, assume the prices p^* now

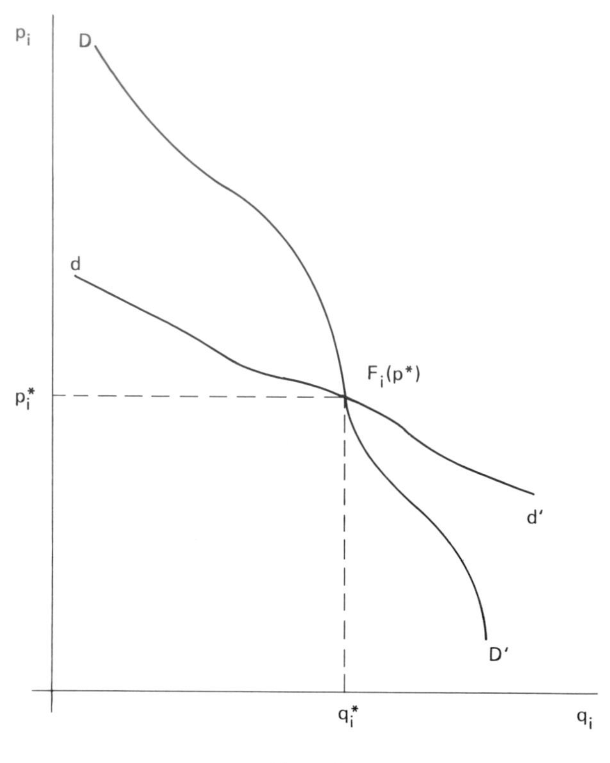

Figure 3.1.

prevail. The firm is presumed to think that if it raises its price, it moves along the *dd'* curve. That is, if it raises its price, all other firms keep their prices constant. If it lowers its price, however, all other firms lower their prices by an equal amount. Thus, the firm moves along the *DD'* curve when it decreases its price. Under these conditions, the firm cannot increase its profit by changing its price if

$$\pi_i^i(p^*) \leqslant 0 \leqslant \sum_{j=1}^{n} \pi_i^j(p^*). \tag{3.4}$$

A *Sweezy equilibrium* obtains if eq. (3.4) holds at p^* for all firms.

A natural question is: Under what circumstances could this equilibrium arise? Clearly, in the single-period, simultaneous decision models examined in Section 2, this equilibrium is nonsensical. A somewhat contrived single-period setting in which it makes some sense is one in which a p^* is given which satisfies eq. (3.4) and in which any firm may choose another price, but if a firm does choose another price this decision is announced before it takes effect and all other firms may alter their price choices in the light of the announcement of the first firm. Intuition suggests that, in a multi-period model, the Sweezy equilibrium could hold if firms wished to maximize discounted profits and there were an adjustment cost associated with changing one's price. Then, even with firms knowing at time t the choices of others through $t-1$, if the adjustment cost were larger than any possible extra one-period gain, Sweezy's equilibrium could hold. In Section 4.2 models of this general character are considered, developed by Marschak and Selten (1978).

A closer look at Sweezy's equilibrium reveals an unsatisfactory element, even given the quick response feature. Consider the whole decision made by one firm. It chooses (a) whether to break away from p^* given that p^* is established and the other firms will behave according to *response functions* which give the prices they adopt when someone deviates from p_i^*, and (b) a response function to adopt for itself. Denote the response function of the ith firm by $\phi_i(p^*, p_j')$, the price which firm i adopts given that p^* has been established and that the jth firm $(j \neq i)$ deviates from p_j^* by choosing p_j'. The Sweezy response function is $\phi_i(p^*, p_j') = p_i^* + (p_j' - p_j^*)$ when $p_j' < p_j^*$ and p_i^* when $p_j' > p_j^*$. (a) and (b) above must be determined for any possible p^*. A natural question is whether the rules of behavior (i.e. strategy) adopted by a firm form a best reply to the rules adopted by the other firms. Consider a Chamberlinian model satisfying Assumptions 2 and 4 and look at a price satisfying eq. (3.4). Say that firm 1 is to deviate from p^* by a price change of $\mathrm{d}p_1$ and that firms $2,\ldots,n$ are to respond by the Sweezy rule that $\mathrm{d}p_j = \mathrm{d}p_1$ if $\mathrm{d}p_1 < 0$ and $\mathrm{d}p_j = 0$ if $\mathrm{d}p_1 > 0$. The profit maximizing response of firm n, where it attains an interior maximum, is characterized by $\pi_n^n(p') = 0$, where $p_1' = p_1^* + \mathrm{d}p_1$ and p_i'. For a small value of $\mathrm{d}p_1$, this is approximated by $\pi_n^n(p_n', \bar{p}_n^*)$. On the margin, the optimal response of firm n does not depend on whether the deviator raises or lowers its price.

The response function equilibrium outlined above is more than Sweezy claims, however. All that he asserts is that, given a p^* which satisfies eq. (3.4) and the response functions which he stipulates, no firm can change its own price and increases its own profit. This is correct even though the various response functions of the firms are not optimal response functions.

Stackelberg, like Sweezy, presents a special form of behavior. He does not consider in a general way what sort of behavior is conceivable given the nature of the model and then look for a reasonable equilibrium. Though Stackelberg does not use an explicitly multi-period model, it is in such a context that his analysis makes sense. He defines two behavioral types, called *follower* and *leader*, and carries out everything for duopoly. A follower is a firm which behaves according to a Cournot reaction function. Assume firm 1 in a Cournot duopoly to be a follower. Then its output choice is given by $q_{1t} = w_1(q_{2,t-1})$ from eq. (3.3). If firm 2 is also a follower, then it chooses its output according to w_2 and, over time, in a stable market the observed outputs converge to q^c. Alternatively, say that firm 2 is a leader. A leader is a firm which believes that its rival is a follower, and it chooses its own output level accordingly. Thus, the profit of firm 2 in period t is $\pi_2(w_1(q_{2,t-1}), q_{2t})$. It is possible to continue to analyze this model as a multi-period model along the lines of the lagged response models of Section 4.1; however, to highlight the Stackelberg leader–follower equilibrium it is sufficient to revert to a single-period model and make the assumption that the leader announces its quantity choice first, then the follower, upon hearing it, makes its.[8] Clearly, the follower would choose $q_1 = w_1(q_2)$ and the leader, taking this into account, would maximize its profits by choosing q_2 to maximize $\pi_2(w_1(q_2), q_2)$. For an interior maximum, this occurs where

$$\pi_2^1(w_1(q_2), q_2)w_1'(q_2) + \pi_2^2(w_1(q_2), q_2) = 0. \tag{3.5}$$

The particular behavioral conventions outlined above, that the leader announces first and the follower second, must be imposed on the model. That is, such conventions have never been shown to arise of themselves in a theoretical model; and it is not plausible to impose them. It is not easy to see how they might arise spontaneously in a world in which the firms may actually make decisions in any fashion they wish. In terms of the multi-period version of the Stackelberg leader–follower model, the situation is that the follower, like a Cournot actor in a multi-period setting, assumes (wrongly) that its rival repeats in the current period the decision it made in the preceding period. Then it maximizes with respect to that incorrect supposition. The leader knows the actual behavior of the follower and maximizes its profit with respect to that knowledge. Thus, the leader's behavior is a best reply to the follower's but the converse is not true.

There is, finally, the possibility that both firms are leaders. This condition leads to what has been named the *Stackelberg disequilibrium*. Not only does each firm make a wrong assumption about how its rival chooses prices but, even if their respective behaviors lead to a steady state pair of prices, they are a pair of prices which neither expects. In terms of the timing of decisions, there is no way to contrive the leader–leader situation.

4. Multi-period models

The models sketched in the present section are all explicitly multi-period. The time structure is, in all cases, made clear, and the equilibria are all of the

[8]Although this timing of decisions is logically consistent, from a behavioral standpoint it is nonsensical. The excuses for making such an assumption here are that it catches the spirit of what Stackelberg was doing and the results are an approximation to the results which obtain in a well-specified multi-period model.

non-cooperative sort. It is not true of each model that the equilibrium is a Nash (1951) non-cooperative equilibrium. That is to say, when the set of actions from which each firm may choose is fully specified, it is not always true that, in equilibrium, each firm is using that action which, among all available to it maximizes its discounted profit given the actions of the others.[9] It is made clear which equilibria do, and which do not, have this desirable best reply property.

The models discussed fall into four distinct categories. The first may be termed *traditional reaction function models*. These models, reviewed in Section 4.1, derive their inspiration from the stability discussion in Cournot (1927). Stackelberg (1934) and Fellner (1949) further this line of investigation, although they do not formalize the role of time. The key elements, stated in terms of an underlying differentiated products model, are that each firm chooses a mode of behavior described by a continuous function, $p_{it} = \psi_i(p_{t-1})$, giving the period t price choice of the firm as a function of the previous period prices of all firms. The firm *reacts* at time t to the choices of $t-1$. The second key element is that each firm chooses its reaction function ψ_i with a view to maximizing a discounted stream of profits. In these models the firms only choose a price policy, there being no investment, advertising, etc. In addition, the profit functions are stationary over time. The models presented are based on the work of Friedman (1977a, ch. 5) and Cyert and deGroot (1970).

Second are *costly price change models*. These models, reviewed in Section 4.2 based on the work of Marschak and Selten (1978), are characterized by firms which are also attempting to maximize the value of a profit stream, however, making price changes is costly. An alternative way to get the same effects as those obtained from the adjustment cost is to assume that in each time period there is a gap of time between when prices are announced and when they take effect. If any firm announces a price different from the one it had used in the period before then the other firms are allowed to change their prices for the current period.

Third are *models formulated as abstract games*, presented in Sections 4.3 and 4.4. They may, clearly, be applied to oligopoly, and they are presented in terms of oligopoly models. In Section 4.3 it is demonstrated that when the firms wish to maximize their discounted profits, it is possible to have equilibria which are, on the one hand, non-cooperative, and which, on the other, yield to the firms Pareto optimal profits. That is to say, the profits associated with collusion and binding agreements may be obtained as profits associated with an equilibrium in the spirit of Cournot.

Fourth is a model having a time-dependent structure in which profits depend on actions in two periods. The model of Section 4.4 differs from the other models

[9]The Nash non-cooperative equilibrium is precisely the Cournot equilibrium for games more general than a one-period oligopoly. For more on this concept, see Nash (1951) or Friedman (1977a, ch. 7).

presented in having the profits received by a firm in period t a function of the firms' actions in both periods t and $t-1$. For example, demand during period t could depend on p_t and p_{t-1}. Investment and capital could be introduced into the model with the firm's cost function depending, in period t, on the amount of capital on hand at the beginning of the period. This amount of capital would depend on the previous level of capital and the amount of investment undertaken in the preceding period.

4.1. *Lagged response reaction function models*

Cyert and deGroot (1970) have followed Cournot quite literally in dealing with a multi-period duopoly. Their model is outlined first, then that of Friedman (1977a). In Cournot's stability discussion, he starts with an arbitrary pair of outputs for a duopoly, (q_1', q_2'). He gets a second by allowing firm 1 to choose its best response to q_2', which is $q_1''=w_1(q_2')$. The third output pair is obtained by allowing firm 2 to choose its best reply to q_1'', which is $q_2''=w_2(q_1'')$. Thus, Cournot describes a processs of alternate decision in which firm 1 chooses a new output, then firm 2, then firm 1, etc. Cyert and deGroot embed this into an explicit multi-period model in which the payoff functions of the firms are

$$\sum_{t=1}^{T} \pi_i(q_{1t}, q_{2t}), \qquad i=1,2. \tag{4.1}$$

With firm 1 choosing a new output level only in periods 1,3,5,..., so that $q_{1t}=q_{1,t-1}$ for $t=2,4,\dots$, and firm 2 choosing a new output level in periods 2,4,6,..., simultaneous decision is removed from the model. Using a model in which the demand function is linear and the total cost function is quadratic yields quadratic profit functions, from which optimal policies for the firms are derived. For firm 1, an optimal policy consists of a sequence of reaction functions of the form

$$q_{1,2T+1}=w_{1,T-2t-1}(q_{2,2t}), \qquad t=1,2,\dots,T_1, \tag{4.2}$$

$$q_{2,2t}=w_{2,T-2t}(q_{1,2t-1}), \qquad t=2,4,\dots,T_2, \tag{4.3}$$

where T_1 is the largest value of t for which $T-2t-1\geqslant 0$, and T_2 is similarly defined. These families of functions have the best reply characteristic. That is, it would be impossible for firm 1, given eq. (4.3), to find a new sequence of reaction functions which would give greater total profits. The same holds for firm 2. Furthermore, under Cyert and deGroot's assumptions the optimal reaction function for a given time period converges as the horizon, T, goes to infinity. That is,

$w_{1,T-2t-1}$ is the optimal reaction function for firm 1 when its time horizon is $T-2t-1$ periods. The optimality of the function does not depend on the actual time, $2t-1$, but only on the number of periods to end of the horizon (i.e. on $T-2t-1$).

The alternate period decision assumption of the model may seem objectionable. If the model were expanded to $n>2$ firms, it would be necessary to modify alternating decision in a way to ensure that in any given time period only one firm could change its output. Such behavior is not consistent with either intuition or casual empiricism. On the other hand, the model does permit a stronger, more acceptable equilibrium than does the reation function model of Friedman. This is made clear below.

Like Cyert and deGroot, Friedman (1977a, ch. 5) uses a model which is explicitly multi-period. It is based on the assumption of differentiated products, rather than being a multi-period extension of a Cournot model. Assumptions 2, 4, 5, and 7, along with a number of additional regularity assumptions which it would be tedious to state are assumed. The firms' discounted profit functions are

$$\sum_{t=1}^{\infty} \alpha_i^{t-1}\pi_i(p_t), \qquad i=1,\dots,n. \tag{4.4}$$

The analysis is aimed at showing the existence and nature of an equilibrium in which the firms all behave according to stationary reaction functions, $p_{it}=\psi_i(p_{t-1})$.

The results are described below in two stages. For the first, the best reply, or discounted profit maximizing behavior, of the ith firm is examined, given that all other firms, j, behave according to stationary reaction functions which are known to firm i. The reaction functions $p_{jt}=\psi_j(p_{t-1})$, $(j\neq i)$, are assumed to obey the following conditions: (a) for any p_{t-1} such that $0\leqslant p_{k,t-1}\leqslant p_k^+$, $k=1,\dots,n$, $0\leqslant p_{jt}=\psi_j(p_{t-1})\leqslant p_j^+$, and (b) $\psi_j(p_{t-1})$ is twice continuously differentiable with $\psi_j^k>0$ and $\sum_{k=1}^n\psi_j^k\leqslant\lambda<1$. The objective of the firm is to maximize eq. (4.4) subject to the conditions $p_{jt}=\psi_j(p_{t-1})$, $(j\neq i)$. This problem is tackled as a dynamic programming problem in which the firm has a finite horizon. It is then shown that optimal behavior under an infinite horizon requires the use of a stationary reaction function, similar to the ψ_j, which is the limit of the reaction functions which are optimal for finite horizons.

In solving the finite horizon dynamic programming problem a sequence of reaction functions is found. Denoting them $p_{it}=\phi_{is}(p_{t-1})$, $s=1,2,\dots,$ the reaction function ϕ_{i1} is optimal when the firm has a horizon of only one period; and ϕ_{i2} is optimal in the second to last period, given that ϕ_{i1} is to be used in the last period. Thus, if the firm has a horizon of T periods, extending from $t=1$ to $t=T$, its discounted profit maximizing plan (i.e. strategy) is to choose $p_{i1}=\phi_{iT}(p_0)$,

$p_{i2}=\phi_{i,T-1}(p_1),\ldots,p_{iT}=\phi_{i1}(p_{T-1})$. When the horizon is allowed to go to infinity, the sequence of reaction functions, ϕ_{is}, converges to ϕ_i. The principal results relating to the ith firm's best reply to $\bar{\psi}_i$ are:

Theorem 7

(a) For $s=1,2,\ldots,\phi_{is}(p)$ exists.
(b) $p_i^+>\phi_{is}(p)>\phi_{i,s-1}(p)\geq 0$ for all p and all $s>1$.
(c) $\lim_{s\to\infty}\phi_{is}=\phi_i$ exists and obeys any Lipschitz condition which is uniformly obeyed by the ϕ_{is}.
(d) If π_i and $\bar{\psi}_i$ are continuously differentiable up to the kth order, then ϕ_{is} is continuously differentiable up to the $k-1$st order and ϕ_i, up to the $k-2$nd.
(e) All first partial derivatives of ϕ_i and of the ϕ_{is} are non-negative and the functions obey the same Lipschitz condition as $\bar{\psi}_i$.
(f) If π_i and $\bar{\phi}_i$ are infinitely continuously differentiable, then so are ϕ_i and the ϕ_{is}.
(g) The functions satisfy

$$\lim_{T\to\infty}\sum_{t=1}^{T}\alpha_i^{t-1}\pi_i\left(\phi_{i,T-t+1}(p_{t-1}),\bar{\psi}_i(p_{t-1})\right)$$
$$=\sum_{t=1}^{\infty}\alpha_i^{t-1}\pi_i\left(\phi_i(p_{t-1}),\bar{\psi}_i(p_{t-1})\right). \tag{4.5}$$

(h) The limiting reaction function, ϕ_i, is optimal for an infinite horizon.

The content of (a) is that, with the horizon finite and fixed, there is a unique best way for the firm to behave; and that best behavior is in the form of a sequence of reaction functions. That is, for a horizon of T, using $\phi_{iT},\ldots,\phi_{i1}$ yields as much profit as any conceivable way of choosing prices. When the firm considers what to do, without being constrained to use reaction functions of the sort the rivals use, it still finds it best to use them. From (b) it is seen that these reaction functions have a monotone property. That is, given p_{t-1}, the farther away is the end of the horizon, the larger is p_{it}. This is illustrated in Figure 4.1. (c) asserts that the reaction function sequence converges, while (d), (e), and (f) give some regularity results. Roughly speaking, (a)–(f) sum up to saying that if the other firms use $\bar{\psi}_i$, the optimal policy of the ith firm is something of the same general sort as the $\psi_j(j\neq i)$. From (g) it is seen that as the horizon goes to infinity, the discounted profit of the firm converges to what it would be if the firm were always to use ϕ_i. This limiting profit, according to (h), is the optimal profit when the firm's horizon is infinite.

A particularly helpful way to summarize some of the results of Theorem 7 is to note that ϕ_i is the best reply of the ith firm to $\bar{\psi}_i$. This may be denoted

$$\phi_i=W_i(\bar{\psi}_i). \tag{4.6}$$

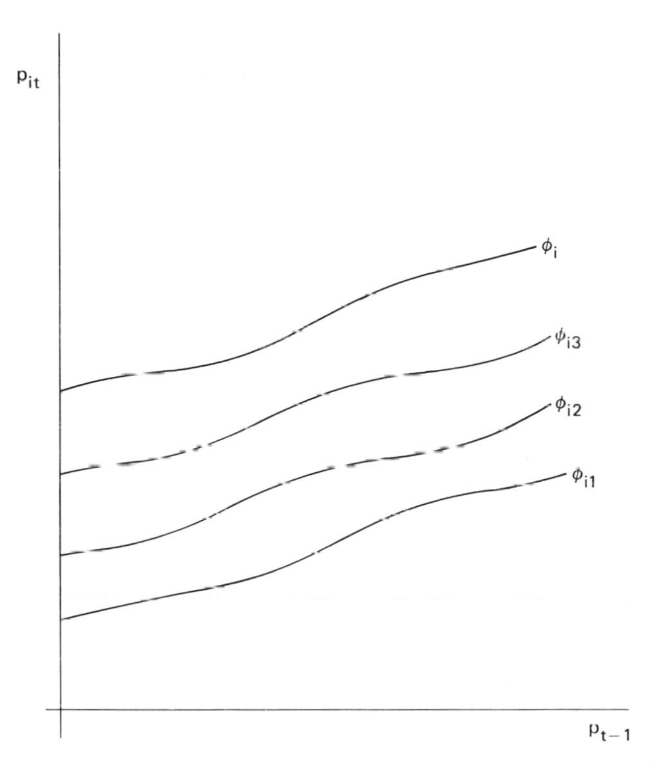

Figure 4.1.

Again, it is important to note that ϕ_i is a member of the same broad family of functions as the ψ_j. They are all monotone increasing differentiable contractions taking p_{t-1} into p_{it} or p_{jt}, as the case may be. The function ϕ_i is in no way constrained to be of this class; the period t choice of firm i could be any function associating $p_{it} \in [0, p_i^+]$ with $(p_0, \dots, p_{t-1})$.

What has been done for the ith firm may, of course, be done for all i, and the best reply function in eq. (4.6) may be used to map one complete set of reaction functions into another:

$$\phi = (\phi_1, \dots, \phi_n) = \big(W_1(\bar{\psi}_1), \dots, W_n(\bar{\psi}_n)\big) = W(\psi). \tag{4.7}$$

It is clear that a fixed point of W is a Nash non-cooperative equilibrium, and it is a natural extension of the Cournot equilibrium to multi-period reaction function models. Is there some fixed point equilibrium $\psi^* = W(\psi^*)$? Except for a trivial case, the existence of such an equilibrium has not been proved. The trivial case consists of the reaction functions $p_{it}=p_i^c$, $i=1,\ldots,n$. That is to say, if all firms always choose the single-period Cournot equilibrium prices no matter what the past history is, then no single firm i can deviate from constantly choosing p_i^c and increase its discounted profit. It would be of interest to show the existence of a reaction function equilibrium different from this trivial one which is a Nash non-cooperative equilibrium. This has not been done. What has been shown is the existence of an approximation to it.

Theorem 8

There exists ψ^* such that (a) $\psi^*(p^r)=p^r$, (b) associated with ψ^* is $\phi^*=W(\psi^*)$ such that, (c) $\phi^*(p^r)=p^r$, and $\psi_i^{*j}(p^r)=\phi_i^{*j}(p^r)$, $i, j=1,\ldots,n$.

Because ψ^* is a contraction which maps the set $[0, p_1^+]\times\cdots\times[0, p_n^+]$ into itself, it follows that, starting from an arbitrary p_0 and generating a sequence of prices $p_t=\psi^*(p_{t-1})$, that p_t must converge to p^r, where p^r is the unique fixed point of ψ^*. In other words, if the firms behave according to the reaction functions ψ^*, they will choose prices which converge to p^r over time. p^r is the only set of prices such that $p_i^r=\psi_i^*(p^r)$, $i=1,\ldots,n$. Theorem 8 states that there is a vector of reaction functions, not necessarily unique, such that the best reply to $\psi^*, \phi^*=W(\psi^*)$, has the same steady-state price vector, p^r, where the two functions ψ^* and ϕ^* are tangent at p^r. This means that if the firms actually use $\psi_1^*,\ldots,\psi_n^*$, then, if the price is actually p^r, no single firm i could better its profit by switching to the best reply, ϕ_i. In addition, at prices close to p^r, even price changes from one period to the next dictated by the ψ_i^* are approximately optimal. As a vector of reaction functions ψ is a mapping from R^n to R^n, even for $n=2$, a diagram can only givc a schematic representation; however, Figure 4.2 gives some idea of the relationship between ψ and ϕ.

The steady-state prices, p^r, occur at levels above p^c, the single-period Cournot prices. Recall that assumptions are made which ensure the uniqueness of p^c. Many prices might be steady-state prices associated with the reaction function equilibrium of Theorem 8. The proof of that theorem proceeds by construction, using a p^r chosen so that $p_i^r-p_i^c=p_j^r-p_j^c>0$, for all i, j. It is clear that p^r is easily chosen so that for all firms, $\pi_i(p^r)>\pi_i(p^c)$. That is necessarily true for p^r, as noted above, which are near to p^c.

Thus, it is clear that the reaction function equilibria can yield profits in excess of those attainable at the single-period Cournot equilibrium. Can Pareto optimal profits be attained? The interested reader, using material from Friedman (1977a, §2), can easily satisfy himself that in a symmetric model (i.e. all demand and cost functions are the same), the price associated with the joint profit maximum may play the role of p^r, with all $\psi_j^{*k}(p^r)$ having the same value. Beyond this established point, nothing has been worked out.

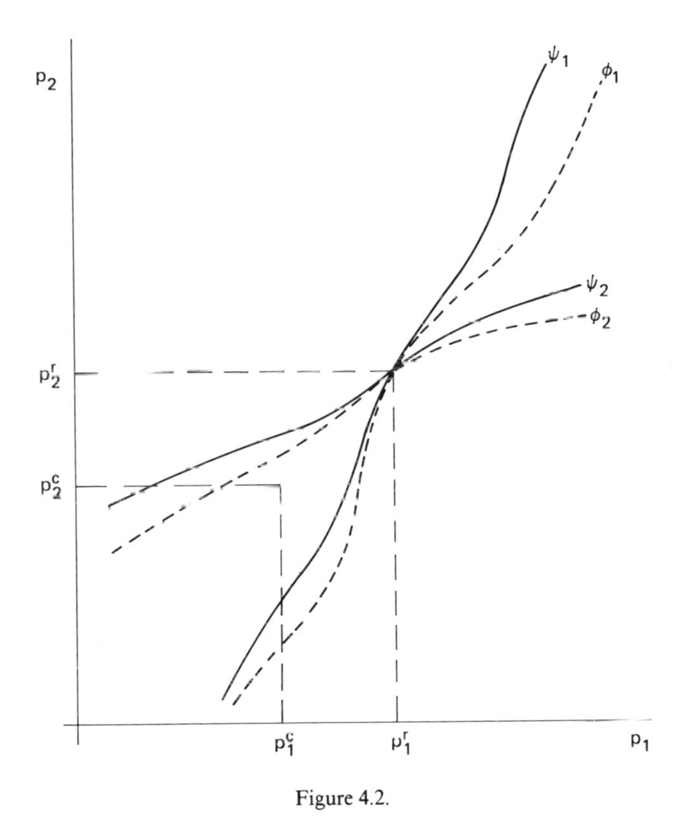

Figure 4.2.

4.2. *Models with quick response or adjustment costs*

Marschak and Selten (1978) developed a pair of parallel models which are outlined below. Their presentation is in the form of an abstract game; however, they make very clear that a prime area for the application of their work is to oligopoly. Indeed, in the later parts of their paper such application is made. In the present section, the main outlines of their models and notion of equilibrium are presented in the framework of the differentiated products price models developed in earlier sections of the present chapter. They present a single-period model in which there are established prices which are used if no firm decides to deviate from its own established price. If some firm does wish to deviate then all other firms have the option of choosing different prices. This is a formalization of the price choosing process discussed in Section 3.2 in conjunction with the Sweezy model. A second model is formulated in which the firms choose prices over an infinite sequence of time periods. The objective function of the firm differs in two ways from eq. (4.4). First, the firm does not discount profit. Instead, it wishes to maximize its average profit per period, over an infinite horizon. Secondly, its profit in a period is given by $\pi_i(p_t)$ only if it does not change its price in the period. If it does change its price then it must pay an additional cost associated with the price adjustment. The equilibria for the two models are shown to be almost equivalent.

For each firm in the single-period model, there are three things to be determined: (a) the status quo prices, p_i^0; (b) the decision as to whether or not to deviate from p_i^0; and (c) the *response function*, $\phi_i(p^0, p_j)$ to use in the event that firm $j(\neq i)$ should decide to deviate from p_j^0. The rules of information stipulate that all firms are informed of p^0 and the ϕ_i. In addition, if a firm deviates from p_j^0 the price it intends to use is announced before it takes effect and in time for all the other firms to change their prices if they wish. The rules also stipulate that no more than one firm may deviate from the status quo prices in a single period. The conditions governing the response functions are given below:

Assumption 8

For each $p^0 \in [0, p_1^+] \times \cdots \times [0, p_n^+]$ and each $p_j \in [0, p_j^+]$, $j=1,\ldots,n$, $\phi_i(p^0, p_j)$ is defined and contained in $[0, p_i^+]$. In addition, $\phi_i(p^0, p_i)=p_i$ and $\phi_i(p^0, p_j^0)=p_i^0$.

The last requirement, that $\phi_i(p^0, p_j^0)=p_i^0$, is merely a consistency condition that when a "deviation" by firm j involves no change on the part of j, then the "response" from i is also no change. Likewise, the condition $\phi_i(p^0, p_i)=p_i$ states that the response of i to its own deviation is the deviation itself. These two consistency requirements allow ϕ_i to be defined for all p^0 and p_j rather than for only those (p^0, p_j) for which $j\neq i$ and $p_j \neq p_j^0$. A joint response function $\phi(p^0, p_j)$ $=(\phi_1(p^0, p_j),\ldots,\phi_n(p^0, p_j))$ is defined in the obvious way. Each firm is also allowed to make a sequence of responses to a sequence of deviations. This is still under the rule that only one firm may deviate; hence, what is involved is a sequence of deviations by one firm, along with a sequence of responses by the others. The response function is used to define an *extended response function*.

Let $\{p_j^1, p_j^2,\ldots, p_j^k\}$ be a sequence of deviations by firm j. The process is that j announces p_j^1, then the others respond with $\phi_i(p^0, p_j^1)=p_i^1$. Next, firm j announces p_j^2 and the others respond with $\phi_i(p^1, p_j^2)=p_i^2$. This continues up to the last deviation, p_j^k. The response to it is $\phi_i(p^{k-1}, p_j^k)$. Denoting the extended response function $\hat{\phi}$,

$$\hat{\phi}\left(p^0, \{p_j^1,\ldots, p_j^k\}\right)=\phi\left(\hat{\phi}\left(p^0, \{p_j^1,\ldots, p_j^{k-1}\}\right), p_j^k\right), \qquad k\geqslant 2, \tag{4.8}$$

$$\hat{\phi}\left(p^0, \{p_j^1\}\right)=\phi\left(p^0, p_j^1\right). \tag{4.9}$$

The prices and response functions (p^0, ϕ) form a non-cooperative equilibrium if the following two conditions are met: (a) the ith firm cannot increase its profit by deviating from p_i^0 given that p^0 are the status quo prices and that the other firms meet deviations with responses dictated by the $\phi_j(j \neq i)$; and (b) the ith firm cannot use a different response function and increase its profits in those instances in which some other firm deviates. That is,

$$\text{(a)} \quad \pi_i(p^0) \geqslant \pi_i(\hat{\phi}(p^0, \{p_i^1, \ldots, p_i^k\})) \tag{4.10}$$

and

$$\text{(b)} \quad \pi_i(\hat{\phi}(p^0, \{p_j^1, \ldots, p_j^k\})) \geqslant \pi_i(\hat{\phi}_i'(p^0, \{p_j^1, \ldots, p^k\}), \hat{\bar{\phi}}_i(p^0, \{p_j^1, \ldots, p_j^k\})), \tag{4.11}$$

for all sequences $\{p_i^1, \ldots, p_i^k\}$ and $\{p_j^1, \ldots, p_j^k\}$, $j \neq i$, and extended response functions $\hat{\phi}_i'$, $i = 1, \ldots, n$.[10] A response function ϕ which, together with status quo prices p^0, forms a non-cooperative equilibrium is called a *convolution*. If a firm i cannot deviate profitably from p^0, given $\bar{\phi}_i$, then it is said that firm i is *stable at* p^0 *with respect to* ϕ. If ϕ_i is a best reply to the deviations of others, and this holds for all i, then ϕ is said to be *restabilizing*.

Finally, a *weak convolution* is defined. This could also be thought of as a weak non-cooperative equilibrium. Let P_i be a subset of $[0, p_1^+] \times \cdots \times [0, p_n^+]$ and let $P = \bigcap_{i=1}^n P_i$ be non-empty. (p^0, ϕ) is a weak non-cooperative equilibrium and ϕ is a weak convolution if $\phi_i(p', p_j'')$ is a best reply for firm i for any $p' \in P_i$ and $p_j'' \in [0, p_j^+]$, $j \neq i$, $i = 1, \ldots, n$, $\phi(p', p_j'') \in P$, and no firm can profitably deviate from p^0.

Consider now a model in which time is discrete and, in each time period, each firm makes a price choice. The information conditions are those of Section 4.1; each firm makes its period t decision simultaneously with all the others, and each knows all price choices made in all previous time periods. The structure of the present multi-period model differs from that of the model in Section 4.1 in two respects. First, the profit received in period t by the ith firm is

$$\pi_i(p_t) - M_i(p_{it}, p_{i,t-1}), \tag{4.12}$$

where $M_i(p_{it}, p_{i,t-1})$ is the cost of making a change in price for firm i. When $p_{it} = p_{i,t-1}$, the cost is zero; however, when $p_{it} \neq p_{i,t-1}$, the cost is larger than any possible one-period gain which can accrue from making the change. Thus, for $p_{it} \neq p_{i,t-1}$

$$M_i(p_{it}, p_{i,t-1}) > \max_{\bar{p}_{it}} (\pi_i(p_{it}, \bar{p}_{it}) - \pi_i(p_{i,t-1}, \bar{p}_{it})). \tag{4.13}$$

[10] Note the continued use of the "bar" notation. $\hat{\bar{\phi}}_i$ is the $n-1$ vector of extended response functions for all firms other than i.

In the model of Section 4.1, as well as the model of Section 4.3, transitory profits play a role in determining the nature of equilibrium, the role being especially prominent in Section 4.3. By a *transitory profit* is meant an additional profit obtained by a firm in a time period during which it uses a new price to which the other firms have not had the opportunity to adjust. Given eq. (4.13), transitory profits can play no role because the cost of changing a price is larger than any possible one-period increase in π_i. As a result, a firm changes its price only if it expects to see its profits higher after its rivals have made their adjustments to the price change. Those adjustments, of course, come after the period in which the deviating firm makes the initial change. Because the firm is not discounting profits, it will make any price change which will permanently increase its profits. That is, its objective function is

$$\liminf_{T\to\infty} \frac{1}{T} \sum_{t=1}^{T} \left[\pi_i(p_t) - M_i(p_{it}, p_{i,t-1})\right], \qquad i=1,\dots,n. \tag{4.14}$$

p_{i0} is taken to be equal to p_{i1}. M_i is bounded above so that it never stands in the way of a price change which leads to a permanent increase in the level of profits.

Now a behavior rule, or reaction function, may be constructed; however, as is apparent below, it is different from the reaction functions reviewed in Section 4.1. Let ψ_i denote the reaction function of the ith firm and $\psi=(\psi_1,\dots,\psi_n)$. Then

$$p_{it} = \psi_i(p_{t-1}, p_{t-2}),$$
$$\text{with} \quad p_{it} = p_{i,t-1} \quad \text{when} \quad p_{t-1} = p_{t-2}, \quad t=2,3,\dots, \quad \text{and} \quad p_0 = p_1. \tag{4.15}$$

The price choice of the ith firm depends on the observed prices in the market during the past two periods and, if no prices have changed from period $t-2$ to period $t-1$, then the ith firm leaves its own price unchanged. Marschak and Selten call this type of behavior *low memory* and *conservative*; it is "low memory" because it only uses information from two past time periods rather than all past periods, and it is "conservative" because it dictates no price change when the most recent past information shows no price change.

In relation to the ψ_i, there is one more notion to define. Imagine a time t and say that up to that time firm i has never deviated from the behavior prescribed by ψ_i. Say also, that in any single period before t, there has been at most one firm j which has deviated from ψ_j. Finally, for the time from t onward, assume that ψ_i is a best reply for firm i to $\bar{\psi}_i$. If there is $\psi=(\psi_1,\dots,\psi_n)$ such that these conditions hold for all i and all t, then ψ is said to be *paraperfect*. A paraperfect vector of reaction functions, ψ, together with an appropriate initial condition, p_0, clearly form a non-cooperative equilibrium. Given the initial condition and the behavior of the others, governed by $\bar{\psi}_i$, firm i can do no better than to use ψ_i. By restricting attention to (p_{t-1}, p_{t-2}) prices for which p_{t-1} and p_{t-2} may differ in, at most, one component, it is possible to define a response function which represents the reaction function ψ_i. Let (p_{t-1}, p_{t-2}) meet the requirement. There are three possibilities to consider: (a) $p_{j,t-1} \neq p_{j,t-2}$ for some $j \neq i$; (b) $p_{i,t-1} \neq p_{i,t-2}$; and (c) $p_{t-1} = p_{t-2}$. The response function for the ith firm is defined as follows:

$$\begin{aligned} &(\text{a}) \quad \phi_i(p_{t-2}, p_{j,t-1}) = \psi_i(p_{t-1}, p_{t-2}), \\ &(\text{b}) \quad \phi_i(p_{t-2}, p_{i,t-1}) = p_{i,t-1}, \\ &(\text{c}) \quad \phi_i(p_{t-2}, p_{i,t-2}) = p_{i,t-2}. \end{aligned} \tag{4.16}$$

While a response function ϕ may be defined from a reaction function ψ, the converse is not quite true. The difficulty here is that a reaction function ψ_i which represents a response function ϕ_i is not completely determined. It is only determined for (p_{t-1}, p_{t-2}) such that p_{t-1} and p_{t-2} differ in at most one component. It is possible to speak of a ψ_i which is consistent with ϕ_i. ψ_i is consistent with ϕ_i if ϕ_i is the response function which is derived from it. The single-period game with instantaneous response possible using response functions has a corresponding multi-period game which is obtained by abandoning the quick response condition and adding, instead, the change cost functions, M_i, and the objective functions, eqs. (4.14). Conversely, a multi-period game with change costs may be associated with a single-period quick response game by dropping the change costs, altering the objective functions, and introducing the response functions. It would not be surprising, therefore, if there were some sort of equivalence between ϕ's which are convolutions and ψ's which are paraperfect.

Theorem 9

Let ψ be a reaction function and ϕ a response function such that ϕ is defined from ψ and ψ is consistent with ϕ. Then if ψ is paraperfect [i.e. if there is p_0 such that (p_0, ψ) is a non-cooperative equilibrium], then ϕ is a weak convolution [i.e. (p_0, ϕ) is a weak non-cooperative equilibrium]. Conversely, if ϕ is a weak convolution, then ψ is paraperfect.

An application of this approach is made to an oligopoly model in which there is a given list of goods which the oligopolists could produce, with each firm able to produce any and all of the goods using the same technology as any other firm. Taking the response function approach and assuming both input prices and total output (equals demand) of each oligopoly good as given, it is shown that there are prices and response functions at which all firms make exactly zero profits and which form a non-cooperative equilibrium. This model is embedded into a model of general equilibrium, though it is possible to conceive it in a partial equilibrium setting also.

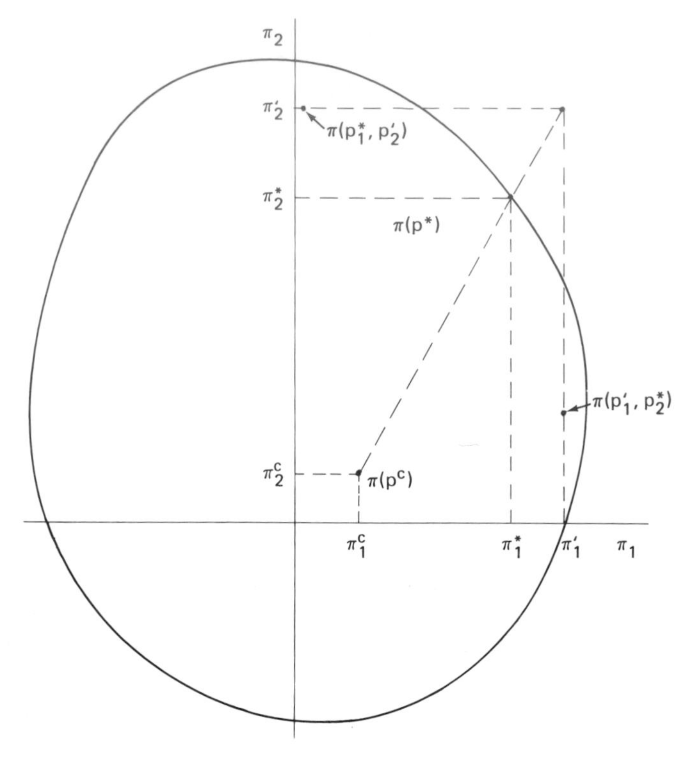

Figure 4.3.

4.3. *A model yielding cooperative equilibria without a cartel*

The equilibrium presented in this section is stated for a market model like that of Section 4.1. Firms seek to maximize a discounted stream of profits, and there are not costs associated with making price changes. There is an element in the equilibrium behavior which is like the reaction functions of Section 4.2; however, transitory profits play a role in the outcome. A more complete and general exposition of these models may be found in Friedman (1977a, ch. 8). The gist of the firm's behavior is this. There is a price vector, such as p^*, which is associated with profits higher for every firm than the profits at the single-period Cournot equilibrium prices p^c. Each firm i chooses in period t the price p_i^* if all other firms have chosen p_j^* in all periods through $t-1$. If any firm has deviated in the past from p_j^*, then the ith firm chooses p_i^c. This is illustrated in Figure 4.3. As long as both firms continue to choose p^*, their profits are $\pi_i^* = \pi_i(p^*)$, $i=1,2$. If, for example, firm 2 deviates in period t from p_2^*, it can obtain $\pi_2' = \pi_2(p_1^*, p_2') = \max_{p_2} \pi_2(p_1^*, p_2)$. However, because firm 1 will commence choosing p_1^c in period

$t+1$, firm 2 can do no better than to choose p_2^c from $t+1$ onward also. The net result is that firm 2 can gain an extra $\pi_2' - \pi_2^*$ in period t in exchange for giving up $\pi_2^* - \pi_2^c$ in each period from $t+1$ onward. Whether the trade is worthwhile depends on the size of the firm's discount parameter and the other profit magnitudes. Formally, the firm's behavior is described by

$$\begin{aligned} &p_{i1} = p_i^*, \quad \text{and for} \quad t \geq 2, \\ &p_{it} = p_i^*, \quad \text{if} \quad p_{j\tau} = p_j^* \quad \text{for} \quad j = 1, \ldots, n \quad \text{and} \quad \tau = 1, \ldots, t-1, \\ &p_{it} = p_i^c, \quad \text{otherwise}, \quad i = 1, \ldots, n. \end{aligned} \tag{4.17}$$

Now assume that all firms except the ith are behaving according to eq. (4.17), and consider what is the best reply to this behavior for firm i. It is immediate that one choice is always to select p_i^*, yielding a discounted profit of

$$\frac{\pi_i(p^*)}{1-\alpha_i}. \tag{4.18}$$

It is also clear that if time t is the first time period in which $p_{it} \neq p_i^*$, then from $t+1$ onward there is no better course than to choose $p_{i\tau} = p_i^c (\tau = t+1, \ldots)$. This is because there is no structural connection between the time periods, and, given the choice of the firms to select p_j^c, there is no way that the price choices of firms in one period affect those of another (after firm i has once departed from p_i^*). Given $p_{it} \neq p_i^*$, the best choice is p_i' which is defined by the condition

$$\pi_i(p_i', \bar{p}_i^*) = \max_{p_i} \pi_i(p_i, \bar{p}_i^*). \tag{4.19}$$

Thus, the discounted profit of the firm is

$$\sum_{\tau=1}^{t-1} \alpha_i^{\tau-1} \pi_i^* + \alpha_i^{t-1} \pi_i' + \sum_{\tau=t+1}^{\infty} \alpha_i^{\tau-1} \pi_i^c = \frac{1-\alpha_i^{t-1}}{1-\alpha_i} \pi_i^* + \alpha_i^{t-1} \pi_i' + \frac{\alpha_i^t}{1-\alpha_i} \pi_i^c. \tag{4.20}$$

Given the stationarity of the model, if discounted profits are increased by choosing p_i' in period t, then they are maximized by making the choice in period 1 instead. The level of the discount parameter is crucial. If

$$\alpha_i > \frac{\pi_i' - \pi_i^*}{\pi_i' - \pi_i^c}, \tag{4.21}$$

the firm does not improve itself by deviating from p_i^*. If the inequality in (4.21) runs the other way, deviating from p_i^* in period 1 is optimal and, finally, if equality holds, then the firm is indifferent. Thus, if (4.21) holds for each firm strictly, the behavior given by eqs. (4.17) describe a non-cooperative equilibrium. If, in addition, $\pi(p^*)$ is Pareto optimal, as in the example of Figure 4.3, the firms are at a non-cooperative equilibrium which corresponds to the kind of qualitative outcome which is expected with collusion. Now, many price vectors may play the role of p^*. All that is required is that (4.21) hold for all firms. For suitably large discount parameters, any p'' such that $\pi(p'')\gg\pi(p^c)$ would be feasible. Among all the possibilities those which are Pareto optimal and which also satisfy

$$\frac{\pi_i' - \pi_i^*}{\pi_i' - \pi_i^c} = \frac{\pi_j' - \pi_j^*}{\pi_j' - \pi_j^c}, \qquad i, j = 1, \dots, n, \tag{4.22}$$

may be singled out as being of special interest. The terms in eq. (4.22) are the ratio of the transitory gain to the sum of the transitory gain and the loss per period. This ratio might be thought of as the "temptation" of the firm to deviate. Eq. (4.22) expresses a *balanced temptation property* and an equilibrium in which firms use the behavior given by eq. (4.17) may be said to satisfy this balanced temptation property. Given the existence of many prices which could play the part of p^*, it is easy to visualize this equilibrium as one which could be chosen by common agreement, but one which requires no formal agreement to keep in effect because the firms are agreeing on a non-cooperative equilibrium, i.e. they are agreeing on something which is self-enforcing. If one firm does deviate from p_i^*, then, given the announced behavior of the others, it is in any one firm's interest to go ahead and do what it had indicated it would (i.e. choose p_j^c). The conditions under which the p_j^c are chosen and the commitment to choose them are in the manner of a threat. At least, they form a deterrent to deviating from p^*.

4.4. *Models with a time-dependent structure*

Time dependence is present in the structure of a model when the payoff received in period t depends on the actions taken in period t and in one or more past periods. A natural way to incorporate time dependence is through adding investment in capital to the firm's decisions. Say that the firm's production cost in period t is a function of its output level, as before, and of the amount of capital inherited from the previous period. Letting the depreciation rate be δ_i and K_{it} be the firm's capital at the end of time t, total costs during period t are

$$C_i(q_{it}, K_{i,t-1}) + (K_{it} - (1-\delta_i)K_{i,t-1}). \tag{4.23}$$

It is also easy to imagine a firm's demand depending on prices in both the current and immediately past periods. This might arise from the consumers' wish to speculate on future prices. A second possibility is that the lagged prices are a proxy for the amount sold in the previous period. That sales in one period affect demand later is natural, especially with durable goods, as well as with goods for which "familiarity breeds contempt" or the reverse. Space does not permit going into detail with models of this sort; however, models of this variety are developed within a game theoretic framework in Friedman (1977a, chs. 9 and 10) and Sobel (1973). An application to oligopoly may be seen in Friedman (1977b). In all these instances the equilibria which are considered are non-cooperative.

5. Oligopoly and game theory

Anyone who has read the present chapter from the beginning to this point cannot fail to see the intimate connection between the material presented and parts of the theory of games. Cournot's equilibrium for the single-period quantity model is the first example, or at the very least, an extraordinarily early example, of a non-cooperative equilibrium in an n-person variable sum game. Surely most of the people working in oligopoly theory now are, and for the past several decades have been, influenced by the fundamental insights of game theory even when game theory is not being brought explicitly to bear. Frequently it *is* brought to bear. In addition, the influences do not all go from game theory to oligopoly (or to other branches of economics); for, among the active contributors to game theory are economists who also work in oligopoly. Often the problems they take up within the theory of games and the way they handle them are influenced by their interest in oligopoly and by the wish to make their results applicable to oligopoly. The reader interested in the present chapter is likely to have an interest in Chapter 7: "Game Theory Models and Methods in Political Economy" by Martin Shubik.

6. Entry and exit in oligopoly models

In the great bulk of the oligopoly literature the number of firms in a market is constant. Firms do not choose whether or not to enter a market. Without a detailed consideration of the mechanisms at work, it is often asserted for competitive markets that zero profits must be the rule in a long-run equilibrium because firms will, over time, leave industries in which profits are negative and enter those in which profits are positive. Whatever the merits of this line of argument for competitive markets, it is untenable for oligopolistic markets. The entry or exit of one firm from such a market is likely to be followed by a large discrete change in the policies being pursued (prices, etc.) by the firms which are active before and after the change. Even if none changed its behavior, amounts demanded of each firm would be much different after the alteration in the number of firms. That is to say, for the competitive firm, it is reasonable for it to suppose that the prices, profitability, etc. of a given market are independent of whether it is in the market. For an oligopolist, however, it is abundantly clear that these variables must depend, in part, on whether or not it is active.

The situation may be summarized by noting that an oligopolistic market is really a game in which the players consist of m active firms who are in the market at the moment and k potential entrants who could come into the market if they wished. It is difficult to see how a useful, insightful and interesting theoretical analysis of entry and exit could be carried out without explicitly taking into account all $m+k$ of these economic agents. Bain (1949), in a seminal article on "entry preventing price", recognizes this in his discussion. Shubik (1959) makes the point much more strongly with his notion of the "firm-in-being", which is the potential entrant.

Summarizing a simple version of the entry preventing price (or limit price) is useful because it has received much attention over the years, another early proponent of it being Harrod (1952). Then it may be seen how the entry preventing price might fit into the modeling of entry and exit. Imagine an industry in which there is one active firm, and assume there is just one potential entrant. According to the simplest notion of limit price there is some p^* such that the potential entrant will not come into the industry if he observes the active firm to charge p^* or less. On the other hand, if he observes a price above p^*, he will come in. The established firm knows how the potential entrant makes his decision and behaves in whatever way he regards as in his best interest.

The problem with this scenario is that it is ad hoc. If, for example, both agents know all three profit functions (the established firm's prior to entry and both firms' after entry), then there is no reason why the pre-entry price of the established firm should have any effect whatever on the decision of the potential entrant. Presumably, the effect of the low price prior to entry is to scare off the potential entrant by fooling him into thinking his profits would be too low to warrant coming in. This could be thought to happen by keeping the established firm's profits down to a dismal level (the Harrod approach), leaving the entrant to think he would fare no better, or by the entrant's believing that the pre-entry price would also prevail after entry. Then the price must be low enough that the entrant concludes he would not fare well enough. Again, in the presence of complete information it is not easy to see how this trickery could be made to work.

A more fruitful approach to justifying a limit price might be sought by assuming the potential entrant does not know what his profit function would be if he entered, and that he assumes the price charged by the established firm conveys some information about it. Some work along these lines may be found in Friedman (1977b). It appears, however, that the entry preventing price survives in this situation only if the profit functions which the entrant believes possible may be unambiguously ordered according to their profitability and, the higher the established firm's price, the more probability weight is shifted to the more profitable functions. These conditions appear rather stringent; however, they are a start at delineating the conditions under which limit prices may be derived in models in which established firms and potential entrants are rational decision-makers.

Very little formal theoretical work has been done on entry into or exit from non-competitive markets. Kamien and Schwartz (1975) deal with a model of oligopoly with the possibility of entry. In that paper and in their earlier work potential entrants are not explicitly modeled; hence, they are not optimizing economic agents within the model. It remains to be seen whether the behavior they attribute to the potential entrants is consistent with the entrant's self-interest in a completely specified model; however, they have certainly stimulated interest among economists in an important and long neglected problem.

7. Bi-lateral monopoly

Bi-lateral monopoly is the name for a market in which a monopolisitic seller faces a monopsonistic buyer. Such a market is, inevitably, a two-person cooperative game. Why it is two person is obvious. That it is cooperative stems from the inability of one firm to function without trading with the other. In a single-period setting, two-person trade represented in an Edgeworth box captures the situation. Though some of the earlier literature is couched in terms of one firm naming the price at which trade takes place with the other naming the amount to be traded [see Fellner (1949, ch. 9)], this is patently artificial since nothing prevents the pair from discussing and agreeing upon any price–quantity combination. At least, this is the case in a one-period model with each firm knowing both profit functions.

As with oligopoly, bi-lateral monopoly theory becomes much more interesting and potentially applicable when behavior over time is taken into account. This was done in the work of Cross (1969) who was interested in analyzing the process of bargaining. He explicitly recognized that as the length of time involved in bargaining to reach agreement is allowed to drag on both parties are likely to receive a smaller payoff in the end than they could have achieved by making an early settlement. An obvious illustration is provided by labor–management negotiations. Not only are resources tied up by both sides in the bargaining process, but also if a strike should ensue, the longer it takes to reach a settlement, the longer both parties go on receiving incomes much below what either would receive under any reasonable agreement. As time goes by, the total pie they have to divide is shrinking.

8. Oligopoly embedded into models of general equilibrium

In the beginning of this chapter it was noted that, in the main, oligopoly is a partial equilibrium study. Considering that the analytical jewel of economic theory is the theory of general competitive equilibrium, it is only natural to wish to treat oligopoly within a general equilibrium framework. Recent years, beginning with Negishi (1961), have seen a number of writings aimed at this goal. They include Farrell (1970), Jaskold-Gabszewicz and Vial (1972), Shitovitz (1973), Marschak and Selten (1974, 1977) Nikaido (1975), Laffont and Laroque (1976), and Roberts and Sonnenschein (1977). The research may be divided into two principal branches – cooperative and non-cooperative. These are briefly sketched in Sections 8.1 and 8.2. In Section 8.3 an impossibility result which is relevant to the non-cooperative approach is discussed.

8.1. *The core theory approach*

Farrell (1970) and Shitovitz (1973) have each put non-cooperative elements into a general equilibrium model of trade in Edgeworth (1881) tradition. Farrell's model is in the spirit of Debreu and Scarf (1963), though he limits himself to two goods and two types of traders, while Shitovitz is in the "continuum" tradition initiated by Aumann (1964) and ably continued in Hildenbrand (1974). The sketch which follows is along the lines of Farrell. Figure 8.1 shows an Edgeworth box for two traders, A and B, whose initial endowments of goods place them at E in the diagram. Any trade which puts them onto a point on the curve GH is Pareto optimal, while any trade which puts them in the shaded region leaves neither trader worse off than he is at E. Thus, efficiency considerations suggest a point on GH and "individual rationality" suggests a point in the shaded region. Any point on the segment DF satisfies both criteria and that segment is the *core* of the two-player trading game under consideration. A point is in the core if there is no subset of players who can trade among themselves in a way which gives each at least as much utility as he would achieve at the given point, with one or more players getting more utility. Clearly, any point outside the shaded area is ruled out by A alone or by B alone; for each has the choice open to him to consume his endowment. Any point inside the shaded region, but not on DF, is ruled out by A

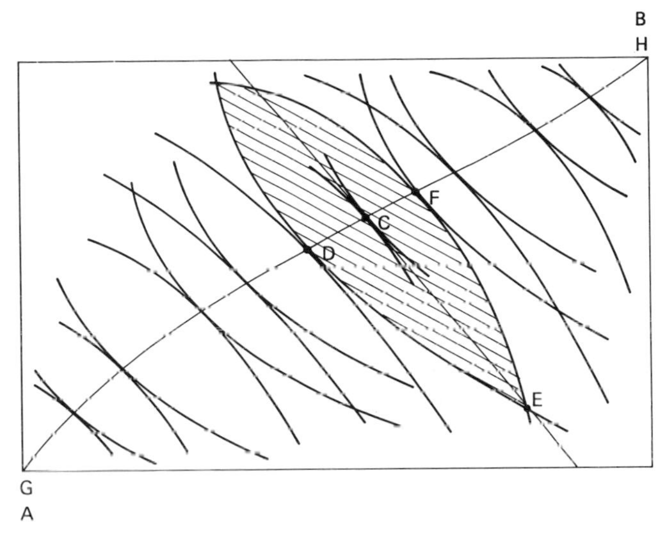

Figure 8.1.

and B together, because they can find a trade on DF which gives more utility to each simultaneously.

Among the results in Debreu and Scarf (1963) are that, if the economy were allowed to grow by replication, then Figure 8.1 could still represent the game, with the core of the larger games still contained in DF. *Growing by replication* means having, say k, traders identical to A in both endowment and preferences and k identical to B. The diagram for a $2k$ trader model still shows only one trader of each type. It is then proved that the core tends to get smaller as k increases, and, in the limit, as $k \to \infty$, the core coincides with the competitive equilibria of the model (i.e. points such as C which are supported by a system of prices).

Farrell's method of including non-competitive traders into the model is to have the number of traders of type A fixed at m, while the number of type B traders is allowed to grow. He then gets a limit result like Debreu and Scarf for $m \geqslant 2$. In the limit, as the number of B traders goes to infinity, the core consists of only the competitive equilibria. For a model in which there is a measure space of traders with the non-competitive traders being atoms in the space, Shitovitz also gets the result that the core and the set of competitive equilibria coincide when there are two or more "large" traders.

These results, which essentially say that non-competitive elements do not matter, may seem surprising. They may stem from the assumption of a lack of organizational costs in coalition formation together with the ability of the small traders to group themselves into whatever number and size coalitions is in their best interest. One is left in the position either that traditional oligopoly theory is a blind alley or that the core models do not get at the essentials of oligopoly.

8.2. *Noncooperative models*

The approach taken in Jaskold-Gabszewicz and Vial (1972), Marschak and Selten (1974, 1977), Nikaido (1975), and Laffont and Laroque (1976) is to model an economy in which collusion and coalition formation are not allowed, and in which some firms are sufficiently large that they are not perfect competitors in the markets for their outputs. To take Laffont and Laroque as a specific example, all consumers are competitors and certain of the goods are produced by competitive firms. The remaining goods are produced by the non-competitive firms, with each firm producing an arbitrary subset of these goods. Each could produce all of them.[11]

The aim of the work is to prove a theorem on existence of equilibrium starting from the kind of "basic" assumptions which are typical of competitive general equilibrium theory. This means, of course, characterizing consumers' tastes by axioms describing their preferences and characterizing the production structures of the firms, etc. Thus, the intent is to start from more basic beginnings than is usual in oligopoly, where the consumers are represented through demand functions which not only are not derived from an underlying preference structure, they are assumed to have many properties which are convenient for analysis, and the production structure makes no explicit appearance.

While an existence theorem is proved, it is necessary to bring in additional assumptions of the sort made in the oligopoly literature.

8.3. *An impossibility theorem*

None of the research using the non-cooperative approach attains an existence theorem based solely on fundamental assumptions on the consumers' preferences, the firms' productions sets, and reasonable assumptions about the way economic agents make decisions. Such assumptions are not enough to ensure that the profit functions of the non-competitive agents are continuous, much less quasi-concave,

[11]These models do imply that the textbook notion of the monopolist, who has no game-theoretic type interactions with any other economic agents, is not fruitful. In Nikaido (1975), for example, each good has only one producer; however, the model then has the characteristics of an n-product, n-firm differentiated products oligopoly. The same holds in Laffont and Laroque (1976) when each "non-competitive" good is assumed to have only one producer.

in the variables under the control of the agent. Such additional conditions have, then, been assumed in order to generate existence theorems. Roberts and Sonnenschein (1977) have produced examples of economies in which equilibrium does not exist, but in which the usual preference and production conditions do obtain. Thus, they show definitively that the goal for which Laffont and Laroque reach unsuccessfully is, in fact, impossible.

It appears beyond hope to obtain results for non-competitive economies which are at the same level of generality as the now standard results for competitive economies. The next task in this line of work would seem to be to see whether the unsatisfactory assumptions of Laffont and Laroque and others could not be replaced with more fundamental assumptions – though these would have to be more restrictive than the assumptions which have been standard in competitive analysis up to this time.

References

Aumann, Robert J. (1964), "Markets with a continuum of traders", Econometrica, 32:39–50.

Bain, Joe S. (1949), "A note on pricing in monopoly and oligopoly", American Economic Review, 39:448–464.

Bertrand, Joseph (1883), "Review of 'Théorie mathématique de la richesse sociale and Recherches sur les principes mathématiques de la théorie des richesess'", Journal des Savants, 499–508.

Bowley, Arthur L. (1924), The mathematical groundwork of economics. Oxford: Oxford University Press.

Chamberlin, Edward H. (1956), The theory of monopolistic competition, 7th edn. Cambridge: Harvard University Press.

Cournot, Augustin (1927), Researches into the mathematical principles of the theory of wealth, translated by N. T. Bacon. New York: Macmillan.

Cross, John G. (1959), The economics of bargaining. New York: Basic Books.

Cyert, Richard M. and Morris de Groot (1970), "Multiperiod decision models with alternating choice as a solution to the duopoly problem", Quarterly Journal of Economics, 84:410–429.

Debreu, Gerard and Herbert Scarf (1963), "A limit theorem on the core of an economy", International Economic Review, 4:235–246.

Edgeworth, Francis Y. (1881), Mathematical psychics. London: Kegan Paul.

Edgeworth, Francis Y. (1925), "The pure theory of monopoly", in: Francis Y. Edgeworth, Papers Relating to Political Economy, Vol. 1. London: Macmillan, 111–142.

Farrell, Michael J. (1970), "Edgeworth bounds for oligopoly prices", Economica, 37:342–361.

Fellner, William J. (1949), Competition among the few. New York: Knopf.

Friedman, James W. (1977a), Oligopoly and the theory of games. Amsterdam: North-Holland Publishing Co.

Friedman, James W. (1977b), "On limit price models of entry", University of Rochester, Department of Economics, April 1977.

Harrod, Roy F. (1952), "Theory of imperfect competition revised", Chapter 8 in: Roy F. Harrod, Economic Essays. London: Macmillan.

Hildenbrand, Werner (1974), Core and equilibria of a large economy. Princeton: Princeton University Press.

Jaskold-Gabszewicz, Jean and Jean-Philippe Vial (1972), "Oligopoly á la Cournot in a general equilibrium analysis", Journal of Economic Theory, 4:381–400.

Kamien, Morton I. and Nancy L. Schwartz (1975), "Cournot oligopoly with uncertain entry", Review of Economic Studies, 48:125–131.
Laffont, Jean-Jacques and Guy Laroque (1976), "Existence d'un equilibre Général de concurrence imparfaite: Une introduction", Econometrica, 44:283–294.
Marschak, Thomas and Reinhard Selten (1974), General equilibrium with price-making firms. Berlin: Springer-Verlag.
Marschak, Thomas and Reinhard Selten (1977), "Oligopolistic economies as games of limited information", Zeitschrift für die Gesamte Staatswissenschaft, 133:385–410.
Marschak, Thomas and Reinhard Selten (1978), "Restabilizing responses, inertia supergames, and oligopolistic equilibria", Quarterly Journal of Economics, 92:71–93.
Nash, John F., Jr. (1951), "Non-cooperative games", Annals of Mathematics, 45:286–295.
Negishi, Takashi (1961), "Monopolistic competition and general equilibrium theory", Review of Economic Studies, 28:196–201.
Neumann, John von and Oskar Morgenstern (1954), Theory of games and economic behavior, 3rd edn. Princeton: Princeton University Press.
Nikaido, Hukukane (1975), Monopolistic competition and effective demand. Princeton: Princeton University Press.
Okuguchi, Koji (1976), Expectations and stability in oligopoly models. Berlin: Springer-Verlag.
Roberts, John and Hugo Sonnenschein (1977), "On the foundations of the theory of monopolistic competition", Econometrica, 45:101–113.
Selten, Reinhard (1970), Preispolitik der Mehrproduktenunternehmung in der Statischen Theorie. Berlin: Springer-Verlag.
Shitovitz, Benyamin (1973), "Oligopoly in market with a continuum of traders", Econometrica, 41:467–501.
Shubik, Martin (1959), Strategy and market structure. New York: Wiley.
Sobel, Matthew J. (1973), "Continuous stochastic games", Journal of Applied Probability, 10:597–604.
Stackelberg, Heinrich von (1934), Marktform and Gleichgewicht. Vienna: Springer.
Sweezy, Paul M. (1939), "Demand under conditions of oligopoly", Journal of Political Economy, 47:568–573.
Szidarovszky, F. and S. Yakowitz (1977), A new proof of the existence and uniqueness of the Cournot equilibrium", International Economic Review, 18:787–789.
Triffin, Robert (1940), Monopolistic competition and general equilibrium theory. Cambridge: Harvard University Press.

Chapter 12

DUALITY APPROACHES TO MICROECONOMIC THEORY*

W. E. DIEWERT

University of British Columbia

1. Introduction

What do we mean when we say that there is a "duality" between cost and production functions? Suppose that a *production function* F is given and that $u=F(x)$, where u is the maximum amount of output that can be produced by the technology during a certain period if the vector of input quantities $x\equiv(x_1, x_2,\dots, x_N)$ is utilized during the period. Thus, the production function F describes the technology of the given firm. On the other hand, the firm's *minimum total cost* of producing at least the output level u given the input prices $(p_1, p_2,\dots, p_N)\equiv p$ is defined as $C(u, p)$, and it is obviously a function of u, p and the given production function F. What is not so obvious is that (under certain regularity conditions) the cost function $C(u, p)$ also completely describes the technology of the given firm; i.e., given the firm's cost function C, it can be used in order to define the firm's production function F. Thus, there is a *duality* between cost and production functions in the sense that either of these functions can describe the technology of the firm equally well in certain circumstances.

In the first part of this chapter we develop this duality between cost and production functions in more detail. In Section 2 we derive the regularity conditions that a cost function C *must* have (irrespective of the functional form or specific regularity properties for the production function F), and we show how a production function can be constructed from a given cost function. In Section 3 we develop this duality between cost and production functions in a more formal manner.

In Section 4 we consider the duality between a (direct) production function F and the corresponding indirect production function G. Given a production

*This work was supported by the Canada Council and National Science Foundation Grant SOC77-11105 at the Institute for Mathematical Studies in the Social Sciences, Stanford University. The author would like to thank C. Blackorby, R. C. Allen, and R. T. Rockafellar for helpful comments and M. D. Intriligator for editorial assistance. A preliminary version of this paper was presented at the New York meetings of the Econometric Society, December 1977.

Handbook of Mathematical Economics, vol. II, edited by K.J. Arrow and M.D. Intriligator

function F, input prices $p \equiv (p_1, p_2, \ldots, p_N)$ and an input budget of y dollars, the *indirect production function* $G(y, p)$ is defined as the maximum output $u = F(x)$ that can be produced, given the budget constraint on input expenditures $p^{\mathrm{T}}x \equiv \sum_{i=1}^{N} p_i x_i \leqslant y$. Thus, the indirect production function $G(y, p)$ is a function of the maximum allowable budget y, the input prices which the producer faces p, and the producer's production function F. Under certain regularity conditions, it turns out that G can also completely describe the technology and thus there is a duality between direct and indirect production functions.

The above dualities between cost, production, and indirect production functions also have an interpretation in the context of consumer theory: simply let F be a consumer's *utility function*, x be a vector of commodity purchases (or rentals), u be the consumer's utility level, and y be the consumer's "income" or expenditure on the N commodities. Then $C(u, p)$ is the minimum cost of achieving utility level u given that the consumer faces the commodity prices p and there is a duality between the consumer's utility function F and the function C, which is often called the *expenditure function* in the context of consumer theory. Similarly, $G(y, p)$ can now be defined as the maximum utility that the consumer can attain given that he faces prices p and has income y to spend on the N commodities. In the consumer context, G is called the consumer's *indirect utility function*.

Thus, each of our duality theorems has two interpretations: one in the producer context and one in the consumer context. In Section 2 we will use the producer theory terminology for the sake of concreteness. However, in subsequent sections we use a more neutral terminology which will cover both the producer and the consumer interpretations: we call a production or utility function F an *aggregator function*, a cost or expenditure function C a *cost function*, and an indirect production or utility function G an *indirect aggregator function*.

In Section 5 the distance function $D(u, x)$ is introduced. The *distance function* provides yet another way in which tastes or technology can be characterized. The main use of the distance function is in constructing the Malmquist (1953) quantity index.

In Section 6 we discuss a variety of other duality theorems: i.e. we discuss other methods for equivalently describing tastes or technology, either locally or globally, in the one-output, N-inputs context. The reader who is primarily interested in applications can skip Sections 3–6.

The mathematical theorems presented in Sections 2–6 may appear to be only theoretical results (of modest mathematical interest perhaps) devoid of practical applications. However, this is not the case. In Sections 7–10 we survey some of the applications of the duality theorems developed earlier. These applications fall in two main categories: (i) the measurement of technology or preferences (Sections 9 and 10) and (ii) the derivation of comparative statics results (Sections 7 and 8).

In Section 10 we consider firms that can produce *many* outputs while utilizing many inputs (whereas earlier we dealt only with the *one* output case). We state some useful duality theorems and then note some applications of these theorems.

Finally, in Sections 11 and 12 we briefly note some of the other areas of economics where duality theory has been applied.

Proofs are omitted for the most part: proofs can be found in the literature referred to or in Diewert (1982).

2. Duality between cost (expenditure) and production (utility) functions: A simplified approach

Suppose we are given an N-input production function F: $u=F(x)$, where u is the amount of output produced during a period and $x=(x_1, x_2,\dots, x_N)\geq 0_N$[1] is a non-negative vector of input quantities utilized during the period. Suppose, further, that the producer can purchase amounts of the inputs at the fixed positive prices $(p_1, p_2,\dots, p_N)\equiv p\gg 0_N$ and that the producer does not attempt to exert any monopsony power on input markets.[2]

The producer's *cost function* C is defined as the solution to the problem of minimizing the cost of producing at least output level u, given that the producer faces the input price vector p:

$$C(u, p)\equiv\min_x\left\{p^{\mathrm{T}}x: F(x)\geq u\right\}. \tag{2.1}$$

In this section it is shown that the cost function C satisfies a surprising number of regularity conditions, *irrespective* of the functional form for the production function F, provided only that solutions to the cost minimization problem (2.1) exist. In a subsequent section, it is shown how these regularity conditions on the cost function may be used in order to prove comparative statics theorems about derived demand functions for inputs [cf. Samuelson (1947, ch. 4)].

Before establishing the properties of the cost function C, it is convenient to place the following minimal regularity condition on the production function F:

Assumption 1 on F

F is continuous from above, i.e. for every $u\in$range F,[3] $L(u)\equiv\{x: x\geq 0_N, F(x)\geq u\}$ is a closed set.

[1]Notation: $x\geq 0_N$ means each component of the N-dimensional vector x is non-negative, $x\gg 0_N$ means that each component is positive, and $x>0_N$ means that $x\geq 0_N$ but $x\neq 0_N$.

[2]In Section 11 of this chapter the assumption of competitive behavior is relaxed.

[3]This simply means that the output u can be produced by the technology. Throughout this section, range F can be replaced with the smallest convex set containing range F.

If F is a continuous function, then of course F will also be continuous from above. Assumption 1 is sufficient to imply that solutions to the cost minimization problem (2.1) exist.

The following seven properties for the cost function C can now be derived, assuming *only* that the production function F satisfies Assumption 1.

Property 1 for C

For every $u \in$ range F and $p \gg 0_N$, $C(u, p) \geqslant 0$, i.e. C is a *non-negative* function.

Proof

$$\begin{aligned} C(u, p) &\equiv \min_x \{p^{\mathrm{T}}x : x \geqslant 0_N, F(x) \geqslant u\} \\ &= p^{\mathrm{T}}x^* \text{ say, where } x^* \geqslant 0_N \text{ and } F(x^*) \geqslant u \\ &\geqslant 0 \text{ since } p \gg 0_N \text{ and } x^* \geqslant 0_N. \quad \text{Q.E.D.} \end{aligned}$$

Property 2 for C

For every $u \in$ range F, if $p \gg 0_N$ and $k > 0$, then $C(u, kp) = kC(u, p)$, i.e. the cost function is (positively) *linearly homogeneous* in input prices for any fixed output level.

Proof

Let $p \gg 0_N$, $k > 0$ and $u \in$ range F. Then

$$\begin{aligned} C(u, kp) &\equiv \min_x \{(kp)^{\mathrm{T}}x : F(x) \geqslant u\} \\ &= k \min_x \{p^{\mathrm{T}}x : F(x) \geqslant u\} = kC(u, p). \quad \text{Q.E.D.} \end{aligned}$$

Property 3 for C

If any combination of input prices increases, then the minimum cost of producing any feasible output level u will not decrease; i.e. if $u \in$ range F and $p^1 > p^0$, then $C(u, p^1) \geqslant C(u, p^0)$.

Proof

$$\begin{aligned} C(u, p^1) &\equiv \min_x \{p^{1\mathrm{T}}x : F(x) \geqslant u\} \\ &= p^{1\mathrm{T}}x^1 \text{ say, where } x^1 \geqslant 0_N \text{ and } F(x^1) \geqslant u \\ &\geqslant p^{0\mathrm{T}}x^1 \text{ since } p^1 > p^0 \text{ and } x^1 \geqslant 0_N \\ &\geqslant \min_x \{p^{0\mathrm{T}}x : F(x) \geqslant u\} \text{ since } x^1 \text{ is feasible for the cost} \\ &\quad \text{minimization problem but not necessarily optimal} \\ &\equiv C(u, p^0). \quad \text{Q.E.D.} \end{aligned}$$

Thus far the properties of the cost function have been intuitively obvious from an economic point of view. However, the following important property is not an intuitively obvious one.

Property 4 for C

For every $u \in$ range F, $C(u, p)$ is a *concave* function[4] of p.

Proof

Let $u \in$ range F, $p^0 \gg 0_N$, $p^1 \gg 0_N$ and $0 \leq \lambda \leq 1$. Then

$$C(u, p^0) \equiv \min_x \{p^{0\mathrm{T}}x : F(x) \geq u\} = p^{0\mathrm{T}}x^0 \text{ say, and}$$

$$C(u, p^1) \equiv \min_x \{p^{1\mathrm{T}}x : F(x) \geq u\} = p^{1\mathrm{T}}x^1 \text{ say. Now}$$

$$\begin{aligned} C(u, \lambda p^0 + (1-\lambda)p^1) &\equiv \min_x \{(\lambda p^0 + (1-\lambda)p^1)^{\mathrm{T}} x : F(x) \geq u\} \\ &= (\lambda p^0 + (1-\lambda)p^1)^{\mathrm{T}} x^{\lambda} \text{ say} \\ &= \lambda p^{0\mathrm{T}} x^{\lambda} + (1-\lambda) p^{1\mathrm{T}} x^{\lambda} \\ &\geq \lambda p^{0\mathrm{T}} x^0 + (1-\lambda) p^{1\mathrm{T}} x^1 \end{aligned}$$

since x^{λ} is feasible for the cost minimization problems associated with the input price vectors p^0 and p^1 but is not necessarily optimal for those problems

$$= \lambda C(u, p^0) + (1-\lambda) C(u, p^1). \quad \text{Q.E.D.}$$

The basic idea in the above proof is used repeatedly in duality theory. Owing to the non-intuitive nature of Property 4, it is perhaps useful to provide a geometric interpretation of it in the two-input case (i.e. $N=2$).

Suppose that the producer must produce the output level u. The u isoquant is drawn in Figure 2.1. Define the set S^0 as the set of non-negative input combinations which are either on or below the optimal isocost line when the producer faces prices p^0; i.e. $S^0 \equiv \{x : p^{0\mathrm{T}}x \leq C(u, p^0), x \geq 0_N\}$, where $C^0 \equiv C(u, p^0) = p^{0\mathrm{T}}x^0$ is the minimum cost of producing output u given that the producer faces input prices $p^0 \gg 0_N$. Note that the vector of inputs x^0 solves the cost minimization problem in this case. Now suppose that the producer faces the input prices $p^1 \gg 0_N$ and define S^1, C^1, and x^1 analogously, i.e. $S^1 \equiv \{x : p^{1\mathrm{T}}x \leq C(u, p^1),$

[4]A function $f(z)$ of n variables defined over a convex set S is *concave* iff $z^1, z^2 \in S$, $0 \leq \lambda \leq 1$ implies $f(\lambda z^1 + (1-\lambda)z^2) \geq \lambda f(z^1) + (1-\lambda) f(z^2)$. A set S is a *convex set* iff $z^1, z^2 \in S$, $0 \leq \lambda \leq 1$ implies $(\lambda z^1 + (1-\lambda)z^2) \in S$.

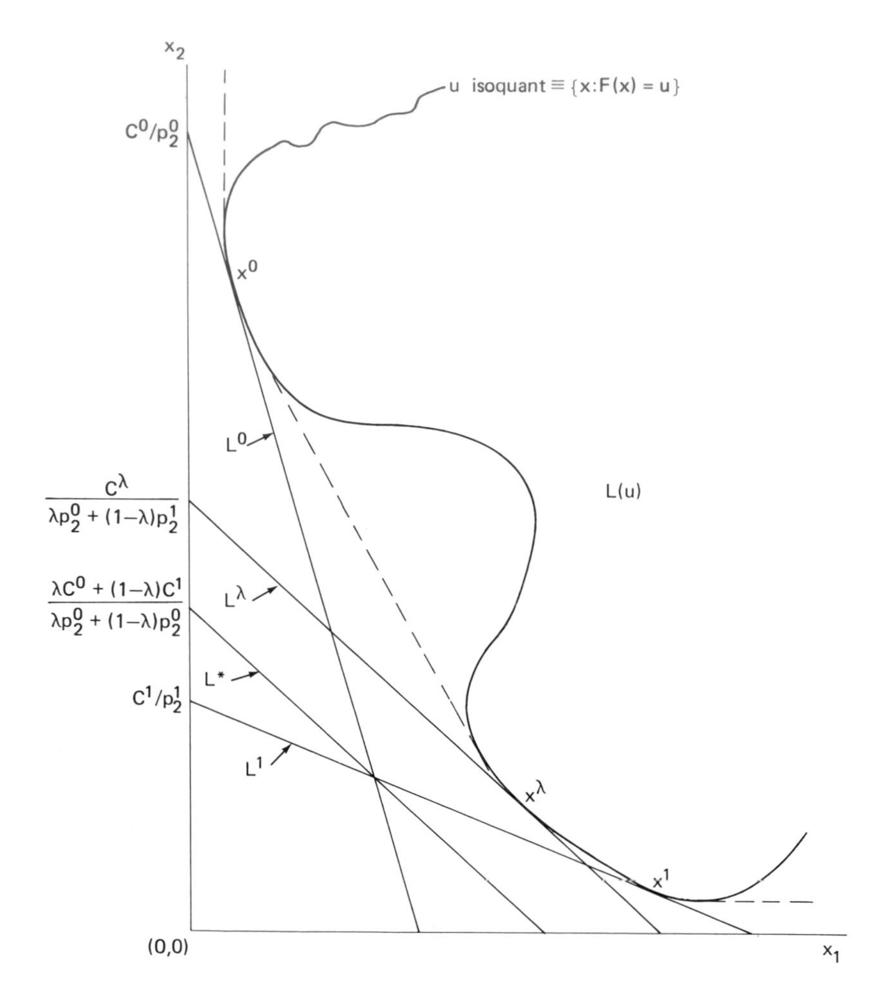

Figure 2.1.

$x \geqslant 0_N\}, C^1 \equiv C(u, p^1) = p^{1\mathrm{T}}x^1$, where the vector of inputs x^1 solves the cost minimization problem when the producer faces prices p^1.

Let $0<\lambda<1$ and suppose now that the producer faces the "average" input prices $\lambda p^0 + (1-\lambda)p^1$. Define S^λ, C^λ, and x^λ as before:

$$S^\lambda \equiv \left\{x : \left(\lambda p^0 + (1-\lambda)p^1\right)^{\mathrm{T}} x \leqslant C\left(u, \lambda p^0 + (1-\lambda)p^1\right), x \geqslant 0_N\right\},$$

$$C^\lambda \equiv C\left(u, \lambda p^0 + (1-\lambda)p^1\right) = \left(\lambda p^0 + (1-\lambda)p^1\right)^{\mathrm{T}} x^\lambda,$$

where x^{λ} solves the cost minimization problem when the producer faces the average prices $\lambda p^0+(1-\lambda)p^1$. Finally, consider the isocost line which would result if the producer spends an "average" of the two initial costs, $\lambda C^0+(1-\lambda)C^1$, facing the "average" input prices, $\lambda p^0+(1-\lambda)p^1$. The set of non-negative input combinations which are either on or below this isocost line is defined as the set $S^*\equiv\{x:(\lambda p^0+(1-\lambda)p^1)^{\mathrm{T}}x\leqslant\lambda C^0+(1-\lambda)C^1, x\geqslant 0_N\}$. In order to show the concavity property for C, we need to show that $C^{\lambda}\geqslant\lambda C^0+(1-\lambda)C^1$ or, equivalently, we need to show that S^{λ} contains the set S^*. It can be shown that the isocost line associated with the set S^*, $L^*\equiv\{x:(\lambda p^0+(1-\lambda)p^1)^{\mathrm{T}}x=\lambda C^0+(1-\lambda)C^1$, passes through the intersection of the isocost lines associated with the sets S^0 and S^1.[5] On the other hand, the isocost line associated with the set S^{λ}, $L^{\lambda}\equiv\{x:(\lambda p^0+(1-\lambda)p^1)^{\mathrm{T}}x=C^{\lambda}\}$ is obviously parallel to L^*. Finally, L^{λ} must be either coincident with or lie above L^*, since if L^{λ} were below L^*, then there would exist a point on the u isoquant which would lie below at least one of the isocost lines $L^0\equiv\{x: p^{0\mathrm{T}}x=C^0\}$ or $L^1\equiv\{x: p^{1\mathrm{T}}x=C^1\}$ which would contradict the cost minimizing nature of x^0 or x^1.[6]

Property 5 for C

For every $u\in$ range F, $C(u,p)$ is *continuous* in p for $p\gg 0_N$. [Proofs of this property are based on results in Fenchel (1953, p. 75) and Rockafellar (1970, p. 82).]

Property 6 for C

$C(u,p)$ is non-decreasing in u for fixed p; i.e. if $p\gg 0_N$, $u^0, u^1\in$ range F, and $u^0\leqslant u^1$, then $C(u^0,p)\leqslant C(u^1,p)$.

Proof

Let $p\gg 0_N$, $u^0, u^1\in$ range F and $u^0\leqslant u^1$. Then

$$
\begin{aligned}
C(u^1,p) &\equiv \min_x \{p^{\mathrm{T}}x: F(x)\geqslant u^1\}\\
&\geqslant \min_x \{p^{\mathrm{T}}x: F(x)\geqslant u^0\} \text{ since if } u^0\leqslant u^1, \text{ then}\\
&\quad \{x: F(x)\geqslant u^1\}\subset\{x: F(x)\geqslant u^0\} \text{ and the minimum}\\
&\quad \text{of } p^{\mathrm{T}}x \text{ over a large set cannot increase}\\
&\equiv C(u^0,p). \quad \text{Q.E.D.}
\end{aligned}
$$

[5] Let $x^*\in L^0\cap L^1$. Then $p^{0\mathrm{T}}x^*=C^0$ and $p^{1\mathrm{T}}x^*=C^1$. Thus, $(\lambda p^0+(1-\lambda)p^1)^{\mathrm{T}}x^*=\lambda C^0+(1-\lambda)C^1$ and $x^*\in S^*$. This also follows from the readily proven proposition that $S^0\cap S^1\subset S^*\subset S^0\cup S^1$ [cf. Diewert (1974a, pp. 157–158)].

[6] It can be seen that the approximating set $S^a\equiv\{x: p^{0\mathrm{T}}x\geqslant C^0, x\geqslant 0_N\}\cap\{x: p^{1\mathrm{T}}x\geqslant C^1, x\geqslant 0_N\}$ *contains* the true technological set $L(u)\equiv\{x: F(x)\geqslant u\}$ and thus the minimum cost associated with S^a will generally be *lower* than the cost associated with $L(u)$.

In contrast to the previous properties for the cost function, the following property requires some heavy mathematical artillery. Since these mathematical results are useful not only in the present section, but also in subsequent sections, we momentarily digress and state these results.

In the following definitions let S denote a subset of R^M, T is a subset of $R^K, \{x^n\}$ a sequence of points of S, and $\{y^n\}$ a sequence of points of T. For a more complete discussion of the following definitions and theorems, see Chapter 1 of the Handbook, by Green and Heller.

Definition

ϕ is a *correspondence* from S into T if for every $x \in S$ there exists a non-empty image set $\phi(x)$ which is a subset of T.

Definitions

A correspondence ϕ is *upper semi-continuous* (or alternatively, *upper hemi-continuous*) at the point $x^0 \in S$ if $\lim_n x^n = x^0$, $y^n \in \phi(x^n)$, $\lim_n y^n = y^0$ implies $y^0 \in \phi(x^0)$. A correspondence ϕ is *lower semi-continuous* at $x^0 \in S$ if $\lim_n x^n = x^0$, $y^0 \in \phi(x^0)$ implies that there exists a sequence $\{y^n\}$ such that $y^n \in \phi(x^n)$ and $\lim_n y^n = y^0$. A correspondence ϕ is *continuous* at $x^0 \in S$ if it is both upper and lower semi-continuous at x^0.

Lemma 2 [Berge (1963, pp. 111–112)]

ϕ is an upper semi-continuous correspondence over S iff graph $\phi \equiv \{(x, y): x \in S, y \in \phi(x)\}$ is a closed set in $S \times T$.[7]

Upper semi-continuous maximum theorem [Berge (1963, p. 116)]

Let f be a continuous from above function[8] defined over $S \times T$, where T is a compact (i.e. closed and bounded) subset of R^K. Suppose that ϕ is a correspondence from S into T and that ϕ is upper semi-continuous over S. Then the function g defined by $g(x) \equiv \max_y \{f(x, y): y \in \phi(x)\}$ is well defined and is continuous from above over S.

Maximum theorem [Debreu (1952, pp. 889–890), (1959, p. 19); Berge (1963, p. 116)]

Let f be a continuous real valued function defined over $S \times T$, where T is a compact subset of R^K. Let ϕ be a correspondence from S into T and let ϕ be

[7] $S \times T$ is the set of (x, y) such that $x \in S$ and $y \in T$.

[8] A real valued function f defined over $S \times T$ is *continuous from above* (or alternatively, is *upper semi-continuous*) at $z^0 \in S \times T$ iff either of the following conditions is satisfied: (i) for every $\varepsilon > 0$ there exists a neighborhood of z^0, $N(z^0)$ such that $z \in N(z^0)$ implies $f(z) < f(z^0) + \varepsilon$, or (ii) if $z^n \in S \times T$, $\lim_n z^n = z^0$, $f(z^n) \geqslant f(z^0)$, then $\lim_n f(z^n) = f(z^0)$. f is *continuous from above over* $S \times T$ if it is continuous from above at each point of $S \times T$. See Chapter 1 of this Handbook, by Heller and Green.

continuous over S. Define the (maximum) function g by $g(x) \equiv \max_y\{f(x, y): y \in \phi(x)\}$ and the (set of maximizers) correspondence ξ by $\xi(x) \equiv \{y: y \in \phi(x)$ and $f(x, y) = g(x)\}$. Then the function g is continuous over S and the correspondence ξ is upper semi-continuous over S.

Property 7 for C

For every $p \gg 0_N$, $C(u, p)$ is continuous from below[9] in u; i.e. if $p^* \gg 0_N$, $u^* \in$ range F, $u^n \in$ range F for all n, $u^1 \leqslant u^2 \leqslant \dots$ and $\lim u^n = u^*$, then $\lim_n C(u^n, p^*) = C(u^*, p^*)$.

A proof of Property 7 can be found in Diewert (1982).

In order to illustrate this property at C, the reader may find it useful to let $N = 1$ and let the production function $F(x)$ be the following (continuous from above) step function [cf. Shephard (1970, p. 89)]:

$$F(x) \equiv \{0, \text{if } 0 \leqslant x < 1; 1, \text{if } 1 \leqslant x < 2; 2, \text{if } 2 \leqslant x < 3, \dots\}.$$

For $p > 0$, the corresponding cost function $C(u, p)$ is the following (continuous from below) step function:

$$C(u, p) \equiv \{0, \text{if } 0 = u;\ p, \text{if } 0 < u \leqslant 1; 2p, \text{if } 1 < u \leqslant 2, \dots\}.$$

The above properties of the cost function have some empirical implications, as we shall see later. However, one implication can be mentioned at this point. Suppose that we can observe cost, input prices, and output for a firm and suppose further that we have econometrically estimated the following linear cost function:[10]

$$C(u, p) = \alpha + \beta^{\mathrm{T}} p + \gamma u, \tag{2.2}$$

where α and γ are constants and β is a vector of constants. Could (2.2) be the firm's true cost function? The answer is *no* if the firm is competitively minimizing costs and if either one of the constants α and γ is non-zero, for in this case C does not satisfy Property 2 (linear homogeneity in input prices).

Suppose now that we have somehow determined the firm's true cost function C, but that we do not know the firm's production function F (except that it satisfies Assumption 1). How can we use the given cost function $C(u, p)$ (satisfying Properties 1–7 above) in order to construct the firm's underlying production function $F(x)$?

[9]A function f is *continuous from below* iff $-f$ is continuous from above.

[10]This type of cost function is often estimated by economists; e.g. see Walters' (1961) survey article on cost and production functions.

Equivalent to the production function $u=F(x)$ are the family of isoproduct surfaces $\{x: F(x)=u\}$ or the family of level sets $L(u)\equiv\{x: F(x)\geqslant u\}$. For any $u\in$range F, the cost function can be used in order to construct an outer approximation to the set $L(u)$ in the following manner. Pick input prices $p^1\gg 0_N$ and graph the isocost surface $\{x: p^{1\mathrm{T}}x=C(u, p^1)\}$. The set $L(u)$ must lie above (and intersect) this set, because $C(u, p^1)\equiv\min_x\{p^{1\mathrm{T}}x: x\in L(u)\}$; i.e. $L(u)\subset\{x: p^{1\mathrm{T}}x\geqslant C(u, p^1)\}$. Pick additional input price vectors $p^2\gg 0_N$, $p^3\gg 0_N,\ldots$, and the graph isocost surfaces $\{x: p^{i\mathrm{T}}x=C(u, p^i)\}$. It is easy to see that $L(u)$ must be a subset of each of the sets $\{x: p^{i\mathrm{T}}x\geqslant C(u, p^i)\}$. Thus,

$$L(u)\subset\bigcap_{p\gg 0_N}\{x: p^{\mathrm{T}}x\geqslant C(u, p)\}\equiv L^*(u), \tag{2.3}$$

i.e. the true production possibilities set $L(u)$ must be contained in the outer approximation production possibilities set $L^*(u)$ which is obtained as the intersection of all of the supporting total cost half spaces to the true technology set $L(u)$. In Figure 2.1, $L^*(u)$ is indicated by dashed lines. Note that the boundary of this set forms an approximation to the true u isoquant and that this approximating isoquant coincides with the true isoquant in part, but it does not have the backward bending and non-convex portions of the true isoquant.

Once the family of approximating production possibilities sets $L^*(u)$ has been constructed, the approximating production function F^* can be defined as

$$\begin{aligned} F^*(x)&\equiv\max_u\{u: x\in L^*(u)\} \\ &=\max_u\{u: p^{\mathrm{T}}x\geqslant C(u, p)\quad\text{for every}\quad p\gg 0_N\}, \end{aligned} \tag{2.4}$$

for $x\geqslant 0_N$. Note that the maximization problem defined by (2.4) has an infinite number of constraints (one constraint for each $p\gg 0_N$). However, (2.4) can be used in order to define the approximating production function F^* given only the cost function C.

It is clear (recall Figure 2.1) that the approximating production function F^* will not in general coincide with the true function F. However, it is also clear that from the viewpoint of observed market behavior, if the producer is competitively cost minimizing, then it does not matter whether the producer is minimizing cost subject to the production function constraint given by F or F^*: observable market data will never allow us to determine whether the producer has the production function F or the approximating function F^*.

It is also clear that if we want the approximating production function F^* to coincide with the true production function F, then it is necessary that F satisfy the following two assumptions:

Assumption 2 on F

F is *non-decreasing*; i.e. if $x^2 \geqslant x^1 \geqslant 0_N$, then $F(x^2) \geqslant F(x^1)$.

Assumption 3 on F

F is a *quasi-concave* function; i.e. for every $u \in$ range F, $L(u) \equiv \{x: F(x) \geqslant u\}$ is convex set.

If F satisfies Assumption 2, then backward bending isoquants cannot occur, while if F satisfies Assumption 3, then non-convex isoquants of the type drawn in Figure 2.1 cannot occur.

It is not too difficult to show that if F satisfies Assumptions 1–3 and the cost function C is computed via (2.1), then the approximating production function F^* computed via (2.4) will coincide with the original production function F; i.e. there is a *duality* between cost functions satisfying Properties 1–7 and production functions satisfying Assumptions 1–3. The first person to prove a formal duality theorem of this type was Shephard (1953).

In the following section we will state a similar duality theorem after placing somewhat stronger conditions on the underlying production function F.

The following result is the basis for most theoretical and empirical applications of duality theory.

Lemma 3 [Hicks (1946, p. 331); Samuelson (1947, p. 68); Karlin (1959, p. 272); and Gorman (1976)]

Suppose that the production function F satisfies Assumption 1 and that the cost function C is defined by (2.1). Let $u^* \in$ range F, $p^* \gg 0_N$ and suppose that x^* is a solution to the problem of minimizing the cost of producing output level u^* when input prices p^* prevail; i.e.

$$C(u^*, p^*) \equiv \min_x \{p^{*\mathrm{T}}x: F(x) \geqslant u^*\} = p^{*\mathrm{T}}x^*. \tag{2.5}$$

If in addition, C is differentiable with respect to input prices at the point (u^*, p^*), then

$$x^* = \nabla_p C(u^*, p^*), \tag{2.6}$$

where

$$\nabla_p C(u^*, p^*) \equiv [\partial C(u^*, p^*_1, \ldots, p^*_N)/\partial p_1, \ldots, \partial C(u^*, p^*_1, \ldots, p^*_N)/\partial p_N]^{\mathrm{T}}$$

is the vector of first-order partial derivatives of C with respect to the components of the input price vector p.

Proof

Given any vector of positive input prices $p \gg 0_N$, x^* is feasible for the cost minimization problem defined by $C(u^*, p)$ but it is not necessarily optimal; i.e. for every $p \gg 0_N$ we have the following inequality:

$$p^T x^* \geq C(u^*, p). \tag{2.7}$$

For $p \gg 0_N$, define the function $g(p) \equiv p^T x^* - C(u^*, p)$. From (2.7), $g(p) \geq 0$ for $p \gg 0_N$ and from (2.5), $g(p^*) = 0$. Thus, $g(p)$ attains a global minimum at $p = p^*$. Since g is differentiable at p^*, the first-order necessary conditions for a local minimum must be satisfied:

$$\nabla_p g(p^*) = x^* - \nabla_p C(u^*, p^*) = 0_N,$$

which implies (2.6). Q.E.D.

Thus, differentiation of the producer's cost function $C(u, p)$ with respect to input prices p yields the producer's system of cost minimizing input demand functions, $x(u, p) = \nabla_p C(u, p)$.

The above lemma should be carefully compared with the following result.

Lemma 4 [Shephard (1953, p. 11)]

If the cost function $C(u, p)$ satisfies Properties 1–7 and, in addition, is differentiable with respect to input prices at the point (u^*, p^*), then

$$x(u^*, p^*) = \nabla_p C(u^*, p^*) \tag{2.8}$$

where $x(u^*, p^*) \equiv [x_1(u^*, p^*), \ldots, x_N(u^*, p^*)]^T$ is the vector of cost minimizing input quantities needed to produce u^* units of output given input prices p^*, where the underlying production function F^* is defined by (2.4), $u^* \in$ range F^* and $p^* \gg 0_N$.

The difference between Lemma 3 and Lemma 4 is that Lemma 3 assumes the existence of the production function F and does not specify the properties of the cost function other than differentiability, while Lemma 4 assumes only the existence of a cost function satisfying the appropriate regularity conditions and the corresponding production function F^* is defined using the given cost function. Thus, from an econometric point of view, Lemma 4 is more useful than Lemma 3: in order to obtain a valid system of input demand functions, all we have to do is postulate a functional form for C which satisfies the appropriate regularity conditions and differentiate C with respect to the components of the input price vector p. It is not necessary to compute the corresponding production function F^* nor is it necessary to endure the sometimes painful algebra involved in deriving the input demand functions from the production function via Lagrangian techniques.[11]

Historical notes

The proposition that there are two or more equivalent ways of representing preferences or technology forms the core of duality theory. The mathematical basis for the economic theory of duality is Minkowski's (1911) Theorem, as presented in Fenchel (1953, pp. 48–50) and Rockafellar (1970, pp. 95–99): every closed convex set can be represented as the intersection of its supporting halfspaces. Thus, under certain conditions, the closed convex set $L(u) \equiv \{x: F(x) \geqslant u, x \geqslant 0_N\}$ can be represented as the intersection of the halfspaces generated by the isocost surfaces tangent to the production possibilities set $L(u)$, $\cap_p \{x: p^{\mathrm{T}}x \geqslant C(u, p)\}$.

If the consumer (or producer) has a budget of $y>0$ to spend on the N commodities (or inputs), then the maximum utility (or output) that he can obtain given that he faces prices $p \gg 0_N$ can generally be obtained by solving the equation $y = C(u, p)$ or by solving

$$1 = C(u, p/y) \tag{2.9}$$

(where we have used the linear homogeneity of C in p) for u as a function of the normalized prices, p/y. Call the resulting function G so that $u = G(p/y)$. Alternatively, G can be defined directly from the utility (or production) function F in the following manner for $p \gg 0_N$, $y>0$:

$$G^*(p, y) \equiv \max_x \{F(x): p^{\mathrm{T}}x \leqslant y, x \geqslant 0_N\} \tag{2.10}$$

or

$$G\left(\frac{p}{y}\right) \equiv \max_x \left\{F(x): \left(\frac{p}{y}\right)^{\mathrm{T}} x \leqslant 1, x \geqslant 0_N\right\}.$$

Houthakker (1951–52, p. 157) called the function G the *indirect utility function* and, like the cost function C, it also can characterize preferences or technology uniquely under certain conditions (cf. Section 4 below). Our reason for introducing it at this point is that historically it was introduced into the economics

[11]For an exposition of the Lagrangian method for deriving demand functions and comparative statics theorems, see Chapter 2 of this Handbook by Intriligator.

literature before the cost function by Antonelli (1971, p. 349) in 1886 and then by Konüs (Konyus) (1924). However, the first paper which recognized that preferences could be equivalently described by a direct or indirect utility function appears to be by Konyus and Byushgens (1926, p. 157) who noted that the equations $u=F(x)$ and $u=G(p/y)$ are equations for the same surface, but in different coordinate systems: the first equation is in pointwise coordinates while the second is in planar or tangential coordinates. Konyus and Byushgens (1926, p. 159) also set up the minimization problem that allows one to derive the direct utility function from the indirect function and, finally, they graphed various preferences in price space for the case of two goods.

The English language literature on duality theory seems to have started with two papers by Hotelling (1932, 1935), who was perhaps the first economist to use the word "duality":

> Just as we have a utility (or profit) function u of the quantities consumed whose derivatives are the prices, there is, dually, a function of the prices whose derivatives are the quantities consumed [Hotelling (1932, p. 594)].

Hotelling (1932, p. 594) also recognized that "the cost function may be represented by surfaces which will be concave upward", i.e. he recognized that the cost function $C(u, p)$ would satisfy a curvature condition in p.

Hotelling (1932, p. 590; 1935, p. 68) also introduced the *profit function* Π which provides yet another way by which a decreasing returns to scale technology can be described. Using our notation, Π is defined as

$$\Pi(p)\equiv\max_{x}\left\{F(x)-p^{\mathrm{T}}x\right\}. \tag{2.11}$$

Hotelling indicated that the profit maximizing demand functions, $x(p)\equiv[x_1(p),\ldots,x_N(p)]^{\mathrm{T}}$, could be obtained by differentiating the profit function, i.e. $x(p)=-\nabla_p\Pi(p)$. Thus, if Π is twice continuously differentiable, one can readily deduce *Hotelling's* (1935, p. 69) *symmetry conditions*:

$$-\frac{\partial x_i}{\partial p_j}(p)=\frac{\partial^2\Pi}{\partial p_i\partial p_j}(p)=-\frac{\partial x_j}{\partial p_i}(p). \tag{2.12}$$

The next important contribution to duality theory was made by Roy who independently recognized that preferences could be represented by pointwise or tangential coordinates:

> Il vient alors tout naturellement à l'esprit d'invoques le principle de dualité qui permet d'utiliser les equations tangentielles au lieu des

> équations ponctuelles; ainsi apparaît-il possible de présenter les équations d'equilibre sous une form nouvelle et susceptible d'interprétations fécondes [Roy (1942, pp. 18–19)].

Roy (1942, p. 20) defined the indirect utility function G^* as in eq. (2.10) above and then he derived the counterpart to Lemma 3 above, which is called *Roy's Identity* (1942, pp.18–19).

$$x\left(\frac{p}{y}\right)=\frac{-\nabla_p G^*(p,y)}{\nabla_y G^*(p,y)}, \tag{2.13}$$

where $x(p/y)\equiv[x_1(p/y),\dots,x_N(p/y)]^{\mathrm{T}}$ is the vector of utility (or output) maximizing demand functions given that the consumer (or producer) faces input prices $p\gg 0_N$ and has a budget $y>0$ to spend. Roy (1942, pp. 24–27) showed that G^* was decreasing in the prices p, increasing in income y, and homogeneous of degree 0 in (p,y); i.e. $G^*(\lambda p,\lambda y)=G^*(p,y)$ for $\lambda>0$. Thus, $G^*(p,y)=G^*(p/y,1)\equiv G(p/y)=G(v)$, where $v\equiv p/y$ is a vector of *normalized prices*. In his 1947 paper, Roy derived the following version of Roy's Identity (1947, p. 219), where the indirect utility function G is used in place of G^*:

$$x_i(v)=\frac{\partial G(v)}{\partial v_i}\bigg/\sum_{j=1}^{N}v_j\frac{\partial G(v)}{\partial v_j},\qquad i=1,2,\dots,N. \tag{2.14}$$

The French mathematician Ville (1951, p. 125) also derived the useful relations (2.14) in 1946, so perhaps (2.14) should be called *Ville's Identity*. Ville (1951, p. 126) also noted that if the direct utility function $F(x)$ is linearly homogeneous, then the indirect function $G(v)\equiv\max_x\{F(x): v^{\mathrm{T}}x\leqslant 1, x\geqslant 0_N\}$ is homogeneous of degree -1, i.e. $G(\lambda v)=\lambda^{-1}G(v)$ for $\lambda>0, v\gg 0_N$, and thus $-G(v)=\sum_{j=1}^{N}v_j(\partial G(v)/\partial v_j)$. Substitution of the last identity into (2.14) yields the simpler equations [see also Samuelson (1972)]:

$$x_i(v)=-\partial\ln G(v)/\partial v_i,\qquad i=1,2,\dots,N. \tag{2.15}$$

At this point it should be mentioned that Antonelli (1971, p. 349) obtained a version of Roy's Identity in 1886 and Konyus and Byushgens (1926, p. 159) *almost* derived it in 1926 in the following manner: they considered the problem of minimizing indirect utility $G(v)$ with respect to the normalized prices v subject to the constraint $v^{\mathrm{T}}x=1$. As Houthakker (1951–52, pp. 157–158) later observed, it turns out that this constrained minimization problem generates the direct utility function, i.e. we have for $x \gg 0_N$:

$$F(x) = \min_v \{G(v): v^T x \leq 1, v \geq 0_N\}. \tag{2.16}$$

Konyus and Byushgens obtained the first-order conditions for the problem (2.16): $\nabla_v G(v) = \mu x$. If the Lagrange multiplier μ is eliminated from this last system of equations using the constraint $v^T x = 1$, we obtain $x = \nabla_v G(v)/v^T \nabla_v G(v)$, which is (2.14) written in vector notation. However, Konyus and Byushgens did not explicitly carry out this last step.

Another notable early paper was written by Wold (1943–44) who defined the indirect utility function $G(v)$ (he called it a "price preference function") and showed that the indifference surfaces of price space were either convex to the origin or possibly linear; i.e. he showed that $G(v)$ was a quasi-convex function[12] in the normalized prices v. Wold's early work is summarized in Wold (1953, pp. 145–148).

Malmquist (1953, p. 212) also defined the indirect utility function $G(v)$ and indicated that it was a quasi-convex function in v.

If the production function F is subject to constant returns to scale (i.e. $F(\lambda x) = \lambda F(x)$ for every $\lambda \geq 0, x \geq 0_N$) in addition to being continuous from above, then the corresponding cost function decomposes in the following manner: let $u > 0$, $p \gg 0_N$; then

$$\begin{aligned} C(u,p) &\equiv \min_x \{p^T x: F(x) \geq u\} \\ &= \min_x \{up^T(x/u): F(x/u) \geq 1\} \\ &= u \min_z \{p^T z: F(z) \geq 1\} \\ &\equiv uC(1,p). \end{aligned} \tag{2.17}$$

(The above proof assumes that there exists at least one $x^* > 0_N$ such that $F(x^*) > 0$ so that the set $\{z: F(z) \geq 1\}$ is not empty.) Samuelson (1953–54) assumed that the production function F was linearly homogeneous and subject to a "generalized law of diminishing returns", $F(x' + x'') \geq F(x') + F(x'')$ (which is equivalent to concavity of F when F is linearly homogeneous). He (p. 15) then defined the unit cost function $C(1,p)$ and indicated that $C(1,p)$ satisfied the same properties in p that F satisfied in x. He (p. 15) also noted that a flat on the unit output production surface (a region of infinite substitutability) would correspond to a corner on the unit cost surface, a point which was also made by Shephard (1953, pp. 27–28).

[12]A function G is *quasi-convex* if and only if $-G$ is quasi-concave.

Shephard's 1953 monograph appears to be the first modern, rigorous treatment of duality theory. Shephard (1953, pp. 13–14) notes that the cost function $C(u, p)$ can be interpreted as the support function for the convex set $\{x: F(x) \geqslant u\}$, and he uses this fact to establish the properties of $C(u, p)$ with respect to p. Shephard (1953, p. 13) also explicitly mentions Minkowski's (1911) Theorem on convex sets and Bonnesen and Fenchel's (1934) monograph on convex sets. It should be mentioned that Shephard did not develop a direct duality between production and cost functions; he developed a duality between production and distance functions, (which we will define in a later section) and then between distance and cost functions.

Shephard (1953, p. 4) defined a production function F to be *homothetic* if it could be written as

$$F(x) = \phi[f(x)],$$

where f is a homogeneous function of degree one and ϕ is a continuous, positive monotone increasing function of f. Let us formally introduce the following additional conditions on F (or f):

Assumption 4 on F

F is (non-negatively) *linearly homogeneous*; i.e. if $x \geqslant 0_N$, $\lambda \geqslant 0$, then $F(\lambda x) = \lambda F(x)$.

Assumption 5 on F

F is *weakly positive*; i.e. for every $x \geqslant 0_N$, $F(x) \geqslant 0$ but $F(x^*) > 0$ for at least one $x^* > 0_N$.

Now let us assume that $\phi(f)$ is a continuous, monotonically increasing function of one variable for $f \geqslant 0$ with $\phi(0) = 0$. Under these conditions the inverse function ϕ^{-1} exists and has the same properties as ϕ, with $\phi^{-1}[\phi(f)] = f$ for all $f \geqslant 0$. If $f(x)$ satisfies Assumptions 1, 4, and 5 above, then the cost function which corresponds to $F(x) \equiv \phi[f(x)]$ decomposes as follows: let $u > 0$, $p \gg 0_N$; then

$$\begin{aligned}
C(u, p) &\equiv \min_x \{p^{\mathrm{T}}x: \phi[f(x)] \geqslant u\} \\
&= \min_x \{p^{\mathrm{T}}x: f(x) \geqslant \phi^{-1}[u]\} \\
&= \phi^{-1}[u] \min_x \{p^{\mathrm{T}}(x/\phi^{-1}[u]): f(x/\phi^{-1}[u]) \geqslant 1\}, \\
&\qquad \text{where } \phi^{-1}[u] > 0 \quad \text{since} \quad u > 0, \\
&= \phi^{-1}[u] c(p),
\end{aligned} \tag{2.18}$$

where $c(p) \equiv \min_z \{p^{\mathrm{T}}z: f(z) \geqslant 1\}$ is the *unit cost function* which corresponds to the linearly homogeneous function f, a non-negative, (positively) linearly homogeneous, non-decreasing, concave and continuous function of p (recall Properties 1–5 above). As usual, we will not be able to derive the original production function $\phi[f(x)]$ from the cost function (2.18) unless f also satisfies Assumptions 2 and 3 above. Shephard (1953, p. 43) obtained the factorization (2.18) for the cost function corresponding to a homothetic production function.

Finally, Shephard (1953, pp. 28–29) noted several practical uses for duality theory: (i) as an aid in aggregating variables, (ii) in econometric studies of production when input data are not available but cost, input price and output data are available, and (iii) as an aid in deriving certain comparative statics results. Thus, Shephard either derived or anticipated many of the theoretical results and practical applications of duality theory.

Turning now to the specific results obtained in this section, McFadden (1966) showed that the minimum in definition (2.1) exists if F satisfies Assumption 1. Property 1 was obtained by Shephard (1953, p. 14), Property 2 by Shephard (1953, p. 14) and Samuelson (1953–54, p. 15), Property 3 by Shephard (1953, p. 14), Property 4 by Shephard (1953, p. 15) [our method of proof is due to McKenzie (1956–57, p. 185)], Properties 5 and 6 by Uzawa (1964, p. 217), and finally Property 7 was obtained by Shephard (1970, p. 83).

The method for constructing the approximating production possibilities sets $L^*(u)$ in terms of the cost function is due to Uzawa (1964).

The very important point that the approximating isoquants do not have any of the backward bending or non-convex parts of the true isoquants was made in the context of consumer theory by Hotelling (1935, p. 74), Wold (1943, p. 231; 1953, p. 146) and Samuelson (1950b, pp. 359–360) and in the context of producer theory by McFadden (1966, 1978a). It is worth quoting Hotelling and Samuelson at some length in order to emphasize this point.

> If indifference curves for purchases be thought of as possessing a wavy character, convex to the origin in some regions and concave in others, we are forced to the conclusion that it is only the portions convex to the origin that can be regarded as possessing any importance, since the others are essentially unobservable. They can be detected only by the discontinuities that may occur in demand with variation in price-ratios, leading to an abrupt jumping of a point of tangency across a chasm when the straight line is rotated. But, while such discontinuities may reveal the existence of chasms, they can never measure their depth. The concave portions of the indifference curves and their many-dimensional generalizations, if they exist, must forever remain in unmeasurable obscurity [Hotelling (1935, p. 74)].

> It will be noted that any point where the indifference curves are convex rather than concave cannot be observed in a competitive market. Such

> points are shrouded in eternal darkness – unless we make our consumer a monopsonist and let him choose between goods lying on a very convex "budget curve" (along which he is affecting the price of what he buys). In this monopsony case, we could still deduce the slope of the man's indifference curve from the slope of the observed constraint at the equilibrium point [Samuelson (1950b, pp. 359–360)].

Our proof of Lemma 3 follows a proof attributed by Diamond and McFadden (1974, p. 4) to W. M. Gorman; however, the same method of proof was also used by Karlin (1959, p. 272). Hicks' and Samuelsons' proof of Lemma 3 assumed differentiability of the production function and utilized the first-order conditions for the cost minimization problem along with the properties of determinants. Our earlier quotation by Hotelling (1932, p. 594) indicates that he also obtained the Hicks (1946, p. 331), Samuelson (1947, p. 68; 1953–54, pp. 15–16) results in a slightly different context.

3. Duality between cost and aggregator (production or utility) functions

In this section we assume that the aggregator function F satisfies the following properties:

Conditions I on F

(i) F is a real valued function of N variables defined over the non-negative orthant $\Omega \equiv \{x: x \geqslant 0_N\}$ and is *continuous* on this domain.
(ii) F is *increasing*; i.e. $x'' \gg x' \geqslant 0_N$ implies $F(x'') > F(x')$.
(iii) F is a *quasi-concave* function.

Note that the Properties (i) and (ii) above are stronger than Assumptions 1 and 2 on F made in the previous section, so that we should be able to deduce somewhat stronger conditions on the cost function $C(u, p)$ which corresponds to an $F(x)$ satisfying Conditions I above.

Let U be the range of F. From I(i) and (ii) it can be seen that $U \equiv \{u: \bar{u} \leqslant u < \bar{\bar{u}}\}$, where $\bar{u} \equiv F(0_N) < \bar{\bar{u}}$. Note that the least upper bound $\bar{\bar{u}}$ could be a finite number or $+\infty$. For consumer theory applications, however, there is no reason to assume that $\bar{u}$ be finite (i.e. $\bar{u}$ could equal $-\infty$) but there is little loss of generality in doing so.

Define the set of positive prices $P \equiv \{p: p \gg 0_N\}$.

Theorem 1

If F satisfies Conditions I, then $C(u, p) \equiv \min_x \{p^{\mathrm{T}}x: F(x) \geqslant u\}$ defined for all $u \in U$ and $p \in P$ satisfies Conditions II below. For a proof see Diewert (1982).

Conditions II on C

(i) $C(u, p)$ is a real valued function of $N+1$ variables defined over $U \times P$ and is jointly *continuous* in (u, p) over this domain.
(ii) $C(\bar{u}, p)=0$ for every $p \in P$.
(iii) $C(u, p)$ is *increasing in u* for every $p \in P$; i.e. if $p \in P$, $u', u'' \in U$, with $u' < u''$, then $C(u', p) < C(u'', p)$.
(iv) $C(\bar{\bar{u}}, p)=+\infty$ for every $p \in P$; i.e. if $p \in P$, $u^n \in U$, $\lim_n u^n = \bar{\bar{u}}$, then $\lim_n C(u^n, p)=+\infty$.
(v) $C(u, p)$ is (positively) *linearly homogeneous in p* for every $u \in U$; i.e. $u \in U$, $\lambda > 0$, $p \in P$ implies $C(u, \lambda p)=\lambda C(u, p)$.
(vi) $C(u, p)$ is *concave in p* for every $u \in U$.
(vii) $C(u, p)$ is *increasing in p* for $u > \bar{u}$ and $u \in U$.
(viii) C is such that the function $F^*(x) \equiv \max_u \{u: p^{\mathrm{T}} x \geqslant C(u, p)$ for every $p \in P$, $u \in U\}$ is continuous for $x \geqslant 0_N$.

Corollary 1.1

If $C(u, p)$ satisfies Conditions II above, then the domain of definition of C can be extended from $U \times P$ to $U \times \Omega$. The extended function C is continuous in p for $p \in \Omega \equiv \{p: p \geqslant 0_N\}$ for each $u \in U$.[13]

Corollary 1.2

For every $x \geqslant 0_N$, $F^*(x)=F(x)$, where F^* is the function defined by the cost function C in part (viii) of Conditions II.

Corollary 1.2 shows that the cost function can completely describe a production function which satisfies Conditions I; i.e. to use McFadden's (1966) terminology, the cost function is a *sufficient statistic* for the production function.

The proof of Theorem 1 is straightforward, with the exception of parts (i) and (viii), the parts that involve the continuity properties of the cost or production function. These continuity complexities appear to be the only difficult concepts associated with duality theory; this is why we tried to avoid them in the previous section as much as possible. For further discussion on continuity problems, see Shephard (1970), Friedman (1972), Diewert (1974a), Blackorby, Primont, and Russell (1978), and Blackorby and Diewert (1979).

Property I(ii), increasingness of F, is required in order to prove the correspondence $L(u)$ continuous and thus that $C(u, p)$ is continuous over $U \times P$.[14] If

[13] However, $C(u, p)$ need not be strictly increasing in u when p is on the boundary of Ω; e.g. consider the function $f(x_1, x_2) \equiv x_1$ which has the dual cost function $C(u, p_1, p_2) \equiv p_1 u$ which is not increasing in u when $p_1 = 0$.

[14] Friedman (1972) shows that I(ii) plus continuity from above (Assumption 1 on F in the previous section) is sufficient to imply the joint continuity of C over $U \times P$. However, unless we assume the additional property on F of continuity from below, we cannot conclude that $C(u, p)$ is increasing in u for $p \in P$, a property which follows from I(i) to I(ii).

Property I(ii) is replaced by a weak monotonicity assumption (such as our old Assumption 2 on F of the previous section), then plateaus on the graph of F ("thick" indifference surfaces to use the language of utility theory) will imply discontinuities in C with respect to u [cf. Friedman (1972, p. 169)].

Note that II(ii) and (iii) imply that $C(u, p)>0$ for $u>\bar{u}$ and $p \gg 0_N$ and that II(vii) is not an independent property of C since it follows from II(ii), (iii), (v), and (vi). Note also that we have not assumed that F be strictly quasi-concave, i.e. that the production possibility sets $L(u) \equiv \{x: F(x) \geqslant u\}$ be strictly convex.

Finally, it is evident that given only a firm's total cost function C, we can use the function F^* defined in terms of the cost function by II(viii) in order to generate the firm's production function. This is formalized in the following theorem.

Theorem 2

If C satisfies Conditions II above, then F^* defined in II(viii) satisfies Conditions I. Moreover, if $C^*(u, p) \equiv \min_x \{p^{\mathrm{T}}x: F^*(x) \geqslant u\}$ is the cost function which is defined by F^*, then $C^* = C$.

Corollary 2.1

The set of supergradients to C with respect to p at the point $(u^*, p^*) \in U \times P$, $\partial C(u^*, p^*)$, is the solution set to the cost minimization problem $\min_x \{p^{*\mathrm{T}}x: F^*(x) \geqslant u^*\}$, where F^* is the aggregator function which corresponds to the given cost function satisfying Conditions II. [The *super gradients* satisfy $x^* \in \partial C(u^*, p^*)$ iff $C(u^*, p) \leqslant C(u^*, p^*) + x^{*\mathrm{T}}(p - p^*)$ for every $p \gg 0_N$.]

Corollary 2.2 [*Shephard's* (1953, p. 11) *Lemma*]

If C satisfies Conditions II and, in addition, is differentiable with respect to input prices at the point $(u^*, p^*) \in U \times P$, then the solution x^* to the cost minimization problem $\min_x \{p^{*\mathrm{T}}x: F(x) \geqslant u^*\}$ is unique and is equal to the vector of partial derivatives of $C(u^*, p^*)$ with respect to the components of the input price vector p; i.e.

$$x^* = \nabla_p C(u^*, p^*). \tag{3.1}$$

The above two theorems provide a version of the Shephard (1953, 1970) Duality Theorem between cost and aggregator functions. The conditions on C which correspond to our Conditions I on F seem to be straightforward with the exception of II(viii), which essentially guarantees the continuity of the aggregator function F^* corresponding to the given cost function C. Condition II(viii) can be dropped if we strengthen II(iii) to $C(u, p)$ increasing in u for every p in $S \equiv \{p: p \geqslant 0_N, 1_N^{\mathrm{T}} p = 1\}$. The resulting F^* can be shown to be continuous [cf. Blackorby, Primont, and Russell (1978)]; however, many useful functional forms do not satisfy the strengthened Condition II(iii).[15] An alternative method of dropping II(viii), which preserves continuity of the direct aggregator function F^* corresponding to a given cost function C, is to develop local duality theorems, i.e. assume that C satisfies Conditions II(i)–(vii) for $(u, p)\in U\times P$, where P is now is restricted to be a *compact*, convex subset of the positive orthant. A (locally) valid continuous F^* can then be defined from C which in turn has C as its cost function over $U\times P$. This approach is pursued in Blackorby and Diewert (1979).

Historical notes

Duality theorems between F and C have been proven under various regularity conditions by Shephard (1953, 1970), McFadden (1962), Chipman (1970), Hanoch (1978), Diewert (1971a, 1974a), Afriat (1973a), and Blackorby, Primont, and Russell (1978).

Duality theorems between C and the level sets of F, $L(u)\equiv\{x\colon F(x)\geqslant u\}$, have been proven by Uzawa (1964), McFadden (1966, 1978a), Shephard (1970), Jacobsen (1970, 1972), Diewert (1971a), Friedman (1972), and Sakai (1973).

4. Duality between direct and indirect aggregator functions

We assume that the direct aggregator (utility or production) function F satisfies Conditions I listed in the previous section. The basic optimization problem that we wish to consider in this section is the problem of maximizing utility (or output) $F(x)$ subject to the budget constraint $p^{\mathrm{T}}x\leqslant y$, where $p\gg 0_N$ is a vector of given commodity (or input) prices and $y>0$ is the amount of money the consumer (or producer) is allowed to spend. Since $y>0$, the constraint $p^{\mathrm{T}}x\leqslant y$ can be replaced with $v^{\mathrm{T}}x\leqslant 1$, where $v\equiv p/y$ is the vector of *normalized* prices. The *indirect aggregator function* $G(v)$ is defined for $v\gg 0_N$ as

$$G(v)\equiv\max_x\left\{F(x)\colon v^{\mathrm{T}}x\leqslant 1,\, x\geqslant 0_N\right\}. \tag{4.1}$$

Theorem 3

If the direct aggregator function F satisfies Conditions I, then the indirect aggregator function G defined by (4.1) satisfies the following conditions:

[15] E.g. consider $C(u, p)\equiv b^{\mathrm{T}}pu$, where $b>0_N$ but b is not $\gg 0_N$. This corresponds to a Leontief or fixed coefficients aggregator function.

Conditions III on G

(i) $G(v)$ is a real valued function of N variables defined over the set of positive normalized prices $V \equiv \{v: v \gg 0_N\}$ and is a *continuous* function over this domain.
(ii) G is *decreasing*; i.e. if $v'' \gg v' \gg 0_N$, then $G(v'') < G(v')$.
(iii) G is *quasi-convex* over V.
(iv) G[16] is such that the function $\hat{F}(x) \equiv \min_v\{G(v): v^{\mathrm{T}}x \leqslant 1, v \geqslant 0_N\}$ defined for $x \gg 0_N$ is continuous over this domain and has a continuous extension[17] to the non-negative orthant $\Omega = \{x: x \geqslant 0_N\}$.

Corollary 3.1

The direct aggregator function F can be recovered from the indirect aggregator function G; i.e. for $x \gg 0_N$, $F(x) = \min_v\{G(v): v^{\mathrm{T}}x \leqslant 1, v \geqslant 0_N\}$.

Corollary 3.2

Let F satisfy Conditions I and let $x^* \gg 0_N$. Define the closed convex set of normalized supporting hyperplanes at the point x^* to the closed convex set $\{x: F(x) \geqslant F(x^*), x \geqslant 0_N\}$, $H(x^*)$.[18] Then: (i) $H(x^*)$ is the solution set to the indirect utility (or production) minimization problem $\min_v\{G(v): v^{\mathrm{T}}x^* \leqslant 1, v \geqslant 0_N\}$, where G is the indirect function which corresponds to F via definition (7.4), and (ii) if $v^* \in H(x^*)$, then x^* is a solution to the direct utility (or production) maximization problem $\max_x\{F(x): v^{*\mathrm{T}}x \leqslant 1, x \geqslant 0_N\}$.

Corollary 3.3 [Hotelling (1935, p. 71); Wold (1944, pp. 69–71; 1953, p. 145) Identity]

If F satisfies Conditions I and in addition is differentiable at $x^* \gg 0_N$ with a non-zero gradient vector $\nabla F(x^*) > 0_N$, then x^* is a solution to the direct (utility or production) maximization problem $\max_x\{F(x): v^{*\mathrm{T}}x \leqslant 1, x \geqslant 0_N\}$ where

$$v^* \equiv \frac{\nabla F(x^*)}{x^{*\mathrm{T}} \nabla F(x^*)}. \tag{4.2}$$

[16] G here is the extension of G to the non-negative orthant that is defined by the Fenchel (1953) closure operation; i.e. define the epigraph of the original G as $\Gamma \equiv \{(u, v): v \gg 0_N, u \geqslant G(v)\}$, define the closure of Γ as $\bar{\Gamma}$, and define the *extended* G as $G(v) \equiv \inf_u\{u: (u, v) \in \bar{\Gamma}\}$ for $v \geqslant 0_N$. The resulting extended G is continuous from below (the sets $\{v: G(v) \leqslant u, v \geqslant 0_N\}$ are closed for all u). If the range of F is $U \equiv \{u: \bar{u} \leqslant u < \bar{\bar{u}}\}$, where $\bar{u} < \bar{\bar{u}}$, then the range of the unextended G is $\{u: \bar{u} < u < \bar{\bar{u}}\}$ and the range of the extended G is $\{u: \bar{u} < u \leqslant \bar{\bar{u}}\}$ so that if $\bar{\bar{u}} = +\infty$, then $G(v) = +\infty$ for $v = 0_N$ and possibly for other points v on the boundary of the nonnegative orthant.

[17] Again F is extended to the non-negative orthant by the Fenchel closure operation: define the hypograph of the original $\hat{F}$ as $\Delta \equiv \{(u, x): x \gg 0_N, u \leqslant \hat{F}(x)\}$, define the closure of Δ as $\bar{\Delta}$, and define the *extended* $\hat{F}$ as $F(x) \equiv \sup_u\{u: (u, x) \in \bar{\Delta}\}$ for $x \geqslant 0_N$. Since the unextended $\hat{F}$ is continuous for $x \gg 0_N$, the extended $\hat{F}$ can easily be shown to be continuous from above for $x \geqslant 0_N$. Condition III(iv) implies that the extended F is continuous from below for $x \geqslant 0_N$ as well.

[18] If $v^* \in H(x^*)$, then $v^{*\mathrm{T}}x^* = 1$, $x^* \geqslant 0_N$ and $F(x) \geqslant F(x^*)$ implies $v^{*\mathrm{T}}x \geqslant v^{*\mathrm{T}}x^* = 1$. The closedness and convexity of $H(x^*)$ is shown in Rockafellar (1970, p. 215).

The system of equations (4.2) is known as the system of *inverse demand functions*; the ith equation

$$p_i/y \equiv v_i^* = \left[\partial F(x^*)/\partial x_i\right] \Big/ \left[\sum_{j=1}^{N} x_j^* \partial F(x^*)/\partial x_j\right]$$

gives the ith commodity price p_i divided by expenditure y as a function of the quantity vector x^* which the producer or consumer would choose if he were maximizing $F(x)$ subject to the budget constraint $v^{*\mathrm{T}}x=1$.

We now assume that a well-behaved indirect aggregator function G is given and we show that it can be used in order to define a well-behaved direct aggregator function $\hat{F}$ which has G as its indirect function.

Theorem 4

Suppose G satisfies Conditions III. Then $\hat{F}(x)$ defined for $x \gg 0_N$ by

$$\hat{F}(x) = \min_v \left\{G(v): v^{\mathrm{T}}x \leqslant 1, v \geqslant 0_N\right\} \tag{4.3}$$

has an extension to $x \geqslant 0_N$ which satisfies Conditions I. Moreover, if we define $G^*(x) \equiv \max_x\{\hat{F}(x): v^{\mathrm{T}}x \leqslant 1, x \geqslant 0_N\}$ for $v \gg 0_N$, then $G^*(v) = G(v)$ for all $v \gg 0_N$.

Corollary 4.1

Let G satisfy Conditions III and let $v^* \gg 0_N$. Define the closed convex set of normalized supporting hyperplanes at the point v^* to the closed convex set $\{v: G(v) \leqslant G(v^*), v \geqslant 0_N\}$, $H^*(v^*)$. Then: (i) $H^*(v^*)$ is the solution set to the direct maximization problem $\max_x\{\hat{F}(x): v^{*\mathrm{T}}x \leqslant 1, x \geqslant 0_N\}$, where $\hat{F}$ is the direct function which corresponds to the given indirect function G via definition (4.3), and (ii) if $x^* \in H(v^*)$, then v^* is a solution to the indirect minimization problem $\min_v\{G(v): v^{\mathrm{T}}x^* \leqslant 1, v \geqslant 0_N\}$.

Corollary 4.2 [Ville (1946, p. 35); Roy (1947, p. 222) Identity]

If G satisfies Conditions III and, in addition, is differentiable at $v^* \gg 0_N$ with a non-zero gradient vector $\nabla G(v^*) < 0_N$, then x^* is the unique solution to the direct maximization problem $\max_x\{F(x): v^{*\mathrm{T}}x \leqslant 1, x \geqslant 0_N\}$, where

$$x^* \equiv \frac{\nabla G(v^*)}{v^{*\mathrm{T}} \nabla G(v^*)}. \tag{4.4}$$

It can be seen that (4.4) provides the counterpart to Shephard's Lemma in the previous section. Shephard's Lemma and Roy's Identity are the basis for a great number of theoretical and empirical applications, as we shall see later.

Finally, we note that although Condition III(iv) appears to be a bit odd, it enables us to derive a continuous direct aggregator function from a given indirect function satisfying Conditions III.[19]

Historical notes

Duality theorems between direct and indirect aggregator functions have been proven by Samuelson (1965, 1969, 1972); Newman (1965, pp. 138–165); Lau (1969); Shephard (1970, pp. 105–113); Hanoch (1978); Weddepohl (1970, ch. 5); Katzner (1970, pp. 59–62); Afriat (1972a, 1973c), and Diewert (1974a).

For closely related work relating assumptions on systems of consumer demand functions to the direct aggregator function F (the integrability problem), see Samuelson (1950b), Hurwicz and Uzawa (1971), Hurwicz (1971), and Afriat (1973a, b).

For a geometric interpretation of Roy's Identity, see Darrough and Southey (1977), and for some extensions, see Weymark (1980).

5. Duality between direct aggregator functions and distance or deflation functions

In this section we consider a *fourth* alternative method of characterizing tastes or technology, a method which proves to be extremely useful for defining a certain class of index number formulae due to Malmquist (1953, p. 232).

As usual, let $F(x)$ be an aggregator function satisfying Conditions I listed in Section 3 above. For u belonging to the interior of the range of F (i.e. $u \in \operatorname{int} U$, where $U \equiv \{u: \bar{u} \leqslant u < \bar{\bar{u}}\}$) and $x \gg 0_N$, define the *distance or deflation function*[20] D as

$$D(u, x) \equiv \max_k \left\{ k: F\left(\frac{x}{k}\right) \geqslant u,\ k>0 \right\}. \tag{5.1}$$

Thus, $D(u^*, x^*)$ is the biggest number which will just deflate (inflate if $F(x^*) < u^*$) the given point $x^* \gg 0_N$ onto the boundary of the utility (or production) possibility set $L(u^*) \equiv \{x: F(x) \geqslant u^*\}$. If $D(u^*, x^*) > 1$, then $x^* \gg 0_N$ produces a higher level of utility or output than the level indexed by u^*.

[19]Without Condition III(iv) we could still deduce continuity of $\hat{F}(x)$ over $x \gg 0_N$ but the resulting F would not necessarily have a continuous extension to $x \geqslant 0_N$ (since $\hat{F}$ is not necessarily concave, but is only quasi-concave over $x \gg 0_N$, its extension is not necessarily continuous). For discussion and examples of these continuity problems, see Diewert (1974a, pp. 121–123).

[20]Shephard (1953, p. 6; 1970, p. 65) introduced the distance function into the economics literature, using the slightly different but equivalent definition: $D(u, x) \equiv 1/\min_\lambda \{\lambda: F(\lambda x) \geqslant u, \lambda > 0\}$. McFadden (1978a) and Blackorby, Primont, and Russell (1978) call D the *transformation* function, while in the mathematics literature [e.g. Rockafellar (1970, p. 28)], D is termed a *guage* function. The term *deflation* function for D would seem to be more descriptive from an economic point of view.

It turns out that the mathematical properties of $D(u,x)$ with respect to x are the same as the properties of $C(u,p)$ with respect to p, but the properties of D with respect to u are reciprocal to the properties of C with respect to u, as the following theorem shows.

Theorem 5

If F satisfies Conditions I, then D defined by (5.1) satisfies Conditions IV below.

Conditions IV on D

(i) $D(u,x)$ is a real valued function of $N+1$ variables defined over $\text{int}\,U \times \text{int}\,\Omega = \{u: \bar{u} < u < \bar{\bar{u}}\} \times \{x: x \gg 0_N\}$ and is *continuous* over this domain.

(ii) $D(\bar{u},x) = +\infty$ for every $x \in \text{int}\,\Omega$; i.e. $u^n \in \text{int}\,U, \lim u^n = \bar{u}, x \in \text{int}\,\Omega$ implies $\lim_n D(u^n,x) = +\infty$.

(iii) $D(u,x)$ is *decreasing in u* for every $x \in \text{int}\,\Omega$; i.e. if $x \in \text{int}\,\Omega, u', u'' \in \text{int}\,U$ with $u' < u''$, then $D(u',x) > D(u'',x)$.

(iv) $D(\bar{\bar{u}},x) = 0$ for every $x \in \text{int}\,\Omega$; i.e. $u^n \in \text{int}\,U, \lim u^n = \bar{\bar{u}}, x \in \text{int}\,\Omega$ implies $\lim_n D(u^n,x) = 0$.

(v) $D(u,x)$ is (positively) *linearly homogeneous in x* for every $u \in \text{int}\,U$; i.e. $u \in \text{int}\,U, \lambda > 0, x \in \text{int}\,\Omega$ implies $D(u,\lambda x) = \lambda\, D(u,x)$.

(vi) $D(u,x)$ is *concave in x* for every $u \in \text{int}\,U$.

(vii) $D(u,x)$ is increasing in x for every $u \in \text{int}\,U$; i.e. $u \in \text{int}\,U, x', x'' \in \text{int}\,\Omega$ implies $D(u, x'+x'') > D(u,x')$.

(viii) D is such that the function

$$\tilde{F}(x) \equiv \{u: u \in \text{int}\,U, D(u,x) = 1\} \tag{5.2}$$

defined for $x \gg 0_N$ has continuous extension to $x \geqslant 0_N$.

Corollary 5.1

$\tilde{F}(x) \equiv \{u: u \in \text{int}\,U, D(u,x) = 1\} = F(x)$ for every $x \gg 0_N$ and thus $\tilde{F} = F$; i.e. the original aggregator function F is recovered from the distance function D via definition (5.2) if F satisfies Conditions I.

As was the case with the cost function $C(u,p)$ studied in Section 3 above, D satisfying Conditions IV over $\text{int}\,U \times \text{int}\,\Omega$ can be uniquely extended to $\text{int}\,U \times \Omega$ using the Fenchel closure operation. It can be verified that the extended D satisfies Conditions IV(v), (vi), and (vii) over $\text{int}\,U \times \Omega$, but the joint continuity Condition IV(i) and the monotonicity conditions in u are no longer necessarily satisfied.[21] It should also be noted that if Condition I(iii) (quasi-concavity of F)

[21] If Conditions IV(i)–(vii) were satisfied by D over $\text{int}\,U \times \Omega$, then we could derive the corresponding continuous F from D without using the somewhat unusual Condition IV(viii).

were dropped, then Theorem 5 would still be valid except that Condition IV(vi) (concavity of D in x) would have to be dropped.

The following theorem shows that the deflation function D can also be used in order to define a continuous aggregator function $\tilde{F}$.

Theorem 6

If D satisfies Conditions IV above, the $\tilde{F}$ defined by (5.2) for $x \in \operatorname{int}\Omega$ has an extension to Ω which satisfies Conditions I. Moreover, if we define the deflation function D^* which corresponds to $\tilde{F}$ by

$$D^*(u,x) \equiv \left\{k: \tilde{F}\left(\frac{x}{k}\right) = u, k > 0\right\}, \tag{5.3}$$

then $D^*(u,x) = D(u,x)$ for $(u,x) \in \operatorname{int} U \times \operatorname{int}\Omega$.

Corollary 6.1 [Shephard[22] (1953, pp. 10–13); Hanoch (1978, p. 7)]

If D satisfies Conditions IV and, in addition, is continuously differentiable at $(u^*, x^*) \in \operatorname{int} U \times \operatorname{int}\Omega$ with $D(u^*, x^*) = 1$ and $\partial D(u^*, x^*)/\partial u < 0$, then x^* is a solution to the direct maximization problem $\max_x\{\tilde{F}(x): v^{*T}x \leqslant 1, x \geqslant 0_N\}$, where $\tilde{F}$ is defined by (5.2) and $v^* > 0_N$ is defined by

$$v^* \equiv \nabla_x D(u^*, x^*). \tag{5.4}$$

Moreover, $\tilde{F}$ is continuously differentiable at x^* with

$$\nabla_x \tilde{F}(x^*) = \frac{-\nabla_x D(u^*, x^*)}{\partial D(u^*, x^*)/\partial u}. \tag{5.5}$$

Thus, the consumer's system of inverse demand functions can be obtained by differentiating the deflation function D satisfying Conditions IV (plus differentiability) with respect to the components of the vector x.

Historical notes

Duality theorems between distance or deflation functions D and aggregator functions $\tilde{F}$ have been proven by Shephard (1953, 1970), Hanoch (1978), McFadden (1978a), and Blackorby, Primont, and Russell (1978).

There are a number of interesting relationships (and further duality theorems) between direct and indirect aggregator, cost, and deflation functions. For exam-

[22] The result can readily be deduced from several separate equations in Shephard but it is explicit in Hanoch's paper.

ple, Malmquist (1953, p. 214) and Shephard (1953, p. 18) showed that the deflation function for the indirect aggregator function, $\max_k\{k: G(v/k)\leqslant u, k>0\}$, equals the cost function, $C(u,v)$. A complete description of these interrelationships and further duality theorems under various regularity conditions may be found in Hanoch (1980) and Blackorby, Primont, and Russell (1978). For some applications see Deaton (1979).

For a local duality theorem between deflation and aggregator functions, see Blackorby and Diewert (1979).

6. Other duality theorems

Concave functions can also be characterized by their *conjugate* functions. Furthermore, it turns out that closed convex sets can also be described by a conjugate function under certain conditions, as discussed in Rockafellar (1970, pp. 102–105) and Karlin (1959, pp. 226–227). Thus, a direct aggregator function F, having convex level sets $L(u)\equiv\{x: F(x)\geqslant u\}$, can also be characterized by its conjugate function as well as by its cost, deflation, or indirect aggregator function. This conjugacy approach was initiated by Hotelling (1932, pp. 36–39; 1960; 1972) and extended by Samuelson (1947, pp. 36–39; 1960; 1972), Lau (1969, 1976, 1978), Jorgenson and Lau (1974a, 1974b), and Blackorby and Russell (1975). We will not review this approach in detail, although in a later section we will review the closely related duality theorems between profit and transformation functions.

Another class of duality theorems [which also has its origin in the work of Hotelling (1935, p. 75) and Samuelson (1960)] is obtained by partitioning the commodity vector $x\geqslant 0_N$ into two vectors, x^1 and x^2 say, and then defining the consumer's *variable indirect aggregator*[23] function g as

$$g(x^1,p^2,y^2)\equiv\max_{x^2}\left\{F(x^1,x^2): p^{2\mathrm{T}}x^2\leqslant y^2, x^2\geqslant 0_{N_2}\right\}, \tag{6.1}$$

where $p^2\gg 0_{N_2}$ is a positive vector of prices the consumer faces for the goods x^2, and $y^2>0$ is the consumer's budget which he has allocated to spend on x^2 goods. The solution set to (6.1), $x^2(x^1,p^2,y^2)$, is the consumer's *conditional* (on x^1) *demand correspondence*. If g satisfies appropriate regularity conditions, conditional demand functions can be generated by applying Roy's Identity (4.4) to the function $G(v^2)\equiv g(x^1,v^2,1)$, where $v^2\equiv p^2/y^2$. For formal duality theorems between direct and variable indirect aggregator functions, see Epstein (1975a), Diewert (1978c) and Blackorby, Primont, and Russell (1977a). For various applications of this duality, see Epstein (1975a) (for applications to consumer

[23]Pollak (1969) uses the alternative terminology, "conditional indirect utility function".

choice under uncertainty) and Pollak (1969) and Diewert (1978c) (estimation of preferences for public goods using market demand functions). Finally, the variable indirect utility function can be used to prove versions of Hicks' (1946, pp. 312–313) composite good theorem – that a group of goods behaves just as if it were a single commodity if the prices of the group of goods change in the same proportion – under less restrictive conditions than were employed by Hicks; see Pollak (1969), Diewert (1978c) and Blackorby, Primont, and Russell (1977a).

We now turn to a brief discussion of a vast literature; i.e. the implications of various special structures on *one* of our many equivalent representations of tastes or technology (such as the direct or indirect aggregator function or the cost function) on the other representations. For example, Shephard (1953) showed that homotheticity of the direct function implied that the cost function factored into $\phi^{-1}(u)c(p)$ [recall eq. (2.18) above]. Another example of a special structure is separability.[24] References which deal with the implications of separability and/or homotheticity include Shephard (1953, 1970), Samuelson (1953–54, 1965, 1969, 1972), Gorman (1959, 1976), Lau (1969, 1978), McFadden (1978), Hanoch (1975, 1978), Pollak (1972), Diewert (1974a), Jorgenson and Lau (1975), Blackorby, Primont, and Russell (1975a, 1975b, 1977a, 1977b, 1977d, 1978), and Blackorby and Russell (1975). For the implications of separability and/or homotheticity on Slutsky coefficients or on partial elasticities of substitution,[25] see Sono (1945), Pearce (1961), Goldman and Uzawa (1964), Geary and Morishima (1973), Berndt and Christensen (1973a), Russell (1975), Diewert (1974a, pp. 150–153), and Blackorby and Russell (1976). For implications of Hicks' (1946, pp. 312–313) Aggregation Theorem on aggregate elasticities of substitution, see Diewert (1974c).

For empirical tests of the separability assumption, see Berndt and Christensen (1973b, 1974), Burgess (1974) and Jorgenson and Lau (1975); for theoretical discussions of these testing procedures, see Blackorby, Primont, and Russell (1977c), Jorgenson and Lau (1975), Lau (1975), Woodland (1978), and Denny and Fuss (1977).

For the implications of assuming *concavity* of the direct aggregator function or of assuming *convexity* of the indirect aggregator function, see Diewert (1978c).

[24]Loosely speaking, $F(x)=F(x^1,x^2,\ldots,x^M)$ is separable in the partition $(x^1,x^2,\ldots,x^M)$ if there exist functions $\hat{F}, F^1,\ldots,F^M$ such that $F(x)=\hat{F}[F^1(x^1),F^2(x^2),\ldots,F^M(x^M)]$ and F is additively separable if there exist functions $F^*, F^1, F^2,\ldots,F^M$ such that $F(x)=F^*[F^1(x^1)+F^2(x^2)+\cdots+F^M(x^M)]$. For historical references and more precise definitions, see Blackorby, Primont, and Russell (1977d, 1978).

[25]Uzawa (1962) observed that the Allen (1938, p. 504) partial elasticity of substitution between inputs i and j, $\sigma_{ij}(u,p)=C(u,p)C_{ij}(u,p)/C_i(u,p)C_j(u,p)$, where $C(u,p)\equiv\min_x\{p^{\mathrm{T}}x\colon F(x)\geqslant u\}$ is the cost function dual to the aggregator function F, and C_i denotes the partial derivative with respect to the ith price in p, p_i, and C_{ij} denotes the second-order partial derivative of C with respect to p_i and p_j. Shephard's Lemma implies that $C_{ij}(u,p)=\partial x_i(u,p)/\partial p_j=\partial x_j(u,p)/\partial p_i=C_{ji}(u,p)$ assuming continuous differentiability of C, where $x_i(u,p)$ and $x_j(u,p)$ are the cost minimizing demand functions. Thus, σ_{ij} can be regarded as a normalization of the response of the cost minimizing x_i to a change in p_j.

The duality theorems referred to above have been "global" in nature. A "local" approach has been initiated by Blackorby and Diewert (1979), where it is assumed that a given cost function $C(u, p)$ satisfies Conditions II above over $U \times P$, where U is a *finite* interval and P is a closed, convex, and *bounded* subset of positive prices. They then construct the corresponding direct aggregator, indirect aggregator, and deflation functions which are dual to the given "locally" valid cost function C. The proofs of these "local" duality theorems turn out to be much simpler than the corresponding "global" duality theorems presented in this paper (and elsewhere), since troublesome continuity problems do not arise due to the assumption that $U \times P$ be compact. These "local" duality theorems are useful in empirical applications, since econometrically estimated cost functions frequently do not satisfy the appropriate regularity conditions for all prices, but the conditions may be satisfied over a smaller subset of prices which is the empirically relevant set of prices.

Epstein has extended duality theory to cover more general maximization problems. In Epstein (1975b), the following utility maximization problem which arises in the context of choice under uncertainty is considered:

$$\max_{x, x^1, x^2} \{F(x, x^1, x^2): x \geq 0_N, x^1 \geq 0_{N_1}, x^2 \geq 0_{N_2}, \\ p^{\mathrm{T}}x + p^{1\mathrm{T}}x^1 \leq y^1, p^{\mathrm{T}}x + p^{2\mathrm{T}}x^2 \leq y^2\}, \tag{6.2}$$

where x represents current consumption, x^i represents consumption in state i $(i=1,2)$, p is the current price vector, p^i is the discounted future price vector which will occur if state i occurs, and $y^i > 0$ is the consumer's discounted income if state i occurs. In Epstein (1981a), the following maximization problem is considered:

$$\max_x \{F(x): x \geq 0_N, c(x, \alpha) \leq 0\}, \tag{6.3}$$

where c is a given constraint function which depends on a vector of parameters α.

We will not attempt to provide a detailed analysis of Epstein's results, but rather we will present a more abstract version of his basic technique which will hopefully capture the essence of duality theory.

The basic maximization problem we study is $\max_x\{F(x): x \in B(v)\}$, where F is a function of N real variables x defined over some set S and $B(v)$ is a constraint set which depends on a vector of M parameters v, which in turn can vary over a set V. Our assumptions on the sets S and V and the constraint set correspondence B are:

(i) S and V are non-empty compact sets in R^N and R^M.

(ii) For every $v \in V$, $B(v)$ is non-empty and $B(v) \subset S$.

(iii) For every $x \in S$, the inverse correspondence[26] $B^{-1}(x)$ is non-empty and $B^{-1}(x) \subset V$.
(iv) The correspondence B is continuous over V.
(v) The correspondence B^{-1} is continuous over S. (6.4)

Our assumptions on the primal function F are:

(i) f is a real valued function of N variables defined over S and is *continuous* over S.
(ii) For every $x^* \in S$, there exists $v^* \in V$ such that $F(x^*) = \max_x\{F(x): x \in B(v^*)\}$. (6.5)

The function G dual to F is defined for $v \in V$ by

$$G(v) \equiv \max_x \{F(x): x \in B(v)\}. \tag{6.6}$$

Theorem 7

If S, V, and B satisfy (6.4) and F satisfies (6.5), then G defined by (6.6) satisfies the following conditions:

(i) G is a real valued function of M variables defined over V and is *continuous* over V.
(ii) For every $v^* \in V$, there exists $x^* \in S$ such that $G(v^*) = \min_v\{G(v): v \in B^{-1}(x^*)\}$. (6.7)

Moreover, if we define the function F^* dual to G for $x \in S$ by

$$F^*(x) \equiv \min_v \{G(v): v \in B^{-1}(x)\}, \tag{6.8}$$

then $F^*(x) = F(x)$ for every $x \in S$.

Corollary 7.1

Let $x^* \in S$ and define $H(x^*)$ to be the set of $v^* \in V$ such that $F(x^*) = \max_x\{F(x): x \in B(v^*)\}$. If $v^* \in H(x^*)$, then x^* is a solution to $\max_x\{F(x): x \in B(v^*)\}$, and v^* is a solution to $\min_v\{G(v): v \in B^{-1}(x^*)\}$.

For proofs, see Diewert (1982).

Notice that property (6.5) (ii) of F is the replacement for our old quasi-concavity assumption in Section 4, and the set $H(x^*)$ defined in Corollary 7.1 replaces the set of normalized supporting hyperplanes which occurred in Corollary 3.2.

Owing to the symmetric nature of our assumptions, it can be seen that the proof of the following theorem is the same as the proof of Theorem 7, except that the inequalities are reversed.

[26] $B^{-1}(x) \equiv \{v: v \in V \text{ and } x \in B(v)\}$.

Theorem 8

If S, V, and B satisfy (6.4) and G satisfies (6.7), then F^* defined by (6.8) satisfies (6.5). Moreover, if we define the function G^* dual to F^* for $v \in V$ by

$$G^*(v) \equiv \max_x \{F^*(x): x \in B(v)\}, \tag{6.9}$$

then $G^*(v) = G(v)$ for every $v \in V$.

Corollary 8.1

Let $v^* \in V$ and define $H^*(v^*)$ to be the set of $x^* \in S$ such that $G(v^*) = \min_v\{G(v): v \in B^{-1}(x^*)\}$. If $x^* \in H^*(v^*)$, then v^* is a solution to $\min_v\{G(v): v \in B^{-1}(x^*)\}$, and x^* is a solution to $\max_x\{F^*(x): x \in B(v^*)\}$.

Note that Condition (6.7) (ii) on G replaces our old quasi-convexity condition on G in Section 4, and the set $H^*(v^*)$ defined in Corollary 8.1 replaces the set of normalized supporting hyperplanes which occurred in Corollary 4.1.

We cannot establish counterparts to Corollary 3.3 (Hotelling–Wold Identity) and Corollary 4.2 (Ville–Roy Identity) since these corollaries made use of the differentiable nature of F or G and the relevant constraint function. Thus, in order to derive counterparts to Corollaries 3.3 and 4.2 in the present context, we need to make additional assumptions on F (or G) and the constraint correspondence B.[27] However, the above theorems (due essentially to Epstein) do illustrate the underlying structure of duality theory. They can also be interpreted as examples of local duality theorems.

7. Cost minimization and the derived demand for inputs

Assume that the technology of a firm can be described by the production function F where $u = F(x)$ is the maximum output that can be produced using the non-negative vector of inputs $x \geqslant 0_N$. Assume that F satisfies Assumption 1 of Section 2 (i.e. the production function is continuous from above). If the firm takes the prices of inputs $p \gg 0_N$ as given (i.e. the firm does not behave monopsonistically with respect to inputs), then we saw in Section 2 that the firm's total cost function $C(u; p) \equiv \min_x\{p^{\mathrm{T}}x: F(x) \geqslant u\}$ was well defined for all $p \gg 0_N$ and $u \in \operatorname{range} F$. Moreover, $C(u, p)$ was linearly homogeneous and concave in prices p for every u and was non-decreasing in u for each fixed p.

Now suppose that C is twice continuously differentiable[28] with respect to its arguments at a point (u^*, p^*) where $u^* \in \operatorname{range} F$ and $p^* \equiv (p_1^*, \ldots, p_N^*) \gg 0_N$.

[27] Epstein (1975b, 1981a) derives counterparts to 4.2 in the context of his specific models.

[28] By this assumption we mean that the second-order partial derivatives of C exist and are continuous functions for a neighborhood around (u^*, p^*).

From Lemma 3 in Section 2, the cost minimizing input demand functions $x_1(u, p),\ldots, x_N(u, p)$ exists at (u^*, p^*) and they are in fact equal to the partial derivatives of the cost function with respect to the N input prices:

$$x_i(u^*, p^*)=\frac{\partial C(u^*, p^*)}{\partial p_i}; \qquad i=1,\ldots, N. \tag{7.1}$$

Thus, the assumption that C be twice continuously differentiable at (u^*, p^*) ensures that the cost minimizing input demand functions $x_i(u, p)$ exist and are once continuously differentiable at (u^*, p^*).

Define $[\partial x_i/\partial p_j]\equiv[\partial x_i(u^*, p^*)/\partial p_j]$ to be the $N\times N$ matrix of derivatives of the N-input demand functions $x_i(u^*, p^*)$ with respect to the N prices p_j^*, $i, j=1,2,\ldots, N$. From (7.1) it follows that

$$\left[\frac{\partial x_i}{\partial p_j}\right]=\nabla^2_{pp}C(u^*, p^*), \tag{7.2}$$

where $\nabla^2_{pp}C(u^*, p^*)\equiv[\partial^2 C(u^*, p^*)/\partial p_i\partial p_j]$ is the Hessian matrix of the cost function with respect to the input prices evaluated at (u^*, p^*). Twice continuous differentiability of C with respect to p at (u^*, p^*) implies (via Young's Theorem) that $\nabla^2_{pp}C(u^*, p^*)$ is a symmetric matrix, so that using (7.2),

$$\left[\frac{\partial x_i}{\partial p_j}\right]=\left[\frac{\partial x_i}{\partial p_j}\right]^{\mathrm{T}}=\left[\frac{\partial x_j}{\partial p_i}\right], \tag{7.3}$$

i.e. $\partial x_i(u^*, p^*)/\partial p_j=\partial x_j(u^*, p^*)/\partial p_i$, for all i and j.

Since C is *concave* in p and is twice continuously differentiable with respect to p around the point (u^*, p^*), it follows from Fenchel (1953, pp. 87–88) or Rockafellar (1970, p. 27) that $\nabla^2 C(u^*, p^*)$ is a negative semi-definite matrix. Thus, by (7.2),

$$z^{\mathrm{T}}\left[\frac{\partial x_i}{\partial p_j}\right]z\leqslant 0, \quad \text{for all vectors } z. \tag{7.4}$$

Thus, in particular, letting $z=e_i$, the ith unit vector, (7.4) implies

$$\frac{\partial x_i(u^*, p^*)}{\partial p_i}\leqslant 0, \qquad i=1,2,\ldots, N, \tag{7.5}$$

i.e. the ith cost minimizing input demand function cannot slope upwards with respect to the ith input price for $i=1,2,\ldots, N$.

Since C is linearly homogeneous in p, we have $C(u^*, \lambda p^*)=\lambda C(u^*, p^*)$ for all $\lambda>0$. Partially differentiating this last equation with respect to p_i for λ close to 1 yields the equation $C_i(u^*, \lambda p^*)\lambda=\lambda C_i(u^*, p^*)$, where $C_i(u^*, p^*)\equiv \partial C(u^*, p^*)/\partial p_i$. Thus, $C_i(u^*, \lambda p^*)=C_i(u^*, p^*)$ and differentiation of this last equation with respect to λ yields (when $\lambda=1$)

$$\sum_{j=1}^{N} p_j^* \partial^2 C(u^*, p^*)/\partial p_i \partial p_j=0, \quad \text{for } i=1,2,\dots, N.$$

Thus, using (7.2), we find that the input demand functions $x_i(u^*, p^*)$ satisfy the following N restrictions:

$$\left[\frac{\partial x_i}{\partial p_j}\right] p^*=\nabla^2_{pp} C(u^*, p^*)p^*=0_N, \tag{7.6}$$

where $p^*\equiv[p_1^*, p_2^*,\dots, p_N^*]^{\mathrm{T}}$.

The final general restriction that we can obtain on the derivatives of the input demand functions is obtained as follows: for λ near 1, differentiate both sides of $C(u^*, \lambda p^*)=\lambda C(u^*, p^*)$ with respect to u and then differentiate the resulting equation with respect to λ. When $\lambda=1$, the last equation becomes

$$\sum_{j=1}^{N} p_j^* \partial^2 C(u^*, p^*)/\partial u \partial p_j=\partial C(u^*, p^*)/\partial u.$$

Note that the twice continuous differentiability of C and (7.1) implies that

$$\begin{aligned}\partial^2 C(u^*, p^*)/\partial u \partial p_j &=\partial^2 C(u^*, p^*)/\partial p_j \partial u \\ &=\partial\left[\partial C(u^*, p^*)/\partial p_j\right]/\partial u=\partial x_j(u^*, p^*)/\partial u.\end{aligned}$$

Thus

$$\begin{aligned}\sum_{j=1}^{N} p_j^* \frac{\partial^2 C(u^*, p^*)}{\partial u \partial p_j} &= \sum_{j=1}^{N} p_j^* \frac{\partial x_j(u^*, p^*)}{\partial u} \\ &= \frac{\partial C(u^*, p^*)}{\partial u} \geqslant 0.\end{aligned} \tag{7.7}$$

The inequality $\partial C(u^*, p^*)/\partial u\geqslant 0$ follows from the non-decreasing in u property of C. The inequality (7.7) tells us that the changes in cost minimizing input demands induced by an increase in output cannot all be negative, i.e. not all inputs can be inferior.

With the additional assumption that F be linearly homogeneous (and there exists $x>0_N$ such that $F(x)>0$), we can deduce (cf. Section 2) that $C(u,p)=uc(p)$, where $c(p)\equiv C(1,p)$. Thus, when F is linearly homogeneous,

$$x_i(u^*,p^*)=u^*\frac{\partial c(p^*)}{\partial p_i}, \qquad i=1,\dots,N, \tag{7.8}$$

and $\partial x_i(u^*,p^*)/\partial u=\partial c(p^*)/\partial p_i$. Thus, if $x_i^*\equiv x_i(u^*,p^*)>0$ for $i=1,2,\dots,N$, using (7.8) we can deduce the additional restrictions

$$\frac{\partial x_i(u^*,p^*)}{\partial u}\frac{u^*}{x_i^*}=\frac{u^*[\partial c(p^*)/\partial p_i]}{x_i^*}=1, \tag{7.9}$$

if F is linearly homogeneous, i.e. all of the input elasticities with respect to output are unity.

For the general two-input case, the general restrictions (7.3)–(7.7) enable us to deduce the following restrictions of the six partial derivatives of the two-input demand functions, $x_1(u^*,p_1^*,p_2^*)$ and $x_2(u^*,p_1^*,p_2^*)$: $\partial x_1/\partial p_1\leqslant 0$, $\partial x_2/\partial p_2\leqslant 0$, $\partial x_1/\partial p_2\geqslant 0$, $\partial x_2/\partial p_1\geqslant 0$ (and if any one of the above inequalities holds strictly, then they all do, since $p_1^*\partial x_1/\partial p_1=-p_2^*\partial x_1/\partial p_2=-p_2^*\partial x_2/\partial p_1=(p_2^*)^2(p_1^*)^{-1}\partial x_2/\partial p_2)$ and $p_1^*\partial x_1/\partial u+p_2^*\partial x_2/\partial u\geqslant 0$. Thus, the signs of $\partial x_1/\partial u$ and $\partial x_2/\partial u$ are ambiguous, but if one is negative, then the other must be positive. For the constant returns to scale two-input case, the ambiguity disappears: we have $\partial x_1(u^*,p^*)/\partial u\geqslant 0$, $\partial x_2(u^*,p^*)/\partial u\geqslant 0$ and at least one of the inequalities must hold strictly if $u^*>F(0_2)$.

An advantage in deriving these well-known comparative statics results using duality theory is that the restrictions (7.2)–(7.7) are valid in cases where the direct production function F is not even differentiable. For example, a Leontief production function has a linear cost function $C(u,p)=ua^{\mathrm{T}}p$, where $a^{\mathrm{T}}\equiv(a_1,a_2,\dots,a_N)>0_N^{\mathrm{T}}$ is a vector of constants. It can be verified that the restrictions (7.2)–(7.7) are valid for this non-differentiable production function.

Historical notes

Analogues to (7.3) and (7.4) in the context of profit functions were obtained by Hotelling (1932, p. 594; 1935, pp. 69–70). Hicks (1946, pp. 311, 331) and Samuelson (1949, p. 69) obtained all of the relations (7.2)–(7.6) and Samuelson (1947, p. 66) also obtained (7.7). All of these authors assumed that the primal function F was differentiable and their proofs used the first-order conditions for the cost minimization (or utility maximization) problem plus the properties of determinants in order to prove their results.

Our proofs of (7.3)–(7.6), using only differentiability of the cost function plus Lemma 3 in Section 2, are due to McKenzie (1956–57, pp. 188–189) and Karlin (1959, p. 273). McFadden (1978a) also provides alternative proofs.

If F is only homothetic rather than being linearly homogeneous, then the relations (7.9) are not longer true. If F is homothetic, then by (2.18), $C(u, p)=\phi^{-1}(u)c(p)$, where ϕ^{-1} is a monotonically increasing function of one variable. Thus, under our differentiability assumptions, $x_i(u^*, p^*)=\phi^{-1}(u^*)\partial c(p^*)/\partial p_i$ and $\partial x_i(u^*, p^*)/\partial u=[d\phi^{-1}(u^*)/du][\partial c(p^*)/\partial p_i]$, so that if $x_i^* \equiv x_i(u^*, p^*)>0$,

$$\frac{\partial x_i(u^*, p^*)}{\partial u}\frac{u^*}{x_i^*}=\frac{u^*\left[d\phi^{-1}(u^*)\right]/du}{\phi^{-1}(u^*)}\equiv\eta(u^*)\geqslant 0, \quad \text{for } i=1,2,\ldots,N. \tag{7.10}$$

Thus, in the case of a homothetic production function, the input elasticities with respect to output are all equal to the same non-negative number independent of the input prices, but dependent in general on the output level u^*. Furthermore, assuming homotheticity of F, we can solve the equation $C(u, p)=\phi^{-1}(u)c(p)=y$ for $u=\phi[y/c(p)]=\phi[1/c(p/y)]\equiv G(p/y)$, where $y>0$ is the producer's allowable expenditure on inputs. If we replace u^* by $\phi(y^*/c(p^*))$ in the system of input demand functions $x_i(u^*, p^*)$, we obtain the system of "market" demand functions

$$\begin{aligned} x_i(\phi[y^*/c(p^*)], p^*)&=\phi^{-1}[(y^*/c(p^*)]\partial c(p^*)/\partial p_i \\ &=[y^*/c(p^*)]\partial c(p^*)/\partial p_i, \qquad i=1,2,\ldots,N. \end{aligned}$$

Thus, if $x_i^* \equiv x_i(u^*, p^*)>0$,

$$\frac{\partial x_i}{\partial y}\left(\phi\left[\frac{y^*}{c}(p^*)\right], p^*\right)\frac{y^*}{x_i^*}=1, \qquad i=1,2,\ldots,N, \tag{7.11}$$

i.e. all inputs have unitary "income" (or expenditure) elasticity of demand if the underlying aggregator function F is homothetic. Note the close resemblance of (7.11) to (7.9). That homotheticity of F implies the relations (7.11) dates back to Frisch (1936, p. 25) at least. For further references, see Chipman (1974a, p. 27).

8. The Slutsky conditions for consumer demand functions

Assume that a consumer has a utility function $F(x)$ defined over $x\geqslant 0_N$ which is continuous from above. Then we have seen in Section 2 that $C(u, p)\equiv\min_x\{p^{\mathrm{T}}x: F(x)\geqslant u\}$ is well defined for $u\in$ range F and $p\gg 0_N$. Moreover, the cost function

C has a number of properties including non-decreasingness in u for each $p \gg 0_N$ and linear homogeneity and concavity in p for each $u \in$ range F.

Assume that the consumer faces prices $p^* \gg 0_N$ and has income $y^* > 0$ to spend on commodities. Then the consumer will wish to choose the largest u such that his cost minimizing expenditure on the goods is less than or equal to his available income. Thus, the consumer's equilibrium utility level will be u^* defined by

$$u^* \equiv \max_u \{u: C(u, p^*) \leqslant y^*, u \in \text{range } F\}.$$

Now assume that C is twice continuously differentiable with respect to its arguments at the point (u^*, p^*) with

$$\frac{\partial C(u^*, p^*)}{\partial u} > 0. \tag{8.1}$$

The fact that C is non-decreasing in u implies that $\partial C(u^*, p^*)/\partial u \geqslant 0$; however, the slightly stronger assumption (8.1) enables us to deduce that the consumer will actually spend all of his income on purchasing (or renting) commodities, i.e. (8.1) implies that

$$C(u^*, p^*) = y^*. \tag{8.2}$$

Furthermore, since C is linearly homogeneous in p, (8.2) implies

$$C\left(u^*, \frac{p^*}{y^*}\right) = 1. \tag{8.3}$$

Our differentiability assumptions plus (8.1) and (8.3) imply (using the Implicit Function Theorem) that (8.3) can be solved for u as a function of p/y in a neighborhood of p^*/y^*. The resulting function $G(p/y)$ is the consumer's *indirect utility function*, which gives the maximum utility level the consumer can attain, given that he faces commodity prices p and has income y to spend on commodities. The Implicit Function Theorem also implies that G will be twice continuously differentiable with respect to its arguments at p^*/y^*. Note that

$$u^* = G\left(\frac{p^*}{y^*}\right). \tag{8.4}$$

The consumer's system of *Hicksian* (1946, p. 33) *or constant real income demand functions*[29] $f_1(u, p), \ldots f_N(u, p)$ is defined as the solution to the expenditure

[29] In the previous section these functions were denoted as $x_1(u, p), \ldots, x_N(u, p)$.

minimization problem $\min_x\{p^Tx: F(x)\geqslant u\}$. Since we have assumed that C is differentiable with respect to p at (u^*, p^*), by Lemma 3 in Section 2

$$f_i(u^*, p^*)=\frac{\partial C(u^*, p^*)}{\partial p_i}, \qquad i=1,\dots, N, \tag{8.5}$$

i.e. the Hicksian demand functions can be obtained by differentiating the cost function with respect to the commodity prices. On the other hand, the consumer's system of *ordinary market demand functions*, $x_1(y, p),\dots, x_N(y, p)$ can be obtained from the Hicksian system (8.5), if we replace u in (8.5) by $G(p/y)$, the maximum utility the consumer can obtain when he has income y and faces prices p. Thus,

$$x_i(y^*, p^*)\equiv f_i\left(G\left(\frac{p^*}{y^*}\right), p^*\right), \qquad i=1,\dots, N. \tag{8.6}$$

Thus, the consumer's system of market demand functions can be obtained from the cost function as well as by using the Ville–Roy Identity (4.4). Finally, it can be seen that if we replace y in the consumer's system of market demand functions by $C(u, p)$, then we should obtain precisely the system of Hicksian demand functions (8.5); i.e. we have

$$x_i(C(u^*, p^*), p^*)=\frac{\partial C(u^*, p^*)}{\partial p_i}, \qquad i=1,\dots, N. \tag{8.7}$$

Differentiating both sides of (8.7) yields:

$$\begin{aligned}\frac{\partial^2 C(u^*, p^*)}{\partial p_i \partial p_j} &= \frac{\partial x_i(y^*, p^*)}{\partial p_j}+\frac{\partial x_i(y^*, p^*)}{\partial y}\frac{\partial C(u^*, p^*)}{\partial p_j}, \quad \text{using (8.2)}\\ &= \frac{\partial x_i(y^*, p^*)}{\partial p_j}+f_j(u^*, p^*)\frac{\partial x_i(y^*, p^*)}{\partial y}, \quad \text{using (8.5)}\\ &= \frac{\partial x_i(y^*, p^*)}{\partial p_j}+x_j(y^*, p^*)\frac{\partial x_i(y^*, p^*)}{\partial y}, \\ &\qquad \text{using (8.4) and (8.6)}\\ &\equiv k^*_{ij}, \qquad i, j=1,2,\dots, N,\end{aligned} \tag{8.8}$$

where k^*_{ij} is known as the ijth *Slutsky coefficient*. Note that the $N\times N$ matrix of these Slutsky coefficients, $K^*\equiv[k^*_{ij}]$, can be calculated from a knowledge of the market demand functions, $x_i(y, p)$, and their first order derivatives at the point (y^*, p^*). (8.8) shows that $K^*\equiv\nabla^2_{pp}C(u^*, p^*)$ and thus [recall eqs. (7.3), (7.4) and

(7.6) of the previous section] K^* satisfies the following *Slutsky–Samuelson–Hicks Conditions*:

$$\begin{aligned} &\text{(i)} \quad K^* = K^{*\mathrm{T}}. \\ &\text{(ii)} \quad z^{\mathrm{T}} K^* z \leqslant 0, \text{ for every } z. \\ &\text{(iii)} \quad K^* p^* = 0_N. \end{aligned} \tag{8.9}$$

Historical notes

Slutsky (1915) deduced (8.9)(i) and part of (8.9)(ii), i.e. that $k_{ii}^* \leqslant 0$. Samuelson (1938, p. 348) and Hicks (1946, p. 311) deduced the entire set of restrictions (8.9) under the assumption that F was twice continuously differentiable at an equilibrium point $x^* \gg 0_N$ and F satisfied the additional property that $v^{\mathrm{T}} \nabla_{xx}^2 F(x^*) v < 0$ for all $v \neq 0_N$ such that $v \neq k \nabla F(x^*)$ for any scalar k. In fact, under these hypotheses, Samuelson and Hicks were able to deduce the following strengthened version of (8.9)(ii): $z^{\mathrm{T}} K^* z < 0$ for every $z \neq 0_N$ such that $z \neq kp^*$ for any scalar k.

Our proof of Conditions (8.9) is due to McKenzie (1956–57) and Karlin (1959, pp. 267–273). See also Arrow and Hahn (1971, p. 105). This method of proof again has the advantage that differentiability of F does not have to be assumed: all that is required is differentiability of the demand functions essentially. Afriat (1972a) makes this point.

For a derivation of Conditions (8.9) which utilizes only the properties of the indirect utility function G, see Diewert (1977a, p. 356).

For a "traditional" derivation of (8.9), see Chapter 2 of this Handbook by Intriligator.

9. Empirical applications using cost or indirect utility functions

Suppose that the technology of an industry can be characterized by a constant returns to scale production function f which has the following properties:[30]

$$f \text{ is a (i) positive, (ii) linearly homogeneous, and (iii) concave function defined over the positive orthant in } R^N. \tag{9.1}$$

It is shown in Samuelson (1953–54) and Diewert (1974a, pp. 110–112) that the cost function which corresponds to f has the following form: for $u \geqslant 0$, $p \gg 0_N$,

$$C(u, p) \equiv \min_x \{ p^{\mathrm{T}} x : f(x) \geqslant u, x \geqslant 0_N \} = uc(p), \tag{9.2}$$

[30] f can be uniquely extended to the non-negative orthant by using the Fenchel closure operation.

where $c(p)\equiv C(1,p)$ is the unit cost function and it also satisfies the three properties listed in (9.1).

The producer's system of input demand functions, $x(u,p)\equiv[x_1(u,p,\ldots, x_N(u,p)]^{\mathrm{T}}$, can be obtained as the set of solutions to the programming problem (9.2) if we are given a functional form for the production function f. Thus, one method for obtaining a system of derived input demand functions that is consistent with the hypothesis of cost minimization is to postulate a (differentiable) functional form for f and then use the usual Lagrangian techniques in order to solve (9.2).

The problem with this first method for obtaining the system of input demand functions $x(u,p)$ is that it is usually very difficult to obtain an algebraic expressions for $x(u,p)$ in terms of the (unknown) parameters which characterize the production function f, particularly if we assume that f is a *flexible*[31] linearly homogeneous functional form.

A second method for obtaining a system of input demand functions $x(u,p)$ makes use of Lemma 4 (Shephard's Lemma): simply postulate a functional form for the cost function $C(u,p)$ which satisfies the appropriate regularity conditions and, in addition, is differentiable with respect to input prices. Then, $x(u,p)=\nabla_p C(u,p)$ and the system of derived demand functions can be obtained by differentiating the cost function with respect to input prices.

For example, suppose that the unit cost function is defined by

$$c(p)=\sum_{i=1}^{N}\sum_{j=1}^{N} b_{ij}p_i^{\frac{1}{2}}p_j^{\frac{1}{2}}, \quad \text{with } b_{ij}=b_{ji}\geqslant 0 \quad \text{for all } i,j.$$

Then if at least one $b_{ij}>0$, the resulting function c satisfies (9.1), and the input demand functions are

$$x_i(u,p)=\sum_{j=1}^{N} b_{ij}\left(\frac{p_j}{p_i}\right)^{\frac{1}{2}}u; \qquad i=1,2,\ldots,N. \tag{9.3}$$

Note that the system of input demand equations (9.3) is linear in the unknown parameters, and thus linear regression techniques can be used in order to estimate

[31] f is a flexible functional form if it can provide a second-order (differential) approximation to an arbitrary twice continuously differentiable function f^* at a point x^*. f differentially approximates f^* at x^* iff (i) $f(x^*)=f^*(x^*)$, (ii) $\nabla f(x^*)=\nabla f^*(x^*)$, and (iii) $\nabla^2 f(x^*)=\nabla^2 f^*(x^*)$, where both f and f^* are assumed to be twice continuously differentiable at x^* (and thus the two-Hessian matrices in (iii) will be symmetric). Thus, a general flexible functional form f must have at least $1+N+N(N+1)/2$ free parameters. If f and f^* are both linearly homogeneous, then $x^{*\mathrm{T}}\nabla f^*(x^*)=f^*(x^*)$ and $\nabla^2 f^*(x^*)x^*=0_N$, and thus a flexible linearly homogeneous functional form f need have only $N+N(N-1)/2=N(N+1)/2$ free parameters. The term "flexible" is due to Diewert (1974a, p. 113) while the term "differential approximation" is due to Lau (1974a, p. 183).

the b_{ij}, if we are given data on output, inputs, and input prices. Note also that b_{ij} in the ith input demand equation should equal b_{ji} in the jth equation for $j \neq i$. These are the Hotelling (1932, p. 594), Hicks (1946, pp. 311, 331), and Samuelson (1947, p. 64) *symmetry restrictions* (7.3), and we can statistically test for their validity. If some of the b_{ij} are negative, then the system of input demand equations can still be locally valid, as shown in Blackorby and Diewert (1979) and Diewert (1974a, pp. 113–114). Finally, note that if $b_{ij}=0$ for $i \neq j$, then (9.3) becomes $x_i(u,p)=b_{ii}u$, $i=1,2,\dots,N$, which is the system of input demand functions that corresponds to the Leontief (1941) production function, $f(x_1,x_2,\dots,x_N) \equiv \min\{x_i/b_{ii}\colon i=1,2,\dots,N]\}$. In the general case, the production function which corresponds to (9.3) is called the *Generalized Leontief Production Function* in Diewert (1971a). It can also be shown that the corresponding unit cost function $\Sigma_i \Sigma_j b_{ij} p_i^{\frac{1}{2}} p_j^{\frac{1}{2}}$, is a flexible linearly homogeneous functional form as shown in Diewert (1974a, p. 115).

As another example of the second method for obtaining input demand functions, consider the following *translog cost function*:

$$\ln C(u,p) \equiv \alpha_0 + \sum_{i=1}^{N} \alpha_i \ln p_i + \tfrac{1}{2} \sum_{i=1}^{N} \sum_{j=1}^{N} \gamma_{ij} \ln p_i \ln p_j + \delta_0 \ln u + \sum_{i=1}^{N} \delta_i \ln p_i \ln u + \tfrac{1}{2}\varepsilon_0 (\ln u)^2, \tag{9.4}$$

where the parameters satisfy the following restrictions:

$$\sum_{i=1}^{N} \alpha_i = 1; \qquad \gamma_{ij} = \gamma_{ji}, \quad \text{for all } i, j;$$

$$\sum_{j=1}^{N} \gamma_{ij} = 0, \quad \text{for } i=1,2,\dots,N; \quad \text{and} \sum_{i=1}^{N} \delta_i = 0. \tag{9.5}$$

The restrictions (9.5) ensure that C defined by (9.4) is linearly homogeneous in p. The additional restrictions

$$\delta_0 = 1; \qquad \delta_i = 0, \quad \text{for } i=1,2,\dots,N; \quad \text{and } \varepsilon_0 = 0 \tag{9.6}$$

ensure that $C(u,p)=uC(1,p)$ so that the corresponding production function is linearly homogeneous. Finally, with the additional restrictions $\gamma_{ij}=0$, for all i, j, and $\alpha_i \geq 0$, for $i=1,2,\dots,N$, C defined by (9.4) reduces to a Cobb–Douglas cost function.

The "translog" functional form defined by (9.4) is due to Christensen, Jorgenson, and Lau (1971), Griliches and Ringstad (1971) (for two inputs), and

Sargan (1971, pp. 154–156) (who calls it the *log quadratic production function*). In general, C defined by (9.4) will not satisfy the appropriate regularity conditions (e.g. Conditions II in Section 3) globally, but Lau (1974, p. 186) shows that it can provide a good *local* approximation to an arbitrary twice differentiable, linearly homogeneous in p, cost function, i.e. the translog functional form (9.4) is flexible.

The cost minimizing input demand functions $x_i(u, p)$ which (9.4) generates via Shephard's Lemma are not linear in the unknown parameters. However, it is easy to verify that the factor share functions

$$s_i(u,p) \equiv p_i x_i(u,p) \Big/ \sum_{k=1}^{N} p_k x_k(u,p) = \frac{p_i x_i(u,p)}{C(u,p)} = \frac{\partial \ln C(u,p)}{\partial \ln p_i}$$

are linear in the unknown parameters:

$$s_i(u,p) = \alpha_i + \sum_{j=1}^{N} \gamma_{ij} \ln p_j + \delta_i \ln u, \qquad i=1,\ldots,N. \tag{9.7}$$

However, since the shares sum to unity, only $N-1$ of the N equations defined by (9.7) can be statistically independent. Moreover, notice that the parameters α_0, δ_0, and ε_0 do not appear in (9.7). However, all of the parameters can be statistically determined, given data on output, inputs, and input prices, if we append equation (9.4) (which is also linear in the unknown parameters) to $N-1$ of the N equations in (9.7).

The above two examples illustrate how simple it is to use the second method for generating systems of input demand functions which are consistent with the hypothesis of cost minimization.

Just as Shephard's Lemma (3.1) can be used to derive systems of cost minimizing input demand functions, Roy's Identity (4.4) can be used to derive systems of utility maximizing commodity demand functions in the context of consumer theory. For example, consider the following *translog indirect utility function*:[32] for $v \equiv p/y \gg 0_N$, define

$$G(v) \equiv \alpha_0 + \sum_{i=1}^{N} \alpha_i \ln v_i + \tfrac{1}{2} \sum_{i=1}^{N} \sum_{j=1}^{N} \gamma_{ij} \ln v_i \ln v_j; \qquad \gamma_{ij} = \gamma_{ji}. \tag{9.8}$$

Roy's Identity (4.4) applied to G defined by (9.8) yields the following system of consumer demand functions where $v \equiv (v_i, \ldots, v_N)^{\mathrm{T}} = p^{\mathrm{T}}/y$, $p^{\mathrm{T}} \equiv (p_1, \ldots, p_N)$ is a

[32]See Jorgenson and Lau (1970) and Christensen, Jorgenson, and Lau (1975). This translog function can locally approximate any twice continuously differentiable indirect utility function. However, $G(v)$ defined by (9.8) will, in general, not satisfy Conditions III globally.

vector of positive commodity prices, and $y>0$ is the consumer's expenditure on the N goods:

$$x_i\left(\frac{p}{y}\right)=\frac{p_i^{-1}y\left(\alpha_i+\sum_{j=1}^{N}\gamma_{ij}\ln p_j-\sum_{j=1}^{N}\gamma_{ij}\ln y\right)}{\sum_{k=1}^{N}\alpha_k+\sum_{k=1}^{N}\sum_{m=1}^{N}\gamma_{km}\ln p_m-\sum_{k=1}^{N}\sum_{m=1}^{N}\gamma_{km}\ln y},\qquad i=1,2,\ldots,N. \tag{9.9}$$

Note that the demand functions are homogeneous of degree zero in all of the parameters taken together. Thus, in order to identify the parameters, a normalization such as

$$\sum_{i=1}^{N}\alpha_i=-1 \tag{9.10}$$

must be appended to equations (9.9). Note also that the parameter α_0 which occurs in (9.8) cannot be identified if we have data only on consumer purchases (rentals in the case of durable goods) x, prices p, and total expenditure y. Moreover, only $N-1$ of the N equations in (9.9) are independent and eq. (9.8) cannot be added to the independent equations in (9.9) to give N independent estimating equations because the left-hand side of (9.8) is the unobservable variable, utility u. Thus, the econometric procedures used to estimate consumer preferences are not entirely analogous to the procedures used to estimate production functions, even though from a theoretical point of view the duality between cost and production functions is entirely isomorphic to the duality between expenditure and utility functions.

The system of commodity demand functions defined by (9.9) is not linear in the unknown parameters and thus non-linear regression techniques will have to be used in order to estimate econometrically the unknown parameters. We generally obtain non-linear demand equations using Roy's Identity if we assume that G is defined by a flexible functional form.[33]

[33] However, if we assume that the direct utility function F is linearly homogeneous, then the corresponding indirect utility function G will be homogeneous of degree -1. A flexible homogeneous of degree -1, G can be obtained by using the translog functional form (9.8) with the additional restrictions $\alpha_0=0$, $\sum_{i=1}^{N}\alpha_i=-1$ and $\sum_{j=1}^{N}\gamma_{ij}=0$ for $i=1,\ldots,N$. In this case, the consumer's system of commodity share equations becomes $s_i\equiv p_ix_i/y=-\alpha_i-\sum_{j=1}^{N}\gamma_{ij}\ln p_j$, $i=1,\ldots,N$, which is linear in the unknown parameters. However, the assumption of linear homogeneity (or even homotheticity) for F is highly implausible in the consumer context, since it leads to unitary income elasticities for all goods [cf. Frisch (1936, p. 25)].

The system of demand equations defined by (9.9) could be utilized given microeconomic data on a single utility maximizing consumer (with constant preferences) or given cross section data on a number of consumers, assuming that each utility maximizing consumer in the sample had the same preferences. However, could we legitimately apply the system (9.9) to market data, i.e. assume that x_i represented total market demand for commodity i divided by the number of independent consuming units, p_i is the price of commodity i, and y is total market expenditure on all goods divided by the number of consuming units? The answer is no in general.[34] However, if we have information on the *distribution* $\phi(y)$ of expenditure y by the different households in the market and we are willing to assume that each household has the same tastes, then the market demand functions X_i can be obtained by integrating over the individual demand functions $x_i(p/y)$:

$$X_i(p)=N^*\int_0^\infty x_i\left(\frac{p}{y}\right)\phi(y)\,\mathrm{d}y, \qquad i=1,\dots,N, \tag{9.11}$$

where N^* is the number of households in the market and $\int_0^\infty\phi(y)\,\mathrm{d}y=1$. The integrations in (9.11) can readily be performed using the $x_i(p/y)$ defined by (9.9) if we impose the following normalizations on the parameters of the translog indirect utility function defined by (9.8): (i) $\alpha_0=0$; (ii) $\sum_{i=1}^N\alpha_i=-1$; and (iii) $\sum_{i=1}^N\sum_{j=1}^N\gamma_{ij}=0$. The effect of these three normalizations is to make G homogeneous of degree -1 along the ray of equal prices, i.e. $G(\lambda 1_N)=\lambda^{-1}G(1_N)$ for all $\lambda>0$, and this in turn is simply a harmless (from a theoretical point of view but not necessarily from an econometric point of view) cardinalization of utility so that utility is proportional to income when the prices the consumer faces are all equal. This approach for obtaining systems of market demand functions consistent with microeconomic theory has been pursued by Diewert (1974a, pp. 127–130) and Berndt, Darrough, and Diewert (1977).

There is a simpler method for obtaining systems of market demand functions consistent with individual utility maximizing behavior which is due to Gorman (1953): assume that each household's preferences can be represented by a cost function of the form

$$C(u,p)=b(p)+uc(p), \tag{9.12}$$

where b and c are unit cost functions which satisfy conditions (9.1), $p\gg 0_N$ and $c(p)u\geqslant y-b(p)\geqslant 0$, where y is household expenditure. Blackorby, Boyce, and

[34] The reader is referred to the considerable body of literature on the implications of microeconomic theory for systems of market (excess) demand functions, which is reviewed by Shafer and Sonnenschein (1980) and Diewert (1981).

Russell (1978) call a functional form for C which has the structure (9.12), a *Gorman polar form*. If $y-b(p)\geqslant 0$, the indirect utility function which corresponds to (9.12) is $G(v)\equiv[1/c(v)]-[b(v)/c(v)]=[y/c(p)]-[b(p)/c(p)]$, where $v\equiv p/y$, and Roy's Identity (4.4) yields the following system of individual household demand functions if the unit cost functions b and c are differentiable:

$$x\left(\frac{p}{y}\right)=\nabla_p b(p)+[c(p)]^{-1}[y-b(p)]\nabla_p c(p); \qquad y\geqslant b(p). \tag{9.13}$$

The interesting thing about the system of consumer demand functions defined by (9.13) is that they are *linear* in the household's income or expenditure y. Thus, if every household in the market under consideration has the same preferences which are dual to C defined by (9.12) and each household has income $y\geqslant b(p)$, then the system of market demand functions $X(p)$ defined by (9.11) is independent of the distribution of income; in fact

$$\frac{X(p)}{N^*}=\nabla_p b(p)+[c(p)]^{-1}[y^*-b(p)]\nabla_p c(p), \tag{9.14}$$

where $X(p)/N^*$ is the per capita market demand vector and $y^*\equiv\int y\phi(y)\,\mathrm{d}y$ is average or per capita expenditure. Comparing (9.14) with (9.13), we see that the per capita market demand system has the same functional form as the individual demand vector for a single decision-making unit. The advantage of this approach over the previous one is that it does not require information on the distribution of expenditure: all that is required is information on market expenditure by commodity, commodity (rental) prices, and the number of consumers or households.[35]

Several flexible functional forms for cost functions have been estimated empirically, using Shephard's Lemma in order to derive systems of input demand functions: see Parks (1971), Denny (1972, 1974), Binswanger (1974), Hudson and Jorgenson (1974), Woodland (1975), Berndt and Wood (1975), Burgess (1974, 1975), and Khaled (1978). Khaled also develops a very general class of functional forms which contains most of the other commonly used functional forms as special cases.

There are also many applications of the above theory to the problem of estimating consumer preferences. For empirical examples, see Lau and Mitchell (1970), Diewert (1974d), Christensen, Jorgenson, and Lau (1975), Jorgenson and Lau (1975), Boyce (1975), Boyce and Primont (1976a), Christensen and Manser

[35]For other generalizations of the Gorman polar form (9.12) which have useful aggregation properties, see Gorman (1959, p. 476), Muellbauer (1975, 1976), and Lau (1977a, 1977b). Diewert (1978c) shows that functional forms of the type (9.12) are flexible. See also Lau (1975).

(1977), Darrough (1977), Blackorby, Boyce, and Russell (1978), Howe, Pollak and Wales (1979), Donovan (1977) and Berndt, Darrough, and Diewert (1977).[36]

10. Profit functions

Up to now, we have considered the case of a firm which produces only a single output, using many inputs. However, in the real world most firms produce a variety of outputs, so that it is now necessary to consider the problems of modelling a multiple output, multiple input firm.

For econometric applications, it is convenient to introduce the concept of a firm's *variable profit function* $\Pi(p, x)$: it simply denotes the maximum revenue minus variable input expenditures that the firm can obtain given that it faces prices $p \gg 0_I$ for variable inputs and outputs and given that another vector of inputs $x \geqslant 0_J$ is held fixed. We denote the variable inputs and outputs by the I-dimensional vector $u \equiv (u_1, u_2, \ldots, u_I)$, the fixed inputs by the J-dimensional vector $-x \equiv (-x_1, \ldots, -x_J)$, and the set of all feasible combinations of inputs and outputs is denoted by T, the firm's *production possibilities set*. Outputs are denoted by positive numbers and inputs are denoted by negative numbers, so if $u_i > 0$, then the ith variable good is an output produced by the firm. Formally, we define Π for $p \gg 0_I$ and $-x \leqslant 0_J$ by

$$\Pi(p, x) \equiv \max_u \{p^{\mathrm{T}} u \colon (u, -x) \in T\}. \tag{10.1}$$

If T is a closed, non-empty, convex cone in Euclidean $I+J$ dimensional space with the additional properties: (i) if $(u, -x) \in T$, then $x \geqslant 0_J$ (the last J goods are always inputs); (ii) if $(u', -x') \in T$, $u'' \leqslant u'$ and $-x'' \leqslant -x'$, then $(u'', -x'') \in T$ (free disposal); and (iii) if $(u, -x) \in T$, then the components of u are bounded from above (bounded outputs for bounded fixed inputs), then Π has the following properties: (i) $\Pi(p, x)$ is a non-negative real valued function defined for every $p \gg 0_I$ and $x \geqslant 0_J$ such that $\Pi(p, x) \leqslant p^{\mathrm{T}} b(x)$ for every $p \gg 0_J$; (ii) for every $x \geqslant 0_J$, $\Pi(p, x)$ is (positively) linearly homogeneous, convex, and continuous in p; and (iii) for every $p \gg 0_I$, $\Pi(p, x)$ is (positively) linearly homogeneous, concave, continuous, and non-decreasing in x. Moreover, Gorman (1968), McFadden (1966), and Diewert (1973a) show that T can be constructed using Π as follows:

$$T = \{(u, -x) \colon p^{\mathrm{T}} u \leqslant \Pi(p, x), \quad \text{for every } p \gg 0_I;\ x \geqslant 0_J\}. \tag{10.2}$$

[36]Lau (1974b) considers the problem of testing for or imposing the various monotonicity and curvature conditions on the cost or indirect utility functions. On the issue of how flexible are flexible functional forms, see Wales (1977) and Byron (1977).

Thus, there is a duality between production possibilities sets T and variable profit functions Π satisfying the above regularity conditions. Moreover, in a manner which is analogous to the proof of Shephard's Lemma (3.1) and Roy's Identity (4.4), the following result can be proven:

Hotelling's Lemma (1932, p. 594)

If a variable profit function Π satisfies the regularity conditions below (10.1) and is in addition differentiable with respect to the variable quantity prices at $p^* \gg 0_I$ and $x^* \geq 0_J$, then $\partial\Pi(p^*, x^*)/\partial p_i = u_i(p^*, x^*)$ for $i = 1, 2, \ldots, I$, where $u_i(p^*, x^*)$ is the profit maximizing amount of net output i (of input i if $\partial\Pi(p^*, x^*)/\partial p_i < 0$) given that the firm faces the vector of variable prices p^* and has the vector x^* of fixed inputs at its disposal.

Hotelling's Lemma can be used in order to derive systems of variable output supply and input demand functions. We need only postulate a functional form for $\Pi(p, x)$ which is consistent with the appropriate regularity conditions for Π and is differentiable with respect to the components of p. For example, consider the translog variable profit function Π defined as:

$$\begin{aligned} \ln \Pi(p, x) \equiv \alpha_0 &+ \sum_{i=1}^{I} \alpha_i \ln p_i + \tfrac{1}{2} \sum_{i=1}^{I} \sum_{h=1}^{I} \gamma_{ih} \ln p_i \ln p_h \\ &+ \sum_{i=1}^{I} \sum_{j=1}^{J} \delta_{ij} \ln p_i \ln x_j + \sum_{j=1}^{J} \beta_j \ln x_j \\ &+ \tfrac{1}{2} \sum_{j=1}^{J} \sum_{k=1}^{J} \phi_{jk} \ln x_j \ln x_k, \end{aligned} \tag{10.3}$$

where $\gamma_{ih} = \gamma_{hi}$ and $\phi_{jk} = \phi_{kj}$. It is easy to see that Π defined by (10.3) is homogeneous of degree one in p if and only if

$$\sum_{i=1}^{I} \alpha_i = 1; \qquad \sum_{i=1}^{I} \delta_{ij} = 0, \quad \text{for } j = 1, \ldots, J;$$

$$\sum_{h=1}^{I} \gamma_{ih} = 0, \quad \text{for } i = 1, \ldots, I. \tag{10.4}$$

Similarly, $\Pi(p,x)$ is homogeneous of degree one in x[37] if and only if

$$\sum_{j=1}^{J} \beta_j = 1; \qquad \sum_{j=1}^{J} \delta_{ij} = 0, \quad \text{for } i=1,\ldots,I;$$
$$\sum_{k=1}^{J} \phi_{jk} = 0, \quad \text{for } j=1,\ldots,J. \tag{10.5}$$

If $\Pi(p,x)>0$, define the ith variable net supply share by $s_i(p,x) \equiv p_i u_i(p,x)/\Pi(p,x)$. Hotelling's Lemma applied to the translog variable profit function defined by (10.3) yields the following system of net supply share functions:

$$s_i(p,x) = \alpha_i + \sum_{h=1}^{I} \gamma_{ih} \ln p_h + \sum_{j=1}^{J} \delta_{ij} \ln x_j; \qquad i=1,\ldots,I. \tag{10.6}$$

Since the shares sum to unity, only $I-1$ of eqs. (10.6) are independent. $I-1$ of eqs. (10.6) plus equation (10.3) can be used in order to estimate the parameters of the translog variable profit function. Note that these equations are linear in the unknown parameters as are the restrictions (10.4) and (10.5) so that modifications of linear regression techniques can be used.

Alternative functional forms for variable profit functions have been suggested by McFadden (1978b), Diewert (1973a), and Lau (1974a). Empirical applications have been made by Kohli (1978), Woodland (1977c), Harris and Appelbaum (1977), and Epstein (1977).

A concept which is closely related to the variable profit function notion is the concept of a *joint cost function*, $C(u,w) \equiv \min_x\{w^{\mathrm{T}}x\colon (u,-x) \in T\}$, where T is the firm's production possibilities set, as before, and $w \gg 0_J$ is a vector of positive input prices. As usual, if $C(u,w)$ is differentiable with respect to input prices w (and satisfies the appropriate regularity conditions), then Shephard's Lemma can be used in order to derive the producer's system of cost minimizing input demand functions $x(u,w)$; i.e. we have

$$x(u,w) = \nabla_w C(u,w). \tag{10.7}$$

Joint cost functions have been empirically estimated by Burgess (1976a) [who utilized a functional form suggested by Hall (1973)], Brown, Caves, and

[37]If we drop the assumption that the production possibilities set T be a cone (so that we no longer assume constant returns to scale in all inputs and outputs), then $\Pi(p,x)$ does not have to be homogeneous of degree one in x. Thus, the restriction (10.5) can be used to test whether T is a cone or not. If we drop the assumption that T be convex, then $\Pi(p,x)$ need not be concave (or even quasi-concave) in x.

Christensen (1975), and Christensen and Greene (1976) [who utilized a translog functional form for $C(u,w)$ analogous to the translog variable profit function defined by (10.3)].

Historical notes

Samuelson (1953–54, p. 20) introduced the concept of the variable profit function[38] and stated some of its properties. Gorman (1968) and McFadden (1966, 1978a) established duality theorems between a production possibilities set satisfying various regularity conditions and the corresponding variable profit function.[39] Alternative duality theorems are due to Shephard (1970), Diewert (1973a, 1974b), Sakai (1974), and Lau (1976). For the special case of a single fixed input, see Shephard (1970, pp. 248–250) or Diewert (1974b).

McFadden (1966, 1978a) introduced the joint cost function, stated its properties, and proved formal duality theorems between it and the firm's production possibilities set T, as did Shephard (1970) and Sakai (1974).

There are also very simple duality theorems between production possibilities sets and transformation functions, which give the maximum amount of one output that the firm can produce (or the minimum amount of input required) given fixed amounts of the remaining inputs and outputs. For examples of these theorems, see Diewert (1973), Jorgenson and Lau (1974a, 1974b), and Lau (1976).

As usual, Hotelling's Lemma can be generalized to cover the case of a non-differentiable variable profit function: the gradient of Π with respect to p is replaced by the set of subgradients. This generalization was first noticed by Gorman (1968, pp. 150–151) and McFadden (1966, p. 11) and repeated by Diewert (1973a, p. 313) and Lau (1976, p. 142).

If $\Pi(p,x)$ is differentiable with respect to the components of the vector of fixed inputs, then $w_j \equiv \partial\Pi(p,x)/\partial x_j$ can be interpreted as the worth to the firm of a marginal unit of the jth fixed input, i.e. it is the "shadow price" for the jth input [cf. Lau (1976, p. 142)]. Moreover, if the firm faces the vector of rental prices $w \gg 0_J$ for the "fixed" inputs, and during some period the "fixed" inputs can be varied, then if the firm minimizes the cost of producing a given amount of variable profits we will have [cf. Diewert (1974a, p. 140)]

$$w = \nabla_x \Pi(p,x), \tag{10.8}$$

and these relations can also be used in econometric applications.

[38]Samuelson called it the *national production function*.

[39]Gorman uses the term "gross profit function" and McFadden uses the term "restricted profit function" to describe $\Pi(p,x)$.

The translog variable profit was independently suggested by Russell and Boyce (1974) and Diewert (1974a, p. 139). Of course, it is a straightforward modification of the translog functional form due to Christensen, Jorgenson, and Lau (1971) and Sargan (1971).

The comparative statics properties of $\Pi(p,x)$ or $C(u,w)$ have been developed by Samuelson (1953–54), McFadden (1966, 1978a), Diewert (1974, pp. 142–146), and Sakai (1974).

In international trade theory it is common to assume the existence of sectoral production functions, fixed domestic resources x, and fixed prices of internationally traded goods p. If we now attempt to maximize the net value of internationally traded goods produced by the economy, we obtain the economy's variable profit function, $\Pi(p,x)$, or Samuelson's (1953–54) *national product function*. If the sectoral production functions are subject to constant returns to scale, $\Pi(p,x)$ will have all the usual properties mentioned above. However, the existence of sectoral technologies will imply additional comparative statics restrictions on the national product function Π: see Chipman (1966, 1972, 1974b). Samuelson (1966), Ethier (1974), Woodland (1977a, 1977b), Diewert and Woodland (1977), and Jones and Scheinkman (1977) and the many references included in these papers.

Finally, note that the properties of $\Pi(p,x)$ with respect to x are precisely the properties that a neoclassical production function possesses. If x is a vector of primary inputs, then $\Pi(p,x)$ can be interpreted as a value added function. If the prices p vary (approximately) in proportion overtime, then $\Pi(p,x)$ can be deflated by the common price trend and the resulting real value added function has all of the properties of a neoclassical production function; see Khang (1971), Bruno (1971) and Diewert (1978c, 1980).

11. Duality and non-competitive approaches to microeconomic theory

Up to now we have assumed that producers and consumers take prices as given and optimize with respect to the quantity variables they control. We indicate in this section how duality theory can be utilized even if there is monopsonistic or monopolistic behavior on the part of consumers or producers. We will not attempt to be comprehensive but will illustrate the techniques involved by means of four examples.

11.1. Approach 1: The monopoly problem

Suppose that a monopolist produces output x_0 by means of the production function $F(x)$, where $x \geq 0_N$ is a vector of variable inputs. Suppose, further, that he faces the (inverse) demand function $p_0 = wD(x_0)$, i.e. $p_0 \geqslant 0$ is the price at which he can sell $x_0 > 0$ units of output, D is a continuous positive function of x_0, and the variable $w > 0$ represents the influence on demand of "other variables". That is to say, if the monopolist is selling to consumers, w might equal disposable income for the period under consideration; if the monopolist is selling to producers, w might be a linearly homogeneous function of the prices that those other producers face.[40] Finally, suppose that the monopolist behaves competitively on input markets, taking as given the vector $p \gg 0_N$ of input prices. The monopolist's profit maximization problem may be written as

$$\begin{aligned}\max_{p_0, x_0, x} &\{p_0 x_0 - p^{\mathrm{T}}x \colon x_0 = F(x), p_0 = wD(x_0), x \geqslant 0_N\} \\ &= \max_x \{wD[F(x)]F(x) - p^{\mathrm{T}}x \colon x \geqslant 0_N\} \\ &- \max_x \{wF^*(x) - p^{\mathrm{T}}x \colon x \geqslant 0_N\} \\ &\equiv \Pi^*(w, p),\end{aligned} \tag{11.1}$$

where $F^*(x) \equiv D[F(x)]F(x) = p_0 x_0 / w$ is the deflated (by w) revenue function or pseudoproduction function and Π^* is the corresponding pseudoprofit function (recall Section 10) which corresponds to F^*.[41] Notice that w plays the role of a price for $F^*(x)$. If F^* is a concave function, then $\Pi^*(1, p/w)$ will be the conjugate function to F^* [recall the Samuelson (1960), Lau (1969, 1978), and Jorgenson and Lau (1974a, 1974b) conjugacy approach to duality theory] and Π^* will be dual to F^* (i.e. F^* can be recovered from Π^*). Even if F^* is not concave, if the maximum in (11.1) exists over the relevant range of (w, p) prices, then Π^* can be used to represent the relevant part of F^* (i.e. the free disposal convex hull of F^* can be recovered from Π^*). Moreover, if Π^* is differentiable at (w^*, p^*) and x_0^*, p_0^*, x^* solve (11.1), then Hotelling's Lemma implies

$$u_0^* \equiv \frac{p_0^* x_0^*}{w^*} = \nabla_w \Pi^*(w^*, p^*) \quad \text{and} \quad -x^* = \nabla_p \Pi^*(w^*, p^*). \tag{11.2}$$

Moreover, if Π^* is twice continuously differentiable at (w^*, p^*), then we can deduce the usual comparative statics results on the derivatives of the deflated sales function $u_0(w^*, p^*) \equiv \nabla_w \Pi^*(w^*, p^*)$ and the input demand functions

[40] If non-quantity variables do not influence the inverse demand function that the monopolist faces for the periods under consideration, then w can be set equal to 1 in each period.

[41] Note that we have suppressed mention of any fixed inputs. We assume sufficient regularity on F and D so that the maximum in (11.1) exists.

$-x(w^*, p^*) \equiv \nabla_p \Pi^*(w^*, p^*)$: namely $\nabla^2 \Pi^*(w^*, p^*)$ is a positive semidefinite symmetric matrix and $[w^*, p^{*\mathrm{T}}]\nabla^2\Pi^*(w^*, p^*) = 0^{\mathrm{T}}_{N+1}$.

Eqs. (11.2) can be used in order to estimate econometrically the parameters of Π^* and hence indirectly of F^*: simply postulate a functional form for Π^*, differentiate Π^*, and then fit (11.2), given a time series of observations on p_0, p, w, x_0, and x. The drawbacks to this method are: (i) we cannot disentangle D from F; (ii) we cannot test whether the producer is in fact behaving competitively on the output market; and (iii) we cannot use our estimated equations to predict output x_0 or selling price p_0 separately.

11.2. Approach 2: The monopsony problem

Consider the problem of a consumer maximizing a utility function $F(x)$ satisfying Conditions I but now we no longer assume that the consumer faces fixed prices for the commodities he purchases. Rather he is able monopsonistically to exploit one or more of the suppliers that he faces. Then in period r he faces a non-linear budget constraint of the form $h_r(x)=0$, where $x \geqslant 0_N$ is his vector of purchases (or rentals). Let $x^r \geqslant 0_N$ be a solution to the period r constrained utility maximization problem, so that

$$\max_x \{F(x): h_r(x)=0, x \geqslant 0_N\} = F(x^r); \qquad r=1,\ldots,T. \tag{11.3}$$

Suppose, further, that the rth budget constraint function h_r is differentiable at x^r with $\nabla_x h(x^r) \gg 0_N$ for each r. Then we may linearize the rth budget constraint around $x=x^r$ by taking a first-order Taylor Series expansion. The linearized rth budget constraint is $h_r(x^r)+[\nabla_x h_r(x^r)]^{\mathrm{T}}(x-x^r)=0$ or $[\nabla h_r(x^r)]^{\mathrm{T}}(x-x^r)=0$ since $h_r(x^r)=0$ using (11.3). It is easy to see that the utility surface $\{x: F(x)=F(x^r), x \geqslant 0_N\}$ is tangent not only to the original non-linear budget surface $\{x: h_r(x)=0, x \geqslant 0_N\}$ at $x=x^r$, but also to the linearized budget constraint surface $\{x: [\nabla h_r(x^r)]^{\mathrm{T}}(x-x^r)=0, x \geqslant 0_N\}$ at $x=x^r$. Since we assume F is quasi-concave, the set $\{x: F(x) \geqslant F(x^r), x \geqslant 0_N\}$ is convex and the linearized budget constraint is a supporting hyperplane to this set, i.e.

$$\max_x \{F(x): p^{r\mathrm{T}}x \leqslant p^{r\mathrm{T}}x^r, x \geqslant 0_N\} = F(x^r), \qquad r=1,\ldots,T, \tag{11.4}$$

where $p^r \equiv \nabla h_r(x^r)$ for $r=1,2,\ldots,T$. But now (11.4) is just a series of aggregator maximization problems of the type we have studied in Section 4 (the rth vector of normalized prices is defined as $v^r \equiv p^r/p^{r\mathrm{T}}x^r$) and the estimation techniques outlined in Section 9 above [recall eqs. (9.9) for example] can be used in order to estimate the parameters of the indirect utility function dual to F.

When we were dealing with linear budget constraints in Section 4, it was irrelevant whether F was quasi-concave or not (recall our discussion and diagram in Section 2). However, now we require the additional assumption that F be quasi-concave in order rigorously to justify the replacement of (11.3) by (11.4). Note also that in order to implement the above procedure, it is necessary to know the vector of derivatives $\nabla_x h_r(x^r)$ for each r; i.e. we have to know the derivatives of the supply functions that the consumer is "exploiting" in each period – information which was not required in Approach 1.

The monopsony model presented here is actually much broader than the classical model of monopsonistic exploitation: prices that the consumer faces can vary with the quantity purchased for a great many reasons, including search and transactions costs, quantity discounts, and the existence of progressive taxes on labor earnings. Most tax systems lead to budget constraints with "kinks" or non-differentiable points. This does not cause any problems with the above procedure unless the consumer's observed consumption–leisure choice falls precisely on a kink in his budget constraint.[42]

11.3. Approach 3: The monopoly problem revisited

Consider again the monopoly problem outlined above. Suppose $x_0^r > 0$, $x^r > 0_N$ is a solution to the period r monopoly profit maximization problem which can be rewritten as

$$\max_{x_0, x} \{ w^r D(x_0) x_0 - p^{rT} x : x_0 = F(x), x \geq 0_N \}$$

$$= w^r D(x_0^r) x_0^r - p^{rT} x^r, \qquad r = 1, 2, \ldots, T, \tag{11.5}$$

where $p_0^r \equiv w^r D(x_0^r) > 0$ is the observed selling price of the output during period $r, w^r D(x_0)$ is the period r inverse demand function, and $p^r \gg 0_N$ is the period r input price vector. If the production function F is continuous and concave (so that the production possibility set $\{(x_0, x): x_0 \leq F(x), x \geq 0_N\}$ is closed and convex) and if the inverse demand function D is differentiable at x_0^r for $r = 1, \ldots, T$, then the objective function for the rth maximization problem in (11.5) can be linearized around (x_0^r, x^r) and this linearized objective function will be tangent to

[42] See Wales (1973) and Wales and Woodland (1976, 1977, 1979) for econometric treatments of this last problem.

the production surface $x_0=F(x)$ at (x_0^r, x^r). Thus,

$$\max_{x_0, x}\{\tilde{p}_0^r x_0 - p^{rT}x: x_0=F(x), x\geq 0_N\}\equiv\Pi(\tilde{p}_0^r, p^r)$$
$$=\tilde{p}_0^r x_0^r - p^{rT}x^r, \qquad r=1,\dots,T, \tag{11.6}$$

where $\tilde{p}_0^r\equiv w^rD(x_0^r)+w^rD'(x_0^r)x_0^r=p_0^r+w^rD'(x_0^r)x_0^r>0$ is the period r *shadow* or *marginal* price of output ($\tilde{p}_0^r<p_0^r$ if $w^r>0$ and $D'(x_0^r)<0$) and Π is the firm's *true* profit function which is dual to the production function F [recall Π^* defined in Approach 1 was dual to the convex hull of $D[F(x)]F(x)\equiv F^*(x)$]. Thus, the true non-linear monopolistic profit maximization problems (11.6) which have the usual structure once the appropriate marginal output prices $\tilde{p}_0^r$ have been calculated so that the usual econometric techniques can be applied [recall eqs. (10.5) in Section 10].[43]

Comparing Approach 3 with Approach 1, it can be seen that Approach 3 requires the extra assumption that the production function be concave (convex technology) and requires additional information, i.e. a knowledge of the slope of the demand curve the monopolist is exploiting is required for each period.

It is easy to see how this approach can be generalized to a multi-product firm which simultaneously exploits several output and input markets: all that is required is the assumption of a convex technology and a (local) knowledge of the demand and supply curves that the firm is exploiting so that the appropriate shadow prices can be calculated.

Of course, the above techniques can also be used in situations where the firm is not behaving monopolistically or monopsonistically in an exploitative sense, but merely faces prices for its outputs or inputs that depend on the quantity sold or purchased for any number of reasons, including transactions costs or quantity discounts.

11.4. Approach 4: The monopoly problem once again

Suppose now that the production function satisfies Conditions I and, as usual, we suppose that $x_0^r>0$, $x^r>0_N$ is the solution to the period r monopolistic profit maximization problem (11.5), which we rewrite as

$$\max_{x_0}\{w^rD(x_0)x_0-C(x_0, p^r): x_0\geq 0\}=w^rD(x_0^r)x_0^r-p^{rT}x^r, \qquad r=1,\dots,T, \tag{11.7}$$

[43] The notation has been changed and we are now holding fixed inputs fixed for all r, so that we can suppress mention of these fixed inputs in (11.5).

where C is the cost function dual to F. If the inverse demand function D is differentiable at $x_0^r>0$ and $\partial C(x_0^r, p^r)/\partial x_0$ exists, then the first-order conditions for the rth maximization problem in (11.7) yield the condition $w^rD(x_0^r)+w^rD'(x_0^r)x_0^r-\partial C(x_0^r, p^r)/\partial x_0=0$ or, recalling that $p_0^r\equiv w^rD(x_0^r)$ is the observed selling price of output in period r,

$$p_0^r=-w^rD'(x_0^r)x_0^r+\frac{\partial C(x_0^r, p^r)}{\partial x_0}, \qquad r=1,\dots,T. \tag{11.8}$$

If the cost function C is differentiable with respect to input prices at (x_0^r, p^r) for each r, then Shephard's Lemma implies the additional equations

$$x^r=\nabla_p C(x_0^r, p^r), \qquad r=1,\dots,T. \tag{11.9}$$

Suppose that the part of the inverse demand function that depends on x_0, $D(x_0)$, can be adequately approximated over the relevant x_0 range by the following function:

$$D(x_0)\equiv\alpha-\beta\ln x_0, \tag{11.10}$$

where $\alpha>0$, $\beta\geq0$ are constants. Substitution of (11.10) into (11.8) yields the equations

$$p_0^r=w^r\beta+\frac{\partial C(x_0^r, p^r)}{\partial x_0}, \qquad r=1,\dots,T. \tag{11.11}$$

Given the observable price and quantity decisions of the firm, p_0^r, p^r, x_0^r, x^r, and data on w^r (we can assume $w^r\equiv1$ if this is appropriate), the system of equations (11.9) and (11.11) can be jointly econometrically estimated once we assume a differentiable functional form for the cost function $C(x_0, p)$. Note that if $\beta=0$ in eq. (11.11), then the producer is behaving competitively, selling output at a price p_0^r equal to marginal cost, $\partial C(x_0^r, p^r)/\partial x_0$. Eq. (11.11) is also consistent with the producer behaving like a "naive" markup monopolist (depending on what w^r is). Thus, we now have the basis for a statistical test of market structure: (i) if $\beta=0$, then the producer's behavior is consistent with competitive price taking behavior; (ii) if $\beta>0$ and $\beta w^r/p_0^r<1$ for $r=1,2,\dots,T$, then we have consistency with classical monopolistic behavior;[44] (iii) if $\beta>0$ but $\beta w^r/p_0^r\geq1$ for

[44]Using (11.10) and $p_0^r\equiv w^rD(x_0^r)$, we find that the first-order conditions (11.2) translate into $w^rD(x_0^r)[1+(D'(x_0^r)x_0^r/D(x_0^r))]=p_0^r[1-(\beta w^r/p_0^r)]=\partial C(x_0^r, p^r)/\partial x_0>0$ or $[1-(\beta w^r/p_0^r)]=(p_0^r)^{-1}\partial C/\partial x_0>0$ which implies $\beta w^r/p_0^r<1$. The second-order necessary conditions for (11.7) require $-\beta\leq(x_0^r/w^r)\partial^2C(x_0^r, p^r)/\partial x_0^2$ which will be satisfied if $\beta\geq0$ and $\partial^2C(x_0^r, p^r)/\partial x_0^2\geq0$ (non-decreasing marginal costs or non-increasing returns to scale).

some r, then we have consistency with markup monopolistic behavior; and (iv) if $\beta<0$, then we have inconsistency with all three of the above types of behavior.[45]

This approach offers several advantages over the previous approaches: (i) we can now statistically test for competitive behavior; (ii) informational requirements are low – we do not require exogenous information on the elasticity of demand (this information is endogenously generated); (iii) we do not have to assume that the production function F is concave so that the model is consistent with an increasing returns to scale production function; and finally (iv) the procedure is particularly simple – just insert the term βw^r into the competitive equation, price equals marginal cost.

11.5. Historical notes

Approach 1 is essentially due to Lau (1974a, pp. 193–194; 1978)[46] but it has its roots in Hotelling (1932, p. 609). Approach 2 is in Diewert (1971b) but it has its roots in the work of Frisch (1936, pp. 14–14). Approach 3 (which is isomorphic to Approach 2) is outlined in Diewert (1974a, p. 155). Approach 4 is due to Appelbaum (1975), who makes somewhat different assumptions on the functional form of the inverse demand function.[47] Appelbaum (1975, 1979) also indicates how his approach can be extended to several monopolistically supplied outputs or monopsonistically demand inputs, and he presents an empirical example based on the U.S. crude petroleum and natural gas industry. Another empirical example of his technique based on Canada–U.S. trade is in Appelbaum and Kohli (1979). Approach 4 has also been applied by Schworm (1980) in the context of investment theory where the price of investment goods purchased by a firm depends on the quantity purchased.

12. Conclusion

We have attempted to give a fairly comprehensive treatment of the foundations of the duality approach to microeconomic theory in Sections 2–6 of this chapter. In Sections 7 and 8 we showed how duality theory could be used in order to derive the usual comparative statics theorems for producer and consumer theory, while

[45]Of course, these tests are conditional on the assumed functional form for C, the assumed functional form for the inverse demand function $w_r D(x_0)$, where D is defined by (11.10), and the assumption of price taking behavior on input markets.

[46]Lau uses a normalized profit function and does not assume that $p_0 = wD(x_0)$, but simply assumes that $p_0 = D(x_0)$.

[47]He models more explicitly the demand function that the producer is exploiting.

in Sections 9 and 10 we showed how duality theory has been used as an aid in the econometric estimation of preferences and technology. Finally, in Section 11 we indicated how duality theory could be applied in certain non-competitive situations.

Unfortunately, the number of papers using duality theory during the last decade is so large that we are unable to review (or even reference) them. Additional topics and references can be found in my earlier survey paper [Diewert (1974a)] and the comments on it [Jacobson (1974), Lau (1974a) and Shephard (1974)] as well as in Fuss and McFadden (1978) which provides a comprehensive treatment of the duality approach to production theory.

We have mentioned the aggregation over consumers problem in Section 9 above but we have not mentioned the corresponding aggregation over producers problem: for results and references to this literature, see Hotelling (1935, pp. 67–70), Gorman (1968), Sato (1975) and Diewert (1980, Part III).

Another area which we have been unable to cover is the economic theory of index numbers: see Konyus (1924), Konyus and Byushgens (1926), Lerner (1935–36), Joseph (1935–36), Frisch (1936), Leontief (1936), Samuelson (1947, 1950a, 1974), Allen (1949), Malmquist (1953), Wold (1953), Kloek (1967), Theil (1967, 1968), Pollak (1971), Afriat (1972b), Samuelson and Swamy (1974), Vartia (1974, 1976a, 1976b), Sato (1976a, 1976b), Diewert (1976a, 1976b, 1978b), and Lau (1979).

Although we have used duality theory to derive several partial equilibrium comparative statics theorems, we have not mentioned the corresponding general equilibrium literature: see Jones (1965, 1972), Diewert (1974e, 1974f, 1978a), Epstein (1974), Woodland (1974), and Burgess (1976b) for various applications. The related literature on optimal taxation often makes use of duality theory: for references to this literature, see Mirrlees (1981) and Deaton (1979).

Finally, some recent references that utilize duality theory in the context of continuous time optimization problems are Lau (1974a, pp. 190–193), Appelbaum (1975), Cooper and McLaren (1977), Epstein (1978, 1981b), McLaren and Cooper (1980), and Schworm (1980).

References

Afriat, S. N. (1972a), "The case of the vanishing Slutsky matrix", Journal of Economic Theory, 5:208–223.

Afriat, S. N. (1972b), "The theory of international comparisons of real income and prices", pp. 13–69 in: D. J. Daly, ed., International comparisons of prices and output. New York: National Bureau of Economic Research.

Afriat, S. N. (1973a), "The maximum hypothesis in demand analysis", Discussion Paper #79. Canada: Department of Economics, University of Waterloo.
Afriat, S. N. (1973b), "On integrability conditions for demand functions", Discussion Paper #82. Canada: Department of Economics, University of Waterloo.
Afriat, S. N. (1973c), "Direct and indirect utility", Discussion Paper #83. Canada: Department of Economics, University of Waterloo.
Allen, R. G. D. (1938), Mathematical analysis for economists. London: Macmillan.
Allen, R. G. D. (1949), "The economic theory of index numbers", Economica N.S., 16:197–203.
Antonelli, G. B. (1971), "On the mathematical theory of political economy", pp. 333–364 in: J. S. Chipman, L. Hurwicz, M. K. Richter and H. F. Sonnenschein, eds., Preferences, utility and demand. New York: Harcourt Brace Jovanovich.
Appelbaum, E. (1975), Essays in the theory and application of duality in economics, Ph.D. thesis. Vancouver: University of British Columbia.
Appelbaum, E. (1979), "Testing price taking behavior", Journal of Econometrics, 9:283–294.
Appelbaum, E. and Ui R. Kohli (1979), "Canada–United States Trade: Test for the small-open-economy hypothesis", Canadian Journal of Economics, 12:1–14.
Arrow, K. J. and F. H. Hahn (1971), General competitive analysis. San Francisco: Holden-Day, Inc.
Berge, C. (1963), Topological spaces. New York: Macmillan.
Berndt, E. R. and L. R. Christensen (1973a), "The internal structure of functional relationships: Separability, substitution, and aggregation", Review of Economic Studies, 40:403–410.
Berndt, E. R. and L. R. Christensen (1973b), "The translog function and the substitution of equipment, structures, and labor in U.S. manufacturing, 1929–1968", Journal of Econometrics, 1:81–114.
Berndt, E. R. and L. R. Christensen (1974), "Testing for the existence of a consistent aggregate index of labor inputs", American Economic Review, 64:391–404.
Berndt, E. R., M. N. Darrough, and W. E. Diewert (1977), "Flexible functional forms and expenditure distributions: An application to Canadian consumer demand functions", International Economic Review, 18:651–676.
Berndt, E. R. and D. O. Wood (1975), "Technology, prices and the derived demand for energy", The Review of Economics and Statistics, 57:259–268.
Binswanger, H. P. (1974), "The measurement of technical change biases with many factors of production", American Economic Review, 64:964–976.
Blackorby, C., R. Boyce, and R. R. Russell (1978), "Estimation of demand system generated by the Gorman polar form: A generalization of the S-branch utility tree", Econometrica, 46:345–363.
Blackorby, C. and W. E. Diewert (1979), "Expenditure functions, local duality and second order approximations", Econometrica, 47:579–601.
Blackorby, C., D. Primont, and R. R. Russell (1975a), "Budgeting, decentralization, and aggregation", Annals of Economic and Social Measurement, 4:23–44.
Blackorby, C., D. Primont, and R. R. Russell (1975b), "Some simple remarks on duality and the structure of utility functions", Journal of Economic Theory, 11:155–160.
Blackorby, C., D. Primont, and R. R. Russell (1977a), "Dual price and quantity aggregation", Journal of Economic Theory, 14:130–148.
Blackorby, C., D. Primont, and R. R. Russell (1977b), "An extension and alternative proof of Gorman's price aggregation theorem", in: W. Eichhorn, ed., Theory and Applications of Economic Indices. Würstburg-Wien: Physica.
Blackorby, C., D. Primont, and R. R. Russell (1977c), "On testing separability restrictions with flexible functional forms", Journal of Econometrics, 5:195–209.
Blackorby, C., D. Primont, and R. R. Russell (1977d), "Separability vs. functional structure: A characterization of their differences", Journal of Economic Theory, 15:135–144.
Blackorby, C., D. Primont, and R. R. Russell (1978), Duality, separability and functional structure: Theory and economic applications. New York: American Elsevier.
Blackorby, C. and R. R. Russell (1975), "Conjugate implicit separability", Discussion Paper 75-12, Review of Economic Studies, forthcoming.
Blackorby, C. and R. R. Russell (1976), "Functional structure and the Allen partial elasticities of substitution: An application of duality theory", Review of Economic Studies, 43:285–292.
Bonnesen, T. and W. Fenchel (1934), Theorie der Konvexen Körper. Berlin: Springer.

Brown, R. S., D. W. Caves, and L. R. Christensen (1979), "Modelling the structure of cost and production for multiproduct firms", Southern Economic Journal, 46:256–270.
Boyce, R. (1975), "Estimation of dynamic Gorman polar form utility functions", Annals of Economic and Social Measurement, 4:103–116.
Boyce, R. and D. Primont (1976), "An econometric test of the representative consumer hypothesis", Discussion Paper #76-31, Department of Economics, University of British Columbia.
Bruno, M. (1978), "The theory of protection, tariff change and the value added function", in: Fuss and McFadden, Eds. (1978).
Burgess, D. F. (1974), "Production theory and the derived demand for imports", Journal of International Economics, 4:103–117.
Burgess, D. F. (1975), "Duality theory and pitfalls in the specification of technologies", Journal of Econometrics, 3:105–121.
Burgess, D. F. (1976a), "Tariffs and income distribution: Some empirical evidence for the United States", Journal of Political Economy, 84:17–45.
Burgess, D. F. (1976b), "The income distributional effects of processing incentives: A general equilibrium analysis", Canadian Journal of Economics, 9:595–612.
Byron, R. P. (1977), "Some Monte Carlo experiments using the translog approximation", mimeo. Washington, D. C.: Development Research Center, World Bank.
Chipman, J. S. (1966), "A survey of the theory of international trade: Part 3, The modern theory", Econometrica, 34:18–76.
Chipman, J. S. (1970), "Lectures on the mathematical foundations of international trade theory: I Duality of cost functions and production functions", mimeo. Vienna: Institute of Advanced Studies.
Chipman, J. S. (1972), "The theory of exploitative trade and investment policies: A reformulation and synthesis", in: L. E. Di Marco ed., International Economics and Development: Essays in Honor of Raul Prebisch. New York: Academic Press.
Chipman, J. S. (1974a), "Homothetic preferences and aggregation", The Journal of Economic Theory, 8:26–38.
Chipman, J. S. (1974b), "The transfer problem once again", pp. 19–78 in: Trade, Stability and Macroeconomics: Essays in Honor of Lloyd A. Metzler. New York: Academic Press.
Christensen, L. R. and W. H. Greene (1976), "Economies of scale in U.S. electric power generation", Journal of Political Economy, 84:655–676.
Christensen, L. R., D. W. Jorgenson, and L. J. Lau (1971), "Conjugate duality and the transcendental logarithmic production function", Econometrica, 39:255–256.
Christensen, L. R., D. W. Jorgenson, and L. J. Lau (1975), "Transcendental logarithmic utility functions", American Economic Review, 65:367–383.
Christensen, L. R. and M. E. Manser (1977), "Estimating U.S. consumer preferences for meat with a flexible utility function", Journal of Econometrics, 6:37–53.
Cooper, R. J. and K. R. McLaren (1977), "Intertemporal duality: Application to consumer theory", Working Paper #10, Department of Econometrics and Operations Research, Monash University, Clayton, Australia.
Darrough, M. N. (1977), "A model of consumption and leisure in an intertemporal framework: A systematic treatment using Japanese data", International Economic Review, 18:677–696.
Darrough, M. N. and C. Southey (1977), "Duality in consumer theory made simple: The revealing of Roy's identity", The Canadian Journal of Economics, 10:307–317.
Deaton, A. S. (1979), "The distance function in consumer behavior with applications to index numbers and optimal taxation", The Review of Economic Studies, 46:391–405.
Debreu, G. (1952), "A social equilibrium existence theorem", Proceedings of the National Academy of Sciences, 38:886–893.
Debreu, G. (1959), Theory of Value. New York: John Wiley and Sons.
Denny, M. (1972), Trade and the production sector: An exploration of models of multiproduct technologies. Ph.D. Thesis, University of California at Berkeley.
Denny, M. (1974), "The relationship between functional forms for the production system", Canadian Journal of Economics, 7:21–31.
Denny, M. and M. Fuss (1977), "The use of approximation analysis to test for separability and the existence of consistent aggregates", American Economic Review, 67:404–418.
Diamond, P. A. and D. L. McFadden (1974), "Some uses of the expenditure function in public

finance", Journal of Public Economics, 3:3–22.
Diewert, W. E. (1971a), "An application of the Shephard duality theorem: A generalized Leontief production function", Journal of Political Economy, 79:481–507.
Diewert, W. E. (1971b), "Choice on labor markets and the theory of the allocation of time", Research Branch, Department of Manpower and Immigration, Ottawa.
Diewert, W. E. (1973), "Functional forms for profit and transformation functions", Journal of Economic Theory, 6:284–316.
Diewert, W. E. (1974a), "Applications of duality theory", in: M. D. Intriligator and D. A. Kendrick, eds., Frontiers of Quantitative Economics, Vol. II. Amsterdam: North-Holland Publishing Company.
Diewert, W. E. (1974b), "Functional forms for revenue and factor requirements functions", International Economic Review, 15:119–130.
Diewert, W. E. (1974c), "A note on aggregation and elasticities of substitution", Canadian Journal of Economics, 7:12–20.
Diewert, W. E. (1974d), "Intertemporal consumer theory and the demand for durables", Econometrica, 42:497–516.
Diewert, W. E. (1974e), "The effects of unionization of wages and employment: A general equilibrium analysis", Journal of Economic Inquiry, 12:319–339.
Diewert, W. E. (1974f), "Unions in a general equilibrium model", Canadian Journal of Economics, 7:475–495.
Diewert, W. E. (1975), "The Samuelson nonsubstitution theorem and the computation of equilibrium prices", Econometrica, 43:57–64.
Diewert, W. E. (1976a), "Exact and superlative index numbers", Journal of Econometrics, 4:115–145.
Diewert, W. E. (1976b), "Harberger's welfare indicator and revealed preference theory", The American Economic Review, 66:143–152.
Diewert, W. E. (1977a), "Generalized Slutsky conditions for aggregate consumer demand functions", Journal of Economic Theory, 15:353–362.
Diewert, W. E. (1978a), "Optimal tax perturbations", Journal of Public Economics, 10:139–177.
Diewert, W. E. (1978b), "Superlative index numbers and consistency in aggregation", Econometrica, 46:863–900.
Diewert, W. E. (1978c), "Hicks' aggregation theorem and the existence of a real value added function", in: Fuss and McFadden, eds. (1978).
Diewert, W. E. (1980), "Aggregation problems in the measurement of capital", in: D. Usher, ed., The Measurement of Capital. Chicago: University of Chicago Press.
Diewert, W. E. (1981), "Symmetry conditions for market demand functions", The Review of Economic Studies, 47:595–601.
Diewert, W. E. (1982), Duality theory in economics. Amsterdam: North-Holland (forthcoming).
Diewert, W. E. and A. D. Woodland (1977), "Frank Knight's theorem in linear programming revisited", Econometrica, 45:375–398.
Donovan, D. J. (1977), Consumption, leisure, and the demand for money and money substitutes. Ph.D. thesis, University of British Columbia, Vancouver, Canada.
Epstein, L. (1974), "Some economic effects of immigration: A general equilibrium analysis", Canadian Journal of Economics, 7:174–190.
Epstein, L. (1975a), "A disaggregate analysis of consumer choice under uncertainty", Econometrica, 43:877–892.
Epstein, L. (1975b), "Duality and consumer choice under uncertainty", Department of Economics, University of British Columbia, April.
Epstein, L. (1977), Essays in the economics of uncertainty. Ph.D. Thesis, University of British Columbia, Vancouver, Canada.
Epstein, L. (1978), "The Le Chatelier principle in optimal control problems", Journal of Economic Theory, 19:103–122.
Epstein, L. G. (1981a), "Generalized duality and integrability", Econometrica, 49:655–678.
Epstein, L. G. (1981b), "Duality theory and functional forms for dynamic factor demand", Review of Economic Studies, 48:81–95.
Ethier, W. (1974), "Some of the theorems of international trade with many goods and factors", Journal of International Economics, 4:199–206.

Fenchel, W. (1953), "Convex cones, sets and functions", Lecture notes, Department of Mathematics, Princeton University.
Friedman, J. W. (1972), "Duality principles in the theory of cost and production revisited", International Economic Review, 13:167–170.
Frisch, R. (1936), "Annual survey of general economic theory: The problem of index numbers", Econometrica, 4:1–39.
Fuss, M. and D. McFadden (1978) (eds.), Production economics: A dual approach to theory and applications. Amsterdam: North-Holland.
Geary, P. T. and M. Morishima (1973), "Demand and supply under separability", in: M. Morishima et al., Theory of demand: Real and monetary. Oxford: Clarendon.
Goldman, S. M. and H. Uzawa (1964), "A note on separability in demand analysis", Econometrica, 32:387–398.
Gorman, W. M. (1953), "Community preference fields", Econometrica, 21:63–80.
Gorman, W. M. (1959), "Separable utility and aggregation", Econometrica, 27:469–481.
Gorman, W. M. (1968), "Measuring the quantities of fixed factors", in: J. N. Wolfe, ed., Value, capital and growth: Papers in honor of Sir John Hicks. Chicago: Aldine Publishing Co.
Gorman, W. M. (1976), "Tricks with utility functions", in: M. Artis and R. Nobay, eds., Essays in economic analysis. Cambridge: Cambridge University Press.
Griliches, Z. and V. Ringstad (1971), Economies of scale and the form of the production function. Amsterdam: North-Holland.
Hall, R. E. (1973), "The specification of technology with several kinds of output", Journal of Political Economy, 81:878–892.
Hanoch, G. (1975), "Production or demand models with direct or indirect implicit additivity", Econometrica, 43:395–420.
Hanoch, G. (1978), "Generation of new production functions through duality", in: Fuss and McFadden, eds. (1978).
Harris, R. and E. Appelbaum (1977), "Estimating technology in an intertemporal framework: A neo-Austrian approach", The Review of Economics and Statistics, 59:161–170.
Heller, W. and J. Green (1980), "Analysis, convexity and topology, with applications to economics", in: K. J. Arrow and M. D. Intriligator, eds., Handbook of mathematical economics. Amsterdam: North-Holland.
Hicks, J. R. (1946), Value and capital. Oxford: Clarendon Press.
Hotelling, H. (1932), "Edgeworth's taxation paradox and the nature of demand and supply functions", Journal of Political Economy, 40:577–616.
Hotelling, H. (1935), "Demand functions with limited budgets", Econometrica, 3:66–78.
Houthakker, H. S. (1951–52), "Compensated changes in quantities and qualities consumed", Review of Economic Studies, 19:155–164.
Howe, H., R. A. Pollak and T. J. Wales (1979), "Theory and time series estimation of the quadratic expenditure system", Econometrica, 47:1231–1247.
Hudson, E. A. and D. W. Jorgenson (1974), "U.S. energy policy and economic growth, 1975–2000", The Bell Journal of Economics and Management Science, 5:461–514.
Hurwicz, L. (1971), "On the problem of integrability of demand functions", pp. 174–214 in: J. S. Chipman, L. Hurwicz, M. K. Richter and H. F. Sonnenschein, eds., Preferences, utility and demand. New York: Harcourt Brace Jovanovich.
Hurwicz, L. and H. Uzawa (1971), "On the integrability of demand functions", pp. 114–148 in: J. S. Chipman, L. Hurwicz, M. K. Richter and H. F. Sonnenchein, eds., Preferences, utility and demand. New York: Harcourt Brace Jovanovich.
Intriligator, M. D. (1980), "Mathematical programming with applications to economics", in: K. J. Arrow and M. D. Intriligator, eds., Handbook of mathematical economics. Amsterdam: North-Holland.
Jacobsen, S. E. (1970), "Production correspondences", Econometrica, 38:754–771.
Jacobsen, S. E. (1972), "On Shephard's duality theory", Journal of Economic Theory, 4:458–464.
Jacobsen, S. E. (1974), "Applications of duality theory: Comments", pp. 171–176 in: M. D. Intriligator and D. A. Kendrick, eds., Frontiers of quantitative economics, Vol. II. Amsterdam: North-Holland.
Jones, R. W. (1965), "The structure of simple general equilibrium models", Journal of Political

Economy, 73:557–572.
Jones, R. W. (1972), "Activity analysis and real incomes: Analogies with production models", Journal of International Economics, 2:277–302.
Jones, R. W. and J. A. Scheinkman (1977), "The relevance of the two-sector production model in trade theory", Journal of Political Economy, 85:909–936.
Jorgenson, D. W. and L. J. Lau (1970), "The 'transcendental logarithmic' utility function and demand analysis", mimeo.
Jorgenson, D. W. and L. J. Lau (1974a), "The duality of technology and economic behavior", The Review of Economic Studies, 41:181–200.
Jorgenson, D. W. and L. J. Lau (1974b), "Duality and differentiability in production", Journal of Economic Theory, 9:23–42.
Jorgenson, D. W. and L. J. Lau (1975), "The structure of consumer preferences", Annals of Economic and Social Measurement, 4:49–101.
Joseph, M. F. W. (1935–6), "Mr. Lerner's supplementary limits for price index numbers", Review of Economic Studies, 3:155–157
Karlin, S. (1959), Mathematical methods and theory in games, programming and economics, Vol. I. Palo Alto, California: Addison-Wesley.
Katzner, D. W. (1970), Static demand theory. New York: Macmillan.
Khaled, M. S. (1978), Productivity analysis and functional specification: A parametric approach. Ph.D. thesis, University of British Columbia, Vancouver, Canada.
Khang, C. (1971), "An isovalue locus involving intermediate goods and its applications to the pure theory of international trade", Journal of International Economics, 1:315–325.
Kloek, T. (1967), "On quadratic approximations of cost of living and real income index numbers", Report #6710, Econometric Institute, Netherlands School of Economics, Rotterdam.
Kohli, U. J. R. (1978), "A gross national product function and the derived demand for imports and supply of exports", Canadian Journal of Economics, 11:167–182.
Konüs, A. A. (1924), "The problem of the true index of the cost of living", translated in Econometrica, 7:1939, 10–29.
Konyus, A. A. and S. S. Byushgens (1926), "K Probleme Pokupatelnoi Cili Deneg", Voprosi Konyunkturi, II: 151–172. (English title: A. A. Conus and S. S. Buscheguennce, "On the problem of the purchasing power of money", The Problems of Economic Conditions (supplement to the Economic Bulletin of the Conjuncture Institute), 2:151–172.)
Lau, L. J. (1969), "Duality and the structure of utility functions", Journal of Economic Theory, 1:374–396.
Lau, L. J. (1974a), "Applications of duality theory: Comments", pp. 176–199 in: M. D. Intriligator and D. A. Kendrick, eds., Frontiers of Quantitative Economics, Vol. II. Amsterdam: North-Holland.
Lau, L. J. (1974b), "Econometrics of monotonicity, convexity and quasi-convexity", Technical Report No. 123, Institute for Mathematical Studies in the Social Sciences (IMSSS), Stanford University, California, forthcoming in Econometrica.
Lau, L. J. (1975), "Complete systems of consumer demand functions through duality", Department of Economics, Stanford University, California.
Lau, L. J. (1976), "A characterization of the normalized restricted profit function", Journal of Economic Theory, 12:131–163.
Lau, L. J. (1977a), "Existence conditions for aggregate demand functions: The case of a single index", Technical Report #248, IMSSS, Stanford University, October.
Lau, L. J. (1977b), "Existence conditions for aggregate demand functions: The case of multiple indexes", Technical Report #249, IMSSS, Stanford University, October.
Lau, L. J. (1978), "Some applications of profit functions", in: Fuss and McFadden, eds. (1978).
Lau, L. J. (1979), "A note on exact index numbers", Review of Economics and Statistics, 61:73–82.
Lau, L. J. and B. M. Mitchell (1970), "A linear logarithmic expenditure system: An application to U.S. data", paper presented at the Second World Congress of the Econometric Society, Cambridge, England, September.
Leontief, W. (1936), "Composite commodities and the problem of index numbers", Econometrica, 4:39–59.

Lerner, A. P. (1935–6), "A note on the theory of price index numbers", Review of Economic Studies, 3:50–56.
Malmquist, S. (1953), "Index numbers and indifference surfaces", Trabajos de Estatistica, 4:209–242.
McFadden, D. (1962), Factor substitutability in the economic analysis of production. Ph.D. Thesis, University of Minnesota.
McFadden, D. (1966), "Cost, revenue and profit functions: A cursory review", Working Paper #86, IBER, University of California at Berkeley, March.
McFadden, D. (1978a), "Cost, revenue and profit functions", in: Fuss and McFadden, eds. (1978).
McFadden, D. (1978b), "The general linear profit function", in: Fuss and McFadden, eds. (1978).
McKenzie, L. W. (1956–7), "Demand theory without a utility index", Review of Economic Studies, 24:185–189.
McLaren, K. R. and R. J. Cooper (1980), "Intertemporal duality: Application to the theory of the firm", Econometrica, 48:1755–1776.
Minkowski, H. (1911), Theorie der konvexen Körper, Gesammelte Abhandlungen II. Leipzig und Berlin: B. G. Teubner.
Mirrlees, J. A. (1981), "The theory of optimal taxation", in: K. J. Arrow and M. D. Intriligator, eds., Handbook of mathematical economics. Amsterdam: North-Holland.
Muellbauer, J. (1975), "Aggregation, income distribution and consumer demand", The Review of Economic Studies, 42:525–543.
Muellbauer, J. (1976), "Community preferences and the representative consumer", Econometrica, 44: 979–999.
Newman, P. (1965), The theory of exchange. Englewood Cliffs: Prentice-Hall.
Parks, R. W. (1971), "Price responsiveness of factor utilization in Swedish manufacturing 1870–1950", Review of Economics and Statistics, 53:129–139.
Pearce, I. F. (1961), "An exact method of consumer demand analysis", Econometrica, 29:499–516.
Pollak, R. A. (1969), "Conditional demand functions and consumption theory", Quarterly Journal of Economics, 83:60–78.
Pollak, R. A. (1971), "The theory of the cost of living index", Research Discussion Paper #11, Office of Prices and Living Conditions, U.S. Bureau of Labor, Statistics, Washington D.C.
Pollak, R. A. (1972), "Generalized separability", Econometrica, 40:431–453.
Rockafellar, R. T. (1970), Convex Analysis. Princeton: Princeton University Press.
Roy, R. (1942), De l'utilite. Paris: Hermann.
Roy, R. (1947), "La distribution du revenu entre les divers biens", Econometrica, 15:205–225.
Rudin, W. (1953), Principles of mathematical analysis. New York: McGraw-Hill.
Russell, R. R. (1975), "Functional separability and partial elasticities of substitution", Review of Economic Studies, 42:79–86.
Russell, R. R. and R. Boyce (1974), "A multilateral model of international trade flows: A theoretical framework and specification of functional forms", Institute for Policy Analysis, La Jolla, California.
Sakai, Y. (1973), "An axiomatic approach to input demand theory", International Economic Review, 14:735–752.
Sakai, Y. (1974), "Substitution and expansion effects in production theory: The case of joint production", Journal of Economic Theory, 9:255–274.
Samuelson, P. A. (1938), "The empirical implications of utility analysis", Econometrica, 6:344–356.
Samuelson, P. A. (1947), Foundations of economic analysis. Cambridge, Mass.: Harvard University Press.
Samuelson, P. A. (1950a), "Evaluation of real national income", Oxford Economic Papers, 2:1–29.
Samuelson, P. A. (1950b), "The problem of integrability in utility theory", Economica, 17:355–385.
Samuelson, P. A. (1953–4), "Prices of factors and goods in general equilibrium", Review of Economic Studies, 21:1–20.
Samuelson, P. A. (1960), "Structure of a minimum equilibrium system", in: Ralph W. Pfouts, ed., Essays in Economics and Econometrics: A volume in Honor of Harold Hotelling. University of North Carolina Press.
Samuelson, P. A. (1965), "Using full duality to show that simultaneously additive direct and indirect utilities implies unitary price elasticity of demand", Econometrica, 33:781–796.

Samuelson, P. A. (1966), "The fundamental singularity theorem for non-joint production", International Economic Review, 7:34–41.
Samuelson, P. A. (1969), "Corrected formulation of direct and indirect additivity", Econometrica, 37:355–359.
Samuelson, P. A. (1972), "Unification theorem for the two basic dualities of homothetic demand theory", Proceedings of the National Academy of Sciences, U.S.A., 69:2673–2674.
Samuelson, P. A. (1974), "Remembrances of frisch", European Economic Review, 5:7–23.
Samuelson, P. A. and S. Swamy (1974), "Invariant economic index numbers and canonical duality: Survey and synthesis", American Economic Review, 64:566–593.
Sargan, J. D. (1971), "Production functions", Part V of P. R. G. Layard, J. D. Sargan, M. E. Ager, and D. J. Jones, Qualified Manpower and Economic Performance. London: Penguin Press.
Sato, K. (1975), Production functions and aggregation. Amsterdam: North-Holland.
Sato, K. (1976a), "The ideal log-change index number", Review of Economics and Statistics, 58:223–228.
Sato, K. (1976b), "the meaning and measurement of the real value added index", The Review of Economics and Statistics, 58:434–442.
Schworm, W. E. (1980), "Financial constraints and capital theory", International Economic Review, 21:643–660.
Shafer, W. and H. Sonnenschein (1980), "Aggregate demand as a function of prices and income", in: K. J. Arrow and M. D. Intriligator, eds., Handbook of mathematical economics. Amsterdam: North-Holland.
Shephard, R. W. (1953), Cost and production functions. Princeton: Princeton University Press.
Shephard, R. W. (1970), Theory of cost and production functions. Princeton: Princeton University Press.
Shephard, R. W. (1974), "Applications of duality theory: Comments", pp. 200–206 in: M. D. Intriligator and D. A. Kendrick eds., Frontiers of Quantitative Economics, Vol. II. Amsterdam: North-Holland.
Slutsky, E. (1915), "Sulla Teoria Del Bilancio Del Consumatore", Giornale Degli Economisti, 51:1–26.
Sono, M. (1945), "The effect of price changes on the demand and supply of separable goods", translation in International Economic Review, 2:239–275 (1961).
Theil, H. (1967), Economics and information theory. Amsterdam: North-Holland.
Theil, H. (1968), "On the geometry and the numerical approximation of cost of living and real income indices", De Economist, 116:677–689.
Uzawa, H. (1962), "Production functions with constant elasticities of substitution", Review of Economic Studies, 29:291–299.
Uzawa, H. (1964), "Duality principles in the theory of cost and production", International Economic Review, 5:216–220.
Vartia, Y. (1974), Relative changes and economic indices. Licensiate Thesis in Statistics, University of Helsinki, June.
Vartia, Y. (1976a), "Ideal log-change index numbers", Scandanavian Journal of Statistics, 3:121–126.
Vartia, Y. (1976b), Relative changes and index numbers. Helsinki: The Research Institute of the Finnish Economy.
Ville, H. (1946), "Sur les conditions d'existence d'une Ophélimité Totale et d'un Indice du Niveau des Prix", Annales de l'Université de Lyon, 9:32–9 (English translation: "The existence conditions of a utility function", Review of Economic Studies, 19 (1951):123–128).
Wales, T. J. (1973), "Estimation of a labor supply curve for self-employed business proprietors", International Economic Review, 14:69–80.
Wales, T. J. (1977), "On the flexibility of flexible functional forms: An empirical approach", Journal of Econometrics, 5:183–193.
Wales, T. J. and A. D. Woodland (1976), "Estimation of household utility functions and labor supply response", International Economic Review, 17:397–410.
Wales, T. J. and A. D. Woodland (1977), "Estimation of the allocation of time for world, leisure and housework", Econometrica, 45:115–132.
Wales, T. J. and A. D. Woodland (1979), "Labor supply and progressive taxes", Review of Economic Studies, 46:83–95.

Walters, A. A. (1961), "Production and cost functions: An econometric survey", Econometrica, 31:1–66.
Weddepohl, H. N. (1970), Axiomatic choice models and duality. Groningen: Rotterdam University Press.
Weymark, J. A. (1980), "Duality results in demand theory", European Economic Review, 14:377–395.
Wold, H. (1943–44), "A synthesis of pure demand analysis", Skandinavisk Aktuarietidskrift, 26:85–144 and 220–275; 27:69–120.
Wold, H. (1953), Demand analysis. New York: John Wiley and Sons.
Woodland, A. D. (1974), "Demand conditions in international trade theory", Australian Economic Papers, 13:209–224.
Woodland, A. D. (1975), "Substitution of structures, equipment and labor in Canadian production", International Economic Review, 16:171–187
Woodland, A. D. (1977a), "A dual approach to equilibrium in the production sector in international trade theory", The Canadian Journal of Economics, 10:50–68.
Woodland, A. D. (1977b), "Joint outputs, intermediate inputs and international trade theory", International Economic Review, 18:517–534
Woodland, A. D. (1977c), "Estimation of a variable profit and planning price functions for Canadian manufacturing, 1947–1970", Canadian Journal of Economics, 10:355–377
Woodland, A. D. (1978), "On testing for weak separability", Journal of Econometrics, 8:383–398.

Chapter 13

ON THE MICROECONOMIC THEORY OF INVESTMENT UNDER UNCERTAINTY*

ROBERT C. MERTON

Massachusetts Institute of Technology

1. Introduction

Investment theory is the study of the individual behavior of households and economic organizations in the allocation of their resources to the available investment opportunities. For the purposes of investment theory, economic organizations are characterized as being members of one of two groups: "business firms" that hold as assets the physical means of production for the economy and finance their production decisions by issuing financial claims or securities; and "financial intermediaries" that hold financial claims as assets and finance these assets by issuing securities. Individuals or households are assumed to invest primarily in securities, and therefore invest only indirectly in physical assets. The markets in which these securities are traded are called the *capital markets*.

The natural starting point for the development of investment theory is to derive the investment behavior of individuals. It is traditional in economic theory to take the existence of households and their tastes as exogenous to the theory. However, this tradition does not extend to economic organizations and institutions. They are regarded as existing primarily because of the functions they serve instead of functioning primarily because they exist. Economic organizations are endogenous to the theory. To derive the functions of these economic organizations, therefore, the investment behavior of individuals must be derived first.

It is convenient to break the investment decision by individuals into two parts: (1) the "*consumption-saving*" *choice* where the individual decides how much of his

*This paper is not a survey of the economics and finance literature on investment theory which is already copious and whose rate of expansion has in recent years accelerated. Rather, the paper is designed to be a self-contained, but cryptic, introduction to some of the major problems in investment theory. In many cases, I have referenced excellent survey articles which themselves contain extensive bibliographies rather than attempt to reference all the important individual contributions to the subject. I gratefully acknowledge financial support from the National Science Foundation for this paper, and thank J. Cox and M. Latham for their helpful comments.

Handbook of Mathematical Economics, vol. II, edited by K.J. Arrow and M.D. Intriligator

wealth to allocate to current consumption and how much to invest for future consumption; and (2) the "*portfolio selection*" *choice* where he decides how to allocate his savings among the available investment opportunities. In general, the two decisions cannot be made independently. However, the format of the paper is to, first, solve the portfolio selection problem taking individual's consumption decisions and firms' production decisions as given, and to derive necessary conditions for financial equilibrium. Second, using these necessary conditions, the optimal production decision rules for firms are derived. Finally, the combined consumption and portfolio selection problem for individuals is solved.

It is, of course, not possible in a single paper to cover all the topics important to investment theory. However, two topics not covered here warrant special mention. First, while some necessary conditions for equilibrium are derived, I have not attempted to integrate these conditions into a general equilibrium theory for the economy. Second, no attempt has been made to make explicit how individuals and firms acquire the information needed to make their decisions, and in particular how they modify their behavior in environments where there are significant differences in the information available to various participants.[1]

2. One-period portfolio selection

The basic investment choice problem for an individual is to determine the optimal allocation of his wealth among the available investment opportunities. The solution to the general problem of choosing the best investment mix is called *portfolio selection theory*. I begin the study of portfolio selection theory with its classic one-period formulation.

There are n different investment opportunities called *securities* and the random variable one-period return per dollar on security j is denoted by Z_j $(j=1,\dots,n)$ where a "dollar" is the "unit of account". Any linear combination of these securities which has a positive market value is called a *portfolio*. It is assumed that the investor chooses at the beginning of a period that feasible portfolio allocation which maximizes the expected value of a von Neumann–Morgenstern utility function[2] for end-of-period wealth. I denote this utility function by $U(W)$, where

[1]An important issue within this topic is the "information efficiency" of the stock market as derived in Fama (1965, 1970a) and Samuelson (1965, 1973). Grossman (1976) and Grossman and Stiglitz (1976) provide some additional insights. Hirshleifer (1973) has shown that, in the absence of production, private sector expenditures on information gathering may be "socially wasteful". Moreover, significant further advances in general equilibrium theory will require explicit recognition of information production and its dissemination, and the development of such an integrated theory is just at its beginning.

[2]von Neumann and Morgenstern (1947). For an axiomatic description, see Herstein and Milnor (1953). Although the original axioms require that U be bounded, the continuity axiom can be extended to allow for unbounded functions. See Samuelson (1977) for a discussion of this and the St. Petersburg Paradox.

W is the end-of-period value of the investor's wealth measured in dollars. It is further assumed that U is an increasing strictly concave function on the range of feasible values for W and that U is twice-continuously differentiable.[3] Because the criterion function for choice depends only on the distribution of end-of-period wealth, the only information about the securities that is relevant to the investor's decision is his subjective joint probability distribution for $(Z_1, \ldots, Z_n)$.

In addition, it is assumed that:

Assumption 1 ("Frictionless" markets)

There are no transactions costs or taxes, and all securities are perfectly divisible.

Assumption 2 ("Price taker")

The investor believes that his actions cannot affect the probability distribution of returns on the available securities. Hence, if w_j is the fraction of the investor's initial wealth, W_0, allocated to security j, then $\{w_1, \ldots, w_n\}$ uniquely determines the probability distribution of his terminal wealth.

A *riskless security* is defined to be a security or feasible portfolio of securities whose return per dollar over the period is known with certainty.

Assumption 3 ("No-arbitrage opportunities")

All riskless securities must have the same return per dollar. This common return will be denoted by R.

Assumption 4 ("No-institutional" restrictions)

Short-sales of all securities, with full use of proceeds, is allowed without restriction. If there exists a riskless security, then the borrowing rate equals the lending rate.[4]

Hence, the only restriction on the choice for the $\{w_j\}$ is the budget constraint that $\sum_1^n w_j = 1$.

Given these assumptions, the portfolio selection problem can be formally stated as

$$\max_{\{w_1, \ldots, w_n\}} \mathrm{E}\left\{U\left(\sum_1^n w_j Z_j W_0\right)\right\}, \tag{2.1}$$

[3]The strict concavity assumption implies that investors are everywhere risk averse. While strictly convex or linear utility functions on the entire range imply behavior that is grossly at variance with observed behavior, the strict concavity assumption also rules out Friedman–Savage type utility functions whose behavioral implications are reasonable. The strict concavity also implies $U'(W) > 0$ which rules out individual satiation.

[4]The "borrowing rate" is the rate on riskless-in-terms-of-default loans. While virtually every individual loan involves some chance of default, the empirical "spread" in the rate on margin loans to individuals suggests that this assumption is not a "bad approximation" for portfolio selection analysis. An explicit analysis of risky loan evaluation is provided in Section 7.

subject to $\sum_1^n w_j = 1$, where E is the expectation operator for the subjective joint probability distribution. If $(w_1^*, \ldots, w_n^*)$ is a solution to (2.1), then it will satisfy the first-order conditions

$$\mathrm{E}\{U'(Z^* W_0) Z_j\} = \lambda, \qquad j = 1, 2, \ldots, n, \tag{2.2}$$

where the prime denotes derivative; $Z^* \equiv \sum_1^n w_j^* Z_j$ is the random variable return per dollar on the optimal portfolio; and λ is the Lagrange multiplier for the budget constraint. Together with the concavity assumptions on U, if the $n \times n$ variance–covariance matrix of the returns $(Z_1, \ldots, Z_n)$ is nonsingular and an interior solution exists, then the solution is unique.[5] This non-singularity condition on the returns distribution eliminates "redundant" securities (i.e. securities whose returns can be expressed as exact linear combinations of the returns on other available securities).[6] It also rules out that any one of the securities is a riskless security.

If a riskless security is added to the menu of available securities (call it the $(n+1)$st security), then it is the convention to express (2.1) as the following unconstrained maximization problem:

$$\max_{\{w_1, \ldots, w_n\}} \mathrm{E}\left\{U\left(\left[\sum_{j=1}^{n} w_j (Z_j - R) + R\right] W_0\right)\right\}, \tag{2.3}$$

where the portfolio allocations to the risky securities are unconstrained because the fraction allocated to the riskless security can always be chosen to satisfy the budget constraint (i.e. $w_{n+1}^* = 1 - \sum_1^n w_j^*$). The first-order conditions can be written as

$$\mathrm{E}\{U'(Z^* W_0)(Z_j - R)\} = 0, \qquad j = 1, 2, \ldots, n, \tag{2.4}$$

where Z^* can be rewritten as $\sum_1^n w_j^*(Z_j - R) + R$. Again, if it is assumed that the variance–covariance matrix of the returns on the risky securities is non-singular and an interior solution exists, then the solution is unique.

[5]The existence of an interior solution is assumed throughout the analyses in this paper. For a complete discussion of the necessary and sufficient conditions for the existence of an interior solution, see Leland (1972) and Bertsekas (1974).

[6]For a trivial example, shares of General Motors with odd serial numbers are technically different from shares of GM with even serial numbers, and are, therefore, technically different securities. However, because their returns are identical, they are perfect substitutes from the point of view of investors. In portfolio theory, securities are operationally defined by their return distributions, and therefore two securities with identical returns are indistinguishable.

In both (2.1) and (2.3), no explicit consideration has been given for the treatment of bankruptcy (i.e. $Z^* < 0$). To rule out bankruptcy, the additional constraint that the probability of $Z^* \geqslant 0$ be one could be imposed on the choices for $(w_1^*, \ldots, w_n^*)$.[7] If the reason for this constraint is to reflect institutional restrictions designed to avoid individual bankruptcy, then it is too weak because the probability assessments on the $\{Z_j\}$ are subjective. A more realistic treatment would be to forbid borrowing and short-selling in conjunction with limited-liability securities where, by law, $Z_j \geqslant 0$. These rules can be formalized as restrictions on the allowable set of $\{w_j\}$ such that $w_j^* \geqslant 0, j=1,2,\ldots,n+1$, and (2.1) or (2.3) can be solved using the methods of Kuhn and Tucker[8] for inequality constraints. However, since the creation of limited-liability securities is itself a bankruptcy rule and only one of many that might be proposed, such explicit restrictions will not be imposed. Rather, it is simply assumed that there exists a bankruptcy law which allows for $U(W)$ to be defined for $W<0$ and that this law is consistent with the continuity and concavity assumptions on U.

The optimal demand functions for risky securities, $\{w_j^* W_0\}$, and the resulting probability distribution for the optimal portfolio will, of course, depend on the risk preferences of the investor, his initial wealth, and the joint distribution for the securities' returns. It is well known that the von Neumann–Morgenstern utility function can only be determined up to a positive affine transformation. Hence, the preference orderings of all choices available to the investor are completely specified by the Pratt–Arrow[9] *absolute risk-aversion function* which can be written as

$$A(W) \equiv \frac{-U''(W)}{U'(W)}, \tag{2.5}$$

and the change in absolute risk aversion with respect to a change in wealth is, therefore, given by

$$\frac{\mathrm{d}A}{\mathrm{d}W} = A'(W) = A(W)\left[A(W) + \frac{U'''(W)}{U''(W)}\right]. \tag{2.6}$$

[7]If U is such that $U'(0)=\infty$, and by extension, $U'(W)=\infty$, $W<0$, then from (2.2) or (2.4), it is easy to show that the probability of $Z^* \leqslant 0$ is a set of measure zero.

[8]Kuhn and Tucker (1951). The analysis of the "no short-sales" case is complicated and leads to virtually no theorems. Although in actual markets there are some restrictions on short sales, these restrictions may not be too important because limited liability securities (e.g. put options) can and have been created that provide essentially the same type of return as a short sale.

[9]The behavior associated with the utility function $V(W) \equiv aU(W)+b$, $a>0$, is indistinguishable from the behavior associated with $U(W)$. *Note:* $A(W)$ is invariant to any positive affine transformation of $U(W)$. See Pratt (1964).

By the assumption that $U(W)$ is increasing and strictly concave, $A(W)$ is positive, and such investors are called *risk-averse*. An alternative, but related, measure of risk aversion is the *relative risk-aversion function* defined to be

$$\mathcal{R}(W) \equiv -\frac{U''(W)W}{U'(W)} = A(W)W, \tag{2.7}$$

and its change with respect to a change in wealth is given by

$$\mathcal{R}'(W) = A'(W)W + A(W). \tag{2.8}$$

The *certainty equivalent end-of-period wealth*, W_c, associated with a given portfolio for end-of-period wealth whose random variable value is denoted by W, is defined to be that value such that

$$U(W_c) = \mathrm{E}\{U(W)\}, \tag{2.9}$$

i.e. W_c is the amount of money such that the investor is indifferent between having this amount of money for certain or the portfolio with random variable outcome W. The term "risk-averse" as applied to investors with concave utility functions is descriptive in the sense that the certainty equivalent end-of-period wealth is always less than the expected value of the associated portfolio, $\mathrm{E}\{W\}$, for all such investors. The proof follows directly by Jensen's Inequality: if U is strictly concave, then

$$U(W_c) = \mathrm{E}\{U(W)\} < U(\mathrm{E}\{W\})$$

whenever W has positive dispersion, and, because U is a non-decreasing function of W, $W_c < \mathrm{E}\{W\}$.

The certainty-equivalent can be used to compare the risk-aversions of two investors. An investor is said to be *more risk averse* than a second investor if for every portfolio, the certainty-equivalent end-of-period wealth for the first investor is less than or equal to the certainty equivalent end-of-period wealth associated with the same portfolio for the second investor with strict inequality holding for at least one portfolio.

While the certainty equivalent provides a natural definition for comparing risk aversions across investors, Rothschild and Stiglitz[10] have attempted in a corresponding fashion to define the meaning of "increasing risk" for a security so that

[10]Rothschild and Stiglitz (1970, 1971). There is an extensive literature, not discussed here, that uses this type of risk measure to determine when one portfolio "stochastically dominates" another. Cf. Hadar and Russell (1970, 1971), Hanoch and Levy (1969), and Bawa (1975).

the "riskiness" of two securities or portfolios can be compared. In comparing two portfolios with the same expected values, the first portfolio with random variable outcome denoted by W_1 is said to be *less risky* than the second portfolio with random variable outcome denoted by W_2 if

$$\mathrm{E}\{U(W_1)\} \geqslant \mathrm{E}\{U(W_2)\} \tag{2.10}$$

for all concave U with strict inequality holding for some concave U. They bolster their argument for this definition by showing its equivalence to the following two other definitions:

> There exists a random variable Z such that W_2 has the same distribution as $W_1 + Z$ where the conditional expectation of Z given the outcome on W_1 is zero (i.e. W_2 is equal in distribution to W_1 plus some "noise"). (2.11)

> If the points of F and G, the distribution functions of W_1 and W_2 are confined to the closed interval $[a, b]$, and $T(y) \equiv \int_a^y [G(x) - F(x)]\,\mathrm{d}x$, then $T(y) \geqslant 0$ and $T(b) = 0$ (i.e. W_2 has more "weight in its tails" than W_1). (2.12)

A feasible portfolio with return per dollar Z will be called an *efficient portfolio* if there exists an increasing, strictly concave function V such that $\mathrm{E}\{V'(Z)(Z_j - R)\} = 0$, $j = 1, 2, \ldots, n$. Using the Rothschild–Stiglitz definition of "less risky", a feasible portfolio will be an efficient portfolio only if there does not exist another feasible portfolio which is less risky than it is. All portfolios that are not efficient are called *inefficient portfolios*.

Proposition 2.1

If the set of available securities contain no redundant securities, then no two efficient portfolios have the same risk in the Rothschild–Stiglitz sense.

Proof

Let Z_e^1 and Z_e^2 denote the random variable returns per dollar on two distinct efficient portfolios. Let E denote the expectation operator over their joint distribution. The proof goes by contradiction. If the two portfolios have the same risk, then $\mathrm{E}(Z_e^1) = \mathrm{E}(Z_e^2)$ and $\mathrm{E}\{U(Z_e^1)\} = \mathrm{E}\{U(Z_e^2)\}$ for every concave U. Consider the following artificial security with return per dollar Z given by $Z = YZ_e^1 + (1 - Y)Z_e^2$, where $Y = 1$ with probability δ $(0 < \delta < 1)$ and $Y = 0$ with probability $(1 - \delta)$ and Y is independent of Z_e^1 and Z_e^2. It follows that $\mathrm{E}_Z(Z) = \mathrm{E}(Z_e^1) = \mathrm{E}(Z_e^2)$

and $E_Z\{U(Z)\}=\delta E\{U(Z_e^1)\}+(1-\delta)E\{U(Z_e^2)\}=E\{U(Z_e^1)\}=E\{U(Z_e^2)\}$, i.e. Z has the same risk as either Z_e^1 or Z_e^2. Define $Z_\delta \equiv E_Y(Z)=\delta Z_e^1+(1-\delta)Z_e^2$. By the independence of Y and Jensen's Inequality, $E\{U(Z_\delta)\}\geq E_Z\{U(Z)\}$ with equality holding if and only if $Z=Z_\delta$ because U can be taken to be strictly concave. If the inequality holds, then Z_δ is less risky than Z because $E(Z_\delta)=E_Z(Z)$. But Z_δ is the random variable return on a feasible portfolio gotten by combining the two portfolios Z_e^1 and Z_e^2 with portfolio weights $[\delta,(1-\delta)]$. Hence, Z_δ is less risky than both Z_e^1 and Z_e^2. But this contradicts the hypothesis that these portfolios are efficient. Hence, $Z=Z_\delta$, and this is possible if and only if $Z_e^1=Z_e^2$. If there are no redundant securities, then $Z_e^1=Z_e^2$ if and only if the two portfolios contain identical holdings, but this contradicts the hypothesis that the two portfolios are distinct.

Hence, from the definition of an efficient portfolio and Proposition 2.1, it follows that no two portfolios in the efficient set can be ordered with respect to one another. From (2.10), it follows immediately that every efficient portfolio is a possible optimal portfolio, i.e. for each efficient portfolio there exists an increasing, concave U and an initial wealth W_0 such that the efficient portfolio is a solution to (2.1) or (2.3). Furthermore, from (2.10), all risk-averse investors will be indifferent between selecting their optimal portfolios from the set of all feasible portfolios or from the set of efficient portfolios. Hence, without loss of generality, I will assume that all optimal portfolios are efficient portfolios.

With these general definitions established, I now turn to the analysis of the optimal demand functions for risky assets and their implications for the distributional characteristics of the underlying securities. A note on notation: the symbol Z_e will be used to denote the random variable return per dollar on an efficient portfolio, and a bar over a random variable (e.g. $\bar{Z}$) will denote the expected value of that random variable.

Theorem 2.1

If Z denotes the random variable return per dollar on any feasible portfolio and if $(Z_e-\bar{Z}_e)$ is riskier than $(Z-\bar{Z})$ in the Rothschild and Stiglitz sense, then $\bar{Z}_e>\bar{Z}$.

Proof

By hypothesis, $E\{U([Z-\bar{Z}]W_0)\}>E\{U([Z_e-\bar{Z}_e]W_0)\}$. If $\bar{Z}\geq\bar{Z}_e$, then trivially, $E\{U(ZW_0)\}>E\{U(Z_eW_0)\}$. But Z is a feasible portfolio and Z_e is an efficient portfolio. Hence, by contradiction, $\bar{Z}_e>\bar{Z}$.

Corollary 2.1.a

If there exists a riskless security with return R, then $\bar{Z}_e\geq R$ with equality holding only if Z_e is a riskless security.

Proof

The riskless security is a feasible portfolio with expected return R. If Z_e is riskless, then by Assumption 3, $\bar{Z}_e = R$. If Z_e is not riskless, then $(Z_e - \bar{Z}_e)$ is riskier than $(R - R)$. Therefore, by Theorem 2.1, $\bar{Z}_e > R$.

Theorem 2.2

The optimal portfolio for a non-satiated, risk-averse investor will be the riskless security (i.e. $w^*_{n+1} = 1$, $w^*_j = 0, j = 1, 2, \ldots, n$) if and only if $\bar{Z}_j = R$ for $j = 1, 2, \ldots, n$.

Proof

From (2.4), $\{w^*_1, \ldots, w^*_n\}$ will satisfy $E\{U'(Z^* W_0)(Z_j - R)\} = 0$, $j = 1, 2, \ldots, n$. If $\bar{Z}_j = R$, $j = 1, 2, \ldots, n$, then $Z^* = R$ will satisfy these first-order conditions. By the strict concavity of U and the non-singularity of the variance–covariance matrix of returns, this solution is unique. This proves the "if" part. If $Z^* = R$ is an optimal solution, then we can rewrite (2.4) as $U'(RW_0)E(Z_j - R) = 0$. By the non-satiation assumption, $U'(RW_0) > 0$. Therefore, for $Z^* = R$ to be an optimal solution, $\bar{Z}_j = R$, $j = 1, 2, \ldots, n$. This proves the "only if" part.

Hence, form Corollary 2.1.a and Theorem 2.2, if a risk-averse investor chooses a risky portfolio, then the expected return on that portfolio exceeds the riskless rate, and a risk-averse investor will choose a risky portfolio if, at least, one available security has an expected return different from the riskless rate.

Define the notation $E(Y | X_1, \ldots, X_q)$ to mean the *conditional expectation of the random variable* Y, conditional on knowing the realizations for the random variables $(X_1, \ldots, X_q)$.

Theorem 2.3

If there exists a feasible portfolio with return Z_p such that for security s, $Z_s = Z_p + \varepsilon_s$, where $E(\varepsilon_s) = E(\varepsilon_s | Z_p, Z_j, j = 1, \ldots, n, j \neq s) = 0$, then the fraction of every efficient portfolio allocated to security s is the same and equal to zero.

Proof

The proof follows by contradiction. Let Z_e be the return on an efficient portfolio with fraction $\delta_s \neq 0$ allocated to security s. Let Z be the return on a portfolio with the same fractional holdings as Z_e except instead of security s, it holds the fraction δ_s in feasible portfolio Z_p. Hence, $Z_e = Z + \delta_s(Z_s - Z_p)$ or $Z_e = Z + \delta_s \varepsilon_s$. By hypothesis, $\bar{Z}_e = \bar{Z}$ and by construction, $E(\varepsilon_s | Z) = 0$. Therefore, for $\delta_s \neq 0$, Z_e is riskier than Z in the Rothschild–Stiglitz sense. But this contradicts the hypothesis that Z_e is an efficient portfolio. Hence, $\delta_s = 0$ for every efficient portfolio.

Corollary 2.3.a

Let ψ denote the set of n securities with returns $(Z_1, \ldots, Z_{s-1}, Z_s, Z_{s+1}, \ldots, Z_n)$ and ψ' denote the same set of securities except Z_s is replaced with $Z_{s'}$. If $Z_{s'} = Z_s + \varepsilon_s$ and $E(\varepsilon_s) = E(\varepsilon_s | Z_1, \ldots, Z_{s-1}, Z_s, Z_{s+1}, \ldots, Z_n) = 0$, then all risk-averse investors would prefer to choose their optimal portfolios from ψ rather than ψ'.

The proof is essentially the same as the proof of Theorem 2.3 with Z_s replacing Z_p. Unless the holdings of Z_s in every efficient portfolio are zero, ψ will be strictly preferred to ψ'.

Theorem 2.3 and its corollary demonstrate that all risk-averse investors would prefer any "unnecessary" uncertainty or "noise" to be eliminated. In particular, by this theorem the existence of lotteries is shown to be inconsistent with strict risk aversion on the part of all investors.[11] While the inconsistency of strict risk aversion with observed behavior such as betting on the numbers can be "explained" by treating lotteries as consumption goods, it is difficult to use this argument to explain other implicit lotteries such as callable, sinking fund bonds where the bonds to be redeemed are selected at random.

As illustrated by the partitioning of the feasible portfolio set into its efficient and inefficient parts and the derived theorems, the Rothschild–Stiglitz definition of increasing risk is quite useful for studying the properties of optimal portfolios. However, it is important to emphasize that these theorems apply only to efficient portfolios and not to individual securities or inefficient portfolios. For example, if $(Z_j - \bar{Z}_j)$ is riskier than $(Z - \bar{Z})$ in the Rothschild–Stiglitz sense and if security j is held in positive amounts in an efficient or optimal portfolio (i.e. $w_j^* > 0$), then it *does not* follow that $\bar{Z}_j$ must equal or exceed $\bar{Z}$. In particular, if $w_j^* > 0$, it does not follow that $\bar{Z}_j$ must equal or exceed R. Hence, to know that one security is riskier than a second security using the Rothschild–Stiglitz definition of increasing risk provides no normative restrictions on holdings of either security in an efficient portfolio. And because this definition of riskier imposes no restrictions on the optimal demands, it cannot be used to derive properties of individual securities' return distributions from observing their relative holdings in an efficient portfolio. To derive these properties, a second definition of risk is required. However, discussion of this measure is delayed until Section 3.

In closing this section, comparative statics results are derived for the optimal risky security demand in the special case of a single risky security and a riskless security (i.e. $n = 1$). Additional results for the general case of many risky securities can be found in Fisher (1972).

[11] I believe that Christian von Weizsäcker proved a similar theorem in unpublished notes some years ago. However, I do not have a reference.

Define κ by

$$\kappa \equiv -\mathrm{E}\{U''(W^*)(Z-R)^2 W_0\} > 0, \tag{2.13}$$

where Z is the random variable return per dollar on the single risky security and $W^* \equiv [w^*(Z-R)+R]W_0$ is the random variable end-of-period wealth for the optimal allocation.

Applying the Implicit Function Theorem to (2.4), the change in the proportion of the optimal portfolio allocated to the risky security with respect to a change in initial wealth can be written as

$$\begin{aligned} \frac{\partial w^*}{\partial W_0} &= \frac{-\mathrm{E}\{w^* \mathcal{R}(W^*)U'(W^*)(W^*-RW_0)\}}{(w^* W_0)^2 \kappa}, \quad \text{for } w^* \neq 0, \\ &= 0 \qquad\qquad\qquad\qquad\qquad\qquad\qquad \text{for } w^* = 0, \end{aligned}$$

where $\mathcal{R}$ is the relative risk-aversion function defined in (2.7). If U is such that $\mathcal{R}$ is a constant, then $\partial w^*/\partial W_0 = 0$ because $\mathrm{E}\{U'(W^*)(W^*-RW_0)\}=0$ from (2.4). By integrating (2.7) twice, the general class of concave utility functions that exhibit constant relative risk aversion can be written as

$$U(W) = C_1\left[\frac{W^\gamma - 1}{\gamma}\right] + C_2, \tag{2.15}$$

where C_1 and C_2 are any constants such that $C_1 > 0$ and $\gamma < 1$. Because von Neumann–Morgenstern utility functions are unique only up to a positive affine transformation, all distinct members of this class are determined by the single parameter γ. The relative risk-aversion function $\mathcal{R}$ equals $(1-\gamma) > 0$. Hence, for investors with utility functions in this class, a larger value of γ implies a smaller value of relative risk aversion. Moreover, for all portfolios that can be ranked,[12] it is straightforward to show that $\partial w^*/\partial \gamma > 0$.

For each possible outcome for W^*, we have by the Mean Value Theorem that there exists a number $\theta = \theta(W^*)$, $0 \leq \theta \leq 1$, such that

$$\mathcal{R}(W^*) = \mathcal{R}(RW_0) + \mathcal{R}'(\eta)(W^* - RW_0), \tag{2.16}$$

where $\eta \equiv RW_0 + \theta(W^* - RW_0)$ and $\mathcal{R}'$ is defined in (2.8). Substituting from

[12] Because the iso-elastic family is not well defined for $W < 0$, portfolios with a positive probability of negative end-of-period wealth cannot be ranked.

(2.16) into (2.14) with $w^* \neq 0$, (2.14) can be rewritten as

$$\frac{\partial w^*}{\partial W_0} = \frac{-\mathrm{E}\{w^* \mathcal{R}'(\eta) U'(W^*)(W^* - RW_0)^2\}}{(w^* W_0)^2 \kappa}. \tag{2.17}$$

From (2.17) a sufficient condition for the optimal proportion in the risky security to decline with an increase in initial wealth is that $\mathcal{R}'w^* > 0$. Similarly, a sufficient condition for the optimal proportion to increase is that $\mathcal{R}'w^* < 0$. Hence, for investors with strictly increasing (decreasing) relative risk aversion, an increase in initial wealth will induce a decrease (increase) in the *absolute* proportion of the optimal portfolio allocated to the risky security, $|w^*|$.

From (2.14) the change in the dollar allocation to the risky security in the optimal portfolio with respect to a change in initial wealth can be written as

$$\begin{aligned} \frac{\partial(w^* W_0)}{\partial W_0} &= \frac{-R\mathrm{E}\{w^* A(W^*) U'(W^*)(W^* - RW_0)\}}{(w^*)^2 \kappa}, & w^* \neq 0, \\ &= 0, & w^* = 0, \end{aligned} \tag{2.18}$$

where A is the absolute risk-aversion function defined in (2.5). If U is such that A is a constant, then $\partial(w^* W_0)/\partial W_0 = 0$. By integrating (2.5) twice, the general class of concave utility functions that exhibit constant absolute risk aversion can be written as

$$U(W) = C_2 - C_1 e^{-\mu W}, \tag{2.19}$$

where C_1 and C_2 are any constants such that $C_1 > 0$ and $\mu > 0$. All distinct members of this class are determined by the single parameter μ which is equal to the absolute risk-aversion function, and $\partial(w^* W_0)/\partial\mu < 0$.

Again, using the Mean Value Theorem, $A(W^*)$ can be written as

$$A(W^*) = A(RW_0) + A'(\eta)(W^* - RW_0). \tag{2.20}$$

Substituting from (2.20) into (2.18) with $w^* \neq 0$, (2.18) can be rewritten as

$$\frac{\partial(w^* W_0)}{\partial W_0} = \frac{-R\mathrm{E}\{w^* A'(\eta) U'(W^*)(W^* - RW_0)^2\}}{(w^*)^2 \kappa}. \tag{2.21}$$

From (2.21) a sufficient condition for the optimal dollar investment in the risky security to decline is that $A'w^* > 0$, and a sufficient condition for it to increase is that $A'w^*<0$. Hence, for investors with strictly increasing (decreasing) absolute risk aversion, an increase in initial wealth will induce a decrease (increase) in the absolute dollar position in the risky security, $|w^*W_0|$.

Although the direction of change in both the proportional and absolute dollar holdings of the risky security depends upon the sign of w^*, the sign of w^* is determined solely by the sign of $(\bar{Z}-R)$. From Theorem 2.2, $w^*=0$ if and only if $\bar{Z}=R$. If $w^*\neq 0$, then $\bar{Z}^*>R$ because an optimal portfolio is an efficient portfolio. But $\bar{Z}^*=R+w^*(\bar{Z}-R)$. Hence, $w^*(\bar{Z}-R)>0$ if $w^*\neq 0$. Therefore, for any non-satiated, risk-averse investor, the sign of w^* will equal the sign of $(\bar{Z}-R)$. *Warning:* this condition need not obtain with respect to an individual security when there is more than one risky security, i.e. the sign of w_j^* need not equal the sign of $(\bar{Z}_j-R)$ for $n>1$.

From (2.14), the change in w^* with respect to a change in the risky security's expected return $\bar{Z}$ can be written as

$$\frac{\partial w^*}{\partial \bar{Z}}=\frac{\mathrm{E}\{U'(W^*)[1-A(W^*)(W^*-RW_0)]\}}{\kappa}. \tag{2.22}$$

Inspection of (2.22) shows that the sign of $\partial w^*/\partial \bar{Z}$ is ambiguous. However, three sufficient conditions for $\partial w^*/\partial \bar{Z}$ to be positive are:

(i) $w^*=0$;

or

(ii) $A'(W)\leqq 0$; (2.23)

or

(iii) $\mathcal{R}(W)\leqq \dfrac{W}{W-RW_0}$, for $RW_0<W\leqslant W^+$,

where W^+ equals the maximum possible outcome for W^*. Conditions (i) and (iii) follow directly from (2.22) and (ii) follows from the substitution for $A(W^*)$ from (2.20) into (2.22).

From (2.14) the change in w^* with respect to a change in R can be written as

$$\frac{\partial w^*}{\partial R}=\frac{-\mathrm{E}\{U'(W^*)[(w^*)^2+A(W^*)w^*(1-w^*)(W^*-RW_0)]\}}{(w^*)^2\kappa}. \tag{2.24}$$

From (2.24), $\partial w^*/\partial R$ can be either positive or negative. Two sufficient conditions for $\partial w^*/\partial R$ to be negative are:

$$\begin{aligned} &\text{(i)} \quad A'(W) \geqq 0 \quad \text{and} \quad w^*(1-w^*) \geqq 0 \\ &\text{or} \\ &\text{(ii)} \quad A'(W) \leqq 0 \quad \text{and} \quad w^*(1-w^*) \leqq 0. \end{aligned} \tag{2.15}$$

These conditions follow from the substitution for $A(W^*)$ from (2.20) into (2.24).

Rothschild and Stiglitz[13] have examined the effect on the optimal demand for the risky security of an increase in that security's riskiness. As with the other comparative statics results derived here, they found that the change in the optimal demand can be either positive or negative. A sufficient condition for an increase in risk to decrease the demand for the risky security is that $f(Z) \equiv (Z-R)U'([w^*(Z-R)+R]W_0)$ be concave in Z. It is straightforward to show by differentiating $f(Z)$ twice with respect to Z that if relative risk aversion is less than or equal to R and non-decreasing and if absolute risk aversion is non-increasing, then $f(Z)$ is strictly concave for $w^*>0$ and strictly convex for $w^*<0$.

In summary, even for the simplest case of one risky security and a riskless security, the sign of the change in the optimal demand for the risky security with respect to changes in its probability distribution is consistently ambiguous unless restrictions are placed on the class of utility functions or on the class of probability distributions for Z.

3. Risk measures for securities and portfolios in the one-period model

In the previous section the Rothschild–Stiglitz measure for the risk of a security was defined, and the comparative statics of an increase in a security's risk was analyzed in the special case of a single risky security and a riskless security. However, in the more-than-one risky security portfolio problem the Rothschild–Stiglitz measure is not a natural definition of risk for a security. In this section a second definition of increasing risk is introduced, and it is argued that this second measure is a more appropriate definition for the risk of a security. Although this second measure will not in general provide the same orderings as the Rothschild–Stiglitz measure, it is further argued that the two measures are not in conflict, and indeed, are complementary.

If Z_e is the random variable return per dollar on an efficient portfolio with positive dispersion, then let $V(Z_e)$ denote an increasing, strictly concave function

[13] Rothschild and Stiglitz (1971, pp. 70–74).

such that

$$\mathrm{E}\{V'(Z_e)(Z_j-R)\}=0, \qquad j=1,2,\ldots,n,$$

i.e. V is a concave utility function such that an investor with initial wealth $W_0=1$ and these preferences would select this efficient portfolio as his optimal portfolio. While such a function V will always exist, it will not be unique. If $\mathrm{cov}[x_1, x_2]$ is the functional notation for the covariance between the random variables x_1 and x_2, then define the random variable, $Y(Z_e)$, by

$$Y(Z_e)=\frac{V'(Z_e)-\mathrm{E}\{V'(Z_e)\}}{\mathrm{cov}[V'(Z_e), Z_e]}. \tag{3.1}$$

$Y(Z_e)$ is well defined as long as Z_e has positive dispersion because $\mathrm{cov}[V'(Z_e), Z_e] < 0$.[14] It is understood that in the following discussion "efficient portfolio" will mean "efficient portfolio with positive dispersion". Let Z_p denote the random variable return per dollar on any feasible portfolio p.

Definition

The *measure of risk of portfolio p relative to efficient portfolio K with random variable return* Z_e^K, b_p^K, is defined by

$$b_p^K \equiv \mathrm{cov}[Y(Z_e^K), Z_p]$$

and portfolio p is said to be *riskier than portfolio p′ relative to efficient portfolio K* if $b_p^K > b_{p'}^K$.

Theorem 3.1

If Z_p is the return on a feasible portfolio p and Z_e^K is the return on efficient portfolio K, then $\bar{Z}_p - R = b_p^K(\bar{Z}_e^K - R)$.

Proof

From the definition of $V(Z_e^K)$, $\mathrm{E}\{V'(Z_e^K)(Z_j-R)\}=0$, $j=1,2,\ldots,n$. Let δ_j be the fraction of portfolio p allocated to security j. Then, $Z_p = \sum_1^n \delta_j(Z_j-R)+R$, and $\sum \delta_j \mathrm{E}\{V'(Z_e)(Z_j-R)\} = \mathrm{E}\{V'(Z_e)(Z_p-R)\}=0$. by a similar argument, $\mathrm{E}\{V'(Z_e^K)(Z_e^K-R)\}=0$. Hence, $\mathrm{cov}[V'(Z_e^K), Z_e^K]=(R-\bar{Z}_e^K)\mathrm{E}\{V'(Z_e^K)\}$ and $\mathrm{cov}[V'(Z_e^K), Z_p]=(R-\bar{Z}_p)\mathrm{E}\{V'(Z_e^K)\}$. By Corollary 2.1.a, $\bar{Z}_e^K > R$. Therefore, $\mathrm{cov}[Y(Z_e^K), Z_p]=(R-\bar{Z}_p)/(R-\bar{Z}_e^K)$.

[14]For a proof, see Theorem 236 in Hardy, Littlewood, and Pölya (1959).

Hence, the expected excess return on portfolio p, $\bar{Z}_p - R$, is in direct proportion to its risk, and because $\bar{Z}_e^K > R$, the larger is its risk, the larger is its expected return. Thus, Theorem 3.1 provides the first argument why b_p^K is a natural measure of risk for individual securities.

A second argument goes as follows. Consider an investor with utility function U and initial wealth W_0 who solves the portfolio selection problem

$$\max_w \mathrm{E}\left\{U\left(\left[wZ_j + (1-w)Z\right]W_0\right)\right\},$$

where Z is the return on a portfolio of securities and Z_j is the return on the security j. The optimal mix, w^*, will satisfy the first-order condition

$$\mathrm{E}\left\{U'\left(\left[w^*Z_j + (1-w^*)Z\right]W_0\right)(Z_j - Z)\right\} = 0. \tag{3.2}$$

If the original portfolio of securities chosen was this investor's optimal portfolio (i.e. $Z = Z^*$), then the solution to (3.2) is $w^* = 0$. However, an optimal portfolio is an efficient portfolio. Therefore, by Theorem 3.1, $\bar{Z}_j - R = b_j^*(\bar{Z}^* - R)$. Hence, the "risk–return tradeoff" provided in Theorem 3.1 is a condition for personal portfolio equilibrium. Indeed, because security j may be contained in the optimal portfolio, w^*W_0 is similar to an excess demand function. b_j^* measures the contribution of security j to the Rothschild–Stiglitz risk of the optimal portfolio in the sense that the investor is just indifferent to a marginal change in the holdings of security j provided that $\bar{Z}_j - R = b_j^*(\bar{Z}^* - R)$. Moreover, by the Implicit Function Theorem, we have from (3.2) that

$$\frac{\partial w^*}{\partial \bar{Z}_j} = \frac{w^*W_0\mathrm{E}\left\{U''(Z - Z_j)\right\} - \mathrm{E}\{U'\}}{\mathrm{E}\left\{U''(Z - Z_j)^2\right\}} > 0, \quad \text{at } w^* = 0. \tag{3.3}$$

Therefore, if $\bar{Z}_j$ lies above the "risk–return" line in the $(\bar{Z} - b^*)$ plane, then the investor would prefer to increase his holdings in security j, and if $\bar{Z}_j$ lies below the line, then he would prefer to reduce his holdings. If the risk of a security increases, then the risk-averse investor must be "compensated" by a corresponding increase in that security's expected return if his current holdings are to remain unchanged.

The Rothschild–Stiglitz measure of risk is clearly different from the b_j^K measure here. The Rothschild–Stiglitz measure provides only for a partial ordering while the b_j^K measure provides a complete ordering. Moreover, they can give different rankings. For example, suppose the return on security j is independent of the return on efficient portfolio K, then $b_j^K = 0$ and $\bar{Z}_j = R$. Trivially, $b_R^K = 0$

for the riskless security. Therefore, by the b_j^K measure, security j and the riskless security have equal risk. However, if security j is not riskless, then by the Rothschild–Stiglitz measure security j is more risky than the riskless security. Despite this, the two measures are not in conflict and, indeed, are complementary. The Rothschild–Stiglitz definition measures the "total risk" of a security in the sense that it compares the expected utility from holding a security *alone* with the expected utility from holding another security *alone*. Hence, it is the appropriate definition for identifying optimal portfolios and determining the efficient portfolio set. However, it is not useful for defining the risk of securities generally because it does not take into account that investors can mix securities together to form portfolios. The b_j^K measure does take this into account because it measures the only part of an individual security's risk which is relevant to an investor: namely, the part that contributes to the total risk of his optimal portfolio. In contrast to the Rothschild–Stiglitz measure of total risk, the b_j^K measures the "systematic risk" of a security (relative to efficient portfolio K). Of course, to determine the b_j^K the efficient portfolio set must be determined. Because the Rothschild–Stiglitz measure does just that, the two measures are complementary.

Other properties of the b_p^K measure of systematic risk are:

Property 1

If a feasible portfolio p has portfolio weights $(\delta_1, \delta_2, \dots, \delta_n)$, then $b_p^K = \sum_1^n \delta_j b_j^K$. The systematic risk of a portfolio is the weighted sum of the systematic risks of its component securities.

Property 1 follows directly from the linearity of the covariance operator, $\text{cov}[x_1, x_2]$, with respect to the random variable x_2.

Let p and p' denote any two feasible portfolios and let K and L denote any two efficient portfolios.

Property 2

If $b_p^K = 0$ for some efficient portfolio K, then $b_p^L = 0$ for any efficient portfolio L.

Property 2 follows from Theorem 3.1 and Corollary 2.1.a. If $b_p^K = 0$, then $\bar{Z}_p = R$ because $\bar{Z}_e^K > R$. But if $\bar{Z}_p = R$ and $\bar{Z}_e^L > R$, then $b_p^L = 0$.

Property 3

$b_p^K \gtreqless b_{p'}^K$ if and only if $b_p^L \gtreqless b_{p'}^L$.

Property 3 follows from Property 2 if $b_p^K = b_{p'}^K = 0$. Suppose $b_p^K \neq 0$, then Property 3 follows from Theorem 3.1 because

$$\frac{b_{p'}^K}{b_p^K} = \frac{\bar{Z}_{p'} - R}{\bar{Z}_p - R} = \frac{b_{p'}^L}{b_p^L}.$$

Property 3 provides a third argument why b_p^K is a natural measure of risk for individual securities, namely the ordering of securities by their systematic risk relative to a given efficient portfolio will be identical to their ordering relative to any other efficient portfolio. Hence, given the set of available securities, there is an unambiguous meaning to the statement "security j is riskier than security i".

Property 4

If portfolio p is an efficient portfolio (call it the Kth one with $Z_p = Z_e^K$), then for any efficient portfolio L, $b_p^L > 0$ and in particular, $b_p^K = 1$. Hence, all efficient portfolios have positive systematic risk relative to any efficient portfolio.

Property 4 follows from Theorem (3.1) and Corollary 2.1.a.

Property 5

If the systematic risk of portfolio p is defined by its expected return, $\bar{Z}_p$, and if portfolio p is said to be riskier than portfolio p' if and only if $\bar{Z}_p > \bar{Z}_{p'}$, then this measure of systematic risk is equivalent to the b_p^K measure.

Property 5 follows directly from Theorem (3.1), Corollary 2.1.a, and Property 3.

Although the expected return of a security provides an equivalent ranking to its b_p^K measure, the b_p^K measure is not vacuous. There exist non-trivial information sets which allow b_p^k to be determined without knowledge of $\bar{Z}_p$. For example, consider a model in which all investors agree on the joint distribution of the returns on securities. Suppose we know the probability distribution of the optimal portfolio, Z^*W_0, and the utility function U for some investor. From (3.1) we therefore know the distribution of $Y(Z^*)$. For security j, define the random variable $\varepsilon_j \equiv Z_j - \bar{Z}_j$. Suppose, furthermore, that we have enough information about the joint distribution of $Y(Z^*)$ and ε_j to compute $\text{cov}[Y(Z^*), \varepsilon_j]$, but do not know $\bar{Z}_j$.[15] By the definition of covariance, $\text{cov}[Y(Z^*), \varepsilon_j] = \text{cov}[Y(Z^*), Z_j] = b_j^*$. However, Theorem 3.1 is a necessary condition for equilibrium in the securities market. Hence, we can deduce the equilibrium expected return on security j from $\bar{Z}_j = R + b_j^*(\bar{Z}^* - R)$. An analysis of the necessary information sets required to deduce the equilibrium structure of security returns is an important part of portfolio theory, and further discussion of this topic is provided in Sections 4 and 5.

[15] A sufficient amount of information would be the joint distribution of Z^* and ε_j. The necessary amount of information will depend upon the functional form of U'. However, in no case will a necessary condition be knowledge of $\bar{Z}_j$.

4. Spanning, separation, and mutual fund theorems

Definition

A set of M feasible portfolios with random variable returns $(X_1,\ldots, X_M)$ are said to *span* the space of portfolios contained in set Ψ if and only if for any portfolio in Ψ with return denoted by Z_p, there exists numbers $(\delta_1,\ldots, \delta_M)$, $\Sigma_1^M \delta_j = 1$, such that $Z_p = \Sigma_1^M \delta_j X_j$. If N is the number of securities available to generate the portfolios in Ψ and if M^* denotes the smallest number of feasible portfolios that span the space of portfolios contained in Ψ, then $M^* \leqq N$.

As was illustrated in Section 2, very little can be derived about the structure of optimal portfolio demand functions unless further restrictions are imposed on the class of investors' utility functions or the class of probability distributions for securities' returns. A particularly fruitful set of such restrictions is the one that provides for a non-trivial (i.e. $M^* < N$) spanning of the feasible portfolio set. Indeed, the spanning property leads to a collection of "mutual fund" or "separation" theorems that are the core of modern financial theory.

A *mutual fund* is a financial intermediary that holds as its assets a portfolio of securities and issues as liabilities shares against these assets. Unlike the portfolio of an individual investor, the portfolio of securities held by a mutual fund need not be an efficient portfolio. The connection between mutual funds and the spanning property can be seen in the following theorem:

Theorem 4.1

If there exist M mutual funds whose portfolios span the portfolio set Ψ, then all investors will be indifferent between selecting their optimal portfolios from Ψ or from portfolio combinations of just the M mutual funds.

The proof of the theorem follows directly from the definition of spanning. If Z^* denotes the return on an optimal portfolio selected from Ψ and if X_j denotes the return on the jth mutual fund's portfolio, then there exist portfolio weights $(\delta_1^*,\ldots, \delta_M^*)$ such that $Z^* = \Sigma_1^M \delta_j^* X_j$. Hence, any investor would be indifferent between the portfolio with return Z^* and the $(\delta^*_1,\ldots, \delta_M^*)$ combination of the mutual fund shares.

Although the theorem states "indifference", if there are information-gathering or other transactions costs and if there are economies of scale, then investors would prefer the mutual funds whenever $M<N$. By a similar argument, one would expect that investors would prefer to have the smallest number of funds necessary to span Ψ. Therefore, the smallest number of such funds, M^*, is a particularly important spanning set. Hence, the spanning property can be used to derive an endogenous theory for the existence of financial intermediaries with the

functional characteristics of a mutual fund. Moreover, from these functional characteristics a theory for their optimal management can be derived.

For the mutual fund theorems to have serious empirical content, the minimum number of funds required for spanning M^* must be significantly smaller than the number of available securities, N. When such spanning obtains, the investor's portfolio selection problem can be separated into two steps: first, individual securities are mixed together to form the M^* mutual funds; second, the investor allocates his wealth among the M^* funds' shares. If the investor knows that the funds span the space of optimal portfolios, then he need only know the joint distribution of $(X_1, \ldots, X_{M^*})$ to determine his optimal portfolio. It is for this reason that the mutual fund theorems are also called "separation" theorems. However, if the M^* funds can be constructed only if the fund managers know the preferences, endowments, and probability beliefs of each investor, then the formal separation property will have little operational significance.

In addition to providing an endogenous theory for mutual funds, the existence of a non-trivial spanning set can be used to deduce equilibrium properties of individual securities' returns and to derive optimal decision rules for business firms making physical investments. Moreover, in virtually every model of portfolio selection in which empirical implications beyond those presented in Sections 2 and 3 are derived, some non-trivial form of the spanning property obtains.

While the determination of conditions under which non-trivial spanning will obtain is, in a broad sense, a subset of the traditional economic theory of aggregation, the first rigorous contributions in portfolio theory were made by Arrow (1964), Markowitz (1959), and Tobin (1958). In each of these papers, and most subsequent papers, the spanning property is derived as an implication of the specific model examined, and therefore such derivations provide only sufficient conditions. In two notable exceptions, Cass and Stiglitz (1970) and Ross (1978) "reverse" the process by deriving necessary conditions for non-trivial spanning to obtain. In this section necessary and sufficient conditions for spanning are developed along the lines of Cass and Stiglitz and Ross, leaving until Section 5 discussion of the specific models of Arrow, Markowitz, and Tobin.

Let Ψ^f denote the set of all feasible portfolios that can be constructed from a riskless security with return R and n risky securities with a given joint probability distribution for their random variable returns $(Z_1, \ldots, Z_n)$. Let Ω denote the $n \times n$ variance–covariance matrix of the returns on the n risky assets.

Theorem 4.2

Necessary conditions for the M feasible portfolios with returns $(X_1, \ldots, X_M)$ to span the portfolio set Ψ^f are (i) that the rank of $\Omega \leqq M$ and (ii) that there exists numbers $(\delta_1, \ldots, \delta_M)$, $\Sigma_1^M \delta_j = 1$, such that the random variable $\Sigma_1^M \delta_j X_j$ has zero variance.

Proof

(i) The set of portfolios Ψ^f defines a $(n+1)$ dimensional vector space. By definition, if $(X_1,\ldots, X_M)$ spans Ψ^f, then each risky security's return can be represented as a linear combination of $(X_1,\ldots, X_M)$. Clearly, this is only possible if the rank of $\Omega \leq M$.

(ii) The riskless security is contained in Ψ^f. Therefore, if $(X_1,\ldots, X_M)$ spans Ψ^f, then there must exist a portfolio combination of $(X_1,\ldots, X_M)$ which is riskless.

Proposition 4.1

If $Z_p = \Sigma_1^n a_j Z_j + b$ is the return on some security or portfolio and if there are no "arbitrage opportunities" (Assumption 3), then (1) $b=[1-\Sigma_1^n a_j]R$ and (2) $Z_p = R + \Sigma_1^n a_j(Z_j - R)$.

Proof

Let $Z^\dagger$ be the return on a portfolio with fraction $\delta_j^\dagger$ allocated to security j, $j=1,2,\ldots, n$; δ_p allocated to the security with return Z_p; $(1-\delta_p - \Sigma_1^n \delta_j^\dagger)$ allocated to the riskless security with return R. If $\delta_j^\dagger$ is chosen such that $\delta_j^\dagger = -\delta_p a_j$, then $Z^\dagger = R + \delta_p(b - R[1-\Sigma_1^n a_j])$. $Z^\dagger$ is a riskless security, and therefore, by Assumption 3, $Z^\dagger = R$. But δ_p can be chosen arbitrarily. Therefore, $b=[1-\Sigma_1^n a_j]R$. Substituting for b, it follows directly that $Z_p = R + \Sigma_1^n a_j(Z_j - R)$.

As long as there are no arbitrage opportunities, from Theorem 4.2 and Proposition 4.1, without loss of generality, it can be assumed that one of the portfolios in any candidate spanning set is the riskless security. If, by convention, $X_M = R$, then in all subsequent analyses the notation $(X_1,\ldots, X_m, R)$ will be used to denote an M-portfolio spanning set where $m \equiv M-1$ is the number of risky portfolios (together with the riskless security) that span Ψ^f.

Theorem 4.3

A necessary and sufficient condition for $(X_1,\ldots, X_m, R)$ to span Ψ^f is that there exist numbers (a_{ij}) such that $Z_j = R + \Sigma_1^m a_{ij}(X_i - R)$, $j=1,2,\ldots, n$.

Proof

If $(X_1,\ldots, X_m, R)$ span Ψ^f, then there exist portfolio weights $(\delta_{1j},\ldots, \delta_{mj})$, $\Sigma_1^M \delta_{ij} = 1$, such that $Z_j = \Sigma_1^M \delta_{ij} X_i$. Noting that $X_M = R$ and substituting $\delta_{Mj} = 1 - \Sigma_1^m \delta_{ij}$, we have that $Z_j = R + \Sigma_1^m \delta_{ij}(X_i - R)$. This proves necessity. If there exist numbers (a_{ij}) such that $Z_j = R + \Sigma_1^m a_{ij}(X_i - R)$, then pick the portfolio weights $\delta_{ij} = a_{ij}$ for $i=1,\ldots, n$, and $\delta_{Mj} = 1 - \Sigma_1^m \delta_{ij}$, from which it follows that $Z_j = \Sigma_1^M \delta_{ij} X_i$. But every portfolio in Ψ^f can be written as a portfolio combination of $(Z_1,\ldots, Z_n)$ and R. Hence, $(X_1,\ldots, X_m, R)$ spans Ψ^{f} and this proves sufficiency.

Let Ω_X be the $m \times m$ variance–covariance matrix of the returns on the m portfolios with returns $(X_1,\ldots, X_m)$

Corollary 4.3.a

A necessary and sufficient condition for $(X_1,\ldots, X_m, R)$ to be the smallest number of feasible portfolios that span (i.e. $M^* = m+1$) is that the rank of Ω equals the rank of $\Omega_X = m$.

Proof

If $(X_1,\ldots,X_m, R)$ span Ψ^{f} and m is the smallest number of risky portfolios that does, then $(X_1,\ldots,X_m)$ must be linearly independent, and therefore rank $\Omega_X = m$. Hence, $(X_1,\ldots,X_m)$ form a basis for the vector space of security returns $(Z_1,\ldots,Z_n)$. Therefore, the rank of Ω must equal Ω_X. This proves necessity. If the rank of $\Omega_X = m$, then $(X_1,\ldots,X_m)$ are linearly independent. Moreover, $(X_1,\ldots,X_m) \in \Psi^{\mathrm{f}}$. Hence, if the rank of $\Omega = m$, then there exist numbers (a_{ij}) such that $Z_j - \bar{Z}_j = \sum_1^m a_{ij}(X_i - \bar{X}_i)$ for $j = 1,2,\ldots,n$. Therefore, $Z_j = b_j + \sum_1^m a_{ij} X_i$, where $b_j \equiv \bar{Z}_j - \sum_1^m a_{ij} \bar{X}_i$. By the same argument used to prove Proposition 4.1, $b_j = [1 - \sum_1^m a_{ij}]R$. Therefore, $Z_j = R + \sum_1^m a_{ij}(X_i - R)$. By Theorem 4.3, $(X_1,\ldots,X_m, R)$ span Ψ^{f}.

It follows from Corollary 4.3.a that a necessary and sufficient condition for non-trivial spanning of Ψ^{f} is that some of the risky securities are redundant securities. Note, however that this condition is sufficient only if securities are priced such that there are no arbitrage opportunities.

In all these derived theorems the only restriction on investors' preferences was that they prefer more to less. In particular, it was not assumed that investors are necessarily risk averse. Although Ψ^{f} was defined in terms of a known joint probability distribution for $(Z_1,\ldots, Z_n)$, which implies homogeneous beliefs among investors, inspection of the proof of Theorem 4.3 shows that this condition can be weakened. If investors agree on a set of portfolios $(X_1,\ldots, X_m, R)$ such that $Z_j = R + \sum_1^m a_{ij}(X_i - R), j = 1,2,\ldots, n$, and if they agree on the numbers (a_{ij}), then by Theorem 4.3 $(X_1,\ldots, X_m, R)$ span Ψ^{f} even if investors *do not* agree on the joint distribution of $(X_1,\ldots, X_m)$. These appear to be the weakest restrictions on preferences and probability beliefs that can produce non-trivial spanning and the corresponding mutual fund theorem. Hence, to derive additional theorems it is now further assumed that all investors are risk averse and that investors have homogeneous probability beliefs.

Define Ψ^{e} to be the set of all efficient portfolios contained in Ψ^{f}.

Proposition 4.2

If Z_e is the return on a portfolio contained in Ψ^e, then any portfolio that combines positive amounts of Z_e with the riskless security is also contained in Ψ^e.

Proof

Let $Z=\delta(Z_e-R)+R$ be the return on a portfolio with positive fraction δ allocated to Z_e and fraction $(1-\delta)$ allocated to the riskless security. Because Z_e is an efficient portfolio, there exists a strictly concave, increasing function V such that $E\{V'(Z_e)(Z_j-R)\}=0$, $j=1,2,\ldots,n$. Define $U(W)\equiv V(aW+b)$, where $a\equiv 1/\delta>0$ and $b\equiv(\delta-1)R/\delta$. Because $a>0$, U is a strictly concave and increasing function. Moreover, $U'(Z)=aV'(Z_e)$. Hence, $E\{U'(Z)(Z_j-R)\}=0, j=1,2,\ldots,n$. Therefore, there exists a utility function such that Z is an optimal portfolio, and thus Z is an efficient portfolio.

Hence, from Proposition 4.2, if $(X_1,\ldots, X_M)$ are the returns on M portfolios that are candidates to span the space of efficient portfolios Ψ^e, then without loss of generality it can be assumed that one of the portfolios is the riskless security.

Theorem 4.4

If $(X_1,\ldots, X_m, R)$ span Ψ^e and if $Z_j=\bar{Z}_j+\varepsilon_j$, where $E(\varepsilon_j)=E(\varepsilon_j|X_1,\ldots, X_m)=0$, then $\bar{Z}_j=R$.

Proof

By hypothesis, $(X_1,\ldots, X_m, R)$ span Ψ^e and therefore, for each efficient portfolio Z_e^k, there exists portfolio weights such that $Z_e^k=R+\Sigma_1^m\delta_i^k(X_i-R)$. Hence, $E(\varepsilon_j|Z_e^k)=0$ for every efficient portfolio. Therefore, the systematic risk of security j with respect to efficient portfolio k, $b_j^k=0$. By Theorem 3.1, $\bar{Z}_j=R$.

Hence, by Theorem 4.4, if in equilibrium the total source of variation in a security's return is noise relative to a set of portfolios that span Ψ^e, then risk-averse investors will be willing to hold the total quantity of that security outstanding even though they are not "compensated" by a positive expected excess return. Moreover, Theorem 4.4 suggests that the equilibrium expected return on a security will depend upon the joint distribution of its return with the set of spanning portfolios.

Theorem 4.5

If $(X_1,\ldots,X_m, R)$ span Ψ^e and if there exist numbers (a_{ij}) such that $Z_j=\bar{Z}_j+\Sigma_1^m a_{ij}(X_i-\bar{X}_i)+\varepsilon_j$, where $E(\varepsilon_j)=E(\varepsilon_j|X_1,\ldots,X_m)=0$, then $\bar{Z}_j=R+\Sigma_1^m a_{ij}(\bar{X}_i-R)$.

Proof

Let $Z^{\dagger}$ be the return on a portfolio with fraction $\delta_i^{\dagger}$ allocated to portfolio X_i, $i=1,\ldots,m$; δ allocated to security j; and $1-\delta-\sum_1^m \delta_i^{\dagger}$ allocated to the riskless security. If $\delta_i^{\dagger}$ is chosen such that $\delta_i^{\dagger}=-\delta a_{ij}$, then $Z^{\dagger}=R+\delta[\bar{Z}_j-R-\sum_1^m a_{ij}(\bar{X}_i-R)]+\delta\varepsilon_j$. By Theorem 4.4, $\bar{Z}^{\dagger}=R$. But δ can be chosen arbitrarily. Therefore, $\bar{Z}_j=R+\sum_1^m a_{ij}(\bar{X}_i-R)$.

Hence, if the return on a security can be written in a linear form relative to the spanning portfolios $(X_1,\ldots,X_m,R)$, then its expected excess return is completely determined by the expected excess returns on the spanning portfolios and the weights a_{ij}.

The following theorem, first proved by Ross (1978) shows that the security returns can always be written in this linear form relative to a set of spanning portfolios.

Theorem 4.6

A necessary and sufficient condition for $(X_1,\ldots,X_m,R)$ to span the set of efficient portfolios Ψ^{e} is that there exist numbers (a_{ij}) such that for $j=1,\ldots,n$, $Z_j=R+\sum_1^m a_{ij}(X_i-R)+\varepsilon_j$, where $\mathrm{E}(\varepsilon_j)=\mathrm{E}(\varepsilon_j\,|\,X_1,\ldots,X_m)=0$.

Proof

The proof of sufficiency is straightforward. Let Z_p be the return on a feasible portfolio with fraction δ_j invested in security j, $j=1,2,\ldots,n$ and $1-\sum_1^n\delta_j$ invested in the riskless security. Then $Z_p=R+\sum_1^n\delta_j(Z_j-R)$. By hypothesis, Z_p can be rewritten as $Z_p=R+\sum_1^m\delta_i'(X_i-R)+\varepsilon_p$ where $\delta_i'\equiv\sum_1^n\delta_j a_{ij}$ and $\varepsilon_p\equiv\sum_1^n\delta_j\varepsilon_j$. Consider the feasible portfolio with return Z_p' constructed by allocating fraction δ_i' to portfolio X_i and fraction $1-\sum_1^m\delta_i'$ to the riskless security. By construction, $Z_p=Z_p'+\varepsilon_p$ where $\mathrm{E}(\varepsilon_p)=\mathrm{E}(\varepsilon_p\,|\,Z_p')=0$. Hence, for $\varepsilon_p\not\equiv 0$, Z_p is riskier than Z_p' in the Rothschild–Stiglitz sense, and hence, Z_p cannot be an efficient portfolio. Therefore, all efficient portfolios can be generated by a portfolio combination of $(X_1,\ldots,X_m,R)$. The proof of necessity is not presented here because it is long and not constructive. The interested reader can find the proof in Ross (1978).

Since Ψ^{e} is contained in Ψ^{f}, any properties proved for portfolios that span Ψ^{e} must be properties of portfolios that span Ψ^{f}. From Theorems 4.3 and 4.6, the essential difference is that to span the efficient portfolio set it is not necessary that linear combinations of the spanning portfolios exactly replicate the return on each available security. Hence, it is not necessary that there exist redundant securities for non-trivial spanning of Ψ^{e} to obtain. Of course, both theorems are empty of any empirical content if the size of the smallest spanning set M^* is equal to $(n+1)$.

As discussed in the introduction to this section, all the important models of portfolio selection exhibit the non-trivial spanning property for the efficient portfolio set. Therefore, for all such models that do not restrict the class of admissible utility functions beyond that of risk aversion, the distribution of individual security returns must be such that $Z_j = R + \sum_1^m a_{ij}(X_i - R) + \varepsilon_j$, where $E(\varepsilon_j | X_1, \ldots, X_m) = 0$ for $j = 1, \ldots, n$. Moreover, given some knowledge of the joint distribution of a set of portfolios that span Ψ^e with $(Z_j - \bar{Z}_j)$, there exists a method for determining the a_{ij} and $\bar{Z}_j$.

Proposition 4.3

If $(X_1, \ldots, X_m, R)$ span Ψ^e with $(X_1, \ldots, X_m)$ linearly independent with finite variances and if the return on security j, Z_j, has a finite variance, then the (a_{ij}), $i = 1, 2, \ldots, m$ in Theorem 4.6 are given by

$$a_{ij} = \sum_1^m v_{ik} \operatorname{cov}[X_k, Z_j],$$

where v_{ik} is the $i-k$th element of Ω_X^{-1}.

The proof of Proposition 4.3 follows directly from the condition that $E(\varepsilon_j | X_k) = 0$, which implies that $\operatorname{cov}[\varepsilon_j, X_k] = 0$, $k = 1, \ldots, m$. The condition that $(X_1, \ldots, X_m)$ be linearly independent is trivial in the sense that knowing the joint distribution of a spanning set one can always choose a linearly independent subset. The only properties of the joint distributions required to compute the a_{ij} are the variances and covariances of the $X_1, \ldots, X_m$ and the covariances between Z_j and $X_1, \ldots, X_m$. In particular, knowledge of $\bar{Z}_j$ is not required because $\operatorname{cov}[X_k, Z_j] = \operatorname{cov}[X_k, Z_j - \bar{Z}_j]$. Hence, for $m < n$ (and especially so for $m \ll n$), there exists a non-trivial information set which allows the a_{ij} to be determined without knowledge of $\bar{Z}_j$. If $\bar{X}_1, \ldots, \bar{X}_m$ are known, then $\bar{Z}_j$ can be computed by the formula in Theorem 4.5. By comparison with the example at the end of Section 3, the information set required there to determine $\bar{Z}_j$ was a utility function and the joint distribution of its associated optimal portfolio with $(Z_j - \bar{Z}_j)$. Here, we must know a complete set of portfolios that span Ψ^e. However, here only the second-moment properties of the joint distribution need be known, and no utility function information other than risk aversion is required.

A special case of no little interest is when a single risky portfolio and the riskless security span the space of efficient portfolios. Indeed, the classic model of Markowitz and Tobin, which is discussed in Section 5, exhibits this strong form of separation. Moreover, most macroeconomic models have highly aggregated financial sectors where investors' portfolio choices are limited to simple combinations of two securities: "bonds" and "stocks". The rigorous microeconomic foundation for such aggregation is precisely that Ψ^e is spanned by a single risky portfolio and the riskless security.

If X denotes the random variable return on a risky portfolio such that (X, R) spans Ψ^e, then the return on any efficient portfolio, Z_e, can be written as if it had been chosen by combining the risky portfolio with return X with the riskless security. Namely, $Z_e = \delta(X-R)+R$, where δ is the fraction allocated to the risky portfolio and $(1-\delta)$ is the fraction allocated to the riskless security. By Corollary 2.1.a, the sign of δ will be the same for every efficient portfolio, and therefore all efficient portfolios will be perfectly positively correlated. If $\bar{X}>R$, then by Proposition 4.2 X will be an efficient portfolio and $\delta>0$ for every efficient portfolio.

Proposition 4.4

If $(Z_1,\ldots,Z_n)$ contain no redundant securities, if δ_j denotes the fraction of portfolio X allocated to security j, and w_j^* denotes the fraction of any risk-averse investor's optimal portfolio allocated to security j, $j=1,\ldots,n$, then for every such risk-averse investor,

$$\frac{w_j^*}{w_k^*} = \frac{\delta_j}{\delta_k}, \qquad j,k=1,2,\ldots,n.$$

The proof follows immediately because every optimal portfolio is an efficient portfolio, and the holdings of risky securities in every efficient portfolio are proportional to the holdings in X. Hence, the relative holdings of risky securities will be the same for all risk-averse investors. Whenever Proposition 4.4 holds and if there exist numbers δ_j^* such that $\delta_j^*/\delta_k^* = \delta_j/\delta_k$, $j,k=1,\ldots,n$ and $\sum_1^n \delta_j^* = 1$, then the portfolio with proportions $(\delta_1^*,\ldots,\delta_n^*)$ is called the *Optimal Combination of Risky Assets*. If such a portfolio exists, then without loss of generality it can always be assumed that $X=\sum_1^n \delta_j^* Z_j$.

Proposition 4.5

If (X, R) spans Ψ^e, then Ψ^e is a convex set.

Proof

Let Z_e^1 and Z_e^2 denote the returns on two distinct efficient portfolios. Because (X, R) spans Ψ^e, $Z_e^1 = \delta_1(X-R)+R$ and $Z_e^2 = \delta_2(X-R)+R$. Because they are distinct, $\delta_1 \neq \delta_2$, and so assume $\delta_1 \neq 0$. Let $Z \equiv \lambda Z_e^1 + (1-\lambda)Z_e^2$ denote the return on a portfolio which allocates fraction λ to Z_e^1 and $(1-\lambda)$ to Z_e^2, where $0 \leqslant \lambda \leqslant 1$. By substitution, the expression for Z can be rewritten as $Z=\delta(Z_e^1-R)+R$, where $\delta \equiv [\lambda + (\delta_2/\delta_1)(1-\lambda)]$. Because Z_e^1 and Z_e^2 are efficient portfolios, the sign of δ_1 is the same as the sign of δ_2. Hence, $\delta \geqslant 0$. Therefore, by Proposition 4.2, Z is an efficient portfolio It follows by induction that for any integer k and numbers λ_i such that $0 \leqslant \lambda_i \leqslant 1$, $i=1,\ldots,k$ and $\Sigma_1^k \lambda_i = 1$, $Z^k \equiv \Sigma_1^k \lambda_i Z_e^i$ is the return on an efficient portfolio. hence, Ψ^e is a convex set.

Definition

A *market portfolio* is defined as a portfolio that holds all available securities in proportion to their market values. To avoid the problems of "double counting" caused by financial intermediaries and inter-household issues of securities, the equilibrium market value of a security for this purpose is defined to be the equilibrium value of the aggregate demand by individuals for the security. In models where all physical assets are held by business firms and business firms hold no financial assets, an equivalent definition is that the market value of a security equals the equilibrium value of the aggregate amount of that security issued by business firms. If V_j denotes the market value of security j and V_R denotes the value of the riskless security, then

$$\delta_j^M = \frac{V_j}{\sum_1^n V_j + V_R}, \qquad j=1,2,\ldots,n,$$

where δ_j^M is the fraction of security j held in a market portfolio.

Theorem 4.7

If Ψ^e is a convex set, and if the securities' market is in equilibrium, then a market portfolio is an efficient portfolio.

Proof

Let there be K risk-averse investors in the economy with the initial wealth of investor k denoted by W_0^k. Define $Z^k \equiv R + \Sigma_1^n w_j^k (Z_j - R)$ to be the return per dollar on investor k's optimal portfolio, where w_j^k is the fraction allocated to security j. In equilibrium, $\Sigma_1^K w_j^k W_0^k = V_j$, $j=1,2,\ldots,n$, and $\Sigma_1^K W_0^k \equiv W_0 = \Sigma_1^n V_j + V_R$. Define $\lambda_k \equiv W_0^k / W_0$, $k=1,\ldots,K$. Clearly, $0 \leqslant \lambda_k \leqslant 1$ and $\Sigma_1^K \lambda_k = 1$. By definition of a market portfolio, $\Sigma_1^K w_j^k \lambda_k = \delta_j^M$, $j=1,2,\ldots,n$. Multiplying by $(Z_j - R)$ and summing over j, it follows that $\Sigma_1^K \lambda_k \Sigma_1^n w_j^k (Z_j - R) = \Sigma_1^K \lambda_k (Z^k - R) = \Sigma_1^n \delta_j^M (Z_j - R) = Z_M - R$, where Z_M is defined to be the return per dollar on the market portfolio. Because $\Sigma_1^K \lambda_k = 1$, $Z_M = \Sigma_1^K \lambda_k Z^k$. But every optimal portfolio is an efficient portfolio. Hence, Z_M is a convex combination of the returns on K efficient portfolios. Therefore, if Ψ^e is convex, then the market portfolio is contained in Ψ^e.

Because a market portfolio can be constructed without the knowledge of preferences, the distribution of wealth, or the joint probability distribution for the outstanding securities, models in which the market portfolio can be shown to be efficient are more likely to produce testable hypotheses. In addition, the efficiency of the market portfolio provides a rigorous microeconomic justification for the use of a "representative man" in aggregated economic models, i.e. if the market portfolio is efficient, then there exists a concave utility function such that maximization of its expected value with initial wealth equal to national wealth, would lead to the market portfolio as the optimal portfolio. Moreover, it is currently fashionable in the real world to advise "passive" investment strategies that simply mix the market portfolio with the riskless security. Provided that the market portfolio is efficient, by Proposition 4.2 no investor following such strategies could ever be convicted of "inefficiency". Unfortunately, necessary and sufficient conditions for the market portfolio to be efficient have not as yet been derived.[16]

However, even if the market portfolio were not efficient, it does have the following important property:

Proposition 4.6

In all portfolio models with homogeneous beliefs and risk-averse investors, the equilibrium expected return on the market portfolio exceeds the return on the riskless security.

The proof follows directly from the proof of Theorem 4.7 and Corollary 2.1.a. Clearly, $\bar{Z}_M - R = \sum_1^K \lambda_k(\bar{Z}^k - R)$. By Corollary 2.1.a, $\bar{Z}^k \geqslant R$ for $k=1,\ldots,K$, with strict inequality holding if Z^k is risky. But, $\lambda_k \geqslant 0$. Hence, $\bar{Z}_M > R$ if any risky securities are held by any investor. Note that using no information other than market prices and quantities of securities outstanding, the market portfolio (and combinations of the market portfolio and the riskless security) is the only risky portfolio where the sign of its equilibrium expected excess return can always be predicted.

Returning to the special case where Ψ^e is spanned by a single risky portfolio and the riskless security, it follows immediately from Proposition 4.5 and Theorem 4.7 that the market portfolio is efficient. Because all efficient portfolios are perfectly positively correlated, it follows that the risky spanning portfolio can always be chosen to be the market portfolio (i.e. $X = Z_M$). Therefore, every efficient portfolio (and hence, every optimal portfolio) can be represented as a simple portfolio combination of the market portfolio and the riskless security with a positive fraction allocated to the market portfolio. If all investors want to hold risky securities in the same relative proportions, then the only way in which

[16] In some unpublished research, C. Baldwin and E. P. Jones, M.I.T., have shown that the market portfolio is efficient for the Arrow–Debreu model and for models with three-fund spanning and a riskless security. S. Ross, Yale, has produced an example where the market portfolio is inefficient.

this is possible is if these relative proportions are identical to those in the market portfolio. Indeed, if there were one best investment strategy, and if this "best" strategy were widely known, then whatever the original statement of the strategy, it must lead to simply this imperative: "hold the market portfolio".

Because for every security $\delta_j^M \geqq 0$, it follows from Proposition 4.4, that in equilibrium, every investor will hold non-negative quantities of risky securities, and therefore, it is never optimal to short-sell risky securities. Hence, in models where $m=1$, the introduction of restrictions against short-sales will not affect the equilibrium.

Theorem 4.8

If (Z_M, R) span Ψ^e, then the equilibrium expected return on security j, can be written as

$$\bar{Z}_j = R + \beta_j(\bar{Z}_M - R),$$

where

$$\beta_j \equiv \frac{\text{cov}[Z_j, Z_M]}{\text{var}(Z_M)}.$$

The proof follows directly from Theorem 4.6 and Proposition 4.3. This relationship, called the *Security Market Line*, was first derived by Sharpe (1964) as a necessary condition for equilibrium in the mean–variance model of Markowitz and Tobin when investors have homogenous beliefs. This relationship has been central to the vast majority of empirical studies of securities' returns published during the last decade. Indeed, the switch in notation from a_{ij} to β_j in this special case reflects the almost universal adoption of the term "the 'beta' of a security" to mean the covariance of that security's return with the market portfolio divided by the variance of the return on the market portfolio.

In the special case of Theorem 4.8, β_j measures the systematic risk of security j relative to the efficient portfolio Z_M (i.e. $\beta_j = b_j^M$ as defined in Section 3), and therefore beta provides a complete ordering of the risk of individual securities. As is often the case in research, useful concepts are derived in a special model first. The term "systematic risk" was first coined by Sharpe and was measured by beta. The definition in Section 3 is a natural generalization. Moreover, unlike the general risk measure of Section 3, β_j can be computed from a simple covariance between Z_j and Z_M. Securities whose returns are positively correlated with the market are called *pro-cyclical*, and will be priced to have positive equilibrium expected excess returns. Securities whose returns are negatively correlated are called *counter-cyclical*, and will have negative equilibrium expected excess returns.

In general, the sign of b_j^k cannot be determined by the sign of the correlation coefficient between Z_j and Z_e^k. However, because $\partial Y(Z_e^k)/\partial Z_e^k > 0$ for each realization of Z_e^k, $b_j^k > 0$ does imply a generalized positive "association" between the return on Z_j and Z_e^k. Similarly, $b_j^k < 0$ implies a negative "association".

Let $\Psi_{\min}$ denote the set of portfolios contained in Ψ^{f} such that there exists no other portfolio in Ψ^{f} with the same expected return and a smaller variance. Let $Z(\mu)$ denote the return on a portfolio contained in $\Psi_{\min}$ such that $\bar{Z}(\mu) = \mu$, and let δ_j^μ denote the fraction of this portfolio allocated to security j, $j = 1, \ldots, n$.

Theorem 4.9

If $(Z_1, \ldots, Z_n)$ contain no redundant securities, then (a) for each value μ, δ_j^μ, $j = 1, \ldots, n$ are unique; (b) there exists a portfolio contained in $\Psi_{\min}$ with return X such that (X, R) span $\Psi_{\min}$; (c) $\bar{Z}_j - R = a_j(\bar{X} - R)$, where $a_j \equiv \mathrm{cov}(Z_j, X)/\mathrm{var}(X)$, $j = 1, 2, \ldots, n$.

Proof

Let σ_{ij} denote the i-jth element of Ω and because $(Z_1, \ldots, Z_n)$ contain no redundant securities, Ω is non-singular. Hence, let v_{ij} denote the i-jth element of Ω^{-1}. All portfolios in $\Psi_{\min}$ with expected return μ must have portfolio weights that are solutions to the problem: minimize $\sum_1^n \sum_1^n \delta_i \delta_j \sigma_{ij}$ subject to the constraint $\bar{Z}(\mu) = \mu$. Trivially, if $\mu = R$, then $Z(R) = R$ and $\delta_j^R = 0$, $j = 1, 2, \ldots, n$. Consider the case when $\mu \neq R$. The n first-order conditions are

$$0 = \sum_1^n \delta_j^\mu \sigma_{ij} - \lambda_\mu (\bar{Z}_i - R), \qquad i = 1, 2, \ldots, n,$$

where λ_μ is the Lagrange multiplier for the constraint. Multiplying by δ_i^μ and summing, we have that $\lambda_\mu = \mathrm{var}[Z(\mu)]/(\mu - R)$. By definition of $\Psi_{\min}$, λ_μ must be the same for all $Z(\mu)$. Because Ω is non-singular, the set of linear equations has the unique solution

$$\delta_j^\mu = \lambda_\mu \sum_1^n v_{ij} (\bar{Z}_i - R), \qquad i = 1, 2, \ldots, n.$$

This proves (a). From this solution, $\delta_j^\mu / \delta_k^\mu$, $j, k = 1, 2, \ldots, n$, are the same for every value of μ. Hence, all portfolios in $\Psi_{\min}$ with $\mu \neq R$ are perfectly correlated. Hence, pick any portfolio in $\Psi_{\min}$ with $\mu \neq R$ and call its return X. Then every $Z(\mu)$ can be written in the form $Z(\mu) = \delta_\mu (X - R) + R$. Hence, (X, R) span $\Psi_{\min}$ which proves (b), and from Theorem 4.6 and Proposition 4.3 (c) follows directly.

From Theorem 4.9, a_k will be equivalent to b_k^K as a measure of a security's systematic risk provided that the $Z(\mu)$ chosen for X is such that $\mu > R$. Like β_k, the only information required to compute a_k are the joint second moments of Z_k, and X. Which of the two equivalent measures will be more useful obviously depends upon the information set that is available. However, as the following theorem demonstrates, the a_k measure is the natural choice in the case when there exists a spanning set for Ψ^e with $m=1$.

Theorem 4.10

If (X, R) span Ψ^e and if X has a finite variance, then Ψ^e is contained in $\Psi_{\min}$.

Proof

Let Z_e be the return on any efficient portfolio. By Theorem 4.6, Z_e can be written as $Z_e = R + a_e(X-R)$. Let Z_p be the return on any portfolio in Ψ^f such that $\bar{Z}_e = \bar{Z}_p$. By Theorem 4.6, Z_p can be written as $Z_p = R + a_p(X-R) + \varepsilon_p$, where $E(\varepsilon_p) = E(\varepsilon_p | X) = 0$. Therefore, $a_p = a_e$ if $\bar{Z}_p = \bar{Z}_e$; $\text{var}(Z_p) = a_p^2 \text{var}(X) + \text{var}(\varepsilon_p) \geq a_p^2 \text{var}(X) = \text{var}(Z_e)$. Hence, Z_e is contained in $\Psi_{\min}$. Moreover, Ψ^e will be the set of all portfolios in $\Psi_{\min}$ such that $\mu \geq R$.

Thus, whenever there exists a spanning set for Ψ^e with $m=1$, the means, variances, and covariances of $(Z_1, \ldots, Z_n)$ are sufficient statistics to completely determine all efficient portfolios. Such a strong set of conclusions suggests that the class of joint probability distributions for $(Z_1, \ldots, Z_n)$ which admit a two-fund separation theorem will be highly specialized. However, as the following theorems demonstrate, the class is not empty.

Theorem 4.11

If $(Z_1, \ldots, Z_n)$ have a joint normal probability distribution, then there exists a portfolio with return X such that (X, R) span Ψ^e.

Proof

Using the procedure applied in the proof of Theorem 4.9, construct a risky portfolio contained in $\Psi_{\min}$, and call its return X. Define the random variables, $\varepsilon_k \equiv Z_k - R - a_k(X-R)$, $k = 1, \ldots, n$. By part (c) of that theorem, $E(\varepsilon_k) = 0$, and by construction, $\text{cov}[\varepsilon_k, X] = 0$. Because $Z_1, \ldots, Z_n$ are normally distributed, X will be normally distributed. Hence, ε_k is normally distributed, and because $\text{cov}[\varepsilon_k, X] = 0$, ε_k and X are independent. Therefore, $E(\varepsilon_k) = E(\varepsilon_k | X) = 0$. From Theorem 4.6, it follows that (X, R) span Ψ^e.

It is straightforward to prove that if $(Z_1, \ldots, Z_n)$ can have *arbitrary* means, variances, and covariances, then a necessary condition for there to exist a portfolio with return X such that (X, R) span Ψ^e is that $(Z_1, \ldots, Z_n)$ be joint normally distributed. However, it is important to emphasize the word "arbitrary". For example, the joint probability density function, $p(Z_1, \ldots, Z_n)$ is called a *symmetric function* if for each set of admissible outcomes for $(Z_1,\ldots,Z_n)$, $p(Z_1,\ldots,Z_n)$ remains unchanged when any two arguments of p are interchanged.

Theorem 4.12

If $p(Z_1,\ldots,Z_n)$ is a symmetric function with respect to all its arguments, then there exists a portfolio with return X such that (X,R) spans Ψ^e.

Proof

By hypothesis, $p(Z_1,\ldots,Z_i,\ldots,Z_n)=p(Z_i,\ldots,Z_1,\ldots,Z_n)$ for each set of given values $(Z_1,\ldots,Z_n)$. Therefore, from the first-order conditions for portfolio selection, (2.4), every risk-averse investor will choose $w_1^*=w_i^*$. But, this is true for $i=1,\ldots,n$. Hence, all investors will hold all risky securities in the same relative proportions. Therefore, if X is the return on a portfolio with an equal dollar investment in each risky security, then (X,R) will span Ψ^e.

Samuelson (1967) was the first to examine this class of density functions in a portfolio context. An obvious example of such a joint distribution is when $Z_1,\ldots,Z_n$ are independently and identically distributed which implies that $p(Z_1,\ldots,Z_n)$ is of the form $p(Z_1)\ldots p(Z_n)$.

A second example comes from the Linear-Factor model developed by Ross (1976). Suppose the returns on securities are generated by

$$Z_j=\bar{Z}_j+\sum_1^m a_{ij}Y_i+\varepsilon_j, \qquad j=1,\ldots,n, \tag{4.1}$$

where $E(\varepsilon_j)=E(\varepsilon_j|Y_1,\ldots,Y_m)=0$ and without loss of generality, $E(Y_i)=0$ and $\text{cov}[Y_i,Y_j]=0$, $i\neq j$. The random variables $\{Y_i\}$ represent common factors that are likely to affect the returns on a significant number of securities. If it is possible to construct a set of m portfolios with returns $(X_1,\ldots,X_m)$ such that X_i and Y_i are perfectly correlated, $i=1,2,\ldots,m$, then the conditions of Theorem 4.6 will be satisfied and $(X_1,\ldots,X_m,R)$ will span Ψ^e.

Although in general it will not be possible to construct such a set, by imposing some mild additional restrictions on $\{\varepsilon_j\}$, Ross (1976) has derived an asymptotic spanning theorem as the number of available securities, n, becomes large. While the rigorous derivation is rather tedious, a rough description goes as follows: let Z_p be the return on a portfolio with fraction δ_j allocated to security j, $j=1,2,\ldots,n$. From (4.1), Z_p can be written as

$$Z_p=\bar{Z}_p+\sum_1^m a_{ip}Y_i+\varepsilon_p, \tag{4.2}$$

where $\bar{Z}_p=R+\Sigma_1^n\delta_j(\bar{Z}_j-R)$; $a_{ip}\equiv\Sigma_1^n\delta_j a_{ij}$; $\varepsilon_p\equiv\Sigma_1^n\delta_j\varepsilon_j$. Consider the set of portfolios (called *well-diversified portfolios*) that have the property $\delta_j \equiv u_j/n$, where $|u_j| \leqslant M_j < \infty$ and M_j is independent of n, $j=1,\ldots,n$. Virtually by the definition of a common factor it is reasonable to assume that for every $n \gg m$, a significantly positive fraction of all securities, λ_i, have $a_{ij} \neq 0$, and this will be true for each common factor i, $i=1,\ldots,m$. Similarly, because the $\{\varepsilon_j\}$ denotes the variations in securities' returns not explained by common factors, it is also reasonable to assume for large n that for each j, ε_j is uncorrelated with virtually all other securities' returns. Hence, if the number of common factors, m, is fixed, then it should be possible to construct a set of well-diversified portfolios $\{X_k\}$ such that for X_k, $a_{ik}=0$, $i=1,\ldots,m$, $i \neq k$ and $a_{kk} \neq 0$ for all $n \gg m$. It follows from (4.2), that X_k can be written as

$$X_k = \bar{X}_k + a_{kk} Y_k + \frac{1}{n} \sum_1^n u_j^k \varepsilon_j, \qquad k=1,\ldots,m.$$

But $|u_j^k|$ is bounded, independently of n, and virtually all the ε_j are uncorrelated. Therefore, by the Law of Large Numbers, as $n \to \infty$, $X_k \to \bar{X}_k + a_{kk} Y_k$. So, as n becomes very large, X_k and Y_k become perfectly correlated, and by Theorem 4.6, asymptotically $(X_1,\ldots,X_m,R)$ will span Ψ^e. In particular, if $m=1$, then asymptotically two-fund separation will obtain independent of any other distributional characteristics of Y_1 or the $\{\varepsilon_j\}$.

It is interesting to note that empirical studies of stock market securities' returns have rarely found more than two or three statistically significant common factors.[17] Given that there are tens of thousands of different corporate liabilities traded in U.S. securities markets, the assumptions used by Ross are not without some empirical foundation. Indeed, whenever non-trivial spanning of Ψ^e obtains and the set of risky spanning portfolios can be identified, much of the structure of individual securities returns can be empirically estimated. For example, if $(X_1,\ldots,X_m,R)$ span Ψ^e, then by Theorem 4.6 and Proposition 4.3, ordinary least squares regression of the realized excess returns on security j, $Z_j - R$, on the realized excess returns of the spanning portfolios, $(X_1 - R,\ldots,X_m - R)$, will always give unbiased estimates of the a_{ij}. Of course, for these estimators to be efficient, further restrictions on the $\{\varepsilon_j\}$ are required to satisfy the Gauss–Markov Theorem.

Although the analyses derived here have been expressed in terms of restrictions on the joint distribution of security returns without explicitly mentioning security

[17] Cf. King (1966), Livingston (1977), Farrar (1962), Feeney and Hester (1967), and Farrell (1974). Unlike standard "factor analysis", the number of common factors here does not depend upon the fraction of total variation in an individual security's return that can be "explained". Rather, what is important is the number of factors necessary to "explain" the *covariation* between pairs of individual securities.

prices, it is obvious that these derived restrictions impose restrictions on prices through the identity that $Z_j \equiv V_j/V_{j0}$, where V_j is the random variable, end-of-period aggregate value of security j and V_{j0} is its initial value. Hence, given the characteristics of any two of these variables, the characteristics of the third are uniquely determined. For the study of equilibrium pricing, the usual format is to derive the equilibrium V_{j0} given the distribution of V_j.

Theorem 4.13

If $(X_1,\ldots, X_m, R)$ span Ψ^e, $M^*=m+1$, and all securities have finite variances, then a necessary condition for equilibrium in the securities' market is that

$$V_{j0}=\frac{\bar{V}_j-\sum_1^m\sum_1^m v_{ik}\mathrm{cov}[X_k,V_j](\bar{X}_i-R)}{R}, \qquad j=1,\ldots,n, \tag{4.3}$$

where v_{ik} is the i-kth element of Ω_x^{-1}.

Proof

Because $M^*=m+1$, Ω_x is non-singular. From the identity $V_j\equiv Z_jV_{j0}$ and Theorem 4.6, $V_j=V_{j0}[R+\sum_1^m a_{ij}(X_i-R)+\varepsilon_j]$, where $\mathrm{E}(\varepsilon_j|X_1,\ldots,X_m)=\mathrm{E}(\varepsilon_j)=0$. Taking expectations $\bar{V}_j=V_{j0}[R+\sum_1^m a_{ij}(\bar{X}_i-R)]$. Noting that $\mathrm{cov}[X_k,V_j]=V_{j0}\mathrm{cov}[X_k,Z_j]$, we have from Proposition 4.3, that $V_{j0}a_{ij}=\sum_1^m v_{ik}\mathrm{cov}[X_k,V_j]$. By substituting for a_{ij} in the $\bar{V}_j$ expression and rearranging terms, the theorem is proved.

Hence, from Theorem 4.13 a sufficient set of information to determine the equilibrium value of security j is the first and second moments for the joint distribution of $(X_1,\ldots, X_m, V_j)$. Moreover, the valuation formula has the following important "linearity" properties:

Corollary 4.13.a

If the hypothesized conditions of Theorem 4.13 hold and if the end-of-period value of some security is given by $V=\sum_1^n\lambda_jV_j$, then in equilibrium

$$V_0=\sum_1^n\lambda_jV_{j0}.$$

The proof of the corollary follows by substitution for V in formula (4.3). This property of formula (4.3) is called "value-additivity".

Corollary 4.13.b

If the hypothesized conditions of Theorem 4.13 hold and if the end-of-period value of some security is given by $V=qV_j+u$, where $\mathrm{E}(u)=\mathrm{E}(u|X_1,\ldots, X_m)=\bar{u}$

and $E(q)=E(q|X_1,\ldots,X_m,V_j)=\bar{q}$, then in equilibrium

$$V_0=\bar{q}V_{j0}+\bar{u}/R.$$

The proof follows by substitution for V in formula (4.3) and by applying the hypothesized conditional expectation conditions to show that $\text{cov}[X_k,V]=\bar{q}\text{cov}[X_k,V_j]$. Hence, to value two securities whose end-of-period values differ only by multiplicative or additive "noise", we can simply substitute the expected values of the noise terms. As will be shown later, both corollaries are central to the theory of investment by business firms.

If non-trivial spanning of Ψ^e is to obtain, the joint probability distribution for securities' returns cannot be arbitrary. How restrictive these conditions are cannot be answered in the abstract. First, the introduction of general equilibrium pricing conditions on securities will impose some restrictions on the joint distribution of returns. Second, the discussed benefits to individuals from having a set of spanning mutual funds may induce the creation of financial intermediaries or additional financial securities which together with pre-existing securities will satisfy the conditions of Theorem 4.6. An important example of the latter is the Arrow model discussed in Section 5.

An alternative approach to the development of non-trivial spanning theorems is to derive a class of utility functions for investors such that for arbitrary joint probability distributions for the available securities, investors within the class can generate their optimal portfolios from the spanning portfolios. Let Ψ^u denote the set of optimal portfolios selected from Ψ^f by investors with strictly concave Von Neumann–Morgenstern utility functions U_i. Cass and Stiglitz (1970) have proved the following theorem:

Theorem 4.14

There exists a portfolio with return X such that (X,R) span Ψ^u if and only if $A_i(W)=1/(a_i+bW)>0$, where A_i is the absolute risk aversion function for investor i in Ψ^u.[18]

The family of utility functions whose absolute risk-aversion functions can be written as $1/(a+bW)>0$ is called the "HARA" (Hyperbolic Absolute Risk Aversion) family.[19] By appropriate choices for a and b, various members of the family will exhibit increasing, decreasing, or constant absolute and relative risk aversion. Hence, if each investor's utility function could be approximated by some member of the HARA family, then it might appear that this alternative

[18]For this family of utility functions, the probability distribution for securities cannot be completely arbitrary without violating the von Neumann–Morgenstern axioms. For example, it is required that for every realization of W, $W>-a/b$ for $b>0$ and $W<-a/b$ for $b<0$. The latter condition is especially restrictive.

[19]Many authors have studied the properties of this family. See Merton (1971, p. 389) for references.

approach would be fruitful. However, it should be emphasized that the b in the statement of Theorem 4.14 does not have a subscript i, and therefore, for separation to obtain, all investors in Ψ^{u} must have virtually the same utility function.[20] Moreover, they must agree on the joint probability distribution for $(Z_1,\ldots, Z_n)$. Hence, the only significant way in which investors can differ is in their endowments of initial wealth.

Cass and Stiglitz also examine the possibilities for general non-trivial spanning ($1 \leqslant m < n$) by restricting the class of utility functions and conclude, "...it is the requirement that there be *any* mutual funds, and not the limitation on the *number* of mutual funds which is the restrictive feature of the property of separability".[21] Hence, the Cass and Stiglitz analysis is essentially a negative report on this approach to developing spanning theorems.

In closing this section, two further points should be made. First, although virtually all the spanning theorems require the generally implausible assumption that all investors agree upon the joint probability distribution for securities, it is not so unreasonable when applied to the theory of financial intermediation and mutual fund management. In a world where the economic concepts of "division of labor" and "comparative advantage" have content, then it is quite reasonable to expect that an efficient allocation of resources would lead to some individuals (the "fund managers") gathering data and actively estimating the joint probability distributions and the rest either buying this information directly or delegating their investment decisions by "agreeing to agree" with the fund managers' estimates. If the distribution of returns is such that non-trivial spanning of Ψ^{e} does not obtain, then there are no gains to financial intermediation over the direct sale of the distribution estimates. However, if non-trivial spanning does obtain and the number of risky spanning portfolios, m, is small, then a significant reduction in redundant information processing and transactions can be produced by the introduction of mutual funds. If a significant coalition of individuals can agree upon a common source for the estimates and if they know that, based on this source, a group of mutual funds offered spans Ψ^{e}, then they need only be provided with the joint distribution for these mutual funds to form their optimal portfolios. On the supply side, if the characteristics of a set of spanning portfolios can be identified, then the mutual fund managers will know how to structure the portfolios of the funds they offer.

The second point concerns the riskless security. It has been assumed throughout that there exists a riskless security. Although some of the specifications will change slightly, virtually all the derived theorems can be shown to be valid in the

[20]As discussed in footnote 18, the range of values for a_i cannot be arbitrary for a given b. Moreover, the sign of b uniquely determines the sign of $A'(W)$.

[21]Cass and Stiglitz (1970, p. 144).

absence of a riskless security.[22] However, the existence of a riskless security vastly simplifies many of the proofs.

5. Two special models of one-period portfolio selection

The two most cited models in the literature of portfolio selection are the *Time–State Preference model* of Arrow (1964) and Debreu (1959) and the *Mean–Variance model* of Markowitz (1959) and Tobin (1958). Because these models have been central to the development of the microeconomic theory of investment, there are already many review and survey articles devoted just to each of these models.[23] Hence, only a cursory description of each model is presented here with specific emphasis on how each model fits within the framework of the analyses presented in the other sections. Moreover, while, under appropriate conditions, both models can be interpreted as multiperiod, intertemporal portfolio selection models, such an interpretation will be delayed until Section 7.

The structure of the Arrow–Debreu model is described as follows. Consider an economy where all possible configurations for the economy at the end of the period can be described in terms of M possible states of nature. The states are mutually exclusive and exhaustive. It is assumed that there are N risk-averse individuals with initial wealth W_0^k and a von Neumann–Morgenstern utility function $U^k(W)$ for investor k, $k=1,\dots,N$. Each individual acts on the basis of subjective probabilities for the states of nature denoted by $\Pi_k(\theta)$, $\theta=1,\dots,M$. While these subjective probabilities can differ across investors, it is assumed for each investor that $0<\Pi_k(\theta)<1$, $\theta=1,\dots,M$. As was assumed in Section 2, there are n risky securities with returns per dollar Z_j and initial market value, V_{j0}, $j=1,\dots,n$, and the "perfect market" assumptions of that section, Assumptions 1–4, are assumed here as well. Moreover, if state θ obtains, then the return on security j will be $Z_j(\theta)$, and all investors agree on the functions $Z_j(\theta)$. Because the set of states is exhaustive, $[Z_j(1),\dots,Z_j(M)]$ describe all the possible outcomes for the returns on security j. In addition, there are available M "*pure*" *securities* with the properties that, $i=1,\dots,M$, one unit (share) of pure security i will be worth \$1 at the end of the period if state i obtains and will be worthless if state i does not obtain. If P_i denotes the price per share of pure security i and if X_i denotes its return per dollar, then for $i=1,\dots,M$, X_i as a function of the states of nature can be written as $X_i(\theta)=1/P_i$ if $\theta=i$ and $X_i(\theta)=0$ if $\theta\neq i$. All investors agree on the functions $\{X_i(\theta)\}$, i, $\theta=1,\dots,M$.

[22] Cf. Ross (1978) for spanning proofs in the absence of a riskless security. Black (1972) and Merton (1972) derive the two-fund theorem for the mean–variance model with no riskless security.

[23] For the Arrow–Debreu model, see Hirshleifer (1965, 1966, 1970), Myers (1968), and Radner (1970). For the mean–variance model, see Jensen (1972), Jensen, Ed. (1972), and Sharpe (1970).

Let $Z = Z(N_1,\ldots, N_M)$ denote the return per dollar on a portfolio of pure securities that holds N_j shares of pure security j, $j=1,\ldots, M$. If $V_0(N_1,\ldots, N_M) \equiv \sum_1^M N_j P_j$ denotes the initial value of this portfolio, then the return per dollar on the portfolio, as a function of the states of nature, can be written as $Z(\theta) = N_\theta / V_0$, $\theta = 1,\ldots, M$.

Proposition 5.1

There exists a riskless security, and its return per dollar R equals $1/\left(\sum_1^M P_j\right)$.

Proof

Consider the pure-security portfolio that holds one share of each pure security ($N_j = 1$, $j=1,\ldots, M$). The return per dollar Z is the same in every state of nature and equals $1/V_0(1,\ldots,1)$. Hence, there exists a riskless security and by Assumption 3 its return R is given by $1/\left(\sum_1^M P_j\right)$.

Proposition 5.2

For each security j with return Z_j, there exists a portfolio of pure securities whose return per dollar exactly replicates Z_j.

Proof

Let $Z^j \equiv Z(Z_j(1),\ldots, Z_j(M))$ denote the return on a portfolio of pure securities with $N_\theta = Z_j(\theta)$, $\theta = 1,\ldots, M$. It follows that $V_0(Z_j(1),\ldots, Z_j(M)) = \sum_1^M P_i Z_j(i)$ and $Z^j(\theta) = Z_j(\theta)/V_0$, $\theta = 1,\ldots, M$. Consider a three-security portfolio with return Z_p where fraction V_0 is invested in Z^j; fraction -1 is invested in Z_j; and fraction $1 - V_0 - (-1) = (2 - V_0)$ is invested in the riskless security. The return per dollar on this portfolio as a function of the states of nature can be written as

$$Z_p(\theta) = (2 - V_0)R + V_0 Z^j(\theta) - Z_j(\theta) = (2 - V_0)R,$$

which is the same for all states. Hence, Z_p is a riskless security, and by Assumption 3 $Z_p(\theta) = R$. Therefore, $V_0 = 1$, and $Z^j(\theta) = Z_j(\theta)$, $\theta = 1,\ldots, M$.

Proposition 5.3

The set of pure securities with returns $(X_1,\ldots, X_M)$ span the set of all feasible portfolios that can be constructed from the M pure securities and the n other securities.

The proof follows immediately from Propositions 5.1 and 5.2. Hence, whenever a complete set of pure securities exists or can be constructed from the available securities, then every feasible portfolio can be replicated by a portfolio of pure securities. Models in which such a set of pure securities exists are called "*complete markets*" *models* in the sense that any additional securities or markets would be redundant. Necessary and sufficient conditions for such a set to be constructed from the available n risky securities alone and therefore, for markets to be complete, are that: $n \geq M$: a riskless asset can be created and Assumption 3 holds; and the rank of the variance–covariance matrix of returns, Ω, equals $M-1$.

The connection between the pure securities of the Arrow–Debreu model and the mutual fund theorems of Section 4 is obvious. To put this model in comparable form, we can choose the alternative spanning set $(X_1, \dots, X_m, R)$ where $m = M-1$. From Theorem 4.3, the returns on the risky securities can be written as

$$Z_j = R + \sum_1^M a_{ij}(X_i - R), \qquad j = 1, \dots, m, \tag{5.1}$$

where the numbers (a_{ij}) are given by Proposition 4.3.

Note that no where in the derivation were the subjective probability assessments of the individual investors required. Hence, individual investors need not agree on the joint distribution for $(X_1, \dots, X_m)$. However, by Theorem 4.3, investors cannot have arbitrary beliefs in the sense that they must agree on the (a_{ij}) in (5.1).

Proposition 5.4

If $V_j(\theta)$ denotes the end-of-period value of security j if state θ obtains, then a necessary condition for equilibrium in the securities' market is that

$$V_{j0} = \sum_1^M P_k V_j(k), \qquad j = 1, \dots, n.$$

The proof follows immediately from the proof of Proposition 5.2. It was shown there that $V_0 = \Sigma_1^M P_k Z_j(k) = 1$. Multiplying both sides by V_{j0} and noting the identity $V_j(k) \equiv V_{j0} Z_j(k)$, it follows that $V_{j0} = \Sigma_1^M P_k V_j(k)$.

However, by Theorem 4.3 and Proposition 5.3, it follows that the $\{V_{j0}\}$ can also be written as

$$V_{j0} = \frac{\bar{V}_j - \sum_1^m \sum_1^m v_{ik} \operatorname{cov}[X_k, V_j](\bar{X}_i - R)}{R}, \qquad j = 1, \dots, n, \tag{5.2}$$

where v_{ik} is the i-kth element of Ω_X^{-1}. Hence, from (5.2) and Proposition 5.4, it follows that the a_{ij} in (5.1) can be written as

$$a_{ij} = [Z_j(i) - R]/[1/P_i - R], \qquad i = 1, \dots, m; j = 1, \dots, n. \tag{5.3}$$

From (5.3), given the prices of the securities $\{P_i\}$ and $\{V_{j0}\}$, the $\{a_{ij}\}$ will be agreed upon by all investors if and only if they agree upon the $\{V_j(i)\}$ functions.

While it is commonly believed that the Arrow–Debreu model is completely general with respect to assumptions about investors' beliefs, the assumption that all investors agree on the $\{V_j(i)\}$ functions can impose non-trivial restrictions on these beliefs. In particular, when there is production, it will in general be inappropriate to define the states, tautologically, by the end-of-period values of the securities, and therefore, investors will at least have to agree on the technologies specified for each firm.[24] However, as discussed in Section 4, it is unlikely that a model without some degree of homogeneity in beliefs (other than agreement on currently observed variables) can produce testable restrictions. Among models that do produce such testable restrictions, the assumptions about investors' beliefs in the Arrow–Debreu model are probably the most general.

Finally, for the purposes of portfolio theory, the Arrow–Debreu model is a special case of the spanning models of Section 4 which serves to illustrate the generality of the linear structure of those models.

The most elementary type of portfolio selection model in which all securities are not perfect substitutes is one where every portfolio can be characterized by two numbers: its "risk" and its "return". The mean–variance portfolio selection model of Markowitz (1959) and Tobin (1958) is such a model. In this model, each investor chooses his optimal portfolio so as to maximize a utility function of the form $H[\mathrm{E}(W), \mathrm{var}(W)]$, subject to his budget constraint, where W is his random variable end-of-period wealth. The investor is said to be "risk averse in a mean–variance sense" if $H_1 > 0$; $H_2 < 0$; $H_{11} < 0$; $H_{22} < 0$, and $H_{11}H_{22} - H_{12}^2 > 0$, where subscripts denote partial derivatives.

In an analogous fashion to the general definition of an efficient portfolio in Section 2, a feasible portfolio will be called a *mean–variance efficient portfolio* if there exists a risk-averse mean–variance utility function such that this feasible portfolio would be preferred to all other feasible portfolios. Let $\Psi_{\mathrm{mv}}^{\mathrm{e}}$ denote the set of mean-variance efficient portfolios. As defined in Section 4, Ψ_{min} is the set of feasible portfolios such that there exists no other portfolio with the same expected return and a smaller variance. For a given initial wealth W_0, every risk-averse

[24] If the states are defined in terms of end-of-period values of the firm in addition to "environmental" factors, then the firms' production decisions will, in general, alter the state space description which violates the assumptions of the model. Moreover, I see no obvious reason why individuals are any more likely to agree upon the $\{V_j(i)\}$ function than upon the probability distributions for the environmental factors. If sufficient information is available to partition the states into fine enough categories to produce agreement on the $\{V_j(i)\}$ functions, then, given this information, it is difficult to imagine how rational individuals would have heterogeneous beliefs about the probability distributions for these states. As with the standard certainty model, agreement on the technologies is necessary for Pareto optimality in this model. However, as Peter Diamond has pointed out to me, it is not sufficient. Sufficiency demands the stronger requirement that everyone be "right" in their assessment of the technologies.

investor would prefer the portfolio with the smallest variance among those portfolios with the same expected return. Hence, Ψ_{mv}^{e} is contained in Ψ_{min}.

Proposition 5.5

If $(Z_1,\ldots,Z_n)$ are the returns on the available risky securities, then there exists a portfolio contained in Ψ_{mv}^{e} with return X such that (X,R) span Ψ_{mv}^{e} and $\bar{Z}_j - R = a_j(\bar{X} - R)$, where $a_j \equiv \text{cov}(Z_j, X)/\text{var}(X), j = 1,2,\ldots,m$.

The proof follows immediately from Theorem 4.9.[25] Hence, all the properties derived in the special case of two-fund spanning ($m=1$) in Section 4 apply to the mean–variance model. Indeed, because all such investors would prefer a higher expected return for the same variance of return, Ψ_{mv}^{e} is the set of all portfolios contained in Ψ_{min} such that their expected returns are equal to or exceed R. Hence, the mean–variance model is also a special case of the spanning models developed in Section 4.

If investors have homogeneous beliefs, then the equilibrium version of the mean–variance model is called the *Capital Asset Pricing Model.*[26] It follows from Proposition 4.5, and Theorem 4.7 that, in equilibrium, the market portfolio can be chosen as the risky spanning portfolio. From Theorem 4.8, the equilibrium structure of expected returns must satisfy the Security Market Line.

Because of the mean–variance model's attractive simplicity and its strong empirical implications, a number of authors[27] have studied the conditions under which such a criterion function is consistent with the expected utility maxim. Like the studies of general spanning properties cited in Section 4, these studies examined the question in two parts. (i) What is the class of probability distributions such that the expected value of an arbitrary concave utility function can be written solely as a function of mean and variance? (ii) What is the class of strictly concave von Neumann–Morgenstern utility functions whose expected value can be written solely as a function of mean and variance for arbitrary distributions? Since the class of distributions in (i) was shown in Section 4 to be equivalent to the class of finite variance distributions that admit two-fund spanning of the efficient set, the analysis will not be repeated here. To answer (ii), it is straightforward to show that a necessary condition is that U be of the form, $W - bW^2$, with $b>0$. This member of the HARA family is called the *quadratic*, and will satisfy the von Neumann axioms only if $W \leqslant 1/2b$, for all possible outcomes for W. Even if U is defined to be $\max[W - bW^2, 1/4b]$ so that U satisfies the axioms for all W, for general distributions its expected value can be written as a function of

[25]In particular, the optimal portfolio demand functions are of the form derived in the proof of Theorem 4.9. For a complete analytic derivation, see Merton (1972).

[26]Sharpe (1964), Lintner (1965), and Mossin (1966) are generally credited with independent derivations of the model. Black (1972) extended the model to include the case of no riskless security.

[27]Cf. Borch (1969), Feldstein (1969), Tobin (1969), and Samuelson (1967).

just $\mathrm{E}(W)$ and $\mathrm{var}(W)$ only if the maximum possible outcome for W is less than $1/2b$.

Although both the Arrow–Debreu and Markowitz–Tobin models were shown to be special cases of the spanning models in Section 4, they deserve special attention because they are unquestionably the genesis of these general models.

6. Investment theory for the firm

In the preceding section the portfolio selection problem for individuals was solved and a set of necessary conditions for financial equilibrium were derived. Neither the current consumption-saving choice by individuals nor the allocation of resources for physical production were explicitly considered. Hence, the preceding analyses are best viewed as a partial equilibrium study of the financial markets taking current consumption and production plans as fixed. In this section the theory of optimal investment in physical assets is presented, and the connection between production theory and the financial markets is made explicit. However, the optimal choice of current consumption by individuals is still taken as given leaving until Section 7 its explicit examination.

There are two essential differences between the portfolio selection problem and the optimal allocation of physical investment problem. First, because it was assumed that individuals behave "competitively" with respect to the securities market (i.e. Assumption 2 in Section 2), only linear allocations of resources are allowed among the available investments in the portfolio selection problem. In general, physical production technologies can be non-linear functions of their inputs. Also, because the available investments in the portfolio problem are securities, it is possible for an individual to invest negative amounts in specific investments. For production technologies, the amount of physical investment must be non-negative.

Second, in developed market economies, most of the physical production is carried on by business firms where the production decisions are made by managers who are generally not the (sole) owners of the firm. Hence, it is important to know the conditions under which an efficient allocation of resources among the available production technologies will obtain when there exists an institutional separation between the owners of the resources and the managers of these resources. In essence, does there exist a set of investment decision rules such that if firm managers follow these rules, the firm will operate "as if" the owners of the firm had made the production decisions directly. Hence, unlike the utility function of an individual which is taken to be exogenous in the portfolio selection problem, the criterion function for production decisions by the firm is derived, and is therefore endogenous.

It is well known in the theory of production under certainty that if firms are competitive and make their production decisions so as to maximize the market value of the firm, then a competitive equilibrium is a Pareto optimum. Arrow (1964) and Debreu (1959) extended these results to uncertainty within the framework of their "complete markets" model. However, the extent to which these results carry over to "incomplete markets" has not as yet been determined. Diamond (1967) and more recently Leland (1974), Eckern and Wilson (1974), Radner (1974), and Hart (1975) have derived conditions under which stockholders will unanimously agree on a production decision. However, the "derived" criterion function for the firm will not, in general, be "to maximize market value". While the non-optimality of the value maximization rule is not surprising when firms do not behave competitively, Jensen and Long (1972), Fama (1972), and Stiglitz (1972) claim that it can be non-optimal even when firms are competitive. Merton and Subrahmanyam (1974) argue that the posited firm behavior in the Jensen–Fama–Stiglitz models is not competitive and, therefore, the findings using their models are not inconsistent with those proven for complete markets. While these issues have not yet been resolved and will not be here, it is hoped that the analysis of this section will shed some light on the controversy.

A firm is defined by the single production technology it owns, and let there be n production technologies where the random variable end-of-period value of firm j, $V_j(I_j; \theta_j)$, can be written as a function of the initial investment in its technology, I_j, and an exogenously specified random variable θ_j. Suppose there exists a set of portfolios with random variable returns per dollar $(X_1, \ldots, X_m, R)$ that span the efficient portfolio set. Let $V_{j0}(I_j)$ be the equilibrium initial value of firm j after I_j has been invested in its technology. Therefore, $V_{j0}(I_j)$ will satisfy the formula (4.3) in Theorem 4.13.

Consider a risk-averse individual with utility function $U(W)$. Prior to investment and trading, his initial endowment contains λ_j fractional ownership of firm j, $j=1, \ldots, n$, in addition to other exogenous assets with market value $\overline{W}_0$. His initial wealth, W_0, can be written as

$$W_0(I_1, \ldots, I_n) = \sum_1^n \lambda_j \left[V_{j0}(I_j) - I_j \right] + \overline{W}_0. \tag{6.1}$$

Consider the expanded portfolio selection problem where the investor chooses his optimal portfolio allocation and the amount of initial investment allocated to those firms in which he has some positive initial ownership. As a natural extension of Assumption 2 in the standard portfolio problem of Section 2, it is assumed that the investor believes that his actions (including his choices for I_j) cannot affect the probability distribution of returns on the set of spanning portfolios $(X_1, \ldots, X_m, R)$.

By hypothesis, $(X_1,\dots, X_m, R)$ span ψ^e. Hence, without loss of generality, we can formulate the problem as

$$\max_{\{w_i, I_j\}} E\left\{U\left(\left[\sum_1^m w_i(X_i - R) + R\right]W_0\right)\right\}, \tag{6.2}$$

where w_i is the fraction of his initial wealth allocated to spanning portfolio X_i, $i=1,\dots, m$; $(1-\Sigma_1^m w_i)$ is the fraction allocated to the riskless security; W_0 is given by (6.1); and E denotes the expectation operator over his subjective probability distribution for $(X_1,\dots, X_m)$.

Define $W^* \equiv \left[\Sigma_1^m w_i^*(X_i - R) + R\right]\left[\Sigma_1^n \lambda_j\left[V_{j0}(I_j^*) - I_j^*\right] + \overline{W}_0\right]$, where $(w_1^*,\dots, w_m^*)$ are his optimal portfolio weights and $(I_1^*,\dots, I_n^*)$ are his optimal choices for initial investment in the n firms. For an interior solution, the optimal choices will satisfy the first-order conditions

$$\mathrm{E}\{U'(W^*)(X_i - R)\} = 0, \qquad i=1,\dots, m \tag{6.3}$$

and

$$\mathrm{E}\left\{U'(W^*)\lambda_j Z^*\left[V_{j0}'(I_j^*) - 1\right]\right\} = 0, \qquad j=1,\dots, n, \tag{6.4}$$

where Z^* is the return per dollar on his optimal portfolio and $V_{j0}'(I_j) \equiv \partial V_{j0}/\partial I_j$. By the first-order conditions (6.3), $\mathrm{E}\{U'(W^*)Z^*\} = R\mathrm{E}\{U'(W^*)\} > 0$ by non-satiation. Hence, for all firms where the individual has positive initial ownership $(\lambda_j > 0)$, the first-order conditions (6.4) can be rewritten as

$$V_{j0}'(I_j^*) - 1 = 0. \tag{6.5}$$

From (6.1), optimality conditions (6.5) simply imply that for a given distribution of the spanning portfolios, the investor would prefer the allocation of physical investment across firms to be the one that maximizes his initial wealth. But the initial wealth function in (6.1) does not depend upon the investor's utility function. Hence, every investor with a positive ownership of firm j will agree that the amount of physical investment in firm j should be chosen so as to maximize its market value. And if the investors agree on the $V_{j0}(I_j)$ function, then they will agree on the optimal choice for I_j. *Note*: This latter condition can be satisfied even if investors do not have homogeneous beliefs about the probability distribution for $(X_1,\dots, X_m)$. For example, for the Arrow–Debreu model in Section 5, it was shown that all investors will agree on the valuation formula (4.3), even though they have heterogeneous beliefs about the probability distribution for the states.

In general, with heterogeneous beliefs, investors will disagree on the amount of investment that the firm should take to maximize its market value. However, as long as each shareholder perceives changes in the firm's investment as having no effect on the return distributions of the efficient set, they will agree that any change in investment which increases the market value of the firm will be preferred to ones that do not.

Theorem 6.1

If changes in the investment made by firm j do not change the distribution of the efficient portfolio set, then all shareholders of firm j will agree that its investment should be chosen to maximize its market value. In addition, if all such shareholders agree on the valuation function $V_{j0}(I_j)$, then the value-maximizing firm will operate "as if" the owners of the firm had made the investment decision directly.

Of course, the critical issue is under what conditions will the hypothesis of Theorem 6.1 obtain? One clear example is if, for every choice I_j, the resulting equilibrium random variable return on firm j is such that it is a "redundant" security (as defined in Sections 2 and 4), then changes in the investment made by firm j will not affect the equilibrium distribution of the efficient portfolio set. This condition will occur whenever, from the point of view of investors, there exist other securities that are perfect substitutes for security j. While "perfect substitutability" is sufficient, it is not necessary as the following analysis demonstrates.

Consider the same risk-averse individual as in the previous analysis, but for simplicity it is assumed that he has an initial endowment of positive fractional ownership of firm j only. Assume that $(X_1,\dots,X_m,Z_j,R)$ span the efficient portfolio set. The investor believes that his actions cannot affect the distribution of returns for $(X_1,\dots,X_m,R)$, and he can only affect the distribution of Z_j through his choice of I_j. Expression (6.2) can be reformulated as

$$\max_{\{w_i,I_j\}} \mathrm{E}\left\{U\left(\left[\sum_1^m w_i(X_i-R)+w_{m+1}(Z_j-R)+R\right]W_0\right)\right\}, \tag{6.6}$$

where w_{m+1} is the fraction of his initial wealth allocated to firm j. The first-order conditions for an interior solution can be written as

$$\begin{aligned}&\mathrm{E}\{U'(W^*)(X_i-R)\}=0, \\ &\qquad\qquad\qquad\qquad i=1,\dots,m \\ &\mathrm{E}\{U'(W^*)(Z_j-R)\}=0,\end{aligned} \tag{6.7}$$

and

$$\mathrm{E}\{U'(W^*)[(V'_{j0}(I_j^*)-1)\lambda_j Z^* + W_0^* w_{m+1}^* Z'_j(I_j^*)]\}=0, \tag{6.8}$$

where $W_0^* \equiv \lambda_j[V_{j0}(I_j^*)-I_j^*]+\overline{W}_0$ and $Z'_j(I_j)$ is shorthand for $\partial[V_j(I_j;\theta_j)/V_{j0}(I_j)]/\partial I_j$ for each given value of the random variable θ_j.

Using (6.7), (6.8) can be rewritten as

$$\lambda_j R\mathrm{E}\{U'(W^*)\}(V'_{j0}(I_j^*)-1)+w_{m+1}^* W_0^* \mathrm{E}\{U'(W^*)Z'_j(I_j^*)\}=0. \tag{6.9}$$

Hence, without further conditions on the distribution of the marginal return on security j, $Z'_j(I_j^*)$, the value-maximizing choice for I_j, will not be optimal. Of course, if the post-investment holdings of security j by the investor are zero, ($w_{m+1}^*=0$), then the value-maximizing choice is optimal for him, but this will not sustain a post-investment equilibrium unless all investors would choose $w_{m+1}^*=0$.

However, if the random variable marginal return can be written in the form $Z'_j(I_j^*)=\gamma_j+\sum_1^m \mu_{ij}(X_i-R)+\eta_j(Z_j-R)+\varepsilon_j$, where $\mathrm{E}(\varepsilon_j)=\mathrm{E}(\varepsilon_j|X_i,\ldots,X_m,Z_j)=0$, then using (6.7), (6.9) can be rewritten as

$$\lambda_j R(V'_{j0}(I_j^*)-1)+w_{m+1}^* W_0^* \gamma_j=0. \tag{6.10}$$

If $\gamma_j=0$, then value maximization is again optimal. Even if $\gamma_j\neq 0$, if the condition is imposed that I_j^* be chosen such that post-investment investors in firm j would not want to change it, then $\lambda_j V_{j0}(I_j^*)=w_{m+1}^* W_0^*$, and (6.10) can be rewritten as

$$R(V'_{j0}(I_j^*)-1)+\gamma_j V_{j0}(I_j^*)=0. \tag{6.11}$$

Hence, if the firm chooses its investment so as to satisfy (6.11), then all (pre- and post-investment) stockholders will agree on the investment chosen by the firms. Eq. (6.11) is an example of stockholder unanimity, and the conditions imposed on Z'_j are the same (except for our including ε_j) as in the proposition proved by Eckern and Wilson (1974, p. 175).

If the available technologies exhibit stochastic constant returns [as in the model of Diamond (1967)] and if the technologies are freely available, then a necessary condition for equilibrium is that the value of firm j be no larger than the value of its factor inputs. In our notation, $V_{j0}(I_j^*)\leq I_j^*$. If the investment process is "reversible" or if ex ante investment decisions are agreed upon by ex post stockholders, then $V_{j0}(I_j^*)=I_j^*$. But stochastic constant returns imply that $V_j(I_j;\theta_j)=I_j\theta_j$ and therefore $Z_j(I_j^*)=\theta_j$. Hence, $Z'_j(I_j^*)=0$, and from (6.9) value maximization is optimal.

In summary, if the choice of investment for each firm taken individually is perceived as having no impact on the return distribution of the efficient portfolio set, then value maximization will produce the same investment allocation as would have been chosen by the stockholders of the firms if they had made the decision directly.

There appears to be a connection between this condition and the conditions for a competitive securities market. Since a change in a firm's investment choice will change its market value and can change its own return distribution, the traditional definition of a competitive market as one in which no single participant's action can affect prices in that market makes no sense for a securities market. However, it seems to me that a reasonable condition for a securities market to be *competitive* is that no single participant's action can affect the distribution of returns on any set of portfolios that span the efficient portfolio set. In the case of a certainty world, this condition implies that no single participant's actions can change the interest rate. In the Arrow–Debreu model this condition implies that no single participant's actions can affect the prices of the pure securities $(P_1,\dots, P_M)$. In the Capital Asset Pricing model it implies that no single participant's actions can affect the return distribution for either the market portfolio or the interest rate. Like the traditional definition, this condition implies that no single participant's actions can affect anything "that matters" to everyone.

While this condition is appealing, there are a number of issues that must be resolved before it could be accepted as a definition. First, unlike prices, return distributions are not observable. In models with homogeneous expectations, this is not a problem because, given a set of prices, the return distributions are unambiguously defined. However, in incomplete markets models with heterogeneous beliefs, there are obvious complications. Second, the definition, existence, and optimality properties of a "competitive" equilibrium would require derivation. Third, while it is already a standard assumption of portfolio theory that the actions of a single individual or financial institution will not affect the distribution of returns on securities, the case for individual firms investment decisions is more subtle. In discussing the value-maximization criterion in incomplete markets, Radner notes that with the exception of the Diamond model: "I have not seen a formulation of this hypothesis that enables the producer to act as a price-taker; i.e., that does not imply that the producer is able to calculate the effect of his actions on the equilibrium prices."[28] In general, his comment would also apply to the assumption that producers take the distribution of the efficient set as given. While this issue merits careful study, my suspicion is that if significant non-trivial spanning of the efficient set obtains, then a producer might not be able to calculate the effect of his actions on the equilibrium return distribution for the efficient set. [See Fama (1972) and Fama and Laffer (1972) for just such a formulation.] If these issues can be satisfactorily decided then much of the current

[28]Radner (1970, p. 460).

controversy surrounding the theory of the firm with incomplete securities markets will be settled.

In discussing the investment decision by firms, it was assumed that each firm had a single production technology and the investment decision was to choose the intensity at which the technology is operated. However, firms can also make investments by buying other firms' technologies. To examine the effects of this type of investment, suppose that the first k technologies are owned by a single firm and so, instead of n, there are only $n-k+1$ firms. If there are no economies of scale in such a consolidation (i.e. no "synergy"), then the end-of-period value for the consolidated firm can be written as

$$V(I_1,\ldots,I_k;\theta_1,\ldots,\theta_k)\equiv\sum_1^k V_j(I_j;\theta_j). \tag{6.12}$$

Theorem 6.2

Suppose $(X_1,\ldots,X_m,R)$ span the efficient portfolio set and (6.12) holds. If the return distribution for $(X_1,\ldots,X_m,R)$ remains the same as it was prior to consolidation, then, post-consolidation, the initial value of the consolidated firm is given by

$$V_0=\sum_1^k V_{j0}(I_j).$$

The proof follows immediately from the "value additivity" property derived in Corollary 4.13.a. Indeed, if the investment decision of any single firm (including the consolidated firm) cannot affect the distribution of $(X_1,\ldots,X_m,R)$ (6.12) holds, and firms make their investment decision according to the value maximization rule, then from Theorem 6.2 the investment allocation chosen by the consolidated firm will be identical to the allocation that would have been chosen by the k individual firms. In this case, consolidation of firms has no effect on individual welfare because both the initial wealth of each individual and the return distribution on his optimal portfolio will remain unchanged. Indeed, if (6.12) holds and the (pre-consolidation) allocation of investment is optimal, then individuals cannot be any better off after consolidation. However, it is possible that, post-consolidation, investors could be worse off. For example, suppose there do not exist "perfect substitute" securities for at least some of the individual firms and pre-consolidation, not all investors chose to hold these k firms in proportion to their market values. Since, post-consolidation, investors can only hold these k firms in the same relative proportions, some pre-consolidation optimal portfolios will not be feasible after consolidation, and therefore some individuals could be worse off.

In an analogous fashion to the individual's portfolio selection behavior, it has been argued that firms acquire other firms to reduce risk. However, unless an acquisition provides a production opportunity otherwise unavailable or unless the acquired firm was not operating its technology efficiently, diversification by the firm cannot improve the welfare of individuals, and in some cases it can reduce it. This result serves as a warning against the indiscriminate use of models that treat firms "like" individuals and ascribe to the firms exogenously given utility functions rather than deriving endogenous criterion functions for ranking their choices.

In the discussion of the investment decision by firms, the method of financing these investments was not made explicit. Implicitly, it was assumed that firms used internally available funds or issued additional financial securities where all financial claims against the firm were of a single type called *equity*. Of course, it is well known that, in addition to equity, firms also issue other types of financial claims (e.g. debt, preferred stock, and convertible bonds). The choice of the menu of financial liabilities is called the firm's *financing decision*. Although, in general, the optimal investment and financing decisions by a firm are determined simultaneously, it is useful to study the financing decision by taking the investment decision as given.

Consider firm j with random variable end-of-period value V^j and q different financial claims. The kth such financial claim is defined by the function $f_k[V^j]$ which describes how the holders of this security will share in the end-of-period value of the firm. The production technology and choice of investment intensity, $V_j(I_j;\,\theta_j)$ and I_j, are taken as given. If it is assumed that the end-of-period value of the firm is independent of its choice of financial liabilities,[29] then $V^j = V_j(I_j;\,\theta_j)$, and $\sum_1^q f_k \equiv V_j(I_j;\,\theta_j)$ for every outcome θ_j.

Suppose when firm j is all equity financed there exists an equilibrium such that the initial value of firm j is given by $V_{j0}(I_j)$.

Theorem 6.3

If firm j is financed by q different claims defined by the functions $f_k(V^j)$, $k=1,\dots,q$, and if there exists an equilibrium such that the return distributions of the efficient portfolio set remains unchanged from the equilibrium in which firm j

[29]By this assumption, I have formally ruled out financial securities that alter the tax liabilities of the firm (e.g. interest deductions) or ones that can induce "outside" costs (e.g. bankruptcy costs). However, by redefining $V_j(I_j;\,\theta_j)$ as the pre-tax and bankruptcy value of the firm and letting one of the f_k represent the government's (tax) claim and another the lawyers' (bankruptcy) claim, then the analysis in the text will be valid for these extended securities as well.

was all equity financed, then

$$\sum_{1}^{q} f_{k0} = V_{j0}(I_j),$$

where f_{k0} is the equilibrium initial value of financial claim k.

Proof

In the equilibrium in which firm j is all equity financed, the end-of-period random variable value of firm j is $V_j(I_j; \theta_j)$ and the initial value, $V_{j0}(I_j)$, is given by formula (4.3) where $(X_1, \ldots, X_m, R)$ span the efficient set. Consider now that firm j is financed by the q different claims. The random variable end-of-period value of firm j, $\Sigma_1^q f_k$, is still given by $V_j(I_j; \theta_j)$. By hypothesis, there exists an equilibrium such that the distribution of the efficient portfolio set remains unchanged, and therefore the distribution of $(X_1, \ldots, X_m, R)$ remains unchanged. By inspection of formula (4.3), the initial value of firm j will remain unchanged, and therefore $\Sigma_1^q f_{k0} = V_{j0}(I_j)$.

Hence, for a given investment policy, the way in which the firm finances this investment will not affect the market value of the firm unless the choice of financial instruments changes the return distributions of the efficient portfolio set. Theorem 6.3 is representative of a class of theorems that describe the impact of financing policy on the market value of a firm when the investment decision is held fixed, and this class is generally referred to as *Modigliani–Miller Theorems* after the pioneering work in this direction by Modigliani and Miller.[30]

Clearly, a sufficient condition for Theorem 6.3 to obtain is that each of the financial claims issued by the firm are "redundant securities" (as defined in Sections 2 and 4). This condition will be satisfied by the subclass of financial securities that provide for *linear* sharing rules, i.e $f_k(V) = a_k V + b_k$, where $\Sigma_1^q a_k = 1$ and $\Sigma_1^q b_k = 0$. They are redundant securities because the investor can replicate exactly the payoff structure of each claim by a portfolio combination of the (all-equity) firm and the riskless security. Hence, in this case Theorem 6.3 will obtain as a special case of Proposition 4.1. Indeed, an example of this subclass is in the original Modigliani–Miller paper where they examined the effect on firm values of borrowing by firms under the assumption that borrowing (either by firms or individuals) is riskless and there is no bankruptcy.

It should be pointed out that the linear structure for firm borrowings only applies if there is no chance of default on the debt. Consider the case of a single homogeneous debt issue where the firm promises to pay B dollars at the end of the period and in the event the firm does not pay (i.e. defaults), then ownership of

[30] Miller and Modigliani (1958). See also Stiglitz (1969, 1974), Fama (1978), and Miller (1977).

the firm is transferred to the debtholders. If the equity of the firm has limited liability, then the payoff function to the debt, f_1, can be written as

$$f_1(V^j)=\min[V^j, B] \tag{6.13}$$

and the payoff function to the equity, f_2, can be written as

$$f_2(V^j)=\max[0, V^j-B]. \tag{6.14}$$

Hence, f_1 and f_2 will have a linear sharing-rule structure only if the probability that $V^j<B$ is zero.

While the linearity of the sharing rules is sufficient, it is not a necessary condition for Theorem 6.3 to obtain as Stiglitz (1969, 1974) has shown for the Arrow–Debreu and Capital Asset Pricing Models, and as will be demonstrated in Section 7.

7. Intertemporal consumption and portfolio selection theory

As with the preceding analyses here most papers on investment theory under uncertainty have assumed that individuals act so as to maximize the expected utility of end-of-period wealth and that intra-period revisions are not allowed. Therefore, all events which take place after next period are irrelevant to their decisions. Of course, individuals, and therefore firms, do care about events beyond "next period", and they can review their allocations periodically. Hence, the one-period, static analyses will only be valid under those conditions such that an intertemporally-maximizing individual acts, each period, as if he were a one-period, expected utility-of-wealth maximizer. In this section the lifetime consumption-portfolio selection problem is solved, and conditions are derived under which the one-period static portfolio problem will be an appropriate "surrogate" for the dynamic, multi-period portfolio problem.

As in the early contributions by Hakansson (1970), Samuelson (1969), and Merton (1969), the problem of choosing optimal portfolio and consumption rules for an individual who lives T years is formulated as follows. The individual chooses his consumption and portfolio allocation for each period so as to maximize[31]

$$E_0\left\{\sum_{0}^{T-1} U[C(t),t]+B[W(T),T]\right\}, \tag{7.1}$$

[31]The additivity of the utility function and the single-consumption-good assumptions are made for analytical simplicity and because the principal topic of this paper is investment allocation and not the individual consumption choice. Fama (1970b) analyzes the problem for non-additive utilities. Although T is treated as known in the text, the analysis is essentially the same for an uncertain lifetime with T a random variable. Cf. Richard (1975) and Merton (1971).

where $C(t)$ is consumption chosen at age t; $W(t)$ is wealth at age t; E_t is the conditional expectation operator conditional on knowing all relevant information available as of time t; the utility function (during life) U is assumed to be strictly concave in C; and the "bequest" function B is also assumed to be concave in W.

It is assumed that there are n risky securities with random variable returns between time t and $t+1$ denoted by $Z_1(t+1),\ldots, Z_n(t+1)$, and there is a riskless security whose return between t and $t+1$, $R(t)$, will be known with certainty as of time t.[32] When the individual "arrives" at date t, he will know the value of his portfolio, $W(t)$. He chooses how much to consume, $C(t)$, and then reallocates the balance of his wealth, $W(t)-C(t)$, among the available securities. Hence, the accumulation equation between t and $t+1$ can be written as[33]

$$W(t+1)=\left[\sum_1^n w_j(t)\left[Z_j(t+1)-R(t)\right]+R(t)\right]\left[W(t)-C(t)\right], \tag{7.2}$$

where $w_j(t)$ is the fraction of his portfolio allocated to security j at date t, $j=1,\ldots, n$. Because the fraction allocated to the riskless security can always be chosen to equal $1-\sum_1^n w_j(t)$, the choices for $w_1(t),\ldots, w_n(t)$ are unconstrained.

It is assumed that there exist m state variables, $\{S_k(t)\}$, such that the stochastic processes for $\{Z_1(t+1),\ldots, Z_n(t+1), S_1(t+1),\ldots, S_m(t+1)\}$ are Markov with respect to $S_1(t),\ldots, S_m(t)$, and $S(t)$ will denote the m-vector of state variable values at time t.[34]

The method of stochastic dynamic programming is used to derive the optimal consumption and portfolio rules. Define the function $J[W(t), S(t), t]$ by

$$J[W(t), S(t), t]\equiv\max E_t\left\{\sum_t^{T-1} U[C(\tau),\tau]+B[W(T),T]\right\}. \tag{7.3}$$

J, therefore, is the (utility) value of the balance of the individual's optimal consumption–investment program from date t forward, and, in this context, is

[32] This definition of a riskless security is purely technical and without normative significance. For example, investing solely in the riskless security will not allow for a certain consumption stream because $R(t)$ will vary stochastically over time. On the other hand, a T-period, riskless-in-terms-of-default coupon bond which allows for a certain consumption stream is not a riskless security because its one-period return is uncertain. For further discussion, see Merton (1973b).

[33] It is assumed that all income comes from investment in securities. The analysis would be the same with wage income provided that investors can sell shares against future income. However, because institutionally this cannot be done, the "non-marketability" of wage income will cause systematic effects on the portfolio and consumption decisions.

[34] Many non-Markov stochastic processes can be transformed to fit the Markov format by expanding the number of state variables. Cf. Cox and Miller (1968, pp. 16–18). To avoid including "surplus" state variables, it is assumed that $\{S(t)\}$ represent the minimum number of variables necessary to make $\{Z_j(t+1)\}$ Markov.

called the "derived" utility of wealth function. By the Principle of Optimality, (7.3) can be rewritten as

$$J[W(t),S(t),t]=\max\{U[C(t),t]+E_t(J[W(t+1),S(t+1),t+1])\}, \tag{7.4}$$

where "max" is over the current decision variables $[C(t),w_1(t),\dots,w_n(t)]$. Substituting for $W(t+1)$ in (7.4) from (7.2) and differentiating with respect to each of the decision variables, we can write the $n+1$ first-order conditions for a regular interior maximum as[35]

$$0=U_C[C^*(t),t]-E_t\left\{J_W[W(t+1),S(t+1),t+1]\left(\sum_1^n w_j^*(Z_j-R)+R\right)\right\} \tag{7.5}$$

and

$$0=E_t\{J_W[W(t+1),S(t+1),t+1](Z_j-R)\},\qquad j=1,2,\dots,n, \tag{7.6}$$

where $U_C\equiv\partial U/\partial C$; $J_W\equiv\partial J/\partial W$; and (C^*,w_j^*) are the optimum values for the decision variables. Henceforth, except where needed for clarity, the time indices will be dropped. Using (7.6), (7.5) can be rewritten as

$$0=U_C[C^*,t]-RE_t\{J_W\}. \tag{7.7}$$

To solve for the complete optimal program, one first solves (7.6) and (7.7) for C^* and w^* as functions of $W(t)$ and $S(t)$ when $t=T-1$. This can be done because $J[W(T),S(T),T]=B[W(T),T]$, a given function. Substituting the solutions for $C^*(T-1)$ and $w^*(T-1)$ in the right-hand side of (7.4), (7.4) becomes an equation and therefore one has $J[W(T-1),S(T-1),T-1]$. Using (7.6), (7.7), and (7.4), one can proceed to solve for the optimal rules in earlier periods in the usual "backwards" recursive fashion of dynamic programming. Having done so, one will have a complete schedule of optimal consumption and portfolio rules for each date expressed as functions of the (then) known state variables $W(t)$, $S(t)$, and t. Moreover, as Samuelson (1969) has shown, the optimal consumption rules will satisfy the "envelope condition" expressed as

$$J_W[W(t),S(t),t]=U_C[C^*(t),t], \tag{7.8}$$

[35]Sufficient conditions for existence are described in Bertsekas (1974). Uniqueness of the solutions are guaranteed by: (1) strict concavity of U and B; (2) no redundant securities; and (3) no arbitrage opportunities. Cf. Dreyfus (1965) for the dynamic programming technique.

i.e. at the optimum, the marginal utility of wealth (future consumption) will just equal the marginal utility of (current) consumption. Moreover, from (7.8) it is straightforward to show that $J_{WW}<0$ because $U_{CC}<0$. Hence, J is a strictly concave function of wealth.

A comparison of the first-order conditions for the static portfolio selection problem, (2.4) in Section 2, with the corresponding conditions (7.6) for the dynamic problem will show that they are formally quite similar. Of course, they do differ in that, for the former case, the utility function of wealth is taken to be exogenous while, in the latter, it is derived. However, the more fundamental difference in terms of derived portfolio selection behavior is that J is not only a function of W but also a function of S. The analogous condition in the static case would be that the end-of-period utility function of wealth is also state dependent.

To see that this difference is not trivial, consider the Rothschild–Stiglitz definition of "riskier" that was used in the one-period analysis to partition the feasible portfolio set into its efficient and inefficient parts. Let W_1 and W_2 be the random variable, end-of-period values of two portfolios with identical expected values. If W_2 is equal in distribution to W_1+Z, where $E(Z|W_1)=0$, then from (2.10) and (2.11), W_2 is riskier than W_1 and every risk-averse maximizer of the expected utility of end-of-period wealth would prefer W_1 to W_2. However, consider an intertemporal maximizer with a strictly concave, derived utility function J. It will not, in general, be true that $E_t\{J[W_1, S(t+1), t+1]\}>E_t\{J[W_2, S(t+1), t+1]\}$. Therefore, although the intertemporal maximizer selects his portfolio for only one period at a time, the optimal portfolio selected may be one that would never be chosen by any risk-averse, one-period maximizer. Hence, the portfolio selection behavior of an intertemporal maximizer will, in general, be operationally distinguishable from the behavior of a static maximizer.

To adapt the Rothschild–Stiglitz definition to the intertemporal case, a stronger condition is required, namely if W_2 is equal in distribution to W_1+Z, where $E[Z|W_1, S(t+1)]=0$, then every risk-averse intertemporal maximizer would prefer to hold W_1 rather than W_2 in the period t to $t+1$. The proof follows immediately from the concavity of J and Jensen's Inequality. Namely, $E_t\{J[W_2, S(t+1), t+1]\} = E_t\{E(J[W_2, S(t+1), t+1]| W_1, S(t+1))\}$. By Jensen's Inequality, $E(J[W_2, S(t+1), t+1]|W_1, S(t+1))< J[E(W_2|W_1, S(t+1)), S(t+1), t+1]= J[W_1, S(t+1), t+1]$, and, therefore, $E_t\{J[W_2, S(t+1), t+1]\}< E_t\{J[W_1, S(t+1), t+1]\}$. Hence, "noise" as denoted by Z must not only be noise relative to W_1 but noise relative to the state variables $S_1(t+1),\ldots,S_m(t+1)$. All the analyses of the preceeding sections can be formally adapted to the intertemporal framework by simply requiring that the "noise" terms there, ε, have the additional property that $E_t(\varepsilon|S(t+1))= E_t(\varepsilon)=0$. Hence, in the absence of further restrictions on the distributions, the resulting efficient portfolio set for intertemporal maximizers will be larger than in the static case.

However, under certain conditions,[36] the portfolio selection behavior of intertemporal maximizers will be "as if" they were one-period maximizers. For example, if $E_t[Z_j(t+1)] \equiv \bar{Z}_j(t+1) = E_t[Z_j(t+1)|S(t+1)]$, $j=1,2,\dots,n$, then the additional requirement that $E_t(\varepsilon|S(t+1))=0$ will automatically be satisfied for any feasible portfolio, and the original Rothschild–Stiglitz "static" definition will be valid. Indeed, in the cited papers by Hakansson, Samuelson, and Merton, it is assumed that the security returns $\{Z_1(t),\dots, Z_n(t)\}$ are serially independent and identically distributed in time which clearly satisfies this condition. Define the *investment opportunity set at time t* to be the joint distribution for $\{Z_1(t+1),\dots, Z_n(t+1)\}$ and the return on the riskless security, $R(t)$. The Hankansson et al. papers assume that the investment opportunity set is constant through time. The condition $\bar{Z}_j(t+1) = E_t[Z_j(t+1)|S(t+1)]$, $j=1,\dots,n$, will also be satisfied if changes in the investment opportunity set are either completely random or time dependent in a non-stochastic fashion. Moreover, with the possible exception of a few perverse cases, these are the only conditions on the investment opportunity set under which $\bar{Z}_j(t+1) = E_t[Z_j(t+1)|S(t+1)]$, $j = 1,\dots,n$. Hence, for arbitrary concave utility functions, the one-period analysis will be a valid surrogate for the intertemporal analysis only if changes in the investment opportunity set satisfy these conditions.

Of course, by inspection of (7.6), if J were of the form $V[W(t),t]+H[S(t),t]$ so that $J_W = V_W$ is only a function of wealth and time, then for arbitrary investment opportunity sets such an intertemporal investor will act "as if" he is a one-period maximizer. Unfortunately, the only concave utility function that will produce such a J function and satisfy the additivity specification in (7.1) is $U[C,t]=a(t)\log[C]$ and $B[W,T]=b(T)\log[W]$, where either $a=0$ and $b>0$ or $a>0$ and $b\geq 0$. While some have argued that this utility function is of special normative significance,[37] any model whose results depend singularly upon all individuals having the same utility function and where, in addition, the utility function must have a specific form, can only be viewed as an example, and not the basis for a theory.

Hence, in general, the one-period, static analysis will not be rich enough to describe the investment behavior of individuals in an intertemporal framework. Indeed, without additional assumptions, the only derived restrictions on optimal demand functions and equilibrium security returns are the ones that rule out arbitrage. Hence, to deduce additional properties, further assumptions about the dynamics of the investment opportunity set will be needed. However, before these assumptions are introduced, I make a brief digression.[38]

[36]See Fama (1970b) for a general discussion of these conditions.

[37]See Latané (1959), Markowitz (1976), and Rubinstein (1976) for arguments in favor of this view, and Samuelson (1971), Goldman (1974), and Merton and Samuelson (1974) for arguments in opposition to this view.

[38]This digression is adapted from Merton (1975a, pp. 662–663).

There are three time intervals or horizons involved in the consumption-portfolio problem. First, there is the *trading horizon* which is the *minimum* length of time between which successive transactions by economic agents can be made in the market. In a sequence-of-markets analysis it is the length of time between successive market openings, and is therefore part of the specification of the structure of markets in the economy. While this structure will depend upon the tradeoff between the costs of operating the market and its benefits, this time scale is not determined by the individual investor, and is the same for all investors in the economy. Second, there is the *decision horizon* which is the length of time between which the investor makes successive decisions, and it is the minimum time between which he would take any action. For example, an investor with a fixed decision interval of one month, who makes a consumption decision and portfolio allocation today, will under no conditions make any new decisions or take any action prior to one month from now. This time scale is determined by the costs to the individual of processing information and making decisions, and is chosen by the individual. Third, there is the *planning horizon* which is the maximum length of time for which the investor gives any weight in his utility function. Typically, this time scale would correspond to the balance of his lifetime and is denoted by T in the formulation (7.1).

The static approach to portfolio selection implicitly assumes that the individual's decision and planning horizons are the same: "one period". While the intertemporal approach distinguishes between the two, when individual demands are aggregated to determine market equilibrium relationships, it is implicitly assumed in both approaches that the decision interval is the same for all investors, and therefore corresponds to the trading interval. If h denotes the length of time in the trading interval, then every solution derived has, as an implicit argument, h. Clearly, if one were to change h, then the derived behavior of investors would change, as indeed would any deduced equilibrium relationships.[39] I might mention, somewhat parenthetically, that empirical researchers almost uniformly neglect to recognize that h is part of a model's specification. For example, in Theorem 4.6 the returns on securities were shown to have a linear relationship to the returns on a set of spanning portfolios. However, because the n-period return on a security is the *product* (and not the sum) of the one-period returns, this linear relationship can only obtain for the single time interval, h. If we define a fourth time interval, the *observation horizon*, to be the length of time between successive observations of the data by the researcher, then the usual empirical practice is implicitly to assume that the decision and trading intervals are equal to the observation interval. This is done whether the observation interval is daily, weekly, monthly, or annually!

[39]If investor behavior were invariant to h, then investors would choose the same portfolio if they were "frozen" into their investments for ten years as they would if they could revise their portfolios every day.

If the "frictionless" markets assumption (Assumption 1) is extended to include no costs of information processing or of operating the markets, then it follows that all investors would prefer to have h as small as physically possible. Indeed, the aforementioned general assumption that all investors have the same decision interval will, in general, only be valid if all such costs are zero. This said, it is natural to consider the limiting case when h tends to zero and trading takes place continuously in time.

Returning from this digression on time scales, consider an economy where the trading interval, h, is sufficiently small such that the state description of the economy can change only "locally" during the interval $(t, t+h)$. Formally, the Markov stochastic processes for the state variables, $S(t)$ are assumed to satisfy the property that one-step transitions are permitted only to the nearest neighboring states. The analogous condition in the limiting case of continuous time is that the sample paths for $S(t)$ are continuous functions of time, i.e. for every realization of $S(t+h)$ except possibly on a set of measure zero, $\lim_{h\to 0}[S_k(t+h)-S_k(t)]=0$, $k=1,\dots,m$. If, however, in the continuous limit the uncertainty of "end-of-period" returns is to be preserved, then an additional requirement is that $\lim_{h\to 0}[S_k(t+h)-S_k(t)]/h$ exists almost nowhere, i.e. even though the sample paths are continuous, the increments to the states are not, and therefore, in particular, "end-of-period" rates of return will not be "predictable" even in the continuous time limit. The class of stochastic processes that satisfy these conditions are called *diffusion processes*.[40]

Although such processes are almost nowhere differentiable in the usual sense, under some mild regularity conditions there is a generalized theory of stochastic differential equations which allows their instantaneous dynamics to be expressed as the solution to the system of equations:[41]

$$\mathrm{d}S_i(t)=G_i(S,t)\,\mathrm{d}t+H_i(S,t)\,\mathrm{d}q_i(t), \qquad i=1,\dots,m, \tag{7.9}$$

where $G_i(S,t)$ is the instantaneous expected change in $S_i(t)$ per unit time at time t; H_i^2 is the instantaneous variance of the change in $S_i(t)$, where it is understood that these statistics are conditional on $S(t)=S$. The $\mathrm{d}q_i(t)$ are Weiner processes with the instantaneous correlation coefficient per unit of time between $\mathrm{d}q_i(t)$ and $\mathrm{d}q_j(t)$ given by the function $\eta_{ij}(S,t)$, i, $j=1,\dots,m$.[42] Moreover, the functions

[40] See Feller (1966), Itô and McKean (1964), and Cox and Miller (1968).

[41] (7.9) is a short-hand expression for the stochastic integral

$$S_i(t)=S_i(0)+\int_0^t G_i(S,\tau)\,\mathrm{d}\tau+\int_0^t H_i(S,\tau)\,\mathrm{d}q_i,$$

where $S_i(t)$ is the solution to (7.9) with probability one. For a general discussion and proofs, see Itô and McKean (1964) McKean (1969), and McShane (1974).

[42] $\int_0^t \mathrm{d}q_i=q_i(t)-q_i(0)$ will be normally distributed with a zero mean and variance equal to t.

$\{G_i, H_i, \eta_{ij}, i,j=1,\ldots,m\}$ completely describe the transition probabilities for $S(t)$ between any two dates.[43]

Under the assumption that the returns on securities can be described by diffusion processes, Merton (1971) has solved the continuous-time analog to the discrete-time formulation in (7.1), namely

$$\max E_0\left\{\int_0^T U[C(t),t]\,\mathrm{d}t + B[W(T),T]\right\}. \tag{7.10}$$

Adapting the notation in that paper,[44] the rate of return dynamics on the security j can be written as

$$\frac{\mathrm{d}P_j}{P_j} = \alpha_j(S,t)\,\mathrm{d}t + \sigma_j(S,t)\,\mathrm{d}Z_j, \qquad j=1,\ldots,n, \tag{7.11}$$

where α_j is the instantaneous conditional expected rate of return per unit time; σ_j^2 is its instantaneous conditional variance per unit time; and $\mathrm{d}Z_j$ are Weiner processes with the instantaneous correlation coefficient per unit time between $\mathrm{d}Z_j(t)$ and $\mathrm{d}Z_k(t)$ given by the function $\rho_{jk}(S,t)$, $j,k=1,\ldots,n$. In addition to the n risky securities, there is a riskless security whose instantaneous rate of return per unit time is the interest rate $r(t)$.[45] To complete the model's dynamics description, define the functions $\mu_{ij}(S,t)$ to be the instantaneous correlation coefficients per unit time between $\mathrm{d}q_i(t)$ and $\mathrm{d}Z_j(t)$, $i=1,\ldots,m; j=1,\ldots,n$.[46]

If J is defined by

$$J[W(t),S(t),t] \equiv \max \mathrm{E}_t\left\{\int_t^T U[C(\tau),\tau]\,\mathrm{d}\tau + B[W(T),T]\right\}, \tag{7.12}$$

then the continuous-time analog to (7.4) can be written as[47]

$$0 = \max\Bigg\{U[C,t] + J_t + J_W\left[\left(\sum_1^n w_j(\alpha_j - r) + r\right)W - C\right] + \sum_1^m J_i G_i + \tfrac{1}{2}J_{WW}\sum_1^n\sum_1^n w_i w_j \sigma_{ij} W^2 + \tfrac{1}{2}\sum_1^m\sum_1^m J_{ij} H_i H_j \eta_{ij} + \sum_1^m\sum_1^n J_{iW} w_j \sigma_j H_i \mu_{ij} W\Bigg\},$$

[43] See Feller (1966, pp. 320–321); Cox and Miller (1968, p. 215). The transition probabilities will satisfy the Kolmogrov or Fokker–Planck partial differential equations.

[44] Merton (1971, p. 377). $\mathrm{d}P_j/P_j$ in continuous time corresponds to $Z_j(t+1)-1$ in the discrete time analysis.

[45] $r(t)$ corresponds to $R(t)-1$ in the discrete time analysis, and is the "force of interest", continuous rate. While the rate earned between t and $(t+\mathrm{d}t)$, $r(t)$, is known with certainty as of time t, $r(t)$ can vary stochastically over time.

[46] Unlike in the Arrow–Debreu model, for example, it is not assumed here that the returns are necessarily completely described by the changes in the state variables, $\mathrm{d}S_i$, $i=1,\ldots,m$, i.e. the $\mathrm{d}Z_j$ need not be instantaneously perfectly correlated with some linear combination of $\mathrm{d}q_1,\ldots,\mathrm{d}q_m$. Rather, it is only assumed that $(\mathrm{d}P_1/P_1,\ldots,\mathrm{d}P_n/P_n, \mathrm{d}S_1,\ldots,\mathrm{d}S_m)$ is Markov in $S(t)$.

[47] See Merton (1971, p. 381) and Kushner (1967, Ch. IV, Theorem 7).

where the subscripts t, W, and i on J denote partial derivatives with respect to the arguments t, W, and S_i, $(i=1,\ldots,m)$ of J, respectively, and $\sigma_{ij}\equiv\sigma_i\sigma_j\rho_{ij}$ is the instantaneous covariance of the returns of security i with security j, $i,j=1,\ldots,n$. As was the case in (7.4), the "max" in (7.13) is over the current decision variables $[C(t), w_1(t),\ldots,w_n(t)]$. If C^* and w^* are the optimum rules, then the $(n+1)$ first-order conditions for (7.13) can be written as

$$0=U_C[C^*,t]-J_W[W,S,t] \tag{7.14}$$

and

$$0=J_W(\alpha_j-r)+J_{WW}\sum_1^n w_i^*\sigma_{ij}W+\sum_1^m J_{iW}\sigma_j H_i\mu_{ij}, \qquad j=1,\ldots,n. \tag{7.15}$$

Eq. (7.14) is identical to the "envelope condition", (7.8), in the discrete time analysis. However, unlike (7.6) in the discrete time analysis, (7.15) is a system of equations which is *linear* in the optimal demands for risky securities. Hence, if none of the risky securities is redundant, then (7.15) can be solved explicitly for the optimal demand functions using standard matrix inversion, i.e.

$$w_j^*(t)W(t)=K\sum_1^n v_{kj}(\alpha_k-r)+\sum_1^m B_i\zeta_{ij}, \qquad j=1,\ldots,n, \tag{7.16}$$

where v_{kj} is the $k-j$th element of the inverse of the instantaneous variance–covariance matrix of returns $[\sigma_{ij}]$;

$$\zeta_{ij}\equiv\sum_1^n v_{kj}\sigma_k H_i\mu_{ik};\ K\equiv -J_W/J_{WW};\ \text{and } B_i\equiv -J_{iW}/J_{WW},\ i=1,\ldots,m.$$

As an immediate consequence of (7.16), we have the following mutual fund theorem:

Theorem 7.1

If the returns dynamics are described by (7.9) and (7.11), then there exist $(m+2)$ mutual funds constructed from linear combinations of the available securities such that, independent of preferences, wealth distribution, or planning horizon, individuals will be indifferent between choosing from linear combinations of just these $(m+2)$ funds or linear combinations of all n risky securities and the riskless security.

Proof

Let mutual fund #1 be the riskless security; let mutual fund #2 hold fraction, $\delta_j\equiv\sum_1^n v_{kj}(\alpha_k-r)$, in security j, $j=1,\ldots,n$, and the balance $(1-\sum_1^n\delta_j)$ in the riskless security; let mutual fund $\#(2+i)$ hold fraction $(\delta_j^i \equiv \zeta_{ij})$ in security j, $j=1,\ldots,n$ and the balance $(1-\sum_1^n \delta_j^i)$ in the riskless security for $i=1,\ldots,m$. Consider a portfolio of these mutual funds which allocates $d_2(t)=K$ dollars to fund #2; $d_{2+i}(t)=B_i$ dollars to fund $(\#2+i)$, $i=1,\ldots,m$; and $d_1(t)=W(t)-\sum_2^{2+m} d_i(t)$ dollars to fund #1. By inspection of (7.16), this portfolio of funds exactly replicates the optimal portfolio holdings chosen from among the original n risky securities and the riskless security. However, the fractional holdings of these securities by the $(m+2)$ funds do not depend upon the preferences, wealth, or planning horizon of the individuals investing in the funds. Hence, every investor can replicate his optimal portfolio by investing in the $(m+2)$ funds.

Of course, as with the mutual fund theorems of Section 4, Theorem 7.1 is vacuous if $m \geqslant n+1$. However, for $m \ll n$, the $(m+2)$ portfolios provide for a non-trivial spanning of the efficient portfolio set, and it is straightforward to show that the instantaneous returns on individual securities will satisfy the same linear specification relative to these spanning portfolios as was derived in Theorem 4.6 for the one-period analysis.

It was shown in the discrete-time analysis that if $E_t[Z_j(t+1)|S(t)]=\bar{Z}_j(t+1)$, $j=1,\ldots,n$, then the intertemporal maximizer will act "as if" he were a static maximizer of the expected utility of end-of-period wealth. The corresponding condition in the continuous-time case translates into $\mu_{ij}=0$, $i=1,\ldots,m$, and $j=1,\ldots,n$. Under these conditions the optimal demand functions in (7.16) can be rewritten as

$$w_j^*(t)W(t)=K\sum_1^n v_{jk}(\alpha_k-r), \qquad j=1,\ldots,n. \tag{7.17}$$

From inspection of (7.17) the relative holdings of risky securities, $w_j^*(t)/w_i^*(t)$, will be the same for all investors and, hence, the efficient portfolio set will be spanned by just two funds: a single risky fund and the riskless security. Moreover, by the procedure used to prove Theorem 4.9 and Theorem 4.10 in the static analysis, the efficient portfolio set here can be shown to be generated by the set of portfolios with minimum (instantaneous) variance for a given expected rate of return. Hence, under these conditions the continuous-time intertemporal maximizer will act "as if" he were a static, Markowitz–Tobin mean–variance maximizer. Although the demand functions are formally identical to those derived from the mean–variance model, the continuous-time analysis is valid for any concave utility function and does not assume that security returns are normally distributed. Indeed, if, for example, the investment opportunity set $\{\alpha_j, r, \sigma_{ij}, i, j=1,\ldots,n\}$ is constant through time, then from (7.11) the return on each risky security will be lognormally distributed, which implies that all securities will have limited liability.[48]

[48]See Merton (1971, p. 384–488). It is also shown there that the return will be long-normal on the risky fund which, together with the riskless security, spans the efficient portfolio set.

In the general case described in Theorem 7.1, the qualitative behavioral differences between an intertemporal maximizer and a static maximizer can be clarified further by analyzing the characteristics of the derived spanning portfolios.

As already shown, fund #1 and fund #2 are the "usual" portfolios that would be mixed to provide an optimal portfolio for a static maximizer. Hence, the intertemporal behavioral differences are characterized by funds #$(2+i)$, $i=1,\dots,m$. At the level of demand functions, the "differential demand" for risky security j, ΔD_j^*, is defined to be the difference between the demand for that security by an intertemporal maximizer at time t and the demand for that security by a static maximizer of the expected utility of "end-of-period" wealth where the absolute risk aversion and current wealth of the two maximizers are the same. Noting that $K\equiv -J_W/J_{WW}$ is the reciprocal of the absolute risk aversion of the derived utility of wealth function, from (7.16) we have that

$$\Delta D_j^* = \sum_1^m B_i \zeta_{ij}, \qquad j=1,\dots,n. \tag{7.18}$$

Lemma 7.1

Define

$$\mathrm{d}Y_i \equiv \mathrm{d}S_i - \left(\sum_1^n \delta_j^\dagger \left(\frac{\mathrm{d}P_j}{P_j} - r\,\mathrm{d}t \right) + r\,\mathrm{d}t \right).$$

The set of portfolio weights $\{\delta_j^\dagger\}$ that minimize the (instantaneous) variance of $\mathrm{d}Y_i$ are given by $\delta_j^\dagger = \zeta_{ij}$, $j=1,\dots,n$ and $i=1,\dots,m$.

Proof

The instantaneous variance of $\mathrm{d}Y_i$ is equal to $[H_i^2 - 2\sum_1^n \delta_j^\dagger H_i \sigma_j \mu_{ij} + \sum_1^n \sum_1^n \delta_j^\dagger \delta_k^\dagger \sigma_{jk}]$. Hence, the minimizing set of $\{\delta_j^\dagger\}$ will satisfy $0 = -H_i \sigma_j \mu_{ij} + \sum_1^n \delta_k^\dagger \sigma_{jk}$, $j=1,\dots,n$. By matrix inversion, $\delta_j^\dagger = \zeta_{ij}$.

The instantaneous rate of return on fund #$(2+i)$ is exactly $[r\,\mathrm{d}t + \sum_1^n \zeta_{ij}(\mathrm{d}P_j/P_j - r\,\mathrm{d}t)]$. Hence, fund #$(2+i)$ can be described as that feasible portfolio whose rate of return most closely replicates the stochastic part of the instantaneous change in state variable $S_i(t)$, and this is true for $i=1,\dots,m$.

Consider the special case where there exist securities that are instantaneously perfectly correlated with changes in each of the state variables. Without loss of generality, assume that the first m securities are the securities such that $\mathrm{d}P_i/P_i$ is perfectly positively correlated with $\mathrm{d}S_i$, $i=1,\dots,m$. In this case[49] the demand

[49] This case is similar in spirit to the Arrow–Debreu complete markets model.

functions (7.16) can be rewritten in the form

$$\begin{aligned} w_i^*(t)W(t) &= K\sum_1^n v_{ik}(\alpha_k - r) + B_i H_i/\sigma_i, \qquad i=1,\ldots,m, \\ &= K\sum_1^n v_{ik}(\alpha_k - r), \qquad i=m+1,\ldots,n. \end{aligned} \tag{7.19}$$

Hence, the relative holdings of securities $m+1$ through n will be the same for all investors, and the differential demand functions can be rewritten as

$$\begin{aligned} \Delta D_i^* &= B_i H_i/\sigma_i, \qquad i=1,\ldots,m, \\ &= 0, \qquad i=m+1,\ldots,n. \end{aligned} \tag{7.20}$$

The composition of fund #$(2+i)$ reduces to a simple combination of security i and the riskless security.

The behavior implied by the demand functions in (7.16) can be more easily interpreted if they are rewritten in terms of the direct utility and optimal consumption functions. The optimal consumption function has the form $C^*(t) = C^*(W, S, t)$, and from (7.14) it follows immediately that

$$K = -U_C[C^*, t]/(U_{CC}[C^*, t]\partial C^*/\partial W), \tag{7.21}$$

$$B_i = -(\partial C^*/\partial S_i)/(\partial C^*/\partial W), \qquad i=1,\ldots,m. \tag{7.22}$$

Because $\partial C^*/\partial W > 0$, it follows that the sign of B_i equals the sign of $(-\partial C^*/\partial S_i)$. An unanticipated change in a state variable is said to be *unfavorable* if, ceteris paribus, such a change would reduce current optimal consumption, e.g. an unanticipated increase in S_i would be unfavorable if $\partial C^*/\partial S_i < 0$. Inspection of (7.20), for example, shows that for such an individual, the differential demand for security i (which is perfectly positively correlated with changes in S_i) will be positive. If there is an unanticipated increase in S_i, then, ceteris paribus, there will be an unanticipated increase in his wealth. Because $\partial C^*/\partial W > 0$, this increase in wealth will tend to offset the negative impact on C^* caused by the increase in S_i, and therefore the unanticipated variation in C^* will be reduced. In effect, by holding more of this security, he expects to be "compensated" by larger wealth in the event that S_i changes in the unfavorable direction. Of course, if $\partial C^*/\partial S_i > 0$, then he will take a differentially short position. However, in all cases investors will allocate their wealth to the funds #$(2+i)$, $i=1,\ldots,m$, so as to "hedge" against unfavorable changes in the state variables $S(t)$.[50]

[50] This behavior will obtain even when the return on fund #$(2+i)$ is not instantaneously perfectly correlated with $\mathrm{d}S_i$.

In the usual static model, such hedging behavior will not be observed because the utility function depends only upon end-of-period wealth, and such a formulation implicitly assumes that $\partial C^*/\partial S_i = 0$, $i = 1, \ldots, m$. Thus, in the intertemporal model, securities have, in addition to their manifest function of providing an "efficient" risk–return tradeoff for end-of-period wealth, a latent function of allowing consumers to "hedge" against other uncertainties.[51]

It was shown in Section 4 that a necessary and sufficient condition for a set of portfolios to span the efficient portfolio set is that the returns on every security can be written as a linear function of the returns on the spanning portfolios plus noise. Moreover, for the spanning property to have operational significance, the size of the minimum spanning set, M^*, must be very much smaller than the number of available securities, $n+1$. This linearity requirement would appear virtually to rule out non-trivial spanning unless the markets are already complete in the Arrow–Debreu sense. To see this consider the following: suppose there exists a non-trivial spanning set $(X_1, \ldots, X_m, R)$ and the return on all-equity financed firm j satisfies the linearity condition. Suppose firm j changes its liabilities structure by issuing debt and retiring some equity where the end-of-period values of the two claims are given by (6.13) and (6.14) in Section 6. Although the payoffs to these claims are perfectly functionally related to the end-of-period value of the firm, this functional relationship is non-linear if there is a positive probability of default. Hence, the returns on either of these claims will not, in general, be expressible as a linear function of the returns on the spanning portfolios plus noise. Therefore, upon such a change in firm j's liability structure, the size of the spanning set would, in general, have to increase. In effect, because the payoff to these liabilities cannot be replicated by a portfolio combination of the previously existing securities, there are no "perfect substitutes" for the created liabilities among these securities. While "perfect substitutability" is not a necessary condition for the spanning set to remain unchanged, without it, it is unlikely that this set will remain unchanged. Because firms can and do issue many different liabilities, it is therefore unlikely that the size of the minimum spanning set will differ significantly from the number of securities outstanding in incomplete market models with arbitrary concave utility functions.

While the issuing of liabilities with non-linear payoffs has this effect in the static and discrete time dynamic models, it does not for the continuous time model described here. To illustrate, consider the following example which uses a type of analysis first presented by Black and Scholes (1973) in the context of option pricing. Let V be the value of some security whose return dynamics are

[51]For further discussion of this analysis, descriptions of specific sources of uncertainty, and extensions to discrete time examples, see Merton (1973b, 1975a, 1975b). In the case of multiple consumption goods with uncertain relative prices, similar behavior obtains. However, C^* is a vector and J_W is the "shadow" price of the "composite" consumption bundle. Hence, the corresponding derived "hedging" behavior is to minimize the unanticipated variations in J_W.

described by the diffusion process

$$\frac{\mathrm{d}V}{V}=\alpha\,\mathrm{d}t+\sigma\,\mathrm{d}Z, \tag{7.23}$$

where α and σ are, at most, functions of V and t so that the process is Markov. Let W be the price of some security whose value as of a specified date in the future, T^*, is given by the function $g[V(T^*)]$, where $g[0]=0$ and $g[V(T^*)]/V(T^*)$ is bounded.

Let $f(V,t)$ be the solution to the partial differential equation[52]

$$\tfrac{1}{2}\sigma^2V^2f_{VV}+rVf_V-rf+f_t=0 \tag{7.24}$$

subject to the boundary conditions

$$\begin{aligned}&\text{(a)}\quad f[0,t]=0,\\ &\text{(b)}\quad f[V,T^*]=g[V],\\ &\text{(c)}\quad f/V \text{ bounded,}\end{aligned} \tag{7.25}$$

where in this example the interest rate r is taken to be constant over time and subscripts denote partial derivatives.

Consider the continuous time portfolio strategy where the investor allocates the fraction $w(t)$ to the security with value $V(t)$ and $[1-w(t)]$ to the riskless security. If $w(t)$ is a right-continuous function and $P(t)$ denotes the value of the portfolio at time t, then the dynamics for P must satisfy

$$\mathrm{d}P=[w(\alpha-r)+r]P\,\mathrm{d}t+w\sigma P\,\mathrm{d}Z. \tag{7.26}$$

Suppose the particular portfolio strategy chosen is $w(t)=f_V[V,t]V(t)/P(t)$. Substituting this strategy into (7.26), we have that

$$\mathrm{d}P=[f_VV(\alpha-r)+rP]\,\mathrm{d}t+f_VV\sigma\,\mathrm{d}Z. \tag{7.27}$$

Since f is twice continuously differentiable, Itô's Lemma[53] can be used to express the stochastic process for f as

$$\mathrm{d}f=[\tfrac{1}{2}\sigma^2V^2f_{VV}+\alpha Vf_V+f_t]\,\mathrm{d}t+f_V\sigma V\,\mathrm{d}Z. \tag{7.28}$$

[52](7.24) is a linear partial differential equation of the parabolic type. If σ^2 is a continuous function and g is piece-wise continuous, then there exists a unique solution that satisfies boundary conditions (7.25). The usual method for solving this equation is Fourier transforms.

[53]Itô's Lemma is for stochastic differentiation the analog to the Fundamental Theorem of the calculus for deterministic differentiation. For a statement of the Lemma and applications in economics, see Merton (1971, 1973a). For its rigorous proof, see McKean (1969, p. 44).

But f satisfies (7.24), and therefore (7.28) can be rewritten as

$$\mathrm{d}f=[f_V V(\alpha-r)+rf]\,\mathrm{d}t+f_V V\sigma\,\mathrm{d}Z. \tag{7.29}$$

Comparing (7.29) with (7.27), if the initial investment in the portfolio, $P(0)$, is chosen such that $P(0)=f[V(0),0]$, then $P(t)=f[V,t]$ for all V and $t\leq T^*$. But $P(t)$ is the value of a feasible portfolio strategy at time t with the properties that $P(t)=0$ if $V(t)=0$ and $P(T^*)=g[V(T^*)]$. Hence, if the value of the other security, $W(t)$, does not equal $P(t)$ for each t, then there will exist an arbitrage opportunity. Therefore, $W(t)=f[V(t),t]$.

Since g can be a non-linear function, it has been shown that there exists a dynamic portfolio strategy that will exactly replicate the payoff to a non-linear function of the price of the security. The application of this procedure to the debt–equity case of (6.13) and (6.14) is immediate. If V denotes the value of the firm, then the debt issue will satisfy (6.24) with $g[V]=\min[V,B]$ and the equity will satisfy the same equation but with $g[V]=\max[0,V-B]$. In addition, this type of analysis can be used to show that the Modigliani–Miller Theorem (Theorem 6.3) will obtain for non-linear sharing rules.[54] Finally, inspection of (7.24) shows that knowledge of the expected return on V, α, is not required to solve for the values of the various claims, and indeed the only non-observable input required is the variance rate, σ^2. Moreover, (7.24) often yields closed-form solutions and in those cases where it does not, there exist efficient numerical solution techniques. Hence, this type of analysis has led to a copious theoretical and empirical literature on the pricing of corporate liabilities and contingent claims generally.[55]

In summary, the combined assumptions of continuous trading and local state variable changes substantially simplifies the analysis required in the discrete time case. Under these conditions, the crucial non-trivial spanning properties will obtain, and the creation of firm-contingent liabilities will not, in general, affect the equilibrium. An in-depth discussion of the mathematical and economic assumptions required for the valid application of the continuous time analysis is beyond the scope of this paper. However, the required assumptions are rather mild when applied in the context of a securities market.[56] Although continuous trading is only a theoretical proposition, the continuous trading solutions will be an asymptotically valid approximation to the discrete time solution as the trading

[54] For a proof and extensions to general contingent claim securities, see Merton (1977).

[55] The literature based on the Black–Scholes type analysis has grown so rapidly that no attempt has been made here to provide a complete set of references. An excellent survey article is Smith (1976). In addition to the pricing of corporate liabilities, it has been applied to the pricing of loan guarantees, insurance contracts, "dual" funds, term-structure bonds, and even tenure for university professors.

[56] Merton (forthcoming) discusses in great detail the economic assumptions required for the continuous time methodology. Moreover, all the necessary mathematical tools for manipulation of these models are derived using only elementary probability theory and the calculus.

interval h becomes small.[57] Indeed, actual securities markets are open virtually every day, and hence the assumption that h is small is not without empirical foundation. In the intertemporal version of the Arrow–Debreu model with complete markets, it is well known that the market need only be open "once" because individuals will have no need for further trade. The continuous trading model is, of course, the opposite extreme. However, the continuous time model appears to have many of the properties of the Arrow–Debreu model without nearly so many securities. Hence, it may be that a good substitute for having so many markets and securities is to have fewer markets and securities but the existing markets open more frequently. The study of this possibility will be left as a topic for future research.

References

Arrow, K. J. (1964), "The role of securities in the optimal allocation of risk bearing", Review of Economic Studies, 31:91–96.

Bawa, V. S. (1975), "Optimal rules for ordering uncertain prospects", Journal of Financial Economics, 2:95–121.

Bertsekas, D. P. (1974), "Necessary and sufficient conditions for existence of an optimal portfolio", Journal of Economic Theory, 8:235–247.

Black, F. (1972), "Capital market equilibrium with restricted borrowing", Journal of Business, 45:444–455.

Black, F. and M. Scholes (1973), "The pricing of options and corporate liabilities", Journal of Political Economy 81:637–654.

Borch, K. (1969), "A note on uncertainty and indifference curves", Review of Economic Studies, 36:1–4.

Cass, D. and J. E. Stiglitz (1970), "The structure of investor preferences and asset returns, and separability in portfolio allocation: A contribution to the pure theory of mutual funds", Journal of Economic Theory, 2:122–160.

Cox, D. A. and H. D. Miller (1968), The theory of stochastic processes. New York: John Wiley & Sons, Inc.

Debreu, G. (1959), Theory of value. New York: John Wiley & Sons, Inc.

Diamond, P. (1967), "The role of a stock market in a general equilibrium model with technological uncertainty", American Economic Review, 57:759–776.

Dreyfus, S. E. (1965), Dynamic programming and the calculus of variations. New York: Academic Press.

Eckern, S. and R. Wilson (1974), "On the theory of the firm in an economy with incomplete markets", Bell Journal of Economics and Management Science, 5:171–180.

Fama, E. (1965), "The behavior of stock market prices", Journal of Business 38:34–105.

Fama, E. (1970a), "Efficient capital markets: A review of theory and empirical work", Journal of Finance, 25:383–417.

Fama, E. (1970b), "Multiperiod consumption–investment decisions", American Economic Review, 60:163–174.

[57]See Samuelson (1970) and Merton and Samuelson (1974). In Merton (1975a, p. 663) there is a brief discussion of the special cases in which the limiting discrete time solutions do not approach the continuous time solutions.

Fama, E. (1972), "Perfect competition and optimal production decisions under uncertainty", Bell Journal of Economics and Management Science, 3:509–530.
Fama, E. (1978), "The effects of a firm's investment and financing decisions on the welfare of its securityholders", American Economic Review, 68.
Fama, E. and A. B. Laffer (1972), "The number of firms and competition", American Economic Review, 62:670–674.
Farrar, D. E. (1962), The investment decision under uncertainty. Englewood Cliffs, New Jersey: Prentice Hall.
Farrell, J. L. (1974), "Analyzing covariation of returns to determine homogeneous stock groupings", Journal of Business, 47:186–207.
Feeney, G. J., and D. Hester (1967), "Stock market indices: A principal components analysis", in: D. Hester and J. Tobin, eds., Risk aversion and portfolio choice. New York: John Wiley & Sons, Inc.
Feldstein, M. S. (1969), "Mean–variance analysis in the theory of liquidity preference and portfolio selection", Review of Economic Studies, 36:5–12.
Feller, W. (1966), An introduction to probability theory and its applications, Vol. 2. New York: John Wiley & Sons, Inc.
Fischer, S. (1972), "Assets, contingent commodities, and the Slutsky equations", Econometrica, 40:371–385.
Friend, I. and J. Bicksler (eds.) (1975), Studies in risk and return, Vols. I & II. Cambridge, Mass: Ballinger.
Goldman, M. B. (1974), "A negative report on the 'near-optimality' of the max expected log policy as applied to bounded utilities for long-lived programs", Journal of Financial Economics, 1:97–103.
Grossman, S. (1976), "On the efficiency of competitive stock markets where traders have diverse information", Journal of Finance, 31:573–585.
Grossman, S. and J. E. Stiglitz (1976), "Information and competitive price systems", American Economic Review, 66:246–253.
Hadar, J. and W. R. Russell (1969), "Rules for ordering uncertain prospects", American Economic Review, 59:25–34.
Hadar, J. and W. R. Russell (1971), "Stochastic dominance and diversification", Journal of Economic Theory, 3:288–305.
Hakansson, N. (1970), "Optimal investment and consumption strategies under risk for a class of utility functions", Econometrica, 38:587–607.
Hanoch, G. and H. Levy (1969), "The efficiency analysis of choices involving risk", Review of Economic Studies, 36:335–346.
Hardy, G. H., J. E. Littlewood, and G. Pölya. (1959), Inequalities. Cambridge: Cambridge University Press.
Hart, O. (1975), "On the optimality of equilibrium when the market structure is incomplete", Journal of Economic Theory, 11:418–443.
Herstein, I. and J. Milnor. (1953), "An axiomatic approach to measurable utility", Econometrica, 21:291–297.
Hirshleifer, J. (1965), "Investment decision under uncertainty: Choice-theoretic approaches", Quarterly Journal of Economics 79:509–536.
Hirshleifer, J. (1966), "Investment decisions under uncertainty: Applications of the state-preference approach", Quarterly Journal of Economics, 80:252–277.
Hirshleifer, J. (1970), Investment, interest and capital. Englewood Cliffs, New Jersey: Prentice-Hall.
Hirschleifer, J. (1973), "Where are we in the theory of information", American Economic Review, 63:31–39.
Itô, K. and H. P. McKean, Jr. (1964), Diffusion processes and their sample paths. New York: Academic Press.
Jensen, M. C. (ed.). (1972), Studies in the theory of capital markets. New York: Praeger Publishers.
Jensen, M. C. (ed.). (1972), "Capital markets: Theory and evidence", Bell Journal of Economics and Management Science, 3:357–398.
Jensen, M. C. (ed.) and J. B. Long, Jr. (1972), "Corporate investment under uncertainty and Pareto optimality in the capital markets", Bell Journal of Economics and Management Science, 3:151–174.
King, B. R. (1966), "Market and industry factors in stock price behavior", Journal of Business, 39 Supplement:139–190.

Kuhn, H. W. and A. W. Tucker. (1951), "Nonlinear programming", in: J. Neyman, ed., Proceedings of the Second Berkeley Symposium of mathematical statistics and probability. Berkeley: University of California Press.
Kushner, H. J. (1967), Stochastic stability and control. New York: Academic Press.
Latane, H. (1959), "Criteria for choice among risky ventures", Journal of Political Economy, 67:144–155.
Leland, H. (1972), "On the existence of optimal policies under uncertainty", Journal of Economic Theory, 4:35–44.
Leland, H. (1974), "Production theory and the stock market", Bell Journal of Economics and Management Science, 5:125–144.
Lintner, J. (1965), "The valuation of risk assets and the selection of risky investments in stock portfolios and capital budgets", Review of Economics and Statistics, 47:13–37.
Livingston, M. (1977), "Industry movements of common stocks", Journal of Finance, 32:861–874.
Markowitz, H. (1959), Portfolio selection: Efficient diversification of investment. New York: John Wiley & Sons, Inc.
Markowitz, H. (1976), "Investment for the long run: New evidence for an old rule", Journal of Finance, 31:1273–1286.
McKean, Jr., H. P. (1969), Stochastic integrals. New York: Academic Press.
McShane, E. J. (1974), Stochastic calculus and stochastic models. New York: Academic Press.
Merton, R. C. (1969), "Lifetime portfolio selection under uncertainty: The continuous-time case", Review of Economics and Statistics, 51:247–257.
Merton, R. C. (1971), "Optimum consumption and portfolio rules in a continuous time model", Journal of Economic Theory, 3:373–413.
Merton, R. C. (1972), "An analytic derivation of the efficient portfolio frontier", Journal of Financial and Quantitative Analysis, 7:1851–1872.
Merton, R. C. (1973a) "Theory of rational option pricing", Bell Journal of Economics and Management Science, 4:141–183.
Merton, R. C. (1973b), "An intertemporal capital asset pricing model", Econometrica, 41:867–887.
Merton, R. C. (1975a), "Theory of finance from the perspective of continuous time", Journal of Financial and Quantitative Analysis, 10:659–674.
Merton, R. C. (1975b), "A reexamination of the capital asset pricing model", in: Friend and Bicksler, eds. (1975).
Merton, R. C. (1977), "On the pricing of contingent claims and the Modigliani–Miller theorem", Journal of Financial Economics, 5:241–249.
Merton, R. C. (forthcoming), "On the mathematics and economic assumptions of continuous-time financial models", in: W. F. Sharpe, ed., Financial Economics: Essays in Honor of Paul Cootner. Englewood Cliffs: Prentice Hall Inc.
Merton, R. C. and P. A. Samuelson (1974), "Fallacy of the log-normal approximation to optimal portfolio decision making over many periods", Journal of Financial Economics, 1:67–94.
Merton, R. C. and M. G. Subrahmanyam (1974), "The optimality of a competitive stock market", Bell Journal of Economics and Management Science, 5:145–170.
Miller, M. H. (1977), "Debt and taxes", Journal of Finance, 32:261–276.
Miller, M. H. and F. Modigliani (1958), "The cost of capital, corporation finance, and the theory of investment", American Economic Review, 48:261–297.
Mossin, J. (1966), "Equilibrium in a capital asset market", Econometrica, 35:768–783.
Myers, S. C. (1968), "A time-state-preference model of security valuation", Journal of Financial and Quantitative Analysis, 3:1–33.
Pratt, J. W. (1964), "Risk aversion in the small and in the large", Econometrica, 32:122–136.
Radner, R. (1970), "Problems in the theory of markets under uncertainty", American Economic Review, 60:454–460.
Radner, R. (1974), "A note on the unanimity of stockholders' preferences among alternative production plans: A reformulation of the Eckern–Wilson model", Bell Journal of Economics and Management Science, 5:181–186.
Richard, S. (1975), "Optimal consumption, portfolio, and life insurance rules for an uncertain lived individual in a continuous-time model", Journal of Financial Economics. 2:187–204.

Ross, S. (1976), "Arbitrage theory of capital asset pricing", Journal of Economic Theory 13:341–360.
Ross, S. (1978), "Mutual fund separation in financial theory: The separating distributions", Journal of Economic Theory. 17:254–286.
Rothschild, M. and J. E. Stiglitz (1970), "Increasing risk I: A definition", Journal of Economic Theory, 2:225–243.
Rothschild, M. and J. E. Stiglitz (1971), "Increasing risk II: Its economic consequences", Journal of Economic Theory, 3:66–84.
Rubinstein, M. (1976), "The strong case for the generalized logarithmic utility model as the premier model of financial markets", Journal of Finance, 31:551–572.
Samuelson, P. A. (1965), "Proof that properly anticipated prices fluctuate randomly", Industrial Management Review, 6:41–49. Reprinted in Samuelson (1972).
Samuelson, P. A. (1967), "General proof that diversification pays", Journal of Financial and Quantitative Analysis, 2:1–13. Reprinted in Samuelson (1972).
Samuelson, P. A. (1969), "Lifetime portfolio selection by dynamic stochastic programming", Review of Economics and Statistics, 51:239–246. Reprinted in Samuelson (1972).
Samuelson, P. A. (1970), "The fundamental approximation theory of portfolio analysis in terms of means, variances, and higher moments", Review of Economic Studies, 37:537–542. Reprinted in Samuelson (1972).
Samuelson, P. A. (1971), "The 'fallacy' of maximizing the geometric mean in long sequences of investing or gambling", Proceedings of the National Academy of Sciences, 68:2493–2496. Reprinted in Samuelson (1972).
Samuelson, P. A. (1972), in: R. C. Merton, ed., The collected scientific papers of Paul A. Sameulson, Vol III. Cambridge: M.I.T. Press.
Samuelson, P. A. (1973), "Proof that properly discounted present values of assets vibrate randomly", Bell Journal of Economics and Management Science, 4:369–374.
Samuelson, P. A. (1977), "St. Petersburg paradoxes: Defanged, dissected, and historically described", Journal of Economic Literature, 15:24–55.
Sharpe, W. (1964), "Capital asset prices: A theory of market equilibrium under conditions of risk", Journal of Finance, 19:425–442.
Sharpe, W. (1970), Portfolio theory and capital markets. New York: McGraw-Hill.
Smith, Jr., C. W. (1976), "Option pricing: A review", Journal of Financial Economics, 3:3–52.
Stiglitz, J. E. (1969), "A re-examination of the Modigliani–Miller theorem", American Economic Review, 59:78–93.
Stiglitz, J. E. (1972), "On the optimality of the stock market allocation of investment", Quarterly Journal of Economics, 86:25–60.
Stiglitz, J. E. (1974), "On the irrelevance of corporate financial policy", American Economic Review, 64:851–866.
Tobin, J. (1958), "Liquidity preference as behavior towards risk", Review of Economic Studies, 25:68–85.
Tobin, J. (1969), "Comment on Borch and Feldstein", Review of Economic Studies, 36:13–14.
Von Neumann, J. and O. Morgenstern (1947), Theory of games and economic behavior, 2nd edn. Princeton: Princeton University Press.

Chapter 14

MARKET DEMAND AND EXCESS DEMAND FUNCTIONS*

WAYNE SHAFER

University of Southern California

and

HUGO SONNENSCHEIN

Princeton University

1. Introduction

Market demand and market excess demand functions are defined by summing the preference-maximizing actions of consumers. These functions are central to most treatments of value theory. Indeed, the (general) equilibrium states of a Walrasian economy are classically defined as the solutions of the equation system "market excess demand equals the zero vector". (Chapter 15 by G. Debreu deals with the existence of such equilibrium states.) For partial equilibrium analysis, in particular for empirical studies of demand, the market demand function plays a central role. Here, demand is a function of prices and the exogenous distribution of income. Given a distribution of income, an equilibrium price is determined by the equality of the partial equilibrium market demand and supply functions.

The demand theory of the individual consumer has been well developed for some time; however, it is only fairly recent that we have achieved a comparable understanding of the properties of market demand functions. Just as consumer demand theory is the study of properties common to demand functions which are generated by the preference-maximization of an individual consumer, market demand theory investigates properties which are shared by market demand functions. (Chapter 9 by A. Barten and V. Böhm deals with the demand theory of an individual consumer.) Since market demand is often the observable and relevant variable for analysis, it is important to see how it is restricted by the utility hypothesis. Our presentation is organized around two complementary themes of market demand theory. First, when preferences are homothetic and the distribution of income (value of wealth) is independent of prices, then the market

*The support of the National Science Foundation is acknowledged.

Handbook of Mathematical Economics, vol. II, edited by K.J. Arrow and M.D. Intriligator

demand function (market excess demand function) has all of the properties of a consumer demand function. This is *Eisenberg's Theorem*. Second, with general (in particular non-homothetic) preferences, even if the distribution of income is fixed, market demand functions need not satisfy in any way the classical restrictions which characterize consumer demand functions. For market excess demand functions this result was developed by Sonnenschein (1973a, 1974), Mantel (1977), and Debreu (1974), and has been extended by several other authors. For market demand functions it was developed by Sonnenschein (1973b) and has been extended by Mantel (1977), Diewert (1977), and Andreu (forthcoming).

The importance of the above results is clear: strong restrictions are needed in order to justify the hypothesis that a market demand function has the characteristics of a consumer demand function. Only in special cases can an economy be expected to act as an "idealized consumer". The utility hypothesis tells us nothing about market demand unless it is augmented by additional requirements.

Following Section 2, which is devoted to notation and preliminaries, Section 3 gives conditions under which market demand and market excess demand functions are consumer demand functions. Section 4 is centered around Debreu's theorem, which says that for every continuous function F which satisfies homogeneity and Walras' Law, and for every compact set of positive prices K, there exists an economy with market excess demand function F on K. Section 5 discusses similar results for demand functions, and Section 6 details some applications of the analysis.

Finally a disclaimer: the choice of topics reflects our personal taste and interests. There is important material on market demand functions which is not surveyed. For example, the reader will find no list of functional forms which are important in applied market demand theory; similarly, we do not consider when suitable income transfers will lead an economy to act as a single consumer. Market demand theory is a broad area. We have chosen a route which allows us to: (a) develop quite deeply two of the important themes; (b) give some interesting applications; and (c) expose a few unsolved problems.

2. Notation and preliminaries

In this section basic notation and definitions are presented, as well as some standard results from individual demand theory. For further discussion of the former see Chapter 1 of this Handbook; for further discussion of the latter, see Chapter 9.

The symbol R^ℓ denotes the *Euclidean ℓ-space* with *inner product* $x \cdot y = \sum_{i=1}^{\ell} x_i y_i$, and *norm* $\|\cdot\|$ given by $\|x\| = (x \cdot x)^{1/2}$. Certain subsets of R^ℓ are frequently considered:

$$R^{\ell}_{+} = \{x \in R^{\ell}: x_i \geqq 0, \quad \text{for all } i\},$$
$$P = \{x \in R^{\ell}: x_i > 0, \quad \text{for all } i\},$$
$$P_{\varepsilon} = \{x \in R^{\ell}: x_i \geqq \varepsilon, \quad \text{for all } i\},$$
$$\bar{S} = \{x \in R^{\ell}_{+}: \|x\| = 1\},$$
$$S = \bar{S} \cap P,$$
$$S_{\varepsilon} = \bar{S} \cap P_{\varepsilon},$$
$$\Lambda = P \times R^{1}_{+}.$$

We consider ℓ-commodity economies, and unless it is explicitly stated to the contrary, the consumption set for each consumer is R^{ℓ}_{+}. The *preference relation* of a consumer is given by a complete preordering $\succsim$ of R^{ℓ}_{+}, with $x \succsim y$ interpreted as *x is at least as good as y*, $x \succ y (x \succsim y$ and not $y \succsim x)$ interpreted as *x is preferred to y*, and $x \sim y (x \succsim y$ and $y \succsim x)$ interpreted as *indifference between x and y*. The preference relation $\succsim$ is: *continuous* if $\{(x, y): x \succsim y\}$ is a closed subset of $R^{2\ell}_{+}$; *convex* if $\{y: y \succsim x\}$ is convex for each $x \in R^{\ell}_{+}$; *strictly convex* if $x \succsim y$, $x \neq y$, and $0 < \alpha < 1$ implies $\alpha x + (1-\alpha) y \succ y$; *monotone* if $x_i > y_i$ for all i implies $x \succ y$; *locally non-satiated* if for each $x \in R^{\ell}_{+}$ and every $\varepsilon > 0$ there exists a y such that $y \succ x$ and $\|y - x\| < \varepsilon$; *representable* if there exists a "utility function" $u: R^{\ell}_{+} \to R$ such that $x \succsim y$ if and only if $u(x) \geqq u(y)$; and *homothetic* if it is representable by a utility function which is homogeneous of degree one.

Let $T \subset \Lambda$. A function $G: T \to R^{\ell}_{+}$ is a *demand function* with domain T if (i) $p \cdot G(p, m) = m$ for all $(p, m) \in T$, and (ii) $G(\lambda p, \lambda m) = G(p, m)$ for all $\lambda > 0, (p, m)$ such that $(p, m) \in T$ and $(\lambda p, \lambda m) \in T$. Here $G(p, m)$ is interpreted as the commodity vector purchased when p is the vector of prices and m is money income. If G is a demand function with domain T and there exists a preference relation $\succsim$ (interpreted as a consumer) such that for each $(p, m) \in T, G(p, m)$ is a singleton defined by $\{x: p \cdot x \leqq m$ and $x \succsim y$ for all y such that $p \cdot y \leqq m\}$, then G is called *consumer demand function*. In this case we say that $\succsim$ generates G. Let G be a demand function, and define a relation V on R^{ℓ}_{+} by xVy if x and y are distinct and there exists $(p, m) \in T$ such that $x = G(p, m)$ and $p \cdot y \leqq m$. The demand function G satisfies the *Weak Axiom of Revealed Preference* (*WARP*) if and only if V is asymmetric, and G satisfies the *Strong Axiom of Revealed Preference* (*SARP*) if and only if V is acyclic. If G is a consumer demand function, then it satisfies SARP, and if G satisfies SARP, then it is a consumer demand function (but not necessarily a representable consumer demand function). If G is a consumer demand function and is differentiable, then at each $(p, m) \in T$, letting $D_pG(p, m)$ denote the $\ell \times \ell$ matrix of the partial derivatives $\partial G_i(p, m)/\partial p_j$, one has (i) $z^T D_pG(p, m)z \leqq 0$ for all $z \in R^\ell$ such that $z \cdot G(p, m) = 0$, and (ii) $\bar{z}^T D_pG(p, m)z = z^T D_pG(p, m)\bar{z}$ for all $\bar{z}, z \in R^\ell$ such that $z \cdot G(p, m) = 0$ and $\bar{z} \cdot G(p, m) = 0$. These are a slightly different form of classical negative semi-definiteness and symmetry conditions on the Slutsky matrix (see Chapter 9 by A. Barten and V. Böhm). Subject to some additional technical conditions, these are the only general restrictions on a differentiable consumer demand function. A demand function G is called a *market demand function* if there exist $n \geqq 1$ positive numbers $\delta_1, \delta_2, \ldots, \delta_n (\sum \delta_i = 1)$ and n consumer demand functions $G^1, G^2, \ldots, G^n$, such that $G(p, m) = \sum_{i=1}^n G^i(p, \delta_i m)$ holds for every (p, m) in the domain of G. Here δ_i is interpreted as the ith consumer's share of community income. Finally, the domain of a demand function G will be taken to be Λ unless stated otherwise.

Let $T \subset P$. A function F: $T \to R^\ell$ is called an *excess demand function* if (i) $p \cdot F(p) = 0$ for all $p \in T$ (Walras' Law), and (ii) $F(tp) = F(p)$ whenever $p, tp \in T$. In view of (ii), it is always assumed that the domain of F is the cone generated by T, excluding 0. F: $T \to R^\ell$ is a *consumer excess demand function* if there exists an "initial endowment vector" $\omega \in R^\ell_+$ and a consumer demand function G: $T \times R^1_+ \to R^\ell_+$ such that $F(p) = G(p, p \cdot \omega) - \omega$ for all $p \in T$. If $\succsim$ generates G, then we say $(\succsim, \omega)$ generates F. Let $T' = \{(p, m) : (p, m) = (p, p \cdot \omega)\}$, then F satisfies WARP (respectively SARP) if G satisfies WARP (respectively SARP) when restricted to T'. The negative semidefiniteness and symmetry conditions on a differentiable G are inherited by F, in the following form. If F is a differentiable consumer excess demand function, then (i) $z^T DF(p)z \leqq 0$ for all $z \in R^\ell$ such that $z \cdot F(p) = 0$ and (ii) $\bar{z}^T DF(p)z = z^T DF(p)\bar{z}$ for all $\bar{z}, z \in R^\ell$ such that $\bar{z} \cdot F(p) = 0$ and $z \cdot F(p) = 0$. A function F: $T \to R^\ell$ is called a *market excess demand function* if there exists $n \geqq 1$ consumer excess demand functions $F^1, F^2, \ldots, F^n$ such that $F = \sum_{i=1}^n F^i$. If for each i, $(\succsim_i, \omega^i)$ generates F^i, then we say the economy $\mathscr{E} = (\succsim_i, \omega^i)_{i=1}^N$ generates F. Finally, if $\mathscr{E}$ generates F and $F(p) = 0$, then p is called an *equilibrium price vector* for $\mathscr{E}$.

3. When the market demand function is a consumer demand function

In this section we consider conditions under which market demand functions and market excess demand functions are of the class generated by a single consumer. The following classic result, due to Antonelli (1886) and later independently discovered by Gorman (1953) and Nataf (1953), gives necessary and sufficient conditions under which market demand functions are independent of both the distribution of income and are consumer demand functions as well.

Theorem 1 (Antonelli)

Market demand $G=\sum G^i$: $\Lambda \to R^{\ell}_{+}$ is independent of the distribution of income and is a consumer demand function if and only if there is a homothetic preference relation $\succsim$ such that each consumer demand function G^i is derived from $\succsim$. In this case, $G=G^i$ for all i.

Proof

Suppose that each G^i is derived from a homothetic preference relation $\succsim$. Let $H=G^i$ denote the common demand function, which must be linear in income. Then, $G(p,m_1,m_2,\ldots,m_n)=\sum_i H(p,m_i)=H(p,1)\sum m_i=H(p,\sum m_i)=G^i(p,\sum m_i)$ for all i and for all $(p,m_1,m_2,\ldots,m_n)$.

Suppose there exists a consumer demand function H such that $\sum G^i(p,m_i)=H(p,\sum m_i)$ for all $(p,m_1,m_2,\ldots,m_n)$. For any i, set $m=m_i$ and $m=0$ for $j\neq i$. Then $H(p,m)=G^i(p,m)$ holds for every p, m and every i, so all G^i are identical to H. To show that H is derived from a homothetic preference relation, it must be shown that H is linear in m. This follows from $H(p,m_1+m_2)=H(p,m_1)+H(p,m_2)$ for all $m_1, m_2 \geqq 0$.

The analog of Theorem 1 for market excess demand functions is easily established.

Theorem 2

Market excess demand $F=\sum F^i$ is independent of the initial distribution $\omega^1,\omega^2,\ldots,\omega^n$ of the aggregate initial endowment ω and is a consumer excess demand function if and only if there is a homothetic preference relation $\succsim$ such that each F^i is derived from $\succsim$. In this case, for every $\omega^1,\omega^2,\ldots,\omega^n$, F is the excess demand function derived from $\succsim$ with endowment vector $\sum\omega^i$.

A pair of examples follow in which market demand depends on the distribution of income. In the first, demand functions are identical but preferences are not homothetic; in the second, preferences are homothetic but are not identical.

Example 1

Let two consumers have identical preferences on R^2_+ which are represented by $u(x,y)=xy+y$ and let prices be $(1,1)$. Then the income distribution of $m_1=1$ and $m_2=1$ leads to a different demand than if $m_1=2$ and $m_2=0$, since the Engel curve for u at $p=(1,1)$ is not linear.

Example 2

Let two consumers have homothetic preferences on R^2_+ which are represented by $u_1(x,y)=x$ and $u_2(x,y)=y$, respectively. Then aggregate demand depends completely on the distribution of income.

The requirement that consumer demand functions be both identical and linear in income is clearly very restrictive. Eisenberg (1961) made a fundamental contribution when he showed that if identical demand is dropped and if the requirement that market demand is independent of the distribution of income is replaced by the hypothesis that the distribution of income remains fixed, then the market demand function is generated by a utility function. In the form presented below, this result is due to Chipman and Moore (1976), as is the idea of the proof.

Theorem 3 (Eisenberg, Chipman, and Moore)

If the preferences of each agent can be represented by a homogeneous of degree one utility function u_i on R^{ℓ}_{+}, and if income shares are fixed (so $m_i=\delta_i m, \delta_i>0, \sum\delta_i=1$), then market demand is generated by the homogeneous of degree one utility function U given by

$$U(x)=\max\prod_{i=1}^{N}\left(u_i(x^i)\right)^{\delta_i}, \quad \text{s.t. } \sum x^i=x.$$

In particular, if individual demand is single-valued then for each $(p,m)\in\Lambda, G(p,m)=\sum G^i(p,\delta_i m)$, where G^i is the demand function generated by u_i and G is the demand function generated by U.

Proof

We will present the proof for $n=2$; the proof for the general case is completely analogous. Let $\delta=\delta_1, 1-\delta=\delta_2$, and consider the following two problems, with p and m fixed:

(I) $\max_{x^i} u_i(x^i)$ s.t. $p\cdot x^i=\delta_i m \qquad (i=1,2)$;

(II) $\max_{x^1,x^2}(u_1(x^1))^{\delta}(u_2(x^2))^{1-\delta}$ s.t. $p\cdot(x^1+x^2)=m$.

Let $(\bar{x}^1,\bar{x}^2)$ be a solution to the problems defined in (I) and $(\hat{x}^1,\hat{x}^2)$ a solution for (II). The essence of the theorem is that both problems have the same solutions; i.e. $(\bar{x}^1,\bar{x}^2)$ is a solution to (II) and $(\hat{x}^1,\hat{x}^2)$ is a solution to (I), and this is what will be shown. The heart of the proof is to show that $p\cdot\hat{x}^1=\delta m$ and $p\cdot\hat{x}^2=(1-\delta)m$. To see this, let $\theta_1=\delta m/p\cdot\hat{x}^1$ and $\theta_2=(1-\delta)m/p\cdot\hat{x}^2$. We will show $\theta_1=\theta_2=1$. Since $(\theta_1\hat{x}^1,\theta_2\hat{x}^2)$ is feasible for (II):

$$\left(u_1(\hat{x}^1)\right)^{\delta}\left(u_2(\hat{x}^2)\right)^{1-\delta}\geqq\left(u_1(\theta_1\hat{x}^1)\right)^{\delta}\left(u_2(\theta_2\hat{x}^2)\right)^{1-\delta},$$

and so

$$\left(u_1(\hat{x}^1)\right)^{\delta}\left(u_2(\hat{x}^2)\right)^{1-\delta}\geqq\theta_1^{\delta}\theta_2^{1-\delta}\left(u_1(\hat{x}^1)\right)^{\delta}\left(u_2(\hat{x}^2)\right)^{1-\delta}.$$

Thus, $\theta_1^{\delta}\theta_2^{1-\delta} \leqq 1$; substituting, this is equivalent to

$$(\delta m/p\cdot\hat{x}^1)^{\delta}((1-\delta)m/p\cdot\hat{x}^2)^{1-\delta} \leqq 1$$

or

$$\delta^{\delta}(1-\delta)^{1-\delta} \leqq (p\cdot\hat{x}^1/m)^{\delta}(p\cdot\hat{x}^2/m)^{1-\delta} = (p\cdot\hat{x}^1/m)^{\delta}(1-(p\cdot\hat{x}^1/m))^{1-\delta}.$$

Consider the function f: $(0,1)\to R$ defined by $f(t)=t^{\delta}(1-t)^{1-\delta}$. It is easy to check that $f''(t)<0$ for $t\in(0,1)$ and that $f'(t)=0$ precisely at $t=\delta$. Thus, f has a unique maximum at δ, so from the inequality above we must have $p\cdot\hat{x}^1/m=\delta$. Therefore, $\theta_1=\theta_2=1$, and so $p\cdot\hat{x}^1=\delta m$ and $p\cdot\hat{x}^2=(1-\delta)m$. Thus, $(\hat{x}^1,\hat{x}^2)$ is feasible for (I), and of course $(\bar{x}^1,\bar{x}^2)$ is feasible for (II). This yields the following inequalities:

$$u_i(\bar{x}^i) \geqq u_i(\hat{x}^i), \qquad i=1,2$$

$$(u_1(\hat{x}^1))^{\delta}(u_2(\hat{x}^2))^{1-\delta} \geqq (u_1(\bar{x}^1))^{\delta}(u_2(\bar{x}^2))^{1-\delta}.$$

Clearly, we must have $u_i(x^i)=u_i(\hat{x}^i)$, $i=1,2$, and this demonstrates what we set out to prove.

Unlike the situation in the Antonelli Theorem, demand here does depend on the distribution of income. It is important for the theorem that the distribution of income is independent of prices, as the following example illustrates.

Example 3

$u_1(x,y)=x$, $u_2(x,y)=y$, $\delta_1=p_x^2/p_x^2+p_y^2$, $\delta_2=p_y^2/p_x^2+p_y^2$. Fix $m=1$. If $p_x=1$ and $p_y=3$, market demand is $(\frac{1}{10},\frac{3}{10})$, while if $p_x=3$ and $p_y=1$, market demand is $(\frac{3}{10},\frac{1}{10})$, which violates WARP.

The following example indicates that a fixed distribution of income, but no restriction on agents preferences is not sufficient to ensure market demand is that of a single agent [see Wold (1953)].

Example 4

Consider Figure 3.1. Individual income is one-half market income. Market budgets for two different price ratios are indicated with dotted lines. The market choices, which fail to satisfy WARP, lie on the dotted budgets. The choices of the first individual are indicated by an "x" and those of the second individual by an "o". Observe that individual demands are consistent with individual utility maximization.

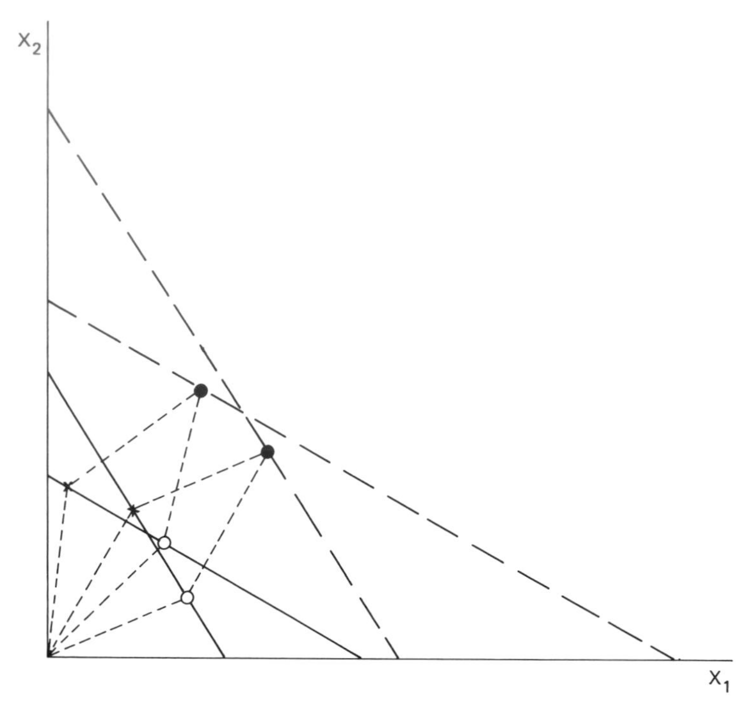

Figure 3.1.

Theorem 4 is the analog of Eisenberg's Theorem for market excess demand functions.

Theorem 4

If preferences of each agent can be represented by a homogeneous of degree one utility function u_i on R^ℓ_+, and, if initial endowments are proportional, so $\omega^i = \delta_i \omega$ ($\delta_i \geqq 0$, $\Sigma \delta_i = 1$), then market excess demand is generated by a single individual with preferences defined by U (as in the Eisenberg Theorem) and initial endowment ω.

With a fixed distribution of income, we have seen that homotheticity of individual preferences is sufficient for the result that market demand is a consumer demand function. However, homotheticity is not necessary. The following problem is open: what are necessary and sufficient conditions for a set of individual demand functions to have the property that all market demand functions obtained by summing members from the set (with a fixed distribution of income) will be consumer demand functions? That homotheticity is not necessary for the case $\ell=2$ was pointed out by Shafer (1977), and can be seen as follows. Suppose each agents' demand function G^i satisfies $D_pG^i(p, m)$ is negative definite. Then $D_pG(p, m)$, for $G=\Sigma G^i$, will have the same property. This implies WARP, and for two commodities ($\ell=2$), WARP implies SARP. An example of a consumer demand function G for which $D_pG(p, m)$ is always negative definite, but which is not linear in income is given by the demand function derived from the utility function $u(x, y)=\ln x+y^{1/2}$.

That homotheticity is not necessary for the case $\ell\geqq 3$ can be seen in the following example.

Example 5

Suppose each agent has a utility function $u_i(x_1, x_2, x_3)=\ln x_1+\ln x_2+\alpha_i x_3^{1/2}$, where $\alpha_i>0$ can be different for different agents. One can easily check that the demand function G^i derived from u_i has the form

$$G^i(p, m_i)=\left(\frac{1}{2p_1}\left(m_i-p_3\varphi_i\left(\frac{m_i}{p_3}\right)\right), \frac{1}{2p_2}\left(m_i-p_3\varphi_i\left(\frac{m_i}{p_3}\right)\right), \varphi_i\left(\frac{m_i}{p_3}\right)\right).$$

If one takes the appropriate derivatives, it can be seen that the above functional form guarantees that the Slutsky matrix is symmetric. Furthermore, with a fixed distribution of income, $G=\Sigma_i G^i$ will have the same functional form as each G^i. Shafer (1977) shows that $D_pG(p, m)$ will be negative definite for each (p, m), so G will have a symmetric negative semidefinite Slutsky matrix for each (p, m).

4. Market excess demand functions

From the above results it is clear that very restrictive conditions are needed to ensure that market demand is the demand of a single agent. The question of whether or not there are any non-trivial restrictions on market demand when no special assumptions are made on individual agents' preferences was articulated by Sonnenschein (1974). For the case of market excess demand, he proved in Sonnenschein (1973a) that on domains of the form S_ε, a dense set of excess demand functions are market excess demand functions. Mantel (1974) subsequently proved that on S_ε any continuously differentiable excess demand function for ℓ commodities is a market demand function of an economy with 2ℓ agents. The "state of the art" result is due to Debreu (1974); it requires only ℓ agents and the continuity of excess demand.[1]

[1]The second half of this section is devoted to a geometric treatment of the results for the case of two commodities. This may be especially useful for the reader who is having his first view of these theorems.

Theorem 5 (Debreu)

Let F be a continuous excess demand function with domain S. For any $\varepsilon>0$, there exists an ℓ consumer exchange economy $\mathscr{E}=(\succsim_i, \omega^i)$ with each $\succsim_i$ continuous, strictly convex, and monotone, such that $\mathscr{E}$ generates F on S_ε.[2]

Debreu's proof that F can be decomposed on S_ε into ℓ demand functions F^i which satisfy SARP is very simple. Since F is continuous, there exists a positive valued continuous function α such that $F(p)+\alpha(p)p$ is a positive vector for every $p\in S_\varepsilon$. The ith agent's excess demand function is defined by $F^i(p)=(F_i(p)+\alpha(p)p_i)(e^i-p_i p)$, where e^i is the ith unit vector. Since F satisfies Walras's Law, each of the F^i satisfy Walras's Law, and they add to F. Also, $p\cdot F^i(q)\leqq 0$ if and only if $p_i/q_i\leqq q\cdot p$; hence, $F(p)$ directly revealed preferred to $F(q)$ implies $p\neq q$, which implies $p_i<q_i$ and guarantees that SARP is satisfied. The difficult part of Debreu's proof is the demonstration that the F^i are generated by preferences with the indicated properties. For the case in which F is continuously differentiable, Geanakoplos (1978) has constructed explicit utility functions for the F^i.

The work of Sonnenschein, Mantel, and Debreu shows that there are no restrictions on market demand functions if no restrictions are placed on agents' preferences. With a slight strengthening of the analytical requirements on F, Mantel (1979) has demonstrated the striking result that even if agents preferences are homothetic, the same conclusion prevails.

Theorem 6 (Mantel)

Let F be a C^1 excess demand function with domain S. For any $\varepsilon>0$ such that F has bounded second-order partial derivatives on S_ε, there exists an ℓ consumer exchange economy $\mathscr{E}=(\succsim_i, \omega^i)_{i=1}^{\ell}$ with each $\succsim_i$ homothetic, convex, monotone, and representable, such that $\mathscr{E}$ generates F on S_ε.

Mantel proved this theorem by exhibiting indirect utility functions for each consumer, and then using Roy's identity to derive the demand functions. Instead we exhibit the demand functions directly and appeal to the integrability theorem, which for homothetic preferences requires the demand function G to satisfy D_pG negative semidefinite and symmetric. See Chipman (1974). If F is the given (candidate) excess demand function, then initial endowments can be $\omega^i=ke^i$ and individual demand functions G^i have the form:[3]

$$G^i(p,1)=-\frac{1}{k}DF^i(p)+\sum_j \frac{a_{ji}}{(p\cdot a^j)}a^j,$$

[2]See Theorem 14 of Section 5 for the case in which there are fewer consumers than commodities.
[3]The approach of writing down explicitly the demand functions associated with Mantel's indirect utility functions and checking the classical symmetry and negative semidefiniteness condition, was suggested to us in private correspondence by M. K. Richter.

where $A=(a_{ij})$ is any positive non-singular $\ell\times\ell$ matrix whose jth row is a^j. The positive number k is chosen big enough so that G^i is positive on S_ε and D_pG^i is negative definite on S_ε. It is easy to establish that the G^i satisfy the symmetry requirement and have associated excess demands which add to F.

The reader should compare this result with Eisenberg's Theorem for excess demand stated above. With homothetic preferences, proportional initial endowments yield a market excess demand which is a consumer demand function. Without proportional initial endowments, the market demand function can be any excess demand function, at least on domains of the type S_ε.[4] Since "translating" the initial endowment and the indifference curves by the same amount does not change the excess demand function, the result can be read another way: Even with proportional initial endowments and preferences which are "translated homothetic", the market excess demand function can be any excess demand function.

The above decomposition results hold on price domains of the form S_ε, for $\varepsilon>0$. It is not known if excess demand functions in general can be decomposed on S. There are some results in this direction. For example, if F is C^2 and can be extended to a C^2 function on $\bar{S}$, then Mantel's proof will give a decomposition on S, since each DF^i and D^2F^i will be bounded. Also, there is a result by McFadden, Mas-Colell, Mantel, and Richter (1974), which states that any excess demand function which is bounded below can be decomposed on S into ℓ excess demand functions, each satisfying SARP. In this case, however, preferences need not be representable. Sonnenschein (1971) presented a result on the equilibrium price set of an economy. If one is concerned about the structure of equilibria of an economy, then the natural question is whether or not any excess demand function F is a market demand function of an economy $\mathscr{E}$ on some S_ε such that $S_\varepsilon\supset F^{-1}(0)$, and $F^{-1}(0)$ is also the set of equilibrium prices for $\mathscr{E}$. Mas-Colell (1977) extended the previous theorems in order to provide the first general answer to this question.

Theorem 7 (Mas-Colell)

Let F be a continuous excess demand function on S which is bounded below and satisfies $\lim_{m\to\infty}\|F(p^m)\|=+\infty$ for $p^m\to p\in\partial S$. For any $\varepsilon>0$ there exists a μ: $0<\mu<\varepsilon$ and a ℓ-consumer exchange economy $\mathscr{E}=(R^\ell_+,\succsim_i,\omega^i)^\ell_{i=1}$ with each $\succsim_i$ continuous, strictly convex, and monotone, such that (i) $\mathscr{E}$ generates F on S_μ and (ii) $F^{-1}(0)\subset S_\mu$ is the set of equilibrium prices of $\mathscr{E}$.

This theorem yields immediately the following result concerning the structure of the set of equilibrium prices of an economy:

[4] Mantel proves a somewhat stronger result as he replaces the e^i by any set of ℓ linearly independent vectors.

Theorem 8 (Mas-Colell)

Let A be a compact non-empty subset of S. Then there exists an economy $\mathscr{E}$ whose agents have continuous, strictly convex, and monotone preferences, such that A is the set of equilibrium prices of $\mathscr{E}$.

Proof

Let $d(p, A)$ denote the distance of p from A; $d(p, A)=0$ if and only if $p \in A$ and $d(\cdot, A)$ is continuous. Pick any $\bar{q} \in A$, and define F by $F(p)= d(p, A)[(\bar{q}_1/p_1, \bar{q}_2/p_2, \ldots, \bar{q}_\ell/p_\ell)-(\bar{q}\cdot e/p\cdot e)e]$. Then it is easy to check that F is an excess demand function satisfying the conditions of the previous theorem, and that $F(p)=0$ if and only if $p \in A$.

This is clearly a very negative result; however, if F is continuously differentiable, then more can be said about $F^{-1}(0)$. Prior to this work on market excess demand, Debreu (1970) had shown that "most" C^1 market excess demand functions satisfying the conditions of Theorem 7 have the property that

$$\det\left[\left(\frac{\partial F_i(p)}{\partial p_j}\right) i, j=1,2,\ldots,\ell-1\right] \neq 0, \quad \text{at every } p \in F^{-1}(0).$$

In particular, the inverse function theorem gives $F^{-1}(0)$ finite if the above property holds. More can be said: E. Dierker (1972) has shown that for such demand functions,[5]

$$\sum_{p \in F^{-1}(0)} \operatorname{sgn} \det\left[\left(\frac{\partial F_i(p)}{\partial p_j}\right) i, j=1,2,\ldots,\ell-1\right]=(-1)^{\ell+1}.$$

This says, in particular, that $F^{-1}(0)$ has an odd number of points. With these considerations in mind, Mas-Colell (1977) showed the following:

Theorem 9 (Mas-Colell)

Suppose $\ell \geqq 3$, and let $A \subset S$ be a finite non-empty set, and i: $A \rightarrow \{-1,+1\}$ which satisfies $\sum_{p \in A} i(p)=(-1)^{\ell+1}$. Then there exists an economy $\mathscr{E}$ whose agents have continuous, monotone, and strictly convex preferences, such that $\mathscr{E}$ generates an excess demand function F satisfying

(i) $F^{-1}(0)=A$,

[5]See Chapter 17 by E. Dierker.

and

(ii) $\operatorname{sgn}\det\left[\left(\frac{\partial F_i(p)}{\partial p_j}\right)i,\ j=1,2,\ldots,\ell-1\right]=i(p),\quad$ for each $p\in A$.

Suppose $\ell=2$, and let $A=\{p^1, p^2,\ldots,p^m\}$ be a finite set with m odd, where the indices are chosen so that the first coordinate of p^j is less than the first coordinate of p^{j+1}, for $j=1,2,\ldots,m-1$. Then there exists an economy $\mathcal{E}$ whose agents have continuous, monotone, and strictly convex preferences, such that $\mathcal{E}$ generates an excess demand function F satisfying

(i) $F^{-1}(0)=A$

and

(ii) $\operatorname{sgn}\frac{\partial F_1}{\partial p_1}(p^j)=(-1)^j$.

A geometric postscript to Section 4

Since the preceding analysis is rather technical and algebraic, the reader may have drawn the mistaken impression that the results depend on some complicated pathology. This is not at all the case, and we offer the following geometric treatment in order to explain why the decomposition theorems are really quite intuitive, at least for the case of two commodities. Let $\ell=2$, and let $D=\{(p_1, p_2)\in R^2_+: p_1+p_2=1,\ 0<p_1<1\}$.

Proposition 10

If F: $D\to R$ is an arbitrary continuously differentiable function, then there exists a two-consumer, two-commodity exchange economy $\mathcal{E}=\{(\succsim_1,\omega_1),(\succsim_2,\omega_2)\}$ such that: (a) for $p\in\operatorname{int} D$, $F(p)$ is the market excess demand generated by $\mathcal{E}$ for the first commodity, and (b) for $p\in D$, $F(p)$ belongs to the market excess demand generated by $\mathcal{E}$ for the first commodity. Furthermore, both consumers can be chosen from the class of agents with homothetic and convex preferences. (The market excess demand for the second commodity is determined by Walras's Identity.)

Proof

Step 1: The line segment $(z_1, z_2)=z$: $D\to R^2$, defined by $p=(p_1, p_2)\to((2p_1+p_2)/(p_1+p_2),\ p_1/(p_1+p_2))$ connecting $(1,0)$ and $(2,1)$, is the price consumption

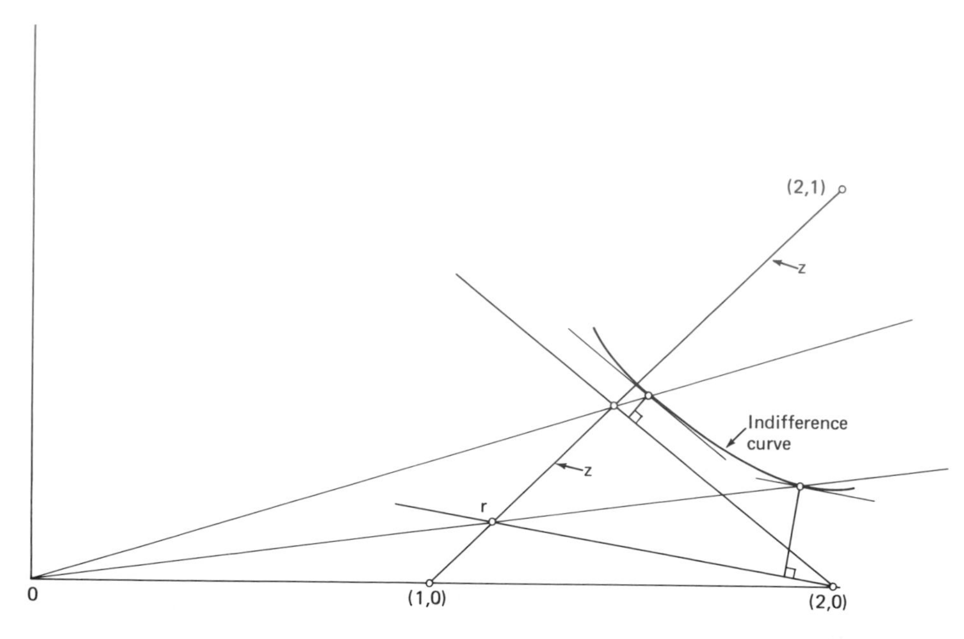

Figure 4.1.

curve of a consumer with a homogeneous utility function and initial endowment (2,0).[6]

Consult Figure 4.1. Preferences which generate the price consumption curve z are obtained by requiring that the marginal rate of substitution along rays such as $\overline{0r}$ (from 0 through r) be the negative of the slope of the line connecting (2,0) and the point where $\overline{0r}$ meets z: if $r=((2p_1+p_2)/(p_1+p_2), p_1/(p_1+p_2))$, then that marginal rate of substitution is p_1/p_2. (Let these preferences be represented by the homogeneous utility function u_1, and observe that they exhibit satiation with respect to the first commodity along the x_1 axis and with respect to the second commodity on and above the ray from the origin with slope one-half.)

Step 2: Sufficiently small smooth perturbations of z are also price consumption curves for consumers with homogeneous utility functions.

The critical feature of the previous construction is that as we move up and to the right along z, then the marginal rate of substitution associated with rays from

[6]For linguistic simplicity the wording will not apply to ∂D. The reader may either verify directly that our construction has the required property on ∂D, or note that this property follows from the upper hemicontinuity of excess demand functions which are generated by utility maximization.

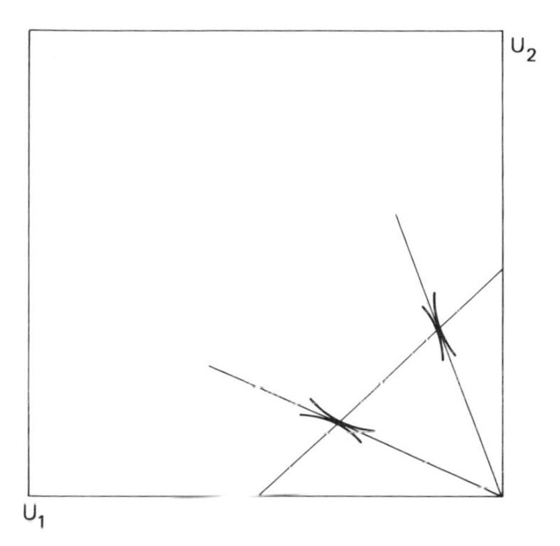

Figure 4.2.

the origin (parameterized by their height at $x_1=1$) is increasing with derivative bounded away from zero. This feature is preserved for sufficiently small smooth perturbations of z. (A smooth perturbation $z+\varepsilon G$ of z is defined by $(p_1, p_2)=((2p_1+p_2)/(p_1+p_2)+\varepsilon G, p_1/(p_1+p_2)-(p_1/p_2)\varepsilon G)$, where $G\colon S^1 \to R$ is continuously differentiable; it is small if ε is small.

Step 3: By adding a (symmetric) consumer with initial endowment (0,2) and preferences represented by $u_2(x_1, x_2)=u_1(x_2, x_1)$, we obtain a two-consumer, two-commodity (Edgeworth Box) economy in which every price is a Walras equilibrium price. (Consult Figure 4.2.)[7]

Step 4: Let F be given which satisfies the hypothesis of the proposition (see Figure 4.3). If the second individual's preferences are fixed at u_2 then the necessary choice of the other consumer [Consumer 1 with endowment (2,0)] for each p is defined by the condition that excess demand for the first commodity be $F(p)$. [For example, in Figure 4.3 the point s represents the choice by Consumer 1 necessary to make the excess demand for the first commodity at $(\frac{1}{2}, \frac{1}{2})$ equal to $F(\frac{1}{2}, \frac{1}{2})$. If F is replaced by $F/2$, the first consumer's choice must be $(s+t)/2$.] The fact that F is continuously differentiable means that the price consumption curve needed for Consumer 1 is a smooth perturbation of z. For ε sufficiently small, Step 2 tells us that $z+\varepsilon F$ is the price consumption curve for a consumer

[7]Some time ago A. Mas-Colell showed us a similar diagram. This, and a perturbation argument used by R. Mantel, are the basis for the current exposition.

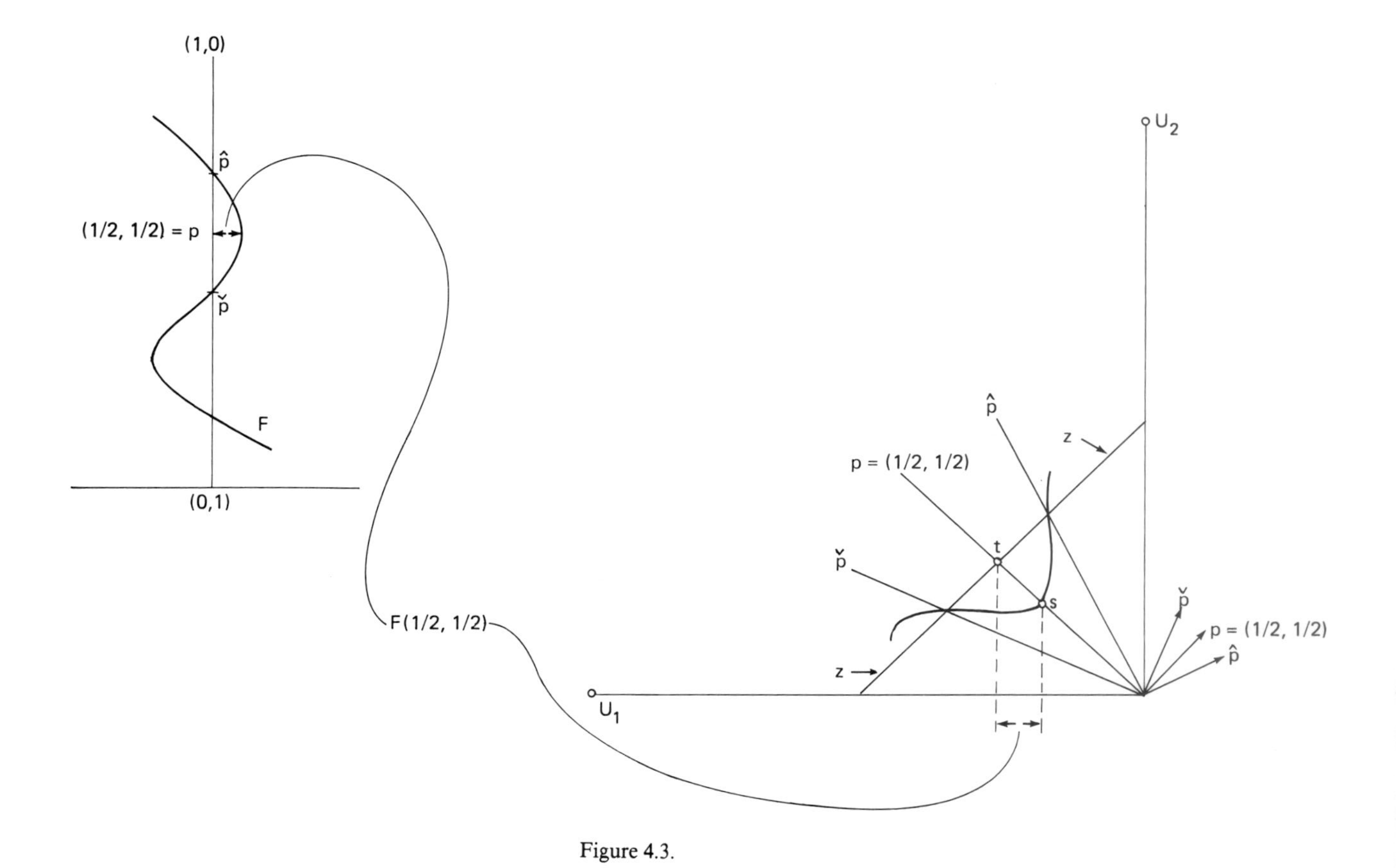

Figure 4.3.

with a homogeneous utility function u_1^ε. The economy $\{(u_1^\varepsilon,(2,0)),(u_2,(0,2))\}$ will generate εF as excess demand for the first commodity. Finally, since multiplying endowments by $1/\varepsilon$ multiplies individual excess demand by $1/\varepsilon$, the economy $\{(u_1^\varepsilon,(2/\varepsilon,0)),(u_2,(0,2/\varepsilon))\}$ generates F as excess demand for the first commodity. This completes the argument.

The analog of Theorem 7 (Mas-Colell) follows.

Proposition 11

Let F: int $D \to R$ be an arbitrary C^1 function such that $\lim_{p_1 \to 0} F(p) = +\infty$ and $\lim_{p_1 \to 0} F(p) < 0$. Then for any $\varepsilon > 0$, there exists a μ, $0 < \mu < \varepsilon$ and an economy with two consumers such that F is the market excess demand function for the first commodity on $D_\mu = \{p \in D\colon \mu \leqq p_1 \leqq 1 - \mu\}$ and such that $F^{-1}(0) \subset D_\mu$ is the set of equilibrium price vectors of the economy. Furthermore, both consumers can be chosen from the class of agents with homothetic and convex preferences.

Proof

Note the boundary and continuity conditions ensure $F^{-1}(0) \neq \varnothing$, by the intermediate value theorem, and that $F^{-1}(0)$ is compact. Pick $0 < \mu < \varepsilon$ such that $F^{-1}(0) \subset D_\mu$. Pick a non-decreasing C^1 function g: $R \to R$ such that $g(t) = 0$ for $t \leqq \mu/2$, $g(t) = 1$ for $t \geqq \mu$, and g is strictly increasing in $(\mu/2, \mu)$. For $p \in D$, let $K(p) = g(p_1)g(p_2)F(p) + g(p_2) - g(p_1)$. Then K is a well-defined C^1 function from D to R, so by the previous proposition K is the market excess demand function for a two-person economy E with homothetic convex preferences. Moreover, $K(p) \equiv F(p)$ if $p \in D_\mu$. If $p \notin D_\mu$, then (a) if $p_1 < \mu$, $F(p) > 0$ and $g(p_2) - g(p_1) > 0$, so $K(p) > 0$, and (b) if $p_2 < \mu$, $F(p) < 0$ and $g(p_2) - g(p_1) < 0$, so $K(p) < 0$. Thus, $F^{-1}(0)$ is the set of equilibrium prices of the economy generating K.

The preceding geometric two-commodity approach also yields analogs of Theorems 8 and 9. A basic result of analysis says that if $A \subset R^2$ is any non-empty closed set, then there exists a C^∞ function F: $R^2 \to R$ such that $F(p) = 0$ if and only if $p \in A$. Simply take F to be the function of Proposition 10. For Theorem 9 ($\ell = 2$), the conditions on the set of prices $A = \{p^1, p^2, \ldots, p^n\}$ make it possible to fit a polynomial excess demand with the desired properties. Note that it has been shown, at least for the case $\ell = 2$, that F is not required to have second derivatives as in Theorem 6 (Mantel) in order to be the excess demand of an economy with homothetic preferences, and that Mas-Colell's results can be obtained using homothetic preferences if excess demand is C^1. Some differentiability is essential: according to Mantel (1977, p. 116), "As noted by Mas-Colell in private communication, the differentiability assumption cannot be totally dispensed with, as can be seen in the two-good case, where homotheticity implies that the ratio of quantities demanded is a monotone function of relative price, and is hence almost everywhere differentiable."

5. Market demand functions

We turn now to the case of market demand functions. Much less is known about restrictions on market demand functions than is known about restrictions on market excess demand functions. The initial investigation of this problem was undertaken by Sonnenschein (1973a). He demonstrated that for $\ell=2$ any demand function is a market demand function with equal income shares, ignoring non-negativity restrictions on commodities. The non-negativity restrictions turn out to be very important: this was observed by Sonnenschein (1973b) and in a related problem by McFadden, Mas-Colell, Mantel, and Richter (1974), who examined restrictions on excess demand if the aggregate demand vector is fixed and consumption sets are R^{ℓ}_{+}. The nature of the non-negativity restriction can be seen as follows: Let G be a consumer demand function with $m=1$. By WARP, $p\cdot G(q)\leqq 1$ implies $q\cdot G(p)\geqq 1$, and since $G\geqq 0$ implies $p\cdot G(q)\geqq 0$ and $q\cdot G(p)\geqq 0$, $p\cdot G(q)+q\cdot G(p)\geqq 1$ must hold for all p, q. This restriction is easily seen to be preserved in aggregation with a fixed distribution of income, and thus it is a property of market demand functions. Consider the demand function G defined by $G(p, m)=mp/p\cdot p$. Since $p\cdot G(q,1)+q\cdot G(p,1)=p\cdot q/q\cdot q+q\cdot p/p\cdot p<1$, provided p and q are chosen close to two distinct unit vectors, G is not a market demand function. Nevertheless, if one examines G and its first derivatives at any point $(p, m)\in\Lambda$, it looks like a market demand function. Precisely, Sonnenschein (1973b) demonstrated the following result:

Theorem 12 (Sonnenschein)

Let G be a C^1 demand function for ℓ commodities. Then for any $(p, m)\in\Lambda$ there exists a market demand function generated by n consumers with demand functions $\{G^i\}$ such that

$$G(p,m)=\sum_{i=1}^{n} G^i(p,m/n) \quad \text{and} \quad \frac{\partial G_k}{\partial p_j}(p,m)=\sum_{i=1}^{n}\frac{\partial G^i_k(p,m/n)}{\partial p_j},$$

for each k, j.

Diewert (1977) showed that there are restrictions on the first derivatives of market demand functions if $n<\ell$. Mantel (1977) showed that these were the only restrictions and in doing so generalized Sonnenschein's result. In particular, the n in Theorem 12 can always be chosen equal to ℓ. Combining Diewert's and

Mantel's work, we have the following result:

Theorem 13 (Diewert, Mantel)

Let G be a C^1 demand function for ℓ commodities. Pick any $(p,m)\in\Lambda$ such that $G(p,m)>0$, and $\delta_1,\delta_2,\ldots,\delta_n$ $(\delta_i>0, \sum\delta_i=1)$, and any linear subspace $L\subset R^\ell$ of dimension $\ell-n\geqq 0$ which satisfies $z\cdot G(p,m)=0$ for all $z\in L$. Then $D_pG(p,m)$ satisfies

(i) $z^T D_pG(p,m)z<0$ for each $z\in L$, $z\neq 0$, and
(ii) $z^T D_pG(p,m)\hat{z}=\hat{z}^T D_pG(p,m)z$ for each z, $\hat{z}\in L$,

if [Diewert (1977)] and only if [Mantel (1977)], there exists n C^1 consumer demand functions $G^1, G^2,\ldots,G^n$ for ℓ commodities, each generated by a C^2 utility function, such that $G^i(p,m)>0$, $i=1,2,\ldots,n$, and

(i) $G(p,m)=\sum_{i=1}^n G^i(p,\delta_i m)$,
(ii) $D_pG(p,m)=\sum_{i=1}^n D_pG^i(p,\delta_i m)$,
(iii) $L\subset\{z\in R^\ell\colon z\cdot G^i(p,\delta_i m)=0 \text{ for each } i=1,2,\ldots,n\}$.

Proof

The special case $n=\ell$ will be shown, based on Mantel's proof. Note that if $n=\ell$, $L=\{0\}$, so that (i) and (ii) are vacuously satisfied by any demand function G. Also, note that for $n=1$ (so $\ell-n=\ell-1$), conditions (i) and (ii) are just the classical restrictions on a consumer demand function which is derived from a C^2 utility function. The theorem for $n=1$ will be assumed known. To prove the theorem for $\ell=n$, first pick ℓ linearly independent positive vectors $x^1, x^2,\ldots,x^\ell\in R^\ell_+$ such that $\sum_i x^i=G(p,m)$ and $p\cdot x^i=\delta_i m$, $i=1,2,\ldots,\ell$. Let X be the $\ell\times\ell$ matrix in which x^i is the ith column vector. Then $Xe=G(p,m)$ and $p^TX=\delta m$, where $\delta=(\delta_1,\delta_2,\ldots,\delta_\ell)$. Choose any $\ell\times\ell$ matrix V of rank $\ell-1$ such that $V\delta=0$. Define a matrix $K=-D_pG(p,m)X^{-1T}-XV^TV$, and let k^i be the ith column vector of K. Note that $p^TK=-p^TD_pG(p,m)X^{-1T}-p^TXV^TV=G(p,m)^TX^{-1T}-m\delta V^TV=e$, so $p\cdot k^i=1$, $i=1,2,\ldots,\ell$. Define $J_i=(-1/\ell)XV^TVX^T-k^ix^{iT}$. Then $p^TJ_i=-x^{iT}$ and $p^TJ_ip=-\delta_i m$, so J_i is the Jacobian $D_p\hat{G}^i(p,\delta_i m)$ of some demand function $\hat{G}^i$ at $(p,\delta_i m)$, since $p^TD_p\hat{G}^i(p,\delta_i m)=-\hat{G}^i(p,\delta_i m)$ is just the derivate version of the budget constraint. Moreover, letting $L_i=\{z\colon\ z\cdot x^i=0\}$, yields $\hat{z}^TJ_iz=-\hat{z}^T(1/\ell)XV^TVX^Tz$ for each $z\in L_i$, and $-XV^TVX^T$ is negative definite symmetric on L_i so (i) and (ii) of the theorem will be satisfied. Thus, applying the theorem for $n=1$ for each i, there is a consumer demand function G^i such that $x^i=G^i(p,\delta_i m)$ and $J_i=D_pG^i(p,\delta_i m)$. Finally, the following holds:

$$\sum_{i=1}^{\ell} J_i=-XV^TVX^T-KX^T=-XV^TVX+D_pG(p,m)X^{-1T}X^T+XV^TVX$$
$$=D_pG(p,m).$$

There is a natural analogue of the above theorem for the case of excess demand. Again, Diewert (1977) worked out restrictions on market excess demand for the case $n<\ell$; Geanakoplos and Polemarchakis (1980) showed that these are only restrictions. Prior to their results, Debreu (1974) had demonstrated the existence of a market excess demand function for ℓ commodities which cannot be decomposed by fewer than ℓ consumers.

Theorem 14 (Diewert, Geanakoplos and Polemarchakis)

Let F: $S \to R^\ell$ be a C^1 excess demand function for ℓ commodities. Pick any $p \in S$ and any linear subspace $L \subset R^\ell$ of dimension $\ell - n - \text{sgn}\|F(p)\| \geqq 0$ which satisfies $z \cdot F(p)=0$ and $z \cdot p=0$ for each $z \in L$. Then, $DF(p)$ satisfies (i) $z^T DF(p) z<0$ for each $z \in L$, $z \neq 0$, and (ii) $\hat{z}^T DF(p) z = \hat{z}^T DF(p) z$ for each $\hat{z}, z \in L$, if [Diewert (1977)] and only if [Geanakoplos and Polemarchakis (1980)] there exists n C^1 consumer excess demand functions $F^1, F^2, \ldots, F^n$, each generated by a C^2 utility function and a positive initial endowment vector, such that

(i) $F(p)=\sum_1^n F^i(p)$,
(ii) $DF(p)=\sum_1^n DF^i(p)$,
(iii) $L \subset \{z \in R^\ell\colon z \cdot F^i(p)=0,\ i=1,2,\ldots,n \text{ and } z \cdot p=0\}$.

Returning to demand functions, we have noted that because of the non-negativity condition not every demand function is a market demand function. Nevertheless, at a given point (p, m) there are no restrictions on the first derivatives of a market demand function other than those implied by homogeneity and the budget equality. There might, however, be restrictions on higher derivatives. In particular, it is not known for $\ell>2$ if every demand function is market demand function in a small neighborhood of a point. Sonnenschein (1973b), posed this question:

True or false? Let G be a C^k demand function for ℓ commodities ($k \geqq 0$). Then for any $(p, m) \in \Lambda$ there exists an open neighborhood N of (p, m) and a market demand function $\bar{G}$ such that $G=\bar{G}$ on N. Furthermore, $\bar{G}$ is the market demand of ℓ consumers.

Note that if $\bar{G}=\sum_{i=1}^{\ell} G^i$ in the above open question, then as long as $G^i(p, \delta_i m) >0$ for each i, the non-negativity restriction on each G^i is irrelevant, since the neighborhood N of (p, m) can always be chosen small enough so that each G^i is positive on N. Thus, one can look at the question by trying to decompose G on Λ into ℓ consumer demand functions G^i which are not required to be positive. Such a result would mean that the only restrictions on market demand would be those which arise because consumer demand is required to be positive. This, as was mentioned above, is what Sonnenschein demonstrated for the case $\ell=2$. Andreu (forthcoming) has recently answered this question on finite domains.

Theorem 15 (Andreu)

Let K be a finite subset of P and G: $P \to R^{\ell}$ satisfy $p \cdot G(p) = \ell$ for all $p \in P$. Then there exist ℓ functions G^i: $P \to R^{\ell}$ such that $p \cdot G^i(p) - 1$ for all $p \in P$ and such that

(i) $\sum_{i=1}^{\ell} G^i(p) = G(p)$ on K, and
(ii) G^i satisfies SARP on K for each i.

Proof

The proof involves a modification of the decomposition given by Debreu in his proof of Theorem 5. Let θ: $R^{\ell} \to R^1$ be arbitrary and define

$$G^i(p) = \left[\theta(p)p_i + G_i(p) - (\ell p_i / p \cdot p)\right]\left[e_i - (p_i p / p \cdot p)\right] + p(p \cdot p).$$

Elementary calculations show that $p \cdot G^i(p) = 1$ for all i and p, and that (i) is satisfied. Andreu shows that (ii) is satisfied for θ sufficiently large.

6. Applications

The theorems of Section 4 on the arbitrary nature of market excess demand functions have had a substantial number of applications; one of these, the arbitrary nature of the set of Walras equilibria, has been explicitly mentioned (Theorem 8). A second application is discussed by G. Debreu in Chapter 15 of this Handbook and concerns an equivalence between the existence of competitive equilibrium and the Brouwer fixed point theorem [see Uzawa (1962)]. The arbitrary nature of market excess demand "combined with Uzawa's observation shows that the proof of existence of a competitive equilibrium requires mathematical tools of the same power as a fixed-point theorem" (Debreu, Chapter 15). As a result, the general problem of computing equilibrium prices (see Chapter 21 by H. Scarf) is equivalent to the general problem of computing fixed points of maps from the n cell into itself.

Two further applications to stability analysis are given by Sonnenschein (1973a); these are even more transparent in the light of Theorem 9 (Mas-Colell): (a) the existence of an economy with an equilibrium that is Hicksian stable but not dynamically stable and (b) the existence of an economy with a single dynamically unstable equilibrium. The first of these results was new and used an example due to Samuelson (1944) of a 3×3 matrix which is Hicksian stable but not dynamically stable. The second result was known, with examples in the literature due to Gale (1963) and Scarf (1960). Also, the theorems of Section 4 are very useful for the construction of examples in which competitive allocations and prices vary in prescribed (and possibly unintuitive) ways with the data of an economy. The results have been applied to show how one can benefit from giving up (either by destruction or transfer) initial endowment (see Guesnerie and Laffont (1978)). Roberts and Sonnenschein (1977) use the results to illustrate the presence of discontinuous reaction functions in the theory of monopolistic competition and show that this can lead to the non-existence of monopolistic competitive equilibrium.

References

Andreu, J. (forthcoming), "Rationalization of market demand on finite domains", Journal of Economic Theory, 22.

Antonelli, G. B. (1886), Sulla Teoria Matemitica della Economia Politica. Pisa: Nella Tipognafia del Folchetto. English translation: J. S. Chipman and A. P. Kirman, in: Chipman, J. S., L. Hurwicz, M. K. Richter, and H. F. Sonnenschein, eds., Preferences, utility, and demand. New York: Harcourt Brace Jovanovich, Inc., 1971, pp. 333–360.

Chipman, J. S. (1974), "Homothetic preferences and aggregation", Journal of Economic Theory, 8:26–38.

Chipman, J. S. and J. Moore (1976), "On the representation and aggregation of homothetic preferences", Manuscript.

Debreu, G. (1970), "Economies with a finite set of equilibria", Econometrica, 38:387–392.

Debreu, G. (1974), "Excess demand functions", Journal of Mathematical Economics, 1:15–23.

Dierker, E. (1972), "Two remarks on the number of equilibria of an economy", Econometrica, 40:951–955.

Diewert, W. E. (1977), "Generalized Slutsky conditions for aggregate consumer demand functions", Journal of Economic Theory, 15:353–336.

Eisenberg, B. (1961), "Aggregation of utility functions", Management Science, 7:337–350.

Gale, D. (1963), "A note on global instability of competitive equilibrium", Naval Research Logistics Quarterly, 10:81–87.

Geanakoplos, J. D. (1978), "Utility functions for Debreu's excess demands", mimeo. Harvard University.

Geanakoplos, J. D. and H. M. Polemarchakis (1980), "On the disaggregation of excess demand functions", Econometrica, 48:315–331.

Gorman, W. M. (1953), "Community preference fields", Econometrica, 21:63–80.

Guesnerie, Roger and Jean-Jacques Laffont (1978) "Advantageous reallocations of initial resources", Econometrica, 46:835–841.

Mantel, R. (1974), "On the characterization of aggregate excess demand", Journal of Economic Theory, 7:348–353.

Mantel, R. (1977), "Implications of microeconomic theory for community excess demand functions", in: M. D. Intriligator, ed., Frontiers of quantitative economics, Vol. IIIA. Amsterdam: North-Holland Publishing Co., pp. 111–126.

Mantel, R. (1979), "Homothetic preferences and community excess demand functions", Journal of Economic Theory, 12:197–201.

Mas-Colell, A. (1977), "On the equilibrium price set of an exchange economy", Journal of Mathematical Economics, 4:117–126.

McFadden, D., A. Mas-Colell, R. Mantel, and M. K. Richter (1974), "A characterization of community excess demand functions", Journal of Economic Theory, 9:361–374.

Nataf, A. (1953), "Sur des questions d'agrégation en économétrie", Publications de l'Institute de Statistique de l'Université de Paris, 2:5–61.

Roberts, John and Hugo Sonnenschein (1977), "On the foundations of the theory of monopolistic competition", Econometrica, 45:101–113.

Samuelson, P. (1944), "The relation between Hicksian stability and true dynamic stability", Econometrica, 12:256–257.
Scarf, H. (1960), "Some examples of global instability of competitive equilibrium", International Economic Review, 1:157–172.
Shafer, W. (1977), "Revealed preference and aggregation", Econometrica, 45:1173–1182.
Sonnenschein, H. (1971), "The numerical construction of economies with given equilibrium price sets", Abstract, Econometrica, 39:22.
Sonnenschein, H. (1973a), "Do Walras' identity and continuity characterize the class of community excess demand functions?" Journal of Economic Theory, 6:345–354.
Sonnenschein, H. (1973b), "The utility hypothesis and market demand theory", Western Economic Journal, 11:404–410.
Sonnenschein, H. (1974), "Market excess demand functions", Econometrica, 40:549–563.
Uzawa, H. (1962), "Competitive equilibrium and fixed point theorems II: Walras' existence theorem and Brouwer's fixed-point theorem", Economic Studies Quarterly, 12:59–62.
Wold, H. (1953), Demand analysis. New York: John Wiley and Sons, Inc., p. 119.

PART 3

MATHEMATICAL APPROACHES TO COMPETITIVE EQUILIBRIUM

Chapter 15

EXISTENCE OF COMPETITIVE EQUILIBRIUM*

GERARD DEBREU

University of California

1. Introduction

The mathematical model of a competitive economy of L. Walras (1874–77) was conceived as an attempt to explain the state of equilibrium reached by a large number of small agents interacting through markets. Walras himself perceived that the theory that he proposed would be vacuous without a mathematical argument in support of the existence of at least one equilibrium state. However, for more than half a century the equality of the number of equations and of the number of unknowns of the Walrasian model remained the only, unconvincing, remark made in favor of the existence of a competitive equilibrium. Study of the existence problem began in the early 1930s when Neisser (1932), Stackelberg (1933), Zeuthen (1933), and Schlesinger (1935) identified several of its basic features and when Wald (1935, 1936a, 1936b) obtained its first solution. After an interruption of some twenty years, the question of existence of an economic equilibrium was taken up again by Arrow and Debreu (1954), McKenzie (1954, 1955), Gale (1955), Debreu (1956), Kuhn (1956a, 1956b), Nikaido (1956), Uzawa (1956), and many other authors whose contributions over the last twenty-five years constitute a bibliography of over three hundred and fifty items.

In this literature four distinct, but closely related, approaches to the existence problem can be recognized. (1) At first, proofs of existence of an economic equilibrium were uniformly obtained by application of a fixed-point theorem of the Brouwer type or of the Kakutani type or by analogous arguments. This approach, which has remained of central importance to the present, is the subject of this chapter. (2) In the last decade, efficient algorithms of a combinatorial nature for the computation of an approximate economic equilibrium were developed. These algorithms, which provide a constructive answer to the existence

*More than forty persons have made suggestions that were incorporated in the text or in the bibliography of this chapter. I wish to express my thanks to each one of them. The support of the National Science Foundation in the preparation of this article is also gratefully acknowledged.

Handbook of Mathematical Economics, vol. II, edited by K.J. Arrow and M.D. Intriligator

question, are surveyed in Chapter 21 of this Handbook by Scarf who played a leading role in this area of research. (3) More recently, the theory of the fixed-point index of a map and the degree theory of maps were used to establish the existence of an economic equilibrium by Dierker (1972, 1974), Mityagin (1972), Smale (1974), Balasko (1975, 1978), Varian (1975a, 1975b), Nishimura (1978), Kalman and Lin (1978a, 1978b), and Kalman, Lin, and Wiesmeth (1978). Chapter 17 of this Handbook by Dierker touches on some of these questions. (4) Finally, in 1976 Smale proposed a differential process whose generic convergence to an economic equilibrium provides an alternative constructive solution of the existence problem [Smale (1976)]. This approach is the subject of Chapter 8 of this Handbook by Smale.

2. The simultaneous optimization approach

Although the first proof of existence of a competitive equilibrium was obtained by Wald (1935, 1936a, 1936b), two articles by von Neumann (1928, 1937) turned out to be of greater importance for the post-1950 development of the subject. In the first of these articles von Neumann established the existence of a pair of equilibrium strategies for a two-person zero-sum game, i.e. of a saddle-point for the utility of either player. In the second, he studied the problem of existence of an equilibrium balanced growth path, transforming it into an equivalent saddle-point problem which was a direct descendant of his earlier contribution to the theory of games. In order to solve this new saddle-point problem, von Neumann proved a topological lemma which, in its reformulation by Kakutani (1941) as a fixed-point theorem for a correspondence, became the most powerful tool for the proof of existence of an economic equilibrium.

Before stating Kakutani's theorem, we introduce the concepts that it concerns. Given two sets, S and T, a *correspondence* φ from S to T associates with every element x of S, a non-empty subset $\varphi(x)$ of T. The *graph* of the correspondence φ is the subset of the Cartesian product $S\times T$ defined by

$$G(\varphi)=\{(x,y)\in S\times T \mid y\in\varphi(x)\}.$$

The correspondence φ is said to be *convex-valued* if T is a real vector space and for every x in S, the set $\varphi(x)$ is convex. A correspondence φ from a subset S of a Euclidean space to a subset T of a Euclidean space is *upper hemicontinuous* (*u.h.c.*) *at a point* x^0 *of* S if there is a neighborhood of x^0 in which φ is bounded, and for every sequence x^q in S converging to x^0 in S, and every sequence y^q in T converging to y^0 in T such that for every q, one has $y^q\in\varphi(x^q)$, then $y^0\in\varphi(x^0)$. More concisely,

$$[x^q\to x^0,\ y^q\to y^0,\ y^q\in\varphi(x^q)]\Rightarrow[y^0\in\varphi(x^0)].$$

This definition of upper hemicontinuity of φ at x^0 clearly implies that $\varphi(x^0)$ is compact. Upper hemicontinuity of φ *on* S is defined as upper hemicontinuity at every point of S. We note that a Cartesian product of u.h.c. correspondences is u.h.c. More precisely, consider for every $i=1,\dots,n$ a correspondence φ_i from S, a subset of a Euclidean space, to T_i, a subset of a Euclidean space, and for every x in S, define $\varphi(x)=\times_{i=1}^{n}\varphi_i(x)$. If every φ_i is u.h.c. at x_0, then φ is clearly u.h.c. at x_0.

Finally, given a correspondence φ from a set S to itself, an element x_0 of S is a *fixed point* of φ if $x^0 \in \varphi(x^0)$.

Theorem 1 (Kakutani)

If S is a non-empty, compact, convex subset of a Euclidean space, and φ is an upper hemicontinuous, convex-valued correspondence from S to S, then φ has a fixed point.

The application of this result made by Nash (1950), whose theorem we now state and prove, shows the suitability of Kakutani's theorem for proofs of existence of an economic equilibrium. Consider an n-person game. The ith player chooses a strategy s_i in the set S_i which is assumed to be a non-empty, compact, convex subset of a Euclidean space. In fact, S_i is the set of probabilities over the finite set of pure strategies of the ith player. The n-list of strategies chosen by the n players is an element $s=(s_1,\dots,s_n)$ of the Cartesian product $S=\times_{i=1}^{n}S_i$. And the resulting utility for the ith player is a real number $f_i(s)$. The utility function f_i is assumed to be continuous on S and to be linear in s_i. In fact, in the game theoretical situation studied by Nash, f_i is linear with respect to each one of the variables $s_1,\dots,s_j,\dots,s_n$. Formally, the game is defined by the n sets $S_1,\dots,S_n$ and the n functions $f_1,\dots,f_n$.

Denote by N the set $\{1,\dots,n\}$ of players and by $N\backslash i$ the set $\{1,\dots,i-1,i+1,\dots,n\}$ of players excluding the ith. An n-list s^* of strategies is a *Cournot–Nash equilibrium* if for every $i\in N$, s_i^* maximizes $f_i(s_i, s_{N\backslash i}^*)$ on S_i, i.e. if every player chooses a strategy that maximizes his utility given the strategies chosen by the other players.

Theorem 2 (Nash)

If, for every $i\in N$, the set S_i is a non-empty, compact, convex subset of a Euclidean space, and f_i is a continuous real-valued function on $S=\times_{j\in N}S_j$ that is linear in its ith variable, then the game $(S_i, f_i)_{i\in N}$ has an equilibrium.

Proof

For each i, define the correspondence μ_i from S to S_i as follows. Given an element s of S, let

$$\mu_i(s)=\left\{x\in S_i \mid f_i(x, s_{N\backslash i})=\max_{y\in S_i} f_i(y, s_{N\backslash i})\right\},$$

i.e. $\mu_i(s)$ is the set of strategies of the ith player that maximize his utility given the strategies of the other players prescribed by s. Actually, $\mu_i(s)$ is independent of s_i. However, it will be more convenient to use μ_i in the form that we have introduced. Note that the maximization operation appearing in the definition of μ_i can be carried out, since f_i is continuous and S_i is compact, non-empty. In particular, $\mu_i(s)$ is non-empty. It is also convex since f_i is linear in s_i, and the maximization operation is carried out on a convex set. Finally, we show that μ_i is upper hemicontinuous. Let (s^q) be a sequence in S converging to s^0, and (x^q) be a sequence in S_i converging to x^0 and such that for every q, one has $x^q \in \mu_i(s^q)$. Consider an arbitrary point y in S_i. For every q, one has $f_i(x^q, s^q_{N\setminus i}) \gtrsim f_i(y, s^q_{N\setminus i})$. In the limit, $f_i(x^0, s^0_{N\setminus i}) \geqslant f_i(y, s^0_{N\setminus i})$. Since this inequality holds for any y in S_i, we have proved that $x^0 \in \mu_i(s^0)$.

Now define the correspondence μ from S to S by

$$\mu(s) = \mathop{\times}_{i=1}^{n} \mu_i(s).$$

A point s^* of S is a Cournot–Nash equilibrium if and only if for every i, $s_i^* \in \mu_i(s^*)$; hence, if and only if $s^* \in \mu(s^*)$ (Figure 2.1).

The concept of a Cournot–Nash equilibrium of the game is therefore equivalent to that of a fixed point of the correspondence μ. The existence of such a fixed

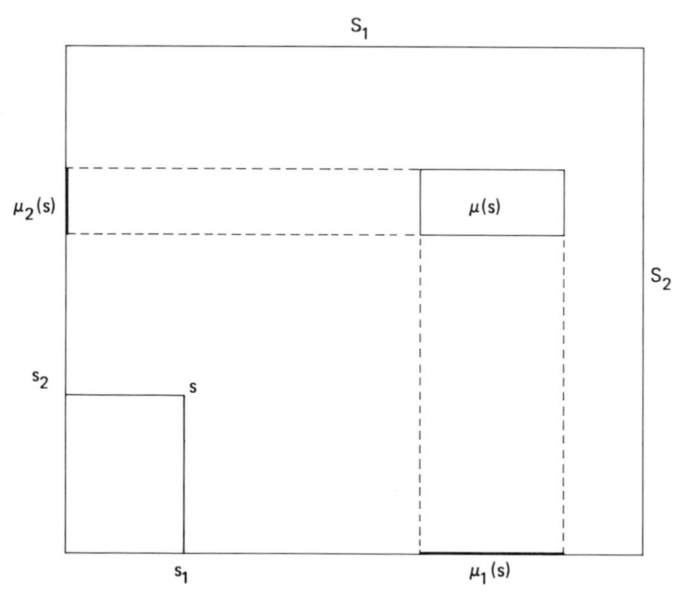

Figure 2.1.

point is proved by an application of Kakutani's theorem whose assumptions are clearly satisfied. Q.E.D.

The preceding concept of equilibrium and Nash's theorem can be generalized [Debreu (1952)] in a manner that makes them applicable to a variety of economic problems. As an introduction to that generalization, we study an economic agent whose environment is described by an element x of a set X. This agent has a set Y of available actions of which he must choose one. However, when his environment is the element x of X, he is restricted in his choice to a non-empty subset $\varphi(x)$ of Y. And if he selects action y in $\varphi(x)$, his utility is $f(x, y)$, where f is a real-valued function defined on $X \times Y$. Given the environment x, the agent strives to choose in $\varphi(x)$ an action that maximizes his utility. We denote the set of his optimal actions by

$$\mu(x)=\left\{y\in\varphi(x)| f(x,y)=\max_{z\in\varphi(x)} f(x,z)\right\}.$$

The proof of existence of an equilibrium at which we are aiming rests on the upper hemicontinuity of the correspondence μ. To establish this property of μ we need two additional concepts of continuity for a correspondence. In a manner symmetric to that in which we defined upper hemicontinuity, we say that a correspondence φ from a subset S of a Euclidean space to a subset T of a Euclidean space is *lower hemicontinuous* (*l.h.c.*) *at a point* x^0 *of* S if, for every sequence (x^q) in S converging to x^0 in S, and every y^0 in $\varphi(x^0)$, there is a sequence (y^q) in T converging to y^0 such that for every q, one has $y^q \in \varphi(x^q)$. More concisely,

$$\left[x^q\to x^0,\ y^0\in\varphi(x^0)\right]\Rightarrow\left[\text{there is } (y^q) \text{ such that } y^q\to y^0,\ y^q\in\varphi(x^q)\right].$$

Lower hemicontinuity on S is defined as lower hemicontinuity at every point of S. And *continuity* of a correspondence at a point or on a set is defined as the conjunction of upper and lower hemicontinuity.

With the help of these concepts we can state the following lemma [Berge (1963, VI.3)].

Lemma 1

Let X and Y be subsets of Euclidean spaces. If the function f and the correspondence φ are continuous, then the correspondence μ is upper hemicontinuous.

Proof

Consider a sequence (x^q) in X converging to x^0 in X, and a sequence y^q in Y converging to y^0 in Y such that for every q, $y^q \in \mu(x^q)$. Since, for every q, $y^q \in \varphi(x^q)$ and φ is upper hemicontinuous, one has $y^0 \in \varphi(x^0)$. On the other hand, let z be an arbitrary point of $\varphi(x^0)$. Since φ is lower hemicontinuous, there is a sequence (z^q) in Y converging to z such that for every q, $z^q \in \varphi(x^q)$. Thus, for every q, one has $f(x^q, y^q) \geqq f(x^q, z^q)$. In the limit, $f(x^0, y^0) \geqq f(x^0, z)$. Since this inequality holds for any z in $\varphi(x^0)$, we have proved that $y^0 \in \mu(x^0)$. Q.E.D.

Let us then study a *social system* composed of n agents. The ith agent must choose an element in the set A_i of his a priori available actions which is assumed to be a non-empty, compact, convex subset of a Euclidean space. However, when the agents other than the ith choose the actions $(a_1, \ldots, a_{i-1}, a_{i+1}, \ldots, a_n)$, the choice of the ith agent is restricted to a non-empty subset of A_i determined by the preceding $(n-1)$-list. Formally, we define a correspondence φ_i from $A = \times_{j=1}^{n} A_j$ to A_i that associates with the generic element a of A, the non-empty subset $\varphi_i(a)$ of A_i to which the choice of the ith agent is restricted. The set $\varphi_i(a)$ is actually independent of the ith component of a, but, as before, we find it more convenient to define φ_i on A rather than on $\times_{j \in N \setminus i} A_j$. The correspondence φ_i is assumed to be continuous and convex-valued.

The utility of the ith agent resulting from the choice of the n-list a of actions is the real number $f_i(a)$, where the function f_i is assumed to be continuous on A and to be quasi-concave relative to a_i. Given a in A, each agent, say, the ith, considers the actions $a_{N \setminus i}$ of the other agents as given and chooses his own action y so as to maximize his utility $f_i(y, a_{N \setminus i})$ in the set $\varphi_i(a)$ to which he is restricted. That is, the ith agent chooses an element of

$$\mu_i(a) = \left\{ x \in \varphi_i(a) \mid f_i(x, a_{N \setminus i}) = \max_{y \in \varphi_i(a)} f_i(y, a_{N \setminus i}) \right\}.$$

An element a^* of A is an *equilibrium* if for every $i \in N$, a_i^* maximizes $f_i(\cdot, a_{N \setminus i}^*)$ in $\varphi_i(a^*)$, i.e. if for every $i \in N$, $a_i^* \in \mu_i(a^*)$. Thus, if we define the correspondence μ from A to A by

$$\mu(a) = \times_{i \in N} \mu_i(a),$$

the element a^* of A is an equilibrium if and only if $a^* \in \mu(a^*)$; in other words, if and only if a^* is a fixed point of the correspondence μ. By an application of Kakutani's theorem, we obtain

Theorem 3

If, for every $i \in N$, the set A_i is a non-empty, compact, convex subset of a Euclidean space, f_i is a continuous real-valued function on $A = \times_{j \in N} A_j$ that is quasi-concave in its ith variable, and φ_i is a continuous, convex-valued correspondence from A to A_i, then the social system $(A_i, f_i, \varphi_i)_{i \in N}$ has an equilibrium.

Proof

Since the correspondence φ_i and the function f_i satisfy the assumptions of Lemma 1, the correspondence μ_i is upper hemicontinuous. Moreover, for every a in A, the set $\mu_i(a)$ is convex, for it is the intersection of the two sets $\varphi_i(a)$ and $\{x \in A_i \mid f_i(x, a_{N \setminus i}) \geqq \max_{y \in \varphi_i(a)} f_i(y, a_{N \setminus i})\}$, both of which are convex. Therefore, for every a in A, the set $\mu(a) = \times_{i \in N} \mu_i(a)$ is convex. The set A is non-empty, compact, and convex. Finally, the correspondence μ is u.h.c. as a Cartesian product of u.h.c. correspondences. Thus, the correspondence μ from A to A has a fixed point. Q.E.D.

The general model of a social system that we have just presented includes as a special case the following economy [Arrow and Debreu (1954)]. The agents of the economy produce, trade, and consume ℓ commodities. Each one of these commodities is a good or a service with specified physical characteristics available at specified date and location [and in a specified event if uncertain exogenous events affect the economy; Arrow (1953), Debreu (1959, ch. 7)]. The different types of human labor are among the commodities.

We distinguish two kinds of agents, namely consumers and producers. The *consumption* of the ith consumer $(i = 1, \ldots, m)$ is an ℓ-list of the quantities of the various commodities that he consumes or that he produces (the latter being typically human labor). His inputs are represented by positive numbers and his outputs by negative numbers. Thus, his consumption is a point x_i in the commodity space R^ℓ. However, his consumption cannot be chosen arbitrarily in R^ℓ. For instance, the combination of a small food input and of a large labor output during a certain period is excluded. We denote by X_i, a non-empty subset of R^ℓ, the set of the possible consumptions of the ith consumer. The tastes of this consumer are described by a complete, reflexive, transitive binary *preference relation* $\precsim_i$ on his consumption set X_i and $x \precsim_i x'$ is read as "x′ is at least as desired by the ith consumer as x". Completeness of this preference relation means that for an arbitrary pair (x, x') of consumptions of the ith consumer, one has $x \precsim_i x'$ and/or $x' \precsim_i x$. Reflexivity means that for an arbitrary consumption x of the ith consumer, one has $x \precsim_i x$. Transitivity means that if x, y, z are three consumptions of the ith consumer, then $[x \precsim_i y$ and $y \precsim_i z]$ implies $[x \precsim_i z]$. The relation of strict preference $x \prec_i x'$ is defined as "$x \precsim_i x'$ and not $x' \precsim_i x$".

Given a price-vector p in R^ℓ listing the prices of all the commodities, the *net value* of the consumption x_i of the ith consumer is the inner product $p \cdot x_i$. It must be at most equal to his wealth w_i. Under this constraint the ith consumer strives to satisfy his preferences $\precsim_i$, i.e. he strives to choose a greatest element for the relation $\precsim_i$ in the set $\{x \in X_i \mid p \cdot x \leqq w_i\}$. To complete the description of the characteristics of the ith consumer, and to explain how his wealth is formed, we specify his *initial endowment* of each one of the commodities as a vector e_i in the commodity space R^ℓ, and his *shares* of the profits of the producers, θ_{ij} being the share of the profit of the jth producer owned by the ith consumer, where $i=1,\dots,m$ and $j=1,\dots,n$. The numbers θ_{ij} are positive or zero, and for every j, $\sum_{i=1}^m \theta_{ij}=1$. Thus, if the profit of the jth producer is r_j, the wealth of the ith consumer is $w_i = p\cdot e_i + \sum_{j=1}^n \theta_{ij} r_j$.

The *production* of the jth producer ($j=1,\dots,n$) is also an ℓ-list of the quantities of the various commodities that he consumes and that he produces, but we represent his inputs by negative numbers and his outputs by positive numbers. Thus, his production is a point y_j in the commodity space R^ℓ, and his technological knowledge defines the non-empty subset Y_j of R^ℓ of his possible productions. Given a price vector p in R^ℓ, the jth producer strives to choose in his production set Y_j a production y_j that maximizes his profit $p\cdot y_j$.

In summary, the *economy* $\mathscr{E}$ is described by the m-list $(X_i, \precsim_i, e_i)$, by the mn shares (θ_{ij}), and by the n-list (Y_j), or, more concisely,

$$\mathscr{E} = ((X_i, \precsim_i, e_i), (\theta_{ij}), (Y_j)).$$

A *state* of the economy $\mathscr{E}$ is an m-list (x_i) of the consumptions of the various consumers, an n-list (y_j) of the productions of the various producers, and a price vector p. The state $((x_i^*), (y_j^*), p^*)$ is an *equilibrium* if

(a) for every i, x_i^* is a best element for $\precsim_i$ of $\{x \in X_i \mid p^*\cdot x \leqq p^*\cdot e_i + \sum_{j=1}^n \theta_{ij} p^*\cdot y_j^*\}$;
(b) for every j, y_j^* maximizes $p^*\cdot y_j$ on Y_j;
(c) $\sum_{i=1}^m x_i^* - \sum_{j=1}^n y_j^* - \sum_{i=1}^m e_i = 0$.

Condition (a) says that every consumer has chosen in his consumption set a consumption that satisfies his preferences best under his budget constraint. Condition (b) says that every producer has maximized his profit in his production set. Condition (c) says that for every commodity the excess of demand over supply is zero. The equilibrium defined by conditions (a), (b), and (c) is competitive in the sense that every agent behaves as if he had no influence on prices and considers them as given when choosing his own action.

However, at first we will study the special concept of a *free disposal-equilibrium* defined as a state of the economy satisfying (a) and (b) as before, but instead of (c),

(d) $z^* = \sum_{i=1}^m x_i^* - \sum_{j=1}^n y_j^* - \sum_{i=1}^m e_i \leqq 0$ with $p^* \geqq 0$ and $p^*\cdot z^* = 0$.

Condition (d) is meaningful if and only if all commodities can be freely disposed of. In that case the price vector p^* must clearly have all its coordinates greater than or equal to zero. For if a commodity had a negative price, by disposing of that commodity the producers could indefinitely increase their total profit and condition (b) would be violated. Equilibrium also requires that on the market for each commodity, supply equals demand, or that supply exceeds demand with the corresponding price equal to zero.

For the statement of the next thoerem, we must also formally define an *attainable state* of the economy by conditions

(a′) for every i, x_i is in X_i;
(b′) for every j, y_j is in Y_j;
(c′) $\sum_{i=1}^m x_i - \sum_{j=1}^n y_j - \sum_{i=1}^m e_i \leqq 0$.

That is to say, the consumption of every consumer is in his consumption set, the production of every producer is in his production set, and for every commodity demand is at most equal to supply. We say that a consumption x_i is attainable for the ith consumer if there is an attainable state of the economy allocating x_i to him. The *attainable consumption set* $\hat{X}_i$ of the ith consumer is the set of his attainable consumptions.

Finally, for two vectors x and y in R^ℓ, "$x<y$" means "$x \leqq y$ and $x \neq y$", while "$x \ll y$" means "$x^h < y^h$ for every coordinate $h=1,\dots,\ell$". The closed positive cone R^ℓ_+ of R^ℓ is $\{x \in R^\ell | x \geqq 0\}$.

Theorem 4

The economy $\mathscr{E}$ has a free disposal-equilibrium if for every i,
X_i is compact and convex,
there is no satiation consumption in $\hat{X}_i$,
the set $\{(x, x') \in X_i \times X_i | x \precsim_i x'\}$ is closed,
if x and x' are two points of X_i such that $x \prec_i x'$, and r is a real number in $]0,1]$, then $x \prec_i (1-r)x + rx'$,
there is x_i^0 in X_i such that $x_i^0 \ll e_i$;
for every j, Y_j is a compact, convex set containing 0.

The insatiability assumption means that for every x in $\hat{X}_i$, there is x' in X_i such that $x \prec_i x'$.

The closedness of the preference relation implies [e.g. Debreu (1959)] that there is a continuous utility function u_i representing $\precsim_i$ in the sense that $x \precsim_i x'$ is equivalent to $u_i(x) \leqq u_i(x')$. Moreover, the convexity assumption on $\precsim_i$ implies that u_i is quasi-concave. To see this, it suffices to note that if x and x' are two points of X_i such that $x \precsim_i x'$ and r is a real number in $[0,1]$, then $x \precsim_i (1-r)x + rx'$. Indeed, if there were x^1 in the straight-line segment $[x, x']$ such that $x^1 \prec_i x$, then by continuity of u_i there would be x^2 in the segment $[x^1, x]$ such that $x^1 \prec_i x^2 \prec_i x$. Thus, $x^2 \prec_i x'$ and x^1 is both between x^2 and x' and different from x^2. By the convexity assumption on preferences made in the statement of the theorem, one would have $x^2 \prec_i x^1$, a contradiction (Figure 2.2).

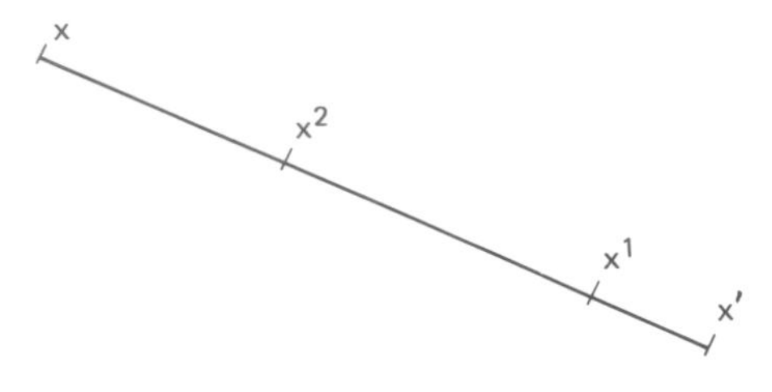

Figure 2.2.

Before proving Theorem 4 we establish two lemmata. Consider a non-empty compact subset S of R^ℓ and define the function $p \mapsto M(p)$ from R^ℓ to R by $M(p) = \operatorname{Max} p \cdot S = \operatorname{Max}_{x \in S} p \cdot x$. As a particular case of Berge (1963, VI.3), we have

Lemma 2

If S is a non-empty, compact subset of R^ℓ, then the function M is continuous.

Proof

(a) *M is lower semicontinuous.* Let p^0 be a point of R^ℓ, x^0 be a point of S such that $p^0 \cdot x^0 = M(p^0)$, and ε be a strictly positive real number. For p close enough to p^0, one has $p \cdot x^0 \geqq p^0 \cdot x^0 - \varepsilon$, hence $M(p) \geqq M(p^0) - \varepsilon$.

(b) *M is upper semicontinuous.* Again let p^0 be a point of R^ℓ, and ε be a strictly positive real number. Consider an arbitrary point x of S. There is an open neighborhood $U(x)$ of p^0 and an open neighborhood $V(x)$ of x such that $[p \in U(x)$ and $y \in V(x)]$ implies $[p \cdot y \leqq p^0 \cdot x + \varepsilon]$. The open neighborhoods $V(x)$ cover the compact set S. Therefore, there is a finite collection $V(x_1), \ldots, V(x_k)$ of them that also covers S. Let $U^0 = \cap_{i=1}^k U(x_i)$. U^0 is an open neighborhood of p^0 of which we consider an arbitrary point p. Let y be any point of S. There is an i in $\{1, \ldots, k\}$ such that $y \in V(x_i)$. Since p is in $U(x_i)$, one has $p \cdot y \leqq p^0 \cdot x_i + \varepsilon$. Therefore, $p \cdot y \leqq M(p^0) + \varepsilon$. Consequently, for any $p \in U^0$, $M(p) \leqq M(p^0) + \varepsilon$. Q.E.D.

Let now p be a price vector in R^ℓ and w be a real number, and define the *budget set* of the ith consumer associated with the pair (p, w) as

$$\beta_i(p, w) = \{x \in X_i \mid p \cdot x \leqq w\}.$$

Assuming that the consumption set X_i is non-empty and compact, the domain of the correspondence β_i is the set $D = \{(p, w) \in R^{l+1} \mid w \geqq \min p \cdot X_i\}$. By Lemma 2, the function $p \mapsto \min p \cdot X_i$ is continuous. Consequently, the set D is closed. Lemma 3 gives conditions under which the correspondence β_i is continuous.

Lemma 3

If X_i is non-empty, compact, and convex, and $w^0 > \min p^0 \cdot X_i$, then the correspondence β_i is continuous at (p^0, w^0).

Proof

The graph of the correspondence β_i is the set

$$\{(p,w,x) \in D \times X_i \mid p \cdot x \leqslant w\}$$

which is closed. Therefore, β_i is upper hemicontinuous.

To show that β_i is lower hemicontinuous at (p^0, w^0), we consider a sequence (p^q, w^q) in D converging to (p^0, w^0) and a point z^0 in $\beta_i(p^0)$.

(i) If $p^0 \cdot z^0 < w^0$, then for q large enough, $p^q \cdot z^0 < w^q$ and the constant sequence $z^q = z^0$ satisfies the conditions appearing in the definition of lower hemicontinuity.

(ii) If $p^0 \cdot z^0 = w^0$, we select a point x^0 in X_i such that $p^0 \cdot x^0 < w^0$. Thus, $p^0 \cdot x^0 < p^0 \cdot z^0$ and for q large enough, the hyperplane $\{z \in R^\ell \mid p^q \cdot z = w^q\}$ intersects the straight line through x^0 and z^0 in a unique point $\bar{z}^q$. Define z^q as $\bar{z}^q$ if $\bar{z}^q$ is between x^0 and z^0 (hence in X_i), as z^0 if $\bar{z}^q$ is not between x^0 and z^0 (hence possibly not in X_i). The sequence z^q satisfies the conditions appearing in the definition of lower hemicontinuity (Figure 2.3). Q.E.D.

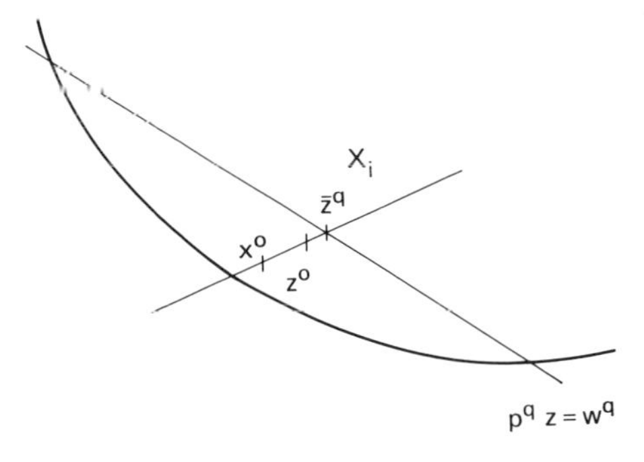

Figure 2.3.

Proof of Theorem 4

We note first that in a free disposal-equilibrium $((x_i^*),(y_j^*), p^*)$ one cannot have $p^* = 0$, for in that case the ith consumer would have no budget constraint and, consequently, would choose a consumption outside $\hat{X}_i$, i.e. an unattainable consumption. In addition, replacing p^* by λp^*, where λ is a strictly positive real number, does not alter any of the equilibrium conditions. Thus, one can restrict the search for an equilibrium price vector $p^* \geqq 0$ to the simplex

$$P = \left\{ p \in R^{\ell}_{+} \middle| \sum_{h=1}^{\ell} p^h = 1 \right\}.$$

Given the price vector p in P, the jth producer maximizes his profit in his production set Y_j. We denote the maximum of his profit relative to p by

$$\pi_j(p) = \max p \cdot Y_j.$$

Similarly, given the price vector p in P, the ith consumer maximizes his utility function u_i in his budget set

$$\beta_i'(p) = \left\{ x \in X_i \middle| p \cdot x \leqq p \cdot e_i + \sum_{j=1}^{n} \theta_{ij} \pi_j(p) \right\}.$$

To cast the economy $\mathscr{E}$ in the form of the general model of a social system presented earlier, we introduce a fictitious market agent whose role is to choose a price vector in P and whose utility function t is defined in the following manner. If the ith consumer ($i = 1, \ldots, m$) chooses the consumption x_i, and the jth producer the production y_j, the resulting excess demand is $\sum_{i=1}^{m} x_i - \sum_{j=1}^{n} y_j - \sum_{i=1}^{m} e_i$. The market agent chooses the price vector p in P so as to make this excess demand as expensive as possible. Thus,

$$t((x_i), (y_j), p) = p \cdot \left[\sum_{i=1}^{m} x_i - \sum_{j=1}^{n} y_j - \sum_{j=1}^{n} e_i \right].$$

The maximization with respect to p of this function agrees with a commonly held view of the way in which prices perform their market-equilibrating role by making commodities with positive excess demand more expensive and commodities with negative excess demand less expensive, thereby increasing the value of excess demand.

The social system can now be fully specified. The agents are the m consumers, the n producers, and the market. Their sets of actions are, respectively, (X_i), (Y_j), and P. The utility function $\tilde{u}_i$ of the ith consumer is defined by

$$\tilde{u}_i((x_i), (y_j), p) = u_i(x_i).$$

The utility function v_j of the jth producer is defined by

$$v_j((x_i),(y_j),p)=p\cdot y_j.$$

And the utility function of the market is t.

It remains to define for each agent the set in which he is restricted to choose his own action when the actions of the other agents are specified. For the jth producer that set is Y_j, and for the market it is P. Thus, for these $n+1$ agents the correspondences φ appearing in the definition of a social system are convex-valued and constant (hence trivially continuous). For the ith consumer, letting

$$w_i(p)=p\cdot e_i+\sum_{j=1}^{n}\theta_{ij}\pi_j(p),$$

we define $\tilde{\beta}_i'$ by

$$\tilde{\beta}_i'((x_i),(y_j),p)=\beta_i'(p)=\{x\in X_i \mid p\cdot x\leqq w_i(p)\}.$$

Since $0\in Y_j$, one has for every p in P, $\pi_j(p)\geqq 0$. Moreover, the inequality $x_i^0\ll e_i$ implies that for every p in P, $p\cdot x_i^0<p\cdot e_i$. Hence

$$\text{for every } p\in P, \min p\cdot X_i<w_i(p). \tag{2.1}$$

Consequently, for every p in P, the set $\beta_i'(p)$ is non-empty. It is also clearly convex. To show that all the assumptions of Theorem 3 are satisfied, it now suffices to check that the correspondence β_i' is continuous. By Lemma 2, the function π_j is continuous for every j. Therefore, the function w_i is continuous. The continuity of the correspondence β_i' now follows from Lemma 3 and from (2.1).

Thus, by Theorem 3, the social system that we have introduced has an equilibrium $((x_i^*),(y_j^*),p^*)$. For every j, y_j^* maximizes profit relative to p^* in Y_j, hence $\pi_j(p^*)=p^*\cdot y_j^*$. And for every i, x_i^* maximizes the function u_i in the set $\beta_i'(p^*)=\{x\in X_i \mid p^*\cdot x\leqq w_i(p^*)\}$, where $w_i(p^*)=p^*\cdot e_i+\sum_{j=1}^{n}\theta_{ij}p^*\cdot y_j^*$.

Since

$$\text{for every } i,\ p^*\cdot x_i^*\leqq p^*\cdot e_i+\sum_{j=1}^{n}\theta_{ij}p^*\cdot y_j^*,$$

we obtain, by summation over i,

$$\sum_{i=1}^{m}p^*\cdot x_i^*\leqq\sum_{i=1}^{m}p^*\cdot e_i+\sum_{i=1}^{m}\sum_{j=1}^{n}\theta_{ij}p^*\cdot y_j^*.$$

Inverting the order of summation in the double sum, and recalling that for every j, $\sum_{i=1}^{m}\theta_{ij}=1$, we have

$$\sum_{i=1}^{m} p^*\cdot x_i^* \leqq \sum_{i=1}^{m} p^*\cdot e_i + \sum_{j=1}^{n} p^*\cdot y_j^*$$

or

$$p^*\cdot\left[\sum_{i=1}^{m} x_i^* - \sum_{i=1}^{m} e_i - \sum_{j=1}^{n} y_j^*\right]\leqq 0.$$

Denoting by z^* the excess demand vector between brackets, we rewrite the last inequality as $p^*\cdot z^*\leqq 0$. However, p^* maximizes the utility function of the market agent in P. Therefore, for every $p\in P$, one has $p\cdot z^*\leqq 0$. This implies that $z^*\leqq 0$, for if the hth coordinate of z^* were strictly positive, it would suffice to take for p the positive unit vector on the hth coordinate axis of R^ℓ to obtain a strictly positive inner product $p\cdot z^*$.

The inequality $z^*\leqq 0$ establishes that the state $((x_i^*),(y_j^*),p^*)$ is attainable. Consequently, for every i, x_i^* is in the attainable consumption set $\hat{X}_i$. By the insatiability assumption, there is x_i' in X_i such that $x_i^* \prec_i x_i'$. This excludes the possibility that $p^*\cdot x_i^* < w_i(p^*)$. For, in that case, one would find a point on the straight-line segment $[x_i^*, x_i']$ different from x_i^*, hence strictly preferred to x_i^*, but close enough to x_i^* to satisfy the budget inequality. But this would contradict the fact that x_i^* maximizes the function u_i in the set $\beta_i'(p^*)$. In conclusion,

$$\text{for every } i,\ p^*\cdot x_i^* = p^*\cdot e_i + \sum_{j=1}^{n}\theta_{ij}p^*\cdot y_j^*.$$

By summation over i, we obtain $p^*\cdot z^*=0$. This equality, together with the inequality $z^*\leqq 0$ and the relation $p^*\in P$, shows that condition (d) in the definition of a free-disposal equilibrium is satisfied. Q.E.D.

However, Theorem 4 is unsatisfactory in several respects. The assumption of free disposal that it makes implicitly should be explicit. And the assumptions that the sets X_i and Y_j are bounded are difficult to justify. In particular, the boundedness of the production sets rules out the standard case of technologies with constant returns to scale. We will therefore prove the following result [Arrow and Debreu (1954) modified according to the suggestion made by Uzawa (1956) to replace the hypothesis that every production set Y_j is convex by the weaker hypothesis that their sum $Y=\sum_{j=1}^{n}Y_j$ is convex].

Theorem 5

The economy $\mathscr{E}$ has an equilibrium if for every i,

X_i is closed, convex, and has a lower bound for $\leqq$,
there is no satiation consumption in $\hat{X}_i$,
the set $\{(x, x') \in X_i \times X_i \mid x \precsim_i x'\}$ is closed,
if x and x' are two points of X_i such that $x \prec_i x'$, and r is a real number in $]0, 1]$, then $x \prec_i (1-r)x + rx'$,
there is x_i^0 in X_i such that $x_i^0 \ll e_i$;

for every j,

$0 \in Y_j$;
Y is closed and convex,
$Y \cap (-Y) = \{0\}$,
$Y \supset (-R_+^\ell)$.

The lower boundedness assumption on the consumption sets is easily justified since the quantity of any commodity that an individual actually consumes is bounded below by zero, while the quantity of any type of labor that he produces during a certain time period is obviously bounded above in absolute value.

The convexity of the total production set Y expresses the idea that the aggregate technology of the economy obeys the law of non-increasing returns to scale. The irreversibility assumption $Y \cap (-Y) = \{0\}$ means that if a total production y different from 0 is possible, then the production $-y$ (in which all the inputs have become outputs and vice versa) is not possible. The last assumption explicitly postulates free disposal of all commodities.

Proof of Theorem 5

The proof consists of constructing an economy $\hat{\mathscr{E}}$ satisfying the assumptions of Theorem 4 and such that a free disposal-equilibrium of $\hat{\mathscr{E}}$ yields an equilibrium of $\mathscr{E}$.

In the first step of this construction we replace in the definition of $\mathscr{E}$, each production set Y_j by the closure $\bar{\dot{Y}}_j$ of its convex hull $\dot{Y}_j$, remarking that

$$\sum_{j=1}^{n} \bar{\dot{Y}}_j = Y.$$

The proof of this equality rests on two facts:

(1) The sum of the convex hulls of a finite collection of sets is equal to the convex hull of their sum. Since $\sum_{j=1}^n Y_j = Y$ has been assumed to be convex, we have $\sum_{j=1}^n \dot{Y}_j = Y$.
(2) The sum of the closures of a finite collection of sets is contained in the closure of their sum. Hence, $\sum_{j=1}^n \bar{\dot{Y}}_j \subset \bar{Y}$. Since Y has been assumed to be closed, we have $\sum_{j=1}^n \bar{\dot{Y}}_j \subset Y$.

However, $Y_j \subset \bar{\dot{Y}}_j$ for every j implies $Y \subset \sum_{j=1}^n \bar{\dot{Y}}_j$, which in conjunction with $\sum_{j=1}^n \bar{\dot{Y}}_j \subset Y$ yields the result.

Let then $\bar{\dot{\mathcal{E}}}$ be the economy obtained by substituting $\bar{\dot{Y}}_j$ for Y_j for every j, and consider an attainable state $((x_i),(y_j), p)$ of $\bar{\dot{\mathcal{E}}}$. We have $\sum_{i=1}^m x_i - \sum_{i=1}^m e_i \leqq \sum_{j=1}^n y_j$. Moreover, each set X_i is bounded below by a vector χ_i. Therefore, $\sum_{j=1}^n y_j \geqq \sum_{i=1}^m \chi_i - \sum_{i=1}^m e_i$. It follows from the basic estimate in Lemma 1, Appendix I of Smale (Chapter 8) that for every j, the set $\hat{Y}_j$ of the attainable productions of the jth producer in the economy $\bar{\dot{\mathcal{E}}}$ is bounded. Now the inequality $\sum_{i=1}^m x_i \leqq \sum_{i=1}^m e_i + \sum_{j=1}^n y_j$ shows that for an attainable state of $\bar{\dot{\mathcal{E}}}$, $\sum_{i=1}^m x_i$ is bounded above. Since each x_i in this sum is bounded below, each x_i is actually bounded. Finally, since the two economies $\mathcal{E}$ and $\bar{\dot{\mathcal{E}}}$ have the same total production set Y, x_i is an attainable consumption of the ith consumer in $\mathcal{E}$ if and only if it is an attainable consumption of the ith consumer in $\bar{\dot{\mathcal{E}}}$. In conclusion, for every i the set $\hat{X}_i$ of the attainable consumptions of the ith consumer (in $\mathcal{E}$ or, equivalently, in $\bar{\dot{\mathcal{E}}}$) is bounded.

To complete the construction of the economy $\widehat{\mathcal{E}}$ we choose a compact, convex set K in R^ℓ containing *in its interior* all the attainable consumption sets $\hat{X}_i$, and all the attainable production sets $\hat{Y}_j$, and we define

$$\widehat{X}_i = X_i \cap K \quad \text{and} \quad \widehat{Y}_j = \bar{\dot{Y}}_j \cap K.$$

The economy $\widehat{\mathcal{E}}$ is obtained from $\bar{\dot{\mathcal{E}}}$ by replacing each X_i by $\widehat{X}_i$ and each $\bar{\dot{Y}}_j$ by $\widehat{Y}_j$.

Note that a state $((x_i),(y_j), p)$ is attainable for $\bar{\dot{\mathcal{E}}}$ if and only if it is attainable for $\widehat{\mathcal{E}}$. Therefore, the attainable consumption set of the ith consumer in $\widehat{\mathcal{E}}$ is again $\hat{X}_i$. Note also that according to the inequality $\sum_{i=1}^m x_i^0 \ll \sum_{i=1}^m e_i$, a state in which the ith consumer consumes $x_i^0 (i=1,\ldots,m)$, and the jth producer produces 0 $(j=1,\ldots,n)$ is attainable in $\mathcal{E}$, hence in $\widehat{\mathcal{E}}$. Consequently, for every i, $x_i^0 \in \widehat{X}_i$, and for every j, $0 \in \hat{Y}_j$.

We now check that the economy $\widehat{\mathcal{E}}$ satisfies all the assumptions of Theorem 4. Actually, the only one for which this is not immediately clear is the insatiability assumption. Let us therefore consider a consumption x in $\hat{X}_i$. By the insatiability assumption of Theorem 5 there is a consumption x' in X_i such that $x \prec_i x'$. Since x is in the interior of K, we can take in the straight-line segment $[x, x']$ a point x'' different from x but close enough to x to be in K. By convexity of X_i, the point x'' is in X_i, hence in $\widehat{X}_i$. By the convexity assumption on preferences of Theorem 5, $x \prec_i x''$. In summary, given any consumption in $\hat{X}_i$, there is a strictly preferred consumption *in* $\widehat{X}_i$.

Thus, we have established that the economy $\widehat{\mathcal{E}}$ has a free-disposal equilibrium $((x_i^*),(y_j), p^*)$. Let

$$z = \sum_{i=1}^m x_i^* - \sum_{j=1}^n y_j - \sum_{i=1}^m e_i$$

be the associated excess demand. According to the definition of a free-disposal equilibrium,

$$z \leqq 0 \quad \text{and} \quad p^* \cdot z = 0.$$

Since $y_j \in \bar{Y}_j$ for every j, the vector $y = \sum_{j=1}^n y_j$ belongs to Y. Moreover, for any λ in $]0, 1]$, the vector $(1/\lambda)z$ belongs to Y since $Y \supset (-R^\ell_+)$. By convexity of Y, the vector $(1-\lambda)y + \lambda((1/\lambda)z)$ is also in Y. Letting λ tend to zero and recalling that Y is closed, we obtain $y + z \in Y$. We now define $y^* = y + z$ and we choose in each Y_j a vector y_j^* such that $\sum_{j=1}^n y_j^* = y^* = y + z$.

The equality

$$\sum_{i=1}^m x_i^* - \sum_{j=1}^n y_j^* - \sum_{i=1}^m e_i = 0$$

implies that the state $((x_i^*), (y_j^*), p^*)$ is attainable for the economy $\mathscr{E}$, hence for the economy $\hat{\mathscr{E}}$. Consequently, for every i, x_i^* belongs to the attainable consumption set $\hat{X}_i$, and for every j, y_j^* belongs to the attainable production set $\hat{Y}_j$. We will prove that the state $((x_i^*), (y_j^*), p^*)$ is in fact an equilibrium of the economy $\mathscr{E}$ by establishing (i) and (ii).

(i) *For every j, y_j^* maximizes profit relative to p^* in Y_j, and $p^* \cdot y_j^* = p^* \cdot y_j$.*

Since $p^* z = 0$, one has $p^* \cdot \sum_{j=1}^n y_j^* = p^* \cdot \sum_{j=1}^n y_j$. However, for every j, y_j and y_j^* are in $\widehat{Y}_j$, and y_j maximizes profit relative to p^* in $\widehat{Y}_j$. Therefore, $\sum_{j=1}^n y_j$ maximizes profit relative to p^* in $\sum_{j=1}^n \widehat{Y}_j$. So does $\sum_{j=1}^n y_j^*$. Consequently, for every j, y_j^* maximizes profit relative to p^* in $\widehat{Y}_j$. This implies that for every j, $p^* \cdot y_j^* = p^* \cdot y_j$.

It remains to prove that for every j, y_j^* maximizes profit relative to p^* in Y_j. Assume to the contrary that there is y_j' in Y_j such that $p^* \cdot y_j' > p^* \cdot y_j^*$. As we noted, y_j^* belongs to $\hat{Y}_j$ which is contained in the interior of K. Therefore, one can find on the straight-line segment $[y_j^*, y_j']$ a point y_j'' different from y_j^* but sufficiently close to y_j^* to be in K. Since the point y_j'' also clearly belongs to $\bar{Y}_j$, it belongs to $\widehat{Y}_j$. But $p^* \cdot y_j'' > p^* \cdot y_j^*$, which contradicts the fact that y_j^* maximizes profit relative to p^* in $\widehat{Y}_j$.

(ii) *For every i, x_i^* is a best element for $\precsim_i$ in $\{x \in X_i \mid p^* \cdot x \leqq p^* \cdot e_i + \sum_{j=1}^n \theta_{ij} p^* \cdot y_j^*\}$.*

According to the definition of a free-disposal equilibrium for the economy $\widehat{\mathscr{E}}$, x_i^* is a best element for $\precsim_i$ in

$$\widehat{\beta}_i'(p^*) = \left\{ x \in \widehat{X}_i \mid p^* \cdot x \leqq p^* \cdot e_i + \sum_{j=1}^n \theta_{ij} p^* \cdot y_j \right\}.$$

And as we have shown, for every j, $p^* \cdot \widetilde{y}_j^* = p^* \cdot y_j$. Assume now that there is x_i' in X_i such that $p^* \cdot x_i' \leqq p^* \cdot e_i + \sum_{j=1}^n \theta_{ij} p^* \cdot y_j^*$ and $x_i' \succ_i x_i^*$. Since x_i^* is in the interior of K, one can find on the straight-line segment $[x_i^*, x_i']$ a point x_i'' different from x_i^* but sufficiently close to x_i^* to be in K. The point x_i'' is clearly also in X_i, hence in $\widehat{X}_i$. Moreover, $x_i'' \succ_i x_i^*$ and $p^* \cdot x_i'' \leqq p^* \cdot e_i + \sum_{j=1}^n \theta_{ij} p^* \cdot y_j^*$ which contradicts the optimality of x_i^* in $\widehat{\beta}_i'(p^*)$. Q.E.D.

In the last few years the preceding three theorems have been extended to models in which the assumptions made on preferences are significantly weakened. Schmeidler (1969) established the existence of a competitive equilibrium for economies with a continuum of agents without assuming completeness of preferences. For economies with a finite set of agents, Sonnenschein (1971) obtained existence without assuming transitivity of preferences, and Mas-Colell (1974) proved a remarkable existence theorem dispensing with completeness as well as with transitivity. Our purpose now is to present the main ideas of Mas-Colell's work and of its continuation by Gale and Mas-Colell (1975), Shafer and Sonnenschein (1975a, 1975b), and several other authors, by focusing on the generalization of Theorem 3 due to Shafer and Sonnenschein.

With the notation of Theorem 3, we consider the correspondence P_i from A to A_i defined by $P_i(a) = \{x \in A_i \mid f_i(a_1, \ldots, a_{i-1}, x, a_{i+1}, \ldots, a_n) > f_i(a)\}$. For every a in A, this correspondence specifies the set of actions that the ith agent strictly prefers to a_i, given the actions $(a_1, \ldots, a_{i-1}, a_{i+1}, \ldots, a_n)$ of the other agents. With the help of this new concept, an *equilibrium* of the social system can be defined as an element a^* of A such that for every $i \in N$, $a_i^* \in \varphi_i(a^*)$ and $P_i(a^*) \cap \varphi_i(a^*) = \varnothing$.

Under the assumptions of Theorem 3, the correspondence P_i has an open graph (by continuity of f_i), and for every $a \in A$, the set $P_i(a)$ is convex (by quasi-concavity of f_i in its ith variable). For every $a \in A$ one also has by the definition of P_i, $a_i \notin P_i(a)$, or equivalently, $a_i \notin \operatorname{co} P_i(a)$, where "co" stands for "convex hull of".

An alternative approach consists of taking as a primitive concept the correspondence P_i from A to A_i and of assuming that P_i has an open graph and for every $a \in A$, $a_i \notin \operatorname{co} P_i(a)$. Theorem 6 asserts that these assumptions, which are substantially weaker than continuity of f_i and quasi-concavity of f_i in its ith variable, suffice to ensure existence of a social equilibrium.

Theorem 6

If, for every $i \in N$, the set A_i is a non-empty, compact, convex subset of a Euclidean space, P_i is an open graph correspondence from A to A_i such that for every $a \in A$, $a_i \notin \operatorname{co} P_i(a)$, and φ_i is a continuous, convex-valued correspondence from A to A_i, then the social system $(A_i, P_i, \varphi_i)_{i \in N}$ has an equilibrium.

Proof

[Shafer–Sonnenschein (1975b)]

$G(P_i)$, the graph of P_i, is an open subset of $A \times A_i$. Therefore, there is a continuous real-valued function g_i on $A \times A_i$ that takes the value zero on the closed set $A \times A_i \backslash G(P_i)$, and strictly positive values on $G(P_i)$. For instance, one can take $g_i(a, x)$ to be the distance from (a, x) to $A \times A_i \backslash G(P_i)$.

Now we define the correspondence μ_i from A to A_i by

$$\mu_i(a) = \left\{ x \in \varphi_i(a) | g_i(a, x) = \max_{y \in \varphi_i(a)} g_i(a, y) \right\},$$

i.e. $\mu_i(a)$ is the set of maximizers of $g_i(a, \cdot)$ in $\varphi_i(a)$. By Lemma 1, the correspondence μ_i is upper hemicontinuous. Consequently, the correspondence $a \mapsto \text{co}\, \mu_i(a)$ from A to A_i is also upper hemicontinuous [Nikaido (1968, Th. 4.8) or Hildenbrand (1974, p. 26)]. This in turn immediately implies that the correspondence from A to A defined by

$$\mu(a) = \mathop{\times}_{i \in N} \text{co}\, \mu_i(a)$$

is upper hemicontinuous. Therefore, by Kakutani's theorem, the correspondence μ has a fixed point a^*.

For every $i \in N$, $a_i^* \in \text{co}\, \mu_i(a^*)$. However, $\mu_i(a^*) \subset \varphi_i(a^*)$, which is convex. Hence, $a_i^* \in \varphi_i(a^*)$. And it remains only to prove that $P_i(a^*) \cap \varphi_i(a^*) = \varnothing$. Assume to the contrary that there is a point y in $P_i(a^*) \cap \varphi_i(a^*)$. The pair (a^*, y) belongs to $G(P_i)$. Consequently, $g_i(a^*, y) > 0$, which implies $g_i(a^*, x) > 0$ for every x in $\mu_i(a^*)$. Therefore, $\mu_i(a^*) \subset P_i(a^*)$. But this implies that $a_i^* \in \text{co}\, P_i(a^*)$, a contradiction. Q.E.D.

3. The excess demand approach

In the previous section we investigated the problem of existence of a competitive equilibrium for an economy by transforming it into the problem of existence of an equilibrium for a social system composed of a finite set of agents simultaneously seeking to maximize their utility functions, or, more generally, to optimize with respect to their preference relations. In the present section we study an alternative approach focusing on the excess demand correspondence of the economy.

Let $\mathcal{E}$ be an economy as in Theorem 4 and consider a vector p in the price simplex P. The jth producer maximizes his profit relative to p in his production set Y_j. We denote by $\eta_j(p)$ the set of his profit-maximizing productions and by $\pi_j(p)$ his maximum profit. The ith consumer maximizes his utility function u_i in his budget set

$$\beta_i'(p) = \left\{ x \in X_i | p \cdot x \leq p \cdot e_i + \sum_{j=1}^{n} \theta_{ij} \pi_j(p) \right\}.$$

We denote by $\xi_i'(p)$ the set of his utility-maximizing consumptions under that constraint. Thus, the ith consumer selects an arbitrary element x_i in $\xi_i'(p)$, and the jth producer selects an arbitrary element y_j in $\eta_j(p)$. The associated excess demand, $\sum_{i=1}^m x_i - \sum_{j=1}^n y_j - \sum_{i=1}^m e_i$, is an arbitrary element of the set

$$\zeta(p) = \sum_{i=1}^{m} \xi_i'(p) - \sum_{j=1}^{n} \eta_j(p) - \sum_{i=1}^{m} e_i.$$

The correspondence ζ from P to R^ℓ so defined is the *excess demand correspondence* of the economy $\mathscr{E}$. And p^* yields a free-disposal equilibrium of $\mathscr{E}$ if and only if there is a vector z^* in $\zeta(p^*)$ such that $z^* \leqq 0$ and $p^* \cdot z^* = 0$. In fact, as we now show, p^* yields a free-disposal equilibrium of $\mathscr{E}$ if and only if $\zeta(p^*)$ intersects $-R^\ell_+$. Indeed, if $\zeta(p^*) \cap (-R^\ell_+) \neq \varnothing$, take a point z^* in that intersection. Since $z^* \in \zeta(p^*)$, for every i, there is x_i^* in $\xi_i'(p^*)$, and for every j, there is y_j^* in $\eta_j(p^*)$ such that

$$z^* = \sum_{i=1}^{m} x_i^* - \sum_{j=1}^{n} y_j^* - \sum_{i=1}^{m} e_i.$$

And since $z^* \leqq 0$, the state $((x_i^*), (y_j^*), p^*)$ is attainable. Consequently, for every i, x_i^* is in $\hat{X}_i$. However, there is no satiation consumption in $\hat{X}_i$, and this implies, as we saw at the end of the proof of Theorem 4, that

$$\text{for every } i,\ p^* \cdot x_i^* = p^* \cdot e_i + \sum_{j=1}^{n} \theta_{ij} p^* \cdot y_j^*.$$

By summation over i, we obtain $p^* z^* = 0$.

We now review the properties of the correspondence ζ. Given any p in P, for every i, the set $\xi_i'(p)$ is convex [since it is the set of maximizers of the quasi-concave utility function u_i on the convex set $\beta_i'(p)$], and for every j, the set $\eta_j(p)$ is convex (since it is the set of maximizers of the linear profit function $y \mapsto p \cdot y$ on the convex set Y_j). Therefore, for every p in P, the set $\zeta(p)$ is convex as a sum of convex sets. According to Lemma 3, for every i, the correspondence β_i' is continuous; hence, by Lemma 1 the correspondence ξ_i' is upper hemicontinuous. Also by Lemma 1, for every j, the correspondence η_j is upper hemicontinuous. Therefore, the correspondence ζ is upper hemicontinuous as a sum of upper hemicontinuous correspondences. Finally, for every p in P, for every i and every x_i in $\xi_i'(p)$, for every j and every y_j in $\eta_j(p)$, one has

$$p \cdot x_i \leqq p \cdot e_i + \sum_{j=1}^{n} \theta_{ij} p \cdot y_j.$$

Hence, by summation over i, for every p in P and every z in $\zeta(p)$, one has $p\cdot z\leqq 0$. More concisely, for every p in P, one has $p\cdot\zeta(p)\leqq 0$. In summary,

the correspondence ζ is upper hemicontinuous and for every p in P, the set $\zeta(p)$ is compact, convex, and satisfies the inequality $p\cdot\zeta(p)\leqq 0$. (3.1)

The existence of a price vector p^* for which $\zeta(p^*)$ intersects $-R^{\ell}_{+}$ will be derived from the following theorem [Gale (1955), Nikaido (1956), and Debreu (1956)]. In its statement we denote the *polar* of a cone C with vertex 0 in R^{ℓ} by $C^0=\{y\in R^{\ell}\,|\,y\cdot x\leqq 0 \text{ for every } x\subset C\}$. We say that a cone with vertex 0 is *degenerate* if it is empty or if it contains only the origin, and that a closed convex cone with vertex 0 is *pointed* if it contains no straight line. S denotes the unit sphere $\{x\in R^{\ell}\,|\,\|x\|=1\}$, i.e. the set of points in R^{ℓ} with unit Euclidean norm (Figure 3.1).

Theorem 7

If C is a non-degenerate pointed closed convex cone with vertex 0 in R^{ℓ} and φ is a convex-valued upper hemicontinuous correspondence from $C\cap S$ to R^{ℓ} such that for every p in $C\cap S$ one has $p\cdot\varphi(p)\leqq 0$, then there is p^* in $C\cap S$ such that $\varphi(p^*)\cap C^0\neq\varnothing$.

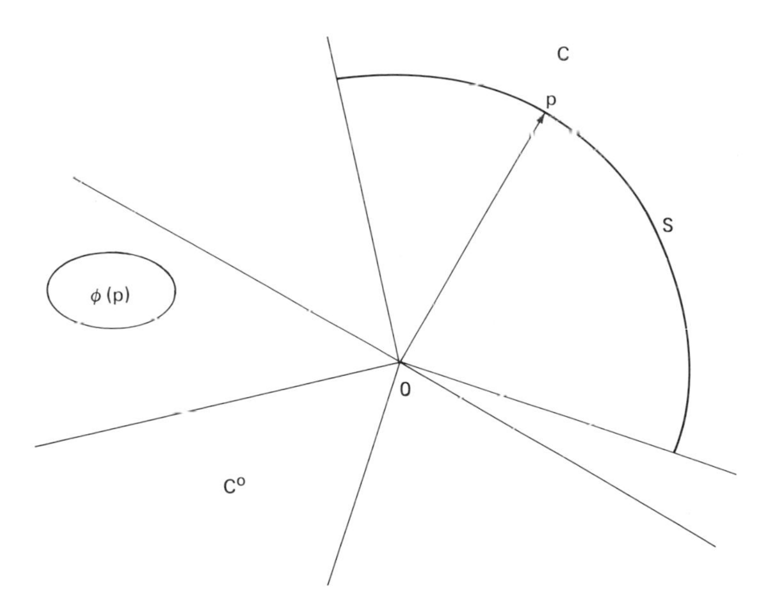

Figure 3.1.

Proof

Note first that the interior of C^0 is not empty. Otherwise, the convex cone C^0 would be contained in a hyperplane H. Consequently, the straight line D orthogonal to H through 0 would be contained in C^{00}, the polar of C^0. However, $C^{00}=C$ [Rockafellar (1970, Section 14)]. Thus, D would be contained in C, a contradiction of the assumption that C is pointed.

We select a vector q in the interior of C^0 and we define

$$\Pi=\{p\in C \mid q\cdot p=-1\}.$$

Let U be a neighborhood of q contained in C^0 and let p be an arbitrary point of C different from 0. For every $y\in U$, one has $y\cdot p\leqq 0$. This implies $q\cdot p<0$. Therefore, every point $p\in C$ different from 0 has a unique projection $p/-q\cdot p$ from 0 into Π.

The set Π is clearly closed and convex. It is also bounded. To see this, assume by contradiction that there is a sequence $p_n\in\Pi$ such that $\|p_n\|\to+\infty$. The sequence $p_n/\|p_n\|$ belongs to the compact set S. Therefore, we can extract a subsequence p'_n of p_n such that $p'_n/\|p'_n\|\to p^0$. However, for every n,

$$q\cdot\frac{p'_n}{\|p'_n\|}=-\frac{1}{\|p'_n\|}.$$

In the limit, $q\cdot p^0=0$, which contradicts the fact that for p^0 in C different from 0, one has $q\cdot p^0<0$.

The projection $p\mapsto p/\|p\|$ from Π into $C\cap S$ is a homeomorphism. Therefore, the correspondence $\hat{\varphi}$ from Π to R^ℓ defined by $\hat{\varphi}(p)=\varphi(p/\|p\|)$ is upper hemicontinuous. It is also convex-valued and for every p in Π, one has $p\cdot\hat{\varphi}(p)\leqq 0$.

Since Π is compact, the upper hemicontinuity of the correspondence $\hat{\varphi}$ on Π implies the existence of a bounded subset Z of R^ℓ such that for every p in Π, $\hat{\varphi}(p)$ is contained in Z. The set Z can clearly be taken to be compact and convex.

Given an arbitrary point z in Z, let then $\mu(z)$ be the set of maximizers of the function $p\mapsto p\cdot z$ from Π to R, i.e.

$$\mu(z)=\{p\in\Pi \mid p\cdot z=\max \Pi\cdot z\}.$$

According to Lemma 1, the correspondence μ is upper hemicontinuous. Moreover, for every z in Z, the set $\mu(z)$ is convex since it is the set of maximizers of a linear function on a convex set.

Consider now the correspondence ψ from $\Pi\times Z$ to itself defined by

$$\psi(p,z)=\mu(z)\times\hat{\varphi}(p).$$

The set $\Pi \times Z$ is non-empty, compact, and convex. The correspondence ψ is upper hemicontinuous and convex-valued. Therefore, by Kakutani's theorem, it has a fixed point (p^*, z^*). Thus,

$$p^* \in \mu(z^*) \quad \text{and} \quad z^* \in \hat{\varphi}(p^*).$$

The first of these two relations implies that for every p in Π, one has $p \cdot z^* \leqq p^* \cdot z^*$. The second implies that $p^* \cdot z^* \leqq 0$. Therefore, for every p in Π (and, consequently, for every p in C), one has $p \cdot z^* \leqq 0$. Hence, $z^* \in C^0$. Q.E.D.

Since the polar of R^{ℓ}_{+} is $-R^{\ell}_{+}$, the existence of a price vector p^* in P such that $\zeta(p^*) \cap (-R^{\ell}_{+}) \neq \varnothing$ immediately follows from the result we have just proved.

Theorem 7 establishes the existence of a price vector yielding a negative or zero excess demand as a direct consequence of a deep mathematical result, the fixed-point theorem of Kakutani. And one must ask whether the preceding proof uses a needlessly powerful tool. This question was answered in the negative by Uzawa (1962a) who showed that Theorem 7 directly implies Kakutani's fixed-point theorem. Let ψ be an upper hemicontinuous convex-valued correspondence from P into itself. We will show that ψ has a fixed point.

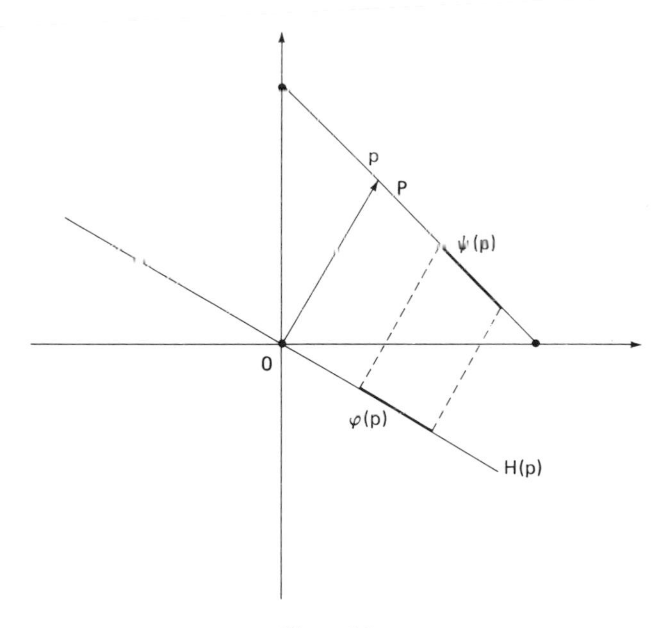

Figure 3.2.

For every p in P, define $H(p)$ as the hyperplane through 0 orthogonal to p and let $\varphi(p)$ be the orthogonal projection of $\psi(p)$ into $H(p)$. The correspondence φ

is upper hemicontinuous, and for every p in P the set $\varphi(p)$ is convex and satisfies $p\cdot\varphi(p)=0$. By Theorem 7, there is a p^* in P such that $\varphi(p^*)\cap(-R^\ell_+)\neq\varnothing$. Let y be a point in that intersection, and let x be the point of P projecting parallel to p^* into y. Since $x-y$ is parallel to p^*, and y is orthogonal to p^*, one has $y\cdot(x-y)=0$. Hence, $y\cdot x=y\cdot y$ (Figure 3.2). However, $y\leqq 0$ and $x\geqq 0$ imply $y\cdot x\leqq 0$. Consequently, $y\cdot y\leqq 0$. Hence, $y\cdot y=0$ and $y=0$. In conclusion, $0\in\varphi(p^*)$, and $p^*\in\psi(p^*)$.

The equivalence of Kakutani's theorem and of Theorem 7 does not suffice to prove that the former (or a result of similar depth) is required to establish the existence of a competitive equilibrium, for the excess demand correspondence generated by an economy might have properties in addition to those listed in (3.1) that would make it possible to obtain the existence of an equilibrium by elementary means. Therefore, the further question is raised whether given a correspondence ζ having properties (3.1), there is an economy generating an excess demand correspondence identical to ζ.

We have repeatedly used the fact that in an economy with insatiable consumers whose preferences satisfy the convexity assumption of Theorems 4 and 5, the budget constraint of each one of them is binding and consequently the value of the excess demand (whenever the latter is defined) is zero. This observation, known as *Walras' Law*, is formally expressed by

for every p in the domain of ζ, *one has* $p\cdot\zeta(p)=0$.

Recent work on the characterization of excess demand functions (Shafer and Sonnenschein, Chapter 14 of this Handbook) provides an affirmative answer to the question raised at the end of the next to last paragraph in the case in which ζ is a continuous *function* obeying Walras' Law. In that case, given any compact subset K of the relative interior of P, there is an exchange economy with ℓ consumers generating an excess demand function that coincides with ζ on K. This result, combined with Uzawa's observation, shows that the proof of existence of a competitive equilibrium requires mathematical tools of the same power as a fixed-point theorem.

In the remainder of this section we study further the simple case of an exchange economy with ℓ commodities and m consumers. The consumption set of the ith consumer ($i=1,\dots,m$) is the closed positive cone R^ℓ_+ of the commodity space R^ℓ and his preference relation $\precsim_i$ satisfies the assumptions of

closedness: the set $\{(x,x')\in R^\ell_+\times R^\ell_+ \mid x\precsim_i x'\}$ is closed; and

monotony: if x and x' are two points of R^ℓ_+ such that $x<x'$, then $x\prec_i x'$.

Given a strictly positive price vector $p\gg 0$ and a strictly positive wealth $w>0$, the budget set

$$\beta_i(p,w)=\{x\in R^\ell_+ \mid p\cdot x\leqq w\}$$

of the ith consumer is compact and non-empty. Moreover, (p, w) has a neighborhood in which β_i is bounded. Consequently, according to Lemma 3 the correspondence β_i is continuous at (p, w).

The closedness of the preference relation implies [e.g. Debreu (1959)] that there is a continuous utility function u_i representing $\precsim_i$ on R^ℓ_+. Maximizing u_i on the budget set $\beta_i(p, w)$ yields the set

$$\xi_i(p,w) = \{x \in \beta_i(p,w) | \text{ for every } y \in \beta_i(p,w), \text{ one has } y \precsim_i x\}$$

of best elements of R^ℓ_+ for $\precsim_i$ under the budget constraint $p \cdot x \leqq w$. By Lemma 1, the demand correspondence ξ_i is upper hemicontinuous.

The monotony of the preference relation implies that the budget constraint is binding, i.e. $p \cdot \xi_i(p, w) = w$. It also implies the following boundary behavior of the correspondence ξ_i [Hildenbrand (1974, p. 103, Corollary 1)]. For a vector x in R^ℓ, we define the norm $|x| = \sum_{h=1}^{\ell} |x^h|$. The distance from the origin 0 to a subset X of R^ℓ is then $d[0, X] = \inf_{x \in X} |x|$. Consider now a sequence (p_q, w_q) in $P \times R$ converging to (p_0, w_0) and such that for every q, p_q belongs to the relative interior of P and $w_q > 0$, while p_0 belongs to the relative boundary ∂P of P and $w_0 > 0$. In other words, for every q, $p_q \gg 0$, while p_0 has some zero coordinates. Under these conditions the sequence $d[0, \xi_i(p_q, w_q)]$ tends to $+\infty$ as we now prove.

Lemma 4

If $\precsim_i$ is a closed, monotone preference relation on R^ℓ_+, $p_q \gg 0$ in P converges to p_0 in ∂P, $w_q > 0$ in R converges to $w_0 > 0$, then $d[0, \xi_i(p_q, w_q)] \to +\infty$.

Proof

Assume that the conclusion does not hold. Then there is a subsequence (p'_q, w'_q) of (p_q, w_q) such that $d[0, \xi_i(p'_q, w'_q)]$ is bounded. For every q, one can select x'_q in $\xi_i(p'_q, w'_q)$ in such a way that the sequence x'_q is bounded. Therefore, one can extract from (p'_q, w'_q, x'_q) a subsequence (p''_q, w''_q, x''_q) converging to (p_0, w_0, x_0). Let B be a closed ball with center 0 containing x_0 in its interior and define $\hat{\beta}_i$ by

$$\hat{\beta}_i(p, w) = B \cap \beta_i(p, w).$$

Since $w_0 > 0$, the correspondence $\hat{\beta}_i$ is continuous at (p_0, w_0) by Lemma 3. By Lemma 1, x_0 is a best element for $\precsim_i$ in $\hat{\beta}_i(p_0, w_0)$. However, the point x_0 is interior to the ball B and p_0 has some zero coordinates. Therefore, there is a point x in $\hat{\beta}_i(p_0, w_0)$ such that $x_0 < x$ and, consequently, $x_0 \prec_i x$, a contradiction of the optimality of x_0 in $\hat{\beta}_i(p_0, w_0)$. Q.E.D.

We now make on the preference relation $\precsim_i$ the additional assumption of

Weak-convexity: If x and x' are two points of R^ℓ_+ such that $x \precsim_i x'$, and r is a real number in [0, 1], then $x \precsim_i (1-r)x + rx'$.

This assumption implies that for every $p \gg 0$ in P and $w > 0$ in R, the set $\xi_i(p, w)$ is convex.

In summary, the demand correspondence ξ_i is defined for every $(p, w) \in P \times R$ such that $p \gg 0$ and $w > 0$; its values are convex subsets of R^ℓ_+; it is upper hemicontinuous; it satisfies the equality $p \cdot \xi_i(p, w) = w$; and it has the boundary property

if (p_q, w_q) is a sequence in $P \times R$ converging to (p_0, w_0) such that (i) for every q, $p_q \gg 0$ and $w_q > 0$, (ii) $p_0 \in \partial P$ and $w_0 > 0$, then $d[0, \xi_i(p_q, w_q)] \to +\infty$.

Instead of taking as a primitive concept the preference relation $\precsim_i$ satisfying the conditions of closedness, monotony, and weak-convexity, one can take the *demand correspondence* ξ_i satisfying the preceding (significantly weaker) conditions. We now do so and define an exchange economy $\mathscr{E}$ by specifying for every $i = 1, \ldots, m$, the demand correspondence ξ_i and the initial endowment $e_i \gg 0$ in R^ℓ of the ith consumer. Formally, $\mathscr{E} = (\xi_i, e_i)_{i=1,\ldots,m}$.

Given a price vector $p \gg 0$ in P, the value of the endowment of the ith consumer is $p \cdot e_i$, his demand set is $\xi_i(p, p \cdot e_i)$, his excess demand set is $\xi_i(p, p \cdot e_i) - e_i$, and the excess demand set of the economy is

$$\zeta(p) = \sum_{i=1}^{m} [\xi_i(p, p \cdot e_i) - e_i].$$

The properties of the excess demand correspondence ζ of the economy are immediately derived from those of the individual demand correspondences. ζ is defined for every $p \gg 0$ in P; its values are convex subsets of R^ℓ; it is bounded below and upper hemicontinuous; it satisfies

Walras' Law: For every $p \gg 0$ in P, one has $p \cdot \zeta(p) = 0$,

and the

Boundary Condition: If $p_q \gg 0$ in P converges to p_0 in ∂P, then $d[0, \zeta(p_q)] \to +\infty$.

These properties of the excess demand correspondence ensure the existence of a price vector p such that $0 \in \zeta(p)$. We state, in the form given by Hildenbrand (1974, p. 150, Lemma 1), this result, some of whose antecedents were McKenzie (1954), Gale (1955), Nikaido (1956), Debreu (1956), Arrow and Hahn (1971), and Dierker (1974, Section 8). The proof that we present is due to Neuefeind (1977).

Theorem 8

If ζ is convex-valued, bounded below, upper hemicontinuous, and satisfies Walras' Law and the Boundary Condition, then there is $p^* \gg 0$ in P such that $0 \in \zeta(p^*)$.

Proof

Let

$$E=\left\{p\in P \mid p\gg 0 \text{ and there is } z\in\zeta(p) \text{ such that } \sum_{h=1}^{\ell} z^h\leqq 0\right\}.$$

Since ζ is bounded below, when p varies in E the distance $d[0,\zeta(p)]$ remains bounded. Therefore, by the Boundary Condition, there cannot be in E a sequence p_q converging to p_0 in ∂P. Consequently, the distance from $p\in E$ to ∂P is bounded below by a strictly positive real number.

Thus, there is a closed convex cone C with vertex 0 in R^ℓ such that $E\subset \operatorname{int} C$ and $C\backslash 0\subset \operatorname{int} R^\ell_+$. We apply Theorem 7 to the cone C and to the correspondence ζ, and we obtain a point $p^*\in C\cap P$ such that $\zeta(p^*)\cap C^0\neq\varnothing$. Let z be a point in that intersection. By Walras' Law the point $(1/\ell,\ldots,1/\ell)$ belongs to E, hence to C. Therefore, $\sum_{h=1}^{\ell} z^h\leqq 0$. Consequently, p^* belongs to E, hence to $\operatorname{int} C$. Moreover, by another application of Walras' Law, one has $p^*\cdot z=0$. This equality with $p^*\in\operatorname{int} C$ and $z\in C^0$ implies that $z=0$. Q.E.D.

We now make on the preference relation $\precsim_i$ the additional assumption of

Strict convexity: If x and x' are two distinct points of R^ℓ_+ such that $x\precsim_i x'$, and r is a real number in $]0,1[$, then $x\prec_i(1-r)x+rx'$.

This assumption implies that the set $\xi_i(p,w)$ has a single element which we denote by $f_i(p,w)$. Thus, the exchange economy $\mathscr{E}$ now specifies for every $i=1,\ldots,m$, the *demand function* f_i and the initial endowment $e_i\gg 0$ in R^ℓ of the ith consumer. The *excess demand function* F of the economy $\mathscr{E}$ is defined for every $p\gg 0$ in P by

$$F(p)=\sum_{i=1}^{m}\left[f_i(p,p\cdot e_i)-e_i\right].$$

Every point (p,w), for which $p\gg 0$ and $w>0$, has a neighborhood in which f_i is bounded. A simple compactness argument shows that in such a neighborhood the upper hemicontinuity of the correspondence $(p,w)\mapsto\{f_i(p,w)\}$ entails the continuity of the function f_i. Therefore, the excess demand function F is continuous at every $p\gg 0$. The other properties of F immediately follow from the properties of the excess demand correspondence ζ. Namely, F is bounded below; it satisfies

Walras' Law: For every $p\gg 0$ in P, one has $p\cdot F(p)=0$,

and the

Boundary Condition: If $p_q\gg 0$ in P converges to p_0 in ∂P, then $|F(p_q)|\to+\infty$.

We formally state as a corollary of Theorem 8 [Dierker (1974, Section 8)] that those properties imply the existence of a price vector p for which $F(p)=0$.

Corollary

If F is continuous, bounded below, and satisfies Walras' Law and the Boundary Condition, then there is $p^* \gg 0$ such that $F(p^*)=0$.

An alternative, intuitively appealing, and direct way of establishing this Corollary rests on a different normalization of the price vector p which we now constrain to have unit Euclidean norm. Thus, p belongs to the strictly positive part of the sphere with center 0 and unit radius,

$$S=\left\{p \in R^{\ell} \mid p \gg 0 \text{ and } \|p\|=1\right\}.$$

By Walras' Law, the excess demand vector $F(p)$ associated with the price vector p in S is orthogonal to p, and can therefore be conceived of as a vector tangent to S at p. In other words, F can be looked upon as defining a vector field on S. However, the Boundary Condition and the assumption that F has a lower bound imply that the vector $F(p)$ points inward near the boundary ∂S of S. A standard argument of degree theory (for which we refer to the bibliographical items listed in Section 1 in connection with the third approach to the existence problem) establishes that there is a p^* in S such that $F(p^*)=0$ (Figure 3.3).

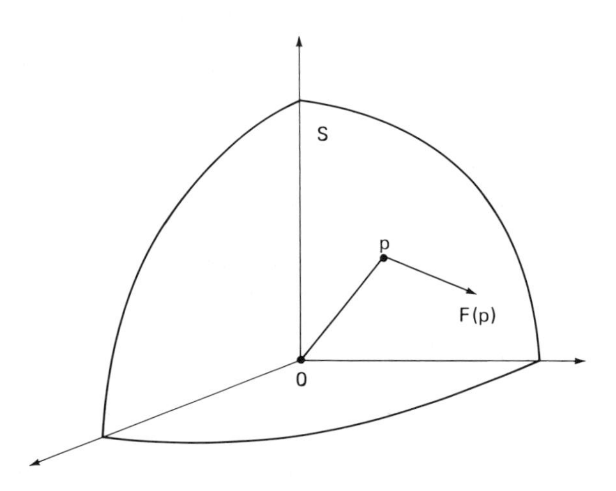

Figure 3.3.

4. Economies with a measure space of agents

The study of the relationship between the core of an economy (Hildenbrand, Chapter 18 of this Handbook) and its set of competitive allocations has required the consideration of sets of economic agents all of whom are negligible, a concept that was formalized in Aumann (1964) by means of measure theory and in Brown and Robinson (1972) by means of non-standard analysis. The theorem that Aumann proved in 1964 about the equality of the core and of the set of competitive allocations of an economy with an atomless measure space of agents would lack content without a parallel result [Aumann (1966)] on the existence of competitive allocations for atomless economies. This final section is devoted to that aspect of the existence problem. (In Chapter 5 of this Handbook Kirman presents the main concepts and results of measure theory used below.)

The *measure space of economic agents* is denoted by $(A, \mathcal{Q}, \nu)$ where A is the non-empty set of economic agents, $\mathcal{Q}$ is the σ-field of *coalitions*, i.e. a non-empty set of subsets of A closed under countable union and under complementation, and ν is a positive, countably additive measure defined on $\mathcal{Q}$ and such that $\nu(A)=1$. The measure $\nu(E)$ of a coalition E is interpreted as the fraction of the totality of agents in E. We assume that the measure space $(A, \mathcal{Q}, \nu)$ is *complete*, i.e. every subset of every set of measure zero belongs to $\mathcal{Q}$.

In an exchange economy the characteristics of an agent $a \in A$ are his preference relation $\precsim_a$ on the closed positive cone R^ℓ_+ of R^ℓ and his initial endowment e_a in R^ℓ_+. We assume that the binary relation $\precsim_a$ is complete, reflexive, transitive, and satisfies the assumptions of

closedness: the set $\{(x, y) \in R^\ell_+ \times R^\ell_+ | x \precsim_a y\}$ is closed; and

monotony: if x and y are two points of R^ℓ_+ such that $x < y$, then $x \prec_a y$.

The *set of preference relations on* R^ℓ_+ satisfying all these assumptions is denoted by $\mathcal{P}$. Thus, the characteristics of agent a are the point $(\precsim_a, e_a)$ in the Cartesian product $\mathcal{P} \times R^\ell_+$. To express formally the idea that the preferences of agent a' are "close" to the preferences of agent a, we introduce a topology on the set $\mathcal{P}$. Since a preference relation in $\mathcal{P}$ is a closed subset of $R^{2\ell}$, it is natural, following Kannai (1970), to endow $\mathcal{P}$ with the *topology of closed convergence* of Hausdorff (1935) which can be characterized [Kannai (1970) and Hildenbrand (1974, p. 96)] as the coarsest separated topology on $\mathcal{P}$ for which the set

$$\{(\precsim, x, y) \in \mathcal{P} \times R^\ell_+ \times R^\ell_+ | x \precsim y\}$$

is closed. The set $\mathcal{P} \times R^\ell_+$ of characteristics is then endowed with the product topology.

The measurable space structure of $\mathcal{P} \times R^\ell_+$ is derived from its topology in a standard manner. It is defined by the *Borel σ-field* of that topology, i.e. by the

σ-field generated by the set of open sets. As we noted, an exchange economy $\mathscr{E}$ specifies for each agent $a \in A$ his characteristics $(\precsim_a, e_a)$ in $\mathscr{P} \times R^\ell_+$. In formal terms, the *economy* $\mathscr{E}$ is a function from A to $\mathscr{P} \times R^\ell_+$. We require that $\mathscr{E}$ be measurable and that the endowment function e from A to R^ℓ_+ be integrable. Its integral $\int_A e\,\mathrm{d}\nu$ is the *mean endowment* of the economy $\mathscr{E}$.

An *allocation* x for the economy $\mathscr{E}$ specifies for each agent $a \in A$ his consumption vector x_a in R^ℓ_+, in other words, x is a function from A to R^ℓ_+ that we require to be integrable. A pair (x^*, p^*) of an allocation x^* and a price vector p^* in R^ℓ is an *equilibrium* if

for a.e. $a \in A$, x^*_a is a best element for $\precsim_a$ of $\{y \in R^\ell_+ \mid p^* \cdot y \leqq p^* \cdot e_a\}$,

$$\int_A x^*\,\mathrm{d}\nu = \int_A e\,\mathrm{d}\nu.$$

The first of these two conditions says that x^*_a is optimal under his budget constraint for almost every (abbreviated to a.e.) agent a in A, i.e. for every agent outside a set of ν-measure zero. The second expresses the equality of mean demand and mean supply.

As before, one must have $p^* \gg 0$ since otherwise, by monotony of preferences, for no agent $a \in A$ could x^*_a be a best element for $\precsim_a$ under a's budget constraint. And replacing p^* by λp^*, where λ is a strictly positive real number, does not alter the equilibrium conditions. Therefore, one can restrict the search for an equilibrium price vector to the relative interior of the simplex

$$P = \left\{ p \in R^\ell_+ \,\middle|\, \sum_{h=1}^{\ell} p^h = 1 \right\}.$$

Given an agent $a \in A$ and a price vector $p \gg 0$ in P, we denote by $\xi'_a(p)$ the set of best-consumption vectors for $\precsim_a$ in the budget set $\beta'_a(p) = \{y \in R^\ell_+ \mid p \cdot y \leqq p \cdot e_a\}$. In order to integrate the correspondence $a \mapsto \xi'_a(p)$ from A to R^ℓ we formally define the *integral of a correspondence* φ from A to R^ℓ over the set A as the subset of R^ℓ:

$\int_A \varphi\,\mathrm{d}\nu = \{y \in R^\ell \mid$ there is an integrable function f from A to R^ℓ such that $\int_A f\,\mathrm{d}\nu = y$ and for a.e. $a \in A$, $f(a) \in \varphi(a)\}$.

By an abuse of notation, we also write occasionally $\int_A \varphi(a)\,\mathrm{d}\nu(a)$ instead of $\int_A \varphi\,\mathrm{d}\nu$. This concept is a natural generalization of the concept of a finite sum of subsets of R^ℓ, in which case

$\sum_{a \in A} \varphi(a) = \{y \in R^\ell \mid$ there is a function f from A to R^ℓ such that $\sum_{a \in A} f(a) = y$ and for every $a \in A$, $f(a) \in \varphi(a)\}$.

Let now x be an allocation for the economy $\mathscr{E}$. If for a.e. $a \in A$, the vector x_a belongs to the set $\xi'_a(p)$, then the mean demand $\int_A x \, d\nu$ belongs to the integral of the correspondence $a \mapsto \xi'_a(p)$ from A to R^ℓ. Thus,

$$\int_A x \, d\nu \in \int_A \xi'_a(p) \, d\nu(a).$$

The correspondence ζ from the relative interior of P to R^ℓ defined by

$$\zeta(p) = \int_A \xi'_a(p) \, d\nu(a) - \int_A e \, d\nu$$

is the mean excess-demand correspondence of the economy $\mathscr{E}$, and p^* is an equilibrium price vector if and only if there is an allocation x^* such that for a.e. $a \in A$, $x^*_a \in \xi'_a(p^*)$ and $\int_A x^* \, d\nu = \int_A e \, d\nu$, in other words, if and only if $\int_A e \, d\nu \in \int_A \xi'_a(p^*) \, d\nu(a)$, or, equivalently,

$$0 \in \zeta(p^*).$$

The proof of existence of an equilibrium that we present follows Schmeidler (1969) and Hildenbrand (1970; 1974, 2.2). It consists in a study of the properties of the correspondence ζ and is based on Theorem 8. First, we note that ζ satisfies Walras' Law,

(i) *For every* $p \gg 0$ *in* P, *one has* $p \cdot \zeta(p) = 0$.

To see this, consider a vector z in the set $\zeta(p)$. By definition of $\zeta(p)$, there is an allocation x such that $\int_A x \, d\nu - \int_A e \, d\nu = z$ and for a.e. $a \in A$, $x_a \in \xi'_a(p)$. However, by monotony of preferences, $x_a \in \xi'_a(p)$ implies $p \cdot x_a = p \cdot e_a$. Therefore, after integrating x and e over A, $p \cdot \int_A x \, d\nu = p \cdot \int_A e \, d\nu$. Hence, $p \cdot z = 0$.

Next, we show that

(ii) *The correspondence* ζ *is bounded below and upper hemicontinuous.*

We must first prove that for every $p \gg 0$ in P, the set $\zeta(p)$ is non-empty.

Given $p \gg 0$ in P, we associate with the pair $(\precsim, w)$, where $\precsim$ is an element of $\mathscr{P}$ and w a positive or zero real number, the set

$$\hat{\xi}(\precsim, w) = \{x \in \beta(p, w) \mid x \text{ is best for } \precsim \text{ in } \beta(p, w)\},$$

i.e. the set of best elements for $\precsim$ in the budget set $\beta(p, w) = \{x \in R^\ell_+ \mid p \cdot x \leqq w\}$. According to Hildenbrand (1974, p. 102, Proposition 2), the correspondence $(\precsim, w) \mapsto \hat{\xi}(\precsim, w)$ from $\mathscr{P} \times R^\ell_+$ to R^ℓ_+ has a measurable graph. Since the function

$a \mapsto (\precsim_a, p \cdot e_a)$ from A to $\mathcal{P} \times R_+$ is measurable, the correspondence $a \mapsto \xi'_a(p) = \hat{\xi}(\precsim_a, p \cdot e_a)$ has a measurable graph by Hildenbrand (1974, p. 59, Proposition 1.b).

Using as before the norm $|x| = \sum_{h=1}^{\ell} |x^h|$ for a vector x in R^ℓ, consider an arbitrary point y of $\xi'_a(p)$. One has $y \geqq 0$. Moreover, $p \cdot y = p \cdot e_a$ implies for every $k = 1, \ldots, \ell$, $p^k y^k \leqq \sum_{h=1}^{\ell} p^h e_a^h \leqq |e_a|$. Hence,

(*) For every $k = 1, \ldots, \ell$, $0 \leqq y^k \leqq |e_a| / p^k$.

Thus, the compact-valued correspondence $a \mapsto \xi'_a(p)$ is bounded below by 0 and above by the integrable function $a \mapsto (|e_a|/p^1, \ldots, |e_a|/p^\ell)$. According to Hildenbrand (1974, p. 62, Theorem 2), $\int_A \xi'_a(p) \mathrm{d}\nu(a) \neq \varnothing$. And, according to Hildenbrand (1974, p. 73, Proposition 7), this set, which is obviously bounded below by 0, is also compact. Therefore, $\zeta(p)$ is non-empty, compact, and bounded below by $-\int_A e \, \mathrm{d}\nu$.

Let now $\bar{p} \gg 0$ be a point of P and let U be a compact neighborhood of $\bar{p}$ contained in the relative interior of P. We claim that for every agent $a \in A$, the correspondence $p \mapsto \xi'_a(p)$ is upper hemicontinuous on U.

(1) If $e_a = 0$, then for every $p \in U$, one has $\xi'_a(p) = \{0\}$ and ξ'_a is trivially upper hemicontinuous on U.

(2) If $e_a > 0$, then for every $p \in U$, one has $p \cdot e_a > 0$. By Lemma 3, this implies that the budget set correspondence β'_a is continuous on U. Consequently, ξ'_a is upper hemicontinuous on U.

Moreover, let

$$\pi(U) = \min_{\substack{k=1,\ldots,\ell \\ p \in U}} p^k,$$

a strictly positive real number. By (*), for every $a \in A$, $p \in U$, and $y \in \xi'_a(p)$, one has $|y| \leqq \ell / \pi(U) |e_a|$. Since the function $a \mapsto |e_a|$ is integrable, the correspondence $p \mapsto \int_A \xi'_a(p) \mathrm{d}\nu(a)$ is upper hemicontinuous on U by Aumann (1965) or (1976), Schmeidler (1970), and especially, Hildenbrand (1974, p. 73, Proposition 8). Therefore, the correspondence ζ is upper hemicontinuous.

(iii) *If ν is atomless, then the correspondence ζ is convex-valued.*

This is a direct consequence of Lyapunov's theorem on the range of an atomless finite-dimensional vector measure [Hildenbrand (1974, p. 62, Theorem 3)].

(iv) *If $\int_A e \, \mathrm{d}\nu \gg 0$ and the sequence $p_q \gg 0$ in P converges to p_0 in ∂P, then $d[0, \zeta(p_q)] \to +\infty$.*

Let

$$f_q(a)=d\left[0, \xi'_a(p_q)\right].$$

According to Hildenbrand (1974, p. 60, Proposition 3), the function f_q is measurable for every q. Actually, f_q is integrable, for

$$f_q(a)\leqq\left(\sum_{h=1}^{\ell}\frac{1}{p_q^h}\right)|e_a|$$

by (*), and the function $a\mapsto|e_a|$ is integrable.

By Fatou's lemma

$$\int_A \liminf_q f_q \mathrm{d}\nu\leqq \liminf_q \int_A f_q \mathrm{d}\nu.$$

However, for a compact subset E of R^ℓ_+, one has $d[0, E]=\min_{x\in E}\sum_{h=1}^{\ell}x^h$. Therefore, by Hildenbrand (1974, p. 63, Proposition 6),

$$d\left[0, \int_A \xi'_a(p_q)\mathrm{d}\nu(a)\right]=\int_A d\left[0, \xi'_a(p_q)\right]\mathrm{d}\nu(a).$$

Consequently,

$$\int_A \liminf_q f_q \mathrm{d}\nu\leqq \liminf_q d\left[0, \int_A \xi'_a(p_q)\mathrm{d}\nu(a)\right].$$

Since $\int_A e\,\mathrm{d}\nu\gg 0$, one has $\int_A p_0\cdot e\,\mathrm{d}\nu>0$, and the set

$$B=\{a\in A|\, p_0\cdot e_a>0\}$$

must have a strictly positive measure. For every $a\in B$, $\lim_q f_q(a)=+\infty$ by Lemma 4. Thus, $\int_A \lim_q\inf f_q \mathrm{d}\nu=+\infty$, and $d\left[0, \int_A\xi'_a(p_q)\mathrm{d}\nu(a)\right]\to+\infty$. So does $d[0, \zeta(p_q)]$.

As an immediate consequence of these properties of the excess demand correspondence, we obtain

Theorem 9

If $\mathscr{E}$ is an atomless exchange economy such that $\int_A e\,\mathrm{d}\nu\gg 0$, then $\mathscr{E}$ has an equilibrium (p^*, x^*) with $p^*\gg 0$.

Proof

According to (i), (ii), (iii), and (iv), the excess demand correspondence ζ satisfies all the conditions of Theorem 8. Therefore, there is $p^* \gg 0$ in P such that $0 \in \zeta(p^*)$. Q.E.D.

References

Allen, B. E. (1978), Generic existence of equilibria for economies with uncertainty when prices convey information. Ph.D. dissertation, University of California, Berkeley.

Allen, B. E. (1980), "Expectations equilibria with dispersed information: Existence, approximate rationality, and convergence in a model with a continuum of agents and finitely many states of the world", Center for Analytic Research in Economics and the Social Sciences, University of Pennsylvania.

Allingham, M. C. (1973), Equilibrium and disequilibrium. Cambridge, MA.: Ballinger.

Allingham, M. C. (1974), "Equilibrium and stability", Econometrica, 42:705–716.

Anderson, R. M. (1977), "A market value approach to approximate equilibria", Department of Mathematics, McMaster University.

Arrow, K. J. (1953), "Le rôle des valeurs boursières pour la répartition la meilleure des risques", Econométrie, Centre National de la Recherche Scientifique, Paris, 41–48. Translated as "The role of securities in the optimal allocation of risk-bearing", Review of Economic Studies, 31 (1964):91–96.

Arrow, K. J. (1968), "Economic equilibrium", International Encyclopedia of the Social Sciences, Vol. 4. Macmillan, 376–389.

Arrow, K. J. (1971), "The firm in general equilibrium theory", in: R. Marris and A. Wood, eds., The Corporate Economy: Growth, Competition and Innovative Potential. Harvard University Press, 68–110.

Arrow, K. J. (1974), "General economic equilibrium: Purpose, analytic techniques, collective choice", American Economic Review, 64:253–272.

Arrow, K. J. and G. Debreu (1954), "Existence of an equilibrium for a competitive economy", Econometrica, 22:265–290.

Arrow, K. J. and F. H. Hahn (1971), General competitive analysis. San Francisco: Holden Day.

Aumann, R. J. (1964), "Market with a continuum of traders", Econometrica, 32:39–50.

Aumann, R. J. (1965), "Integrals of set-valued functions", Journal of Mathematical Analysis and Applications, 12:1–12.

Aumann, R. J. (1966), "Existence of competitive equilibria in markets with a continuum of traders", Econometrica, 34:1–17.

Aumann, R. J. (1976), "An elementary proof that integration preserves upper-semicontinuity", Journal of Mathematical Economics, 3:15–18.

Balasko, Y. (1975), "Some results on uniqueness and on stability of equilibrium in general equilibrium theory", Journal of Mathematical Economics, 2:95–118.

Balasko, Y. (1978), "Economie et théorie des catastrophes", Cahiers du Séminaire d'Econométrie, Vol. 19. Centre National de la Recherche Scientifique, Paris, 105–120. English version: "Economic equilibrium and catastrophe theory: An introduction", Econometrica, 46:557–569.

Balasko, Y. and K. Shell (1979), "The overlapping-generations model", Center for Analytic Research in Economics and the Social Sciences, University of Pennsylvania.

Barkuki, R. A. (1977), "The existence of an equilibrium in economic structures with a Banach space of commodities", Akad. Nauk Azerbaijan SSR Dokl., 33:8–12.

Baudier, E. (1973), "Competitive equilibrium in a game", Econometrica, 41:1049–1068.

Benassy, J. P. (1973), Disequilibrium Theory. Ph.D. dissertation, University of California, Berkeley.

Benassy, J. P. (1975), "Disequilibrium exchange in barter and monetary economies", Economic Enquiry, 13:131–156.

Benassy, J. P. (1975), "Neo-Keynesian disequilibrium theory in a monetary economy", Review of Economic Studies, 42:503–523.
Benassy, J. P. (1976), "The disequilibrium approach to monopolistic price setting and general monopolistic equilibrium", Review of Economic Studies, 43:69–81.
Benassy, J. P. (1976), "Théorie néo-Keynesienne du Déséquilibre dans une économie monétaire", Cahiers du Séminaire d'Econométrie, No. 17. Paris: Centre National de la Recherche Scientifique, 81–113.
Benninga, S. (1979), "Competitive equilibrium with bankruptcy in a sequence of markets", International Economic Review, 20:557–575.
Berge, C. (1959), Espaces Topologiques. Dunod. Translated as Topological Spaces, Macmillan (1963).
Bergstrom, T. C. (1970), "A 'Scandinavian consensus' solution for efficient income distribution among non-malevolent consumers", Journal of Economic Theory, 2:383–398.
Bergstrom, T. C. (1971), "On the existence and optimality of competitive equilibrium for a slave economy", Review of Economic Studies, 38:23–36.
Bergstrom, T. C. (1973), "Competitive equilibrium without transitivity, monotonicity or local non-satiation", Department of Economics, Washington University.
Bergstrom, T. C. (1973), "Equilibrium in a finite economy with bounded non-convexity".
Bergstrom, T. C. (1975), "The existence of maximal elements and equilibria in the absence of transitivity", Department of Economics, University of Michigan.
Bergstrom, T. C. (1976), "How to discard 'free disposability' at no cost", Journal of Mathematical Economics, 3:131–134.
Bergstrom, T. C. (1976), "Collective choice and the Lindahl allocation method", in: S. A. Y. Lin, ed., Theory and Measurement of Economic Externalities. New York: Academic Press.
Bewley, T. F. (1970), Equilibrium theory with an infinite dimensional commodity space. Ph.D. dissertation, University of California, Berkeley.
Bewley, T. F. (1972), "Existence of equilibria in economies with infinitely many commodities", Journal of Economic Theory, 4:514–540.
Bewley, T. F. (1978), "General equilibrium theory with market frictions", Center on Decision and Conflict in Complex Organizations, Harvard University.
Bewley, T. F. (1979), "An integration of equilibrium theory and turnpike theory", Center for Mathematical Studies in Economics and Management Science, Northwestern University.
Bhattacharya, R. N. and M. Majumdar (1973), "Random exchange economies", Journal of Economic Theory, 6:37–67.
Bidard, C. (1975), Equilibre général et spécialisation internationale. Thèse, Université de Paris I.
Blume, L. E. (1977), The stationarity of Markov processes of temporary equilibrium. Ph.D. dissertation, University of California, Berkeley.
Blume, L. E. (1978), "The ergodic behavior of stochastic processes of economic equilibria", Econometrica, forthcoming.
Blume, L. E. (1978), "New techniques for the study of stochastic equilibrium processes", Department of Economics, University of Michigan.
Blume, L. E. (1979), "Consistent expectations", Department of Economics, University of Michigan.
Boehm, V. H. (1972), Stable firm structures and the core of an economy with production. Ph.D. dissertation, University of California, Berkeley.
Boehm, V. H. (1973), "Firms and market equilibria in a private ownership economy", Zeitschrift für Nationalökonomie, 33:87–102.
Boehm, V. and P. Lévine (1979), "Temporary equilibria with quantity rationing", Review of Economic Studies, 46:361–377.
Bojan, P. (1974), "A generalization of theorems on the existence of competitive economic equilibrium to the case of infinitely many commodities", Mathematica Balkanica, 4:491–494.
Borch, K. (1962), "Equilibrium in a reinsurance market", Econometrica, 30:424–444.
Borglin, A. (1973), "Price characterization of stable allocations in exchange economies with externalities", Journal of Economic Theory, 6:483–494.
Borglin, A. (1976), "The Theory of Lindahl equilibria derived from the theory of uncertainty", Department of Economics, University of Lund.

Borglin, A. and H. Keiding (1976), "Existence of equilibrium actions and of equilibrium", Journal of Mathematical Economics, 3:313–316.
Broome, J. (1972), "Approximate equilibrium in economies with indivisible commodities", Journal of Economic Theory, 5:224–249.
Brown, D. J. (1976), "Existence of a competitive equilibrium in a non-standard exchange economy", Econometrica, 44:537–546.
Brown, D. J. and G. Heal (1976), "Existence of a market equilibrium in an economy with increasing returns to scale", Cowles Foundation Discussion Paper.
Brown, D. J. and A. Robinson (1972), "A limit theorem on the cores of large standard exchange economies", Proceedings of the National Academy of Sciences of the U.S.A., 69:1258–1260.
Burger, E. (1959), Einführung in die Theorie der Spiele. Berlin: W. de Gruyter; translated as: Introduction to the theory of games. Prentice-Hall (1963).
Champsaur, P. (1976), "Symmetry and continuity properties of Lindahl equilibria", Journal of Mathematical Economics, 3:19–36.
Chetty, V. K. and D. Dasgupta (1978), "Temporary competitive equilibrium in a monetary economy with uncertain technology and many planning periods", Journal of Mathematical Economics, 5:23–42.
Chichilnisky, G. (1976), Manifolds of preferences and equilibria. Ph.D. dissertation, University of California, Berkeley.
Chichilnisky, G. and P. Kalman (1978), "Rothe's fixed point theorem and the existence of equilibria in monetary economies", Journal of Mathematical Analysis and Applications, 65:56–65.
Chipman, J. S. (1965), "A survey of the theory of international trade, Part 1, The classical theory, Econometrica, 33:477–519; Part 2, The neo classical theory", Econometrica, 33:685–760.
Christiansen, D. S. and M. Majumdar (1977), "On shifting temporary equilibrium", Journal of Economic Theory, 16:1–9.
Cornwall, R. R. (1978), "Marketing costs and imperfect competition in general equilibrium", in: G. Schwödiauer ed., Equilibrium and disequilibrium in economic theory. D. Reidel, 239–254.
Cournot, A. (1838), Recherches sur les Principes Mathématiques de la Théorie des Richesses. Paris: Hachette. Translated as: Researches into the Mathematical Principles of the Theory of Wealth. New York: Kelley (1960).
Dantzig, G. B., B. C. Eaves, and David Gale (1981), "An algorithm for a piecewise linear model of trade and production with negative prices and bankruptcy", Mathematical Programming, forthcoming.
Dasgupta, P. and E. Maskin (1977), "The existence of economic equilibria: Continuity and mixed strategies", Technical Report No. 252, Institute for Mathematical Studies in the Social Sciences, Stanford University.
Debreu, G. (1952), "A social equilibrium existence theorem", Proceedings of the National Academy of Sciences of the U.S.A., 38:886–893.
Debreu, G. (1956), "Market equilibrium", Proceedings of the National Academy of Sciences of the U.S.A., 42:876–878.
Debreu, G. (1959), Theory of value. Wiley.
Debreu, G. (1962), "New concepts and techniques for equilibrium analysis", International Economic Review, 3:257–273.
Debreu, G. (1970), "Economies with a finite set of equilibria", Econometrica, 38:387–392.
Debreu, G. (1974), "Four aspects of the mathematical theory of economic equilibrium", Proceedings of the International Congress of Mathematicians, Vancouver, 1974, Vol. I, 65–77.
Delbaen, F. (1974), "Stochastic preferences and general equilibrium theory", in: J. H. Drèze, ed., Allocation Under Uncertainty: Equilibrium and Optimality. New York: Wiley, 98–109.
Dierker, E. (1971), "Equilibrium analysis of exchange economies with indivisible commodities", Econometrica, 39:997–1008.
Dierker, E. (1972), "Two remarks on the number of equilibria of an economy", Econometrica, 40:951–953.
Dierker, E. (1974), Topological methods in Walrasian economics. Springer-Verlag.
Dierker, E. (1980), "Regular economies", Chapter 17 in this Handbook.

Diewert, W. E. (1973), "On a theorem of Negishi", Metroeconomica, 25:119–135.
Diewert, W. E. (1978), "Walras' theory of capital formation and the existence of a temporary equilibrium", in: G. Schwödiauer, ed., Equilibrium and Disequilibrium in Economic Theory. D. Reidel, 73–126.
Dorfman, R., P. A. Samuelson, and R. M. Solow (1958), Linear programming and economic analysis. New York: McGraw-Hill.
Drandakis, E. M. (1966), "On the competitive equilibrium in a monetary economy", International Economic Review, 7:304–328.
Drèze, J. (1974), "Investment under private ownership: Optimality, equilibrium, and stability", in: J. M. Drèze, ed., Allocation under uncertainty: Equilibrium and optimality. New York: Wiley, 129–166.
Drèze, J. (1975), "Existence of an exchange equilibrium under price rigidities", International Economic Review, 16:301–320.
Drèze, J. (1978), "Public goods with exclusion", CORE Discussion Paper.
Drèze, J. and H. Müller (1979), "Optimality properties of rationing schemes", CORE Discussion Paper.
Dubey, P. and M. Shubik (1977), "A closed economic system with production and exchange modelled as a game of strategy", Journal of Mathematical Economics, 4:253–287.
Dubey, P. and M. Shubik (1977), "A closed economy with exogenous uncertainty, different levels of information, money, futures and spot markets", International Journal of Game Theory, 6:231–248.
Dubey, P. and M. Shubik (1977), "Trade and prices in a closed economy with exogenous uncertainty, different levels of information, money and compound future markets", Econometrica, 45:1657–1680.
Eaves, B. C. (1976), "A finite algorithm for the linear exchange model", Journal of Mathematical Economics, 3:197–203.
Eisenberg, E. (1961), "Aggregation of utility functions", Management Science, 7:337–350.
Ellickson, B. (1979), "Competitive equilibrium with local public goods", Journal of Economic Theory, 21:46–61.
Evers, J. J. M. (1975), "Invariant competitive equilibrium in a dynamic economy with negotiable shares", Research Memorandum, Tilburg University.
Fabre-Sender, F. and R. Guesnerie (1973), "Etude des structures de revenus associées aux optima Parétiens d'une economie convexe", Bulletin de Mathématiques Economiques, 10:25–66.
Farrell, M. J. (1970), "Edgeworth bounds for oligopoly prices", Economica, 37:342–361.
Feltenstein, A. (1979), "Market equilibrium in a model of a planned economy of the Soviet type: A proof of existence and results of numerical simulations", Review of Economic Studies, 46:631–652.
FitzRoy, F. R. (1974), "Monopolistic equilibrium, non convexity and inverse demand", Journal of Economic Theory, 7:1–16.
Foley, D. K. (1967), "Resource allocation and the public sector", Yale Economic Essays, 7:42–98.
Foley, D. K. (1970), "Economic equilibrium with costly marketing", Journal of Economic Theory, 2:276–291.
Foley, D. K. (1970), "Lindahl's solution and the core of an economy with public goods", Econometrica, 38:66–72.
Föllmer, H. (1974), "Random economies with many interacting agents", Journal of Mathematical Economics, 1:51–62.
Fourgeaud, C. (1969), "Contribution à l'étude du rôle des administrations dans la théorie mathématique de l'équilibre et de l'optimum", Econometrica, 37:307–323.
Fourgeaud, C., B. Lenclud, and P. Sentis (1974), "Equilibre, optimum et décentralisation dans un cas de rendement croissant", Cahiers du Séminaire d'Econométrie, Centre National de la Recherche Scientifique, Paris, 15:29–46.
Frank, C. R. and R. E. Quandt (1963), "On the existence of Cournot equilibrium", International Economic Review, 4:92–96.
Friedman, J. W. (1974), "Non-cooperative equilibria in time-dependent supergames", Econometrica, 42:221–237.
Friedman, J. W. (1976), "Reaction functions as Nash equilibria", Review of Economic Studies, 43:83–90.

Friedman, J. W. (1977), Oligopoly and the theory of games. Amsterdam: North-Holland.
Friesen, P. H. (1979), "The Arrow–Debreu model extended to financial markets", Econometrica, 47:689–707.
Fuchs, G. (1974), "Private ownership economies with a finite number of equilibria", Journal of Mathematical Economics, 1:141–158.
Fuchs, G. and R. Guesnerie (1979), "Structure of tax equilibria", Laboratoire d'Econométrie, Ecole Polytechnique.
Gabszewicz, J. J. and J. P. Vial (1972), "Oligopoly 'a la Cournot' in a general equilibrium analysis", Journal of Economic Theory, 4:381–401.
Gale, David (1955), "The law of supply and demand", Mathematica Scandinavica, 3:155–169.
Gale, David (1957), "General equilibrium for linear models", The RAND Corporation.
Gale, David (1976), "The linear exchange model", Journal of Mathematical Economics, 3:205–209.
Gale, David and A. Mas-Colell (1975), "An equilibrium existence theorem for a general model without ordered preferences", Journal of Mathematical Economics, 2:9–15.
Gale, David and A. Mas-Colell (1978), "On the role of complete, transitive preferences in equilibrium theory", in: G. Schwödiauer, ed., Equilibrium and Disequilibrium in Economic Theory. D. Reidel, 7–14.
Gale, David and A. Mas-Colell (1979), "Corrections to an equilibrium existence theorem for a general model without ordered preferences", Journal of Mathematical Economics, 6:297–298.
Gale, D. and H. Nikaido (1965), "The Jacobian matrix and global univalence of mappings", Mathematische Annalen, 159:81–93.
Gale, Douglas (1978), "A note on conjectural equilibria", Review of Economic Studies, 45:33–38.
Gale, Douglas (1979), "Large economies with trading uncertainty", Review of Economic Studies, 46:319–338.
Gay, A. (1978), "The exchange rates approach to general economic equilibrium", Economie Appliquée, 31:159–174.
Geistdoerfer, M. (1978), "Le lemme de Gale–Nikaido–Debreu et l'existence de l'équilibre transitif en l'absence d'hypothèse de libre disposition", C.E.P.R.E.M.A.P., Paris.
Glustoff, E. (1968), "On the existence of a Keynesian equilibrium", Review of Economic Studies, 35:327–334.
Grandmont, J. M. (1970), On the temporary competitive equilibrium. Ph.D. dissertation, University of California, Berkeley.
Grandmont, J. M. (1974), "On the short-run equilibrium in a monetary economy", in: J. Drèze, ed., Allocation under uncertainty: Equilibrium and optimality. New York: Wiley, 213–228.
Grandmont, J. M. (1977), "Temporary general equilibrium theory", Econometrica, 45:535–572.
Grandmont, J. M. and W. Hildenbrand (1974), "Stochastic processes of temporary equilibria", Journal of Mathematical Economics, 1:247–277.
Grandmont, J. M. and G. Laroque (1973), "Money in the pure consumption loan model", Journal of Economic Theory, 6:382–395.
Grandmont, J. M. and G. Laroque (1975), "On money and banking", Review of Economic Studies, 42:207–236.
Grandmont, J. M. and G. Laroque (1976), "On temporary Keynesian equilibria", Review of Economic Studies, 43:53–67.
Grandmont, J. M. and Y. Younès (1972), "On the role of money and the existence of a monetary equilibrium", Review of Economic Studies, 39:355–372.
Green, J. R. (1973), "Temporary general equilibrium in a sequential trading model with spot and futures transactions", Econometrica, 41:1103–1123.
Green, J. R. (1974), "Preexisting contracts and temporary general equilibrium", in: M. Balch, D. L. McFadden, and S. Y. Wu, eds., Essays on economic behavior under uncertainty. Amsterdam: North-Holland, 263–299.
Green, J. R. and M. Majumdar (1975), "The nature of stochastic equilibria", Econometrica, 43:647–660.
Greenberg, J. (1977), "An elementary proof of the existence of a competitive equilibrium with weak gross substitutes", Quarterly Journal of Economics, 91:513–516.

Greenberg, J. (1977), "Quasi-equilibrium in abstract economies without ordered preferences", Journal of Mathematical Economics, 4:163–165.

Greenberg, J. (1977), "Existence of an equilibrium with arbitrary tax schemes for financing local public goods", Journal of Economic Theory, 16:137–150.

Greenberg, J. (1979), "Existence and optimality of equilibrium in labor managed economies", Review of Economic Studies, 46:419–433.

Greenberg, J. and H. Müller (1977), "Equilibrium under price rigidities and externalities", in: O. Moeschlin, ed., Game theory and related topics. Amsterdam: North-Holland.

Greenberg, J., B. Shitovitz, and A. Wieczorek, (1979), "Existence of equilibria in atomless production economies with price dependent preferences", Journal of Mathematical Economics, 6:31–41.

Grossman, S. J. and O. D. Hart (1979), "A theory of competitive equilibrium in stock market economies", Econometrica, 47:293–329.

Groves, T. and J. O. Ledyard (1978), "The existence of efficient and incentive compatible equilibria with public goods", Technical Report No. 257, Institute for Mathematical Studies in the Social Sciences, Stanford University.

Guesnerie, R. (1979), "Financing public goods with commodity taxes: The tax reform viewpoint", Econometrica, 47:393–421.

Hahn, F. H. (1965), "On some problems of proving the existence of an equilibrium in a monetary economy", in: F. H. Hahn and F. P. R. Brechling, eds., The theory of interest rates. New York: Macmillan, 126–135.

Hahn, F. H. (1968), "On some propositions of general equilibrium analysis", Economica, 35:424–430.

Hahn, F. H. (1971), "Equilibrium with transaction costs", Econometrica, 39:417–439.

Hahn, F. H. (1973), "On transaction costs, inessential sequence economies, and money", Review of Economic Studies, 40:449–462.

Hahn, F. H. (1978), "On non Walrasian equilibria", Review of Economic Studies, 45:1–17.

Hammond, P. J. (1979), "Straightforward individual incentive compatibility in large economies", Review of Economic Studies, 46:263–282.

Hart, O. D. (1974), "On the existence of equilibrium in a securities model", Journal of Economic Theory, 9:293–311.

Hart, O. D. and H. W. Kuhn (1975), "A proof of the existence of equilibrium without the free disposal assumption", Journal of Mathematical Economics, 2:335–343.

Hausdorff, F. (1935), Mengenlehre. Dover. Translated as: Set theory. Chelsea (1957).

Hayashi, T. (1974), "The non-Pareto efficiency of initial allocation of commodities and monetary equilibrium: An inside money economy", Journal of Economic Theory, 7:173–187.

Hayashi, T. (1976), "Monetary equilibrium in two classes of stationary economies", Review of Economic Studies, 43:269–284.

Heller, W. P. (1972), "Transaction with set-up costs", Journal of Economic Theory, 4:465–478.

Heller, W. P. (1974), "The holding of money balances in general equilibrium", Journal of Economic Theory, 7:93–108.

Heller, W. P. (1977), "Oligopoly equilibrium with discontinuous reaction functions", Department of Economics, University of California, San Diego.

Heller, W. P. (1978), "Continuity in general nonconvex economies (with applications to the convex case)", in: G. Schwödiauer, ed., Equilibrium and Disequilibrium in Economic Theory. D. Reidel, 27–38.

Heller, W. P. and R. M. Starr (1976), "Equilibrium with non-convex transaction costs: Monetary and non-monetary economies", Review of Economic Studies, 43:195–215.

Heller, W. P. and R. M. Starr (1979), "Unemployment equilibrium with myopic complete information", Review of Economic Studies, 46:339–359.

Helpman, E. and J. J. Laffont (1975), "On moral hazard in general equilibrium theory", Journal of Economic Theory, 10:8–23.

Henry, C. (1970), "Indivisibilités dans une économie d'échanges", Econometrica, 38:542–558.

Hildenbrand, K. (1972), "Continuity of the equilibrium set correspondence", Journal of Economic Theory, 5:152–162.

Hildenbrand, W. (1970), "Existence of equilibria for economies with production and a measure space

of consumers", Econometrica, 38:608–623.
Hildenbrand, W. (1971), "Random preferences and equilibrium analysis", Journal of Economic Theory, 3:414–429.
Hildenbrand, W. (1972), "Metric measure spaces of economic agents", in: L. LeCam et al., eds., Proceedings of the Sixth Berkeley Symposium on Mathematical Statistics and Probability, Vol. II. University of California Press, 81–95.
Hildenbrand, W. (1974), Core and equilibria of a large economy. Princeton University Press.
Hildenbrand, W. (1975), "Distributions of agents' characteristics", Journal of Mathematical Economics, 2:129–138.
Hildenbrand, W. (1977), "Limit theorems on the core of an economy", in: M. D. Intriligator, ed., Frontiers of quantitative economics, Vol. IIIA. Amsterdam: North-Holland, 149–165.
Hildenbrand, W. (1980), "Core of an economy", Chapter 18 in this Handbook.
Hildenbrand, W. and A. P. Kirman (1976), Introduction to equilibrium analysis. Amsterdam: North-Holland.
Hildenbrand, W., D. Schmeidler, and S. Zamir (1973), "Existence of approximate equilibria and cores", Econometrica, 41:1159–1166.
Honkapohja, S. (1978), "On the efficiency of a competitive monetary equilibrium with transaction costs", Review of Economic Studies, 45:405–415.
Hool, B. (1974), Money, financial assets, and general equilibrium in a sequential market economy. Ph.D. dissertation, University of California, Berkeley.
Hool, B. (1976), "Money, expectations and the existence of a temporary equilibrium", Review of Economic Studies, 43:439–445.
Hool, B. (1979), "Liquidity, speculation, and the demand for money", Journal of Economic Theory, 21:73–87.
Hurwicz, L. (1979), "On allocations attainable through Nash equilibria", Journal of Economic Theory, 21:140–165.
Ichiishi, T. (1974), Some generic properties of smooth economies. Ph.D. dissertation, University of California, Berkeley.
Ichiishi, T. (1977), "Coalition structure in a labor-managed market economy", Econometrica, 45:341–360.
Ichiishi, T. (1978), "Management versus ownership", Graduate School of Industrial Administration, Carnegie-Mellon University.
Ichiishi, T. (1979), "A social coalitional equilibrium existence theorem", Graduate School of Industrial Administration, Carnegie-Mellon University.
Ichiishi, T. and S. Weber (1978), "Some theorems on the core of a non-sidepayment game with a measure space of players", International Journal of Game Theory: 7, 95–112.
Intriligator, M. D. (1971), Mathematical optimization and economic theory. Prentice-Hall.
Isard, W. (1960), "General interregional equilibrium", Journal of Regional Science, 2:67–74.
Isard, W. and D. J. Ostroff (1958), "Existence of a competitive interregional equilibrium", Papers and Proceedings of the Regional Science Association, 4:49–76.
Isard, W., T. E. Smith et al. (1969), General theory: Social, political, economic, and regional. Cambridge, Mass.: M.I.T. Press.
Jamison, D. T. and L. J. Lau (1977), "The nature of equilibrium with semiordered preferences", Econometrica, 45:1595–1605.
Jordan, J. S. (1976), "Temporary competitive equilibrium and the existence of self-fulfilling expectations", Journal of Economic Theory, 12:455–471.
Kakutani, S. (1941), "A generalization of Brouwer's fixed point theorem", Duke Mathematical Journal, 8:457–459.
Kalman, P. J. and K-P. Lin (1978), "Application of Thom's transversality theory and Brouwer degree to economics", in: J. Green, ed., Some aspects of the foundations of general equilibrium theory: The posthumous papers of Peter J. Kalman. Springer-Verlag, 27–48.
Kalman, P. J. and K-P. Lin (1978), "On some properties of short-run monetary equilibrium with uncertain expectations", in: J. Green, ed., Some aspects of the foundations of general equilibrium theory: The posthumous papers of Peter J. Kalman. Springer-Verlag, 77–114.

Kalman, P. J., K-P. Lin, and H. Wiesmeth (1978), "Equilibrium theory in Veblen–Scitovsky economies: Local uniqueness, stability and existence", in: J. Green, ed., Some aspects of the foundations of general equilibrium theory: The posthumous papers of Peter J. Kalman. Springer-Verlag, 131–149. Also Journal of Mathematical Economics, 6 (1979):263–276.

Kannai, Y. (1970), "Continuity properties of the core of a market", Econometrica, 38:791–815.

Karlin, S. (1959), Mathematical methods and theory in games programming and economics, Vol. I. Massachusetts: Addison-Wesley.

Kehoe, T. J. (1980), "An index theorem for general equilibrium models with production", Econometrica, 48:1211–1232.

Kemp, M. C. and H. Y. Wan (1972), "The gains from free trade", International Economic Review, 13:509–522.

Khan, M. A. (1975), "Some approximate equilibria", Journal of Mathematical Economics, 2:63–86.

Khan, M. A. and A. Yamazaki (1979), "On the cores of economies with indivisible commodities and a continuum of traders", Department of Political Economy, The Johns Hopkins University.

Kirman, A. P. (1980), "Measure theory and its application to economics", Chapter 5 in this Handbook.

Koopmans, T. C. (1957), Three essays on the state of economic science. McGraw-Hill.

Koopmans, T. C., and A. F. Bausch (1959), "Selected topics in economics involving mathematical reasoning", SIAM Review, 1:79–148.

Krass, Y. (1976), Mathematical models of economic dynamics. Moscow: Sovetskovo Radio.

Kuenne, R. E. (1963), The theory of general economic equilibrium. Princeton University Press.

Kuga, K. (1965), "Weak gross substitutability and the existence of competitive equilibrium", Econometrica, 33:593–599.

Kuga, K. (1973), "Tariff retaliation and policy equilibrium", Journal of International Economics, 3:351–366.

Kuhn, H. W. (1956), "On a theorem of Wald", in: H. W. Kuhn and A. W. Tucker, eds., Linear inequalities and related systems, Annals of Mathematical Studies, Vol. 38. Princeton University Press, 265–273.

Kuhn, H. W. (1956), "A note on 'The law of supply and demand'", Mathematica Scandinavica, 4:143–146.

Kurz, M. (1974), "Equilibrium in a finite sequence of markets with transaction cost", Econometrica, 42:1–20.

Kurz, M. (1974), "Equilibrium with transaction cost and money in a single market exchange economy", Journal of Economic Theory, 7:418–452.

Kurz, M. (1974), "Arrow–Debreu equilibrium of an exchange economy with transaction cost", International Economic Review, 15:699–717.

Kurz, M. (1980), "Unemployment equilibrium in an economy with linked prices", Institute for Mathematical Studies in the Social Sciences, Stanford University.

Laan, G. van der (1980), "Equilibrium under rigid prices with compensation for the consumers", International Economic Review, 21:63–73.

Laffont, J. J. (1972), "Une note sur la compatibilité entre rendements croissants et concurrence parfaite", Revue d'Economie Politique, 82:1188–1193.

Laffont, J. J. (1975), "Optimism and experts against adverse selection in a competitive economy", Journal of Economic Theory, 10:284–308.

Laffont, J. J. and G. Laroque (1972), "Effets externes et théorie de l'équilibre général", Cahiers du Séminaire d'Econométrie, No. 14. Paris: Centre National de la Recherche Scientifique, 25–48.

Laffont, J. J. and G. Laroque (1976), "Existence d'un équilibre général de concurrence imparfaite: une introduction", Econometrica, 44:283–294.

Lancaster, K. (1968), Mathematical economics. New York: Macmillan.

Laroque, G. and H. Polemarchakis (1978), "On the structure of the set of fixed price equilibria", Journal of Mathematical Economics, 5:53–69.

Lewis, L. M. (1977), Essays on purely competitive intertemporal exchange. Ph.D. dissertation, Yale University.

McKenzie, L. W. (1954), "On equilibrium in Graham's model of world trade and other competitive

systems", Econometrica, 22:147–161.
McKenzie, L. W. (1955), "Competitive equilibrium with dependent consumer preferences", in: H. A. Antosiewicz, ed., Proceedings of the Second Symposium in Linear Programming, 277–294.
McKenzie, L. W. (1959), "On the existence of general equilibrium for a competitive market", Econometrica, 27:54–71.
McKenzie, L. W. (1961), "On the existence of general equilibrium: Some corrections", Econometrica, 29:247–248.
McKenzie, L. W. (1980), "The classical theorem on existence of competitive equilibrium", Econometrica, forthcoming.
Makarov, V. L. and A. M. Rubinov (1977), Mathematical theory of economic dynamics and equilibria. Springer-Verlag.
Makowski, L. (1979), "Value theory with personalized trading", Journal of Economic Theory, 20:194–212.
Malinvaud, E. (1972), Lectures on microeconomic theory. Amsterdam: North-Holland.
Manne, A. S., H. Chao, and R. Wilson (1979), "Computation of competitive equilibria by a sequence of linear programs", Department of Operations Research, Stanford University.
Mantel, R. (1968), "Toward a constructive proof of the existence of equilibrium in a competitive economy", Yale Economic Essays, 8:155–196.
Mantel, R. (1975), "General equilibrium and optimal taxes", Journal of Mathematical Economics, 2:187–200.
Marschak, T. and R. Selten (1974), General equilibrium with price-making firms. Springer-Verlag.
Marschak, T. and R. Selten (1977), "Oligopolistic economies as games of limited information", Zeitschrift für die gesamte Staatswissenschaft, 133:385–410.
Marschak, T. and R. Selten (1978), "Restabilizing responses, inertia supergames, and oligopolistic equilibria", Quarterly Journal of Economics, 92:71–93.
Mas-Colell, A. (1974), "An equilibrium existence theorem without complete or transitive preferences", Journal of Mathematical Economics, 1:237–246.
Mas-Colell, A. (1975), "A model of equilibrium with differentiated commodities", Journal of Mathematical Economics, 2:263–295.
Mas-Colell, A. (1977), "Indivisible commodities and general equilibrium theory", Journal of Economic Theory, 16:443–456.
Mas-Colell, A. and W. Neuefeind (1977), "Some generic properties of aggregate excess demand and an application", Econometrica, 45:591–599.
Milleron, J. C. (1972), "Theory of value with public goods: A survey article", Journal of Economic Theory, 5:419–477.
Milne, F. (1976), "Default risk in a general equilibrium asset economy with incomplete markets", International Economic Review, 17:613–625.
Mityagin, B. S. (1972), "Remarks on mathematical economics", Russian Mathematical Surveys, 27, No. 3:1–19.
Montbrial, T. de (1971), Intertemporal general equilibrium and interest rates theory, Ph.D. dissertation, University of California, Berkeley.
Montbrial, T. de (1971), Economie théorique. Paris: Presses Universitaires de France.
Moore, J. C. (1975), "The existence of 'compensated equilibrium' and the structure of the Pareto efficiency frontier", International Economic Review, 16:267–300.
Morishima, M. (1960), "Existence of solution to the Walrasian system of capital formation and credit", Zeitschrift für Nationalökonomie, 20:238–243.
Müller, H. and U. Schweizer (1978), "Temporary equilibrium in a money economy", Journal of Economic Theory, 19:267–286.
Nash, J. F. (1950), "Equilibrium points in N-person games", Proceedings of the National Academy of Sciences, U.S.A., 36:48–49.
Negishi, T. (1960), "Welfare economics and existence of an equilibrium for a competitive economy", Metroeconomica, 12:92–97.
Negishi, T. (1961), "Monopolistic competition and general equilibrium theory", Review of Economic Studies, 28:196–201.
Negishi, T. (1966), "General equilibrium in a monetary economy – existence and dichotomy", Keizaigaku Ronshū (The Journal of Economics), 32:27–53 [in Japanese].

Negishi, T. (1972), General equilibrium theory and international trade. Amsterdam: North-Holland.

Neisser, H. (1932), "Lohnhöhe und Beschäftigungsgrad im Marktgleichgewicht", Weltwirtschaftliches Archiv, 36:415–455.

Neuefeind, W. (1975), "Increasing returns to scale, Part II, A general model", Center for Research in Management Science, University of California, Berkeley.

Neuefeind, W. (1977), Non-convexities and decentralization in general equilibrium analysis. Habilitationsschrift Universität Bonn.

Neuefeind, W. (1980), "Notes on existence of equilibrium proofs and the boundary behavior of supply", Econometrica, 48:1831–1837.

Von Neumann, J. (1928), "Zur Theorie der Gesellschaftsspiele", Mathematische Annalen, 100:295–320. Translated in: A. W. Tucker and R. D. Luce, eds., Contributions to the Theory of Games IV. Princeton (1959).

Von Neumann, J. (1937), "Über ein ökonomisches Gleichungssystem und eine Verallgemeinerung des Brouwerschen Fixpunktsatzes", Ergebnisse eines mathematischen Kolloquiums, No. 8:73–83. Translated in: Review of Economic Studies, 13 (1945):1–9.

Nikaido, H. (1956), "On the classical multilateral exchange problem", Metroeconomica, 8:135–145.

Nikaido, H. (1957), "A supplementary note to 'On the classical multilateral exchange problem'", Metroeconomica, 9:209–210.

Nikaido, H. (1959), "Coincidence and some systems of inequalities", Journal of the Mathematical Society of Japan, 11:354–373.

Nikaido, H. (1962), "A technical note on the existence proof for competitive equilibrium", The Economic Studies Quarterly, 8:54–58.

Nikaido, H. (1964), "Generalized gross substitutability and extremization", in: M. Dresher, L. S. Shapley, and A. W. Tucker, eds., Advances in game theory. Princeton University Press, 55–68.

Nikaido, H. (1968), Convex structures and economic theory. Academic Press, New York.

Nikaido, H. (1975), Monopolistic competition and effective demand. Princeton University Press.

Nishimura, K. (1978), "A further remark on the number of equilibria of an economy", International Economic Review, 19:679–685.

Nishino, H. (1971), "On the occurrence and the existence of competitive equilibria", Keio Economic Studies, 8:33–67.

Novshek, W. (1978), Essays on equilibrium with free entry. Ph.D. dissertation, Northwestern University.

Novshek, W. (1980), "Cournot equilibrium with free entry", Review of Economic Studies, 47:473–486.

Novshek, W. and H. Sonnenschein (1978), "Cournot and Walras equilibrium", Journal of Economic Theory, 19:223–266.

Novshek, W. and H. Sonnenschein (1980), "Small efficient scale as a foundation for Walrasian equilibrium", Journal of Economic Theory, 22:243–255.

Oddou, C. (1976), "Teorèmes d'existence et d'equivalence pour des économies avec production", Econometrica, 44:265–281.

Okuno, M. (1976), "General equilibrium with money: Indeterminacy of price level and efficiency", Journal of Economic Theory, 12:402–415.

Osana, H. (1973), "On the boundedness of an economy with externalities", Review of Economic Studies, 40:321–331.

Osana, H. (1977), "Competitive equilibrium with Marshallian externalities", Keio Economic Studies, 14:31–40.

Paquin, L. T. (1978), "Monetary and value theory: An attempt at integration – the stationary equilibrium case", Université Laval.

Pazner, E. A. and D. Schmeidler (1975), "Competitive analysis under complete ignorance", International Economic Review, 16:246–257.

Pearce, I. F. and J. Wise (1973), "On the uniqueness of competitive equilibrium: Part I. Unbounded demand", Econometrica, 41:817–828.

Pearce, I. F. and J. Wise (1974), "On the uniqueness of competitive equilibrium: Part II. Bounded demand", Econometria, 42:921–932.

Peleg, B. and M. E. Yaari (1970), "Markets with countably many commodities", International Economic Review, 11:369–377.

Pethig, R. (1979), "Environmental management in general equilibrium: A new incentive compatible

approach", International Economic Review, 20:1–27.
Postlewaite, A. and D. Schmeidler (1978), "Approximate efficiency of non-Walrasian Nash equilibria", Econometria, 46:127–135.
Quirk, J. and R. Saposnik (1968), Introduction to general equilibrium theory and welfare economics. New York: McGraw-Hill.
Rader, T. (1964), "Edgeworth exchange and general economic equilibrium", Yale Economic Essays, 4:133–180.
Rader, T. (1972), Theory of general economic equilibrium. New York: Academic Press.
Rader, T. and J. Little (1979), "Tiebout–Lindahl equilibrium and the optimal location of population", Department of Economics, Washington University.
Radner, R. (1968), "Competitive equilibrium under uncertainty", Econometrica, 36:31–58.
Radner, R. (1972), "Existence of equilibrium of plans, prices, and price expectations in a sequence of markets", Econometrica, 40:289–303.
Radner, R. (1974), "Market equilibrium and uncertainty: Concepts and problems", in: M. D. Intriligator and D. A. Kendrick, eds., Frontiers of quantitative economics, Vol. II. Amsterdam: North-Holland, 43–105.
Radner, R. (1979), "Rational expectations equilibrium: Generic existence and the information revealed by prices", Econometrica, 47:655–678.
Rashid, S. (1978), "Existence of equilibrium in infinite economies with production", Econometrica, 46:1155–1164.
Richter, D. K. (1975), "Existence of general equilibrium in multi-regional economies with public goods", International Economic Review, 16:201–221.
Richter, D. K. (1978), "Existence and computation of a Tiebout general equilibrium", Econometrica, 46:779–805.
Roberts, D. J. (1973), "Existence of Lindahl equilibrium with a measure space of consumers", Journal of Economic Theory, 6:355–381.
Roberts, J. and A. Postlewaite, (1976), "The incentives for price-taking behavior in large exchange economies", Econometrica, 44:115–128.
Roberts, J. and H. Sonnenschein (1977), "On the foundations of the theory of monopolistic competition", Econometrica, 45:101–114.
Rockafellar, R. T. (1970), Convex analysis. Princeton University Press.
Roemer, J. E. (1980), "A general equilibrium approach to Marxian economics", Econometrica, 48:505–530.
Rosen, J. B. (1965), "Existence and uniqueness of equilibrium points for concave N-person games", Econometrica, 33:520–534.
Ruys, P. H. M. (1971), "A procedure for an economy with collective goods only", Research Memorandum, Tilburg University.
Ruys, P. H. M. (1972), "On the existence of an equilibrium for an economy with public goods only", Zeitschrift für Nationalökonomie, 32:189–202.
Ruys, P. H. M. (1974), Public goods and decentralization. Tilburg University Press.
Ruys, P. H. M. and H. N. Weddepohl (1978), "Economic theory and duality", in: Convex analysis and mathematical economics. Research Memorandum, Tilburg University.
Saito, M. (1961), "Professor Debreu on 'Theory of value': A review article", International Economic Review, 2:231–237.
Scarf, H. E. (1967), "On the computation of equilibrium prices", in: W. J. Fellner, ed., Ten Economic Studies in the tradition of Irving Fisher. Wiley, 207–230.
Scarf, H. E. (1967), "The approximation of fixed points of a continuous mapping", SIAM Journal of Applied Mathematics, 15:1328–1343.
Scarf, H. E. (1980), "The computation of equilibrium prices: An exposition", Chapter 21 in this Handbook.
Scarf, H. E. (with the collaboration of T. Hansen) (1973), The computation of economic equilibria. New Haven: Yale University Press.
Schlesinger, K. (1935), "Über die Produktionsgleichungen der ökonomischen Wertlehre", Ergebnisse eines mathematischen Kolloquiums, No. 6:10–11.
Schmeidler, D. (1969), "Competitive equilibria in markets with a continuum of traders and incomplete preferences", Econometrica, 37:578–585.

Schmeidler, D. (1970), "Fatou's lemma in several dimensions", Proceedings of the American Mathematical Society, 24:300–306.

Schweizer, U., P. Varaiya, and J. Hartwick (1976), "General equilibrium and location theory", Journal of Urban Economics, 3:285–303.

Shafer, W. J. (1976), "Equilibrium in economies without ordered preferences or free disposal", Journal of Mathematical Economics, 3:135–137.

Shafer, W. J. and H. F. Sonnenschein (1975), "Some theorems on the existence of competitive equilibrium", Journal of Economic Theory, 11:83–93.

Shafer, W. J. and H. F. Sonnenschein (1975), "Equilibrium in abstract economies without ordered preferences", Journal of Mathematical Economics, 2:345–348.

Schafer, W. J. and H. F. Sonnenschein (1976), "Equilibrium with externalities, commodity taxation, and lump sum transfers", International Economic Review, 17:601–611.

Schafer, W. J. and H. F. Sonnenschein (1980), "Aggregate demand functions", Chapter 14 in this Handbook.

Shaked, A. (1976), "Absolute approximations to equilibrium in markets with non-convex preferences", Journal of Mathematical Economics, 3:185–196.

Shapley, L. and H. Scarf (1974), "On cores and indivisibility", Journal of Mathematical Economics, 1:23–37.

Shitovitz, B. (1973), "Oligopoly in markets with a continuum of traders", Econometrica, 41:467–501.

Shoven, J. B. (1974), "A proof of the existence of a general equilibrium with ad valorem commodity taxes", Journal of Economic Theory, 8:1–25.

Shoven, J. B. and J. Whalley (1973), "General equilibrium with taxes: A computational procedure and an existence proof", Review of Economic Studies, 40:475–489.

Silvestre, J. (1977), "General monopolistic equilibrium under non-convexities", International Economic Review, 18:425–434.

Silvestre, J. (1977), "A model of general equilibrium with monopolistic behavior", Journal of Economic Theory, 16:425–442.

Silvestre, J. (1978), "Fixprice analysis: A synopsis of three solution concepts", Center for Research in Management Science, University of California, Berkeley.

Silvestre, J. (1978), "Fixprice analysis: The classification of disequilibrium prices", Center for Research in Management Science, University of California, Berkeley.

Silvestre, J. (1978), "Increasing returns in general non-competitive analysis", Econometrica, 46:397–402.

Slutsky, S. (1977), "A voting model for the allocation of public goods: Existence of an equilibrium", Journal of Economic Theory, 14:299–325.

Smale, S. (1974), "Global analysis and economics IIA", Journal of Mathematical Economics, 1:1–14.

Smale, S. (1976), "A convergent process of price adjustment and global Newton methods", Journal of Mathematical Economics, 3:107–120.

Smale, S. (1977), "Convergent process of price adjustment and global Newton methods", in: M. D. Intriligator, ed., Frontiers of Quantitative Economics, Vol. IIIA. Amsterdam: North-Holland, 191–205.

Smale, S. (1980), "Global analysis and economics", Chapter 8 in this Handbook.

Sondermann, D. (1974), "Economies of scale and equilibria in coalition production economies", Journal of Economic Theory, 8:259–291.

Sondermann, D. (1974), "Temporary competitive equilibrium under uncertainty", in: J. Drèze, ed., Allocation under uncertainty: Equilibrium and optimality. New York: Wiley, 229–253.

Sonnenschein, H. F. (1971), "Demand theory without transitive preferences, with applications to the theory of competitive equilibrium", in: J. Chipman, L. Hurwicz, M. K. Richter, and H. F. Sonnenschein, eds., Preferences, Utility, and Demand. Harcourt-Brace-Jovanovich, 215–223.

Sonnenschein, H. F. (1977), "Some recent results on the existence of equilibrium in finite purely competitive economies", in: M. D. Intriligator, ed., Frontiers of quantitative economics, Vol. IIIA. Amsterdam: North-Holland, 103–109.

Sontheimer, K. C. (1971), "An existence theorem for the second best", Journal of Economic Theory, 3:1–22.

Sontheimer, K. C. (1971), "The existence of international trade equilibrium with trade tax-subsidy distortions", Econometria, 39:1015–1035.

Sontheimer, K. C. (1972), "On the determination of money prices", Journal of Money, Credit, and Banking, 4:489–508.
Stackelberg, H. von (1933), "Zwei kritische Bemerkungen zur Preisthorie Gustav Cassels", Zeitschrift für Nationalökonomie, 4:456–472.
Starr, R. M. (1969), "Quasi-equilibria in markets with non-convex preferences", Econometrica, 37:25–38.
Starr, R. M. (1974), "The price of money in a pure exchange monetary economy with taxation", Econometrica, 42:45–54.
Starrett, D. (1973), "Inefficiency and the demand for 'money' in a sequence economy", Review of Economic Studies, 40:437–448.
Stigum, B. P. (1969), "Competitive equilibria under uncertainty", Quarterly Journal of Economics, 83:533–561.
Stigum, B. P. (1972), "Resource allocation under uncertainty", International Economic Review, 13:431–459.
Stigum, B. P. (1972), "Competitive equilibria with infinitely many commodities", Metroeconomica, 24:221–244.
Stigum, B. P. (1973), "Competitive equilibria with infinitely many commodities (II)", Journal of Economic Theory, 6:415–445.
Stigum, B. P. (1974), "Competitive resource allocation over time under uncertainty", in: M. S. Balch, D. L. McFadden, and S. Y. Wu, eds., Essays on economic behavior under uncertainty. Amsterdam: North-Holland, 301–331.
Takayama, A. (1974), Mathematical Economics. Dryden.
Takayama, A. and M. El-Hodiri (1968), "Programming, Pareto optimum and the existence of competitive equilibria", Metroeconomica, 20:1–10.
Tal, Y. (1978), Public goods and cooperative solutions in large economies. Ph.D. dissertation, The Johns Hopkins University.
Todd, M. J. (1977), "Existence and computation of economic equilibria", CORE Discussion Paper No. 7725.
Todd, M. J. (1979), "A note on computing equilibria in economies with activity analysis models of production", Journal of Mathematical Economics, 6:135–144.
Townsend, R. M. (1979), "Optimal contracts and competitive markets with costly state verification", Journal of Economic Theory, 21:265–293.
Ulph, A. and D. Ulph (1978), "Efficiency, inessentiality and the 'Debreu property' of prices", in: G. Schwödiauer, ed., Equilibrium and disequilibrium in economic theory. D. Reidel, 337–359.
Uzawa, H. (1956), "Note on the existence of an equilibrium for a competitive economy", Department of Economics, Stanford University.
Uzawa, H. (1960), "Walras' tatonnement in the theory of exchange", Review of Economic Studies, 27:182–194.
Uzawa, H. (1962), "Walras' existence theorem and Brouwer's fixed point theorem", Economic Studies Quarterly, 8:59–62.
Uzawa, H. (1962), "Aggregative convexity and the existence of competitive equilibrium", Economic Studies Quarterly, 12:52–60.
Varian, H. R. (1975), "On persistent disequilibrium", Journal of Economic Theory, 10:218–228.
Varian, H. R. (1975), "A third remark on the number of equilibria of an economy", Econometrica, 43:985–986.
Varian, H. R. (1976), "On balanced inflation", Economic Inquiry, 14:45–51.
Varian, H. R. (1977), "Non-Walrasian equilibria", Econometrica, 45:573–590.
Varian, H. R. (1978), Microeconomic analysis. Norton.
Wagstaff, P. (1975), "A uniqueness theorem", International Economic Review, 16:521–524.
Wald, A. (1935), "Über die eindeutige positive Lösbarkeit der neuen Produktionsgleichungen", Ergebnisse eines mathematischen Kolloquiums, No. 6:12–20.
Wald, A. (1936), "Über die Produktionsgleichungen der ökonomischen Wertlehre", Ergebnisse eines mathematischen Kolloquiums, No. 7:1–6.
Wald, A. (1936), "Über einige Gleichungssysteme der mathematischen Ökonomie", Zeitschrift für Nationalökonomie, 7:637–670. Translated as: "On some systems of equations of mathematical economics", Econometrica, 19 (1951): 368–403.

Walras, L. (1874-7), Eléments d'économie politique pure. Lausanne: Corbaz. Translated as: Elements of pure economics. Chicago: Irwin (1954).

Wan, H. Y. (1965), "An elementary proof of the existence and uniqueness of competitive equilibrium in Graham's model of world trade", Econometrica, 33:238–240.

Wan, H. Y. (1970), "The existence of competitive equilibrium with externalities", Department of Economics, Cornell University.

Wan, H. Y. (1972), "A note on trading gains and externalities", Journal of International Economics, 2:173–180.

Weddepohl, H. N. (1972), "Duality and equilibrium", Zeitschrift für Nationalökonomie, 32:163–187.

Weddepohl, H. N. (1973), "Dual sets and dual correspondences, and their application to equilibrium theory", Research Memorandum, Tilburg University.

Weddepohl, H. N. (1979), "An equilibrium model with fixed labor time", Econometrica, 47:921–938.

Weddepohl, H. N (1978), "Equilibrium in a market with incomplete preferences where the number of consumers may be finite", in: G. Schwödiauer, ed., Equilibrium and disequilibrium in economic theory. D. Reidel, 15–26.

Westhoff, F. (1977), "Existence of equilibria in economies with a local public good", Journal of Economic Theory, 14:84–112.

Wieczorek, A. (1977), "Equilibria in an exchange economy with many agents: A game-theoretical method for proving the existence" in: M. Aoki and A. Marzollo, eds., New trends in dynamic system theory and economics. Academic Press.

Wilson, R. (1978), "The bilinear complementarity problem and competitive equilibria of piecewise linear economic models", Econometrica, 46:87–103.

Wilson, R. (1978), "Competitive exchange", Econometrica, 46:577–585.

Wooders, M. (1978), "Equilibria, the core and jurisdiction structures in economies with a local public good", Journal of Economic Theory, 18:328–348.

Yamazaki, A. (1976), "Production economy with competitive firm structures", Department of Economics, University of Rochester.

Yamazaki, A. (1977), General Equilibrium Analysis of Large Nonconvex Economies. Ph.D. dissertation, University of Rochester.

Yamazaki, A. (1978), "An equilibrium existence theorem without convexity assumptions", Econometrica, 46:541–555.

Yamazaki, A. (1979), "Regularizing effect of aggregation on mean demand and existence of competitive equilibria in large nonconvex economies", Econometrica, forthcoming.

Yamazaki, A. (1979), "Diversified consumption characteristics and conditionally dispersed endowment distribution: Regularizing effect and existence of equilibria", Department of Economics, University of Illinois at Chicago Circle.

Younès, Y. (1975), "On the role of money in the process of exchange and the existence of a non-Walrasian equilibrium", Review of Economic Studies, 42:489–501.

Zeuthen, F. (1933), "Das Prinzip der Knappheit, technische Kombination, und ökonomishe Qualität", Zeitschrift für Nationalökonomie, 4:1–24.

Chapter 16

STABILITY

FRANK HAHN*

Cambridge University

0. Introduction

We are concerned with endogenous processes operating in an economy which may bring about an equilibrium. Indeed, the latter is often implicitly or explicitly defined as a stationary point of such a process. The most famous and most discussed is "the law of demand and supply". If at prevailing market signals there is a positive excess demand in a market price there will rise; if a negative excess demand, it will fall. In a partial equilibrium framework when all other prices are held fixed, it seemed that this process would bring about equilibrium provided only that the excess demand in this market was a diminishing function of its price. But even this modest conclusion was at risk until the actual dynamic process was properly specified [Samuelson (1941, 1942)]. For instance, it matters to the conclusion whether it is written as a differential or a difference equation. In a general equilibrium framework, i.e. in a situation where many prices are changing simultaneously, the analysis of the "law of demand and supply" becomes more complex and convergence to equilibrium more problematical. The main available results are discussed below. We shall have to reach the conclusion that there is a large class of economies which have unstable equilibria under the most popular form of the price mechanism. This conclusion is a by-product of recent theorems [Sonnenschein (1972, 1973), Debreu (1974), Mantel (1974)] which, roughly summarised, show that any arbitrary set of excess demand functions continuous on the interior of the simplex and satisfying Walras' Law, can be generated by utility maximising behavior of agents for some utility functions and endowments (see below).

The most popular form of modelling the price mechanism was, until recently, the *tâtonnement*. The device is resorted to for two reasons: the assumption of perfect competition (under which agents believe that they can trade what they want to at current prices) left no room for any actual agent to change price and so

*Kenneth Arrow, Franklin Fisher and Michael Intriligator made very useful comments on a first draft and I am grateful to them.

Handbook of Mathematical Economics, vol. II, edited by K.J. Arrow and M.D. Intriligator

the fictitious auctioneer appears. Secondly, it seemed difficult to make the transition from planned trades and production to actual trades and production when plans were inconsistent. Moreover, to incorporate this transition into the formal analysis is very hard. Hence, no actual trading or production at "false" prices was allowed in the process. While these reasons justify the tâtonnement as a first approach to a complex problem, it is obvious that it is incapable of providing a satisfactory answer to the stability question in most actual economies.

The *tâtonnement* is easiest to modify or abandon altogether in pure exchange economies. If one retains the auctioneer but introduces the assumption that at each date markets are *orderly* [Hahn and Negishi (1962b)], one can show convergence to a competitive equilibrium whatever the exact form of the excess demand functions. Markets are orderly if no agent is restrained in his planned demand (supply) of a good when that good is in aggregate excess supply (demand). Convergence is proved by noting that the mechanism acts like a gradient process. It will be discussed below (Section 3). Recently this approach has been extended to allow for production [Fisher (1976b)].

One does without the auctioneer altogether by considering economies without prices and assuming that exchange will take place as long as there are two agents who can do so to mutual advantage and only then [Uzawa (1962), Smale (1976)]. Once again this *Edgeworth process* is a gradient process, and its asymptotic state is a Pareto efficient allocation. The process can be formulated stochastically and, with certain assumptions, production can be incorporated [Hurwicz, Radner and Reiter (1975)]. This approach is discussed below.

Neither of these approaches is very satisfactory. The assumption that markets are at all times orderly and the retention of the auctioneer cast doubt on the realism and relevance of the first approach. It is true that Fisher (1970, 1972) has gone some way to allow prices to be set by firms but he needs some strong assumptions. In a single market context Diamond (1971) has studied adjustments as a search process with firms setting prices. But it does not seem possible to extend this approach to the multi-market case. In the second approach, we study adjustment without prices (markets). This entails a considerable loss in descriptive power and makes the results more interesting for the theory of planning than for the functioning of an actual economy. While some special other models exist, we shall have to conclude that we still lack a satisfactory descriptive theory of the invisible hand.

This conclusion will be reinforced when we look more carefully at the specification of the economy in the context of which adjustment processes have often been studied. This is the Arrow–Debreu specification in which goods are distinguished by physical characteristics, location, date of delivery and state of nature. It is a commonplace that many of these goods lack markets in actual economies. It is also then clear that many planned excess demands are not public knowledge and that one cannot sensibly retain the traditional price adjustment schema. An alternative is to study a sequence of short period equilibria – a procedure familiar from the theory of growth. Typically one now abandons the attempt to explain how short period equilibrium is established. Even so considerable technical problems remain which arise because short period equilibrium is not generally unique nor is the set of such equilibria convex. One must now rely on being able to make a continuous selection from the short period equilibrium correspondence. The aim is to study convergence to long-run equilibrium whether in a deterministic or in a stochastic setting. Expectations and their formation are central to this kind of approach once durable goods and production lags are included. One is then concerned with convergence to a rational expectation equilibrium. This approach has, as yet, borne little fruit and we do not discuss it in detail.

The conclusion of the ensuing survey will be this: a great deal of skilled and sophisticated work has gone into the study of processes by which an economy could attain an equilibrium. Some of the (mainly) technical work will surely remain valuable in the future. But the whole subject has a distressing ad hoc aspect. There is at present no satisfactory axiomatic foundation on which to build a theory of learning, of adjusting to errors and of delay times in each of these. It may be that in some intrinsic sense such a theory is impossible. But without it this branch of the subject can aspire to no more than the study of a series of suggestive examples.

There is a vast variety of dynamic and adjustment models in the literature. It is not possible to consider them all or indeed a large majority. The following pages are mainly concerned with work in the context of Walrasian General Equilibrium models. Thus, there will be, for instance, no discussion of the large growth literature. Excluded also is the large planning literature [Heal (1973), Drèze and Vallee Poussin (1971)] although an account will be given of certain abstract adjustment mechanisms.

1. Technical preliminaries

Much of the literature has modelled the dynamics of the economy by means of differential or difference equations. Recently there have been extensions to differential and difference correspondences [e.g. Henry (1972, 1973, 1974)]. Also, some progress has been made by the use of Global Analysis, which is surveyed in Chapter 8 of this Handbook by Smale and will not be discussed here.

Let $x \in R^n$ and let a dot over a symbol denote differentiation with respect to time. One is interested in the behavior of

$$\dot{x}=F(x) \tag{1.1}$$

where F is an n-dimensional vector valued function which is continuous on R^n. One call $x^0 \in R^n$ a *critical point* or *equilibrium point* of (1.1) if $F(x^0)=0$. We think of time (t) as being in the interval $(0, \infty)$. A solution of (1.1) is a function from $I \times R^n \to R^n$:

$$x(t, x(0))$$

where $x(0)$ is an element of R^n at $t=0$.

Definition D.1.1

$F(x)$ is said to be *Lipschitzian* (obey a Lipschitz condition) if there exists a constant L such that

$$|F(x)-F(x')| \leqq L|x-x'|, \quad \text{for all } x, x' \in R^n.$$

Theorem T.1.1

If $F(x)$ is Lipschitzian, then the solution $x(t, x(0))$ is unique and continuous in $x(0)$.

Proof

Any text, e.g. Sansom and Conti (1964).

We shall first discuss matters on the hypothesis that $F(x)$ is Lipschitzian. This is not in fact the case in many economic applications, and the issue is taken up again below.

Let $\bar{x}(0)$ be a critical or equilibrium point of $F(x)$. Then one has

Definition D.1.2

The solution $x(t, \bar{x}(0))$ is said to be *stable* if for every $\varepsilon > 0$ there exists a δ (depending on ε), such that for all solutions for which

$$|x(0)-\bar{x}(0)| \leqq \delta,$$

one has

$$|x(t, x(0))-x(t, \bar{x}(0))| \leqq \varepsilon, \quad \text{all } t \geqq 0.$$

Definition D.1.3

The solution $x(t, \bar{x}(0))$ is said to be *asymptotically stable* if (a) it is stable and (b) for all $x(0)$ such that $|x(0)-\bar{x}(0)| \leqq \alpha > 0$ one has

$$\lim_{t \to \infty} |x(t, x(0))-x(t, \bar{x}(0))| = 0. \tag{1.2}$$

Definition D.1.4

The solution $x(t, \bar{x}(0))$ is *globally (asymptotically) stable* if (a) it is stable and (b) (1.2) holds for any $x(0)$ in the domain of definition of F.

These definition are concerned with the stability of the particular solution originating at a critical point of $F(x)$. But we shall also be interested in what we shall call the stability of the process (1.1). To motivate this, consider the case

$$\lim_{t\to\infty} x(t, x(0)) = x^*,$$

where x^* is finite. Then it is easy to see that x^* is a critical point of $F(x)$. For suppose not and that say $F_k(x^*)>0$, where F_k is the kth coordinate of F. Since F is continuous,

$$\lim_{t\to\infty} F_k(x(t, x(0))) = F_k\Big(\lim_{t\to\infty} x(t, x(0))\Big) = F_k(x^*) = F_k^* \text{ say.}$$

By our hypothesis given $\varepsilon>0$ such that $F_k^* - \varepsilon > 0$ there is a $\bar{t}$, depending on ε, such that

$$|F_k(x(t, x(0))) - F_k^*| < \varepsilon, \quad \text{all } t \geqq \bar{t},$$

or

$$F_k(x(t, x(0))) > F_k^* - \varepsilon, \quad \text{all } t \geqq \bar{t}, \tag{1.3}$$

or

$$\dot{x}_k > F_k^* - \varepsilon.$$

Integrating (1.3) from $\bar{t}$ to t', $t' > \bar{t}$, yields

$$x_k(t') = x_k(\bar{t}) + (F_k^* - \varepsilon)(t' - \bar{t}),$$

whence $x_k(t')$ grows without bound as $t' \to \infty$. This contradicts the convergence of the solution to a finite limit. The case $F_k(x^*)<0$ is treated analogously. We have therefore established:

Theorem T.1.2

If the solution $x(t, x(0))$ converges as $t \to \infty$ then it converges on a critical point of $F(x)$.

This then motivates:

Definition D.1.5

The process (1.1) is *globally stable* if for all $x(0)$ in the domain of definition of F the solution $x(t, x(0))$ converges.

This still leaves at least once case of possible interest. To introduce it, assume that all solutions of (1.1) stay in a compact cube. Then $x(t, x(0))$ has a convergent subsequence:

$$\lim_{\lambda\to\infty} x(t_\lambda, x(0)) = x^0 \text{ say.}$$

Then we call

$$x^0(y) = x(t, x^0) \tag{1.4}$$

a *limit path* of $x(t, x(0))$. Assuming $F(x)$ Lipschitzian and using T.1.1 we can write

$$\begin{aligned} x^0(t) &= x(t, x^0) = x\left(t, \lim_{\lambda\to\infty} x(t_\lambda, x(0))\right) \\ &= \lim_{\lambda\to\infty} x(t, x(t_\lambda, x(0))) = \lim_{\lambda\to\infty} x(t+t_\lambda, x(0)). \end{aligned}$$

From this we see that every point of (1.4) is a limit point of $x(t, x(0))$. We now have

Definition D.1.6

The process (1.1) is *quasi-globally stable* if for all $x(0)$ in the domain of definition of F all limit points $x(t, x(0))$ are critical (equilibrium) points of $F(\cdot)$. [All points of the limit path of $x(t, x(0))$ are equilibria of F.]

If the process is quasi-globally stable then every solution approaches arbitrarily closely to the set of equilibria of F as $t\to\infty$. This can only occur if these equilibria are not isolated. Suppose that $F(x^*) = F(x^{**}) = 0$, $x^* \neq x^{**}$, where x^* and x^{**} are limit points. Then x^* and x^{**} are isolated if there are small closed neighbourhoods $N(x^*)$ and $N(x^{**})$ of x^* and x^{**} in R^n such that (i) $N(x^*) \cap N(x^{**}) = \varnothing$ and (ii) x^* is the only equilibrium in $N(x^*)$ and x^{**} is the only equilibrium in $N(x^{**})$. By definition there are infinitely many points of the solution path which are in $N(x^*)$ and infinitely many which are in $N(x^{**})$. It follows that there are infinitely many points on the boundary of $N(x^*)$. The latter is bounded whence there is a convergent subsequence to x^{***} on the boundary. But then x^{***} is a limit point of the solution and by quasi-global stability an equilibrium of F. This contradicts the assumption that the equilibria are isolated.

Theorem T.1.3

If $F(x)$ has isolated equilibria then if the process (1.1) is quasi-globally stable it is globally stable.

It now follows that in economic applications one will only be interested in quasi-global stability of the process when economies are not regular (Debreu). For a discussion of regular economies see Chapter 17 of this Handbook by Dierker.

The basic tool of stability analysis in economics has been Lyapounov's second method [Lyapounov (1947)]. For the case of (1.1) and the Lipschitzian assumption the method is quite elementary.

Definition D.1.7

A function $V: R^n \to R$ is called a *Lyapounov Function* if: (i) it is continuous; (ii) $V(x(t, x(0)))$ converges for all $x(0) \in C$ some $C \subset R^n$; and (iii) it is constant if and only if x is a critical point of $F(x)$.

Suppose that a Lyapounov Function exists for (1.1) and that all the solutions of (1.1) are in a bounded cube $C \subset R^n$. Then by (ii) of D.1.7

$$\lim_{t \to \infty} V(x(t, x(0))) - V^*, \quad \text{for all } x(0) \in C.$$

By property (i) of D.1.7 and the boundedness of the solutions and the Lipschitzian assumption,

$$\begin{aligned}\lim_{t \to \infty} V(x(t, x(0))) &= \lim_{\lambda \to \infty} V(x(t + t_\lambda, x(0))) \\ &= V\Big(\lim_{\lambda \to \infty} x(t + t_\lambda, x(0))\Big) = V\big(x(t, x^0)\big),\end{aligned}$$

where $x(t, x^0)$ is the limit path. We now have

$$V\big(x(t, x^0)\big) = V^*(x(0))$$

and so, by property (iii) of D.1.7, every point of the limit path is an equilibrium. Hence, taking account of our earlier discussion we can state:

Theorem T.1.4

Let the solutions of (1.1) all lie in $C \subset R^n$ for all t where C is bounded. Let $F(\cdot)$ satisfy a Lipschitz condition. If there exists a Lyapounov function for (1.1) then the process (1.1) is a quasi-globally stable. If, in addition, the equilibria of $F(\cdot)$ are isolated then the process is globally stable.

Now let x^* be a critical point of $F(x)$ and, without loss of generality, take $x^*=0$. We shall say that $V(x)$ is *positive definite* if

$$V(x) \geqq a(x), \quad \text{all } x \in C,$$

where $a(0)=0$ and $a(|x|)$ is continuous and increasing. Assume that $W(x)$ is well defined by

$$\lim_{h\to 0}\left[V(x(t+h,x(0)))-V(x(t,x(0)))\right]/h=W(x(t,x(0))).$$

We can state

Theorem T.1.5

Let $x^*=0$ be a critical point of $F(x)$. Then if there exists a positive definite Lyapounov function for (1.1) with $W(x)<0$ when $x\neq 0$, then the solution $x(t,x^*)$ is globally asymptotically stable.

Proof

Use Definition D.1.4 and that of positive definiteness and note that $V(x(t,x(0)))$ converges on zero as $t\to\infty$.

These theorems are, as we shall see, repeatedly used in the literature. They are appropriate when we are dealing with a simultaneous system of non-linear differential equations where one can only look for qualitative results. When the equations are linear then more classical theorems are available. Thus, for the system

$$\dot{x}=Ax, \tag{1.5}$$

where $x\in R^n$, A is $(n\times n)$ we have the classical result:

Theorem T.1.6

The solution $x(t,0)$ is globally asymptotically stable if and only if the real parts of the eigenvalues of A are negative.

In what follows we list some of the main stability theorems for A. Note that if S is non-singular, then $B=S^{-1}AS$ has the same eigenvalues as A. Hence, in the following results if the properties listed apply to B rather than to A the result still holds.

Theorem T.1.7

The real parts of the eigenvalues of A are negative if

(a) A is *negative definite*, ($x'Ax<0$ for all $x\neq 0$) or if A is *quasi-negative definite* ($x'(A+A')x<0$ for $x\neq 0$) [Hicks (1939), Samuelson (1941, 1942) and Arrow and Hurwicz (1958a)].

(b) The off-diagonal elements of A are positive, the diagonal elements are negative (A is a *Metzler matrix*) and there is $h \in R^n_+$ such that

$$Ah < 0 \tag{1.6}$$

[Metzler (1945), Hahn (1958) and Negishi (1958)].

(c) Let A^* be the matrix derived from A by substituting $|a_{ij}|$ for a_{ij} for all $i \neq j$. Then if there is $h \in R^n_+$ such that

$$A^*h \ll 0 \tag{1.7}$$

the assertion of the theorem holds. A is said to have *diagonal dominance* [McKenzie (1960(b)].

(d) Let the indices be divided into two disjoint groups J and J'. Assume that $a_{rs} > 0$ if $r \neq s$ and both r and s belong to either J or J' and that $a_{rs} < 0$ if $r \neq s$ and r and s belong to different groups. Let $S = \begin{bmatrix} I & 0 \\ 0 & -I \end{bmatrix}$ where each identity matrix is of the order of the cardinality of J and J', respectively. Then for some permutation of the rows and columns of A: $SAS = B$ with B a Metzler matrix. (Note $S = S^{-1}$.) Hence, if (1.6) holds for B the assertion of the theorem is true by (b) [Morishima (1952), Karlin (1959) and Mukherji (1972)].

(e) If A is a Metzler matrix satisfying (1.6) and $\hat{k}$ is a diagonal matrix then $\hat{k}A$ has all roots with negative real parts if and only if all elements of k are positive [Enthoven and Arrow (1956)].

(f) Let A_i be the left-hand upper principal minor of A of order i ($i = 1, \ldots, n$). Then there exists $\varepsilon > 0$ such that if the diagonal matrix $\hat{k}$ has the properties (i) $|k_i|/|k_{i-1}| < \varepsilon$, $i = 2 \ldots n$, and (ii) $k_i A_i / A_{i-1} < 0$, $i = 1 \ldots n$ ($A_0 = 1$), then $\hat{k}A$ has distinct real negative roots [Fisher and Fuller (1958), McFadden (1968) and Fisher (1972)].

To this we may add the following remarks:

Remark 1.1

(a) If a Metzler matrix satisfies (1.6) then its principal minors oscillate in sign. A matrix with this property is called *Hicksian*.

(b) If A is a Metzler matrix which satisfies (1.6) then $Ay > 0$ has no solution $y > 0$. From this one can establish that $-A^{-1}$ is a positive matrix [(Arrow–Hahn (1971)].

(c) The conditions listed in T.1.7 are those which in economic application can be given a proper interpretation. The mathematically general conditions, called the *Routh–Hurwicz* (1877, 1895) *conditions*, do not lend themselves to economic interpretation.

Finally, in this section let us return to the case when $F(x)$ in (1.1) may fail to be Lipschitzian. In the cases we are interested in this arises when one restricts $x(t)>0$, all t, and $F(\cdot)$ is not of the form that $F_i(x)>0$ whenever $x_i=0$. Accordingly, one imposes a different regime at such a boundary point, namely

$$\dot{x}_i=0, \quad \text{if } x_i=0 \text{ and } F_i(x)<0.$$

The process as a whole will now fail to be Lipschitzian at these boundary points.

We owe it to Henry (1972, 1973, 1974) that we now know that a solution continuous in the initial conditions nonetheless exists in the applications which concern us. Champsaur, Drèze and Henry (1977) have also studied the case of differential correspondences.

The mathematics of this argument is rather technical and appeals to an existence theorem by Attouch and Damlavian (1972). The interested reader should consult the cited literature.

2. Tâtonnement

2.1. Local results

The economy has ℓ goods, H households and F firms. Define $x^h \in R^\ell$ as the *net trade vector* of household h, i.e. its vector of demand minus the vector of its endowment; $x=\sum_i x^h$. Also, $y^f \in R^\ell$ is an *activity* of firm f where, as usual, positive components denote outputs and negative ones, inputs; $y=\sum y^f$. Let z the *aggregate excess demand vector* and s the *aggregate excess supply vector* by defined by

$$x=\sum_h x^h-\sum_f y^f=x-y \quad \text{and} \quad s=-z.$$

Of course z and s are elements of R^ℓ. It will also be convenient to define Z, X, Y and S, all in $R^{\ell-1}$, as the vectors z, x, y and s with their first component deleted. Also, $p\in R^\ell_+$ is a *price vector* and $P\in R^{\ell-1}_+$ is the vector $(1/p_1)p$ with its first component deleted (of course it is only well defined for $p_1>0$). The *endowment* of h is written as $\omega^h \in R^\ell{}_+$ and one defines

$$\omega=\sum \omega^h \quad \text{and} \quad \hat{\omega}=(\omega^1\ldots\omega^H).$$

We are here concerned with economies which have continuously differentiable excess supply (demand) functions. With some violence to notation, I write $s(p,\hat{\omega})$ and $z(p,\hat{\omega})$ as the excess supply and demand functions, respectively. We know that they are homogeneous of degree zero in p, i.e.

$$s(p,\hat{\omega})=s(1,P,\hat{\omega}) \tag{H}$$

and that they obey Walras' Law, i.e.

$$p\cdot s(p,\hat{\omega})=0, \quad \text{all } p\in R^{\ell}_{+}. \tag{W}$$

Let

$$\Delta=\{p\,|\,p\gg 0, \textstyle\sum p_i=1\}$$

and let $\delta\Delta$ denote the boundary of Δ. The following assumptions will often prove useful:

Assumption A.2.1 (Desirability)

Let p^{λ} be a sequence in Δ which converges to $\bar{p}\in\delta\Delta$. Then there exists an i in $(1\ldots\ell)$ such that $p_i^{\lambda}\to 0$ and $\overline{\lim}_{\lambda\to\infty} s_i(p^{\lambda})=-\infty$.

There are various forms of A.2.1 one can use [the above is based on Dierker (1972)]. Its virtue is that (a) equilibrium prices are strictly positive and (b) in the price adjustment process to be described the motions take boundary points into Δ.

The following is written down for the sake of completeness:

Definition D.2.1 (Equilibrium)

$p^0\in\Delta/\delta\Delta$ is an *equilibrium* if for each i

$$s_i(p^0,\hat{\omega})\geqq 0 \quad\text{and}\quad p_i^0 s_i(p^0,\hat{\omega})=0.$$

As already noted, A.2.1 implies $p^0\in\Delta$.

To proceed, we must place a restriction on the set of equilibria which is achieved by postulating a regular economy [Debreu, (1970); see also Chapter 17 of this Handbook by Dierker]:

Definition Δ*.2.2* (Regularity)

Let $DS(P)$ be the Jacobian of $S(P)$ (where the constant elements $1,\hat{\omega}$ have been omitted as arguments from the excess supply function). Then the economy is *regular* if for every equilibrium P^0 there is an open neighbourhood N of P^0 in $R^{\ell-1}_{+}$ such that $\det DS(P)\neq 0$, all P in N.

Debreu has shown that almost all economies with continuously differentiable excess demand functions are regular. The equilibria of a regular economy are isolated (see Section 1).

We first consider the simplest tâtonnement given by the differential equation system

$$\dot{P}_i = -S_i(P), \qquad i=2\ldots\ell \text{ for } P \gg 0. \tag{2.1}$$

Notice that $\hat{\omega}$ has been omitted as an argument because it is taken to be constant throughout. No actual trade or production takes place as long as $S(P) \neq 0$.

Assume that the economy is regular and let E be the set of equilibria $\bar{P}$.

Theorem T.2.1 [Varian (1975)]

Let the economy be regular and satisfy desirability. Then if for all $\bar{P} \in E$ one has det $DS(\bar{P}) > 0$, equilibrium is unique. E has only one member.

Proof (Sketch)

For each $\bar{P}$ in E define an index $I(\bar{P})$ as plus one if det $DS(\bar{P}) > 0$ and minus one if $DS(\bar{P}) < 0$. By the Poincaré–Hopf Index Theorem [e.g. Milnor (1972)] one has $\Sigma_E I(\bar{P}) = +1$. The theorem now follows.

Theorem T.2.1 [Dierker (1972)]

A regular economy with desirability has an odd number of equilibria.

Proof

Use the Poincaré–Hopf theorem. If e is the number of equilibria e must be odd since 1 is the sum of e terms of absolute value one.

Since the economy is regular and by desirability $\bar{P} \gg 0$, we may express (2.1) in some small neighbourhood of $\bar{P}$ by the linearisation

$$\dot{P}_i = -DS(\bar{P})(P - \bar{P}). \tag{2.2}$$

This is indeed the form in which most of tâtonnement stability problems were studied to begin with [e.g. Samuelson (1941)]. One now has

Corollary 2.2 [Dierker (1972)]

Let the economy be regular and satisfy desirability. Then if for all $\bar{P} \in E$ and $P(0)$ in a small neighbourhood of $\bar{P}$, the solution to (2.2): $P(t, P(0)) \to \bar{P}$ as $t \to \infty$ then E is a singleton (equilibrium is unique).

Proof

A necessary condition for $P(t, P(0)) \to \bar{P}$ as $t \to \infty$ is det $DS(\bar{P}) > 0$. Now apply Theorem T.2.1.

There is thus a close connection between uniqueness of equilibrium and local asymptotic stability of ((2.1). The economic conditions which in the sixties were found for the stability of (2.1) were thus also sufficient for the uniqueness of equilibrium. Since (2.2) is a linear system conditions listed in T.1.7 suffice for local asymptotic stability. The matrix A of that theorem is here $DZ(P)$. However, in now giving economic examples, it will sometimes be more convenient to work with $S(P)$.

2.1.1. *DS(P) is positive definite*

This case was first studied by Hicks (1939). It arises when all households have parallel linear Engel curves which pass though the origin. This in turn is a sufficient condition for the existence of community preference fields [Gorman (1953)] and allows one to treat the economy as if it consisted of one household only. If $\bar{p} \gg 0$ is an equilibrium, then from D.2.1 and (W):

$$p \cdot s(\bar{p}) \geqq p \cdot s(p). \tag{2.3}$$

From profit maximisation: $p \cdot y(p) \geqq p \cdot y(\bar{p})$ and so

$$p(x(p)-x(\bar{p})) \geqq 0.$$

Then from revealed preference for the "as if" single household,

$$\bar{p}(x(p)-x(\bar{p}))>0,$$

so using profit maximisation again: $\bar{p}\bar{y}(p) \geqq \bar{p}y(p)$ and hence

$$ps(p)>\bar{p}s(p). \tag{2.4}$$

Combining (2.3) and (2.4) and setting $p_1 = \bar{p}_1 = 1$, we obtain

$$(P-\bar{P})[S(P)-S(\bar{P})]>0$$

or, for all P close enough to $\bar{P}$,

$$(P-\bar{P})\, DS(\bar{P})(P-\bar{P})>0. \tag{2.5}$$

Hence, $DS(\bar{P})$ is quasi-positive definite [$DZ(\bar{P})$ is quasi-negative definite].

This can perhaps be better understood by considering the income term in the usual Slutsky equation. If μ_i^h is h's marginal propensity to consume i and θ_f^h is the share of h in the profits of firm f then

$$\left[\sum_f \theta_f^h y_j^f(\bar{P}) - x_j^h(\bar{P})\right]\mu_i^h \tag{2.6}$$

is the income term[1] in $\partial x_i^h/\partial p_j$. The Gorman hypothesis is $\mu_i^h = \mu_i$, all h, so that summing (2.6) over h and using $\Sigma_h \theta_f^h = 1$ yields $S_j(\bar{P})\mu_i = 0$ since $S_j(\bar{P}) = 0$. So in this case there are no income effects on aggregate demand and one is left with substitution effects which are known to give a negative definite matrix $[\partial d_i(P)/\partial P_j]$.

Now rewrite (2.6), summed over h, as

$$\sum_h \left[\sum_f \theta_f^h y_j^f(\bar{P}) - x_j^h(\bar{P})\right](\mu_i^h - \bar{\mu}_i),$$

where $\bar{\mu}_i$ is the average propensity to consume good i. This expression is a covariance between the marginal propensity to consume good i and the wealth effect of the change in the price of good j. This covariance may, in large economies, be zero even if the Gorman conditions are not met.

2.1.2. $DZ(\bar{P})$ is a Metzler matrix

The leading case of this arises when all goods are *gross substitutes*. That is, if $s_{ij} = \partial s_i/\partial p_j$ one has

$$s_i(p) \text{ finite implies } s_{ij}(p) < 0, \quad \text{all } i \neq j, \text{ and } s_{ii}(p) > 0. \tag{G.S}$$

By (H) one has $\Sigma_{j\neq 1} s_{ij}(\bar{p})\bar{p}_j = -s_{i1}\bar{p}_1 > 0$, all i, so setting $h = \bar{p}$:

$$DS(\bar{P})h \gg 0$$

(or equivalently $DZ(\bar{P})h \ll 0$). In a pure exchange economy in which households have Cobb–Douglas utility functions, (G.S.) holds. [But see Fisher (1972c).] It is also easy to prove (e.g. Arrow–Hahn) that

$$p_i = 0 \text{ implies } s_i(p) = -\infty, \tag{2.7}$$

so that in equilibrium one must have $\bar{p} \gg 0$.

[1]Recall the property of the partial derivatives of firm f's profit function $\pi^f(P)$ and h's budget constraint: $p \cdot x^h \leqq \Sigma_f \theta_f^h \pi^f(P)$.

2.1.3. $DZ(\bar{P})$ has diagonal dominance

Clearly (G.S.) implies diagonal dominance, but not vice versa. The interpretation to be put on the condition is this: at every equilibrium $\bar{p}\gg 0$ there exists units in which to measure goods such that the absolute value of the own price effect exceeds the sum of the absolute values of the non-numeraire cross effects. There seem to be no examples in the literature of utility and production functions which yield diagonal dominance other than of course the G.S. cases.

2.1.4. $DZ(\bar{P})$ is a Morishima matrix

The interpretation is that goods can be divided into two groups such that any two goods in the same group are gross substitutes and two goods of different groups are gross complements. However, care must be taken to check that the remainder of the condition of T.1.7(d), i.e. on the principal minors of A, are consistent with micro theory [Arrow–Hurwicz (1958), Kennedy (1970) and Morishima (1970)]. It turns out that this requires us to exclude the numeraire good from the Morishima conditions. For suppose not. Then by homogeneity:

$$\sum_{j\in J} z_{rj}p_j = -\sum_{j\in J'} s_{rj}p_j > 0, \quad \text{all } r\in J.$$

The inequality follows from $r\notin J'$ and the definitions. (If r and j are gross complements then $s_{rj}=\partial s_r/\partial p_j>0$.) In matrix notation

$$[z_{rj}]p(J)>0,$$

where $p(J)$ is the price sub-vector with all subscripts in J and $[z_{rj}]$ the Jacobian whose elements have both subscripts in J. Now $[z_{rj}]$ is a Metzler matrix but it cannot satisfy (1.6). For if it did, the above inequality would have no solution $p(J)>0$. Hence $[z_{rj}]$ will not satisfy the conditions on the principal minors of A. For if A is Hicksian $[z_{rj}]$ satisfies (1.6).

Morishima (1970) has shown that if his conditions apply to all goods other than the numeraire and if another (economically not very meaningful) condition replaces the postulate on the principal minors of A, that stability can be established.

Notice now that in view of T.2.1 one need only postulate any one of these conditions at every $\bar{P}\in E$ to deduce that E is a singleton which is locally asymptotically stable. [Caution: except for (G.S.) one needs to postulate desirability directly.]

An obvious application of T.1.7(e) occurs if we generalise (2.1) to

$$\dot{P}_i = -H_i(S_i(P)), \qquad i=2\ldots l,\ P\gg 0, \tag{2.1*}$$

where $H_i(0)=0$, $H_i'>0$, all P, and $S_i(P)H_i(S_i(P))>0$ for $S_i(P)\neq 0$. For local analysis, one obtains

$$\dot{P}_i = k_i DS_i(\bar{P})(P-\bar{P}); \qquad k_i = H_i'(S_i(\bar{P})), \qquad i=2\ldots\ell. \tag{2.2*}$$

One is now interested in the matrix $\hat{k}DS(\bar{P})$ and hence T.1.7(e) tells us that the G.S. case will still yield stability. It is easy to show that if $DS(\bar{P})$ is positive definite that local symptotic stability will continue to hold for the system (2.1*). To do so, let

$$V(P)=\tfrac{1}{2}\sum_{i\neq 1}(P_i-\bar{P}_i)^2/k_i. \tag{2.8}$$

Then for $\bar{P}\gg 0$ and P in a close enough neighbourhood of $\bar{P}$ one has, using (2.2*),

$$\dot{V}(P) = -(P-\bar{P})DS(\bar{P})(P-\bar{P})<0, \quad \text{for} \quad P\neq\bar{P}.$$

Since now $P(t)$ is clearly bounded, we see that $V(P)$ is a Lyapounov function and so by T.1.5 the equilibrium is (locally) stable under (2.1*). Finally, it is clear that the matrix $\hat{k}DS(\bar{P})$ preserves diagonal dominance.

A somewhat more interesting application of T.1.7(e) is the Enthoven-Arrow (1956) problem. It is assumed that the excess supply functions are given by

$$S_i(P, P_i'), \qquad i=2\ldots\ell,$$

where P_i' is the price of good i expected to rule in equilibrium. (The expectational assumptions of this model sit uncomfortably in a tâtonnement.) In equilibrium one must have $P=P'$.

Expectations are formed extrapolatively:

$$P_i' = P_i + \alpha_i\dot{P}_i, \qquad i=2\ldots\ell. \tag{2.9}$$

Lastly the price adjustment rule is

$$\dot{P}_i = -k_i S_i(P, P_i'), \qquad i=2\ldots\ell, \tag{2.10}$$

which, since only local analysis is involved, can be taken as derived from (2.1*). Notice that only the expected price of i enters the excess supply function for i. Let $S_{ii}'=\partial S_i/\partial P_i'$ and δ_{ij} be the Kronecker delta. Then taking a Taylor expansion of (2.10) and using (2.9), one obtains:

$$\dot{P}_i = -k_i\sum_j(S_{ij}+\delta_{ij}S_{ii}')(P_j-\bar{P}_j)-k_i\alpha_i S_{ii}'\dot{P}_i, \qquad i=2\ldots1. \tag{2.11}$$

If now T is the matrix with typical element $k_i(S_{ij}+\delta_{ij}S'_{ij})$, then one can write (2.11) as

$$\dot{P}=-\hat{m}T(P-\bar{P}), \tag{2.12}$$

where $\hat{m}$ is the diagonal matrix with elements $(1-k_i\alpha_i S'_{ii})$ on the diagonal. We know from T.1.7(e) that if T is a Metzler matrix such that for $h\gg 0$ one has $Th\gg 0$, then (2.12) is stable if and only if

$$1-k_i\alpha_i S'_{ii}>0, \quad \text{all } i.$$

This is the Arrow–Enthoven Theorem, and it shows that in this model the conditions of stability cannot be specified solely in terms of the elasticity of expectations [Hicks (1939)]. On the other hand, it is not easy to give economic meaning to expectation in a tâtonnement.

A different model with adaptive expectations has been investigated by Arrow and Nerlove (1958). Here, in the previous notation, the expectational hypothesis is given by

$$\dot{P}'_i=\beta_i(P_i-P'_i), \qquad \beta_i>0, \qquad i=2\ldots\ell, \tag{2.13}$$

and the excess demand now depends on P and P' so (2.10) becomes

$$\dot{P}_i=k_i Z_i(P,P'), \qquad i=2\ldots\ell. \tag{2.14}$$

Let $Z_{ij}=\partial Z_i/\partial P_j$ and $Z'_{ij}=\partial Z_i/\partial P'_j$. Expanding (2.14) around the equilibrium $P=P'=\bar{P}$, yields

$$\dot{P}_i=k_i\sum Z_{ij}(P_j-\bar{P}_j)+k_i\sum Z'_{ij}(P'_j-\bar{P}_j), \qquad i=2\ldots\ell, \tag{2.15}$$

and expanding (2.13):

$$\dot{P}'_i=\beta_i(P_i-\bar{P}_i)-\beta_i(P'_i-\bar{P}_i), \qquad i=2\ldots\ell. \tag{2.16}$$

So if q is the $2l$ vector $(p-\bar{p},\, p'-\bar{p})$ we may write (2.15) and (2.16) as

$$\dot{q}=Mq, \tag{2.17}$$

where M is the matrix of the partial differential coefficients in (2.15) and (2.16). Suppose now that M is a Metzler matrix (all goods, including future goods, are gross substitutes). Then by homogeneity one has

$$M\bar{q}=\{-k_2Z_{21}\ldots-k_\ell Z_{\ell 1},0\ldots 0\}<\{0\},$$

since $Z_{i1}>0$, all $i=2\ldots\ell$. [Here $\bar{q}=(\bar{P},\bar{P})$.] Hence, by T.1.7(b) the system is locally asymptotically stable. (Arrow and Enthoven did not use the homogeneity property, and their result was worded differently.)

Next let us consider an application of T.1.7(f). Recall that Hicks (1939) proposed the notion of *perfect stability* which required that $\mathrm{d}Z_i(\bar{P})/\mathrm{d}P_i<0$, when it is evaluated when all other prices are constant, when the price in one other market adjusts to preserve equilibrium there, when the prices in two other markets so adjust, and so on. More formally, consider $i=2$. Then one wants $Z_{22}<0$ when all prices other than P_2 are fixed. When P_3 is allowed to adjust one has

$$\begin{bmatrix} Z_{22} & Z_{23} \\ Z_{32} & Z_{33} \end{bmatrix}\begin{bmatrix} 1 \\ \mathrm{d}P_3/\mathrm{d}P_2 \end{bmatrix}\begin{bmatrix} 1 \\ \mathrm{d}P_3/\mathrm{d}P_2 \end{bmatrix}=\begin{bmatrix} \mathrm{d}Z_2/\mathrm{d}P_2 \\ 0 \end{bmatrix}. \tag{2.18}$$

So for $\mathrm{d}Z_2/\mathrm{d}P_2<0$ one wants $Z_{33}/\Delta<0$, where Δ is the determinant of the matrix in (2.18). Proceeding in this way, one verifies that perfect stability requires the Jacobian of $Z(\bar{P})$ to have left-hand principal minors which oscillate in sign. Samuelson (1941) pointed out that Hicks had given no dynamic story to justify this condition. However, as T.1.7 easily shows, gross substitutability or the existence of community preference fields suffice not only for local stability but also for perfect stability.

McFadden (1968), however, has proposed an alternative ingenious way of linking perfect stability to dynamic stability. To do this he formulates a difference equation version of price adjustment:

$$P_i(t+1)=P_i(t)+k_iZ_i(P), \qquad i=2\ldots\ell. \tag{2.19}$$

One is once again concerned with local stability and so with the Taylor expansion of (2.19) about $\bar{P}$. Now note that if the Jacobian of $Z(P)$ satisfies perfect stability then condition (ii) of T.1.7(f) leads one to look for $k_i>0$, all i, which can be ranked as required, i.e. $k_i/k_{i-1}<\varepsilon$, some ε. The interpretation now is that markets can be ranked by the speed of their price adjustment. So the dynamic story one can tell to motivate the idea of perfect stability is this: markets are ranked by their speed of adjustment. The first market adjusts first to approximate equilibrium and stays there; then the second market adjusts and so on.

It is assumed that fixing any subset of prices yields a unique equilibrium vector of prices for the markets in which prices are not fixed and that $Z(\bar{P})$ satisfies perfect stability. On these assumptions one has:

Theorem T.2.3 [McFadden (1968)]

There exists $\varepsilon>0$ such that if (a) $k_i<\varepsilon$ and $k_i/k_{i-1}<\varepsilon$, for $i=2\ldots\ell$, and (b) $|P(0)-\bar{P}|<\varepsilon$, then (2.19) is a locally asymptotically stable process.

The proof depends on a corollary to T.1.7(f) to take in the case where A is a continuous function on the compact set (of prices). By taking a Taylor expansion of (2.19) and noting that continuity ensures the preservation of the perfect stability property for P close to $\bar{P}$, one shows that the norm $[P(t)-\bar{P}]^2$ is a Lyapounov function, i.e.

$$|P(t+1)-\bar{P}|^2-|P(t)-\bar{P}|^2<0.$$

This follows at once from the application of T.1.7(f). McFadden also proves a global stability result.

Lastly there are two warnings. The work of Sonnenschein (1972, 1973), Debreu (1974) and Mantel (1974) has established that there are economies based on the actions of utility maximising agents which have equilibria that are not locally asymptotically stable. This is so because we can always choose a set of continuous excess demand functions which obey Walras' Law and which have Jacobians with roots not all of which have negative real parts. We can do this without violating the axioms of rational choice for some preferences and distribution of endowments. (I assume that there are more households than there are different goods.)

The second warning is this: stability may not be independent of the choice of numeraire. This is obvious, for instance, if the particular numeraire good does not exhibit gross substitutability while all the other goods do [see Arrow–Hahn (1971)]. Also, the adjustment process is different if no numeraire is stipulated, i.e. if one does not set $\dot{p}_i\equiv 0$, some i.

2.2. *Global results*

For global analysis when the price of some good may become zero the specification of the process (2.1) must be amplified to

$$\begin{aligned} &\dot{P}_i=0, && \text{if } P_i=0 \text{ and } S_i(P)>0, \\ &\dot{P}_i=-H_i(S_i(P)), && \text{otherwise.} \end{aligned} \qquad i=2\ldots\ell. \tag{2.1.N}$$

Here (2.1.N) will be called the *numeraire process* since $\dot{p}_1\equiv 0$. (It is assumed that the excess demand for good one is unbounded large when $p_1=0$.) It is to be contrasted with the case where all prices respond to excess supply:

$$\begin{aligned} &\dot{p}_i=0, && \text{if } p_i=0 \text{ and } s_i(p)>0, \\ &\dot{p}_i=-H_i(s_i(p)), && \text{otherwise.} \end{aligned} \qquad i=2\ldots\ell. \tag{2.1.A}$$

In both these processes $H_i(\cdot)$ has the same properties as in (2.1*). It is clear that both (2.1.N) and (2.1.A) will fail to satisfy a Lipschitz condition at a boundary of the price space where some price(s) may be zero. Henry (1972), however, has shown that in both systems with $s(p)$ Lipschitzian there are unique solutions determined by $P(0)$ [or $p(0)$] and that they are continuous in these.

A leading case which has been much studied arises when

$$H_i(s_i)=k_is_i, \qquad k_i>0. \tag{2.20}$$

It has the advantage that it leads to

Theorem T.2.4

The solution to (2.1.A) under (2.20) is bounded.

Proof

Differentiate $\sum p_i^2/k_i$ with respect to time. Using (2.1.A) and (2.20) as well as Walras' Law verify that this differential is zero.

In general, for (2.1.N) we need special assumptions to ensure the boundedness of the solution. For instance, one may stipulate that for some $k>0$ one has $S_1(P)<0$ whenever $\sum_2 P_i \geqq K$. Then using (2.1.N) and Walras' Law one has

$$\frac{\mathrm{d}}{\mathrm{d}t}\sum_2 P_i^2/k_i<0, \quad \text{for } \sum_{i=2} P_i \geqq K.$$

When we do not use the specialisation to (2.20), further assumptions on $H_i(\cdot)$ or $S_i(\cdot)$ are needed to ensure boundedness.

If an equilibrium is to be globally asymptotically stable it evidently must be unique. It is not surprising that the leading special cases for uniqueness also suffice for global stability, although of course the latter does not imply the former. It is natural to study the Lyapounov function

$$V(p)=\sum(p_i-\bar{p}_i)^2/k_i$$

for (2.1.A) specialised to (2.20). One finds

$$\dot{V}(p)=2\bar{p}s(p), \tag{2.21}$$

where one uses (2.1.A), (2.20) and Walras' Law. One is interested in the cases for which one can prove $\bar{p}\cdot s(p)<0$ for $p\neq k\bar{p}$, $k>0$, for then we may apply T.1.4.

(i) When the Gorman conditions for the existence of concave preference fields hold, then we know from earlier discussion that $\bar{p}s(p)<0$ for $\bar{p}\gg 0$ and $p\neq k\bar{p}$.

(ii) Suppose all goods are gross substitutes. We know $\bar{p} \gg 0$ and $s(p)$ is bounded above, hence $\bar{p}s(p)$ attains a maximum at some p. In view of (2.7), this p is strictly positive. One verifies that the necessary conditions for a maximum of $\bar{p}s(p)$ are satisfied at $p=\bar{p}$ when of course $\bar{p}s(\bar{p})=0$. So we now show that this maximising solution is unique (up to a scalar multiple). If not, let $\bar{\bar{p}} \neq k\bar{p}$ be another solution. Let

$$h=\bar{\bar{p}}_r/\bar{p}_r \geq \bar{\bar{p}}_i/\bar{p}_i, \quad \text{all } i.$$

Notice that by assumption one of the inequalities must be strict. By homogeneity and gross substitutability:[2]

$$0=s_r(\bar{p})=s_r(h\bar{p})<s_r(\bar{\bar{p}}). \tag{2.22}$$

But then Walras' Law implies

$$\sum \bar{p}_i s_{ir}(\bar{\bar{p}})=-s_r(\bar{\bar{p}})<0 \qquad (s_{ir}=\partial s_i/\partial p_r)$$

and the necessary condition for a maximum of $\bar{p}s(p)$ is violated at $\bar{\bar{p}}$. Hence $\bar{p}s(p)<0$, all $p \neq kp$. So we have

Theorem T.2.5

If community preference fields exist or if all goods are gross substitutes, then the unique equilibrium $\bar{p}$ is globally asymptotically stable under (2.1.A) specialised to (2.20). (Assume that in the Gorman case $\bar{p} \gg 0$.)

Now consider (2.1.N) specialized to (2.21). The natural Lyapounov candidate is

$$W(P)=\sum_{i=j}(P_i-\bar{P}_i)^2/k_i$$

so

$$\dot{W}(P)=-2(P-\bar{P})S(P).$$

When community preference fields exist then the revealed preference argument already noted gives $(\bar{P}-P)S(P)<0$, for $P \neq \bar{P}$, so once again global stability can be deduced.

[2] To pass from $h\bar{p}$ to $\bar{\bar{p}}$ no change is required in the rth entry and none of the other entries are increased while at least one is reduced.

For the gross substitute case choose units in which to measure goods such that $\bar{p}=1$, the unit vector. The Lyapounov function candidate is

$$L(p)=\max_i (p_i-1)^2.$$

This has well-defined right- and left-hand derivatives with respect to t. Under (2.1.N) we have $p_1(t)=1$ identically in t. Suppose $\mathrm{d}L(p)/\mathrm{d}t=2(p_r-1)\dot{p}_r$. Using (2.1.N) and (2.22) we deduce that if $p_r>1$ we have $s_r(P)>0$ and so $L(p)$ is declining. If $p_r<1$, a parallel argument gives $s_r(P)<0$ and once again $L(p)$ is declining. So once again we can deduce global stability. But now notice that this proof makes no use of the specialisation to (2.20). So we can do without it in particular as boundedness of the solution is ensured by the fact that the price which exceeds the equilibrium price in the greatest proportion must be falling and vice versa, for the price which falls short of the equilibrium price in the greatest proportion. So we have

Theorem T.2.6

If community preference fields exist and (2.20) and the solution path is bounded or if all goods are gross substitutes, equilibrium is globally asymptotically stable under (2.1.N).

The remaining leading case of diagonal dominance is not completely settled. Recall that in this case there is $h=(h_2\dots h_n)\gg 0$ such that

$$h_i|s_{ii}(P)|>\sum_{j\neq 1}|s_{ij}(P)|h_j, \qquad i=2\dots\ell,$$

and $s_{ii}(P)>0$, all i. In principle h may depend on P. Although I know of no counterexamples, global stability has not been established for this case. However, when h is independent of P one has

Theorem T.2.7

Let the Jacobian of the excess supply function have diagonal dominance at all P for weights which are independent of P. Let the solutions of (A.2.1.N) be bounded. Then equilibrium is globally stable under (2.1.N).

Proof

Show that $\max_i[-H_i(s_i(P))/h_i]$ is a Lyapounov function.

There are a good number of other special cases. For instance, the adaptive expectation model of Section 2.1 can be shown to be globally stable under gross substitutability [Arrow–Hurwicz (1962) and Arrow–Hahn (1971)]. But it is clear that the adjustment processes so far considered cannot, for reasons already given, be globally stable in general. Accordingly some attempts have been made to find adjustment processes which possess global stability properties independently of the form of the excess demand functions.

Arrow and Hahn (1971) proposed a process based on Newton's method of solving non-linear equations. Thus, for any P define $Q \in R^{n-1}$ by

$$S(P)+DS(P)(Q-P)=0,$$

and consider the process $\dot{P}=(Q-P)$. Make the usual assumptions which ensure that equilibrium prices are strictly positive. Assume further that (i) $DS(P)$ is non-singular for all $P \gg 0$ and (ii) that $DS(P)^{-1}S(P)$ is continuously differentiable. Then this process has a solution which converges to an equilibrium for all $P(0) \gg 0$. However (i) is strong enough to ensure that equilibrium is unique which shows that it is too strong for the general result we are now chasing. Smale (1976) provides the complete answer: making assumptions which keep prices away from the boundary and letting (P) be a real valued function of P where sign $\lambda(P)=(-1)^{n-1}$ sign det $DZ(P)$, he shows that the process

$$DZ(P)P=-\lambda(P)Z(P)$$

always converges to an equilibrium. (See Chapter 8 of this Handbook.) Obviously these results are interesting as algorithms and not as models of the invisible hand.

While these results are of interest as algorithms they have the drawback that it does not seem possible to give them an economic motivation. In particular the behaviour of the price in any one market now depends in principle on what is happening in all other markets. It is therefore of interest to ask whether one can find a process which does as well as the Newton method but does not need to make the price change in any one market depend on the properties of the excess demand in all other markets. This question has been studied by Saari and Simon (1978) with negative results.

From the Newton method [taking $|DS(P)| \neq 0$] one has

$$\dot{P}=-\;DS(P)^{-1}S(P),$$

which one can write as

$$\dot{P}=F(S(P),Y(P)),$$

where Y is an n-vector assuming there to be $(n+1)$ goods. Saari and Simon say that $F(\cdot)$ is *locally effective* if for any P such that $S(P)=0$, P is an attractor and they say that it is *effective* if the solution asymptotically tends to a zero of S. (Both definitions are augmented by technical regularity conditions and some restriction on the set of admissible initial conditions.) They say that a co-ordinate y_{ij} of F is *ignorable* if $\partial F/\partial y_{ij} \equiv 0$. The question is how many (and what kind of) ignorable co-ordinates F can one have and still be locally effective or effective? They prove the following discouraging theorem where they make use of the fact that in a neighbourhood of $S=0$ one can write the above differential equation as

$$\dot{P}=A(S,Y)S,$$

where $A(S,Y)$ is a matrix.

Theorem 2.8 (Saari and Simon)

(a) If there are no fewer than three goods and if F has some ignorable co-ordinate y_{ij} in some neighbourhood of $S=0$, then F cannot be locally effective.
(b) If there are no fewer than three goods and y_{ij} and y_{hk} are ignorable co-ordinates of F in some neighbourhood of $S=0$ where $i \neq h$ and $j \neq k$, then F cannot be effective provided that a_{ji} or a_{kh} is not identically zero on $\{0\} \times R^{n^2}$. [a_{ji} is an element of $A(S,Y)$.]
(c) If the jth column of Y and some y_{ik} for $k \neq j$ are ignorable co-ordinates for F, then F cannot be effective.

The proof of this theorem is far too long and complex to be given here. In reading the theorem it should be recalled that one is here investigating stability *without* any restrictions on the forms of the excess demand function. That is to say, one is looking for a process which has desirable properties (is effective). As noted, the result is discouraging for the hopes of finding such a process which at least to some extent preserves the rule that the price change in any one market should depend only on what is happening in the market. Thus (c) of the theorem tells us that "an effective price mechanism cannot fail to take into consideration how two commodities affect the rate of change of demand for all the commodities" [Saari and Simon (1978, p. 1106)].

2.3. Discrete time

Consider the adjustment process

$$P_i(t+1)=\max[0, P_i(t)+k_i Z_i(P)], \qquad i=2\ldots\ell. \tag{2.23}$$

To obtain stability results now we must do more than restrict the form of the excess demand functions; restrictions must also be placed on the adjustment speeds. [Compare with (2.1) and the discussion of McFadden's theorem there.]

That this is so can be seen in a simple way by considering the gross substitute case and taking the linear expansion of (2.23) about $\bar{P}$. Choose units so that $\bar{P}=e$, the unit vector. We shall then be interested in the Jacobian

$$[I+\hat{k}\,DZ(e)],$$

where $\hat{k}=\text{diag}(k_2\ldots k_n)$. If now k_i is small enough for all i this is a positive matrix. By homogeneity $1+k_i\sum_{j\neq 1}z_{ij}(\bar{e})=1-k_iz_{i1}(e)<1$, all i, and so by a well-known theorem on positive matrices local stability is assured. If, on the other hand, the k_i's are large enough so that $(n-1)+\sum_{i\neq 1}k_iz_{ii}(e)<1-n$, then there will be roots with absolute value greater than one and the necessary condition for local asymptotic stability will not hold. What is true of local results is *a fortiori* true for global ones.

Results for (2.23) thus turn on particular adjustment speeds [Uzawa (1961)]. This is fundamentally a negative conclusion since we have no theory to help in this matter. Indeed, if the theory is to be taken as in some way descriptive we could not exclude mixed difference–differential equations for price adjustment and theory would yield few qualitative results.

However the following technical result is available for the process:

$$x_{t+1}=f(x_t),\ \text{where}\ f\colon R^n\to R^n,\ f(0)=0\ \text{and}\ f\in C^1.$$

Theorem T.2.9 [Brock and Scheinkman (1975)]

Suppose there exists a twice continuously differentiable function V from R^n to R_+ with the following properties:

(i) $V(x)>0$, for $x\neq 0$, $V(0)=0$ and $\lim_{|x|\to\infty}V(x)=\infty$.
(ii) If $V(f(x))=V(x)$ and $x\neq 0$, then $V'(f(x))J(x)x-V'(x)x<0$, where $J(x)$ is the Jacobian of f at x and $V'(x)$ is the gradient vector of V.
(iii) $V'(0)J(0)x-V'(0)x<0$, if $V'(0)\neq 0$.
(iv) If $V'(0)=0$, then $[V''(0)J(0)x]J(0)x-xV''(0)x<0$, for $x\neq 0$, where V'' is the Hessian of V.

Then $x=0$ is a globally asymptotically critical (fixed) point of f.

The proof of this theorem turns on showing that V with the above properties is a Lyapounov function, in particular,

$$V(f(x))-V(x)<0,\quad \text{for}\ x\neq 0.$$

This is achieved by considering a function g: $[0,1]-R$, where

$$g(\lambda)=\lambda[V(f(\lambda x))-V(\lambda x)].$$

Evidently $g(0)=g'(0)=0$. On the other hand by (iii) $g''(0)<0$ if $V'(0)\neq 0$ and $g'''(0)<0$ if $V'(0)=0$. Hence $g(\lambda)<0$ for λ small. If $g(\lambda)$ changes sign at $\bar{\lambda}$ then $g(\bar{\lambda})=0$. But by (ii) $g'(\bar{\lambda})<0$. So evidently $g(1)<0$ for $x\neq 0$. Lastly by (i) $\lim_{|x|\to\infty}V(x)=\infty$, but by what has just been shown $V(f(x_0))<V(x_0)$, hence $|x_0|$ is bounded and the usual Lyapounov theorem applies.

This result has an interesting corollary:

Corollary 2.8 [Brock–Scheinkman (1975)]

If $|J(0)x|<|x|$ for $x\neq 0$ and if $\theta\neq|x|=|f(x)|$ implies $|J(x)x|<x$, then $x=0$ is a globally stable critical point of f.

This is proved by checking that $V(x)=|x|^2$ satisfies the conditions of T.2.8 when the conditions of the corollary hold.

These two results when applied to the tâtonnement confirm our previous local conclusion that the adjustment speeds play an important role in stability.

2.4. *A tâtonnement for the core: A digression*

In a pioneering paper Green (1974) has investigated a tâtonnement process without prices which converge, in probability, to the core. Below we shall give an account of a non-tâtonnement process with similar properties [Graham and Weintraub (1975)].

The idea behind the Green process is very simple, although it is surprisingly hard to prove the desired result. Consider any allocation $\hat{u}=(u_1\ldots u_H)$ of utilities. Let $V(S)$ be the utility allocations a coalition $S\subset H$ could attain on its own and $\bar{V}(S)$ be the Pareto efficient subset of $V(S)$. One writes $B_s(\hat{u})$ as the subset of $\bar{V}(S)$ that could block $\hat{u}$. Now think of $\hat{u}$ as an allocation proposal. We can now define the set of all possible blocking coalitions for $\hat{u}$: $B(\hat{u})=\{S|B_s(\hat{u})\neq\varnothing\}$. One now assumes that over each collection of possible blocking coalitions there is a strictly positive probability distribution according to which the coalition which will block a given allocation is decided. The blocking coalition chooses an allocation proposal in $B_s(\hat{u})$ according to a probability distribution (see below). Lastly the complementary coalition selects an allocation with a certain probability from among those allocations which are Pareto efficient for itself. In this (random) way a new proposal $\hat{u}'$ is generated and the story starts again. It should be noted that any given new proposal may make some members of the complementary coalition worse off so no simple gradient process is involved. One has

Theorem T.2.10 [Green (1974)]

Under Assumption A (below) and the given process starting with an arbitrary proposal $\hat{u}(0)$ the probability that $\hat{u}(t)$ is not in the core approaches zero with t.

The assumptions are:

Assumption A

(1) For every coalition greater than one there are allocations which give a member less than he could achieve by himself.
(2) All coalitions can improve the positions of all their members.
(3) Let $V^*(S)$ be a subset of $V(S)$ such that if $\hat{u}_s \in V^*(S)$ every agent in S does at least as well in $\hat{u}_s$ as he could by himself. Suppose $\hat{u}_s \in V^*(S)$ and also $\hat{u}_s(\varepsilon, i) \in V(S)$, where $\hat{u}_s(\varepsilon, i)$ is the vector $\hat{u}_s$ with $\varepsilon > 0$ added to its ith component. Then there exists an $|s|$-dimensional vector $k \gg 0$ depending only on S and ε such that $\hat{u}_s + k \in V(S)$.
(4) $V(S)$ is closed for all S, exhibits free disposal [i.e. $\hat{u} \in V(S)$ and $\hat{v} \leqslant \hat{u}$ implies $\hat{v} \in V(S)$], and the subset of $V(S)$ in which each agent does at least as well as he could on his own is non-empty and bounded.
(5) There is an allocation in the core which for any S assigns $\hat{u}_s$ which S could not have attained on its own (superadditivity).
(6) For every collection of coalitions that could be a set of possible blocking coalitions for some proposal, the probability that any one coalition in this set will form is strictly positive and depends only on the collection of coalitions and that particular coalition.
(7) Let E be a Lebesgue measurable subset of $B_s(\hat{u})$ and $\mu_b^s(E|\hat{u})$ the probability that coalition S will select an allocation in E given that S blocks $\hat{u}$. Let $\lambda^s(E)$ be the Lebesgue measure of E. Then (i) $\mu_b^s(E|\hat{u}) = 0$ if $\lambda^s(E) = 0$, (ii) $\lambda_b^s(B_s(\hat{u})|\hat{u}) = 1$ and (iii) $\mu_b^s(E|\hat{u}) > \varepsilon\lambda^s(EB_s(\hat{u}))/\lambda^s(B_s(\hat{u}))$ for any measurable E, where $\varepsilon > 0$ is a constant independent of $\hat{u}$ and E.
(8) Make exactly similar assumptions concerning $\mu_c^{s_c}(F)$, the probability that the complementary coalition will select a point in F where $F \subset \bar{V}(S^e)$.

The proof of the theorem is too long and complex to be given here. But something can be said concerning its strategy. Suppose that the only blocking coalition for $\hat{u}(0)$ is the whole-player coalition H. This coalition selects some $\hat{u}(1)$ which is Pareto efficient. But $\hat{u}(1)$ must also be in the core. For if not then there is some S such that $\hat{u}_s(1) \in \text{int } V(S)$. Since $\hat{u}_s(1) \geqslant \hat{u}_s(0)$ and there is free disposal it now follows that S could have blocked $\hat{u}(0)$ contrary to assumption. Green now shows that for all possible cases there is a finite probability that H will at some finite time from zero be a blocking coalition and with finite probability propose the core allocation. That is to say there is a number δ and date T such that the probability of $\hat{u}_t$ not being in the core is $(1-\delta)^{t/T}$. This probability goes to zero with t. Unfortunately this result requires an exhaustive treatment of a number of possible cases.

3. Trading out of equilibrium

We now consider situations where trade can take place at each t whatever the terms on which such trade can be carried out. We shall report on a variety of approaches to this problem but one can allocate them all to one of two categories. The first of these does not insist on market prices at which exchange must be conducted [Uzawa (1962), Smale (1976) and Hurwicz, Radner and Reiter (1975)] while the second considers exchange at given market prices and models changes in the latter [Hahn–Negishi (1962b), Hahn (1962a), Arrow–Hahn (1971), Diamond (1971) and Fisher (1972, 1974, 1975b, 1978)]. The first category really considers economies without markets in which individuals meet and carry out exchange (or production for exchange). The second category, the descriptively more interesting one, takes account of market prices.

3.1. Exchange without prices

The easiest case to consider here is what Uzawa (1962) called the *Edgeworth process*. This process rests on the following:

Assumption A.3.1

Trade will take place if and only if it is feasible and there exists a Pareto improving trade.

This, of course, is a very strong assumption, and it can be somewhat relaxed by a stochastic formulation so the probability of trade taking place is positive if and only if there exists a Pareto improving trade. Even this version is quite demanding and leads to a somewhat differently formulated result than does the deterministic one.

Let $\bar{X}(t)$ be the endowment matrix at t, i.e. in its kth column one has $\bar{x}_k(t)$, the endowment vector of k at t. Suppose $\bar{X}(t, \bar{X}(0))$ is a continuous path for $X(t)$ starting at $\bar{X}(0)$ with the following properties:

(i) $\bar{x}_k(t)\in R^{\ell}_{+}$, all k and t.
(ii) $\Sigma\bar{x}_k(t)=\text{constant}$, all t.
(iii) If $\bar{X}(t+\varepsilon)\neq\bar{X}(t)$, then

$$U_k(\bar{x}_k(t+\varepsilon))\geqq U_k(\bar{x}_k(t)), \quad \text{all } k,$$
$$U_k(\bar{x}_k(t+\varepsilon))>U_k(\bar{x}_k(t)), \quad \text{some } k.$$

Then consider the function

$$V(\bar{X}(t))=\sum U_k(\bar{x}_k(t)).$$

It is clear that $V(\bar{X}(t))$ is increasing as long as trade takes place. Since $\bar{X}(t, \bar{X}(0))$ is bounded we may take it that V is bounded above. So $\lim_{t\to\infty} V(\bar{X}(t, \bar{X}(0)) = V^*$, say. From the boundedness of the allocation path we new deduce that $\bar{X}(t, \bar{X}(0)$ has a subsequence converging on $\bar{X}^*$ so that

$$V^* = V\big(\bar{X}(t, \bar{X}^*)\big).$$

If preferences are strictly convex one finds $V^* = V(\bar{X}^*)$ so that the process converges on a particular allocation. The latter is Pareto efficient.

One is now interested in specifying more precisely a process which generates an endowment path with these agreeable properties. A simple example is the following.

Assume that the marginal utility of any good to an agent becomes unboundedly large as the holdings of that good approach zero. Take $\bar{X}(0)$ to be a strictly positive matrix and assume all utility functions are concave and differentiable. Let μ_{ki} be the marginal rate of substitution between good i and good 1 for agent k and let μ_i be the average (over agents) of this marginal substitution rate. Consider the process

$$\dot{\bar{x}}_{ki} = (\mu_{ki} - \mu_i), \qquad i = 2\ldots l, \text{ all } k,$$
$$\dot{\bar{x}}_{k1} = -\sum_{i\neq 1} \mu_i(\mu_{ki} - \mu_i), \quad \text{all } k.$$

One readily verifies $\Sigma_k \dot{\bar{x}}_{ki} = 0$, all i, and $\bar{x}_{ki}(t) > 0$, all t. Next, differentiating $u_k(\bar{x}_k)$ with respect to time we find

$$\frac{\mathrm{d}}{\mathrm{d}t} U_k(\bar{x}_k(t)) = \frac{\partial U_k}{\partial x_{k1}} \sum_i (\mu_{ki} - \mu_i)^2,$$

which is positive as long as agent k has some marginal rate of substitution different from the average, i.e. as long as there is trade. Making the usual continuity assumptions, the above process has a solution $\bar{X}(t, \bar{X}(0))$ which satisfies conditions (i)–(iii) above and so converges to a Pareto efficient allocation.

However, it is not easy to tell a really good story to motivate this process. For a more satisfactory procedure, one goes to the work of Hurwicz, Radner and Reiter (1975). On the other hand, the process they suggest still needs a kind of auctioneer.

The idea is as follows. Let $\bar{x}_k(t)$ again be k's endowment at t. Let $g(\bar{x}_k(t))$ be the set of all trades which lead to an endowment at least as high in k's preferences as does $\bar{x}_k(t)$. Let $F(\bar{x}_k(t))$ be the set of trades which are admissible for k, i.e. lead to an outcome in k's consumption set and do not require the supply of any one good in excess of his endowment. Then $g(\bar{x}_k(t)) \subset F(\bar{x}_k(t))$ is the set from which k selects *bids* to trade. This he does according to a probability distribution which of course depends on $\bar{x}_k(t)$. When goods are indivisible (so that bids are integers) no more need be said on this score. When goods are divisible the story must be modified otherwise there might never be any trade since the probability of different agents making *compatible* bids is of measure zero. Bids are compatible if the offered trades in each good sum to zero. In order to avoid this problem, the agent is assumed to choose a bid according to his probability distribution as before. This bid is called the *central* bid. The agent, however, announces a set of bids. This is defined as the intersection of a cube with the central bid at its centre and the set $g(\bar{x}_k(t)) \subset F(\bar{x}_k(t))$. The size of the cube is small and fixed once and for all. It is called a *bidding cube*.

If it is possible to select from each bidding cube a bid such that the total is compatible, the trade takes place in accordance with the bids. If no such compatible trades exist, then we wait for the next throw of the dice to generate a new central bid and bidding cube. It is here that one must think of an auctioneer-like agent who checks for compatibility. Under certain assumptions one shows that if the allocation is not Pareto efficient a compatible trial will occur with probability one at some date.

One must now recognise that bids will continue for ever even when the allocation of goods is Pareto efficient. Accordingly one defines an *equilibrium bidding distribution* (which is the vector of bidding distribution of all agents) as one which does not depend on t. (Recall that while exchange is taking place, the upper contour set and so the probability distribution of bids of the agent is changing.) Stability is now defined as the almost sure convergence of the bidding distribution. In terms of utilities, this means that $\lim_{t\to\infty} U_k(\bar{x}_k(t))$ exists with probability one. It is really immediately obvious that if the allocation is Pareto efficient then the bidding distribution generated by that allocation is an equilibrium distribution (Theorem 3.1, Hurwicz, Radner and Reiter).

The case of indivisible goods is very simple. One assumes that the unconstrained probability of making any admissible bid is positive. This means that if a compatible Pareto improving bid is possible it will occur almost surely. (To make this precise one thinks of the bidding set as denumerable and the set of compatible bids as finite.) One has here a stochastic version of the idea discussed at the outset of this section.

Theorem T.3.1 [Hurwicz–Radner–Reiter (1975)]

Under the assumptions of the previous paragraph, the Hurwicz–Radner–Reiter process for indivisible goods has the property that with probability one: (i) for some finite t the bidding distribution (allocation) is Pareto efficient, and (ii) the bidding distribution remains constant for all $t' > t$.

Notice that because of the finiteness assumption convergence to the Pareto efficient allocation occurs in finite time. The result is agreeable to common sense.

The case of divisible goods is a great deal harder and a number of technical assumptions are required to establish convergence to a Pareto efficient allocation. One postulates: (i) that for each agent the admissible set of bids is closed; (ii) that there are no externalities so that the economy's feasible set of bids is the intersection of the product of the agent's bidding set with the set of compatible bids; (iii) that the economy's set of feasible bids is bounded; (iv) that there is a vector of compatible bids each component of which is in the interior of each agents' bidding set; and (v) that an agent's preferences are representable by a continuous real valued utility function. Assumption (iv) can be given an alternative form but we do not pursue this. To these five assumptions a sixth and crucial one is added. It (or some version of it) is needed to ensure that at non-Pareto efficient allocations there is a trade which almost surely takes place. (Of course this is a Pareto improving trade.)

One version of this assumption which is called the *openness* assumption goes as follows. (vi) For any $\varepsilon>0$ consider an allocation such that there is no Pareto efficient allocation which would give each agent a utility which exceeds the given utility by no more than ε. Then there exists (an open) feasible cube which (a) is wholly contained in the set of Pareto improving allocations and (b) which has a non-empty intersection with the set of allocations for each element of which there exists a Pareto efficient allocation giving utility to each agent which do not exceed the given utility by more than ε.

Hurwicz, Radner and Reiter show that given (i)–(v) then (vi) is satisfied for a pure exchange economy if the agent's consumption set is the positive orthant and if utility functions are strictly increasing. It is clear that something like assumption (vi) is needed if there is to be a positive probability of a Pareto improving trade whenever the given allocation is Pareto inefficient. In any case one has:

Theorem T.3.2

Given (i)–(vi) the Hurwicz–Radner–Reiter bidding process converges almost surely to a Pareto efficient allocation as $t\to\infty$.

One advantage of this model is that it need not be confined to the case of pure exchange. Production is introduced through agents who are indifferent amongst all elements of their production sets. Another advantage is that we can dispense with the usual convexity assumptions. While in part technically hard to prove, the results themselves are not surprising. They are entirely in the spirit of the opening paragraphs of this section. The utility–improving mechanism – even when formulated stochastically – is very powerful in bringing about convergence to a Pareto efficient allocation. It has a natural upper bound and as long as one ensures that when a Pareto improving exchange is feasible that it will occur with positive probability, the result is intuitively in the bag.

It needs now to be stressed that the process we have discussed is not to be taken descriptively and is best thought of as a planning algorithm.

As a last example we consider a contribution by Graham and Weintraub (1975). This is to be compared with Green (1974) where no exchange takes place. The idea here is to ensure that Pareto inefficient allocations will be blocked by some coalition. The ensuing reallocations are such that a finite fraction of the remaining potential gains from trade are realised with a high probability. This then ensures that in the limit all gains from trade will be exhausted with probability one, i.e. that there is convergence to a Pareto efficient allocation.

To be more precise, let e be the vector of aggregate endowments and let $\hat{e}=(e_1\dots e_H)$ be an allocation of the endowment amongst the H agents. Define

$$\lambda(\hat{e})=\inf\left\{\lambda\in R_+ | x_h \underset{h}{\geqslant} e_h,\ \text{all } h,\ \lambda\hat{e}=\sum x_h\right\},$$

where $(x_1\dots x_H)$ is an allocation. Then $1-\lambda(\hat{e})$ is an (arbitrary) measure of the potential gains from trade at the allocation $\hat{e}$. Also, $\lambda(\hat{e}')-\lambda(\hat{e})$ can be thought of as the gains from trade captured by a move from $\hat{e}$ to $\hat{e}'$. If $\hat{e}(t)$ is the allocation at t, write $\lambda_t=\lambda(\hat{e}(t))$.

The allocation process is as follows. Let $C(t)$ be a coalition which blocks $\hat{e}(t-1)$ with $\{x_h(e)\}_{h\in C(t)}$. That is, $x_h(t)>_h e_h(t-1)$, all $h\in C(t)$, and $\sum_{c(t)}x_h(t) =\sum_{c(t)}e_h(t-1)$. Then

$$\begin{aligned} e_h(t)&=x_h(t), && h\in C(t),\\ e_h(t)&=e_h(t-1), && h\notin C(t). \end{aligned}$$

It follows that along this sequence λ_t is non-decreasing (and since $\lambda\in[0,1]$ it has a limit). The sequence is called a *blocking sequence*.

Suppose

$$\lambda_{t+1}-\lambda_t<\varepsilon(1-\lambda_t), \qquad \varepsilon>0.$$

Then when $\lambda_t<1$ this can be interpreted as saying that the gains of the $(t+1)$th iteration as a fraction of the potential gains at t are less than ε. Subtracting $(1-\lambda_t)$ from both sides of the inequality we can write it as

$$(1-\lambda_{t+1})>(1-\varepsilon)(1-\lambda_t).$$

Assumption A

For any $\delta>0$ there is $\varepsilon>0$ such that

$$\text{prob}\left[(1-\lambda_{t+1})>(1-\varepsilon)(1-\lambda_t)\right]<\delta, \quad \text{for all } t.$$

This ensures that very small gains from trade at each step have very low probability. One now has

Theorem T.3.3 [Graham and Weintraub (1975)]

Assume all utility functions monotone, continuous and strictly quasi-concave. Then under Assumption A the blocking sequence converges to a Pareto efficient allocation with probability one.

Proof

We know that $\lambda^* = \lim_{t\to\infty}\lambda_t$ exists. We need to prove that $\text{prob}(\lambda^* < 1) = 0$. Suppose $0 < k < 1$. Then if $\lambda_t \leqq k$, the inequality

$$1 - \lambda_{t+1} \leqq (1-\varepsilon)(1-\lambda_t)$$

can hold at most $r = \ln(1-k)/\ln(1-\varepsilon)$ times between $t=0$ and $t=T-1$. There can be at most r occasions in which gains increase by ε or more. Hence, there are $T-r$ terms where the increase is less than ε and so by Assumption A for $\delta < 1$

$$\text{prob}(\lambda_T \leqq k) < \delta^{T-r},$$

and this probability goes to zero as $T \to \infty$. But

$$\text{prob}(\lambda^* < 1) = \lim_{k\to 1} \text{prob}(\lambda^* < k) = \lim_{k\to 1}\lim_{T\to\infty} \text{prob}(\lambda_T \leqq k) = 0.$$

Since the set of possible allocations is compact, the blocking sequence has at least one limit point, $\hat{e}^*$. One easily verifies that $\lambda(\hat{e}_t)$ is continuous whence

$$\lim_{t\to\infty} \lambda(\hat{e}_t) = \lambda\left(\lim_{t\to\infty} \hat{e}_t\right) = \lambda(\hat{e}^*) = \lambda^*$$

along some subsequence. Hence as just shown, $\lambda(\hat{e}^*) = 1$ with probability one. It is routine to verify that $\hat{e}^*$ must be Pareto efficient. The strict convexity of preferences ensures that $\hat{e}^*$ is the unique limit of the blocking sequences so that $\hat{e}^*$ is the limit of $\hat{e}_t$ as $\lambda_t \to 1$.

3.2. Exchange with prices

3.2.1 Pure exchange

It is now assumed that exchange which takes place at any time t takes place at a given set of prices ruling at t. This is the first point of departure from the discussion in Section 3.1. The second one is that exchange takes place even when "bids" are not compatible. The third point of departure is that an auctioneer is reintroduced to bring about changes in price.

Let us assume strictly convex preferences representable by continuous, real valued and differentiable utility functions. As before, let $\bar{X}(t)$ be the endowment matrix at t. The excess demand functions now explicitly carry $\bar{X}(t)$ as an argument, i.e.

$$Z(t)=Z(p(t),\bar{X}(t)).$$

We stipulate the price adjustment rule (2.1.A). One now also needs some rule of the form

$$\dot{\bar{X}}(t)=F(\bar{X}(t),p(t)). \tag{3.1}$$

This rule must obey two restrictions:

$$\sum_h \dot{\bar{x}}_h(t)=0, \tag{3.2}$$

$$\sum_i p_i(t)\dot{\bar{x}}_{hi}(t)=0. \tag{3.3}$$

Here (3.2) simply stipulates pure exchange and (3.2) quid pro quo.

In the spirit of the previous section, one may now wish to add the further restriction that trade takes place if and only if there is a Pareto improving trade which is budget feasible for each agent. (One could almost certainly proceed with a stochastic formulation à la Hurwicz, Radner and Reiter, of this rule.) Uzawa (1962) has shown that at least one process satisfying these three restrictions exists. In addition, one wants (2.1.A) and (3.1) to have a unique continuous solution. By assuming $\bar{x}_h(0)\gg 0$, all k, and postulating unboundedly high marginal utility for any good of which the agent has a zero stock, one can ensure $x_h(t)\gg 0$, all t. By postulating strong desirability and $p(0)\gg 0$, one can ensure $p(t)\gg 0$, all t. That being so, one can, without running into difficulties, require (2.1.A) and (3.1) to satisfy a Lipschitz condition.

Now it is perfectly possible for $\bar{X}(0)$ to be Pareto inefficient and yet for there to be no Pareto improving trade at $p(0)$. One need only think of the usual box diagram with the ray of prices outside the ellipse formed by the indifference curves through $\bar{X}(0)$ to see this. Take $p(0)\gg 0$ and consider the numeraire price vector $P(0)\gg 0$. Assume utility functions differentiable and define $U_i^h=\partial U^h/\partial x_i^h$ and

$$\mu_{hi}=\frac{1}{P_i}\frac{U_i^h}{U_1^h}-1.$$

Then it is elementary to prove

Theorem T.3.4 [Arrow–Hahn (1971)]

Let $\bar{X}(0) \gg 0$, $p(0) \gg 0$. Then a necessary condition for no Pareto improving trade to be possible at these prices is

(a) For all i, μ_{hi} is of the same sign all h.
(b) For all possible pairs i and j, $\mu_{hi} - \mu_{hj}$ has the same sign all h.

If preferences are strictly convex, one can also establish:

Lemma 3.1

Let $\alpha_{hi} = P_i \mu_{hi}$, $\alpha_i = \sum_h \alpha_{hi}$. Let $\bar{X} \gg 0$, $p \gg 0$ and Z_h the excess demand vector of h at $(\bar{X}, p)$, $Z = \sum_h Z_h$. Then if $Z_h \neq 0$, some h, and no Pareto improving change is possible at the given prices

$$\sum_i \alpha_i Z_i > 0. \tag{3.4}$$

Proof

Let $Z_h \neq 0$. Then by convexity of preferences and $0 < \lambda < 1$

$$U^h[\lambda(\bar{x}_h + Z_h) + (1-\lambda)\bar{x}_h] = U^h(\bar{x}_h + \lambda Z_h) > U^h(\bar{x}_h)$$

and for λ small enough

$$0 < U^h(\bar{x}_h + \lambda Z_h) - U^h(\bar{x}^h) \sim \sum U_i^h \lambda Z_{hi}. \tag{3.5}$$

Subtract $\lambda P Z_h = 0$ from the r.h.s. of (3.5) and divide by U_1^h to obtain

$$\lambda \sum \alpha_{hi} Z_{hi} > 0. \tag{3.6}$$

(The partial differential coefficients are taken at $\bar{x}_h$.) Now consider household $k \neq h$. Since $\bar{x}_k \gg 0$ this household can, for λ small enough, satisfy the trade λZ_h of h. Note also that $p \cdot \bar{x}_k - p\lambda Z_h = p \cdot \bar{x}_k$. His utility change would be

$$U^k(\bar{x}_k - \lambda Z_h) - U^k(\bar{x}_k)$$

and this must be negative or else exchange would be possible. Proceeding as before one verifies that

$$-\lambda \sum_i U_i^k Z_{hi} < 0$$

or

$$\lambda \sum_i \alpha_{ki} Z_{hi} > 0.$$

Since this must be true for all $k \neq h$ we have in conjunction with (3.6)

$$\lambda \sum_i \alpha_i Z_{hi} > 0.$$

But this must be true for all h with $Z_h \neq 0$. So, summing over h, we obtain (3.4.)

The following assumption will now be maintained throughout:

Assumption 3.2.1 [Fisher (1974)]

Divide the commodity space into any two non-intersecting sets S and S' with $S \cup S'$ the commodity space. Then there exists a pair of goods $i \in S$, $j \in S'$ such that they are both demanded by some h.

Now consider (2.1.N) of the form

$$\dot{P}_i = Z_i(P, \bar{X}), \qquad i \neq 1. \tag{3.7}$$

Then we have

Theorem T.3.5 [Arrow–Hahn (1971)]

Assume (i) $\bar{X}(0) \gg 0$, $P(0) \gg 0$, (ii) for all i, $Z_i(P, \bar{X}(0)) > 0$ if $P_i = 0$, and (iii) Pareto improving change is not possible at any $P(t)$ generated by (3.7) [i.e. $\bar{X}(t) = \bar{X}(0)$ all $t \geqslant 0$]. Then (3.7) is a globally stable process.

Proof

(a) If P is not an equilibrium, then for some h, $Z_h(P, \bar{x}_h(0)) \neq 0$ and so $\alpha_i(P, \bar{X}(0)) \neq 0$, some i. Let

$$V\big(P, \bar{X}(0)\big) = \sum \alpha_i^2.$$

Certainly $V(P, \bar{X}(0)) \geqq 0$, all P. Also, $V(P, \bar{X}(0)) = 0$ iff P is an equilibrium. Note that by assumption $P(t) \gg 0$, all $t \geqq 0$. It can also be shown [Arrow – Hahn (1971, p. 333) and references there] that $P(t)$ is bounded.

(b) For $Z \neq 0$: $\dot{V}(P, \bar{X}(0)) = -2\alpha_i Z_i < 0$ by Lemma 3.1. Hence $V(\cdot)$ is a Lyapounov function and so, if P^* is a limit point of $P(t, P(0))$, it must be an equilibrium. Since $P^* \gg 0$ one has $Z(P^*, \bar{X}(0)) = 0$ at each such limit point.

(c) Suppose $Z_h(P^*, \bar{X}(0)) \neq 0$ some h. Then $\alpha_{hi} \neq 0$ some i. By Theorem T.3.4 since no exchange is possible, $\alpha_i \neq 0$. But this leads to a contradiction of (3.4) of Lemma 3.1. Hence $Z_h(P^*, \bar{X}(0))=0$, all h, whence $\bar{X}(0)$ is Pareto efficient. But by our assumption $\alpha_i(P, \bar{X}^0)=0$, all i, can have only one solution. Hence P^* is the only limit point and the theorem is proved.

One can now continue this result with the Pareto improving property of the exchange process to prove:

Theorem T.3.6 [Arrow–Hahn (1971)]

Assume (i) $\bar{X}(t)\gg 0$, $P(t)\gg 0$, all $t\geqq 0$, and (ii) utility functions are strictly monotone and strictly quasi-concave everywhere. Then (3.7) together with (3.1) is a globally stable process.

Proof

By arguments already given $\sum U^h(\bar{x}_h(t))$ is a Lyapounov function and $\bar{X}(t)\to\bar{X}^*$ which must be a unique limit point because of the strict quasi-concavity of U^h all h. If P^* is any limit point of $P(t)$ we may now consider the path $P(t, P^*)$ with $\bar{X}^*$ fixed. By Theorem 3.3, it converges to P^{**}, an equilibrium, and $\bar{X}^*$ is Pareto efficient.

These results are of some, but limited, interest. The reason is two-fold: We still retain the auctioneer and the assumption $\bar{X}(t)\gg 0$, all t, is very strong.

An alternative approach which retains the auctioneer is the following. Stipulate that the exchange process satisfies (3.2) and (3.3) but that it need no longer be Pareto improving. Instead it satisfies the further condition:

$$Z_{hi}(t)Z_i(t)\geqq 0, \quad \text{all } i \text{ and } h \text{ and } t\geqq 0 \tag{3.8}$$

The interpretation is this. Exchange is now with an anonymous market. At all times (including $t=0$) markets are *orderly*: no agent has an excess demand (supply) for a good which is in aggregate excess supply (demand). The idea is that there is very good information in all markets so that (instantaneously) agents know of all supplies and demands. It is not a very convincing assumption but it has lately become popular in rationing models (e.g. Drèze (1975)).

Theorem T.3.7

Let the exchange process satisfy (3.2), (3.3) and (3.8) and assume it to satisfy a Lipschitz condition and to be differentiable. Postulate: (a) $Z_i(p(i), \bar{X})>0$ for all $p(i)=\{p_1\dots p_{i-1}0\,p_{i+1}\dots p_n\}$ and $\bar{X}>0$; and (b) strictly quasi-concave utility functions $U^h(x_h)$ which are differentiable and for $x_{hi}=0$ one has $\partial U^h/\partial x_{hi}=\infty$. Then if the price process is given by $\dot{p}_i=Z_i(p, \bar{X})$, $i=1\dots n$, it and the exchange process converge to an equilibrium for $p(0)\gg 0$, $\bar{X}(0)\gg 0$.

Proof

Let $v^h(p, p\cdot\bar{x}_h)$ be the indirect utility function of h, $v_i^h = \partial v^h/\partial p_i$, $v_w^h = \partial u^h/\partial(p\cdot x_h)$. Then

$$\dot{v}^h = \sum_i \left(v_i^h + v_w^h \bar{x}_{hi}\right)\dot{p}_i + v_w^h \overline{px_h}^{\,\cdot}$$
$$= -v_w^h \sum_i Z_{hi}\dot{p}_i = -v_w^h \sum_i Z_{hi} Z_i \leqq 0. \tag{3.9}$$

The second equality utilises a well-known theorem on the partial derivatives of indirect utility functions and (3.3), while the next equality utilises the rule for $\dot{p}$. The inequality follows from (3.8). Let $V = \sum v^h(p, p\cdot\bar{x}_h)$. By (3.9): $\dot{V} < 0$ iff $Z \neq 0$ and $\dot{V} = 0$ if $Z = 0$. Our assumptions ensure $p(t) \gg 0$, all t, and $p(t)\bar{x}_h(t) > 0$, all t. We can take V as bounded below and so V is a Lyapounov function. We have

$$\lim_{t\to\infty} \sum v^h(p(t), p(t)\bar{x}_h(t)) = \sum v^h(p^*, p^*\bar{x}_h^*) = \sum v^{*h}, \text{ say.}$$

Suppose $(p^{**}, p^{**}\bar{x}_h^*)$ is another set of limit points. Then

$$v^h(p^*, p^*\bar{x}_h^*) = v^h(p^{**}, p^{**}\bar{x}_h^{**}) = v^{*h}.$$

If x_h^* is the demand vector at $(p^*, p^*\bar{x}^*)$ and x_h^{**} that at $(p^{**}, p^{**}\bar{x}_h^{**})$ it follows from strict quasi-concavity of u^h that

$$p^*(x_h^* - x_h^{**}) < 0.$$

Summing over h and noting that $\sum x_h^* = \sum x_h^{**} = \sum \bar{x}_h$ leads to a contradiction. Hence the process is globally stable.

One must now show that the stipulations of Theorem T.3.5 do not make it vacuous, by showing that an exchange process satisfying the requirements exist. This was done by Hahn–Negishi (1962b). However, the example was not very satisfactory because the process was consistent with agents being made worse off by exchange (not just by the changing prices). It was also entirely artificial. Notice also that unlike the earlier cases discussed we have chosen the maximal planned ("target") utilities as our Lyapounov function and that these utilities are declining out of equilibrium.

A somewhat more satisfactory approach takes account of the restraint that in actual economies most exchange must be against money. So we now introduce money into the story and write $\bar{m}_h(t)$ as h's stock of money at t. Let Z_t^+ be the vector with components $\max(0, Z_{hi})$. Then we now suppose that the agent faces two constraints at t, where p is the price vector in terms of money:

$$p \cdot x_h(t) + m_h(t) \leqq p \cdot \bar{x}_h(t) + \bar{m}_h(t), \tag{3.10}$$

$$p \cdot Z_h^+(t) \leqq \bar{m}_h(t). \tag{3.11}$$

In (3.10) $m_h(t)$ is the money demanded at t and (3.10) is just the ordinary budget constraint. In (3.11) we insist that the agent spend no more on purchases at t than his money stock at t.

Let $Z_h(t)$ be the excess demand vector of h at t if it is not constrained by (3.11). Call it h's *target excess demand vector*. If now $Z_h(t)$ violates (3.11) we think of the household as planning to achieve its target by a sequence of trades. In this it may follow a rule of thumb and reduce its purchases from their target values in the same proportion. More formally, let a_h be called h's *active excess demand* and define it by:

$$\text{for } Z_{hi}(t)>0: \quad a_{hi}(t) = k_h Z_{hi}(t), \quad \text{where } k_h = \min\left(1, \frac{\bar{m}_h(t)}{p \cdot Z_h^+(t)}\right),$$

$$Z_{hi}(t) \leqq 0; \qquad a_{hi}(t) = Z_{hi}(t). \tag{3.12}$$

One now lets price adjustments depend on active excess demands:

$$\dot{p}_i = 0, \text{ if } p_i = 0 \quad \text{and} \quad a_{hi} < 0, \dot{p}_i = \sum a_{hi}, \text{ otherwise.} \tag{3.13}$$

Trade proceeds on the basis of active excess demands. We, however, want an analogue to (3.8), i.e. we want, when $a_i = \Sigma_h a_{hi}$:

$$a_{hi}(t) a_i(t) \geqq 0, \quad \text{all } i, h \text{ and } t. \tag{3.14}$$

Assuming that (3.14) holds at $t=0$ and that $\bar{m}_h(t)>0$, all h and t, Arrow and Hahn (1971) describe a process of proportional rationing which ensures (3.14) for $t>0$. To make this process continuous, one must think of the economy as large. The process satisfies

$$\sum_h \dot{\bar{x}}_h(t) = 0; \qquad \sum_h \dot{\bar{m}}_h(t) = 0, \tag{3.15}$$

and

$$p \cdot \dot{\bar{x}}_h(t) + \dot{\bar{m}}_h(t) = 0,$$

which are simply (3.2) and (3.3) appropriate to the present case.

Now by (3.12) and $\bar{m}_h(t)>0$: $Z_{hi}(t)a_{ki}(t) \geqq 0$, all h and i, and so by (3.14) $Z_{hi}(t)a_i(t) \geqq 0$, all h and i, whence by (3.13): $Z_{hi}(t)\dot{p}_i \geqq 0$. This makes it clear that once again we shall be able to show declining target utilities for some agents as long as $a_i(t)>0$, $p_i(t)>0$, some i. Notice, on the other hand, that $a_i(t)=0$, all i, implies $Z_i(t)=0$, all i. This is so since in that case $a_{hi}(t)=0$, all h and i, and since $m_h(t)>0$, all h. We ensure the boundedness of the price solution by stipulating

$$Z_m(p, \bar{X})>0, \quad \text{for } \| p \| \geqq k, \tag{3.16}$$

where k is large. For the adjustment process (3.13) one easily checks that this suffices for bounded price paths.

Theorem T.3.8 [Arrow–Hahn (1971)]

Let the exchange process satisfy (3.14) and (3.15) and the price process satisfy (3.13), where $a_{hi}(t)$, $i=1\ldots n$, satisfies (3.12) each h. Assume (a) $Z_i(p(i), \bar{X})>0$ for all $p(i)$ and $\bar{X}>0$; (b) strictly quasi-concave differentiable utility functions; (c) (3.11); and (d) $m_h(t)>0$, all t and h. Then from any $p(0) \gg 0$, $\bar{X}(0)>0$, the price and exchange process converge to an equilibrium.

The proof of T.3.8 is similar to that for T.3.7. Full details are in Arrow–Hahn (1971).

There is some advance in realism in this account but many questions remain. One of these is the following. In the process just described, agents experience quantity constraints but do not take them into account when making their plans. This is unsatisfactory. Fisher (1978) has taken note of this shortcoming and presents a model in which a household which perceives quantity constraints recalculates its optimum active excess demand in the light of these constraints. The story is not easy to follow and in any case is restricted to situations close to a Walrasian equilibrium. But it seems that we start with something like (3.12). Households then express their "first stage" active excess demands. In the light of these they discover whether or not they will be rationed (on the usual assumptions). Given this information, households now recalculate their active excess demands. These recalculations do not reverse the sign of the first stage aggregate active effective demand. The outcome is $Z_{hi}(t)a_i(t) \geqq 0$, all h and i, which in our earlier discussion was the outcome of (3.12) and (3.14). This whole matter still requires further study, and it may well be that possible divergences between expected and actual quantity constraints should play a role.

But it is clear that it will be hard to obtain very general results. This can be seen in this outline of a rather natural story.

Let $P(t)$ be the price vector at t in terms of good 1. Take it as fixed and for a given rationing scheme calculate a Drèze equilibrium (1975). This may not be unique. However, let us make the rather strong assumption that it is and indeed that the Drèze equilibrium rations are continuous functions of $P(t)$. Next, let

$$\dot{P}_i = (\mu_i - P_i)$$

where, as before, μ_i is the average (over households) of the marginal rate of substitution of good i for good one. If rationing is orderly, then the prices of goods where buyers are constrained will be rising and those where sellers are constrained, falling. There is no intrinsic reason why in a Drèze equilibrium it should not be public knowledge which side is constrained in which market. If one now wants to utilise the earlier approach in which target utilities can serve as a Lyapounov norm, one will have to ensure that an agent with target excess demand (supply) at $P(t)$ is demand (supply) constrained in a Drèze equilibrium. But that step is extremely hard. It is one (in a different set up) taken by Fisher (1978) but there is no micro theory to justify it.

3.2.2. Production

When both trading and production at "false prices" takes place the problem becomes very much harder. The main contribution here is by Fisher [1974, 1976a, 1976b].

The simplest case arises when one allows production commitments but no actual production as long as the economy is out of equilibrium. Trade in commitments at "false" prices is permitted. For the purpose of this account, let us assume that target and active commitments of firms always coincide. That is, firm's demands are not restricted by a lack of cash. [A full account is in Fisher (1974).]

Let $\bar{y}_f(t)$ be the vector of actual commitments of firm f at t. One takes $\bar{y}_{fi}(t)>0$ to be a commitment to sell i and $\bar{y}_{fi}(t)<0$ a commitment to buy i. Desired commitments are given by the vector $y_f(t)$, which must be in the production set which is taken to be compact and strictly convex. The usual profit function $\pi_f(p)$ is well defined and we recall that $\partial\pi_f(p)/\partial p_i = y_{fi}$. If $\bar{\pi}_f(t)$ are profits at t arising from all actually traded commitments up to t, then

$$\bar{\pi}_f(t) = \bar{\pi}_f(0) + \int_0^t p(u)\dot{\bar{y}}_f(u)\,\mathrm{d}u.$$

The firm's target profits $\pi_f^*(p(t))$ at t then are

$$\pi_f^*(p(t)) = \pi_f(p(t)) - p(t)\bar{y}_f(t) + \bar{\pi}_f(t).$$

Using the definition of $\bar{\pi}_f(t)$ one finds

$$\dot{\pi}_f^*(p(t)) = \dot{p}\left[y_f(t) - \bar{y}_f(t)\right] \equiv \dot{p}a_f(t), \text{ say.} \tag{3.17}$$

Fisher now postulates a trading rule which is of the kind which led to (3.14). But now we must include trade in commitments so

$$a_i(t) = \sum_h a_{hi}(t) + \sum_f a_{fi}(t)$$

and the analogue of (3.14) for firms i,

$$a_{fi}(t)a_i(t) \leqq 0, \quad \text{all } i \text{ and } f, \tag{3.18}$$

since sales commitments are taken positively. But now from (3.17) and (3.18) together with the pricing rule (3.13) it follows that

$$\dot{\pi}_f^*(p(t)) < 0, \quad \text{unless } a_{fi}(t) \geqq 0, \quad \text{all } i \text{ with } p_i(t)a_{fi}(t) = 0, \quad \text{all } i. \tag{3.19}$$

The rest is now straightforward if one accepts the key assumption that $\sum \pi_f^*$ appears as part of the wealth of households through their share ownership of firms. Fisher has a fairly complicated story involving both money holdings by firms and dividend payments during the adjustment process, but he does not really explain why the target profits of firms are counted as wealth by households nor indeed how they come to know what these target profits are. In any case, once this is accepted, it is easy to see from (3.19) that the demonstration of declining target utilities during the process can proceed as before. Commitments are bounded because they are assumed to be in a compact set (there is no speculation) and one ensures that prices are bounded by an assumption like (3.15). The target utilities can now, as before, be made into Lyapounov functions and on the hypothesis of strictly concave utility functions convergence to an equilibrium results. It is clear that in such an equilibrium desired and actual commitments coincide. Production of commitments takes place in equilibrium.

This story is told by Fisher with much circumstantial detail, but it cannot be taken as descriptively very satisfying. In particular, one would like actual production (and consumption) to take place during the adjustment process. In Fisher (1976b) an analysis of this more difficult case is attempted. The model is rather complex and, in this account, I concentrate on a number of crucial points only.

Firms and households now make plans over the infinite future. They always assume that current prices will continue. A further crucial assumption is the following: "stocks of goods (actual not merely promised) held by participants plus the deliveries made to them are always sufficient for *current* production and consumption at time t" (p. 916). Thus, while trades may be frustrated at t, production and consumption are not. This is very strong. Otherwise the derivation of active from target trades and the rationing rule is just as we have described it before.

The crucial step is once again to show that target profits are declining during the process when production is permitted. To do this, Fisher introduces the notion of *replannable profits*. At any time after t, the replannable profit is the maximum which can be achieved given all constraints and the history of inputs and outputs up to t. This is to be contrasted with actual target profits. At any time t' after t they are the maximum which can be achieved given all constraints and the history of inputs and outputs *up to* t'. It follows that since at t the two profit notions coincide, that at t' replannable profit cannot be less than actual target profit is at t'. For it is assumed that production possibilities at any date depend on the history of inputs and outputs up to that date. In calculating replannable profits at t', there are thus fewer constraints on the firm than there are when calculating actual target profits.

It is assumed that there is no discounting of future receipts and expenditures. There is also no uncertainty. If $\pi_f^{**}(t)$ are replannable profits at t then by usual envelope arguments and the assumption that current prices persist into the indefinite future, one finds

$$\dot{\pi}_f^{**}(t)=\dot{p}\int_t^{\infty} y_f(\theta)\,\mathrm{d}\theta, \tag{3.20}$$

where $y_f(\theta)$ is the vector of planned net sales at θ. To sign (3.20), we now need a further assumption called the *present action postulate*. This is that

$$\operatorname{sign} y_{fi}(t)=\operatorname{sign}\int_t^{\infty} y_{fi}(\theta)\,\mathrm{d}\theta, \quad \text{all } i \text{ and } f. \tag{3.21}$$

Once it is granted and the usual postulate $a_{fi}(t)y_{fi}(t)\geqq 0$, $a_{fi}(t)a_i(t)\geqq 0$ is made, we deduce $\dot{\pi}_f^{**}(t)<0$ out of equilibrium when f is trading. But now for h small one has

$$\pi_f^{*}(t+h)\leqslant\pi_f^{**}(t+h)<\pi_f^{**}(t)=\pi_f^{*}(t),$$

and so actual target profits are declining.

Needless to say, various further technical assumptions are needed to clinch convergence. One must ensure boundedness of prices and commitments. The latter, when the infinite future is involved, is now more difficult to do convincingly. One must also assume that the infinite programmes of firms and households have unique continuous solutions. One needs a present action postulate for households and an argument by "replannable utilities" similar to that of replannable profits. For actual consumption of the past may affect current utility from current consumption. Lastly, since there is money (which I have ignored in presenting the arguments on the firm), one must postulate that agents "never run out of money".

The model is ingenious but the results are bought at considerable costs. Most of these, like for instance the postulate that current prices are taken as permanent, are obvious. The present action postulate is, however, the most arbitrary and unconvincing. Fisher gives a good account of why he felt obliged to make it. As he notes, when trading, etc. out of equilibrium is permitted, difficulties arise with dated goods. As the process goes through actual time, goods of particular date disappear. Since the economy is out of equilibrium, this may "suddenly" force an agent to conclude that he cannot complete his transaction. Moreover, there is no reason to suppose that prices will be such as to leave him indifferent to that discovery. There are thus continuity problems which, in part at least, are due to the hypothesis that agents always believe that they can complete their transactions. Accordingly, he did not allow trade in future dated goods. But then the present action postulate is peculiarly unconvincing. It is clear that something is wrong not just with Fisher's approach, but with the whole modelling of a process in actual time where trading, etc. is permitted at false prices.

4. Doing without the auctioneer

When one allows actual agents to change prices, one must endow them with some, possibly transient, monopoly power. Such power may stem from the circumstance that customers, while they may know the distribution of prices over firms, do not know which firm is charging the lowest prices. Accordingly, such customers must engage in search. Diamond (1971) was the first to study such a model in the context of price adjustments. He showed that in a single industry model his process converged on the monopoly price. To reach this result he had to ensure that each firm faced the same negatively inclined demand curve. Fisher (1970, 1972) makes almost the reverse assumption, namely that firms believe that they are facing perfectly elastic demand curves although they do not know with certainty the position of this curve. Recently, Hey (1974) has produced a synthesis and an account of his model and results are as follows.

We are to think of an industry with many firms. Each firm knows that it faces a downward sloping demand curve which has to be estimated and the position of which depends on the relation of that firm's price to the prices charged by other firms.

Customers sample k firms, where k is taken as given, and buy from the cheapest firm. If $f(p)$ is the density function of the distribution of prices let $F(p)$ be the distribution function. Then one calculates:

$$g(p|k)=kf(p)[1-F(p)]^{k-1}, \tag{4.1}$$

where $g(p|k)$ is the density function of the lowest price firm in a sample of k

firms. It is assumed that all customers are alike so that k is independent of the customer as is the demand curve according to which he buys from the cheapest firm. There are very many customers relative to sellers (the ratio is denoted by N), so that (4.1) gives the mean distribution of customers by the minimum price which they find in a sample of k. The ratio of customers to firms at p is $Ng(p|k)/f(p)$ so using (4.1) and writing this ratio as $R(p|k,N)$, we have

$$R(p|k,N)=Nk[1-F(p)]^{k-1}. \tag{4.2}$$

Notice now that R can be affected by a price change only if this changes the position of the firm in the distribution of the firm by price. This is because N and k are taken to be independent of price. Let $\mu(p)$ be the elasticity of R with respect to price; it is zero if the relative position of the firm is constant. If $x(p)$ is the customer's demand function, then $X(p)$, the demand function facing the firm, is $Rx(p)$. Let $-\varepsilon(p)$ be the elasticity of $x(p)$ and $\theta(p)$ that of $X(p)$. Then

$$\theta(p)=-\varepsilon(p)+\mu(p).$$

Hey now considers the special (Pareto) case

$$F(p)=1-\alpha^{\beta}p^{-\beta}$$

and assumes $x(p)$ to be of constant elasticity ε. Then easily

$$\theta=-\varepsilon+(k-1)\beta.$$

This is called the cross-section elasticity of demand because it is calculated on the assumption that the relative position of the firm changes. When this is not the case, the elasticity of demand is just $-\varepsilon$.

Now the relative position of the firm will not change if all firms follow the pricing rule:

$$p_{t+1}=a_t p_t^{b} \qquad (a_t, b_t>0). \tag{4.3}$$

For then the parameters of the distribution obey

$$\alpha_{t+1}=a_t\alpha_t^{b_t}, \tag{4.4}$$

$$\beta_{t+1}=\beta_t/b_t, \tag{4.5}$$

so that

$$1-F_{t+1}(p_{t+1})=1-F_t(p_t). \tag{4.6}$$

When will firms follow (4.3)? Suppose average costs are constant at c so that a profit maximising firm will charge a price $c\eta$, where $\eta=\varepsilon/(\varepsilon-1)$, and then assumes $\varepsilon>1$. Assume that the firm estimates η and denote this estimated by η^*. Hey assumes that the true value of η is observed by the firm but that it up-dates its estimate adaptively according to

$$\eta^*_{t+1}=(\eta^*_t)^{\lambda}\eta^{1-\lambda}. \tag{4.7}$$

From this and profit maximisation

$$p_{t+1}=c\eta^*_{t+1}=c\eta^{*\lambda}_t\eta^{1-\lambda}=(c\eta)^{1-\lambda}p_t\lambda, \tag{4.8}$$

since $c\eta^*_t=p_t$. But now (4.8) is of the same form as (4.3) with $a_t=(c\eta)^{1-\lambda}$ and $b_t=\lambda$. The assumptions are strong but the example shows that (4.3) can be generated by profit maximisation. Using these values of a_t and b_t, it now follows from (4.4) and (4.5) that the distribution converges on the degenerate distribution at the monopoly price $c\eta$. It should now be noted that if firms acted on the basis of the cross-section elasticity of demand, then the process would approach the competitive solution (since $\beta_t\to\infty$). The results can be generalised to other distributions. In particular, convergence to the monopoly solution depends only on firms throughout maintaining their relative position in the price distribution.

Fisher (1973) has a more elaborate analysis of the single industry case, the outcome of which is convergence to the competitive equilibrium. The model is too elaborate to be given a detailed account here. It hinges on a number of crucial assumptions such as that in a situation of excess demand at least some dealer will behave as the auctioneer would have done. Also, customer search is sufficiently efficient to lead to "low" price firms facing perfectly elastic demand curves over a range which is such that marginal cost=price is a maximising solution. The search process is not directly modelled but is appealed to in justifying some of the assumptions. A particularly important one is the following: if the minimum price charged by firms is bounded above by some p^0 for long enough, then the maximum priced firms will not charge above p^0 for ever.

It is clear that in this area the assumptions are everything and there is a lack of primitive postulates from which the behaviour of the system could be deduced. Fisher has certainly shown that it is possible to have search models with temporary monopoly power which converge to the competitive solution. Diamond and Hey, on the other hand, have demonstrated that monopoly power may continue in the limit. No attempt has as yet been made to link these stories to the theory of large economies, leave alone to elaborate them to permit phenomena such as entry into the industry. It is also a drawback of the present state of play that agents behave myopically and in particular that expectations play no role.

Fisher (1972) is also the only one to study price movements determined by actual agents in a general equilibrium context. This is a model of pure exchange in which certain agents are dealers, i.e. agents who set prices. This they do by guessing the market price at which they think that they can transact what they wish. If there is excess demand, they conclude that they have guessed a price which is too low and raise it, and vice versa when there is excess supply. Search by customers once again is sufficiently good to effect all arbitrage as time goes to infinity. A crucial assumption here is that a low price dealer faces a higher excess demand than does a high priced one. Money is introduced very much in the manner of Arrow–Hahn (1971). Certain assumptions are made to avoid discontinuities which might occur as customers change dealers. Convergence to the competitive equilibrium is established by showing that target utilities can serve as a Lyapounov norm when a "Hahn-process" postulate is used.

References

Arrow, K. J., H. D. Block, and L. Hurwicz (1959), "On the stability of the competitive equilibrium II", Econometrica, 27:82–109.

Arrow, K. J. and L. Hurwicz (1958a) "On the stability of the competitive equilibrium I", Econometrica, 26:522–52.

Arrow, K. J. and L. Hurwicz (1960), "Competitive stability under weak gross substitutability: The 'Euclidean distance' approach", International Economic Review, 1:38–49.

Arrow, K. J. and L. Hurwicz (1962), "Competitive equilibrium under weak gross substitutability: Non-linear price adjustment and adaptive expectations", International Economic Review, 3:233–255.

Arrow, K. J. and N. Nerlove (1958), "A note on expectation and stability", Econometrica, 26:297–305.

Arrow, K. J. and F. H. Hahn (1971), General competitive analysis. San Francisco: Holden-Day and Edinburgh: Oliver and Boyd.

Attouchett and A. Damlauian (1972), "On multivalued evolution equations in Hilbert spaces", Israel Journal of Mathematics, 12:373–390.

Brock, W. A. and J. Scheinkman (1975), "Some results of global asymptotic stability of difference equations", Journal of Economic Theory, 10.

Castaing, C. and M. Valadier (1969), "Equations differentielle multivoque dans les espaces localement conveses", Revue Francaise d'Informatique et de Recherche Operationelle, 16:3–16.

Champsaur, P., J. Drèze and C. Henry (1977), "Stability theorems with economic applications", Econometrica, 45:273–294.

Debreu, G. and J. N. Herstein (1953), "Non-negative square matrices", Econometrica, 21:597–607.

Debreu, G. (1959), Theory of value. New York: Wiley.

Debreu, G. (1970), "Economies with a finite set of equilibria", Econometrica, 38:387–392.

Debreu, G. (1974), "Excess demand functions", Journal of Mathematical Economics, 1:15–23.

Diamond, P. (1971), "A model of price adjustment", Journal of Economic Theory, 3:156–168.

Dierker, E. and H. Dierker (1972), "The local uniqueness of equilibria", Econometrica, 40:867–881.

Dierker, E. (1972), "Two remarks on the number of equilibria of an economy", Econometrica, 40:951–955.

Drèze, J. H. and D. de la Vallee Poussin (1971), "A tâtonnement process for public goods", Review of Economic Studies, 38:133–150.

Drèze, J. H. (1975), "Existence of an exchange equilibrium under price rigidities", International Economic Review, 16:301–320.

Enthoven, A. C. and K. J. Arrow, (1956), "A theorem on expectations and the stability of equilibrium", Econometrica, 24:288–293.

Fisher, F. (1970), "Quasi-competitive price adjustment by individual firms: A preliminary paper", Journal of Economic Theory, 2:195–206.
Fisher, F. (1972a), "On price adjustment without an auctioneer", Review of Economic Studies, 39:1–16.
Fisher, F. (1972b), "A simple proof of the Fisher–Fuller theorem", Proceedings of the Cambridge Philosophical Society.
Fisher, F. (1972c), "Gross-substitutes and the utility function", Journal of Economic Theory, Vol. 4, No. 1:82–87.
Fisher, F. (1974), "The Hahn process with firms but no production", Econometrica, 42:471–486.
Fisher, F. (1976a), "The stability of general equilibrium: Results and problems", in: Artis and Nobey, eds., Essays in economic analysis. Cambridge: Cambridge University Press.
Fisher, F. (1976b), "A non-tâtonnement model with production and consumption", Econometrica, 44:907–938.
Fisher, F. (1978), "Quantity constraints, spillovers and the Hahn process", Review of Economic Studies, 45:19–31.
Fisher, M. E. and A. T. Fuller (1958), "On the stabilisation of matrices and the convergence of linear iteration processes", Proceedings of the Cambridge Philosophical Society, 54:417–425.
Feldman, A. M. (1974), "Recontracting stability", Econometrica, 42:35–44.
Gorman, W. M. (1953), "Community preference fields", Econometrica, 21:63–80.
Graham, D. A. and E. R. Weintraub (1975), "On convergence to Pareto allocations", Review of Economic Studies, 42:469–472.
Green, J. R. (1974), "The stability of Edgeworth's recontracting process", Econometrica, 42:21–34.
Hahn, F. H. (1958), "Gross substitutability and the dynamic stability of general equilibrium", Econometrica 26:169–170.
Hahn, F. H. (1962a), "On the stability of pure exchange equilibrium", International Economic Review, 3:206–213.
Hahn, F. H. and T. Negishi (1962b), "A theorem on non-tâtonnement stability", Econometrica, 30:463–469.
Hahn, F. H. (1968), "On some propositions of general equilibrium analysis", Economica, 35:424–430.
Heal, G. M. (1973), The theory of economic planning. Amsterdam: North-Holland.
Henry, C. (1972), "Differential equations with discontinuous right hand side for planning procedures", Journal of Economic Theory, 4:545–551.
Henry, C. (1973), "An existence theorem for a class of differential equations with multivalued right-hand side", Journal of Mathematical Analysis and Applications, 41:179–186.
Henry, C. (1974), "Problèmes d'existence et de stabilité pour des processes dynamiques considerés en economie mathematique", Compte Rendue de l'Academie des Sciences, Paris, 278 (Series A): 97–100.
Hey, John D. (1974), "Price adjustment in an atomistic market", Journal of Economic Theory 8:483–499.
Hicks, J. (1939), Value and capital. Oxford: Clarendon Press.
Hurwicz, L., R. Radner and S. Reiter (1975), "A stochastic decentralised resource allocation process I and II", Econometrica, 43:187–221, 363–393.
Hurwitz, A. (1895), "Ueber die Bedingungen unter Welchen eine Gleichung nur Wurzeln mit negativen reellen Theilen besitzt", Mathematishche Annalen, XLVI.
Karlin, S. (1959), Mathematical methods and theory in games, programming and economics. Reading, Mass: Addison-Wesley.
Kennedy, C. (1970), "The stability of the Morishima system", Review of Economic Studies, 37:173–175.
La Salle J. and S. Lefschetz (1961), Stability by Lyapounov's direct method with applications. New York: Academic Press.
Lyapounov, A. (1947), Problème général de la stabilité du mouvement, Annals of Mathematics Study, Vol. 17. Princeton: Princeton University Press.
Mantel, R. (1974), "On the characterisation of aggregate excess demand", Journal of Economic Theory, 7:348–353.
McFadden, D. (1968), "On Hicksian stability", in: J. N. Wolfe, ed., Value, capital and growth. Edinburgh: Edinburgh University Press.

McKenzie, L. (1960a) "Stability of equilibrium and the value of positive excess demand", Econometrica, 28:606–617.
McKenzie, L. (1960b), "Matrices with dominant diagonal and economic theory", in: K. J. Arrow, S. Karlin and P. Suppes, eds., Mathematical methods in the social sciences. Stanford: Stanford University Press.
Metzler, L. (1945), "The stability of multiple markets: The Hicks conditions", Econometrica, 13:277–292.
Milnor, J. (1972), Topology from the differentiable viewpoint. Charlottesville: University Press of Virginia.
Morishima, M. (1952), "On the law of change of price-system in an economy which contains complementary commodities", Osaka Economic Papers, 1:101–113.
Morishima, M. (1957), "Notes on the theory of stability of multiple exchange", Review of Economic Studies, 24:203–208.
Morishima, M. (1962), "The stability of exchange equilibrium: an alternative approach", International Economic Review, 3:214–217.
Morishima, M. (1964), Equilibrium stability and growth. Oxford: Clarendon Press.
Morishima, M. (1970), "A generalisation of the gross substitute system", Review of Economic Studies, 37:177–186.
Mukherji, A. (1972), "On complementarity and stability", Journal of Economic Theory, 4:442–457.
Negishi, T. (1958), "A note on the stability of an economy where all goods are gross substitutes", Econometrica, 26:445–457.
Negishi, T. (1961), "On the formation of prices", International Economic Review, 2:122–126.
Negishi, T. (1962), "The stability of a competitive economy: A survey article", Econometrica, 30:635–669.
Negishi, T. (1964), "Stability of exchange and adaptive expectations", International Economic Review, 5:104–111.
Nemetskii, V. V. and V. V. Stepanov (1960), Qualitative theory of differential equations. Princeton: Princeton University Press.
Rothschild, M. (1973), "Models of market organisation with imperfect information: A survey", Journal of Political Economy, 81:1283–1308.
Routh, E. J. (1877), Stability of given state of motion. London.
Saaris, D. G. and C. T. Simon (1978), "Effective price mechanisms", Econometrica, 46, No. 5:1097–1125.
Samuelson, P. A. (1941, 1942), "The stability of equilibrium", Econometrica, 9:97–120, and 10:1–75.
Samuelson, P. A. (1947), Foundations of economic analysis. Cambridge, Mass: Harvard University Press.
Sansome, G. and Conti, R. (1964), Non-linear differential equations. London: Pergamon Press.
Smale, S. (1974), "Global analysis and economics IV; Finiteness and stability of equilibria with general consumption sets and production", Journal of Mathematical Economics, 1:119–127.
Smale, S. (1976), "A convergent process of price adjustments and global Newton Methods", Journal of Mathematical Economics, 3:1–14.
Sonnenschein, H. (1972), "Market excess demand functions", Econometrica, 40:549–563.
Sonnenschein, H. (1973), "Do Walras identity and continuity characterize the class of community excess demand functions?", Journal of Economic Theory, 6:345–354.
Uzawa, H. (1959), "Walras' tâtonnement in the theory of exchange", Review of Economic Studies, 27:182–194.
Uzawa, H. (1961), "The stability of dynamic processes", Econometrica 29:617–631.
Uzawa, H. (1962), "On the stability of Edgeworth's barter process", International Economic Review, 3:218–232.
Varian, R. (1975), "A third remark on the number of equilibria of an economy", Econometrica, 13:985–986.

Chapter 17

REGULAR ECONOMIES

EGBERT DIERKER*

University of Bonn

1. Introduction

The main task of Walrasian equilibrium theory is to explore how prices help to coordinate individual decisions. For a given economy one would like to have a well-determined price system which, if known to each individual agent, would induce the agent to make his economic decision independently of the decisions of all other agents in such a way that all individual decisions happen to be compatible with each other.

There are several basic problems that arise immediately. First, the coordination of individual decisions by prices seems to require that individual decisions are uniquely determined by prices. However, there is no reason to assume that this requirement is fulfilled for every agent. If an agent is a consumer with non-convex preferences then his demand need not be uniquely specified for every price system. One may hope, though, that those consumers whose demand at a given price system does not consist of a unique commodity bundle form a very small subset of all consumers so that aggregate demand can reasonably well be described by a function of prices. On the production side, however, the situation is worse. There is empirical evidence that constant returns to scale play an important role. To take an extreme case, consider a production sector that can be described by a linear model such as a Leontief model. A producer in such a model will be inactive or have no profit maximizing production plan at all unless the prevailing price system leads to a continuum of profit maximizing production plans. It is the rule here rather than the exception that production plans are far from being uniquely determined by equilibrium prices. The importance of constant returns to scale is a reason not to expect that the aggregate supply of all producers can be described by a function of prices. As far as we shall use the concept of an excess demand function in this chapter, we shall therefore restrict

*This chapter is based on my survey article on regular economies that appeared in *Frontiers of Quantitative Economics*, vol. III A, edited by M. D. Intriligator, Amsterdam, 1977, 167–189. This research has been supported in part by the Deutsche Forschungsgemeinschaft, Sonderforschüngsbereich, 21.

ourselves to the case of pure exchange. Also, the implicit assumption that no agent can influence prices fits far better to consumers than to producers.

Furthermore, assume now that the decisions of almost all individual agents are uniquely determined by prices. The aggregate excess demand function then can still be highly sensitive with respect to small price variations, e.g. it can have extremely steep slopes. Although there will exist a precise equilibrium price system, one can hardly say that the use of prices is a viable method to achieve the coherence of decentralized individual decisions if very small price variations can lead to considerable changes of economic plans. Therefore, to keep the slope of excess demand bounded in a neighborhood of equilibrium prices, it is useful to deal with continuously differentiable demand functions. Empirical observations seem to support the view that aggregate demand can be described by a smooth function.

Finally, assume that individual decisions are not only unique but also depend smoothly on prices. Assume, furthermore, that all markets clear at price system p. Then p cannot be considered a well-determined equilibrium price system if it happens that markets also clear at price systems very close to p. The occurrence of equilibrium price systems that are not isolated may cause severe difficulties, e.g. for purposes of comparative statics. Thus, one would like to have aggregate economic plans not being too insensitive with respect to price changes. The concept of a regular economy captures the desirable properties of determinacy and robustness of equilibrium price systems. These properties are intimately related to the sensitivity of excess demand with respect to small price variations.

The concept of a regular economy was introduced into economic theory by Debreu (1970). Two questions were studied in Debreu's seminal paper.

(1) What does the set of equilibrium prices of an economy typically look like, and
(2) how does the set of equilibrium prices of an economy E vary when the characteristic data of E vary?

These problems, which are of basic importance for the explanatory value of equilibrium theory, have to be posed in this generality because the uniqueness of equilibrium prices can only be derived from highly restrictive assumptions [cf. Arrow and Hahn (1971, ch. 9)]. It is the consumption sector of an economy rather than the production sector which may cause the multiplicity of equilibrium prices. It is a difficult open problem to formulate reasonable assumptions on the distribution of consumers' characteristics which entail the uniqueness of equilibrium prices. Therefore, being aware of the possibility that multiple equilibria may occur, one wants to establish the local uniqueness of each equilibrium price system and its continuous dependence on the characteristic data of the economy.

The continuous dependence of an equilibrium price system on the economic data is particularly desirable because data such as consumers' preferences can certainly not be observed precisely. However, even in simple examples of economies, called "critical", this continuity requirement is not satisfied. Such examples need not have many equilibrium price systems, but they may have a whole continuum in extreme cases. The remarkable fact about critical economics is that, loosely speaking, they form a set of very small size. Most economies are regular and their equilibrium price systems depend continuously on the economic data. The purpose of this chapter is, essentially, to give this statement a precise form and to indicate why it is true. This is done in models of different generality. For a concise survey of regular economies, see Debreu (1976).

Although critical economies can be considered atypical or even negligible for the purpose of explaining the observed state of a given economy in a static framework, they are of interest to explain sudden changes in an economy. An illustrative example of an economic model involving sudden changes at critical states is Zeeman's (1974) model of a stock market where longer periods of relatively slow movements are disrupted from time to time by short periods of relatively fast change. In a way similar to Zeeman's one may associate various qualitative changes with certain types of critical phenomena, in particular with elementary catastrophes in the sense of Thom's catastrophe theory [cf. Thom (1972), Bröcker (1975) and Zeeman (1977)]. Critical economies also make us aware of the fact that comparative statics between two economies may very much depend on the path that is chosen to transform one economy into the other. That is to say, if p^* is an equilibrium price system of an economy, E_0, and if there are two continuous paths connecting E_0 with another economy, E_1, and if to each of these paths there corresponds a well-defined continuous path of equilibrium price systems starting at p^*, then it may very well happen that the paths lead to completely different equilibrium price systems of E_1. The study of critical economies is not far developed at present, see however Balasko (1978b) and (1979b). In this article emphasis lies on regular, not on critical, economies.

The theory of regular economies is useful for studies of the local uniqueness of equilibria, of the continuity of the equilibrium price correspondence, and of the possibility of comparative statics. The theory of regular economies also helps give a better understanding of the core correspondence, since core allocations in a large economy can be approximately decentralized through prices; see Hildenbrand's contribution to this Handbook. Furthermore, there is an intimate relationship between the theory of regular economies and the new methods of computing economic equilibria. The link between these two branches of equilibrium theory becomes apparent from the work of Eaves and Scarf (1976) and Smale (1976); see Scarf's and Smale's contributions to this Handbook.

Regularity studies play a role, too, in fields related to economic equilibrium theory. For instance, the general complementarity problem in mathematical programming has been investigated by regularity methods; see Saigal and Simon (1973). Another field where regularity methods can be applied is game theory. In this chapter we restrict ourselves to the Arrow–Debreu equilibrium theory. Temporary equilibria, as studied in Fuchs and Laroque (1976), or fixed price equilibria, as studied in Laroque and Polemarchakis (1978) and in Wiesmeth (1979), are not considered here; see Grandmont's contribution to this Handbook. Other problems arising in this context, such as the basic problem of existence of price equilibria or the problem of how equilibrium prices are reached, are discussed in various other chapters of this Handbook.

The investigation of regular equilibria has led to a revival of calculus methods in equilibrium theory. Mas-Colell (1976) presents a valuable comprehensive study of many important questions of equilibrium theory in a differentiable framework. A detailed mathematical study focusing on price equilibria is Balasko (1976).

2. Excess demand approach: Debreu's theorem

We want to study the problem of decentralization of economic decisions by means of prices under the assumption that individual decisions are uniquely determined by prices and vary smoothly with the price system. It seems appropriate for that purpose to focus attention on the concept of a smooth excess demand function. An equilibrium price system of an economy will be a zero of its excess demand function, and the local study of equilibria will amount to a local study of the excess demand function near its zeros.

We would like to know what the equilibrium set of a given economy looks like and how it depends on the parameters describing the economy. To formulate the problem in a precise way we have to have a space of economies. As excess demand functions make far more sense in exchange economies than in economies with production, we concentrate upon the case of pure exchange. A pure exchange *economy* consists of a fixed number of consumers, $i=1,\ldots,m$, exchanging commodities $h=1,\ldots,\ell$. A *consumer* is described by an initial endowment and by a smooth demand function. For simplicity's sake we keep the demand functions fixed in this section so that initial endowments become the only variable economic parameters. The space of economies can thus be identified with a subset of a finite dimensional vector space, an economy E being an m-tuple $(e_1,\ldots,e_m)$ of initial endowment vectors.

The *commodity space* is the ℓ-dimensional Euclidean space, $\boldsymbol{R}^\ell$. The *price space* is

$$S=\left\{p=(p_1,\ldots,p_\ell)\gg 0 \,|\, \|p\|=\left(\sum_{h=1}^{\ell} p_h^2\right)^{\frac{1}{2}}=1\right\}.^1$$

We consider individual *initial endowments* in $P=\operatorname{int}\boldsymbol{R}^\ell_+$. The *wealth* of consumer

[1] We use $\gg$ to denote strict inequality in each component, $\geqslant$ to denote weak inequality in each component, and avoid other forms of vector inequalities.

i at price system $p \in S$ and endowment $e_i \in P$ is $w_i = pe_i \in]0, \infty[$. Consumer i has a C^1 *demand function* f_i: $S \times]0, \infty[\to \mathbf{R}^\ell_+ (i = 1, \ldots, m)$, $\mathbf{R}^\ell_+$ being i's consumption set. The demand functions are supposed to fulfill $pf_i(p, w) - w = 0$ for any price–wealth pair $(p, w) \in S \times]0, \infty[$.

The *space of economies* is $\mathfrak{S} = P^m$. The *excess demand* of economy $E = (e_1, \ldots, e_m) \in \mathfrak{S}$ at price system p is

$$Z_E(p) = \sum_{i=1}^{m} (f_i(p, pe_i)) - e_i).$$

We normalize prices with respect to the Euclidean norm because *Walras' Law*, i.e. $pZ_E(p) = 0$ for all p, then allows us to interpret $Z_E(p)$ as a vector tangent to S at p.

We assume that all *goods are desirable* in the following sense:

(D) If the sequence $(p^q, w_1^q, \ldots, w_m^q)$ in $S \times]0, \infty[^m$ converges to $(p^0, w_1^0, \ldots, w_m^0) \in (\bar{S} \setminus S) \times]0, \infty[^m$, then $\sum_{i=1}^m \| f_i(p^q, w_i^q) \|$ converges to infinity.

This says in particular that the demand for commodity h becomes arbitrarily large if p_h is the only variable approaching zero. If demand functions are derived from preferences, (D) follows from strict monotonicity of preferences.

An equilibrium price system of E is a zero of the excess demand Z_E. *If* (D) *holds, then every* $E \in \mathfrak{S}$ *has at least one equilibrium*, as can be seen from a degree-type argument or, equivalently, from a fixed point argument. For rigorous proofs of the existence of equilibria, see Debreu's and Smale's contributions to this Handbook. Also, compare the remarks on indices in Section 3. An informal existence argument goes as follows. Let g_0 be a continuous vector field on S which points radially inwards to the center $(\ell^{-\frac{1}{2}}, \ldots, \ell^{-\frac{1}{2}})$ of S. By definition g_0 has a zero at $(\ell^{-\frac{1}{2}}, \ldots, \ell^{-\frac{1}{2}})$. We want to use this fact and assumption (D) to conclude that Z_E must have a zero. For that purpose we transform g_0 continuously into $Z_E = g_1$ by putting

$$g_t = (1-t) g_0 + t Z_E, \qquad 0 \leq t \leq 1.$$

Degree theory, or fixed point theory, says that g_t must have a zero for all $t \in [0, 1]$ provided $C = \bigcup_{t=0}^{1} \{p \in S \mid g_t(p) = 0\}$ stays away from the boundary of S, i.e. provided C is compact ["homotopy invariance"; see Dold (1972, p. 204)]. In particular, $g_1 = Z_E$ has a zero if C is compact. Now (D) implies that $Z_E(p)$ does not point radially away from the center of S provided p is close enough to the boundary of S. Therefore $Z_E(p)$ and $g_0(p)$ do not point into opposite directions outside a compact subset of S. Hence no convex combination of Z_E and g_0 can vanish outside a compact subset of S. Thus, C is compact.

The *equilibrium price correspondence* $\Pi: \mathfrak{E} \to S$ assigns the set $\{p \in S \mid Z_E(p)=0\}$ to $E \in \mathfrak{E}$. To describe the set of equilibrium prices of an economy E and its dependence on E, we study the graph Γ of Π. Defining $Z: \mathfrak{E} \times S \to \boldsymbol{R}^\ell$ by $Z(E, p)=Z_E(p)$ we obtain that

$$\text{graph}(\Pi)=\Gamma=Z^{-1}(0)$$

is closed, since Z is continuous. This is a weak continuity property of Π. Moreover, *if* (D) *holds, then the equilibrium price correspondence is upper hemicontinuous* (*u.h.c.*) *and compact-valued*. That is to say, if (E_n) is a sequence in $\mathfrak{E}$ converging to $E \in \mathfrak{E}$ and if $p_n \in \Pi(E_n)$, n a natural number, then (p_n) has a subsequence converging to an element of $\Pi(E)$. This property follows directly from the definitions. Hildebrand and Mertens (1972) derived, in a rather general model that Π is u.h.c.

Because of Walras' Law all markets clear at p iff any $\ell - 1$ markets clear at p. Let

$$\hat{Z}: \mathfrak{E} \times S \to \boldsymbol{R}^{\ell-1}$$

be obtained from Z by deleting, say, the last component. Let $p \in \Pi(E)$, i.e. $\hat{Z}(E, p)=0$. The following simple calculation shows that the derivative $D\hat{Z}(E, p)$ is surjective, i.e. the derivative of $\hat{Z}$ at (E, p) can be represented by a matrix of maximal rank $\ell - 1$. Let $e_i=(e_{i1},\ldots,e_{i\ell}) \in P$ be the endowment of consumer i. Partial differentiation of the excess demand for commodity k,

$$Z_k(E, p)= \sum_{i=1}^{m} (f_{ik}(p, pe_i)-e_{ik}),$$

with respect to $e_{ih}(h, k \in \{1,\ldots, \ell-1\})$ yields

$$\frac{\partial Z_k}{\partial e_{ih}}(E, p)=\frac{\partial f_{ik}}{\partial w}(p, pe_i)\cdot p_h - \delta_{hk},$$

where $\delta_{hk}=0$ for $h \neq k$ and $\delta_{hk}=1$ for $h=k$. Also,

$$\frac{\partial Z_k}{\partial e_{i\ell}}(E, p)=\frac{\partial f_{ik}}{\partial w}(p, pe_i)\cdot p_\ell.$$

Hence the mapping $\hat{Z}$ has rank $\ell-1$ at (E, p). By the implicit function theorem *the graph of the equilibrium price correspondence*, $\Gamma=\hat{Z}^{-1}(0)$, *is a differentiable manifold of dimension* $m\ell$, which is called an *equilibrium price manifold*. This means that each point in Γ has a neighborhood in Γ which is diffeomorphic to an open subset of $\boldsymbol{R}^{\ell m}$. Figure 2.1 gives a low-dimensional picture of the equilibrium

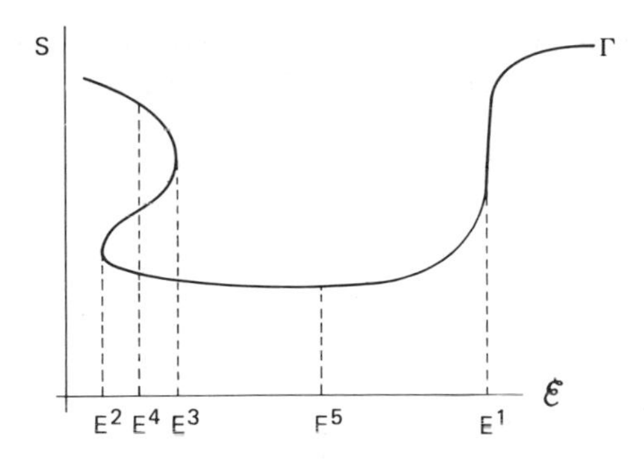

Figure 2.1.

price manifold. A reader not familiar with the concepts of a differentiable manifold, a differentiable mapping, etc. is referred to Milnor (1965).

The equilibrium price correspondence drawn in Figure 2.1 is continuous except at a few points, E^1, E^2, and E^3. Economy E^1 represents an extremely bad case. Although E^1 appears very artificial, it can easily be represented in an Edgeworth box. E^1 has a whole continuum of equilibria, but arbitrarily close there are economies with a unique equilibrium. No equilibrium price p of E^1 has any robustness at all. By a slight parameter change one obtains an economy with its unique equilibrium far away from p. Whenever one happens to cross E^1, equilibrium prices are forced to a sudden change. At E^2 (resp. E^3) the situation is somewhat better. If one crosses E^2 (resp. E^3) one may be forced to jump. But if so, one jumps to another equilibrium price system which is well determined in the sense that no sudden change takes place if one crosses E^2 (resp. E^3) again. One may interpret the situation as an economy coming to rest in a new position after a radical change.

If one considers E^4 then all three equilibrium price systems are robust in the following sense. There is a neighborhood U of E^4 such that to every movement along a smooth path starting at E^4 and staying in U and to every equilibrium price system p of E^4 there corresponds a movement of equilibrium prices along a well-determined smooth path on Γ which starts at (E^4, p). In particular, the equilibrium price correspondence is continuous at E^4. The situation is similar at E^5 where the equilibrium is unique.

A two-dimensional drawing like Figure 2.1 is suitable only to illustrate economies which are either well-behaved like E^4 or E^5 or to illustrate jumps caused by the fact that Γ is folded above economies like E^2 and E^3. Above E^1 the shape of Γ is very artificial. In contrast to the folds the shape of Γ is very sensitive here with respect to disturbances. A higher dimensional version of Figure 2.1 would allow us to illustrate more complicated phenomena than folds occurring in a natural way. For drawings of this kind see books on catastrophe theory or on singularity theory, e.g. Bröcker (1975) and, in particular, Woodcock and Poston (1974).

In our analysis here we shall distinguish only two classes of economies: regular economies like E^4 and E^5, where there are a finite number of equilibria each responding smoothly to small parameter changes, and critical economies like E^1, E^2, and E^3, where there is a finite or, in extreme cases, infinite number of equilibria, and where at least one equilibrium price system does not respond smoothly to a small parameter change. We do not distinguish between different types of critical economies although this might be useful for the study of some dynamic phenomena. For an example see Zeeman (1974).

We proceed to give a first definition of a regular (resp. critical) economy that is motivated by Figure 2.1. Let $pr = \text{proj}_1|\Gamma$: $\Gamma \to \mathfrak{E}$ project the equilibrium price manifold to the parameter space $\mathfrak{E}$. The mapping pr is a left inverse of the equilibrium price correspondence Π, i.e. $pr \circ \Pi = id$. At a regular economy E the space of vectors tangent to Γ at any equilibrium of E is projected onto the tangent space of $\mathfrak{E}$ at E. A critical economy possesses an equilibrium price system where the projection of the tangent space of Γ is not surjective, i.e. has rank less than ℓm.

Definition

If f: $M \to N$ is a differentiable mapping then $m \in M$ is called a *critical point* of f if the derivative $Df(m)$ of f at m is not surjective. A *critical value* of f is the image $f(m)$ of a critical point m of f. A *regular value* of f is a point in N which is not a critical value.

Note that a regular value of f need not be a value of f.

Definition

$E \in \mathfrak{E}$ is a *regular* (resp. *critical*) *economy* iff E is a regular (resp. critical) value of the projection pr: $\Gamma \to \mathfrak{E}$.

To extend this definition of a regular economy to some larger space of economic parameters, the parameter space and the graph of the equilibrium price correspondence have to be differentiable manifolds of possibly infinite dimensions. The above definition therefore does not lend itself to the case where the consumption sector of an economy with many agents is described by a measure and the space of these measures is endowed with the weak topology. To deal with the case where differentiable manifolds are not available we shall present an alternative definition in Section 3.

The above definition of a regular economy is particularly well suited to yield, by Sard's theorem, that "almost all" economies are regular. Here "almost all" has a probabilistic meaning. $\mathfrak{E}$ being a subset of a Euclidean space, the concept of a

Lebesgue null-set is available. A set $A \subset \boldsymbol{R}^n$ is a *null set* iff for every $\varepsilon > 0$ there exists a covering of A by a countable collection of cubes in $\boldsymbol{R}^n$, the volumes of which add up to less than ε. Since a countable union of null sets is null, the concept of a null set in $\boldsymbol{R}^n$ is a local one. One checks easily that diffeomorphisms transfer null sets into null sets so that the concept of a null set is available for finite dimensional manifolds. The following theorem applied to $pr\colon \Gamma \to \mathfrak{E}$ shows that the set of critical economies is null.

Theorem

Let M and N be finite dimensional differentiable manifolds, $\dim M = \dim N$. The set of critical values of a C^1 mapping $f\colon M \to N$ is null.

Corollary

The set of critical economies is null.

The proof of this theorem, which presents a special case of *Sard's theorem*, relies on the mean value theorem for differentiable functions; see Sternberg (1964, p. 48f). An extension to spaces of finite but not necessarily equal dimensions can be found in Milnor (1965, §3). For infinite dimensional spaces see Smale (1965) or Abraham and Robbin (1967, §16). The concept of a null set, which is not naturally available on infinite dimensional manifolds, is replaced by the topological concept of a set of first category, i.e. a countable union of nowhere dense sets. Under suitable compactness assumptions nowhere dense sets can be used instead of sets of first category.

Smallness in the topological sense and smallness in the probabilistic sense are two very different concepts. In the case of a finite dimensional parameter space both notions are available. One may ask whether the set of critical economies is small in the topological sense.

Lemma

If the demand functions $f_1, \ldots, f_m$ are C^1 and if (D) holds, then the set of critical economies is closed and nowhere dense in $\mathfrak{E}$.

Proof

Let B be a compact ball in $\mathfrak{E}$. We want to show that the set of critical economies in B is compact and nowhere dense. Because of (D) the equilibrium price correspondence is u.h.c. and compact-valued. Hence $pr\colon \Gamma \to \mathfrak{E}$ is proper and $pr^{-1}(B)$ is compact. Let C denote the set of critical points of $pr\colon \Gamma \to \mathfrak{E}$. C is closed by definition. Therefore $C \cap pr^{-1}(B)$ is compact. It follows that $pr(C \cap pr^{-1}(B))$ is compact. By Sard's theorem this set is also null. Being a compact null set the set $pr(C \cap pr^{-1}(B))$ of critical economies in B is nowhere dense. Q.E.D.

Summarizing we obtain the following theorem which is due to Debreu (1970).

Theorem

Given m continuously differentiable demand functions $(f_1,\ldots,f_m)$, if (D) is satisfied, then the set of critical economies in $\mathscr{E}=P^m$ is a closed null set.

It has also been observed by Debreu in his original paper (1970) that regular economies have a finite number of equilibria and this number remains locally constant when parameters are changed.

Proposition

Under the assumptions of the theorem, if E is a regular economy, then there is a neighborhood V of E in $\mathscr{E}=P^m$ and there are r C^1 functions $g_1,\ldots, g_r$: $V\to S$ such that for every $E'\in V$, the set $\Pi(E')$ consists of the r distinct elements $g_1(E'),\ldots, g_r(E')$.

Proof

If E is a regular economy and if p is an equilibrium price system of E, then there is, according to the inverse function theorem, a neighborhood U of (E, p) in Γ such that $pr|U$ is a diffeomorphism. In particular, equilibrium prices are isolated. Since the equilibrium price correspondence Π: $\mathscr{E}\to S$ is compact-valued, E has a finite set of equilibrium prices, $p_1,\ldots,p_r$. To each equilibrium price there corresponds a local diffeomorphism describing the dependence of that equilibrium price on the economic parameters. Q.E.D.

3. Excess demand approach: Some extensions

We proceed to give another definition of a regular economy. Whether an economy is regular or critical can be seen from the behavior of its excess demand function. The space of economic parameters is not involved in the second definition.

To illustrate the definition we look at economies with two commodities. We identify the price space S with its projection to the first coordinate axis. Let $\zeta_1(p_1)$ be the excess demand for commodity 1 if the price for commodity 1 equals $p_1\in]0,1[$. The excess demand for commodity 2 is given by Walras' Law and need not be considered explicitly. We draw excess demand functions ζ_1 corresponding to the economies E^1, E^2, E^3, E^4, and E^5, which we discussed in connection with Figure 2.1.

The excess demand of economy E^1 may appear as in Figure 3.1. Economies E^2 or E^3 correspond to an excess demand function as plotted in Figure 3.2. Figure 3.3 represents the excess demand of E^4 and Figure 3.4 that of E^5.

The equilibria which are not robust with respect to parameter changes correspond to zeros of the excess demand where the derivative $\zeta_1'(p)$ vanishes. The regular economies E^4 and E^5 do not have such equilibria.

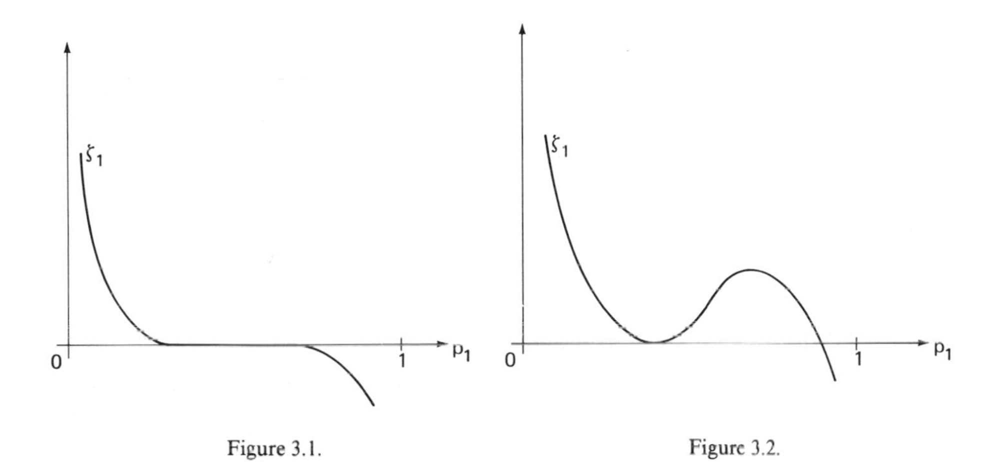

Figure 3.1.

Figure 3.2.

The robustness of an equilibrium price system will be captured by the following definition.

Definition

Let ζ be the excess demand of economy E. A price system $p \in S$ is a *regular equilibrium price system* iff $\zeta(p)=0$ and $[(\partial\zeta_1,\ldots,\partial\zeta_{\ell-1}/\partial p_1,\ldots,\partial p_{\ell-1})(p)]$ has full rank.

Using the concept of the Hessian of a vector field at a zero [see Abraham and Robbin (1967, §22)] this definition can be formulated without the use of local coordinates: *$p \in S$ is a regular equilibrium price system iff p is a non-degenerate zero of the vector field ζ on S*, that is to say, iff $\zeta(p)=0$ and the Hessian of ζ at p has all eigenvalues different from zero. Since the eigenvalues of the Hessian do not

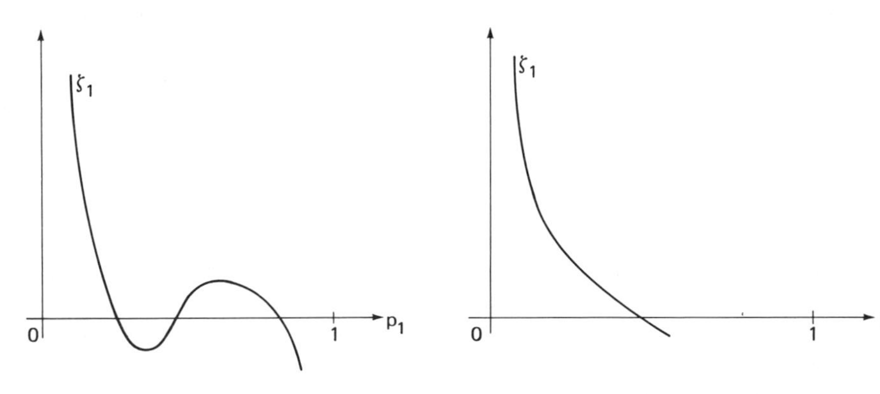

Figure 3.3.

Figure 3.4.

depend on the choice of the local coordinates, the way commodities are indexed does not play any role. Note that regular equilibria are isolated.

As suggested by the drawings, the definition of a regular economy and the definition of a regular equilibrium price system fit together in the following way.

Theorem

Given m continuously differentiable demand functions, let the space of economies be $\mathfrak{S}=P^m$. The economy $E\in\mathfrak{S}$ is regular iff all equilibrium prices of E are regular.

In other words, E is a regular economy iff 0 is a regular value of $\hat{Z}_E$: $S\to \boldsymbol{R}^{\ell-1}$.

The proof of this theorem relies on the following remarkable principle. Assume there is a differentiable manifold A, which plays the role of a parameter space. Assume there are two other differentiable manifolds, M and N, of finite dimension and for each $a\in A$ there is a map f_a: $M\to N$. Let f: $A\times M\to N$, defined by $f(a, m)= f_a(m)$, be sufficiently differentiable and let $n\in N$ be a regular value of f. Then $f^{-1}(n)$ is a differentiable manifold Γ (possibly empty). Now the principle states that the set of regular values of the projection of Γ to the parameter space A coincides with $A_n=\{a\in A|n$ is a regular value of $f_a\}$. This implies in particular by Sard's theorem that A_n contains almost all of A. For a more precise and also more general statement of this principle including a proof, see the Transversal Density Theorem in Abraham and Robbin (1967, §19).

In our situation, f: $A\times M\to N$ corresponds to $\hat{Z}$: $\mathfrak{S}\times S\to \boldsymbol{R}^{\ell-1}$, $n\in N$ corresponds to $0\in \boldsymbol{R}^{\ell-1}$. Thus one obtains the above theorem. A direct proof of this argument can easily be extracted from Abraham–Robbin (1967, §19).

The Transversal Density Theorem yields immediately that almost all economies $E\in\mathfrak{S}=P^m$ are regular. As regularity of an equilibrium is defined by a full-rank-condition, we also obtain that the set of regular economies is open, when we take account of the fact that the equilibrium price correspondence is u.h.c. and compact-valued. In this way, Debreu's theorem, which states that the set of critical economies is a closed null-set, can be derived from the second characterization of a regular economy.

Knowing that most economies are regular, one wants to find out what properties the equilibrium set of a regular economy has to fulfill. Consider for the moment economies with two commodities. In Figure 3.4 there is a unique equilibrium which happens to be stable with respect to the classical tâtonnement process, where the price of a commodity is adjusted proportionally to the excess demand for that commodity. In Figure 3.3 there are three equilibria, two are locally stable and one is locally unstable. In a similar way one can draw, for any natural number k, the excess demand of an economy with $2k-1$ equilibria, k of which are locally stable and $k-1$ of which are locally unstable.

To obtain a statement that generalizes to higher dimensions we have to replace the words "stable" and "unstable". Instead of distinguishing between unstable and stable equilibria we distinguish between equilibria where the excess demand vector field preserves orientation and those where it reverses orientation, that means we distinguish equilibria p according to sign $\det[(\partial\zeta_1,\ldots,\partial\zeta_{\ell-1}/\partial p_1,\ldots,\partial p_{\ell-1})(p)]$. More precisely, let p be a regular equilibrium price system and define

$$\begin{aligned} \textit{index}\ (p) &= \operatorname{sign}\det\left(\frac{-\partial\zeta_1,\ldots,-\partial\zeta_{\ell-1}}{\partial p_1,\ldots,\partial p_{\ell-1}}(p)\right) \\ &= (-1)^{\ell-1}\operatorname{sign}\det\left(\frac{\partial\zeta_1,\ldots,\partial\zeta_{\ell-1}}{\partial p_1,\ldots,\partial p_{\ell-1}}(p)\right). \end{aligned}$$

The index at p depends on the eigenvalues of the Hessian of ζ at p and is thus defined invariantly. In the above definition the sign is chosen such that locally stable equilibria get index $+1$. Other regular equilibria have index $+1$ or -1. A reader familiar with Scarf's example of an economy with a unique completely unstable equilibrium will notice that the equilibrium there also has index $+1$; see Scarf (1960) and Hahn's contribution to this Handbook. The index of isolated equilibria, which may be regular or critical, is defined and illustrated in Milnor (1965, §6). More general, but perhaps less geometric, ways to introduce indices can be found in books on algebraic topology; see Dold (1972, ch. VII, §5). In recent years indices have begun to play an important role in the computation of economic equilibria; see Scarf's contribution to this Handbook.

A most remarkable fact about indices of zeros of vector fields on compact manifolds is that they add up to a number which does not depend on the vector field itself but only on the underlying manifold. In our case excess demand is a vector field on the price space S which is not compact. However, as a substitute for compactness we made the desirability assumption (D), which restricts the behavior of the excess demand near the boundary of the price space in such a way that the invariance of the index sum is still guaranteed. A short proof of this fact is contained in Dierker (1972), where, however, knowledge of elementary homology theory is assumed. An intuitive presentation of the invariance of the index sum is given in Milnor (1965; see the Poincaré–Hopf Theorem, p. 35). Also, see Varian (1975), Nishimura (1978), and Varian's contribution to this Handbook.

The index sum equals the Euler characteristic $\chi(S)$ *of* S, *if* (D) *holds*. Since S is contractible to a point, $\chi(S)=\chi(\text{point})=1$.

Theorem

Given m continuously differentiable demand functions $(f_1,\ldots,f_m)$, if (D) is satisfied, then any regular economy $E\in\mathcal{E}=P^m$ has $2k-1$ equilibria for some $k\in\{1,2,\ldots\}$, k of which have index $+1$ and $k-1$ of which have index -1.

Note that this implies the existence of an equilibrium because $2k-1$ never vanishes for integral k. The existence of equilibria in critical economies can be obtained in a similar way.

The index assigned to a critical equilibrium can be different from ± 1. Index zero, for example, is assigned to the critical equilibrium of Figure 3.2. This example shows that counting critical equilibria in the way of the index sum formula need not be informative.

One may ask whether there are restrictions on the set of equilibria of a regular economy in addition to the index sum requirement. Mas-Colell (1977a) has shown that there is, essentially, no further condition that the equilibrium set has to fulfill. The index sum condition is purely mathematical and *there is no restriction on the equilibrium set which arises from the economic source of the problem.*

Balasko (1979a) pointed out that, loosely speaking, not too many economies can have very many equilibria. More precisely, the following inequality is true. *Let parameters be initial endowments. Then there exists, for any compact set K of economies, a constant $c(K)$ such that the Lebesgue measure of all economies in K with at least n equilibria does not, for any n, exceed $(1/n)c(K)$.*

This can be explained as follows. Imagine the graph Γ of the equilibrium price correspondence as an ℓm-dimensional manifold embedded in $\mathcal{E} \times \boldsymbol{R}^{\ell-1}$, where $\boldsymbol{R}^{\ell-1}$ represents the price space. The embedding defines a Riemannian structure on Γ. In particular, the ℓm-dimensional area can be measured on Γ. This measure on Γ, denoted by μ, is related to the Lebesgue measure λ on $\mathcal{E}$ through the projection pr: projecting a set here does not increase the measure. This relation between λ and μ is independent of the particular way in which the price space is identified with $\boldsymbol{R}^{\ell-1}$.

Since K is compact and pr is proper, $\mu(pr^{-1}(K))$ is finite. Call this number $c(K)$.

Let E be a regular economy with at least n equilibria. Then E has a neighborhood U in K such that $pr^{-1}(U)$ consists of at least n layers, each contributing at least $\lambda(U)$ to $\mu(pr^{-1}(U))$. Let the set of economies in K with at least n equilibria be covered, up to a null set, by a countable, disjoint union of such sets U_j. Then $\sum_j n\lambda(U_j) \leqslant c(K)$, which amounts to the desired inequality.

Next we want to extend Debreu's theorem on regular economies to the case where demand functions are allowed to vary in addition to initial endowments. Since demand functions do not fit into a finite dimensional space, our aim is to show that regular economies are open and dense with respect to a suitable topology and that the equilibria of each economy in a neighborhood of a regular economy can be described by a fixed number of continuous functions.

Let D^1 be the set of all C^1 demand functions f: $S \times]0,\infty[\to \boldsymbol{R}^{\ell}_{+}$ which fulfill the following desirability assumption [cf. Debreu (1970, p. 388)]:

(A) If the sequence (p^q, w^q) in $S \times]0,\infty[$ converges to (p^0, w^0) in $(\bar{S} \setminus S) \times]0,\infty[$, then $\| f(p^q, w^q)|$ converges to infinity.

If (A) holds for at least one agent then (D) is fulfilled.

The topology to be used on D^1 must take the first derivative into account. We use here the topology of uniform C^1 convergence. The C^1 distance between two C^1 functions with common domain and with values in the same Euclidean space is measured by comparing the values and also the first derivatives of both functions at all points of their domain in a uniform way. The space of economic parameters is now $\mathfrak{E}=(D^1\times P)^m$.

One could employ some coarser or finer C^1 topologies instead of the uniform C^1 topology. What we actually need are the following two properties of the topology: first, the equilibrium price correspondence is u.h.c; second, the derivatives of the excess demand functions of two neighboring economies are uniformly close to each other on compact sets

It may be natural in this context to work with Banach manifolds. This is done by Fuchs (1974) in a framework including production functions. But, as density of regular economies in our set-up follows immediately from Debreu's theorem, it suffices to show the openness of the set of regular economies in an elementary way. To avoid the use of infinite dimensional manifolds, it is convenient to think of regular economies as being characterized by their individual excess demand functions without recourse to the space of economic parameters.

Theorem

The set $\mathfrak{R}$ of regular economies is dense and open in $\mathfrak{E}=(D^1\times P)^m$. The projection pr: $\{(E,\Pi(E))|E\in\mathfrak{R}\}\to\mathfrak{R}$ is a covering map: for every $E\in\mathfrak{R}$ there exists a neighborhood U in $\mathfrak{R}$ such that $pr^{-1}(U)$ is the disjoint union of finitely many sets homeomorphic to U. In particular, the equilibrium price correspondence Π is continuous on $\mathfrak{R}$ and the number of equilibria is locally constant on $\mathfrak{R}$.

Proof

We outline a proof along the lines of Dierker and Dierker (1972). To show the openness of $\mathfrak{R}$ let $E=(f_1,\ldots,f_m,e_1,\ldots,e_m)\in(D^1\times P)^m$ be a regular economy and let Z_E be its excess demand function. The set $Z_E^{-1}(0)$ consists of a finite number of equilibrium price systems $p_1,\ldots,p_r$. Each p_k, $k=1,\ldots,r$, has a neighborhood which is mapped diffeomorphically by $\hat{Z}_E$ onto a neighborhood of the origin in $\boldsymbol{R}^{\ell-1}$.

Now we make use of the following fact [cf. E. Dierker (1974, p. 98, lemma 10.1)]. If z maps a neighborhood of $p\in S$ diffeomorphically onto a neighborhood of $0\in\boldsymbol{R}^{\ell-1}$ and if $z(p)=0$, then there exist a neighborhood V of p and $\varepsilon>0$ such that every C^1 function $\tilde{z}$: $V\to\boldsymbol{R}^{\ell-1}$ maps V diffeomorphically onto some neighborhood of 0 provided the C^1 distance of $\tilde{z}$ and $z|V$ is less than ε.

It follows that each equilibrium price system p_k has a neighborhood V_k in S such that $\tilde{z}$: $V_k\to\boldsymbol{R}^{\ell-1}$ is a diffeomorphism onto a neighborhood of the origin in

$\boldsymbol{R}^{\ell-1}$, if $\tilde{z}_k$ is C^1 close enough to $\hat{Z}_E|V_k$. Let each V_k be so small that its closure lies in S and that $V_k \cap V_{k'} = \varnothing$ for $k \neq k'$. The set $K = \cup_{k=1}^r \text{cl}(V_k)$ is a compact subset of S.

Next observe that two economies in $\mathcal{E}$ have their excess demand functions C^1 close on K provided the initial endowments of corresponding agents are close and their demand functions are C^1 close. Therefore, if $\tilde{E}$ is close enough to E, then $\hat{Z}_{\tilde{E}}|V_k$ maps V_k diffeomorphically onto some neighborhood of the origin in $\boldsymbol{R}^{\ell-1}$ for $k=1,\ldots,r$.

Furthermore, the topology used in $\mathcal{E}$ is fine enough to ensure the upper hemi-continuity of the equilibrium price correspondence Π. Hence $\Pi(\tilde{E}) \subset \cup_{k=1}^r V_k$ if $\tilde{E}$ is close enough to E. Summarizing, an economy $\tilde{E}$ close enough to E has exactly r equilibrium price systems and all of them are regular. This shows that the set of regular economies is open in $(D^1 \times P)^m$ and that the number of equilibria is locally constant.

Moreover, the correspondences $\tilde{E} \mapsto \Pi(\tilde{E}) \cap V_k$, $k=1,\ldots,r$, can be considered as single-valued functions in a neighborhood of E. They are continuous since Π is u.h.c. Therefore $\tilde{E} \mapsto (\tilde{E}, \Pi(\tilde{E}) \cap V_k)$, $k=1,\ldots,r$, defines local homeomorphisms near the regular economy E. That is to say, E has a neighborhood U in $\mathfrak{R}$ such that $pr^{-1}(U)$ is homeomorphic to r copies of U, which was to be shown. Q.E.D.

To incorporate *production* into the theory of regular economies one can use the above excess demand approach if one is willing to stipulate the existence of producers' supply functions. The existence of supply functions, however, excludes in particular the economically important case of constant returns to scale. Furthermore, three difficulties arising in the context of supply functions should be pointed out.

First, desirability assumptions such as (A) or (D) do not make sense in an economy with production because there are many primary or intermediate goods involved in the production process which no consumer desires. The boundary behavior of the excess demand function, therefore, also depends on properties of the production sector. For a study of regular economies in the context of supply functions, see Fuchs (1974).

Second, the domain of a supply function depends on the asymptotic cone of the production set. This leads to the problem of how to express when functions with different domains are close to one another and leads to technical difficulties in the subsequent treatment of the dependence of equilibrium prices on supply functions. Fuchs and Laroque (1974) extend the results on regular exchange economies to economies with supply functions with variable domains.

Third, in the presence of production, individuals will not have a strictly positive vector of initial endowments.

Complementary to the treatment of production by the use of supply functions is the work of Mas-Colell (1975), who concentrates upon the case of constant returns to scale. Mas-Colell (1977c) and Kehoe (1980) present the theory of regular equilibria, including the theorem on the index sum for a model in which Koopman's activity analysis is used to describe production.

The most natural way to study regularity properties of economies with production is taken by Smale who uses the production sets themselves. We shall describe Smale's approach in Section 6.

4. Structure of the equilibrium manifold

In this section we keep demand functions fixed and vary initial endowments. We have seen in Section 2 that the graph Γ of the equilibrium price correspondence is a differentiable manifold. We are now going to study the question: Which global properties does Γ possess? For that purpose we place ourselves in the framework of Section 2 and assume in addition that the given demand functions $f_1,\ldots, f_m$ take their values in $P = \operatorname{int} \boldsymbol{R}^\ell_+$. The following result is due to Balasko (1975a).

Proposition

Let the demand functions f_i: $S \times]0, \infty[\to P$, $i=1,\ldots, m$, be continuous. Then the graph Γ of the equilibrium price correspondence Π: $\mathfrak{E} = P^m \to S$ is contractible.

A set is contractible iff it possesses one of its points as a deformation retract. In particular, a contractible set has no "holes". For precise explanations the reader is referred to Dold (1972, ch. III, §5).

Proof

Consider the subset of Γ that consists of all no trade equilibria. A point $(e_1,\ldots, e_m, p) \in \Gamma$ is a *no trade equilibrium* iff, for every agent i, the demand $f_i(p, pe_i)$ equals the initial endowment e_i. The proof consists of showing that the set of no trade equilibria is contractible and that it is a deformation retract of Γ.

The following argument shows that the set of no trade equilibria is contractible. Let

$$B :=]0, \infty[^m \times S$$

denote the space of budget situations for the agents $i=1,\ldots, m$. A point $(w_1,\ldots, w_m, p) \in B$ describes the wealth of every agent and a price system common to all agents. Define f: $B \to \Gamma$ by assigning $(f_1(p, w_1),\ldots, f_m(p, w_m), p)$ to $(w_1,\ldots, w_m, p) \in B$. The set of no trade equilibria equals $f(B)$. Associate $(pe_1,\ldots, pe_m, p) \in B$ with $(e_1,\ldots, e_m, p) \in f(B)$ to see that f establishes a C^r isomorphism between the space B of budget situations and the space $f(B)$ of no trade equilibria, if f is C^r, $r \geq 0$. Since B is contractible, $f(B)$ is contractible.

It remains to show that $f(B)$ is a deformation retract of Γ. Let $(e_1,\dots,e_m,p)\in\Gamma$. To $(e_1,\dots,e_m,p)$ there corresponds the no trade equilibrium $(f_1(p,pe_1),\dots,f_m(p,pe_m),p)$. The desired deformation retraction is defined by continuously transforming any $(e_1,\dots,e_m,p)\in\Gamma$ into the corresponding no trade equilibrium keeping the budget situation fixed. Formally, for $0\leqslant t\leqslant 1$, define $H_t\colon \Gamma\to\Gamma$ by

$$H_t(e_1,\dots,e_m,p)=(tf_1(p,pe_1)+(1-t)e_1,\dots,tf_m(p,pe_m)+(1-t)e_m,p).$$

We have $H_0=id_\Gamma$, $H_1(\Gamma)=f(B)$, and $H_t|f(B)$ is the inclusion of $f(B)$ into Γ for $0\leqslant t\leqslant 1$. Therefore $f(B)$ is a (strong) deformation retract of Γ and the proposition is proven. Q.E.D.

Observe that the proof immediately above does not rely on the differentiability assumption which we make throughout the paper to study regular economies.

It may be useful to illustrate the argument of the above proof by an Edgeworth box, although one loses some degrees of freedom by keeping the size of the box, i.e. the total endowments, fixed. Assume demand is derived from smooth, convex preferences. A no trade equilibrium corresponds to a Pareto efficient allocation. In Figure 4.1 we have drawn the set of those initial allocations which correspond to a given Pareto efficient allocation. This set is represented by an open segment of a straight line which is tangent to the indifference curves of both agents at the given Pareto efficient allocation. The deformation retraction H_t moves every point on the line segment along the line segment with constant speed towards the Pareto efficient allocation. The set of Pareto efficient points in the Edgeworth box is contractible.

If $(\ell,m)=(2,2)$, Figure 4.7 suggests that one may think of the graph of the equilibrium price correspondence as a family of straight line segments indexed by the no trade equilibria.

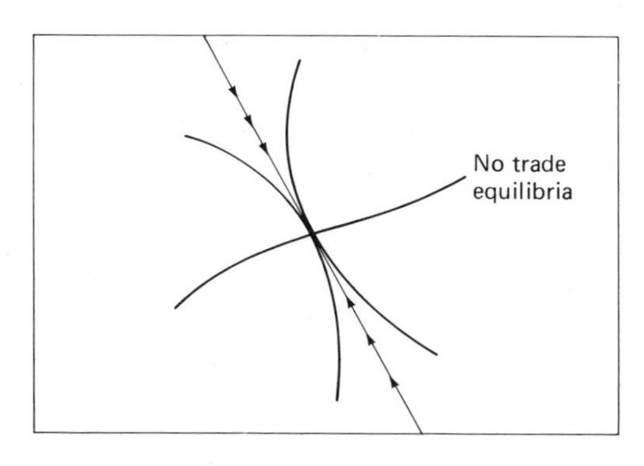

Figure 4.1.

In the description of Γ prices appear explicitly so that different straight line segments have empty intersections. If there are m traders and ℓ commodities, the open line segments have to be replaced by open convex $(\ell-1)(m-1)$-dimensional sets. One expects Γ to be homeomorphic or even diffeomorphic to the Cartesian product of $f(B)$ and a Euclidean space.

Theorem

Let the demand functions f_i: $S\times]0,\infty[\to P$ be C^r, $r\geq 0$, $i=1,\dots,m$. Then the graph Γ of the equilibrium price correspondence Π: $\mathcal{E}=P^m\to S$ is C^r equivalent to $\boldsymbol{R}^{\ell m}$.

For $r=0$, this theorem is shown in Balasko (1975b). Using a result of Stallings on the uniqueness of the differentiable structure of Euclidean space in dimensions other than four, Balasko (1975b) obtained the above theorem for $(\ell, m)\neq(2,2)$. A direct proof of the fact that Γ is C^r diffeomorphic to $\boldsymbol{R}^{\ell m}$, given demand functions are C^r, $r\geq 1$, was given by Schecter (1979).

If one neglects all positivity constraints in the above theorem, i.e. if the consumption set of an agent is $\boldsymbol{R}^{\ell}$ and if the initial endowments of an agent are allowed to vary within $\boldsymbol{R}^{\ell}$, then the proof of the C^r equivalence of Γ and $\boldsymbol{R}^{\ell m}$ is shortened considerably. Following Balasko (1978a) we shall now admit negative amounts of commodities.

Define φ: $\boldsymbol{R}\times\boldsymbol{R}^{\ell(m-1)}\times S\to\Gamma$ by

$$\varphi(w_1,e_2,\dots,e_m,p)=\left(f_1(p,w_1)+\sum_{i=2}^{m}(f_i(p,pe_i)-e_i),e_2,\dots,e_m,p\right).$$

Of course, we now define

$$\Gamma=\left\{(e_1,\dots,e_m,p)\in\boldsymbol{R}^{\ell m}\times S\;\middle|\;\sum_{i=1}^{m}(f_i(p,pe_i)-e_i)=0\right\}.$$

The mapping φ assigns to agent 1 that initial endowment vector that brings the economy into equilibrium without changing the wealth of agent 1, the endowments of all others, and the price system. To see that φ is invertible, let ψ: $\Gamma\to\boldsymbol{R}\times\boldsymbol{R}^{\ell(m-1)}\times S$ be given by

$$\psi(e_1,e_2,\dots,e_m,p)=(pe_1,e_2,\dots,e_m,p).$$

It follows from the definitions that $\psi\circ\varphi=id_{\boldsymbol{R}\times\boldsymbol{R}^{\ell(m-1)}\times S}$ and that $\varphi\circ\psi=id_E$.

Therefore φ establishes a C^r isomorphism between $\boldsymbol{R}\times\boldsymbol{R}^{\ell(m-1)}\times S$ and Γ if all demand functions are of class C^r, $r\geqslant 0$. As S is C^∞ equivalent to $\boldsymbol{R}^{\ell-1}$, we have shown the analog to the above theorem without positivity constraints.

Taking $\boldsymbol{R}^\ell$ as the consumption set of a consumer may appear unusual or even doubtful on economic grounds. This practice, however, appears useful in the study of critical equilibria. If negative amounts of commodities are admissible, a one-parameter family of straight lines instead of a one-parameter family of straight line segments is obtained in the above Edgeworth box situation. An economy (e_1, e_2) is critical iff there are two "infinitesimally close" straight lines in the family that pass through (e_1, e_2). For a precise statement, see Balasko (1978b, theorem 1). The study of critical economies here amounts to the study of the envelope of the one-parameter family of straight lines. A brief, intuitive introduction to the use of envelope theory is given in Balasko (1978a). An extensive mathematical treatment is in Balasko (1978b).

The global structure of the graph of the equilibrium price correspondence being well understood, one is led to ask more specific questions. For instance, reconsidering the classical uniqueness problem one would like to know what properties the set of regular economies with a unique equilibrium price system has. From the point of view of envelopes Balasko (1978b) [or, alternatively, Balasko (1978a)], obtains the result that the set of regular economies with a unique equilibrium price system is arcwise connected if there are only two agents. The following complementary result [theorem 17 in Balasko (1978b)] is independent of the number of agents: "If initial endowments 'at infinity' are suitably added, the set of economies having several equilibria is arcconnected."

Assume now that demand is derived from preferences which fulfill the classical assumptions, such as strict convexity, monotonicity, and smoothness. If $(e_1,\ldots, e_m, p)$ is a no trade equilibrium, then the economy $(e_1,\ldots, e_m)$ has p as its unique equilibrium price system. Furthermore, it follows from the negative definiteness of the Slutsky matrix that the equilibrium price system p is locally stable with respect to the tâtonnement process where price adjustment equals excess demand. Balasko (1978c) shows that the set of asymptotically stable equilibria is arcwise connected and contains the set of no trade equilibria.

Our understanding of the singularities of pr: $\Gamma\to\mathcal{E}$ is rather incomplete and much research has to be done in this direction. A detailed study of the singularities of pr, however, seems to require a considerable amount of mathematical sophistication. It appears desirable that, in addition to a general, abstract, mathematical study of the singularities of pr, some particular types of singularities be examined in their relation to specific economic phenomena. It is a difficult task, though, to relate types of singularities to types of economic events, since empirical observations are often biased by an oversimplified conception of the world in which there is little room for any kind of irregular behavior.

5. Large economies

Until now the number of agents in an economy has been considered a fixed datum. In a large economy, however, the number of agents cannot be precisely observed. On the other hand, one intuitively expects that dropping an average person from a large economy hardly affects the set of equilibrium prices. It is the *distribution of agents' characteristics* that is expected to be essential for the set of equilibrium prices. If an economy is large, the distribution of agents' characteristics remains nearly unaltered when an average person is dropped.

The idea of pure competition fits perfectly well to the conception of a consumption sector with infinitely many agents, each individual agent being negligibly small. Atomless measures are natural tools to describe a purely competitive world. Atomless measures have been introduced into economic theory by Aumann. For the use of measure theory in economics, see W. Hildenbrand's book (1974) and W. Hildenbrand's and A. Kirman's contributions to this Handbook.

The concept of an excess demand function remains available if the distribution of agents' characteristics is described by a probability measure. The excess demand approach therefore lends itself to regularity studies of exchange economies with many agents. We now follow Delbaen (1971) and, in particular, K. Hildenbrand (1974) in the treatment of large, regular exchange economies.

Consider the set D^1 of all C^1 demand functions f: $S\times]0,\infty[\to \boldsymbol{R}^{\ell}_{+}$ which fulfill the desirability assumption (A). The topology of uniform C^1 convergence on D^1 is too fine to yield seperability. Since we shall restrict ourselves later to compact subsets of D^1, we prefer the use of a coarser topology than the topology of uniform C^1 convergence. As observed in Section 3, important aspects for the choice of the topology on D^1 are the uniform C^1 convergence on compact sets and the upper hemi-continuity of the equilibrium price correspondence.

To obtain a suitable topology take the one-point compactification $\boldsymbol{R}^{\ell}_{+}\cup\{\infty\}$ of the consumption set. A C^0 demand function f: $S\times]0,\infty[\to \boldsymbol{R}^{\ell}_{+}$ has a C^0 extension $(\bar{S}/\bar{S}\setminus S)\times]0,\infty[\to \boldsymbol{R}^{\ell}_{+}\cup\{\infty\}$. The extension maps $(\bar{S}\setminus S)\times]0,\infty[$ to ∞. Define two elements of D^1 to be close to each other iff their C^0 extensions are close to each other with respect to the topology of uniform convergence of compact subsets of $(\bar{S}/\bar{S}\setminus S)\times]0,\infty[$ and, moreover, their derivatives are uniformly close to each other on compact subsets of $S\times]0,\infty[$.

This defines a topology on D^1 which is fine enough to ensure the upper hemi-continuity of the equilibrium price correspondence, because $(\bar{S}/\bar{S}\setminus S)$ is a compact space of prices. Furthermore, the uniform convergence on compact sets entails separability. The topology is also metrizable. To describe such a metric in explicit terms, let f^1 and f^2 be in D^1, let $\bar{\delta}$ be any metric belonging to the one-point compactification of $\boldsymbol{R}^{\ell}_{+}$, and let δ be any bounded metric belonging to the usual topology on the linear maps $\boldsymbol{R}^{\ell}\to\boldsymbol{R}^{\ell}$, i.e. on $\boldsymbol{R}^{\ell\cdot\ell}$. Then a metric on D^1

can be defined by

$$d(f^1, f^2) = \sum_{n=1}^{\infty} 2^{-n} \sup_{(p,w)\in S\times[1/n,n]} \bar{\delta}\left(f^1(p,w), f^2(p,w)\right)$$
$$+ \sum_{n=1}^{\infty} 2^{-n} \sup_{(p,w)\in S_n\times[1/n,n]} \delta\left(Df^1(p,w), Df^2(p,w)\right),$$

where the S_n are compact subsets of S with $\cup_{n=1}^{\infty} S_n = S$.

The above way to define a separable metrizable topology on a space of demand functions is quite natural, provided every demand function fulfills (A). If the demand functions are derived from classical preferences, then the above topology is derived from a topology on the space of preferences, see K. Hildenbrand (1974, p. 170).

Let K be a compact subset of D^1 topologized as explained above. We consider agents who have demand functions in K and endowments in $[\alpha, \beta]^{\ell}, 0<\alpha<\beta<\infty$. The *space of traders* is $T = K\times[\alpha, \beta]^{\ell}$. *An economy is described by a probability measure on* $(T, \mathfrak{B}(T))$, where $\mathfrak{B}(T)$ denotes the system of Borel sets of T. *The space* $\mathcal{E}$ *of economies is the space* $\mathfrak{M}_T$ *of probability measures on* $(T, \mathfrak{B}(T))$ *endowed with the topology of weak convergence*. For the reasonableness of this topology, see W. Hildenbrand (1974). $\mathfrak{M}_T$ is a compact, separable metric space.

To every economy $\mu \in \mathfrak{M}_T$ there corresponds a mean excess demand function Z_μ: $S \to \boldsymbol{R}^{\ell}$. Here $Z_\mu(p)$ is defined as the mean with respect to μ of the difference between individual demand at p and individual endowment. Mean excess demand is interpreted as aggregate excess demand per capita. In the simpler framework of the previous sections it was unnecessary to measure excess demand per capita.

Mean excess demand Z_μ is a C^1 function. This follows from the fact that all individual demand functions are assumed to be C^1. Since it is a common observation that the mean tends to behave more smoothly than the corresponding individual entities, it would be desirable to dispense with smoothness on the individual level. It is not clear at present whether this can be done in a general way. We shall comment on this problem in the final part of this section.

Definition

An *equilibrium price system* of μ is a zero of Z_μ. The equilibrium price correspondence Π: $\mathcal{E} \to S$ is defined by $\Pi(\mu) = Z_\mu^{-1}(0)$. An *equilibrium price system is regular* iff it is a non-degenerate zero of Z_μ. The *economy* μ *is regular* iff all its equilibrium price systems are regular.

Theorem

The set $\mathfrak{R}$ of regular economies is open and dense in $\mathcal{E} = \mathfrak{M}_T$. The projection *pr*: $\{(\mu, \Pi(\mu)) | \mu \in \mathfrak{R}\} \to \mathfrak{R}$ is a covering map. In particular, the equilibrium price correspondence is continuous on $\mathcal{R}$ and the number of equilibria, which is finite, is locally constant on $\mathcal{R}$.

To prove the theorem one verifies that Z: $\mathfrak{M}_T \times S \to \boldsymbol{R}^\ell$, defined by $(\mu, p) \to Z_\mu(p)$, is continuous. Therefore, and because assumption (A) holds for every agent in T, Π is u.h.c. and compact-valued. The desirability assumption also ensures that $\Pi(\mu) \neq \varnothing$ for every $\mu \in \mathfrak{M}_T$.

Π being compact-valued, the number of regular and therefore isolated equilibria of an economy is finite. The openness and the covering property are shown in the same way as in Section 3.

The density of $\mathcal{R}$ in $\mathfrak{M}_T$ actually follows from Debreu's theorem (see Section 2), because the set of economies with finitely many traders having characteristics in T is dense in $\mathfrak{M}_T$. It is simple, though, to give a direct argument that shows the density of $\mathcal{R}$ in $\mathfrak{M}_T$. For instance, one can proceed as follows. The set $\mathfrak{M}'_T$ of measures with support in $K \times]\alpha, \beta[^\ell$ is dense in $\mathfrak{M}_T$. Let μ have support in $K \times]\alpha + \varepsilon, \beta - \varepsilon[^\ell$ for some $\varepsilon > 0$. Perturb the initial endowment of all agents in support (μ) by the same vector of length less than ε. Sard's theorem then yields that almost all these small perturbations lead to a regular economy.

The requirement that T is compact is unnecessarily restrictive. It serves to present the arguments of a precise proof in a simple manner. For details, we refer to K. Hildenbrand (1974). This ends the discussion of this particular model of a large economy.

For some economic questions, for instance for questions concerning the core of an economy, preferences must be used as a primitive concept instead of demand functions. This leads one to ask when preferences give rise to C^1 demand functions.

To answer this question it is useful to look at the inverse of a demand function. Consider the consumption set $P = \operatorname{int} \boldsymbol{R}^\ell_+$ and let u: $P \to \boldsymbol{R}$ be a strictly quasi-concave C^2 utility function satisfying $Du(x) \gg 0$, for every $x \in P$. Define g: $P \to S$ by $g(x) = Du(x)/\|Du(x)\|$. As u is C^2, g is C^1. The commodity bundle $x \in P$ will be demanded at the price system $p = g(x)$ and the wealth $w = g(x) \cdot x$. Assigning $(g(x), g(x) \cdot x)$ to $x \in P$ defines a C^1 mapping φ: $P \to S \times]0, \infty[$. Assume that the closure of each indifference surface lies in P. Then φ is surjective. Furthermore φ is injective as u is strictly quasi-concave. Therefore φ has an inverse, which is the demand function $f = \varphi^{-1}$: $S \times]0, \infty[\to P$. The demand function f is C^1 iff the derivative $D\varphi(x)$ is invertible at every $x \in P$.

Considering the local invertibility of φ at $\bar{x} \in P$ in the two-commodity case one sees that $g(x)$ must vary on the indifference curve I through $\bar{x}$ as u is strictly quasi-concave, but $g|_I$ may vary at $\bar{x}$ with zero speed. For instance, if the indifference curve is as flat at $\bar{x}$ as the graph of $y \to y^4$ at the origin, then the change of $g|_I$ at $\bar{x}$ is not of first but of higher order.

More generally, the indifference hypersurface I through $\bar{x}\in P$ has *non-zero Gaussian curvature* at $\bar{x}$ iff $g|_I$: $I\to S$ has full rank $\ell-1$ at $\bar{x}$. In the present framework, where the preference pattern is convex, the indifference surface I has a non-zero Gaussian curvature at $\bar{x}$ iff the utility function u restricted to the budget hyperplane through $\bar{x}$ with normal $g(\bar{x})$ has a non-degenerate maximum at $\bar{x}$. Thus, one can express convexity in the very strict form that requires non-zero Gaussian curvature everywhere on P as follows. If u: $P\to \boldsymbol{R}$ is a C^2 utility function with non-vanishing derivative, then, for all $x\in P$ and all $y\in \boldsymbol{R}^\ell$ such that $y\neq 0$ and $Du(x)y=0$, it is required that $D^2u(x)(y,y)<0$.

We have seen that $D\varphi(\bar{x})$ is not invertible if the indifference hypersurface I through $\bar{x}$ has zero Gaussian curvature at $\bar{x}$. On the other hand, suppose I has a non-zero Gaussian curvature at $\bar{x}$. Then it is clear that $D\varphi(\bar{x})$ is invertible because $g|_I$ has full rank and the wealth $g(x)\cdot x$ varies at $\bar{x}$ with non-zero speed in the direction of the Engel curve defined by $g(x)=g(\bar{x})$. Therefore, *the demand function* f: $S\times]0,\infty[\to P$ *is* C^1 *if all indifference hypersurfaces have everywhere a non-zero Gaussian curvature*. A formal proof of this fact is given in Debreu (1972, pp. 612, 613).

The usefulness of starting with the function g as a basic concept to describe a consumer has been established by Debreu (1972). More precisely, it can be shown that the preferences of a consumer can be given via a C^1 function g: $P\to S$ which satisfies the following *condition of local integrability*.

For all i, j, k and for all $\bar{x}\in P$

$$g_i(\bar{x})\left[\frac{\partial g_k}{\partial x_j}(\bar{x})-\frac{\partial g_j}{\partial x_k}(\bar{x})\right]+g_j(\bar{x})\left[\frac{\partial g_i}{\partial x_k}(\bar{x})-\frac{\partial g_k}{\partial x_i}(\bar{x})\right]+g_k(\bar{x})\left[\frac{\partial g_j}{\partial x_i}(\bar{x})-\frac{\partial g_i}{\partial x_j}(\bar{x})\right]=0.$$

This condition is necessarily fulfilled if g is the normalized gradient of a C^2 utility function. On the other hand, using the Frobenius theorem on involutive distributions, Debreu (1972) derives from the local integrability condition that for each point $\bar{x}\in P$ there exist an open neighborhood V of $\bar{x}$ and a C^1 utility function u on V such that $g(x)=Du(x)/\|Du(x)\|$ for all $x\in V$. Furthermore, Debreu (1972) deduces global integrability from local integrability provided the monotonicity condition $g(x)\in S$ for $x\in P$ holds.

Thus, it becomes possible to deal with exchange economies with many agents who are described by their preferences. Let G be the *space of all preference relations* expressed by C^1 functions g: $P\to S$ that are locally integrable and give rise to weakly convex systems of C^1 indifference hypersurfaces with closures contained in P. Two such preference relations, g_1 and g_2, are considered close iff

g^1 and g^2 are C^1 uniformly close on compact subsets of P. This defines a topology which makes G a separable metrizable space. The *space of consumers' characteristics* is $G \times P$ endowed with the product topology. An *economy* now is a probability measure with compact support on $(G \times P, \mathcal{B}(G \times P))$, where $\mathcal{B}$ denotes Borel sets. The set of all economies is denoted by $\mathcal{M}^*(G \times P)$.

To study regularity questions in this framework H. Dierker (1975a) introduces the following topology on $\mathcal{M}^*(G \times P)$. If ρ denotes the Prohorov metric inducing the weak topology, if e denotes initial endowments, and if δ denotes the Hausdorff distance on the space of compact subsets of $G \times P$, then a metric η on $\mathcal{M}^*(G \times P)$ is defined by the formula

$$\eta(\mu_1, \mu_2) = \rho(\mu_1, \mu_2) + |\int e \, d\mu_1 - \int e \, d\mu_2| + \delta(\text{support}(\mu_1), \text{support}(\mu_2))$$

for $\mu_1, \mu_2 \in \mathcal{M}^*(G \times P)$. Convergence in $\mathcal{M}^*(G \times P)$ means weak convergence, convergence of mean endowments, and convergence of supports. The convergence of supports is included into the definition of the topology to ensure that regular economies form an open set. Observe that individual demand in this framework need not be a function because only weak convexity is assumed. If supports were neglected in the definition of η, then differentiability and hence regularity of an economy could be destroyed by giving an arbitrarily small, positive weight to agents whose individual demand is not a C^1 function. The *space* $\mathcal{E} = \mathcal{M}^*(G \times P)$ *of economies*, endowed with the topology induced by η, is separable.

Mean excess demand and the equilibrium price correspondence are defined in the usual way. An *economy* μ is called *regular* iff the following two conditions are satisfied:

(i) there is an open set B in S with $\Pi(\mu) \subset B \subset \bar{B} \subset S$ such that the Gaussian curvature of an indifference surface of any agent in support(μ) at a commodity bundle which is demanded by the agent at a price system $p \in \bar{B}$ is non-zero, and
(ii) each equilibrium price system is regular.

Condition (i) guarantees that mean demand is C^1 in a neighborhood of $\Pi(\mu)$. Hence condition (ii) is meaningful. Let $\mathcal{R}$ denote the set of regular economies in $\mathcal{M}^*(G \times P)$.

Theorem

The set $\mathcal{R}$ of regular economies is open and dense in $\mathcal{E} = \mathcal{M}^*(G \times P)$. The projection pr: $\{(\mu, \Pi(\mu)) | \mu \in \mathcal{R}\} \to \mathcal{R}$ is a covering map.

For a proof see H. Dierker (1975a). A related model is contained in Ichiishi (1974).

Regular economies are useful for studying the relation between the *core* and the set of price equilibrium allocations (see Section 7 of W. Hildenbrand's contribution to this Handbook). The Edgeworth conjecture says that a core allocation in a large, but finite, competitive economy can be approximately decentralized through a price system. Such a price system, however, need not be close to an equilibrium price system of the economy, the reason being that the equilibrium price correspondence need not be continuous. The continuity of Π has been studied above under differentiability assumptions. Since differentiability assumptions are not common in core theory, one uses approximation theorems which state that continuous preferences can be approximated by smooth preferences [see Mas-Colell (1974)]. Thus, core theory is linked with the theory of regular equilibria. H. Dierker (1975b, theorem 3) states a result of the following kind in a non-differentiable framework. The core of a large class of exchange economies with sufficiently many agents is contained in a finite number of small balls centered at equilibrium allocations of these economies (cf. Theorem 7 of W. Hildenbrand's contribution to this Handbook).

The *rate of convergence* of the core towards the set of price equilibrium allocations depends on the Gaussian curvature of the indifference hypersurfaces of the agents and on the regularity of the limit economy. Debreu (1975) and Grodal (1975) use the concept of a regular economy to show that the distance between the core and the set of Walras allocations tends to zero at least as fast as the inverse of the number of agents (cf. W. Hildenbrand's contribution to this Handbook).

Until now we have used either demand functions or convex preferences as basic concepts to study the decentralization of economic decisions. This is an unsatisfactory, fundamental restriction. Indeed, Debreu (1972, p. 614), speaking about an atomless measure ν on a space A of economic agents remarks: "*One expects that if the measure ν is suitably diffused over the space A, integration over A of the demand correspondences of the agents will yield a total demand function, possibly even a total demand function of class C^1.*" This expectation, originally referring to a framework of weakly convex preferences, is, of course, of particular interest in the context of consumption sectors without any convexity assumption on preferences. Weakly convex preferences in an economy represented by an atomless measure appear intrinsically as convexifications of originally non-convex preferences. In what follows we shall concentrate upon the case of non-convex preferences on convex consumption sets. For non-convex consumption sets, see Mas-Colell (1977d) and Yamazaki (1978). For a treatment of Debreu's conjecture in a convex framework, see Ichiishi (1976).

Regularity of exchange economies with many agents whose preferences need not be convex is studied in Mas-Colell (1977b). This study is based on the following observation. Even if preferences are not convex, the demand of an agent at a given price system will very often be a unique point in a neighborhood of which preferences are strictly convex. If prices or the agent's characteristics are varied slightly, then the demand will behave as if preferences were convex.

A consumer in Mas-Colell (1977b) is characterized by his initial endowment $e \in P$ and by his *preferences* which are representable by a C^2 utility function u: $P \to \boldsymbol{R}$ such that $Du(x) \gg 0$ for $x \in P$ and such that the closure in $\boldsymbol{R}^\ell$ of each indifference hypersurface lies in P. The topology used is that induced by the uniform C^1 convergence of the normalized gradients on compact sets. The space of consumers' characteristics is separable and metrizable. An *economy* is a probability measure with compact support on the space of consumers' characteristics. At an equilibrium price system the value of the mean excess demand correspondence contains zero.

An *economy* ν of this kind is called *regular* iff the following two conditions are satisfied:

(1) Every agent in support (ν) has, at every equilibrium price system, a unique, non-degenerate maximal element in his budget set.

Non-degeneracy of the maximal element amounts to local convexity of the preference pattern and non-zero Gaussian curvature of the indifference hypersurface at the maximal element. Condition (1) is an analog in the non-convex case of condition (i) above. If (1) holds, the mean demand of the economy is a C^1 function in a neighborhood of the equilibrium price set and the following condition can be stated:

(2) Each equilibrium price system is regular.

Two metrizable and separable topologies, called the α- (resp. β-) topology, are considered on the set of economies. Convergence with respect to the α-topology means weak convergence and convergence of mean endowments. Convergence with respect to the β-topology means convergence with respect to the α-topology and convergence of supports. Mas-Colell (1977b) obtains the following result:

Theorem

The set $\mathfrak{R}$ of regular economies is dense in the space of all economies endowed with the α-topology. $\mathfrak{R}$ is open with respect to the β-topology.

Furthermore, an example is given which demonstrates that $\mathfrak{R}$ is not dense with respect to the β-topology. Obviously $\mathfrak{R}$ is not open with respect to the α-topology.

The fact that all critical behavior of demand arising in a non-convex framework is excluded by the above definition of a regular economy is reflected by the discrepancy between the two topologies yielding the density (resp. openness) of the set of regular economies. Although budget situations with multiple utility maximizers or with a degenerate maximum are rare, they cannot freely be neglected. The above approach has to be contrasted with that of Sondermann (1975, 1976, 1980) where multiple utility maximizers are explicitly taken into account. Sondermann's approach is based on the idea that such critical situations vary from consumer to consumer. If consumers' characteristics are dispersed, then the critical situations are dispersed so that one can hope to obtain a nice mean demand function.

In particular one expects that, at a given price system, the demand of all but a null set of consumers is a unique point, provided consumers' characteristics are sufficiently dispersed. Mean demand then is a continuous function under the usual conditions which guarantee the upper hemi-continuity of the mean demand correspondence. To obtain a continuous mean demand function the null sets of consumers whose demand at given price systems is not unique can be neglected. However, these null sets must be taken into account when price variations are considered in order to define the derivative of the mean demand function.

To extend the theory of regular economies to large exchange economies with preference non-convexities, one has to find conditions under which mean demand is a C^1 function. This rather difficult task has not been solved satisfactorily until now. It is not clear, at present, whether the standard results on regular economies carry over to large economies with individual non-convexities without provisions. To obtain a C^1 mean demand function one has to study critical phenomena such as the presence of several utility maximizers in a budget set. Moreover, one is confronted with the basic conceptual problem of how to express dispersion of preferences.

The notion of dispersion is commonly associated with some properties of a measure, in particular with properties of the Lebesgue measure in the case of a Euclidean space. There seems to be no obviously natural candidate for such a measure on the space of all preferences. Motivated by the perception that specific classes of preferences are often described by utility functions with finitely many parameters, one is led to stipulate that the space of admissible preferences is a finite dimensional manifold. Thus, the concept of a Lebesgue null set becomes available. A measure on a finite dimensional space of agents' characteristics is called *dispersed* in Sondermann (1975, 1976, 1980) iff no Lebesgue null set receives positive weight. If the distribution of consumers' characteristics is dispersed in this sense, then mean demand need not be a function because the finite dimensional space of admissible preferences may be an awkward subset of the space of all preferences. Therefore, a regularity condition is imposed upon the space of admissible preferences by Sondermann (1975, 1976) to the effect that any admissible indifference surface can be twisted around a little bit either by a little change of wealth or by a little change within the space of admissible preferences. It is shown in Sondermann (1975) and, by more powerful methods, in Sondermann

(1976) how this condition leads to a continuous mean demand function. An extension of this result to a wide class of maximization problems is given by Araujo and Mas-Colell (1978), theorem 1. The proof there is surprisingly simple. Sondermann (1980), in a sequel to Araujo and Mas-Colell (1978), presents a result on the generic occurrence of continuous mean demand functions in the presence of non-convexities.

Furthermore, without using Sondermann's parametric approach and regularity condition imposed on the space of admissible preferences, Araujo and Mas-Colell (1978) obtain the result that the set of economies with a C^∞ mean demand function is dense with respect to the α-topology. This extends a result of Mas-Colell and Neuefeind (1977). Denseness of C^∞ mean demand is also shown in Neuefeind (1978). These denseness results refer to rather weak topologies. Stability of smoothness with respect to perturbations is not achieved. These results, therefore, do not provide a basis sufficient to derive the usual results on regular economies.

Sondermann (1976), exploiting the possibilities of the parametric approach and regularity condition, aims at the differentiability of mean demand for a large class of economies which are represented by dispersed measures. Dispersed measures have local densities with respect to the Lebesgue measure. Under the assumption that these densities are continuous, Sondermann (1976) obtains mean demand functions that are C^1 except on a closed null set of prices.

An alternative to the parametric approach is to carry out aggregation in two steps. The first step is aggregation with respect to wealth keeping preferences fixed. Dispersion with respect to wealth is easy to express, since Lebesgue measure can be used to describe wealth distribution. The irregularities remaining after aggregation with respect to wealth are considerably smaller than those of individual demand [Dierker, Dierker, and Trockel (1980a, b)]. These irregularities may be further reduced by aggregation with respect to preferences.

We do not have, at present, a satisfactory theory of large regular economies with preference non-convexities. However, in the next section we shall see that regularity in the presence of non-convexities can very well be handled provided there is a well-determined, finite number of agents.

6. Smale's approach

Demand or supply functions need not exist in economies with a finite number of agents who do not necessarily have strictly convex preference patterns or strictly convex technology sets. Dispensing with convexity and with the concept of an excess demand function, Smale (1974a, b) describes an equilibrium state in the following disaggregated way. In an equilibrium state the marginal rates of substitution of each individual agent are compatible with the prevailing price system and each consumer obeys his budget restriction; and, furthermore, an equilibrium allocation is attainable for the economy.

These conditions for an equilibrium state allow one very well to deal with production, whatever the returns to scale are. It is assumed, though, as usual, that competition prevails in the sense that no agent tries to exert an influence on the price system. As the above requirements for an equilibrium state associate explicitly with each individual agent a set of equations, the number of agents plays a certain role here, whereas the mean excess demand function in a large economy hardly depends on a few agents.

As convexity is not assumed, the compatibility of the marginal rates of substitution and the price system does not entail utility maximization. Therefore the compatibility of the marginal rates of substitution and the price system, together with the budget conditions and the feasibility condition, is a requirement too weak to characterize equilibrium states. However, this set of conditions turns out to be restrictive enough to describe states of an economy which are, in most cases, locally unique and vary continuously with the economic data. Therefore, it suffices for the present purpose to concentrate upon this set, called the set of *extended price equilibria*, instead of the set of price equilibria itself.

The local uniqueness and robustness of extended price equilibria is most easily studied in exchange economies. We now follow Smale (1974a). Also see Smale's contribution to this Handbook. There are m consumers. The consumption space of each consumer is the non-negative orthant $\boldsymbol{R}^{\ell}_{+}$. The preferences of consumer i are represented by a C^2 utility function u_i: $\boldsymbol{R}^{\ell}_{+} \to \boldsymbol{R}$ with a nowhere vanishing derivative. The marginal rates of substitution of consumer i at $x \in \boldsymbol{R}^{\ell}_{+}$ can be described by

$$g_i(x) = Du_i(x)/\|Du_i(x)\|.$$

The following analysis does not depend directly on the utility functions u_i but on the functions g_i. Therefore it is unimportant which particular utility function is selected to represent a given preference ordering.

The space

$$U = \left\{ u: \boldsymbol{R}^{\ell}_{+} \to \boldsymbol{R} \,|\, u \text{ is } C^2, \ \forall_{x \in \boldsymbol{R}^{\ell}_{+}} \ Du(x) \neq 0 \right\}$$

is endowed with the C^2 Whitney topology. A neighborhood basis for $u \in U$ in the C^2 Whitney topology is obtained by taking, for any continuous mapping δ: $\boldsymbol{R}^{\ell}_{+} \to]0, \infty[$, the set of those elements of U for which, at every $x \in \boldsymbol{R}^{\ell}_{+}$, the C^2 distance to u is smaller than $\delta(x)$. The Whitney topology has the merits of being very fine and of making U a Baire space. For an extensive discussion of the Whitney topology, see Golubitsky and Guillemin (1973, §5).

The *space of economies* is $\mathcal{E}=(\boldsymbol{R}^\ell_+)^m \times U^m$. The price space is $\bar{S}=\{p\in\boldsymbol{R}^\ell \mid p\geqslant 0, \|p\|=1\}$. The *space of states* is the space $\mathcal{S}=(\boldsymbol{R}^\ell_+)^m\times\bar{S}$ of allocation–price pairs.

Definition

An *extended price equilibrium* of economy $(e_1,\ldots,e_m,u_1,\ldots,u_m)\in\mathcal{E}$ is a state $(x_1,\ldots,x_m,p)\in\mathcal{S}$ such that x_i belongs to the budget set $B_{i,p}=\{x\in\boldsymbol{R}^\ell_+ \mid px=pe_i\}$ and x_i is a critical point of $u_i|B_{i,p}$ $(i=1,\ldots,m)$, and such that $\sum_{i=1}^m x_i=\sum_{i=1}^m e_i$.

To obtain a large set of economies which have a finite number of extended price equilibria, all of them robust with respect to changes of endowments and utilities, a small set of utility functions has first to be excluded. If $(x_1,\ldots,x_m)$ is an equilibrium allocation with corresponding price vector p and if all $x_i\in$ int $\boldsymbol{R}^\ell_+$, then

$$g_1(x_1)=g_2(x_2)=\cdots=g_m(x_m)=p,$$

i.e. the point $(g_1(x_1),\ldots,g_m(x_m),p)$ lies on the diagonal Δ of the $(m+1)$-fold product Σ^{m+1} of the $(\ell-1)$-sphere Σ. For $u=(u_1,\ldots,u_m)\in U^m$, define Ψ_u: $\mathcal{S}\to\Sigma^{m+1}$ by

$$\Psi_u(x_1,\ldots,x_m,p)=(g_1(x_1),\ldots,g_m(x_m),p).$$

Exclude those m-tuples u of utility functions for which Ψ_u does not intersect Δ transversely.

A mapping f: $M\to N$ between differentiable manifolds is said to *intersect* a submanifold $\tilde{N}$ of N *transversely* iff, for every $m\in M$ with $f(m)\in\tilde{N}$, the image of the tangent space of M at m together with the tangent space of $\tilde{N}$ at $f(m)$ spans the tangent space of N at $f(m)$.

It follows directly from this definition that Ψ_u intersects Δ transversely if the Gaussian curvature of the indifference surfaces of all agents is non-zero everywhere. The above transversality requirement on Ψ_u can thus be considered an extension to the non-convex case of the requirement that the preferences of all agents give rise to C^1 demand functions (cf. Section 5). The condition that Ψ_u intersects Δ transversely is hardly restrictive since it is fulfilled for all u in a dense and open subset of U^m. For the results on transversality theory which are relevant here, see Abraham and Robbin (1967, ch. 4).

Now consider a fixed m-tuple u of utility functions such that Ψ_u intersects Δ transversely. An exchange economy then is completely described by an initial allocation $(e_1,\ldots,e_m)\in(\boldsymbol{R}^\ell_+)^m$. Thus, one is in a situation similar to that of Debreu (1970) and can proceed in a manner similar to that given in Section 2. The graph of the extended price equilibrium correspondence for the given utility functions $u=(u_1,\ldots,u_m)$ now is the set

$$\Gamma_u=\left\{(e,x,p)\in(\boldsymbol{R}^\ell_+)^m\times\mathbb{S}\,\middle|\,\sum_{i=1}^m x_i=\sum_{i=1}^m e_i,\right.$$

$$\left.\forall_{i=1,\ldots,m-1}\; px_i=pe_i,\Psi_u(x,p)\in\Delta\right\}.$$

The budget condition for the last agent is omitted since it follows from the other conditions.

Γ_u has a manifold structure as one can see in the following way. The condition that an extended price equilibrium is an attainable state in which all agents satisfy their budget restrictions leads to the definition of Φ: $(\boldsymbol{R}^\ell_+)^m\times\mathbb{S}\to\boldsymbol{R}^{\ell+m-1}$ as:

$$\Phi(e,x,p)=\left(\sum_{i=1}^m x_i-\sum_{i=1}^m e_i,\, px_1-pe_1,\ldots,px_{m-1}-pe_{m-1}\right).$$

A short computation yields that zero is a regular value of Φ. Therefore $\Phi^{-1}(0)$ is the intersection of a $(2\ell-1)$ m-dimensional differentiable manifold with $(\boldsymbol{R}^\ell_+)^m\times\mathbb{S}$. Furthermore, the projection of $\Phi^{-1}(0)$ to $\mathbb{S}$ has a surjective derivative everywhere in $\Phi^{-1}(0)$. Composing this projection with Ψ_u one therefore obtains a mapping transversally intersecting Δ. This mapping sends a point of $\Phi^{-1}(0)$ to Δ iff this point represents an extended price equilibrium. Hence, Γ_u is the intersection of a differentiable manifold with $(\boldsymbol{R}^\ell_+)^m\times\mathbb{S}$. The codimension of Δ in Σ^{m+1} being $(\ell-1)m$, it follows that Γ_u has dimension $(2\ell-1)m-(\ell-1)m=\ell m$.

The *economy* described by the initial allocation e and the given m-tuple u of utility functions is defined to be *regular* iff e is a regular value of the projection of Γ_u to the space $(\boldsymbol{R}^\ell_+)^m$ of initial endowments. Sard's theorem is used to show that the set of critical economies, given u, is a closed subset of $(\boldsymbol{R}^\ell_+)^m$ of measure zero. Each regular economy has a finite set of extended price equilibria. This set contains those price equilibria (x,p) in the usual sense for which all $x_i\in\operatorname{int}\boldsymbol{R}^\ell_+$. To summarize, we have obtained Smale's (1974a) extension to the non-convex case of Debreu's theorem:

Theorem

There exists an open dense subset of the space U^m of m-tuples of utility functions such that, for each u in this set, every pair (e,u) is a regular economy for all initial allocations $e\in(\boldsymbol{R}^\ell_+)^m$ except for e in a closed null set. The open dense subset of U^m contains the set of all m-tuples of utility functions which define C^1 demand functions.

Smale's analysis sketched above can be modified so as to permit utility functions to have critical points. As the normalized gradients $g_i(x)$ then are no longer available, Lagrange multipliers are used explicitly in order to relate the $Du_i(x)$ and p. This extension of Smale (1974a) has been carried out by van Geldrop (1978).

In an economy with production a consumption set will, in general, have less than full dimension ℓ because many goods involved in the production process are not suitable for consumption. Therefore, to prepare for the introduction of a production sector, consumption sets more general than $\boldsymbol{R}^\ell_+$ will be considered.

Assume for the moment that i's consumption set X_i equals the non-negative orthant of some, possibly lower dimensional, coordinate subspace of the commodity space $\boldsymbol{R}^\ell (i=1,\dots,m)$. If $x_i \in X_i$ is i's consumption plan, let α_i be the smallest coordinate subspace of $\boldsymbol{R}^\ell$ that contains x_i. To facilitate notation think of all u_i as having the same domain $\boldsymbol{R}^\ell_+$. Assume that the restriction of i's utility function u_i to the non-negative orthant of α_i has no critical point. If x_i is chosen at price system p, then the compatibility of i's marginal rates of substitution with p requires that $Du_i(x_i)|\alpha_i$ and $p|\alpha_i$ are equal when normalized or, more precisely, that there is a number $\lambda_i > 0$ such that $\lambda_i Du_i(x_i)|\alpha_i = p|\alpha_i$. This condition generalizes the condition $g_i(x_i) = p$ used above. There are only finitely many subspaces of $\boldsymbol{R}^\ell$ that may play the role of α_i for any consumption plan $x_i \in X_i$. This fact allows one to extend the above density and openness result on regular economies to economies in which consumers need not consume positive amounts of all commodities [see Smale (1974b)].

More generally, let i's *consumption set* X_i be of the following kind.

There are real-valued C^1 functions $f_i^1,\dots, f_i^{M_i}$ and $f_i^{M_i+1},\dots, f_i^{N_i}$ defined on the commodity space such that

$$X_i = \left\{ x \in \boldsymbol{R}^\ell \;\middle|\; \begin{array}{ll} f_i^k(x) \geq 0, & \text{for } k=1,\dots, M_i \\ f_i^k(x) = 0, & \text{for } k = M_i+1,\dots, N_i \end{array} \right\}.$$

Assume that if $x_i \in X_i$ and $B_i = \{k \mid f_i^k(x_i) = 0\}$, then the vectors $Df_i^k(x_i)$, $k \in B_i$, are linearly independent.

The last condition is well-known from the Kuhn–Tucker theory of constrained optimization where it is called "constraint qualification." The consumption sets are considered as fixed. Compatibility of marginal rates of substitution and of a price system can be expressed in this framework, and Smale (1974b) has carried out a study of the regularity of extended price equilibria.

Define a *production set* in the following manner.

Let $h = (h^1,\dots, h^N)$ be an N-tuple of C^1 functions from $\boldsymbol{R}^\ell$ to $\boldsymbol{R}$. Assume that if $y \in \boldsymbol{R}^\ell$, $A = \{k \mid h^k(y) = 0\}$, then the vectors $Dh^k(y)$, $k \in A$, are linearly independent. The production set described by h is

$$Y_h = \left\{ y \in \boldsymbol{R}^\ell \mid h^k(y) \leq 0, \quad \text{for } k = 1,\dots, N \right\}.$$

The production sets will not be considered as fixed. If two C^1 functions, h and h', defining production sets Y_h and $Y_{h'}$ are close with respect to the C^1 Whitney topology, then Y_h and $Y_{h'}$ are considered close to each other. In this way a topological space of production sets is obtained which is denoted by $\tilde{T}$.

There are n firms in the economy which are privately owned. Consumer i's share of firm j is a fixed number Θ_{ij}, with $0 \leq \Theta_{ij} \leq 1, \sum_{i=1}^{m} \Theta_{ij} = 1$. Given all fixed characteristics, an *economy* is described by the initial allocation, the utility functions, and by n production sets. For simplicity the initial allocation is allowed to vary through $(\boldsymbol{R}^{\ell})^m$. The *space of economies* then is $\mathfrak{S} = (\boldsymbol{R}^{\ell})^m \times U^m \times \tilde{T}^n$.

Theorem

There is a dense open set $\mathcal{O}$ in $\mathfrak{S}$ with the following property: if $E \in \mathcal{O}$, then each price equilibrium is locally unique and the set of price equilibria is continuous in $E \in \mathcal{O}$. Finally, if $E \in \mathcal{O}$ and the set of states attainable for E is compact, then the set of price equilibria is finite.

This theorem of Smale (1974b) describes the price equilibria of production economies with a well-defined, finite number of agents under very general conditions in a quite satisfactory way. Smale's emphasis on every agent's rates of substitution and their compatibility with the decentralizing market price system seems to be the natural way to incorporate producers into the study of regular equilibria. Smale's approach seems to be less suitable, though, for the treatment of a large consumption sector where the distribution of agents has to play the essential role. One may hope that the aggregation of demand over a large number of consumers will be developed under assumptions on individuals comparable to those of Smale so that the merits of both approaches can be integrated into a model of an economy with production and with a measure space of consumers.

References

Abraham, R. and J. Robbin (1967), Transversal mappings and flows. New York: Benjamin.
Arrow, K. J. and F. H. Hahn (1971), General competitive analysis. San Francisco: Holden Day.
Araujo, A. and A. Mas-Colell (1978), "Notes on the smoothing of aggregate demand", Journal of Mathematical Economics, 5:113–127.
Balasko, Y. (1975a), "The graph of the Walras correspondence", Econometrica, 43:907–912.
Balasko, Y. (1975b), "Some results on uniqueness and on stability of equilibrium in general equilibrium theory", Journal of Mathematical Economics, 2:95–118.
Balasko, Y. (1976), L'équilibre économique du point de vue différentiel. Thesis, Université Paris IX-Dauphine.
Balasko, Y. (1978a), "Economic equilibrium and catastrophe theory: an introduction", Econometrica, 46:557–569.
Balasko, Y. (1978b), "Equilibrium analysis and envelope theory", Journal of Mathematical Economics, 5:153–172.
Balasko, Y. (1978c), "Connectedness of the set of stable equilibria", SIAM Journal of Applied Mathematics, 35:722–728.
Balasko, Y. (1979a), "Economics with a finite but large number of equilibria", Journal of Mathemati-

cal Economics, 6:145–147.

Balasko, Y. (1979b), "A geometric approach to equilibrium analysis", Journal of Mathematical Economics, 6:217–228.

Bröcker, Th. (1975), Differentiable germs and catastrophes. Cambridge: Cambridge University Press.

Debreu, G. (1970), "Economies with a finite set of equilibria", Econometrica, 38:387–392.

Debreu, G. (1972), "Smooth preferences", Econometrica, 40:603–615 and "Corrigendum", Econometrica, 44:831–832.

Debreu, G. (1975), "The rate of convergence of the core of an economy", Journal of Mathematical Economics, 2:1–7.

Debreu, G. (1976), "Regular differentiable economies", The American Economic Review, 66:280–287.

Delbaen, F. (1971), Lower and upper hemi-continuity of the Walras correspondence. Doctoral dissertion, Free University of Brussels.

Dierker, E. (1972), "Two remarks on the number of equilibria of an economy", Econometrica, 40:951–953.

Dierker, E. (1974), Topological Methods in Walrasian Economics, Lecture Notes in Economics and Mathematical Systems, Vol. 92. Berlin: Springer.

Dierker, E. and H. Dierker (1972), "The local uniqueness of equilibria", Econometrica, 40:867–881.

Dierker, E., H. Dierker, and W. Trockel (1980a), "Continuous mean demand functions derived from nonconvex preferences", Journal of Mathematical Economics, 7:27–33.

Dierker, E., H. Dierker, and W. Trockel (1980b), "Smoothing demand by aggregation with respect to wealth", Journal of Mathematical Economics, 7:227–247.

Dierker, H. (1975a), "Smooth preferences and the regularity of equilibria", Journal of Mathematical Economics, 2:43–62.

Dierker, H. (1975b), "Equilibria and core of large economies", Journal of Mathematical Economics, 2:155–169.

Dold, A. (1972), Lectures on algebraic topology. Berlin: Springer.

Eaves, C. and H. Scarf (1976), "The solution of systems of piecewise linear equations", Mathematics of Operations Research, 1:1–27.

Fuchs, G. (1974), "Private ownership economies with a finite number of equilibria", Journal of Mathematical Economics, 1:141–158.

Fuchs, G. and G. Laroque (1974), "Continuity of equilibria for economies with vanishing external effects", Journal of Economic Theory, 9:1–22.

Fuchs, G. and G. Laroque (1976), "Dynamics of temporary equilibria and expectations", Econometrica, 44:1157–1178.

van Geldrop, J. H. (1978), "Extension of a theorem of Smale on equilibria for pure exchange economies", Journal of Mathematical Economics, 5:245–253.

Golubitsky, M. and V. Guillemin (1973), Stable mappings and their singularities. New York: Springer.

Grodal, B. (1975), "The rate of convergence of the core for a purely competitive sequence of economies", Journal of Mathematical Economics, 2:171–186.

Hildenbrand, K. (1974), "Finiteness of $\Pi(\mathcal{E})$ and continuity of Π",: W. Hildenbrand (1974), Appendix to ch. 2.

Hildenbrand, W. (1974), Core and equilibria of a large economy. Princeton, NJ: Princeton University Press.

Hildenbrand, W. and J. F. Mertens (1972), "Upper hemi-continuity of the equilibrium-set correspondence for pure exchange economies", Econometrica, 40:99–108.

Ichiishi, T. (1974), Some generic properties of smooth economies. Doctoral dissertation, University of California, Berkeley.

Ichiishi, T. (1976), "Economies with a mean demand function", Journal of Mathematical Economics, 3:167–171.

Kehoe, T. (1980), "An index theorem for general equilibrium models with production", Econometrica, 48:1211–1232.

Laroque, G. and H. Polemarchakis (1978), "On the structure of the set of fixed price equilibria", Journal of Mathematical Economics, 5:53–69.

Mas-Colell, A. (1974), "Continuous and smooth consumers: Approximation theorems", Journal of Economic Theory, 8:305–336.

Mas-Colell, A. (1975), "On the continuity of equilibrium prices in constant-returns production economies", Journal of Mathematical Economics, 2:21–33.

Mas-Colell, A. (1976), "The theory of economic equilibrium from the differentiable point of view", course notes, Berkeley.
Mas-Colell, A. (1977a), "On the equilibrium price set of an exchange economy", Journal of Mathematical Economics, 4:117–126.
Mas-Colell, A. (1977b), "Regular, nonconvex economies", Econometrica, 45:1387–1407.
Mas-Colell, A. (1977c), "An introduction to the differentiable approach in the theory of economic equilibrium", to appear in: S. Reiter, ed., Studies in mathematical economics.
Mas-Colell, A. (1977d), "Indivisible commodities and general equilibrium theory", Journal of Economic Theory, 16:443–456.
Mas-Colell, A. and W. Neuefeind (1977), "Some generic properties of aggregate excess demand and an application", Econometrica, 45:591–599.
Milnor, J. (1965), Topology from the differentiable viewpoint. Charlottesville: University of Virginia Press.
Neuefeind, W. (1978), "A note on the denseness of differentiable demand", Journal of Mathematical Economics, 5:129–131.
Nishimura, K. (1978), "A further remark on the number of equilibria of an economy", International Economic Review, 19:679–685.
Saigal, R. and C. P. Simon (1973), "Generic properties of the complementarity problem," Mathematical Programming, 4:324–335.
Scarf, H. (1960), "Some examples of global instability of the competitive equilibrium", International Economic Review, 1:157–172.
Schecter, S. (1979), "On the structure of the equilibrium manifold", Journal of Mathematical Economics, 6:1–5.
Smale, S. (1965), "An infinite dimensional version of Sard's theorem", American Journal of Mathematics, 87:861–866.
Smale, S. (1974a), "Global analysis and economics IIA; extension of a theorem of Debreu", Journal of Mathematical Economics, 1:1–14.
Smale, S. (1974b), "Global analysis and economics IV: Finiteness and stability of equilibria with general consumption sets and production", Journal of Mathematical Economics, 1:119–127.
Smale, S. (1976), "Convergent process of price adjustment and global Newton methods", Journal of Mathematical Economics, 3:107–120.
Sondermann, D. (1975), "Smoothing demand by aggregation", Journal of Mathematical Economics, 2:201–223.
Sondermann, D. (1976), "On a measure theoretical problem in mathematical economic", Springer Lecture Notes in Mathematics, Vol. 541. Berlin: Springer.
Sondermann, D. (1980), "Uniqueness of mean maximizers and continuity of aggregate demand", Journal of Mathematical Economics, 7:135–144.
Sternberg, S. (1964), Lectures on differentiable topology. Englewood Cliffs, NJ: Prentice Hall.
Thom, R. (1972), Stabilité structurelle et morphogenèse. New York: Benjamin.
Varian, H. (1975), "A third remark on the number of equilibria of an economy", Econometrica, 43:985–986.
Wiesmeth, H. (1979), "Regular competitive equilibria in disequilibrium economics", Journal of Mathematical Economics, 6:23–29.
Woodcock, A. E. R. and T. Poston (1974), A geometrical study of the elementary catastrophes, Lecture Notes in Mathematics, Vol. 373. Berlin: Springer.
Yamazaki, A. (1978). "An equilibrium existence theorem without convexity assumptions", Econometrica, 46:541–555.
Zeeman, E. C. (1974), "On the unstable behavior of stock exchanges", Journal of Mathematical Economics, 1:39–49.
Zeeman, E. C. (1977), Catastrophe Theory, Addison–Wesley: Reading, Massachusetts.

Chapter 18

CORE OF AN ECONOMY*

WERNER HILDENBRAND†

University of Bonn

1. Introduction

The *core* of an economy consists of those states of the economy which no group of agents can "improve upon". A group of agents can *improve upon* a state of the economy if the group, by using the means available to it, can make each member of that group better off, regardless of the actions of the agents outside that group. Nothing is said in this definition of how a state in the core actually is reached. The actual process of economic transactions is not considered explicitly.

Let us apply this general definition to a situation where the terms "state", "means available", "regardless of the actions", etc. have some specific meaning. Consider the simplest form of economic activity: the exchange of commodities. Thus, we consider a finite set A of economic agents, each of whom, say $a \in A$, is described by his preference relation $\succ_a$ and his initial endowment e_a. The outcome of any exchange, i.e. a *state* (f_a) of the exchange economy, is simply a redistribution of the total endowments, i.e. $\sum_{a \in A} f_a = \sum_{a \in A} e_a$. A coalition of agents, say $S \subset A$, can *improve upon* a redistribution (f_a), if that coalition S, by using the endowments available to it, i.e. $\sum_{a \in S} e_a$, can make each member of that coalition better off, regardless of the actions of the agents not belonging to the coalition S. Now, and this is an important point, if one assumes that the welfare (utility, preference) of every agent depends *only* on the commodity bundle that is allocated to him, that is to say, if all externalities in consumption are excluded[1] and if there are no public goods,[2] then a coalition can disregard the actions of all agents outside the coalition. In this case, a coalition S *can improve upon* a state (f_a) if there is a redistribution, say g_a, $a \in S$, of the endowments $\sum_{a \in S} e_a$ of the coalition S such that g_a is preferred to f_a for every a in S. We emphasize that in

*Sections 1, 2, 4 and 7 do not require any knowledge of measure theory.

†This research was supported in part by the Deutsche Forschungsgemeinschaft, Sonderforschungsbereich 21, and the National Science Foundation, grant SOC76-19700A01 to the University of California, Berkeley, administered through the Centre for Research in Management Science.

[1] For an explicit treatment of externalities we refer to Shapley and Shubik (1969).

[2] The core and public goods was analysed first in Foley (1970).

Handbook of Mathematical Economics, vol. II, edited by K.J. Arrow and M.D. Intriligator

the analysis presented here no group of agents is a priori excluded from forming a coalition.

Restricting attention to states in the core implicitly assumes that information and communication, needed in forming a coalition, are freely available. Under this assumption any allocation not belonging to the core can be considered as "unstable".

The core of an economy might be empty. But if an allocation in the core exists at all then there are, in general, many such allocations. Any two different redistributions, say f and g, in the core are equally plausible outcomes of the exchange process. On the basis of individual preferences and given distribution of initial endowments they cannot be discriminated. Of course, there will always be some agents who rather prefer the situation in f than the one in g; and the other way around. But once a state f or g belonging to the core is proposed, no agent will be able to form a coalition which can improve upon the situation in f or g. The outcome of exchange according to the theory of the core is thus indeterminate.

The core is a rather theoretical equilibrium concept. It will be used to give a theoretical foundation of a more operational equilibrium concept, the *competitive equilibrium* which, in fact, is a very different notion of equilibrium. The allocation process is organized through markets; there is a price for every commodity. All economic agents take the price system as given and make their decisions independently of each other. The equilibrium price system coordinates these independent decisions in such a way that all markets are simultaneously balanced.

It is easy to show (see Proposition 1) that a competitive equilibrium belongs to the core. Thus, a state of the economy which is decentralized by an equilibrium price system cannot be improved upon by cooperation. This proposition strengthens the well-known Proposition of Welfare Economics (i.e. a competitive equilibrium is Pareto-efficient) to the case where ownership in initial endowments is not collective but individual.

However, there are other states in the core than these of competitive equilibria. Consequently, given the distribution of endowments, a competitive equilibrium necessarily discriminates or favours certain agents. Some agents will find the price system "unfair" in comparison with other allocations in the core. Thus, given the distribution of endowments, a generally agreed justification for organizing the allocation process through markets, that is to say, for concentrating the analysis on competitive equilibria, would therefore require the core and the set of competitive equilibria to coincide.

We shall show in Section 3 that, indeed, the core and the set of competitive equilibria coincide for an economy where the influence of every agent is strictly negligible. For economies with a finite number of participants, the influence of every agent is not strictly negligible and the core is, in general, larger than the set of competitive equilibria. Then we shall consider a sequence $(\mathscr{E}_n)$ of economies where the number of participants tends to infinity, and shall show that under certain assumptions on this sequence $(\mathscr{E}_n)$ the "difference" between the core and the set of competitive equilibria of the economy $\mathscr{E}_n$ becomes arbitrarily small.

2. Definitions: Economy and core

2.1. The core of an exchange economy

We consider ℓ distinguishable commodities. Hence, a commodity bundle (consumption plan) can be represented by a vector $x \in R^{\ell}_{+}$, the closed positive orthant of R^{ℓ}. Let $\mathscr{P}$ denote the set of all preference relations on R^{ℓ}_{+}. A generic element of $\mathscr{P}$ is denoted by $\succ$, and $x \succ y$ is interpreted as "x is preferred to y". For the moment, we just assume that every element $\succ \in \mathscr{P}$ is a binary irreflexive relation.

An economic agent in an exchange economy is characterized by his preference relation and his initial endowment, hence by an element in $\mathscr{P} \times R^{\ell}_{+}$.

Definition 1

An *exchange economy* is a mapping $\mathscr{E}$ of a finite set A of economic agents into the set $\mathscr{P} \times R^{\ell}_{+}$ of agents' characteristics.

If we consider just one economy we shall write $(\succ_a, e_a)$ for $\mathscr{E}(a)$.

An *allocation* for the economy $\mathscr{E}$ is a mapping f of A into R^{ℓ}_{+}. An allocation for $\mathscr{E}$ is said to be *attainable* (or a *state* of $\mathscr{E}$) if

$$\sum_{a \in A} f_a - \sum_{a \in A} e_a.$$

Any non-empty subset of A we call a *coalition*.

Definition 2

Let f be an allocation for the economy $\mathscr{E}$: $A \rightarrow \mathscr{P} \times R^{\ell}_{+}$. The coalition S can *improve upon*[3] the allocation f if there exists an allocation g for $\mathscr{E}$ such that

(i) $g_a \succ_a f_a$, for every $a \in S$.

(ii) $\sum_{a \in S} g_a = \sum_{a \in S} e_a$.

The set of all attainable allocations for the economy $\mathscr{E}$ that no coalition can improve upon is called the *core of the economy* and is denoted by $\boldsymbol{C}(\mathscr{E})$.

[3] The core expresses what coalitions can do for themselves, not what they can do to their opponents. Therefore, we use the term "to improve upon" and not "to block", which is often used in the literature (compare Chapter 7 of this Handbook by Shubik).

The core, as a plausible equilibrium concept for an exchange economy, has been discussed at length in the introduction. Edgeworth (1845–1926) in his *Mathematical Psychics* (1881) introduced the concepts of improving and core under the terms "recontracting" and "contract curve". The fundamental contribution of Edgeworth received little attention and was not further developed for almost 80 years.

In 1953 Shapley and Gillies analysed von Neumann–Morgenstern solutions for games in characteristic function forms with transferable utility. In this connection they formulated a new solution concept for such games which they called the *core*. The connection between the Edgeworth contract curve of an exchange economy and the core of a game was perceived by Shubik (1959) who developed and partly extended Edgeworth's ideas in the framework of game theory with transferable utility.

Inspired by these important contributions Scarf (1962) presented a stimulating analysis of the core and competitive equilibria for general exchange economies without the assumption of transferable utility. From there on the theory developed in two connected but somewhat different branches. On the one hand, the core was analysed for general exchange economies; the present chapter gives an account of this development. On the other hand, the analysis was formulated in terms of games in characteristic function forms, often with transferable utility. The main contributions to this development are due to Shapley and Shubik; they are discussed in Chapter 7 of this Handbook by Shubik.

For a two-person and two-commodity exchange economy the core can be represented geometrically in the well-known Edgeworth Box[4] if we assume that the preference relations $\precsim_1$ and $\precsim_2$ of agents 1 and 2 can be represented by their indifference curves. The size of the box is determined by the total endowment $e_1+e_2=\bar{e}$ (Figure 2.1). Every point (x, y) in the box represents a state f of the economy; indeed, $f_1=(x, y)$ and $f_2=\bar{e}-(x, y)$. Some indifference curves for both agents among different states of the economy are drawn. The solid lines describe the indifference curves for the first agent. The dotted lines describe the indifference curves of states of $\mathfrak{E}$ for the second agent. Clearly, the first agent can improve upon every point in the box which is below his indifference curve through e_1, and the second agent can improve upon every point in the box which is above his (dotted) indifference curve through e_1. Thus, we remain with the "shadowed lens" in the box. Some points in the "shadowed lens" can be improved upon by the two agents acting together, for example e_1 or h in Figure 2.1. A point z in the box cannot be improved upon by the coalition of the two

[4]Actually, Edgeworth used a somewhat different graphical representation [see Edgeworth (1881, figure 1, p. 28), where the origin (0,0) represents the initial distribution of endowments]. The "Edgeworth Box", as we use it today, was introduced by Pareto (1906), *Manuel d'Economie Politique*, e.g. fig. 16, ch. III or fig. 50, ch. VI.

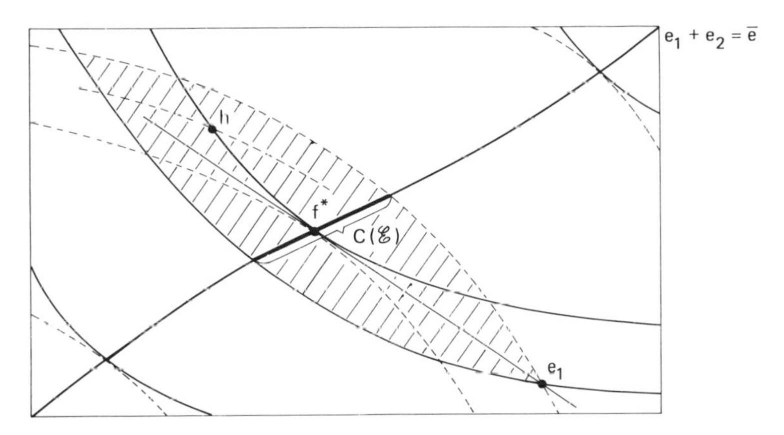

Figure 2.1.

agents, if the set of points which are above the first agent's indifference curve through z and the set of points which are below the second agent's indifference curve through z, have an empty intersection. The locus of these points having this property (called the Pareto-efficient allocations) is, in Figure 2.1, the line from the origin to $\bar{e}$. Thus, the core is the intersection of this line with the shadowed lens.

There is a special point f^* in the core in Figure 2.1; the straight line through e_1 and f^* separates the set of points which are above the first agent's indifference curve through f^* and the set of points which are below the second agent's indifference curve through f^*. Such a state is called a Walrasian or competitive equilibrium (see Chapter 15 in this Handbook by Debreu). For completeness we recall the definition

Definition 3

An allocation f for the economy $\mathscr{E}$: $A \to \mathscr{P} \times R^{\ell}_{+}$ is called a *Walras allocation* (or a *competitive equilibrium*) if there exists a price vector $p \in R^{\ell}$ such that

(α) f_a is a maximal element for $\succ_a$ in the budget-set $\{x \in R^{\ell}_{+} \mid p \cdot x \leq p \cdot e_a\}$ for every agent a in A, and

(β) $\Sigma_A f_a = \Sigma_A e_a$, i.e. total demand equals total supply.

The set of all Walras allocations for the economy $\mathscr{E}$ is denoted by $W(\mathscr{E})$.

A price vector $p \in R^{\ell}$ is called an *equilibrium price vector* for the economy $\mathscr{E}$ if there exists an allocation f for $\mathscr{E}$ with properties (α) and (β). The set of all equilibrium price vectors for the economy $\mathscr{E}$ is denoted by $\Pi(\mathscr{E})$.

We recall that one can easily draw an Edgeworth Box with two, or even infinitely many, Walras allocations. The following result is mathematically trivial but its economic interpretation is important.

Proposition 1

$\boldsymbol{W}(\mathcal{E}) \subset \boldsymbol{C}(\mathcal{E})$ for every economy $\mathcal{E}$.

Proof

Let $f \in \boldsymbol{W}(\mathcal{E})$ but $f \notin \boldsymbol{C}(\mathcal{E})$. Thus, there is a coalition S and an allocation g such that (i) $g_a \succ_a f_a$ for every $a \in S$ and (ii) $\sum_S g_a = \sum_S e_a$. By property (α) of the definition of a Walras allocation we obtain $p \cdot e_a < p \cdot g_a$ for every $a \in S$, where p denotes an equilibrium price vector associated with f. Hence

$$p \cdot \sum_S e_a < p \cdot \sum_S g_a,$$

which contradicts (ii). Q.E.D.

Proposition 1 and the existence theorems for Walras allocations (see Chapter 15 in this Handbook by Debreu) immediately yield sufficient conditions for the existence of allocations in the core of an exchange economy.

Theorem

The core is non-empty for every exchange economy $\mathcal{E}$ with continuous, convex and monotonic preferences.

We emphasize that preferences are not assumed to be complete or transitive. Continuity means that the set $\{(x, y) \in R^\ell_+ \times R^\ell_+ \mid x \succ y\}$ is open relative to $R^\ell_+ \times R^\ell_+$. The preference relation $\succ$ is convex if the set $\{x \in R^\ell_+ \mid x \succ z\}$ is convex for every $z \in R^\ell_+$ and monotonic if $x \geqslant y$ and $x \neq y$ implies $x \succ y$.

It is worthwhile to remark that there are examples of economies $\mathcal{E}$ with $\boldsymbol{C}(\mathcal{E}) \neq \varnothing$ but $\boldsymbol{W}(\mathcal{E}) = \varnothing$. Furthermore, for an economy $\mathcal{E}$ with only few participants the concept of Walras equilibrium is questionable, since the underlying behavioral assumption that all economic agents take the price system as given (beyond their influence) is not justified. The emphasis in this chapter is laid, however, on economies with a large number of participants. As explained in the introduction, the theoretical (non-operational) concept of the core of an economy is then used to justify concentrating our attention on the more operational concept of Walras equilibria. Hence, our main goal is to show that for large competitive economies every allocation in the core of that economy can be "approximately" decentralized by a suitably chosen price system, that is to say, for every $f \in \boldsymbol{C}(\mathcal{E})$ there exists a price vector p such that if this price vector p were to prevail, no (or at least only few) agents a in A would have a strong incentive (given preference $\succ_a$ and income $p \cdot e_a$) to deviate from the allocation $f(a)$. In the following sections we shall make this statement precise in different ways.

2.2. *The core of a coalition production economy*

In an exchange economy the only activity the agents of a coalition can carry out is to redistribute their initial endowments. Now we extend this model by considering for every coalition of agents its productive capabilities as well. There are no a priori given firms or producers. The productive capabilities of every coalition S of economic agents are described by a set of input–output combinations – the production possibility set Y_S for the coalition S. The production possibility set Y_S can be described by a subset in the commodity space R^ℓ (which now also includes various types of labour) if we use the following convention: $y \in Y_S$ means that the coalition S can produce the output vector $\max(0, y)$ provided it has at its disposal the input vector $\max(0, -y)$. Thus, input quantities are represented by negative numbers, and output quantities by positive numbers. The production possibility set Y_S is determined by purely technological as well as institutional conditions. It can also depend on factors or capabilities possessed by members of the coalition, but which do not belong to the list of commodities which defines the commodity space R^ℓ.

In this new context an economic agent a is described by his *consumption set* X_a, his *preference relation* $\succ_a$, and his *initial endowment* e_a. The consumption set X_a now typically is different from R^ℓ_+, since the amount of a certain type of labor that agent a wants to supply is represented by a negative number.

Definition 1′

A *coalition production economy* $(\mathfrak{E}, \boldsymbol{Y})$ is defined by a mapping $\mathfrak{E}$ which associates with every agent a of a finite set A, a consumption set X_a, a preference relation $\succ_a$ on X_a and an initial endowment vector e_a, and by a production set correspondence $\boldsymbol{Y}$ which associates with every coalition $S \subset A$ a production possibility set Y_S.

An *allocation* is a mapping f of A into R^ℓ such that $f_a \in X_a$ for every $a \in A$. An allocation is called *attainable* (or a *state* of $(\mathfrak{E}, \boldsymbol{Y})$) if $\sum_A f_a \in \sum_A e_a + Y_A$.

The definition of the core for a coalition production economy[5] is now straightforward.

Definition 2′

Let f be an allocation for the coalition production economy $(\mathfrak{E}, \boldsymbol{Y})$. The coalition S can *improve upon* the allocation f if there exists an allocation g such that

(i) $g_a \succ_a f_a$ for every $a \in S$, and
(ii) $\sum_{a \in S} g_a \in \sum_{a \in S} e_a + Y_S$.

[5] The concept of a coalition production economy has also been used in the theory of labour managed economies; see, for example, Drèze (1976).

The set of all attainable allocations for $(\mathscr{E}, \boldsymbol{Y})$ which no coalition can improve upon is called the *core* of the economy $(\mathscr{E}, \boldsymbol{Y})$, and is denoted by $\boldsymbol{C}(\mathscr{E}, \boldsymbol{Y})$.

The existence of allocations in the core of $(\mathscr{E}, \boldsymbol{Y})$ surely requires certain assumptions on the production set correspondence $\boldsymbol{Y}$. The correspondence $\boldsymbol{Y}$ is called *balanced* if for every balanced family $\mathfrak{B}$ of coalitions and associated weights δ_S it follows that $\sum_{S\in\mathfrak{B}}\delta_S Y_S \subset Y_A$. (Recall, a family $\mathfrak{B}$ of coalitions is balanced if there exist non-negative weights δ_S for every $S\in\mathfrak{B}$ with $\sum_{S\in\mathfrak{B},\, S\ni a}\delta_S=1$ for every $a\in A$.) It has been shown by Böhm (1974) that $\boldsymbol{C}(\mathscr{E}, \boldsymbol{Y})$ is non-empty if for every agent a: (1) the consumption set X_a is closed, convex and bounded from below, and $e_a \in X_a$; (2) the preference indifference relation $\precsim_a$ is convex; and (3) the production set correspondence $\boldsymbol{Y}$ is balanced, Y_S is closed and contains 0, and asymptotic cone of $Y_A \cap R^\ell_+ = \{0\}$.

In order to extend the concept of a Walras allocation to a coalition production economy, one has to specify the income of every agent, particularly the way in which total profit is distributed. Given a production correspondence $\boldsymbol{Y}$ and a price vector p, the profit distribution $\pi(a, p)$ is called *stable* if $\sup\{p\cdot y \mid y\in Y_A\} = \sum_A \pi(a, p)$ and $\sup\{p\cdot y \mid y\in Y_S\} \leqslant \sum_S \pi(a, p)$ for every coalition S. If total profit $\sup p\cdot Y_A$ is distributed in this way, then, taking the price vector p as given, the best that every individual a in A can do is to accept his profit share $\pi(a, p)$. In other words, a stable profit distribution is defined as an element in the core of the side payment game $v_p(S)$, defined by $v_p(S)=\sup\{p\cdot y \mid y\in Y_S\}$. The core of such a game is non-empty if and only if the game is balanced (see Chapter 7 in this Handbook by Shubik). With this concept of a stable profit distribution which has been proposed, independently, by Böhm (1974), Sondermann (1974) and Oddou (1976) – of course, one can think of other profit distributions – we obtain the following equilibrium concept.

Definition 3′

Given a coalition production economy $(\mathscr{E}, \boldsymbol{Y})$, an allocation f and a total production plan $y\in Y_A$ is called a *price equilibrium* if there exists a price vector $p\in R^\ell$, $p\neq 0$, and if there is a stable profit distribution $\pi(a, p)$ with respect to $\boldsymbol{Y}$ and p such that

(α) f_a is a maximal element for $\succ_a$ in the budget set $\{x\in X_a \mid p\cdot x \leqslant p\cdot e_a + \pi(a, p)\}$ for every agent $a\in A$;
(β) $p\cdot y = \max p\cdot Y_A$; and
(γ) $\sum_A f_a = \sum_A e_a + y$, i.e. total demand equals total supply.

Proposition 1′

Every price equilibrium for a coalition production economy belongs to the core.

Proof

Let (f, y) be a price equilibrium with respect to the price vector p and profit distribution $\pi(a, p)$. Assume that f can be improved upon by the coalition S, i.e. there exists an allocation g such that

(i) $g_a \succ_a f_a$ for every $a \in S$, and
(ii) $\sum_S g_a \in \sum_S e_a + Y_S$.

Property (i) implies $p \cdot g_a > p \cdot e_a + \pi(a, p)$ for every $a \in S$. Hence, $p \cdot \sum_S g_a > p \cdot \sum_S e_a + \sum_S \pi(a, p) \geqslant p \cdot \sum_S e_a + \sup p \cdot Y_S$, which is a contradiction with (ii). Q.E.D.

Sufficient conditions for the existence of price equilibria for a coalition production economy can be found in Böhm (1974) and Sondermann (1974).

Appendix: The space of preferences[6]

Let $\mathcal{P}$ denote the set of continuous and irreflexive binary relations on R^ℓ_+. Thus, every element $\succ \in \mathcal{P}$ is a subset of $R^\ell_+ \times R^\ell_+$ which is open relative to $R^\ell_+ \times R^\ell_+$ and $(x, x) \notin \succ$ for every $x \in R^\ell_+$. Instead of $(x, y) \in \succ$ we shall use the more suggestive notation $x \succ y$. Thus, $x \not\succ y$ means $(x, y) \notin \succ$. A relation $\succ$ in $\mathcal{P}$ is called

transitive, if $x \succ y$, and $y \succ z$ imply $x \succ z$;
monotonic, if $0 \leqslant x \leqslant y$ and $x \neq y$ imply $y \succ x$;
convex, if $\{x \in R^\ell_+ | x \succ z\}$ is convex for every $z \in R^\ell_+$.

A *preference–indifference relation* $\precsim$ on R^ℓ_+ is a reflexive ($x \precsim x$, $x \in R^\ell_+$), complete ($x \precsim y$ or $y \precsim x$ for every $x, y \in R^\ell_+$), transitive ($x \precsim y$ and $y \precsim z$ implies $x \precsim z$), and continuous ($\{(x, y) \in R^\ell_+ \times R^\ell_+ | x \precsim y\}$ is closed) binary relation. Clearly, every preference–indifference relation $\precsim$ can be considered as an element in $\mathcal{P}$ by defining $x \succ y$ as $y \precsim x$, but not $x \precsim y$. The set of all preference–indifference relations on R^ℓ_+ is denoted by $\mathcal{P}^*$. A relation $\precsim$ in $\mathcal{P}^*$ is called *strongly convex* if $x \sim y$ (i.e. $x \precsim y$ and $y \precsim x$) and $x \neq y$ imply $\lambda x + (1-\lambda) y \succ x$ for all λ with $0 < \lambda < 1$.

The tastes of economic agents are described by preference relations. The intuitive concept of "similar" tastes is therefore made precise mathematically by a topology on the set $\mathcal{P}$ of preferences. From the economic point of view, tastes can be qualified to be similar if they give rise to similar demand in similar wealth–price situations. This is certainly a necessary condition for any economically meaningful and operational concept of similarity of tastes.

[6]The material of this Appendix is needed for a full understanding of the theory presented in this chapter. However, essential use of the topological structure on $\mathcal{P}$ is only made in Sections 6, 7 and 8.

Surely there is always a topology on the set $\mathcal{P}$ of preference relations (e.g. the discrete topology) such that the demand correspondence will have continuity properties. However, we want a topology which is *metrizable* and *compact*. Such a topology exists[7] [for details see Hildenbrand (1974, ch. 1.2, p. 96)] and is easily characterized: it is the coarsest topology on $\mathcal{P}$ such that the set

$$(*) \quad \{(\succ, x, y) \in \mathcal{P} \times R^{\ell}_{+} \times R^{\ell}_{+} \mid x \succ y\}$$

is closed. This topology is called the *topology of closed convergence*. We do not need an explicit definition here, its characteristic property $(*)$ and the following properties are sufficient for our purpose. *If the set $\mathcal{P}$ is endowed with the topology of closed convergence, then $\mathcal{P}$ becomes compact and metrizable. The subsets of transitive and convex preference relations are both compact. The subset $\mathcal{P}_{\mathrm{mo}}$ of monotonic preference relations in $\mathcal{P}$ and the subset $\mathcal{P}^{*}_{\mathrm{sco}}$ of strongly convex preference–indifference relations in $\mathcal{P}$ are both countable intersections of open sets in* $\mathcal{P}$ (hence Borel subsets of $\mathcal{P}$). For more details on the topological structure of $\mathcal{P}$ and relevant subsets of preferences, we refer to Grodal (1974). The main justification of the topology of closed convergence is the following result. Let $\varphi(\succ, p, w)$ denote the set of maximal elements for $\succ$ in the budget set $\{x \in R^{\ell}_{+} \mid p \cdot x \leqslant w\}$. If $\varphi(\succ, p, w) \neq \varnothing$, $p \gg 0$ and $w \geqslant 0$, then the demand correspondence is compact-valued and upper hemi-continuous at $(\succ, p, w)$. In particular, on the space $\mathcal{P}^{*}_{\mathrm{sco}}$ of strongly convex preferences the demand $\varphi(\precsim, p, w)$ is a continuous function in $(\precsim, p, w)$. For a proof we refer to Hildenbrand (1974, Theorem 2, p. 99).[8] We remark that for a preference relation $\succ \in \mathcal{P}$ the demand set $\varphi(\succ, p, w)$ is non-empty for $p \gg 0$ and $w \geqslant 0$ if $\succ$ is either transitive or convex. In the latter case, a fixed point argument is used to establish the existence of maximal elements [Sonnenschein (1971) and Mas-Colell (1974)].

Smooth preferences

A *smooth* preference relation is a binary relation on $P = \{x \in R^{\ell} \mid x \gg 0\}$ which is (a) reflexive, complete, transitive, monotonic, continuous (i.e. $\{(x, y) \in P \times P \mid x \precsim y\}$ is closed in $P \times P$); (b) of class C^2, i.e. the set $\mathcal{I} = \{(x, y) \in P \times P \mid x \sim y\}$ is a C^2-hypersurface in $R^{2\ell}$; and (c) the closure of every indifference hypersurface is contained in P.[9] Equivalently, we can define a smooth preference relation by a

[7] This topology on the space $\mathcal{P}^{*}_{\mathrm{mo}}$ was first introduced by Kannai (1970), who gives an explicit expression for a corresponding metric.

[8] The upper hemicontinuity of the demand correspondence was first shown by Debreu (1969) in the case where $\mathcal{P}^{*}$ is endowed with the Hausdorff distance.

[9] We remark that usually this assumption is not required in the definition of a smooth preference relation. Whenever we use smooth preferences in this chapter we need, however, the boundary assumption (c). Therefore we find it convenient to define smooth preferences by including property (c).

real-valued utility function u on P of class C^2, satisfying $Du(x)>0$ for every $x\in P$ [see Debreu (1972, p. 610)].

The set of smooth preferences can be endowed with a metric in the following way. Every smooth preference relation can be represented by a C^1-function g on P, where $g(x)$ is the unit normal to the indifference hypersurface through x oriented in the direction of preferences. Let G be the set of functions g obtained in this way. We now endow G with the topology of uniform C^1 convergence on compact sets. Thus, G becomes a separable metric space. For details we refer to H. Dierker (1975a) or Chapter 17 in this Handbook by E. Dierker. We remark that this topology is finer than the topology of closed convergence.

3. Economies with a continuum of agents

In this section we present a mathematical model of an exchange economy in which the influence of every individual agent is negligible. We shall show that in this model the two equilibrium concepts, the core and the set of Walras allocations, coincide. It will turn out that an "atomless measure space of economic agents" is the proper mathematical formulation of the traditional economic concept of perfect competition. The latter concept is, in fact, as abstract as the former, which has the decisive advantage of being mathematically well defined.

A measure space[10] $(A,\mathcal{Q},\nu)$ is called *atomless* if for every $E\in\mathcal{Q}$ with $\nu(E)>0$ there is a $S\subset E$ with $S\in\mathcal{Q}$, $0<\nu(S)<\nu(E)$. A standard example is the unit interval with Lebesgue measure λ.

Definition 4

An *atomless exchange economy* $\mathcal{E}$ is a measurable mapping of an atomless measure space $(A,\mathcal{Q},\nu)$ with $\nu(A)=1$ into the space $\mathcal{P}\times R^{\ell}_{+}$ of agents' characteristics such that the vector of mean endowments, $\int \text{proj}_2\circ\mathcal{E}\,\text{d}\nu$, is finite.

An *allocation* for the economy $\mathcal{E}$ is an integrable function f of A into R^{ℓ}_{+}. An allocation for $\mathcal{E}$ is called *attainable* or a *state* of $\mathcal{E}$ if $\int f\,\text{d}\nu=\int \text{proj}_2\circ\mathcal{E}\,\text{d}\nu$.[11]

Remark

The basic result of this section, Theorem 1, can be formulated and proved without referring explicitly to the topological structure on the space of preferences $\mathcal{P}$. Instead of assuming that the mapping $\mathcal{E}$ is Borel-measurable, as in Definition 4, one simply assumes what is needed in the proof, namely: for every $z\in R^{\ell}$, the set

[10] We refer to Chapter 5 in this Handbook by Kirman or Billingsley (1968, ch. I) for definitions of measure-theoretic concepts used in the sequel.

[11] If there is no danger of ambiguity we shall use the simplifying notation $(\succ_a, e(a))$ instead of $(\text{proj}_1\circ\mathcal{E}(a), \text{proj}_2\circ\mathcal{E}(a))$.

$\{a \in A | z + e(a) \succ_a f(a)\}$ is measurable, i.e. belongs to $\mathcal{A}$. In any case, measurability is a technical assumption without any economic interpretation.

An atomless exchange economy is, in fact, a quite abstract concept. The interpretation relies on analogy to economies as introduced in Section 2. Any such economy $\mathcal{E}: A \to \mathcal{P} \times R^\ell_+$ can be viewed as a mapping of the simple measure space $(A, \mathcal{A}, \nu)$ into $\mathcal{P} \times R^\ell_+$, where $\mathcal{A}$ is the set of *all* subsets of the finite set A, and ν is the normalized counting measure, i.e. $\nu(E) = \#E/\#A$. The number $\nu(E)$ is interpreted as the fraction of the totality of agents belonging to the coalition E. The σ-algebra $\mathcal{A}$ of coalitions is introduced only for technical (measure-theoretic) reasons. As in the case of economies with finitely many participants, there is conceptually no a priori restriction on possible coalitions. Since for an atomless measure space $(A, \mathcal{A}, \nu)$ the set A of agents must be uncountably infinite, we shall speak of the set of participants as a "continuum of agents". Given an allocation f, then $f(a)$ denotes the commodity vector allocated to agent a. We emphasize that $\int_E f \mathrm{d}\nu$ does not mean the commodity vector allocated to the coalition E; it should be interpreted as the commodity vector per capita, since its analogy for an economy with a finite set A is $(1/\#A)\sum_{a \in E} f(a)$.

There are no "infinitesimally small" agents in our model. There are "many" agents of normal size, and hence infinite amounts of commodities. Yet the per capita – i.e. mean – amounts are finite.

We shall now define the core and the set of Walras allocations for an atomless exchange economy by rewriting the definitions of Section 2 in the new context.

Definition 5

Let f be an allocation for the atomless economy $\mathcal{E}: (A, \mathcal{A}, \nu) \to \mathcal{P} \times R^\ell_+$. The coalition $S \in \mathcal{A}$ can improve upon the allocation f if there exists an allocation g such that

(i) $g(a) \succ_a f(a)$, a.e. in S, and
(ii) $\nu(S) > 0$ and $\int_S g \mathrm{d}\nu = \int_S e \mathrm{d}\nu$.

The set of all attainable allocations for the economy $\mathcal{E}$ that no coalition in $\mathcal{A}$ can improve upon is called the *core* of the economy $\mathcal{E}$ and is denoted by $\boldsymbol{C}(\mathcal{E})$.

Definition 6

An allocation f for the economy $\mathcal{E}$ is called a *Walras allocation for* $\mathcal{E}$ if there exists a price vector $p \in R^\ell$ such that
(i) a.e. in A, $f(a)$ is a maximal element for $\succ_a$ in the budget set $\{x \in X_a | p \cdot x \leqslant p \cdot e(a)\}$, and
(ii) $\int f \mathrm{d}\nu = \int e \mathrm{d}\nu$,

that is, *mean demand equals mean supply*. The set of all Walras allocations for $\mathscr{E}$ is denoted by $\boldsymbol{W}(\mathscr{E})$.[12]

The following result, duc to Aumann (1964), shows that for atomless economies the two equilibrium concepts, the core and the set of Walras allocations, coincide.

Theorem 1

Let $\mathscr{E}\colon (A,\mathscr{A},\nu)\to\mathscr{P}_{\mathrm{mo}}\times R^{\ell}_{+}$ be an atomless economy with $\int e\,\mathrm{d}\nu\gg 0$. Then

$$\boldsymbol{W}(\mathscr{E})=\boldsymbol{C}(\mathscr{E}).$$

Proof

The inclusion $\boldsymbol{W}(\mathscr{E})\subset\boldsymbol{C}(\mathscr{E})$ is shown analogously as Proposition 1. Let $f\in\boldsymbol{C}(\mathscr{E})$. We have to show that there is a price vector $p\in R^{\ell}$ such that ν-a.e., $p\cdot f(a)=p\cdot e(a)$ and $z+e(a)\succ_a f(a)$ implies $p\cdot z>0$. Hence, the intuitive idea of the proof is to find a supporting hyperplane for a set $\bigcup_{a\in A'}\{z\in R^{\ell}\,|\,z+e(a)\succ_a f(a)\}$, where $\nu(A')=1$.

Since it might happen that for some z the set of agents $\{a\in A\,|\,z+e(a)\succ_a f(a)\}$ has measure zero, which creates a difficulty in our proof, we shall proceed as follows.

Let Q^{ℓ} denote the set of all vectors in R^{ℓ} with rational coordinates. For every $z\subset Q^{\ell}$, consider the measurable set of agents

$$A_z=\{a\in A\,|\,z+e(a)\succ_a f(a)\}.$$

Define

$$A'=A\setminus\bigcup_{\substack{z\in Q^{\ell}\\ \nu(A_z)=0}}A_z.$$

In other words, if $z\in Q^{\ell}$ and $\nu(A_z)=0$, then $A'\cap A_z=\varnothing$. Since Q^{ℓ} is countable we have $\nu(A')=1$.

Consider now the set

$$C=\mathrm{co}\bigcup_{a\in A'}\{z\in Q^{\ell}\,|\,z+e(a)\succ_a f(a)\}.$$

[12] For the existence of Walras allocations for an economy with a measure space of agents, we refer to Theorem 18 in Chapter 15 of this Handbook by Debreu.

We shall show that $0 \notin C$. Assume to the contrary that $0 \in C$. Then we can write 0 in the form

$$0 = \sum_{i=1}^{r} \alpha_i z_i,$$

where $\alpha_i > 0$, $\sum_{i=1}^{r} \alpha_i = 1$ and $z_i + e(a_i) \succ_{a_i} f(a_i)$ for at least one agent $a_i \in A'$, i.e. $A'_{z_i} = A_{z_i} \cap A' \neq \emptyset$. By definition of A' we therefore obtain $\nu(A'_{z_i}) > 0$ for every $i = 1, \ldots, r$.

Since the measure space $(A, \mathscr{A}, \nu)$ is atomless, we know by Liapunov's theorem (i.e. $\{\nu(B') | B' \subset B,\ B' \in \mathscr{A}\} = [0, \nu(B)]$) that for every measurable subset B and every number $0 \leqslant \delta \leqslant \nu(B)$ there exists a measurable subset $B' \subset B$ with $\nu(B') = \delta$. Consequently, for λ sufficiently small [e.g. $\lambda = (1/r)\min_i \nu(A_{z_i})$], there are disjoint subsets S_i, where $S_i \subset A_{z_i}$ and $\nu(S_i) = \lambda\alpha_i$, $i = 1, \ldots, r$. (See Figure 3.1.) But now the coalition $S = \bigcup_{i=1}^{r} S_i$ can improve upon the allocation f by using the redistribution $g(a) = z_i + e(a)$, for $a \in S_i$ $(i = 1, \ldots, r)$.

Since we have shown that 0 does not belong to the convex set C there is, by Minkowski's theorem, a hyperplane through 0 supporting the set C. Hence, there exists a price vector $p \neq 0$ such that $p \cdot z \geqslant 0$ for every $z \in C$. Consequently, for every $a \in A'$, $z \in Q^{\ell}$ and $z + e(a) \succ_a f(a)$ imply $p \cdot z \geqslant 0$. Hence, for every $a \in A'$, if $x \in R_{+}^{\ell}$ and $x \succ_a f(a)$ then $p \cdot x \geqslant p \cdot e(a)$. Since preferences are monotonic we obtain $p \geqslant 0$, and by standard arguments one now verifies that (f, p) is a Walras equilibrium. Indeed, it follows that $p \cdot e(a) \leqslant p \cdot f(a)$ for every $a \in A'$, since $f(a)$ is limit of a sequence (x_n) with $x_n \succ_a f(a)$.

Since $\int p \cdot f \,\mathrm{d}\nu = \int p \cdot e \,\mathrm{d}\nu$ it follows that $p \cdot e(a) = p \cdot f(a)$ for ν-almost all $a \in A$. By assumption, $\int e \,\mathrm{d}\nu \gg 0$; therefore there is an agent $a \in A'$ with $p \cdot f(a) = p \cdot e(a) > 0$. For this agent it then follows that $f(a)$ is a maximal element in the budget set

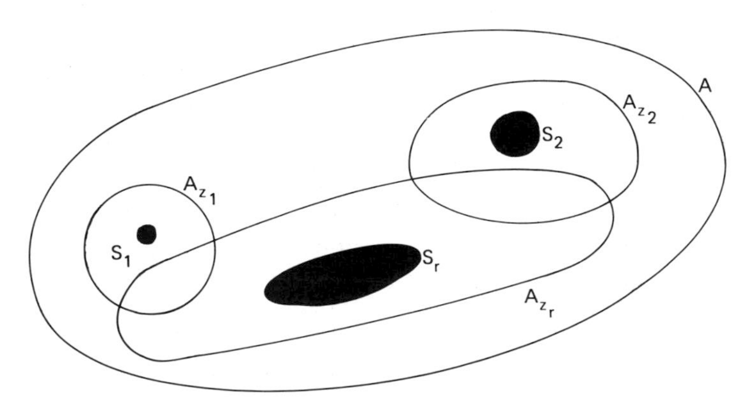

Figure 3.1.

$\{x \in R^{\ell}_{+} | p \cdot x \leqslant p \cdot e(a)\}$. Indeed, for $x \in R^{\ell}_{+}$ with $p \cdot x < p \cdot e(a)$ it follows for every $a \in A'$ that $x \not\succ_a f(a)$. Since in the case $p \cdot e(a) > 0$ every $x \in R^{\ell}_{+}$ with $p \cdot x = p \cdot e(a)$ is limit of a sequence (x_n), with $x_n \in R^{\ell}_{+}$ and $p \cdot x_n < p \cdot e(a)$, the continuity of the preference relation $\succ_a$ implies $x \not\succ_a f(a)$. Thus, $f(a)$ is a maximal element in the budget set. This, together with the monotonicity of preferences, implies that $p \gg 0$. Hence, for ν-almost every agent a in A, $f(a)$ is a maximal element in the budget set, even in the case $p \cdot e(a) = 0$, since $f(a)$ belongs to the budget set for ν-almost every a in A and in this case the budget set is equal to $\{0\}$. This completes the proof that (f, p) is a Walras equilibrium. Q.E.D.

Remark 1

In the above proof we made no explicit use of convexity or transitivity properties of the preference relations. The monotonicity of preferences has been used to obtain $p \gg 0$. For an alternative proof of Theorem 1 and generalizations on it, we refer to Hildenbrand (1974). If one drops the monotonicity assumption and allows that the consumption set X_a depends on a, then the above proof yields the following result. Let $\mathcal{P}_{\mathrm{lns}}$ denote the set of continuous and locally non-saturated preference relations (i.e. $x \in X_a$ implies $x \in$ closure $\{z \in X_a | z \succ_a x\}$) defined on convex consumption sets. If f belongs to the core of an atomless economy $\mathfrak{E}$: $(A, \mathcal{Q}, \nu) \to \mathcal{P}_{\mathrm{lns}} \times R^{\ell}_{+}$, then f is a quasi-Walras equilibrium, i.e. there exists a price vector $p \neq 0$ such that ν-a.e. in A, $p \cdot f(a) = p \cdot e(a)$ and that $\inf p \cdot X_a < p \cdot e(a)$ implies that $f(a)$ is a maximal element in the budget set $\{x \in X_a | p \cdot x \leqslant p \cdot e(a)\}$.

Remark 2

The coalition S in the proof can be chosen such that $\nu(S) \leqslant \varepsilon$ for any given $\varepsilon > 0$. Hence, if one restricts in the definition of the core of an *atomless* economy the set of admissible coalitions to those with $\nu(S) \leqslant \varepsilon$, then one still obtains the identity of the core and $W(\mathfrak{E})$. Thus the core is not enlarged by allowing only "small" coalitions. One can even show [Grodal (1972), Schmeidler (1972) and Vind (1972)] that this is still true if for any $0 < \varepsilon < 1$ the admissible coalitions are restricted to those with $\nu(S) = \varepsilon$.

Furthermore, the proof also shows that if $f \notin C(\mathfrak{E})$ then there is a coalition S and an allocation g such that $g(a) \succ_a f(a)$, $a \in S$, and $\int_S g \, d\nu \ll \int_S e \, d\nu$.

The fundamental contribution of Aumann (1964) has stimulated research in various directions. For an alternative formulation of the content of Theorem 1 we refer to Vind (1964). Vind considers coalitions rather than individual agents as the primitive concept of a perfectly competitive economy. Vind's results have been extended by Cornwall (1969) and Richter (1971).

Non-Standard Analysis. In order to define a mathematical model of an economy where the influence of every individual economic agent is strictly negligible, one can use the theory of Non-Standard Analysis instead of measure theory. In this framework one obtains essentially the same results. The basic references are Brown and Robinson (1972, 1975) and Khan (1974a, b).

Measure spaces of economic agents with atoms. The author's opinion on the significance of an *exchange* economy with atoms has been expressed in Hildenbrand (1974, p. 126):

> From a formal point of view one might also consider a measure space $(A, \mathcal{A}, \nu)$ with atoms and an atomless part. In terms of the interpretation given earlier, one could consider an atom as a syndicate of traders, that is a group of agents which cannot be split up; either all of them join a coalition or none does so. Note, however, that the mapping $\mathfrak{E}$, and also every allocation f, must be constant on an atom. This means that all agents in the atom must have identical characteristics and must all receive the same bundle in an allocation. This is so special a case that it makes the interpretation of atoms as syndicates of little economic significance.
>
> An alternative approach is to consider an atom as a "big" agent. In the framework of the model under consideration "big" can only mean "big" in terms of endowment. Thus, an atom would be an agent who has infinitely more endowment than any agent in the atomless part. Now the measure $\nu(S)$ has a different interpretation. Formerly it expressed the relative number of agents in the coalition S, here it expresses something like the relative size of the endowment of S. Moreover the allocation $f(a)$ and the preferences $\succ_a$ must be reinterpreted. The net result is far from clear.
>
> No doubt a rigorous interpretation of a model with "big" and "small" traders is possible. However, in a model with pure exchange, where there is no production, where all goods have to be consumed, and where there is no accumulation, it is difficult to see the economic content of such an interpretation.

Here are some basic references: the equivalence between the core and the set of Walras allocations still holds if the atoms are not "too big" [Gabszewicz–Mertens (1971)] or if there are at least two atoms, all with the same characteristics, Shitovitz (1973, 1974). Exchange economies with a measure space of economic agents where the equivalence theorem does not hold have been analysed by Aumann (1973), Champsaur and Laroque (1973, 1976), Drèze and Gabszewicz (1971), Drèze, Gabszewicz and Postlewaite (1977), and Drèze, Gabszewicz, Schmeidler and Vind (1972).

Economies with production. Analogously as in Section 2 one can define a coalition production economy for an atomless measure space of economic agents. If the production set correspondence is additive, then the core and the set of price equilibria coincide [Hildenbrand (1974, ch. 4)]. Although the assumption of additivity is fairly strong, it cannot be dispensed with altogether, see Böhm (1973) and Oddou (1976).

We close this section with a historical note. In a stimulating paper, Scarf (1962) proved an equivalence theorem of the core and the set of competitive equilibria for an economy with *countably* many agents (a countably infinite replica economy). The conceptual difficulty with this model is the definition of *attainable* allocation, i.e. to give a meaning to the statement "total demand equals total endowments".[13] Debreu (1963) gave a very simple proof of Scarf's equivalence theorem, and it was this proof which stimulated Aumann to introduce an economy with a continuum of agents. Indeed, Debreu considered the set $C = \mathrm{co} \bigcup_{a \in A} \{z \in R^{\ell} \mid z + e(a) \succsim_a f(a)\}$ for the particular economy considered by Scarf and applied a separating hyperplane argument as in the proof of Theorem 1.

4. Core and prices of a large but finite economy

In this section we shall show that for a large class of economies with a finite but large number of participants *one can associate with every allocation f in the core of the economy a price vector p such that only "few" agents would have a "strong incentive" to deviate from the core-allocation f if the price vector p were to prevail.* Before we discuss the economic content of this result, we shall formulate it more precisely.

Let Δ denote the price simplex, i.e. $\Delta = \{p \in R^{\ell}_{+} \mid \sum_{h=1}^{\ell} p_h = 1\}$. For every economy $\mathscr{E}$: $A \to \mathscr{P} \times R^{\ell}_{+}$ let $B(\mathscr{E}) = \sup\{e^h(a_1) + \cdots + e^h(a_\ell) \mid h = 1, \ldots, l;\ a_1 \in A, \ldots, a_\ell \in A\}$.

Lemma

For every economy $\mathscr{E}$: $A \to \mathscr{P}_{\mathrm{mo}} \times R^{\ell}_{+}$ and every allocation $f \in \boldsymbol{C}(\mathscr{E})$ there exists a price vector $p \in \Delta$ such that

(i) $\sum_{a \in A} |p \cdot f(a) - p \cdot e(a)| \leqslant 2B(\mathscr{E})$, and
(ii) $\sum_{a \in A} |\inf\{p \cdot x \mid x \succ_a f(a)\} - p \cdot e(a)| \leqslant 2B(\mathscr{E})$.

[13] Recently Weiss (1981) proved an equivalence theorem for a finitely additive and "atomless' measure space which covers the case of countably many agents.

The Lemma applied to an economy, where the fraction $B(\mathscr{E})/\#A$ is small, says that there is a price vector such that the *average deviation in budget* $(1/\#A)\Sigma_A|p\cdot f(a)-p\cdot e(a)|$ and *the average deviation in compensated expenditure* $(1/\#A)\Sigma_A|\inf\{p\cdot x|x\succ_a f(a)\}-p\cdot e(a)|$ is small. Thus, in particular, for "most" agents a in A the deviations in budget, $p\cdot f(a)-p\cdot e(a)$, and in compensated expenditure, $\inf\{p\cdot x|x\succ_a f(a)\}-p\cdot e(a)$, will be small. We emphasize, however, that this does not imply that the commodity bundle $f(a)$ in the commodity space is near to the demand $\varphi(a, p)$ of agent a at the price p (Figure 4.1).

The total demand $\Sigma_A\varphi(a, p)$ at this price system p might be quite different from the total supply $\Sigma_A e(a)$. What then is the operational meaning of this "core-allocation price system"? If this price system were to prevail then, indeed, only few agents would have a "strong incentive" to deviate from the core-allocation f. But one cannot conclude that the price system "approximately decentralizes" the core-allocation f; for this one should require in addition to the properties of the lemma that the price vector leads to consistent actions, that is to say, $\Sigma_A e(a)\in\Sigma_A\varphi(a, p)$, i.e. $p\in\Pi(\mathscr{E})$, or at least that total (or mean) excess demand is small. The lemma should therefore be only considered as a first step; it will later be used to prove much stronger results.

For example, if preferences are strongly convex then nearness in budget and in compensated expenditure implies nearness to competitive demand. More precisely, let K be a compact subset of agents' characteristics in $\mathscr{P}^*_{\substack{\text{mo}\\ \text{sco}}}\times R^\ell_+$ and B a bounded subset in R^ℓ_+. Then one easily verifies that for every $\varepsilon>0$ there exists $\delta>0$ such that for every $(\precsim, e)\in K$, $x\in B$ and $p\in\Delta$,

$$|p\cdot x-p\cdot e|<\varepsilon \quad \text{and} \quad |\inf\{p\cdot z|z\succ x\}-p\cdot e|<\varepsilon$$

imply that $|x-\varphi(\precsim, e, p)|<\delta$.

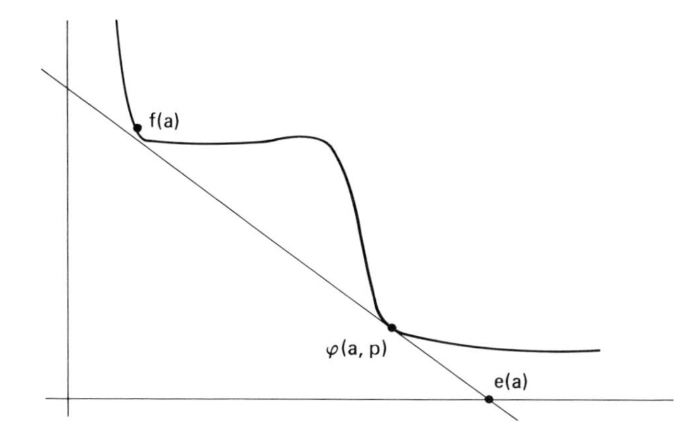

Figure 4.1.

This observation leads to the following result.

Theorem 2

Let K be a compact subset of agents' characteristics in $\mathcal{P}^*_{\substack{\text{mo}\\ \text{sco}}} \times R^\ell_+$ and let δ be any positive number. Then for every $\varepsilon > 0$ there exists an integer N with the following property: for every economy $\mathcal{E}: A \to K$ with $\#A \geqslant N$ and $\sum_{a\in A} e(a) \geqslant \delta$ and every allocation f in $C(\mathcal{E})$, there exists a price vector $p_f \in \Delta$ such that

$$\frac{1}{\#A} \sum_{a\in A} |f(a) - \varphi(a, p_f)| \leqslant \varepsilon .$$

We remark that this result implies that mean excess demand is small, i.e.

$$\left| \frac{1}{\#A} \sum_{a\in A} (\varphi(a, p_f) - e(a)) \right| \leqslant \varepsilon .$$

However, we emphasize that excess demand is not zero; that is to say, it cannot be shown that p_f is an equilibrium price vector for $\mathcal{E}$ (see the example in Section 7). Furthermore, it is not possible to strengthen the conclusion of Theorem 2 to a uniform statement: $|f(a) - \varphi(a, p_f)| \leqslant \varepsilon$ for *every* agent $a \in A$ [see Bewley (1973, example 5, p. 439)]. For such a stronger result it does not suffice just to have a large number of agents in the economy with characteristics in a compact set. Largeness alone is not sufficient. If the economy becomes large one needs an assumption which ensures that for every agent, say a^*, in the economy there are "many" other agents who have similar characteristics as agent a^*. That is to say, no agent becomes "isolated" in the space of characteristics. Such an assumption is most easily formulated by considering sequences of economies. We return to this in Section 6. Theorem 2 was first proved by Bewley (1973). We outline here a much simpler proof by using the fundamental lemma.

In order to prove Theorem 2 it suffices to show the following. Let $(\mathcal{E}_n)$ be any sequence of economies with $\mathcal{E}_n: A_n \to K$, $\sum_{a\in A_n} e(a) \geqslant \delta$ and $\#A_n \to \infty$. Then for every allocation f_n in $C(\mathcal{E}_n)$ there exists a price vector $p_n \in \Delta$ such that the sequence $(d_n)n = 1, \ldots,$

$$d_n = \frac{1}{\#A_n} \sum_{a\in A} |f_n(a) - \varphi(a, p_n)|$$

has a subsequence (d_{n_q}) which converges to zero. This result follows quite easily from the lemma. We sketch a proof. Apply the lemma to every $\mathcal{E}_n$ and $f_n \in C(\mathcal{E}_n)$. Since K is compact it follows that the sequence $B(\mathcal{E})$ is bounded. Consequently it follows from the lemma that there is a subsequence of $(\mathcal{E}_n)$, which we denote again by $(\mathscr{E}_n)$, and a subset E_n of A_n with $\#E_n/\#A_n \to 1$ such that $p_n \to p \in \Delta$ and

(i) $p_n \cdot (f_n(a) - e(a)) \to 0$ uniformly on E_n, and

(ii) $\inf\{p_n \cdot z \mid z \succ_a f_n(a)\} - p_n \cdot e(a) \to 0$ uniformly on E_n.

Since K is a compact set, one can show (see, for example, Proposition 5) that the sequence (f_n) of core-allocations is bounded. Thus we can apply the above remark and hence properties (i) and (ii) imply that

(iii) $|f_n(a) - \varphi(a, p_n)| \to 0$ uniformly on E_n.

Since the sequence (f_n) is bounded and $\#E_n/\#A_n \to 1$ it follows from property (iii) that there is a subsequence for which d_n converges to zero. Q.E.D.

The fundamental Lemma which improves a result of Vind (1965) was first proved with additional assumptions by Arrow–Hahn (1971, Theorem 2, p. 189). E. Dierker (1975) gave a proof for monotonic and transitive preference relations. Keiding (1974) observed that a separating hyperplane theorem can be used instead of the fixed point theorem in Dierker's proof. The simple proof given here is due to Anderson (1978).

Proof of the lemma

For every $a \in A$, let $\Psi(a) = \{x - e(a) \in R^\ell \mid x \succ_a f(a)\} \cup \{0\}$, and define

$$\Psi_A = \frac{1}{\#A} \sum_{a \in A} \Psi(a).$$

We now claim that $\Psi_A \cap R^\ell_- = \{0\}$. Assume to the contrary that there is a function $h\colon A \to R^\ell$ with $h(a) \in \Psi(a)$ and $(1/\#A)\Sigma_A h(a) < 0$. Then the coalition $S = \{a \in A \mid h(a) \neq 0\}$ can improve upon the allocation f with the allocation

$$g(a) = h(a) + e(a) - \sum_A h(a)/\#S, \qquad a \in S.$$

Indeed, by monotonicity of preferences $g(a) \succ_a f(a)$ for every $a \in S$ and $\Sigma_{a \in S} g(a) = \Sigma_{a \in S} e(a)$.

The set ψ_A which contains no strictly negative vector, however, need not be convex. Hence, its convex hull $\mathrm{co}\,\psi_A$ might contain strictly negative vectors. But we shall show that

$$\mathrm{co}\,\psi_A \cap \{x \in R^\ell \mid x \ll -b\} = \varnothing, \quad \text{where } b = \frac{B(\mathscr{E})}{\#A} \cdot (1, \ldots, 1).$$

This is illustrated in Figure 4.2.

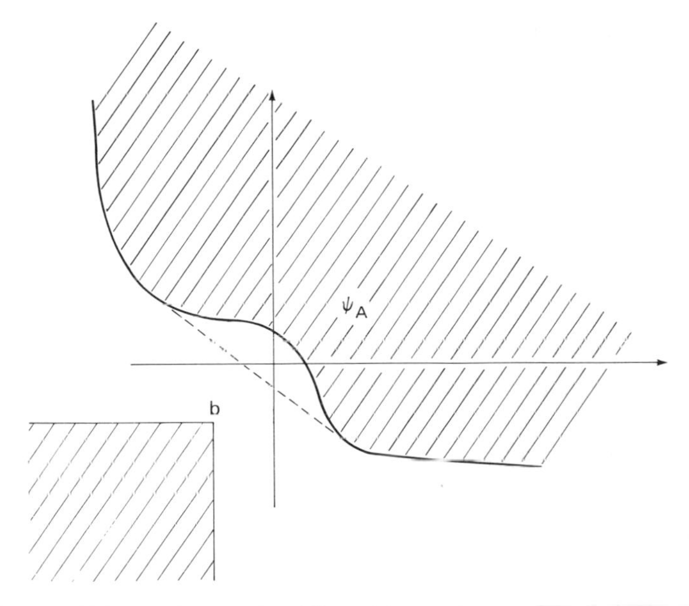

Figure 4.2.

Suppose there is a vector $x \in \operatorname{co}\psi_A$ with $x \ll -b$. Then we shall show that there would be a strictly negative vector y in ψ_A, which is a contradiction. By the Shapley–Folkman Theorem (cf. Chapter 1 of this Handbook) we can write x in the form $x = (1/\#A)\sum_{a\in A} g(a)$, where $g(a) \in \operatorname{co}\psi(a)$ for every $a \in A$ and $\#\{a \in A \mid g(a) \notin \psi(a)\} \leqslant \ell$. Let E denote those agents in A. Define $y = (1/\#A)\sum_{a\in A\setminus E} g(a)$. Since $0 \in \psi(a)$ we obtain $y \in \psi_A$. Clearly, $\psi(a) \geqslant -e(a)$, hence $\operatorname{co}\psi(a) \geqslant -e(a)$. Now,

$$y = x - \frac{1}{\#A}\sum_{a\in E} g(a) \leqslant x + \frac{1}{\#A}\sum_{a\in E} e(a) \leqslant x + b \ll 0.$$

By Minkowski's Theorem there exists a hyperplane separating the convex sets $\operatorname{co}\psi_A$ and $\{x \in R^\ell \mid x \ll -b\}$. Thus, there exists a price vector $p \in \Delta$ such that

$$\inf\{p\cdot x \mid x \in \psi_A\} \geqslant -p\cdot b = -\frac{B(\mathscr{E})}{\#A}.$$

Since

$$\inf\{p\cdot x \mid x \in \psi_A\} = \frac{1}{\#A}\sum_{a\in A} \inf\{p\cdot x \mid x \in \psi(a)\}$$

and by definition $0 \in \psi(a)$ we obtain

$$-B(\mathscr{E}) \leqslant \sum_{a \in A} \inf\{p \cdot x \mid x \in \psi(a)\} \leqslant 0.$$

Let $S = \{a \in A \mid p \cdot (f(a) - e(a)) < 0\}$. Since preferences are locally non-saturated, we have $p \cdot (f(a) - e(a)) \geqslant \inf p \cdot \psi(a) \leqslant 0$. It follows that

$$\sum_{a \in S} |p \cdot (f(a) - e(a))| \leqslant \sum_{a \in S} -\inf p \cdot \psi(a) \leqslant \sum_{a \in A} -\inf p \cdot \psi(a) \leqslant B(\mathscr{E}).$$

Clearly, $\sum_{a \in A} p \cdot (f(a) - e(a)) = 0$, which implies that

$$\sum_{a \in A} |p \cdot (f(a) - e(a))| = 2 \sum_{a \in S} |p \cdot (f(a) - e(a))| \leqslant 2B(\mathscr{E}).$$

This proves property (i) of the lemma.

Let $d(a) = \inf\{p \cdot x \mid x \succ_a f(a)\} - p \cdot e(a)$. Since $d(a) \leqslant p \cdot (f(a) - e(a))$ and $d(a) \geqslant \inf p \cdot \psi(a) \leqslant 0$, $a \in A$, we obtain

$$\begin{aligned}\sum_{a \in A} |d(a)| &= \sum_{a \in A} \max(0, d(a)) + \sum_{a \in A} \max(0, -d(a)) \\ &\leqslant \sum_{a \notin S} p \cdot (f(a) - e(a)) + \sum_{a \in A} -\inf p \cdot \psi(a) \\ &\leqslant B(\mathscr{E}) + B(\mathscr{E}). \quad \text{Q.E.D.}\end{aligned}$$

Remark 1

A careful reader will have observed that we did not use convexity and continuity of preferences. Also, transitivity and monotonicity can be slightly weakened; we just need local non-satiation and $x \gg y$, $y \succ z$ implies $x \succ z$.

Remark 2

The above proof shows that an allocation $f \in \boldsymbol{C}(\mathscr{E})$ is in fact an equilibrium if the corresponding set ψ_A is convex. This set is indeed convex for an atomless economy, since the integral of any correspondence with respect to an atomless measure space is a convex set. This result is a direct consequence of Liapunov's Theorem for vector-valued measures. Theorem 1 of the previous section has been proved in this way in Hildenbrand (1974, Theorem 1, p. 133). In the above proof of the fundamental lemma Liapunov's Theorem has been replaced by Shapley–Folkman's Theorem.

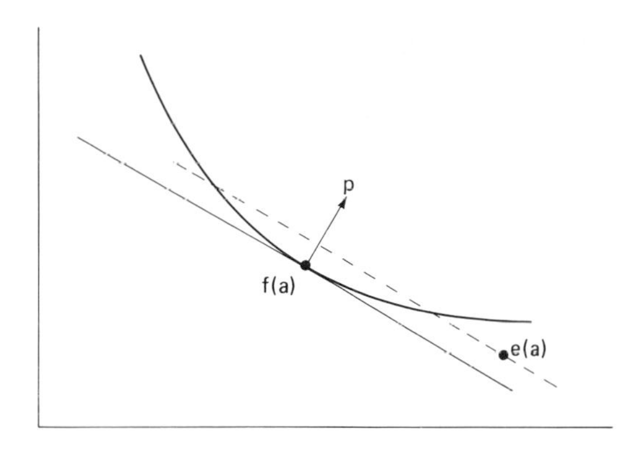

Figure 4.3.

Smooth preferences: Pareto prices

In the framework of smooth preferences the conclusion of the Lemma can be considerably strengthened. If all agents of the economy $\mathscr{E}$ have smooth preferences, then with every Pareto-efficient allocation (i.e. the coalition $S=A$ cannot improve upon f) we can associate the uniquely determined "Pareto price vector" p, that is to say, the price vector p has the property: $x \succsim_a f(a)$ implies $p \cdot x \geqslant p \cdot f(a)$ for every $a \in A$ (Figure 4.3).

We remark that for a Pareto price vector, assertion (ii) of the lemma reduces to assertion (i). The following result shows that for a large economy with smooth *preferences* and for every allocation f in the core *the deviation* in budget, $p \cdot f(a) - p \cdot e(a)$, with respect to the Pareto price vectors, is small for *every* agent.

Let Q denote a compact[14] set of agents' characteristics, where for every $(\precsim, e) \in Q$ the preference relation $\precsim$ is smooth and the endowment $e \gg 0$.

Proposition 2

Given a compact set Q of smooth characteristics, there exists a constant q such that for every economy $\mathscr{E}$: $A \to Q$ and every allocation $f \in \boldsymbol{C}(\mathscr{E})$ the associated Pareto price vector has the property

$$|p \cdot f(a) - p \cdot e(a)| \leqslant q / \# A, \quad \text{for every } a \in A.$$

[14] Compact, of course, with respect to the topology on the space of smooth preferences; see Appendix of Section 2.

For a proof we refer to Grodal (1975) who generalizes a result of Debreu (1975). We remark that, as in the lemma, the total excess demand $z(p,\mathcal{E})=\Sigma_A(\varphi(a,p)-e(a))$ at the Pareto price vector p need not be small in general. However, with the assumptions of Proposition 2, we shall show later that $(1/\#A)z(p,\mathcal{E})$ becomes arbitrarily small if $\#A$ is large enough. One can even show (Theorem 4) that for certain sequences of economies $(\mathcal{E}_n)$, with $\#A_n\to\infty$ the total excess demand $z(p_n,\mathcal{E}_n)$ remains bounded.

We close this section with a discussion of a very interesting result due to Mas-Colell (1978). Let $(\mathcal{E}_n)$ be a sequence of economies with characteristics in the compact set Q and let $\#A_n\to\infty$. Consider in every economy $\mathcal{E}_n$ a Pareto-efficient allocation f_n. Let p_n denote the Pareto price vector associated with f_n.

Assume now that the sequence (f_n) is bounded and that the average deviation in budget

$$b_n=\frac{1}{\#A_n}\sum_{A_n}|p_n\cdot(f_n(a)-\mathrm{proj}_2\circ\mathcal{E}_n(a))|$$

does not converge to zero. To simplify, assume that $b_n\geqslant\delta>0$ for all n. Then it follows from Proposition 2 that for n large enough, say $n\geqslant\bar{n}$, the allocation f_n does not belong to $C(\mathcal{E}_n)$. Consequently, there are coalitions in the economy $\mathcal{E}_n$ that can improve upon the allocation f_n. The question now is: How many such coalitions are there? It follows from Mas-Colell (1978) that *the fraction of those coalitions which can improve upon the allocation f_n tends to $\frac{1}{2}$ with $n\to\infty$*. Clearly $\frac{1}{2}$ is an upper bound, since for a Pareto-efficient allocation of $\mathcal{E}_n$ either the coalition S or its complement $A_n\setminus S$ can improve upon f_n. Thus, if the average deviation in budget b_n with respect to the Pareto price vector of a Pareto-efficient allocation does not become arbitrarily small, then there are many coalitions which can improve upon the allocation. In other words, the probability that a coalition S or its complement $A_n\setminus S$ can improve f_n becomes arbitrarily close to one, if any partition $(S,A_n\setminus S)$ of A_n into two sets is considered as equally likely.

5. Replica economies

In this section we prove the well-known limit theorem on the core for a sequence of replica economies. This result has first been proved by Edgeworth (1881) in the case of two commodities and two types of agents. An exposition in modern terminology and a rigorous treatment of the relevant parts in *Mathematical Psychics* (pp. 34–38) can be found in Debreu and Scarf (1972). Shubik (1959) analysed Edgeworth's limit theorem in the framework of game theory with transferable utility. A generalization of Edgeworth's theorem to the case of arbitrary finite numbers of commodities and types of agents with strongly convex preferences, Theorem 3 of this section, is due to Debreu and Scarf (1963).

In order to prove this result we shall use the same method of proof as for the more general limit theorems, where measure-theoretic tools are needed. In proceeding in this way, instead of presenting the beautiful original proof of Debreu and Scarf (1963), we hope that the more complex proof in the general case (Section 8) becomes more transparent.

Let $\mathcal{E}: A \to \mathcal{P} \times R^{\ell}_{+}$ be an exchange economy. We consider its *n-replica* $\mathcal{E}_n$, i.e. an economy in which each of the original agents appears n times. More formally,

$$\mathcal{E}_n: A \times \{1, \ldots, n\} \to \mathcal{P} \times R^{\ell}_{+},$$

where $\precsim_{(a,i)} = \precsim_a$ and $e(a,i) = e(a)$, $a \in A$, $1 \leqslant i \leqslant n$.

Let $\delta(\boldsymbol{C}(\mathcal{E}), \boldsymbol{W}(\mathcal{E}))$ be the smallest number δ such that for every allocation f in the core of $\mathcal{E}$ there exists a Walras allocation f^* of $\mathcal{E}$ such that

$$|f(a) - f^*(a)| \leqslant \delta, \quad \text{for every agent } a \text{ in } \mathcal{E}.$$

Clearly $\delta(\boldsymbol{C}(\mathcal{E}), \boldsymbol{W}(\mathcal{E}))$ measures the "difference" between the core and the set of Walras allocations of an economy.

Theorem 3

Let $\mathcal{E}$ be an economy with monotonic and strongly convex preferences and strictly positive total endowments. If $(\mathcal{E}_n)$ denotes the sequence of replica economies of $\mathcal{E}$, then

$$\delta(\boldsymbol{C}(\mathcal{E}_n), \boldsymbol{W}(\mathcal{E}_n)) \to 0.$$

The first step in the proof consists in showing that for an allocation $f \in \boldsymbol{C}(\mathcal{E}_n)$ all agents with the same characteristics receive the same commodity bundle.

Proposition 3 (equal treatment)[15]

Let $f \in \boldsymbol{C}(\mathcal{E}_n)$ then $f(a,i) = f(a,j)$ for all $a \in A$ and $1 \leqslant i, j \leqslant n$.

Proof

If Proposition 3 did not hold then there would be $f \in \boldsymbol{C}(\mathcal{E}_n)$, $a \in A$, say a^*, and $i \neq j$ such that

$$f(a^*, i) \neq f(a^*, j).$$

[15] It has been shown by Green (1972) that core-allocations in non-replica economies have the equal treatment property only in exceptional cases; see also Khan and Polemarchakis (1978). The unequalness, however, becomes small if the economy is large, see Hildenbrand and Kirman (1973).

For every $a \in A$ there is a least desired commodity bundle among $f(a,1)$, $f(a,2),\ldots, f(a,n)$ according to the common preference $\precsim_a$. Without loss of generality we can assume that $f(a,1) \precsim_a f(a,i)$, $1 \leq i \leq n$. Since preferences are strongly convex we obtain

$$\frac{1}{n}\sum_{i=1}^{n} f(a^*,i) \succ_{a^*} f(a^*,1) \quad \text{and} \quad \frac{1}{n}\sum_{i=1}^{n} f(a,i) \succsim_a f(a,1), \qquad a \in A.$$

Hence, the coalition S consisting of agent $(a,1)$ for every a in A can improve upon the allocation f, by allocating

$$g(a,1) = \frac{1}{n}\sum_{i=1}^{n} f(a,i), \qquad a \in A.$$

Indeed,

$$\sum_{a \in A}(g(a,1)) = \sum_{a \in A}\frac{1}{n}\sum_{i=1}^{n} f(a,i) = \frac{1}{n}\sum_{a \in A}\sum_{i=1}^{n} f(a,i)$$

$$= \frac{1}{n}\sum_{a \in A}\sum_{i=1}^{n} e(a,i) = \sum_{a \in A} e(a,1).$$

Thus, the allocation $g(a,1)$, $a \in A$, is attainable for the coalition S. Now, if agent $(a^*,1)$ gives a sufficiently small amount of his commodities to the other agents of the coalition S then we obtain $g'(a,1) \succ_a f(a,1)$ for every $a \in A$, since preferences are monotonic. Q.E.D.

Proof of Theorem 3

By the equal treatment property an allocation $f_n \in \boldsymbol{C}(\mathscr{E}_n)$ is completely described by $f_n(a,1)$, $a \in A$. Consequently, in order to prove the theorem, it suffices to prove the following *assertion*: *for every sequence* (f_n), *where* $f_n \in \boldsymbol{C}(\mathscr{E}_n)$, *there exists a subsequence* $Q \subset \boldsymbol{N}$ *such that* $f_n(a,1) \vec{Q} f(a)$, *for every* $a \in A$, *and* $f \in \boldsymbol{W}(\mathscr{E})$.

Clearly, by the equal treatment property, the sequence $(f_n(a,1))$ is bounded for every $a \in A$. Hence, there is a subsequence $Q \subset \boldsymbol{N}$ such that for every $a \in A$ $f_n(a,1) \vec{Q} f(a)$. Since $\sum_{a \in A} f_n(a,1) = \sum_{a \in A} e(a)$ we obtain $\sum_{a \in A} f(a) = \sum_{a \in A} e(a)$. It remains to show that f can be decentralized by a suitably chosen price system.

By the fundamental lemma of the last section it follows that for every $f_n \in \boldsymbol{C}(\mathscr{E}_n)$ there exists a price vector $p_n \in \Delta$ such that

(i) $n\sum_{a \in A} |p_n \cdot (f_n(a,1) - e(a))| \leq 2B(\mathscr{E})$

and

(ii) $n\sum_{a\in A}|\inf\{p_n\cdot x|x\succ_a f_n(a,1)\}-p_n\cdot e(a)|\leqslant 2B(\mathscr{E})$.

Consequently, we obtain for every $a\in A$ that

(i′) $p_n\cdot(f_n(a,1)-e(a))\to 0$

and

(ii′) $\inf\{p_n\cdot x|x\succ_a f_n(a,1)\}-p_n\cdot e(a)\to 0$.

(We remark that the convergence is as fast as $1/n$.)

Without loss of generality we can assume that $p_n\to p$. Hence we obtain for every $a\in A$

$$p\cdot f(a)=p\cdot e(a).$$

By (ii′) we know that $\lim_Q[\inf\{p_n\cdot x|x\succ_a f_n(a,1)\}]=p\cdot e(a)$. Note, the function $(p,z)\to\inf\{p\cdot x|x\succ_a z\}$ may not be continuous. Still we can show that $p\cdot e(a)=\inf\{p\cdot x|x\succ_a f(a)\}=\xi$. Indeed, since we have shown already that $p\cdot f(a)=p\cdot e(a)$, it follows that $p\cdot e(a)\geqslant\xi$. We now show that $p\cdot e(a)\leqslant\xi$.[16] For every $\varepsilon>0$ there exists $x_\varepsilon\succ_a f(a)$ such that $p\cdot x_\varepsilon-\varepsilon\leqslant\xi$. For n large enough we have $x_\varepsilon\succ_a f_n(a,1)$ since preferences are continuous. Thus, $\inf\{p_n\cdot x|x\succ_a f_n(a,1)\}\leqslant p_n\cdot x_\varepsilon$. Consequently, $p\cdot e(a)\leqslant p\cdot x_\varepsilon\leqslant\xi+\varepsilon$. Since this holds for every $\varepsilon>0$, we obtain $p\cdot e(a)\leqslant\xi$.

Hence, we have shown that for every $a\in A$,

$$\inf\{p\cdot x|x\succsim_a f(a)\}=p\cdot e(a).$$

Thus, if $p\cdot e(a)>0$, then $f(a)$ is a best element in the budget set. Since there is at least one $a\in A$ with $p\cdot e(a)>0$, it follows that $p\gg 0$, since preferences are monotonic. But this then implies that $f(a)$ is a best element in the budget for every agent; hence f is a Walras allocation for $\mathscr{E}$. Q.E.D.

The following characterization of $\boldsymbol{W}(\mathscr{E})$ – which is essentially the way in which Debreu and Scarf have formulated their result – follows immediately from the method of proof used for Theorem 1.

Proposition 4

Let $\mathscr{E}: A\to\mathscr{P}_{\mathrm{mo}}\times R^\ell_+$ be an economy with $\#A<\infty$ and strictly positive total endowments. Then $f\in W(\mathscr{E})$ if and only if $f_n\in\boldsymbol{C}(\mathscr{E}_n)$, $n=1,\ldots,$ where $\mathscr{E}_n$ denotes the n-replica economy of $\mathscr{E}$ and f_n the n-replica allocation of f, respectively.

[16]This follows directly from the fact that $(p,z)\mapsto\inf\{p\cdot x|x\succ z\}$ is an upper semi-continuous function [e.g. Hildenbrand (1974, Proposition 11, p. 28)].

Indeed, it is easily shown that the convex set

$$C = \operatorname{co} \bigcup_{a \in A} \{z \in R^{\ell} \mid z + e(a) \succ_a f(a)\},$$

defined in an analogous way as in the proof of Theorem 1, does not contain the origin in its interior. Thus, there exists a separating hyperplane which implies by standard arguments that $f \in \boldsymbol{W}(\mathcal{E})$. This sketches the method of proof of Debreu and Scarf (1963).

Proposition 4, together with the equal treatment property Proposition 3, then implies Theorem 3.

In order to understand why for a sequence of replica economies with strictly convex preferences the strong conclusion $\delta(\boldsymbol{W}(\mathcal{E}_n), \boldsymbol{C}(\mathcal{E}_n)) \to 0$ holds, and is so simple to prove, it is useful to look at the method of proof in a more general situation.

The distinguishing property of a replica sequence $(\mathcal{E}_n)$ is that the distributions of agents' characteristics remain fixed, i.e.

$$\frac{1}{\# A_n} \# \mathcal{E}_n^{-1}(B) = \mu_n(B) = \mu(B), \quad \text{for every } B \subset \mathcal{P} \times R^{\ell}_{+}.$$

If, however, one considers (non-replica) sequences $(\mathcal{E}_n)$ of economies where $\# A_n \to \infty$ and the sequence (μ_n) of characteristic distributions is convergent, say $\mu_n \to \mu$, then the situation becomes much more complicated. We shall show later that one can associate with such a sequence $(\mathcal{E}_n)$ a "limit economy" $\mathcal{E}_\infty$, which is an economy with an atomless measure space of economic agents having μ as characteristic distribution.

To prove the assertion $\delta(\boldsymbol{W}(\mathcal{E}_n), \boldsymbol{C}(\mathcal{E}_n)) \to 0$, we proceed in three steps, the first and third are trivial for replica sequences due to the equal treatment property. We start with a sequence (f_n), with $f_n \in \boldsymbol{C}(\mathcal{E}_n)$. Then

(1) We show that there is a "convergent" subsequence, say $(f_{n_q}) \to f$. For replica economies this could be made precise by the assertion $f_{n_q}(a, 1) \to f(a)$ and the equal treatment property. In general, however, f_n is a function on the set A_n of agents in the economy $\mathcal{E}_n$ and $\# A_n \to \infty$. As we shall show later, the natural concept of convergence then becomes convergence in distribution.

(2) We show that the "limit" f is a Walras allocation for the limit economy $\mathcal{E}_\infty$. For replica economies the limit economy need not be made explicit; the original economy $\mathcal{E}$ could be used instead.

(3) We show that there are Walras allocations $f^*_{n_q}$ for the economy $\mathcal{E}_{n_q}$ such that $f^*_{n_q} \to f$. For replica economies this is trivial since an n-replicated Walras allocation for $\mathcal{E}$ is a Walras allocation for $\mathcal{E}_n$. In general, however, it might

happen that there are Walras allocations for the limit economy which cannot be obtained as a limit of Walras allocations for $\mathscr{E}_n$. To overcome this "discontinuity" of the set of Walras equilibria we shall need the concept of regular economies (see Chapter 17 of this Handbook).

To illustrate the usefulness of associating a "limit economy" with a sequence $(\mathscr{E}_n)$ of (replica) economies we now give an alternative proof for Theorem 3 by explicitly using the limit economy and the fact (Theorem 1) that for this economy the core and the set of Walras allocations coincide.

Alternative proof of Theorem 3

We partition the unit interval [0, 1] into $\#A$ subintervals of equal length, and associate with every agent $a \in A$ one and only one subinterval I_a. The limit economy $\mathscr{E}_\infty$: $[0,1] \to \mathscr{P}_{mo} \times R^\ell_+$ is now defined by $\mathscr{E}_\infty(t) = (\precsim_a, e(a))$ if $t \in I_a$. With respect to the Lebesgue measure λ on [0, 1], $\mathscr{E}_\infty$ is clearly an economy with an atomless measure space of economic agents and Theorem 1 can be applied.

Now, by the equal treatment property, every allocation $f_n \in C(\mathscr{E}_n)$ can be represented by a step function $\tilde{f}_n$, as illustrated in Figure 5.1. We know that a subsequence of $(\tilde{f}_n)$ converges to a step function $\tilde{f}$. We shall show that $\tilde{f} \in C(\mathscr{E}_\infty)$. Then by Theorem 1 we know that $\tilde{f} \in W(\mathscr{E}_\infty)$. Hence, if we define $f_n^*(a, i) = \tilde{f}(t)$, $t \in I_a$, $1 \leq i \leq n$, we obtain $f_n^* \in W(\mathscr{E}_n)$, which implies that $\delta(W(\mathscr{E}_n), C(\mathscr{E}_n)) \to 0$.

Assume that $\tilde{f} \notin C(\mathscr{E}_\infty)$. Then (see Section 3, Remark 2) there is a coalition $E \subset [0,1]$ and an allocation g such that

(i) $g(t) \succ_t \tilde{f}(t)$, $t \in E$, and
(ii) $\int_E g \, d\lambda \ll \int_E e \, d\lambda$.

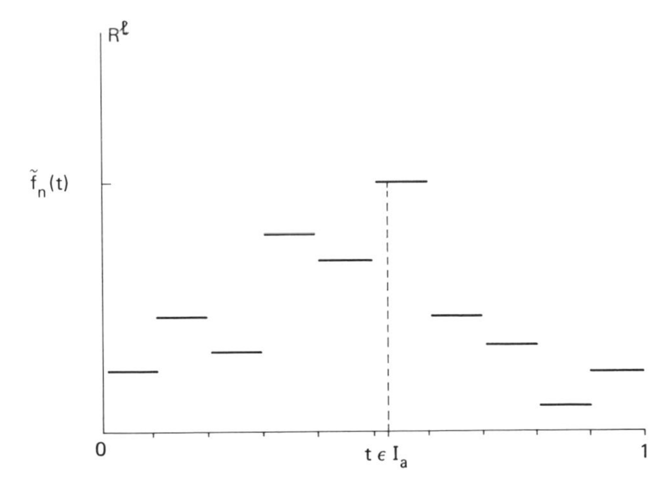

Figure 5.1.

Without loss of generality we can assume that $\lambda(E \cap I_a)$ is rational, say r_a/q, and that g is constant on every interval $E \cap I_a$ (if not, replace g on $E \cap I_a$ by the average $[1/\lambda(E \cap I_a)] \cdot \int_{E \cap I_a} g \,\mathrm{d}\lambda$).

Now, if n is large enough, the coalition E_n consisting of r_a agents with characteristics $(\precsim_a, e(a))$, $a \in A$, can improve upon the allocation f_n by using $g(a) = g(t)$ if $t \in E \cap I_a$, a contradiction.

Rate of convergence of $\delta(\boldsymbol{W}(\mathcal{E}_n), \boldsymbol{C}(\mathcal{E}_n))$

Knowing the convergence $\delta(\boldsymbol{W}(\mathcal{E}_n), \boldsymbol{C}(\mathcal{E}_n)) \to 0$ a natural question to ask is how fast does the distance $\delta(\boldsymbol{W}(\mathcal{E}_n), \boldsymbol{C}(\mathcal{E}_n))$ converge to zero? Shapley (1975) provided examples of replicated economies with C^1-utility functions whose cores converge arbitrarily slowly. It has been shown by Aumann (1979) that arbitrarily slow convergence can occur even if the utility functions are infinitely differentiable. The following result, due to Debreu (1975), shows however that for economies with smooth preferences and differentiable demand functions the examples of slow convergence are "exceptional cases".

The demand function derived from a smooth preference relation $\precsim_a$ is continuously differentiable if and only if every indifference hypersurface has everywhere a non-zero Gaussian curvature [see Debreu (1972, p. 612)]. An economy $\mathcal{E}$ with smooth preferences and a differentiable excess demand function $z(p, \mathcal{E})$ is called *regular* if for every p with $z(p, \mathcal{E}) = 0$ the Jacobian matrix $(\partial z_h / \partial p_i)_{\substack{h=1,\dots,\ell-1 \\ i=1,\dots,\ell-1}}$ is regular (i.e. the rank of the matrix is $\ell - 1$). (For alternative definitions Chapter 17 of this Handbook by Dierker.) Given the preferences $\precsim_a$, $a \in A$, with the above properties, then it is known [Debreu (1970) or Chapter 17 of this Handbook] that the set of initial endowment allocations $(e_a)_{a \in A} \in R^{\ell \cdot \# A}$ for which the economy $(\precsim_a, e_a)_{a \in A}$ is not regular is contained in a closed subset of $R^{\ell \cdot \# A}$ with Lebesgue measure zero. Thus, given the preferences $\precsim_a$, only in "exceptional cases" is the economy $(\precsim_a, e_a)_{a \in A}$ not regular.

With these concepts we can now formulate the following

Theorem 4

Let $\mathcal{E}$ be an economy with smooth preferences $\precsim_a$, strictly positive individual endowments e_a and a continuously differentiable excess demand function $z(\cdot, \mathcal{E})$. If the economy $\mathcal{E}$ is regular then the distance $\delta(\boldsymbol{W}(\mathcal{E}_n), \boldsymbol{C}(\mathcal{E}_n))$ for the n-replica of $\mathcal{E}$ converges to zero at least as fast as the inverse of the number of agents, that is to say,

$$\delta(\boldsymbol{W}(\mathcal{E}_n), \boldsymbol{C}(\mathcal{E}_n)) = 0\left(\frac{1}{n}\right).$$

We shall give no proof for Theorem 4. We just want to explain the role of the two assumptions, namely smoothness of preferences and regularity of the economy.

The smoothness assumption on preferences allows us to apply Proposition 2. Thus, if p_n denotes the Pareto price vector associated with the allocation $f_n \in \boldsymbol{C}(\mathcal{E}_n)$ we obtain:

$$(\alpha) \quad p_n \cdot (f_n(a,1) - e(a)) = 0\left(\frac{1}{n}\right).$$

The regularity of the economy $\mathcal{E}$ is used to show that for every p_n there exists an equilibrium price vector p_n^* of $\mathcal{E}$ such that

$$(\beta) \quad p_n - p_n^* = 0\left(\frac{1}{n}\right).$$

Propositions (α) and (β) now immediately yield the result. Indeed, they imply

$$p_n \cdot f_n(a,1) - p_n^* \cdot e(a) = 0\left(\frac{1}{n}\right).$$

Since the demand function φ is continuously differentiable one obtains from the mean value theorem of calculus that

$$\varphi(\precsim_a, p_n, p_n \cdot f_n(a,1)) - \varphi(\precsim_a, p_n^*, p_n^* \cdot e(a)) = 0\left(\frac{1}{n}\right).$$

Since $f_n(a,1) = \varphi(\precsim_a, p_n, p_n \cdot f_n(a,1))$ and since p_n^* is an equilibrium price vector for $\mathcal{E}$, it follows that

$$\delta(\boldsymbol{W}(\mathcal{E}_n), \boldsymbol{C}(\mathcal{E}_n)) = 0\left(\frac{1}{n}\right).$$

6. Competitive sequences of economies

In an economy with a finite but large number of participants the influence of an individual agent – provided he has no "singular market position" – will be small, and will become negligible if the number of participants increases indefinitely. Thus we consider a sequence

$$\mathcal{E}_n : A_n \to \mathcal{P} \times R^{\ell}_+, \qquad n = 1, \ldots,$$

of economies and ask what property of the *sequence* would correspond to the idea that the influence of every agent becomes negligible.

Obviously, a first condition is that the number of participants tends to infinity, thus:

(i) $\#A_n \to \infty$.

Secondly, the preference–endowment distribution μ_n of the economy $\mathcal{E}_n$, i.e.

$$\mu_n(B) = \frac{1}{\#A_n} \cdot \#\{a \in A_n | \mathcal{E}_n(a) \in B\}, \qquad B \subset \mathcal{P} \times R^\ell_+,$$

should be approximately the same since we wish to speak of the "same" economy but with an increasing degree of competition. Formally:

(ii) the sequence (μ_n) of preference-endowment distributions is weakly convergent.[17]

These two conditions, however, do not yet exclude that some few agents may have a "singular market position". In order to simplify we shall require more than is really needed. For a more general treatment we refer to Trockel (1976). First, we shall assume that the sequence $(1/\#A_n)\Sigma_{a \in A_n} \mathrm{proj}_2 \circ \mathcal{E}_n(a)$ of mean endowments converges and that the limit is strictly positive. Then we require that small groups of agents hold per capita small amounts of endowments. More precisely:

(iii) If $E_n \subset A_n$ and $\#E_n / \#A_n \to 0$, then $(1/\#A_n)\Sigma_{a \in \mathcal{E}_n} \mathrm{proj}_2 \circ \mathcal{E}_n(a) \to 0$.

Obviously, property (iii) follows from (ii) if all endowments belong to a bounded set. It can be shown that (iii) is equivalent with $\int \mathrm{proj}_2 \, \mathrm{d}\mu_n \to \int \mathrm{proj}_2 \, \mathrm{d}\mu$.

This discussion motivates the following:

Definition 7

Let T be any subset of $\mathcal{P} \times R^\ell_+$. A sequence $(\mathcal{E}_n)$ of economies is said to be *competitive* on T if

(0) $\mathcal{E}_n: A_n \to T$, for all n.
(i) $\#A_n \to \infty$.
(ii) $\mu_n \to \mu$ on T.
(iii) $\int \mathrm{proj}_2 \, \mathrm{d}\mu_n \to \int \mathrm{proj}_2 \, \mathrm{d}\mu \gg 0$.

Examples

1. *Sequence of replica economies*. In this case $\#A_n = n \cdot \#A_1$ and $\mu_n = \mu_1$, for all n.

[17]For a definition and details see Billingsley (1968, ch. 1) or Chapter 5 in this Handbook by Kirman.

2. *Sequence of type economies.* The preference–endowment distributions μ_n have all the same finite support, say T. If $\lim(1/\#A_n)\#\{a\in A_n|\mathcal{E}_n(a)=(\succ, e)\}$ exists for every $(\succ, e)\in T$, then $\mu_n \to \mu$.
3. *Sequence of sample economies.* Let μ be any measure on $\mathcal{P}\times R^\ell_+$ such that $0\ll\int \text{proj}_2 \mathrm{d}\mu \ll\infty$. Draw an independent sample of size n from the distribution μ. This defines an economy $\mathcal{E}_n$ with n participants. The sequence $(\mathcal{E}_n)$ of "sample economies" is competitive with probability one (Glivenko–Cantelli's Theorem).
4. An example of a sequence $(\mathcal{E}_n)$ with properties (i) and (ii) but not (iii) is given by: $\mathcal{E}_n$ is an economy with n agents, $n-1$ of whom have the endowment $(0,1)$ and one agent possesses $(n,0)$. The preferences, for simplicity, are all identical. The per capita endowment $(1/n)(n,0)=(1,0)$ of the last agent does not tend to zero. This agent has a "singular market position". An example of a competitive sequence with unbounded initial endowments is the following: $\#A_n=2n+1$; n agents have the endowment $(1,0)$; one agent has the endowment $(\sqrt{n},0)$, and the remaining n agents have the endowments $(0,1)$.

For a detailed discussion of competitive sequences, see Hildenbrand (1974, ch. 2.1) and Trockel (1976).

It seems natural to consider as a "*limit economy* $\mathcal{E}_\infty$" of a competitive sequence $(\mathcal{E}_n)$ any economy with an atomless measure space of agents; the only condition would be that the preference–endowment distribution $\mu_{\mathcal{E}_\infty}$ of this limit economy is equal to $\lim_n \mu_{\mathcal{E}_n}$. If one takes this view, that is to say the *measure space* $(A,\mathcal{Q},\nu)$ *of agents has no intrinsic significance*, then one has to show that two different atomless economies with the same distribution have the same set of equilibria. Clearly the set Π of equilibrium price vectors depends only on the preference–endowment distribution μ, since the mean excess demand – which determines the set of equilibrium price vectors – depends only on the distribution of the *atomless* limit economy. Therefore the notation $\Pi(\mu)$ is justified even though the limit economy is not fully specified. Since allocations are defined as functions on the space of agents, one cannot directly compare the allocations for different spaces of agents. We therefore consider their distributions on the commodity space R^ℓ, i.e.

$$(\mathcal{D}f)(B)=\nu\left(f^{-1}(B)\right), \qquad B\in\mathcal{B}(R^\ell).$$

It is known [Kannai (1970)] that for two atomless economies $\mathcal{E}$ and $\mathcal{E}'$ with the same preference–endowment distribution the sets $\mathcal{D}\boldsymbol{W}(\mathcal{E})$ and $\mathcal{D}\boldsymbol{W}(\mathcal{E}')$ of distributions of Walras allocations are not necessarily identical. However, the difference between $\mathcal{D}\boldsymbol{W}(\mathcal{E})$ and $\mathcal{D}\boldsymbol{W}(\mathcal{E}')$ is not substantial; the two sets have the same closure. [For details see Hart, Hildenbrand and Kohlberg (1974) or Hildenbrand (1974, Proposition 5, p. 155)].

We now state a limit theorem on the core for a competitive sequence with strongly convex preferences. A limit theorem without convexity of preferences is given in Section 8.

Theorem 5

Let the sequence $(\mathscr{E}_n)$ be competitive on $\mathscr{P}^*_{\text{mo, sco}} \times R^{\ell}_{+}$. For every $f_n \in \boldsymbol{C}(\mathscr{E}_n)$ there exists a price vector $p_n \in \Delta$ such that

(1) $$\frac{1}{\#A_n} \sum_{A_n} | f_n(a) - \varphi(\mathscr{E}_n(a), p_n)| \underset{n\to\infty}{\to} 0,$$

(2) every limit point of the sequence (p_n) belongs to $\Pi(\mu)$, and consequently the mean excess demand of the economy $\mathscr{E}_n$ at the price vector p_n tends to zero, i.e.

$$\frac{1}{\#A_n} \sum_{A_n} (\varphi(\mathscr{E}_n(a, p_n) - \text{proj}_2 \mathscr{E}_n(a)) \to 0.$$

Furthermore, if $\text{supp}(\mu_n) \to \text{supp}(\mu)$[18] and if $\text{supp}(\mu)$ is compact then (1) can be strengthened to uniform convergence, i.e.

(3) $$\sup_{A_n} | f_n(a) - \varphi(\mathscr{E}_n(a), p_n)| \to 0.$$

We emphasize that in the above theorem one cannot assume that the price vector p_n is an equilibrium price vector for the economy $\mathscr{E}_n$; in fact, p_n might be quite apart from the set $\Pi(\mathscr{E}_n)$. Therefore, the above statement does not allow comparing the core $\boldsymbol{C}(\mathscr{E}_n)$ and the set of Walras allocations $\boldsymbol{W}(\mathscr{E}_n)$ of the economy $\mathscr{E}_n$, even if n is large enough. Limit theorems on the core in the stronger formulation $\delta(\boldsymbol{W}(\mathscr{E}_n), \boldsymbol{C}(\mathscr{E}_n))$ will be discussed in Section 7.

Clearly Theorem 5 (1) implies immediately Theorem 2 of Section 4, since every sequence $(\mathscr{E}_n)$ of economies with $\mathscr{E}_n$: $A_n \to K$, $\sum_{a\in A_n} e(a) \geqslant \delta$ and $\#A_n \to \infty$ contains a competitive subsequence. In Theorem 5, assertions (1) and (2), it is not assumed however that agents' characteristics belong to a compact set.

The additional assumption for assertion (3) needs a comment. We mentioned above, in discussing Theorem 2, that in order to obtain assertion (3) one has to require that no individual agent becomes "isolated" in the space of agents' characteristics when the economy becomes large. This can be expressed mathematically by assuming that every limit point of agents' characteristics in the

[18] $\text{supp}(\mu)$ denotes the support of the distribution μ, i.e. the smallest closed set with full measure. The convergence is with respect to the Hausdorff distance.

sequence $(\mathcal{E}_n)$ is contained in the support of the limit distribution $\mu = \lim \mu_n$, i.e.

$$\limsup(\mathrm{supp}(\mu_n)) \subset \mathrm{supp}(\mu).$$

Now, this assumption together with the convergence $\mu_n \to \mu$ implies that the support of the distribution μ_n converges (in the Hausdorff distance) to the support of the limit distribution μ.

The proof of assertions (1) and (2) is based on Proposition 5. One either adapts the proof of Theorem 2 to the more general case [where the sequence (f_n) of core allocations is only uniformly integrable and not bounded] or one proceeds as in the general proof of Theorem 9 in Section 8 (see the remark at the end of Section 9). In order to obtain the uniform convergence, assertion (3), from the convergence in the mean, assertion (1), one has to make essential use of the compactness of the support of the limit distribution μ and the convergence $\mathrm{supp}(\mu_n) \to \mathrm{supp}(\mu)$. The proof is due to Bewley (1973), but is too lengthy to be given here; for details see Bewley (1973) or Hildenbrand (1974, Theorem 2, p. 192). Cheng (1980) has recently given an alternative proof based on Steinitz's Theorem.

The analysis of the core of a sequence of replica economies is quite simple due to the equal treatment property, Proposition 3. The equal treatment property allows us to make the following conclusions: let $f_n \in \boldsymbol{C}(\mathcal{E}_n)$, $n = 1, \ldots,$ then the sequence (f_n) is bounded (i.e. $\sup\{|f_n(a)|, a \in A_n, n = 1, \ldots\} < \infty$), and there exists a converging subsequence of (f_n). The following result acts as a substitute for the equal treatment property in the general case of a competitive sequence of economies.

Proposition 5

Let $(\mathcal{E}_n)$ be a competitive sequence of economies on $\mathcal{P}_{\mathrm{mo}} \times R^{\ell}_{+}$ and let $f_n \in \boldsymbol{C}(\mathcal{E}_n)$. Then

(α) the sequence $(f_n)_{n=1,\ldots}$ is uniformly integrable,[19] and
(β) the sequence $(f_n)_{n=1,\ldots}$ is bounded if the sequence of initial endowments is bounded.

Assertion (β) was first proved by Bewley (1973) for a compact set of agents' characteristics.

We shall show that Proposition 5 follows quite easily from the fundamental lemma in Section 4; it can be proved directly, but in a more elaborate way without using core prices [Hildenbrand (1974, Proposition 1, p. 181, and Proposition 2, p. 193)]. One easily verifies that Proposition 5 follows directly from the following.

[19]For a definition see, for example, Billingsley (1968, p. 32).

Proposition 6

Let the sequence $(\mathscr{E}_n)$ be competitive on $\mathscr{P}_{mo} \times R^{\ell}_{+}$. For every $f_n \in C(\mathscr{E}_n)$ let p_n be a price vector with the properties of Proposition 2. Then every limit point of the sequence (p_n) is strictly positive.

We just sketch the main idea of the proof. Assume that a limit point p is not strictly positive; without loss of generality we can assume $p=(0, \pi_2, \ldots)$ with $\pi_2 > 0$.

We claim: there is a subsequence $Q \subset N$ and there are agents $a_n \in A_n$, $n \in Q$, such that

(0) $p_n \to p$,
(i) $p_n \cdot (f_n(a_n) - e(a_n)) \to 0$,
(ii) $\inf\{p_n \cdot z \mid z \succ_{a_n} f_n(a_n)\} - p_n \cdot e(a_n) \to 0$,
(iii) $(\succ_{a_n}, e(a_n)) \to (\succ, e) \in \mathscr{P}_{mo} \times R^{\ell}_{+}$ with $e_2 > 0$, and
(iv) $f_n(a_n) \to x \in R^{\ell}_{+}$.

This leads immediately to a contradiction. Indeed, by (0), (i), (iii), and (iv) we obtain

(1) $p \cdot x = p \cdot e > 0.$

The function $(p, \succ, x) \to \inf\{p \cdot z \mid z \succ x\}$ is upper semi-continuous [e.g. Hildenbrand (1974, Proposition 11, p. 28)]. Hence, we obtain from (ii) and (iv)

(2) $\inf\{p \cdot z \mid z \succ x\} \geqslant p \cdot e.$

But (1) and (2) cannot hold for a monotonic preference relation $\succ$. (See Figure 6.1.)

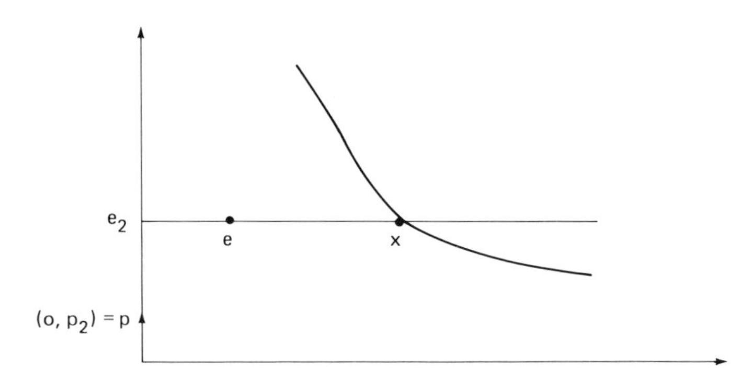

Figure 6.1.

In order to prove the above claim we shall apply the lemma of Section 4. It follows that there is a subsequence Q_1 of N such that $p_n \underset{Q_1}{\to} p$ and $p_n \cdot (f_n(a) - e(a)) \underset{Q_1}{\to} 0$ *uniformly on* $\mathcal{E}_n \subset A_n$

and

$$\inf\{p_n \cdot z \mid z \succ_a f_n(a)\} - p_n \cdot e(a) \underset{Q_1}{\to} 0 \text{ uniformly on } E_n \subset A_n$$

with $\# E_n / \# A_n \underset{Q_1}{\to} 1$.

By again choosing subsequences of Q_1 one can show that there are agents $a_n \in E_n$ $(n=1,\dots)$ such that properties (iii) and (iv) hold.

Proposition 6 has been generalized by Cheng (1980) to more general sequences $(\mathcal{E}_n)$ than the competitive ones on $\mathcal{P}_{\text{mo}} \times R^{\ell}_{+}$.

Appendix: Equilibrium distributions

The arbitrariness in the choice of the atomless measure space of agents for the limit economy strongly suggests that one should define the equilibria for the limit economy only in terms of the limit preference–endowment distribution $\lim \mu_{\mathcal{E}_n}$. In the limit the individual agent has, so to speak, lost his identity and therefore it seems artificial to keep the individualistic description of an economy in the limit.

We now define an equilibrium distribution for an arbitrary preference–endowment distribution μ. This concept is then used in Section 8 to state a general limit theorem on the core.

Definition 8

An equilibrium distribution τ for the distribution μ of agents' characteristics is a measure on $\mathcal{P} \times R^{\ell}_{+} \times R^{\ell}_{+}$ with the properties:

(i) the marginal distribution of τ on the characteristic space $T = \mathcal{P} \times R^{\ell}_{+}$ is equal to the given measure μ, i.e. $\mu = \tau \circ \text{proj}_T^{-1}$;
(ii) mean supply equals mean demand, i.e. $\int \text{proj}_2 \, d\tau = \int \text{proj}_3 \, d\tau$; and
(iii) there exists a price vector $p \in R^{\ell}$, $p \neq 0$, such that $\tau\{(\succ, e, x) \in \mathcal{P} \times R^{\ell}_{+} \times R^{\ell}_{+} \mid x \in \varphi(\succ, e, p)\} = 1$.

The set of all equilibrium distributions for μ is denoted by $ED(\mu)$. By $\Pi(\mu)$ we denote the set of equilibrium price vectors for μ, i.e. $p \in \Pi(\mu)$ if there exists $\tau \in ED(\mu)$ where p and τ have property (iii).

Clearly the definition of an equilibrium distribution τ is simply a reformulation of the usual concept of Walras equilibrium in terms of distributions. If $\mathcal{E}: A \to \mathcal{P} \times R^{\ell}_{+}$ is any economy with characteristic distribution $\mu_{\mathcal{E}}$ and if $f \in W(\mathcal{E})$, then the distribution of the mapping $(\mathcal{E}, f)$ is an equilibrium distribution for $\mu_{\mathcal{E}}$.

However, if τ is an equilibrium distribution for the characteristic distribution $\mu_{\mathcal{E}}$ and preferences are not convex, then there is not necessarily a Walras allocation f for the economy $\mathcal{E}$ such that the distribution of $(\mathcal{E}, f)$ is equal to τ. Thus, in general, one has to distinguish the two sets of measures $ED(\mu_{\mathcal{E}})$ and $\boldsymbol{W}_{\mathfrak{D}}(\mathcal{E}) = \{\mathfrak{D}(\mathcal{E}, f) | f \in \boldsymbol{W}(\mathcal{E})\}$. The reason for this is that the space A of agents may be too small (even if it is atomless). One can assert only that for every $\tau \in ED(\mu)$ there exists a suitably chosen economy $\mathcal{E}$ with characteristic distribution μ and an $f \in \boldsymbol{W}(\mathcal{E})$ such that τ is the distribution of $(\mathcal{E}, f)$. [For details see Hildenbrand (1974, pp. 155–159).]

Theorem 6

For every distribution μ of agents' characteristics in $\mathcal{P}_{\mathrm{mo}} \times R^{\ell}_{+}$ with strictly positive mean endowments $\int \mathrm{proj}_2 \mathrm{d}\mu$ there exists an equilibrium distribution τ. The correspondence ED: $\mu \to ED(\mu)$ has the following continuity property: if the sequence (μ_n) is weakly convergent to μ on $\mathcal{P}_{\mathrm{mo}} \times R^{\ell}_{+}$, with

$$\lim \int \mathrm{proj}_2 \mathrm{d}\mu_n = \int \mathrm{proj}_2 \mathrm{d}\mu \gg 0,$$

then for every sequence (τ_n), $\tau_n \in ED(\mu_n)$, there exists a weakly convergent subsequence whose limit belongs to $ED(\mu)$.

If we restrict all characteristic distributions to a compact subset T of $\mathcal{P}_{\mathrm{mo}} \times R^{\ell}_{+}$, then Theorem 6 means that the correspondence $\mu \mapsto ED(\mu)$ is compact-valued and u.h.c. at every characteristic distribution μ on T with strictly positive mean endowments. For a proof we refer to Hildenbrand (1977).

7. Many agents are not enough

In this section we want to explain why the following "desirable" assertion cannot be true. Given a "nice" (say compact or even finite) set of "nice" (say strongly convex or even smooth) preferences and given a uniform bound on the amount of endowments an individual agent may ever possess, denote this universe of agents' characteristics by T. Then it is not true that the distance $\delta(\boldsymbol{W}(\mathcal{E}), \boldsymbol{C}(\mathcal{E}))$ becomes as small as we like provided only that the economy $\mathcal{E}$ with characteristics in T is large enough. More formally: *it is not true that for every* $\varepsilon > 0$ *there exists an integer* N *such that any economy* $\mathcal{E}$: $A \to T$ *with* $\# A > N$ *has the property* $\delta(\boldsymbol{W}(\mathcal{E}), \boldsymbol{C}(\mathcal{E})) < \varepsilon$.

Since the space T of agents' characteristics is assumed to be compact we can restrict our analysis to a competitive sequence of economies. Now, if for a competitive sequence $(\mathcal{E}_n)$ we want to obtain the same strong conclusion as for replica sequences, that is to say

$$\delta(\boldsymbol{W}(\mathcal{E}_n), \boldsymbol{C}(\mathcal{E}_n)) \to 0,$$

then we have to make sure that the set $\Pi(\mathcal{E}_n)$ of equilibrium prices for the economy $\mathcal{E}_n$ converges[20] to the set $\Pi(\mu)$ of equilibrium prices associated with the limit distribution μ. We remark here that the equilibrium price correspondence Π is always upper-hemi-continuous [Hildenbrand (1974, Proposition 4, p. 152)], i.e., for every neighborhood U of $\Pi(\mu)$, we have $\Pi(\mathcal{E}_n) \subset U$ for n large enough. But, in general, not every $p \in \Pi(\mu)$ can be obtained as a limit of a sequence (p_n) with $p_n \in \Pi(\mathcal{E}_n)$; thus the lower-hemi-continuity of Π is lacking. It is, in fact, the *continuity of the set* $\Pi(\mu)$ *of equilibrium prices at the limit distribution* μ which lies behind the desired "shrinking" of the core to the set of Walras allocations. As we shall show, the essential property of a replica sequence is not the particular symmetric structure of the sequence – surely this simplifies the exposition since it implies the equal treatment property – but the fact that $\Pi(\mathcal{E}_n) = \Pi(\mu)$. The following example shows that preferences might enjoy all "nice" properties, and that for every economy in the sequence the equal treatment property may hold, yet still we do not obtain the conclusion $\delta(\boldsymbol{W}(\mathcal{E}_n), \boldsymbol{C}(\mathcal{E}_n)) \to 0$, due to the fact that $\Pi(\mathcal{E}_n)$ does not converge to $\Pi(\mu)$.

Example[21]

We shall define a competitive sequence $(\mathcal{E}_n)$, where the characteristic distributions μ_n and μ are concentrated with equal weights on two characteristics only: $t_1^n = (\precsim_1, e_1^n)$, $t_2^n = (\precsim_2, e_2^n)$, for μ_n, and $t_1 = (\precsim_1, e_1)$, $t_2 = (\precsim_2, e_2)$, for μ, where $e_1 = \lim e_1^n$ and $e_2 = \lim e_2^n$ and $e_1^n + e_2^n = e_1 + e_2$.

The economy $\mathcal{E}_n$ will be defined by choosing k_n agents with characteristics t_1^n, and k_n agents with characteristics t_2^n. Since for these replica economies we have the equal treatment property for allocations in the core we can represent the economy and its core in the Edgeworth Box (Figure 7.1).

The two preference relations $\precsim_1$ and $\precsim_2$ are such that for every point on the Pareto line above f^*, the common tangent is above e_1. For the points h^* and f^* the common tangent goes through e_1 and for every point below f^* but different from h^* the common tangent is below e_1. Clearly in this example we have $\boldsymbol{W}(\mathcal{E}_n) = f_n^*$ and $\boldsymbol{W}(\mathcal{E}_\infty) = \{f^*, h^*\}$; hence $\Pi(\mathcal{E}_n)$ does not converge to $\Pi(\mu)$.

We now claim that there exist integers k_n with $\#k_n \to \infty$ such that $h^* \in \boldsymbol{C}(\mathcal{E}_n)$, for all n. Indeed, consider the point x_n, the intersection of the line through h^* and e_1^n with the indifference curve through h^*. Thus $x_n = \alpha_n h^* + (1-\alpha_n) e_1^n$. Clearly, if $e_1^n \to e_1$, then $\alpha_n \to 1$. By a well-known argument it follows that the allocation h^* can be improved upon in the k_n-replica economy $\mathcal{E}_n$ if and only if $(k_n - 1)/k_n > \alpha_n$.

[20] We remark that convergence for sets always means with respect to the Hausdorff distance.
[21] This and other interesting examples can be found in Bewley (1973).

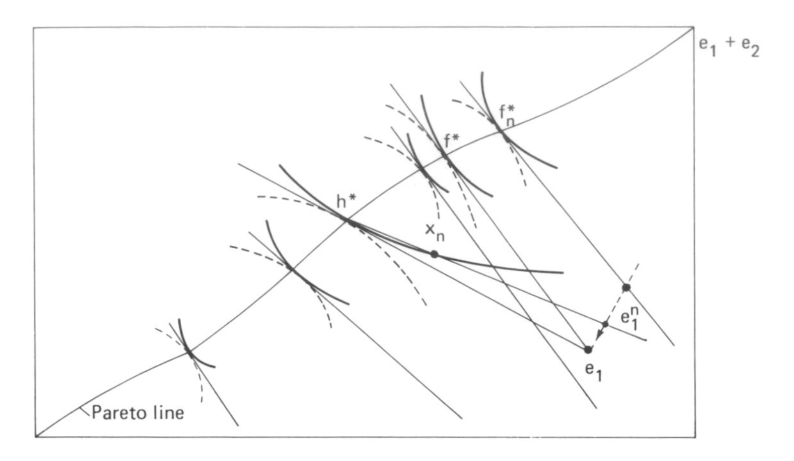

Figure 7.1.

Thus, we obtained a competitive sequence $(\mathscr{E}_n)$ with $h^* \in \boldsymbol{C}(\mathscr{E}_n)$ and the competitive equilibria $\boldsymbol{W}(\mathscr{E}_n)$ lie on the Pareto line above f^*; hence the distance $\delta(\boldsymbol{W}(\mathscr{E}_n), \boldsymbol{C}(\mathscr{E}_n))$ does not converge to zero.

We emphasize that in this example the particular position of the initial endowments in the Edgeworth Box is crucial. Given the preferences a small change in the initial endowments would yield the desired convergence. If preferences are smoothly drawn such that they yield a differentiable excess demand function, then the distribution μ would not be regular. The example makes clear that one has to require the convergence of $\Pi(\mathscr{E}_n)$ to $\Pi(\mu)$.

Let K be a compact subset of $\mathscr{P}_{\mathrm{mo}} \times R^{\ell}_{+}$, and let $\mathfrak{M}$ denote the set of all distributions on K with $\int \mathrm{proj}_2 \mathrm{d}\mu \gg 0$, endowed with the topology of weak convergence. According to Theorem 6, the equilibrium price correspondence Π of $\mathfrak{M}$ into Δ is upper-hemi-continuous. Hence, we can apply Fort's Theorem. For every $\varepsilon > 0$ we define the set $C(\Pi, \varepsilon)$ of "up to ε-continuity" points by

$$C(\Pi, \varepsilon) = \left\{ \mu \in \mathfrak{M} \,\middle|\, \begin{array}{l} \text{there exists a neighborhood } \mathscr{U}_\mu \text{ of } \mu \text{ such} \\ \text{that } \delta(\Pi(\mu), \Pi(\mu')) < \varepsilon, \text{ for every } \mu' \in \mathscr{U}_\mu \end{array} \right\}.$$

With this notation Fort's Theorem says [Fort (1949) or Hildenbrand (1974, p. 31)] that for every $\varepsilon > 0$ the set $C(\Pi, \varepsilon)$ of ε-continuity points is an *open and dense* subset in $\mathfrak{M}$. Clearly, a distribution μ is a continuity point of Π if it belongs to every set $C(\Pi, \varepsilon)$; thus, the set $C(\Pi)$ of continuity points is a set of second category in $\mathfrak{M}$. The most popular continuity points are the *regular* distributions μ, i.e. the corresponding excess demand is a differentiable function $z(\mu, \cdot)$ such that the Jacobian matrix, introduced above, is regular. Whether the set $C(\Pi)$ of continuity points of Π can be viewed as a large set (i.e. open and dense) depends on the "universe of preferences" from which one looks at them. For example, in the setting of smooth preferences the continuity points of Π form an open and dense set [for a precise statement see H. Dierker (1975b) or the remark after Theorem 8].

Let K be a compact subset of agents' characteristics with monotonic and strongly convex preferences and strictly positive initial endowments. Let $\mathfrak{M}(K)$ denote the set of all distributions on K endowed with the topology of weak convergence.

Theorem 7[22]

For every $\varepsilon>0$ there exists an open and dense subset $\mathcal{G}$ of $\mathfrak{M}(K)$ such that for every competitive sequence $(\mathcal{E}_n)$ on K with $\text{supp}(\mu_n)\to\text{supp}(\mu)$ and $\mu\in\mathcal{G}$ it follows that

$$\delta(\boldsymbol{W}(\mathcal{E}_n),\boldsymbol{C}(\mathcal{E}_n))\leqslant\varepsilon,\quad \text{for } n \text{ large enough.}$$

In particular, if μ is a continuity point for Π, then

$$\delta(\boldsymbol{W}(\mathcal{E}_n),\boldsymbol{C}(\mathcal{E}_n))\to 0.$$

Proof

According to Theorem 6 the equilibrium price correspondence Π of $\mathfrak{M}_K$ into Δ is compact-valued and u.h.c. Consequently, the set $E=\cup\{\Pi(\mu)|\mu\in\mathfrak{M}\}$ is compact. Since the demand function $\varphi(\precsim,e,p)$ is continuous, it is uniformly continuous on the compact set $K\times E$. Hence, given $\varepsilon>0$, there exists $\delta>0$ such that

$$|\varphi(\precsim,e,p')-\varphi(\precsim,e,p)|\leqslant e/2,\quad \text{for every } (\precsim,e)\in K \quad\text{and}\quad |p'-p|\leqslant\delta.$$

We now apply Fort's Theorem to the correspondence Π. Thus, there exists an open and dense subset $\mathcal{G}$ in $\mathfrak{M}$ such that $\mu\in\mathcal{G}$ implies that $\delta(\Pi(\mu_n),\Pi(\mu))<\delta$ for n large enough, i.e. if $p\in\Pi(\mu)$ there exists a sequence (p_n), $p_n\in\Pi(\mu_n)$, such that $|p-p_n|<\delta$ for n large enough.

Theorem 7 now easily follows from Theorem 5, assertion (3). Indeed, let $f_n\in\boldsymbol{C}(\mathcal{E}_n)$, $(n=1,\dots)$. Then there exists a subsequence of (f_n), which we denote again by (f_n), and there is a price vector $p\in\Pi(\mu)$ such that for n large enough we have

$$|f_n(a)-\varphi(\mathcal{E}_n(a),p)|\leqslant\varepsilon/2,\quad \text{for every } a\in A_n.$$

[22] For a slightly stronger assertion (making use of regular distributions) see H. Dierker (1975b).

Hence, for every $a \in A_n$ we obtain

$$|f_n(a)-\varphi(\mathcal{E}_n(a),p_n)| \leqslant |f_n(a)-\varphi(\mathcal{E}_n(a),p)| \\ +|\varphi(\mathcal{E}_n(a),p)-\varphi(\mathcal{E}_n(a),p_n)| \leqslant \varepsilon,$$

if n is large enough.

Summing up, we have shown that for every $\varepsilon>0$ there is an open and dense subset $\mathcal{G}$ in $\mathfrak{M}$ with the property: for every competitive sequence $(\mathcal{E}_n)$ with $\operatorname{supp}(\mu_n) \to \operatorname{supp}(\mu)$ and $\mu \in \mathcal{G}$ and every sequence (f_n), where $f_n \in \boldsymbol{C}(\mathcal{E}_n)$, there exists a subsequence (f_{n_q}) and $f^*_{n_q} \in \boldsymbol{W}(\mathcal{E}_{n_q})$ such that for q large enough we have $|f_{n_q}(a)-f^*_{n_q}(a)| \leqslant \varepsilon$, for every $a \in A_{n_q}$. Since $\boldsymbol{W}(\mathcal{E}_n) \subset \boldsymbol{C}(\mathcal{E}_n)$, this clearly implies that the Hausdorff distance $\delta(\boldsymbol{W}(\mathcal{E}_{n_q}), \boldsymbol{C}(\mathcal{E}_{n_q})) \leqslant \varepsilon$, for q large enough. One easily verifies that this result implies Theorem 7.

Remark

The restriction to strictly positive endowments in Theorem 7 has been made only for convenience. In the proof we need that the set $\mathfrak{M}$ of distributions be compact and that for every distribution in $\mathfrak{M}$ the mean endowment vector is strictly positive. Note, however, that the distributions μ_n of the competitive sequence also have to belong to $\mathfrak{M}$.

It should by now be clear that limit theorems on the core in the form of $\delta(\boldsymbol{W}(\mathcal{E}_n), \boldsymbol{C}(\mathcal{E}_n)) \to 0$ are closely related to the theory of regular distributions of agents' characteristics. As in the case of replica sequences, one can obtain, within the framework of smooth preferences and regular characteristic distributions, a result even on the *rate* of convergence [Grodal (1975)].

We state this result for completeness but give no proof since the proof makes essential use of the theory of regular economies which is not the subject of this chapter (see Chapter 17 in this Handbook by Dierker).

Let $\mathcal{P}_0$ denote the set of smooth preference relations on $P=\{x \in R^\ell \mid x \gg 0\}$ such that every indifference hypersurface has everywhere a non-zero Gaussian curvature. The set $\mathcal{P}_0$ is endowed with a metric (see Appendix to Section 2). We remark that this topology on $\mathcal{P}_0$ is finer than Hausdorff's topology of closed convergence which has been used up to now.

Let Q be a compact subset in $\mathcal{P}_0 \times P$.

Theorem 8

Let the sequence $(\mathcal{E}_n)$ of economies be competitive on Q. If the limit distribution $\mu=\lim \mu_n$ is regular then

$$\delta(\boldsymbol{W}(\mathcal{E}_n), \boldsymbol{C}(\mathcal{E}_n))=0\left(\frac{1}{\# A_n}\right).$$

Thus, the distance between the core of $\mathscr{E}_n$ and its set of Walras allocations converges to zero at least as fast as the inverse of the number of agents.

We remark that the set of regular distributions is open in the set $\mathfrak{M}(Q)$ of all distributions on Q [Grodal (1974)], and it is dense in $\mathfrak{M}(Q)$, if the interior of $\text{proj}_2 Q$ is dense in $\text{proj}_2 Q$ [H. Dierker (1975b)].

8. A limit theorem for competitive sequences

In this section we prove a limit theorem on the core for an arbitrary competitive sequence $(\mathscr{E}_n)_{n=1,\dots}$ on $\mathscr{P}_{\text{mo}} \times R^{\ell}_{+}$.

Let $ED(\mu)$ denote the set of equilibrium distributions for the preference–endowment distribution $\mu = \lim \mu_{\mathscr{E}_n}$. If $f \in \boldsymbol{C}(\mathscr{E})$, we denote by $\mathscr{D}(\mathscr{E}, f)$ the distribution of the mapping $a \mapsto (\mathscr{E}(a), f(a))$, i.e.

$$\mathscr{D}(\mathscr{E}, f)(B) = 1/\#_A \cdot \# \{a \in A | (\mathscr{E}(a), f(a)) \in B\},$$

and let $\boldsymbol{C}_{\mathscr{D}}(\mathscr{E}) = \{\mathscr{D}(\mathscr{E}, f) | f \in \boldsymbol{C}(\mathscr{E})\}$. The set of measures on $\mathscr{P}_{\text{mo}} \times R^{\ell}_{+} \times R^{\ell}_{+}$ is endowed with the weak topology.

Theorem 9

Let the sequence $\mathscr{E}_n$ of economies be competitive on $\mathscr{P}_{\text{mo}} \times R^{\ell}_{+}$ and let U be a neighborhood of $ED(\mu)$. Then for n large enough

$$\boldsymbol{C}_{\mathscr{D}}(\mathscr{E}_n) \subset U.$$

Proof

In order to prove the theorem it suffices to show the following assertion: for every sequence (f_n), where $f_n \in \boldsymbol{C}(\mathscr{E}_n)$, there exists a subsequence $Q \subset N$ such that

$$\tau_n = \mathscr{D}(\mathscr{E}_n, f_n) \underset{Q}{\to} \tau \in ED(\mu).$$

By Proposition 5 the sequence (f_n) is uniformly integrable, hence the sequence of its distributions is tight. Consequently, there exists a converging subsequence of (τ_n). Indeed, the sequence (τ_n) is tight – and hence contains a converging subsequence – if and only if the two sequences of its marginal distributions (μ_n) and $(\mathscr{D}f_n)$ are tight. But the sequence (μ_n) is tight since $\mu_n \to \mu$ and the limit measure μ is tight.

Since a subsequence of $(\mathscr{E}_n, f_n)$ converges in distribution we can apply

Skorokhod's Theorem[23] to this subsequence which we denote again by $(\mathscr{E}_n, f_n)$. Thus, there exists a "continuous representation" of the sequence $(\mathscr{E}_n, f_n)$, that is to say, there exists an atomless economy $\mathscr{E}$: $(A, \mathscr{Q}, \nu) \to \mathscr{P}_{mo} \times R^{\ell}_{+}$ with characteristic distribution μ and an allocation f for this economy; furthermore, there are measurable mappings α_n: $A \to A_n$ such that

(1) $\nu(\alpha_n^{-1}(S)) = \#S/\#A_n$, for every $S \subset A_n$, and
(2) $(\tilde{\mathscr{E}}_n, \tilde{f}_n) \to (\mathscr{E}, f)$ ν-a.e. on A, where $\tilde{\mathscr{E}}_n = \mathscr{E}_n \circ \alpha_n$ and $\tilde{f}_n = f_n \circ \alpha_n$.

Since the sequence $(\tilde{f}_n)$ is uniformly integrable, it follows from (2) that $\int \tilde{f}_n \mathrm{d}\nu \to \int f \mathrm{d}\nu$. Thus, f is an attainable allocation for the limit economy $\mathscr{E}$.

It remains to show that $f \in W(\mathscr{E})$. But this follows easily from the lemma in Section 4. Let p_n be the price vector associated with f_n according to the lemma. Hence, we obtain:

(i) $\int_A | p_n \cdot (\tilde{f}_n - \mathrm{proj}_2 \circ \tilde{\mathscr{E}}_n) | \mathrm{d}\nu =$
$1/\#A_n \sum_{A_n} | p_n \cdot (f_n(a) - \mathrm{proj}_2 \cdot \mathscr{E}_n(a)) | \leq 2B(\mathscr{E}_n)/\#A_n$ and
(ii) $\int_A | \inf\{ p_n \cdot x | x \succ_a \tilde{f}_n(a)\} - p_n \cdot \mathrm{proj}_2 \circ \tilde{\mathscr{E}}_n(a) | \mathrm{d}\nu =$
$1/\#A_n \sum_{A_n} | \inf\{ p_n \cdot x | x \succ_a f_n(a)\} - p_n \cdot e_n(a) | \leq 2B(\mathscr{E}_n)/\#A_n$.

For a competitive sequence $(\mathscr{E}_n)$ it follows that $2B(\mathscr{E}_n)/\#A_n \to 0$. We can assume that the sequence (p_n) is convergent, say $p_n \to p$. Since $0 \leq \inf\{ p_n \cdot x | x \succ_a \tilde{f}_n(a)\} \leq p_n \cdot \tilde{f}_n(a)$ and since the sequences $(\tilde{f}_n)$ and $(\mathrm{proj}_2 \circ \tilde{\mathscr{E}}_n)$ are uniformly integrable, we obtain from (i), (ii) and (2) that ν-a.e. on A,

(3) $0 = \lim[p_n \cdot (\tilde{f}_n(a) - \mathrm{proj}_2 \circ \tilde{\mathscr{E}}_n(a))] = p \cdot (f(a) - \mathrm{proj}_2 \circ \mathscr{E}(a))$, i.e. $f(a)$ belongs to the budget set of agent a, and
(4) $0 = \lim[\inf\{ p_n \cdot x | x \succ_a \tilde{f}_n(a)\} - p_n \cdot \mathrm{proj}_2 \circ \tilde{\mathscr{E}}_n(a)] \leq \inf\{ p \cdot x | x \succ_{\mathscr{E}(a)} f(a)\} -$
$p \cdot \mathrm{proj}_2 \circ \mathscr{E}(a)$.

The inequality in (4) is due to the fact that the function $(p, \succ, x) \mapsto \inf\{ p \cdot z | z \succ x\}$ is upper-semi-continuous [e.g. Hildenbrand (1974, Proposition 11, p. 28)].

Now properties (3) and (4) imply

(5) ν-a.e. on A, $p \cdot x \geq p \cdot \mathrm{proj}_2 \circ \mathscr{E}(a)$, for every $x \succ_{\mathscr{E}(a)} f(a)$.

Since by Proposition 6 we have $p \gg 0$, properties (3) and (5) imply that f is a Walras allocation for the limit economy $\mathscr{E}$, and hence $\mathscr{D}(\mathscr{E}, f) = \tau \in ED(\mu)$. Q.E.D.

Remark

The above proof implies assertions (1) and (2) of Theorem 5. Indeed, we have shown above that every limit point p of the sequence (p_n) belongs to $\Pi(\mu)$. This

[23] Skorokhod (1965) or Hildenbrand (1974, p. 50 and p. 139).

implies assertion (2) of Theorem 5 since preferences are assumed to be strongly convex. Again, by the strong convexity of preferences, the function f in the above proof is equal to $\varphi(\mathscr{E}(\cdot), p)$. Since $\tilde{f}_n \to f$ in L_1 and $\varphi(\tilde{\mathscr{E}}_n(\cdot), p_n) \to \varphi(\mathscr{E}(\cdot), p)$ in L_1, it follows that $\tilde{f}_n - \varphi(\tilde{\mathscr{E}}_n(\cdot), p_n)$ converges in L_1 to zero, which implies assertion (1) of Theorem 5.

References

Anderson, R. M. (1978), "An elementary core equivalence theorem", Econometrica 46:1483–1487.
Arrow, K. J. and F. M. Hahn (1971), General competitive analysis. San Francisco: Holden Day.
Aumann, R. J. (1964), "Markets with a continuum of traders", Econometrica 32:39–50.
Aumann, R. J. (1973), "Disadvantageous monopolies", Journal of Economic Theory, 6:1–11.
Aumann, R. J. (1979), "On the rate of convergence of the core", International Economic Review 20:349–357.
Bewley, T. F. (1973), "Edgeworth's conjecture", Econometrica, 41:425–454.
Billingsley, P. (1968), Convergence of probability measures. New York: Wiley.
Böhm, V. (1973), "On Cores and equilibria of productive economies with a measure space of consumers – an example", Journal of Economic Theory, 6:409–412.
Böhm, V. (1974), "The core of an economy with production", The Review of Economic Studies, 41:429–436.
Brown, D. F. and A. Robinson (1972), "A limit theorem on the cores of large standard exchange economics", Proceedings of the National Academy of Sciences of the U.S.A., 69:1258–1260, Correction 69:3068.
Brown, D. F. and A. Robinson (1975), "Non-standard exchange economies", Econometrica, 43:41–55.
Champsaur, P. and G. Laroque (1973), "The equivalence between the core and the set of competitive equilibria: A new approach", I.N.S.E.E., Paris.
Champsaur, P. and G. Laroque (1976), "A note on the core of economies with atoms or syndicates", Journal of Economic Theory, 13:458–471.
Cheng, H.-C. (1980), "Edgeworth conjecture revisited", IP-292, University of Berkeley, Center for Research in Management.
Cornwall, R. (1969), "The use of prices to characterize the core of an economy", Journal of Economic Theory, 1:353–373.
Debreu, G. (1963), "On a theorem of Scarf", Review of Economic Studies, 30:177–180.
Debreu, G. and H. Scarf (1963), "A limit theorem on the core of an economy", International Economic Review, 4:235–246.
Debreu, G. (1969), "Neighboring economic agents", La Décision. Paris: CNRS, 85–90.
Debreu, G. (1970), "Economies with a finite set of equilibria", Econometrica, 38:387–392.
Debreu, G. (1972), "Smooth preferences", Econometrica, 40:603–615, and "Smooth Preferences: A Corrigendum", Econometrica, 44 (1976):831–832.
Debreu, G. and H. Scarf (1972), "The limit of the core of an economy", in: C. B. McGuire and R. Radner, eds., Decision and Organization. Amsterdam: North-Holland.
Debreu, G. (1975), "The rate of convergence of the core of an economy", Journal of Mathematical Economics, 2:1–8.
Dierker, E. (1975), "Gains and losses at core allocations", Journal of Mathematical Economics, 2:119–128.
Dierker, H. (1975a), "Smooth preferences and the regularity of equilibria", Journal of Mathematical Economics, 2:43–62.
Dierker, H. (1975b), "Equilibria and core of large economies", Journal of Mathematical Economics, 2:155–169.
Drèze, J. and J. Gabszewicz (1971), "Syndicates of traders in an exchange economy", in: H. Kuhn and G. Szegö, eds., Differential Games and Related Topics, Amsterdam: North Holland, 399–414.

Drèze, J., J. Gabszewicz, D. Schmeidler, and K. Vind (1972), "Cores and prices in an exchange economy with an atomless sector", Econometrica, 40:1091–1108.
Drèze, J. (1976), "Some theory of labor management and participation", Econometrica, 44:1125–1139.
Drèze, J., J. Gabszewicz, and A. Postlewaite (1977), "Disadvantageous monopolies and disadvantageous endowments", Journal of Economic Theory, 16:116–121.
Edgeworth, F. Y. (1881), Mathematical psychics. London: Paul Kegan.
Foley, D. K. (1970), "Lindahl's solution and the core of an economy with public goods", Econometrica, 38:66–72.
Fort, M. K. (1949), "A unified theory of semi-continuity", Duke Mathematical Journal, 16:237–246.
Gabszewicz, J. and J. F. Mertens (1971), "An equivalence theorem for the core of an economy, whose atoms are not 'too' big", Econometrica, 39:713–721.
Green, J. R. (1972), "On the inequitable nature of core allocations", Journal of Economic Theory, 4:132–143.
Grodal, B. (1972), "A second remark on the core of an atomless economy", Econometrica, 40:581–583.
Grodal, B. (1974), "A note on the space of preference relations", Journal of Mathematical Economics, 1:279–294.
Grodal, B. (1975), ""The rate of convergence on the core for a purely competitive sequence of economies", Journal of Mathematical Economics, 2:171–186.
Hart, S., W. Hildenbrand, and E. Kohlberg (1974), "On equilibrium allocations as distributions on the commodity space", Journal of Mathematical Economics, 1:159–166.
Hildenbrand, W. and A. P. Kirman (1973), "Size removes inequity", The Review of Economic Studies, 30:305–314.
Hildenbrand, W. (1974), Core and equilibria of a large economy. Princeton: Princeton University Press.
Hildenbrand, W. (1977), "Limit theorems on the core of an economy", in: M. D. Intriligator, ed., Frontiers of Quantitative Economics. Amsterdam: North Holland, ch. 4b.
Kannai, Y. (1970), "Continuity properties of the core of a market", Econometrica, 38:791–815.
Keiding, H. (1974), "A limit theorem on the core of large but finite economies, Preprint, University of Copenhagen, April 1974.
Khan, M. A. (1974a), "Some equivalence theorem", The Review of Economic Studies, 41:549–565.
Khan, M. A. (1974b), "Some remarks on the core of a 'large economy'", Econometrica, 42:633.
Khan, M. A. and H. M. Polemarchakis (1978), "Unequal treatment in the core", Econometrica, 46:1475–1481.
Mas-Colell, A. (1974), "An equilibrium existence theorem without complete or transitive preferences", Journal of Mathematical Economics, 1:237–246.
Mas-Colell, A. (1978), "A note on the core equivalence theorem: How many blocking coalitions are there?", Journal of Mathematical Economics, 5:207–215.
Muench, T. J. (1972), "The core and the Lindahl equilibrium of an economy with public goods: An example", Journal of Economic Theory, 4:241–255.
Nishino, H. (1971), "On the occurrence and the existence of competitive equilibria", Keio Economic Studies, 8:33–47.
Oddou, C. (1976), "Coalition production economies with productive factors", Econometrica, 44:265–281.
Pareto, V. (1906), Manuel d'économie politique. Paris: Marcel Giard.
Richter, M. K. (1971), "Coalition, core and competition", Journal of Economic Theory, 3:323–334.
Scarf, H. (1962), "An analysis of markets with a large number of participants", Recent Advances in Game Theory (The Princeton University Conference).
Schmeidler, D. (1972), "A remark on the core of an atomless economy", Econometrica, 40:579–580.
Shapley, L. S. (1975), "An example of a slow-converging core", International Economic Review, 16:345–351.
Shapley, L. S. and M. Shubik (1969), "On the core of an economic system with externalities", Econometrica, 59:678–684.
Shitovitz, B. (1973), "Oligopoly in markets with a continuum of traders", Econometrica, 41:467–502.
Shitovitz, B. (1974), "On some problems arising in markets with some large traders and a continuum of small traders", Journal of Economic Theory, 8:458–470.

Shubik, M. (1959), "Edgeworth market games", in: R. D. Luce and A. W. Tucker, eds., Contributions to the theory of games, IV, Annals of Mathematical Studies, vol. 40. Princeton: Princeton University Press, 267–278.

Skorokhod, A. V. (1965), Studies in the theory of random proccsses. Reading, Massachusetts: Addison-Wesley Publishing Company.

Sondermann, D. (1974), "Economies of scale and equilibria in coalition production economies", Journal of Economic Theory, 8:259–291.

Sonnenschein, H. (1971), "Demand theory without transitive preferences, with applications to the theory of competitive equilibrium", in: J. Chipman et al., eds., Preferences, Utility and Demand. New York: Harcourt-Brace-Jovanovich, ch. 10.

Trockel, W. (1976), "A limit theorem on the core", Journal of Mathematical Economics, 3:247–264.

Vind, K. (1964), "Edgeworth-allocations in an exchange economy with many traders", International Economic Review, 5:165–177.

Vind, K. (1965), "A theorem on the core of an economy", The Review of Economic Studies, 5:165–177.

Vind, K. (1972), "A third remark on the core of an atomless economy", Econometrica, 40:585–586.

Weiss, E. A. (1981), "Finitely additive exchange economies", Journal of Mathematical Economics, 8.

Chapter 19

TEMPORARY GENERAL EQUILIBRIUM THEORY

JEAN-MICHEL GRANDMONT*

Cepremap

1. Introduction

Traditional general equilibrium theory [Arrow and Hahn (1971), Debreu (1959)] has for a long time studied an economic model which is essentially static, where the agents' expectations are self-fulfilling, and where the equilibrium is achieved solely by the price system. While this model is a very useful framework of reference, it relies upon extreme assumptions which make it a poor tool for the description of the world in which we live. Among others, it is known that the Arrow–Debreu model cannot account for money or the existence of active stock markets. In practice, economic agents have only a very imperfect knowledge of economic laws, and moreover their computing abilities are so limited that they cannot correctly forecast future economic events. Finally, in the real world, some adjustments take place to some extent, and for some time by means of quantitative rationing (excess supply or demand).

These issues, which have been ignored for a long time by traditional general equilibrium theory, stand of course among the important problems which are studied in many other parts of economic science. The idea that agents are unable to correctly forecast their future environment and that accordingly the future overshadows the present through the agents' expectations, is one of the main ideas of Keynesian theorising and more generally of macroeconomic theory. The study of the generation of expectations is indeed one of the important topics of theoretical and applied econometrics. Keynesian models are a major attempt at analysing the functionings of an economy where equilibrium is achieved at least partially by quantitative adjustments (rationing). Important domains of economics science have thus displayed somewhat divergent developments. General equilibrium theorists have tried recently however to fill this gap and to move closer to economic reality. The idea underlying this effort is not new and can be found in Hicks's writings under the label *temporary equilibrium*. According to this view-

*This chapter is an extension of a survey that I made at the World Meetings of the Econometric Society held in Toronto in 1975 [Grandmont (1977)].

Handbook of Mathematical Economics, vol. II, edited by K.J. Arrow and M.D. Intriligator

point, trade takes place sequentially over time, and at each date. In order to make a decision, every agent must forecast what will be his future environment as a function of his information concerning current and past states of the economy. One can then study the state of the economy in every period either by assuming that prices move fast enough in the short run so as to achieve equilibrium on every market (*temporary competitive equilibrium*), or by postulating that at each date, adjustments take place at least partially by quantity rationing (*temporary equilibrium with quantity rationing*). The goal of this chapter is to appraise the state of the field.

This approach takes into account a "disequilibrium" phenomenon which seems important in practice, i.e. the fact that, at a given moment, the agents' plans for the future are not coordinated and therefore may not be compatible. This approach, which allows economic agents to make incorrect forecasts, must accordingly be contrasted with the *Perfect Foresight* approach, which postulates that trade indeed takes place sequentially over time but that all agents forecast correctly their future environment [Hahn (1971, 1973), Radner (1967, 1968, 1970, 1972)]. The perfect foresight approach is very useful as a tool for indicative planning or for the description of stationary states, where it seems natural to assume that agents correctly forecast their future environment. It can be employed also in order to check the validity of an economic proposition when agents do not make mistakes. However, it is surely an improper tool for the description of actual economies, for the reasons mentioned already. For an appraisal of this line of thought, the reader is referred to Radner's (1974) excellent survey.

The recent interest in formal temporary equilibrium models relies upon a number of interesting earlier studies. For the temporary competitive equilibrium approach, one can cite the work of Morishima (1964) with particular emphasis on durable goods, the attempt made by Drandakis (1966) to integrate money in the theory of value, and the study of temporary competitive equilibrium under certainty of Arrow and Hahn (1971). The first systematic attempt to study temporary competitive equilibrium under uncertainty appears to be due to Stigum (1969, 1972) and to Grandmont (1970, 1974). On the other hand, the recent research on temporary equilibrium with quantity rationing has been greatly influenced by the reappraisal of Keynesian theorising undertaken by Clower (1965), Leijonhufvud (1968), Patinkin (1965) and more recently by Barro and Grossman (1971). The first systematic studies of this concept in a general equilibrium framework appear to be due to Glustoff (1968), Younès (1970a, 1970b, 1975), Drèze (1975) and Benassy (1973).

The use of this approach permits one, in principle, to construct and to study models which are similar to those which are employed by theoretical and applied macroeconomists. The subject has not yet reached, however, a state which allows a general and concise presentation, owing to the bewildering variety of new themes that have been studied within this framework. Thus, rather than attempting to make a summary of the whole literature, I chose to discuss in detail a few pieces of research from a unified viewpoint. Each section is supplemented accordingly with a short bibliographical note which should enable the reader to pursue his own inquiry.

This chapter is organised in the following way. In Section 2 the conceptual framework underlying temporary equilibrium models is presented. I show there the central role played by an agent's expectations in his decision-making process. This model is the microeconomic foundation of the whole theory. In Section 3 a few models using the temporary competitive equilibrium method are studied. It is postulated there that prices move fast enough in each period to match supply and demand. The first problem studied is the issue of the existence of an equilibrium in asset markets which is so important in any theory of financial markets (Section 3.2). It is shown there, in the framework of a simple model that a necessary and sufficient condition for the existence of a competitive equilibrium in asset markets is some partial agreement among economic agents concerning expected future spot prices of these assets. The method is then applied to monetary phenomena in Section 3.3, within the framework of a simple model involving a banking institution which can influence the money stock by open market operations on the bond market. I study there such issues as the existence of an equilibrium involving a positive price for money, the possibility for the bank to control the interest rate or the money stock, the meaning and validity of the *quantity theory of money*, and the existence of a *liquidity trap*. In Section 3.4 some of the problems raised by the introduction of production in such models are reviewed.

Section 4 is devoted to the study of the temporary equilibrium with the quantity rationing method and to the possible contributions of this approach to the theory of employment and inflation. The basic axiom of this approach is that in the short run, quantities adjust infinitely faster than prices. One is thus led to study a model where prices are temporarily fixed in each period and where equilibrium is achieved by quantity rationing. In Section 4.1 I proceed to a unified presentation and to an evaluation of the equilibrium concepts used in this area. In Section 4.2 this method is applied to a simple "macroeconomic" model involving three aggregate commodities (output, labour, and money), and three aggregated agents (a firm, a consumer, and a government). This example illustrates the power of the method in unifying macroeconomic theories which appeared fundamentally different beforehand, since it allows one to obtain within the same model very different regimes depending upon the characteristics of the economy, such as Keynesian unemployment, stagflation or repressed inflation, where the economic policies aimed at stimulating economic activities (increase of public spending, changes of prices or wages) may have very different or even contradictory consequences on the level of employment.

2. Individual decision making

The goal of this section is to study an agent's decision-making process in temporary equilibrium theory. An important feature of the analysis is the fact that, at a given date, an agent has to make a decision while ignoring his future environment, and thus make a *forecast* of this environment in order to make a choice. Therefore, an important concept enters the analysis, namely that of an *expectation function* which describes the dependence of an agent's forecast of the future upon his information at the date under consideration.

2.1 General formulation

Assume that time is divided into discrete periods. Typically, an agent receives *signals* about his environment in every period, which describe his information about the variables which characterise the state of the economic system in that period (prices and interest rates, endowments, parameters describing the available technology, exogenous states of nature, statistics published by the government, etc.).

The agent's problem at date t is the following one: given the signals received up to date t and his past decisions, he must choose an *action* in period t. Typically, an action will involve a commitment to deliver or receive currently available commodities, as well as commitments to exchange commodities to be delivered at later dates (forward contracts). In the particular and important case where the commodity to be delivered later on is money, an action involves the decision to hold cash balances or securities of various maturity dates, or to borrow.

Since the choice of an action in a given period involves commitments for future periods, it cannot be considered independently of what the agent will do in subsequent periods. Thus, the agent must not only choose an action at date t, but also *forecast* what will be his future environment and *plan* which actions he will take at later dates, depending upon this future environment. Any action and any plan will determine in turn some *consequences* for the agent in the current and future periods. The action and the plan chosen by the agent will be such that the associated consequences are most preferred by him subject to the perceived constraints.

In order to make precise formally the agent's decision-making process, I focus on a simple case where the agent is planning only one period ahead. This is by no means essential: the analysis which follows is valid for an arbitrary finite or infinite planning horizon.

Consider therefore an agent at date 1. The signals he has received before that date are given, as well as his past decisions, and cannot be changed by current decisions or events. Although they evidently influence the agent's current situation (e.g. his current wealth) as well as his forecast of the future, I will not mention them explicitly, since they are fixed in the analysis. Let S_t be the set of all possible signals s_t that the agent may receive in period t ($t=1, 2$). Similarly, A_t is the set of all possible actions a_t that the agent can take at date t ($t=1, 2$). In many economic applications, S_t and A_t are subspaces of finite Euclidean spaces: *I will also make this assumption here* in order to simplify the exposition, although the analysis which follows could be conducted at a more general and more abstract level.

Given a signal $s_1 \in S_1$, the agent must choose an *action* a_1 in A_1 which may be constrained to lie in some subset, say $\beta_1(s_1)$, of A_1. He must also choose a *plan*, represented by a measurable function $a_2(\cdot)$ from S_2 into A_2, which describes the choice $a_2(s_2)$ which he intends to make at date 2 if he perceives the signal s_2 at that date. The constraints which the agent will face in period 2 may depend upon his current and future environment, and upon the action chosen at date 1, and are represented by $a_2(s_2) \in \beta_2(a_1, s_1, s_2)$.

The *consequences* for the agent of the choices he makes in the two periods may depend upon his environment at the two dates and thus upon (s_1, s_2). For simplicity of exposition, however, I will make these consequences depend only on (a_1, a_2) and will represent them by $\gamma(a_1, a_2)$, an element of the space of consequences C, which I take again to be a *subspace of a finite Euclidean space*.

Since the agent's choice is made under uncertainty about the future, the *preferences* of the agent will be described by a complete preordering $\succsim$ on the space of *probability distributions* on C, say $\mathcal{M}(C)$.

The agent's *forecast* of his future environment takes the form of a probability distribution of S_2. It is in general a function of all the observations made by the agent in the past and of the current signal s_1. In some cases (e.g. imperfect competition), it will be a function of the action currently chosen. In order to simplify the exposition, I will single out the dependence upon the current signal, and I will describe the agent's *expectation function* by a mapping ψ taking S_1 into the space of probability distributions on S_2, $\mathcal{M}(S_2)$.

It is now easy to describe how the agent makes his choice. Given $s_1 \in S_1$, each action $a_1 \in A_1$ and each plan $a_2(\cdot)$ defines a *random variable* $\gamma(a_1, a_2(\cdot))$ defined on the space S_2 endowed with the subjective probability $\psi(s_1)$ and taking values in the space of consequences C. This random variable induces a probability distribution C, say $\mu(s_1, a_1, a_2(\cdot))$. Given s_1, the agent chooses an action a_1 and a corresponding plan $a_2(\cdot)$ subject to the constraints $a_1 \in \beta_1(s_1)$ and $a_2(s_2) \in \beta_2(a_1, s_1, s_2)$ for every s_2 in S_2, so that the associated probability distribution μ is most preferred. The resulting optimal actions form a set $\alpha(s_1)$. The relation α describes the *decision rule* adopted by the agent.

It is usually assumed that the agent's preferences satisfy the *Expected Utility Hypothesis*, in which case the agent's decision problem can be described in a simple way.

Assumption 1 (Expected Utility Hypothesis)

There exists a von Neumann–Morgenstern utility function u: $C \to R$ which is continuous, bounded, such that the mapping v: $\mathfrak{M}(C) \to R$ defined by $[v(\mu) = \int_C u \mathrm{d}\mu, \mu \in \mathfrak{M}(C)]$ is a representation of the preferences $\succsim$.

In such a case, the agent's problem amounts to maximising the expected utility of $\mu(s_1, a_1, a_2(\cdot))$,

$$\int_{S_2} u(\gamma(a_1, a_2(\cdot)) \mathrm{d}\psi(s_1) = \int_C u \mathrm{d}\mu(s_1, a_1, a_2(\cdot)),$$

subject to $a_1 \in \beta_1(s_1)$ and $a_2(s_2) \in \beta_2(a_1, s_1, s_2)$ for every s_2 in S_2.

2.2. *Preferences among actions*

The agent's decision problem was up to now described as the simultaneous choice of an action in the current period and a plan for the future. It is possible, by using a well-known backward dynamic procedure, to reduce the problem to a single period.

Given an arbitrary signal s_1 in S_1, consider an action a_1 in A_1. Let u be a von Neumann–Morgenstern utility as in Assumption 1. If the correspondence $\beta_2(a_1, s_1, \cdot)$ from S_2 into A_2 is compact-valued and continuous, and if the mapping γ: $A_1 \times A_2 \to C$ is continuous too, then for every s_2, the maximum value $u^*(a_1, s_1, s_2)$ of $u(\gamma(a_1, a_2))$ when a_2 is constrained by $a_2 \in \beta_2(a_1, s_1, s_2)$ exists and is a continuous function of s_2. Now let

$$v(a_1, s_1) = \int_{S_2} u^*(a_1, s_1, \cdot) \mathrm{d}\psi(s_1). \tag{2.1}$$

Clearly, $v(a_1, s_1)$ can be interpreted as the maximum expected utility that can be achieved if s_1 is observed and a_1 is chosen in the current period. It is therefore tempting to say that *a_1' is preferred or indifferent to a_1'' when s_1 is currently observed if and only if $v(a_1', s_1) \geqq v(a_1'', s_1)$*. Indeed, this definition is consistent for, as the reader will easily check, the preferences among actions so defined for each s_1 depend only on the preferences $\succsim$ on consequences and not on the particular representation u which was used in the definition. The definition is further justified by the fact, which is easy to verify, that for every s_1, *$a_1^* \in \alpha(s_1)$ if and only if a_1^* maximises $v(a_1, s_1)$ subject to $a_1 \in \beta_1(s_1)$*.

This technique is useful since it permits viewing a complex problem involving several periods as a single-period decision problem. But it has interesting economic implications too. Indeed, the above procedure allows the introduction of money, bonds and financial assets into the utility function $v(a_1, s_1)$, although they may have no intrinsic value. The interesting point is that the current signal s_1 (and past observations) enter the utility function by their influence on the agent's forecast of the future. Since perceived signals involve in general prices, this gives a sound foundation to discuss a few important issues in monetary theory, such as the absence of money illusion, the quantity theory of money, and so on. I will return to these issues in the section on money.

2.3. *Properties of the decision rule*

The goal of this section is to state and briefly prove a few results about the expected utility v defined in (2.1) of Section 2.2, and about the decision rule α, which are repeatedly used in temporary equilibrium models. I state first a few common continuity assumptions, which will guarantee the continuity of the expected utility v and of the decision rule α.

Assumption 2 (Continuity of expectations)

The Expectation Function ψ: $S_1 \to \mathcal{M}(S_2)$ is continuous when $\mathcal{M}(S_2)$ is endowed with the topology of weak convergence of probability measures.

For the definition of the weak convergence of probability measures, see Billingsley (1968), Parthasarathy (1967), or Chapter 5 in this Handbook.

Assumption 3 (Continuity of the constraints)

The correspondences β_1 and β_2 are compact-valued and continuous.

Assumption 4

The mapping γ: $A_1 \times A_2 \to C$ is continuous.

In many applications, one needs the convexity of the set of optimal actions $\alpha(s_1)$. Here is an assumption which is satisfied in many temporary equilibrium models, which guarantees this property provided that any von Neumann–Morgenstern utility is concave.

Assumption 5

A_t $(t=1,2)$ and C are convex. In addition,

(i) The correspondences β_1 and β_2 are convex-valued.
(ii) Whenever $a_2' \in \beta_2(a_1', s_1, s_2)$, $a_2'' \in \beta_2(a_1'', s_1, s_2)$, then $a_2 \in \beta_2(a_1, s_1, s_2)$ for every $a_1 = \lambda a_1' + (1-\lambda)a_1''$, $a_2 = \lambda a_2' + (1-\lambda)a_2''$, where $0 \leqq \lambda \leqq 1$.
(iii) γ: $A_1 \times A_2 \to C$ is linear.

Parts (ii) and (iii) of Assumption 5 state essentially the "linearity" of β_2 with respect to a_1 and the linearity of γ with respect to (a_1, a_2). Then, it is straightforward to check:

Proposition 1

Assume that S_t, A_t $(t=1,2)$ and C are subspaces of Euclidean spaces. Then under Assumptions 1, 2, 3 and 4, the expected utility v: $A_1 \times S_1 \rightarrow R$ defined in (2.1) is continuous and bounded, and the correspondence α: $S_1 \rightarrow A_1$ is compact-valued and upper hemi-continuous. If, in addition, Assumption 5 is satisfied and every von Neumann–Morgenstern utility is concave, then v is concave with respect to a_1, and α is convex-valued.

Proof

In order to prove the first part of Proposition 1 it suffices to check that v is continuous (it is obviously bounded). Consider a sequence (a_1^k, s_1^k) which converges to (a_1, s_1) in $A_1 \times S_1$. Now, the function u^* is jointly continuous in all its arguments, under Assumptions 1, 3 and 4. Assumption 2 means that the sequence $\psi(s_1^k)$ converges weakly to $\psi(s_1)$. Thus, all the conditions of Billingsley (1968, Theorem 5.5) are satisfied, which proves that

$$\lim \int_{S_2} u^*(a_1^k, s_1^k, \cdot)\, \mathrm{d}\psi(s_1^k) = \int_{S_2} u^*(a_1, s_1, \cdot)\, \mathrm{d}\psi(s_1)$$

and therefore the continuity of v.

To prove the last part, it is enough to show that v is concave in a_1. But Assumption 5 and the concavity of u imply that $u^*(a_1, s_1, s_2)$ is concave in a_1, which proves the claim. Q.E.D.

2.4. *Bibliographical note*

The dynamic programming technique employed in Section 2.2 is well known. It was first used in the context of a temporary equilibrium theory by Arrow and Hahn (1971) and by Stigum (1969a, 1969b). The use of the concept of weak convergence of probability measures to describe the continuity of expectations originated in Grandmont (1970, 1972). Most of the material of this section can be found in Grandmont (1972). The most general treatment can be found in Jordan (1977). The continuity of the decision rule α with respect to the expectation function ψ has been investigated by Christiansen (1974) and by Christiansen and Majumdar (1977), by making use of the topology of uniform convergence on compacta.

3. Temporary competitive equilibrium

The aim of temporary equilibrium theory is to study the interaction through markets of different individuals in a given period, and to analyse the behaviour over time of the sequence of these equilibria. One possible starting assumption is to postulate that in each period the actions of the agents are made compatible exclusively by price adjustments: prices and interest rates are assumed to move fast enough in each period to match supply and demand in spot markets for currently available commodities and in asset markets (forward contracts). The evolution of the economic system is then viewed as a succession of competitive equilibria. This viewpoint is quite restrictive since it prevents the occurrence of disequilibrium phenomena like unemployment, which will be studied in Section 4. However, the equilibrium reached in each period is temporary since only the current *actions* of the agents are coordinated by the price system. Indeed, the *plans* for the future made by the agents at a given date are not coordinated, and may well be incompatible if the agents have incorrect expectations. This approach is thus to be contrasted with the *rational expectations* approach [Hicks (1946), Radner (1972)], where by definition such a disequilibrium phenomenon is impossible.

One important problem in temporary competitive equilibrium models is to study the existence and properties of a market equilibrium reached in a given period, especially when there are asset markets involved. It is clear beforehand that the properties of the agents' expectations will play a central role in the analysis. This point will indeed be forcefully illustrated in the models that I will study in this section. First, I shall recall a standard result on the existence of a market equilibrium which will be repeatedly used below (Section 3.1). Next, I shall look at the general problem of the existence of an equilibrium in forward markets (Section 3.2). The approach will be then applied to the case of financial assets in Section 3.3. I shall study there such issues as the existence of an equilibrium with a positive price for money, the quantity theory of money, the meaning and the existence of a liquidity trap, and the regulation of the money supply by a banking institution. Finally, I shall give some remarks in Section 3.4 on the problem of the introduction of firms' activities in such models, which is still open, owing to the well-known difficulty of modelling satisfactorily a firm's behaviour under uncertainty.

3.1. Market equilibrium

Let D be a convex subset of the unit simplex of some finite Euclidean space; D should be interpreted as the set of admissible price systems p of the contracts exchanged in the markets under consideration. To every p in D is associated a set of aggregate excess demands $\zeta(p)$ for these contracts, which satisfies Walras' Law $p\cdot\zeta(p)=0$. Now, if the projection on D of an excess demand vector has the property to point toward the interior of D when prices are near the boundary of D, then, by continuity, there should exist p in D such that $0\in\zeta(p)$. Formally, we have the following lemma:

Lemma 1 (Market Equilibrium Lemma)

Let D be an open and convex subset of the unit simplex of R^ℓ. Let ζ: $D\to R^\ell$ be a correspondence which is compact and convex-valued, upper hemi-continuous and satisfies Walras' Law, $p\cdot z=0$ for $z\in\zeta(p)$. Assume that for every sequence p^k in D which tends to some p in the boundary ∂D of D, and every sequence $z^k\in\zeta(p^k)$, there exists $\bar{p}$ in D such that $\bar{p}\cdot z^k>0$ for infinitely many k. Then, there exists p^* in D such that $0\in\zeta(p^*)$.

For the sake of completeness, I give a short proof of this statement which is adapted from known arguments [Debreu (1956)].

Proof

Consider an increasing sequence of compact, convex subsets D^k of D such that D is contained in the union of all the D^k. For each k there exists a compact convex set Z^k which contains $\zeta(p)$ for every p in D^k. For any z in Z^k, let $\mu^k(z)$ be the set of prices p of D^k which maximize $p\cdot z$. To each couple (p,z) of $D^k\times Z^k$, associate the set $\mu^k(z)\times\zeta(p)$. This correspondence has a fixed point. That is, there exists p^k in D^k and $z^k\in\zeta(p^k)$ such that $0=p^k\cdot z^k\geqq p\cdot z^k$, for all p in D^k. It follows that the sequence p^k must be bounded away from the boundary ∂D of D; otherwise, one could contradict the boundary assumption of the lemma. There is thus a subsequence (same notation) such that (p^k) converges to $p^*\in D$ and such that (z^k) converges to $z^*\in\zeta(p^*)$. By continuity, $p\cdot z^*\leqq 0$ for all p in D. Since D is open, this implies $z^*=0$, which proves the lemma. Q.E.D.

3.2. *Temporary competitive equilibrium in forward markets*

The problem considered in this section arises from the fact that, in temporary equilibrium models, an agent typically exchanges contracts at a given date that implies the delivery or the receipt of commodities (goods and services, money) at later dates. In many models, spot markets are assumed to be active at the date of the delivery of these commodities. The possibility of *arbitrage* on forward markets is therefore open to the agent.

This possibility raises new and interesting problems compared to traditional equilibrium theory. The first point to note is that in a competitive framework the size of the contracts on futures markets is not limited by the availability of resources at future dates. It is then intuitively clear that for some patterns of expectations, an agent may find it profitable to buy and sell forward different commodities at given prices and that there may be no bound to these *arbitrage* operations. Accordingly, the existence of a competitive equilibrium in forward markets is not warranted for all expectations patterns.

This raises the question: for which expectations patterns does there exist a competitive equilibrium in forward markets? An answer to this question was provided by Green (1973, 1974): *there must exist some partial agreement between the agents' expectations concerning future spot prices.* Qualitatively similar results were obtained by Hart (1974) in his study of markets for financial securities of the Lintner–Sharpe type. The methods used in these studies, which appear to be of wide applicability, will be analysed in this section.

Green (1973) studies an exchange economy which lasts two periods, 1 and 2. There are ℓ_1 perishable consumption goods available in period 1, and ℓ_2 in period 2, with $\ell=\ell_1+\ell_2$. In period 1, each agent knows his endowment of current deliverable goods, $e_1\in R^{\ell_1}_+$, but does not know yet what will be his endowment at date 2. In period 1, there are ℓ_1 spot markets for current goods, and ℓ_2 forward markets for sure delivery of the ℓ_2 goods at date 2. Each trader knows that in period 2 there will be spot markets for goods available at that date, and that spot prices at that date may differ from prices for forward commitments in period 1. The problem is to find sufficient conditions for the existence of a competitive equilibrium in period 1.

Consider a representative agent in period 1. His decision-making problem fits exactly into the framework described in Section 2. Indeed, the signal received by this agent at date 1 is $s_1=(p_1,q_1)$. It represents the prices $p_1\in R^{\ell_1}_+$ of current goods and the prices $q_1\in R^{\ell_2}_+$ of forward purchases of goods to be delivered in the second period. Thus, $S_1=\Delta^\ell$, the unit simplex of R^ℓ. In period 2 the agent will receive another signal $s_2=(p_2,e_2)$, describing spot prices $p_2\in R^{\ell_2}_+$ and his endowment $e_2\in R^{\ell_2}_+$ at that date. Hence $S_2=\Delta^{\ell_2}\times R^{\ell_2}_+$. At date 1 the agent forms expectations about s_2, which are summarized by a map ψ: $S_1\to\mathfrak{M}(S_2)$, as in Section 2.1.

An action $a_1=(x_1,b_1)$ in period 1 describes the agent's consumption $x_1\in R^{\ell_1}_+$ and his forward purchases $b_1\in R^{\ell_2}$ (b_1 is not bounded below, for unlimited short selling is allowed). An action a_2 in period 2 represents the agent's consumption $x_2\in R^{\ell_2}_+$ at that date. The consequence of a couple of actions (a_1,a_2) is simply the associated consumption stream. Therefore, $A_1=R^{\ell_1}_+\times R^{\ell_2}$, $A_2=R^{\ell_2}_+$ and $C=R^\ell_+$. As for the agent's preferences, they are defined on $\mathfrak{M}(C)$, as in Section 2.1.

It remains to describe the constraints faced by the agent in every period. These are essentially budget constraints. An action a_2 will be feasible in period 2 if its value $p_2\cdot a_2$ does not exceed the agent's wealth at that time, $p_2\cdot(b_1+e_2)$; if his wealth is negative, the agent is bankrupt and is allocated a zero consumption bundle. Thus,

$$\beta_2(a_1,s_2)=\{a_2\in A_2\,|\,p_2\cdot a_2\leqq\max(0,p_2\cdot(b_1+e_2))\}.$$

An action a_1 at date 1 is feasible if, again, its value $s_1 \cdot a_1$ is not greater than the agent's wealth at that date, $p_1 \cdot e_1$. It is convenient to make the additional assumption, which is by no means essential, that the agent never plans to be bankrupt, even with a small probability. This is described by the additional constraint: $p_2 \cdot (b_1 + e_2) \geqq 0$, for all $s_2 = (p_2, e_2)$ in the support of $\psi(s_1)$. Formally,

$$\beta_1(s_1) = \{a_1 \in A_1 | s_1 \cdot a_1 \leqq p_1 \cdot e_1 \text{ and } p_2 \cdot (b_1 + e_2) \geqq 0, \text{ for all } s_2 \in \operatorname{supp} \psi(s_1)\}.$$

The trader's decision problem is therefore identical to the one which was presented in Section 2. For each s_1 in S_1, one can define a set of optimal actions $\alpha(s_1)$, which may be empty at this stage, and a corresponding set of excess demands by:

$$\zeta(s_1) = \{z | z = (x_1 - e_1, b_1), a_1 = (x_1, b_1) \in \alpha(s_1)\}.$$

I state now the assumptions made on the agent's characteristics:

Assumption 6

(1) The preferences on $\mathfrak{M}(C)$ satisfy the expected utility hypothesis (Assumption 1 of 2.1). In addition, any von Neumann–Morgenstern utility u is concave and monotone.
(2) The expectation function ψ satisfies the continuity assumption (Assumption 2) of Section 2.1.
(3) For every s_1 in S_1, $\psi(s_1)$ assigns probability one to the set of $s_2 = (p_2, e_2)$ in S_2 such that $p_2 \gg 0$.
(4) The support of $\psi(s_1)$ is independent of s_1. The convex hull of the projection of $\operatorname{supp} \psi(s_1)$ on Δ^{ℓ_2} has a non-empty relative interior, denoted by Π.

Parts (1) and (2) of the assumption restate the continuity assumptions made in Section 2.1 and require no comment. Part (3) rules out expectations of zero prices at date 2 and is justified by the assumption of monotonic preferences. It allows us to assume that in fact $S_2 = \operatorname{int} \Delta^{\ell_2} \times R^{\ell_2}_{+}$, which will be done in the sequel. The first part of (4), which says that $\operatorname{supp} \psi(s_1)$ is not influenced by s_1, is made only for convenience. It could be replaced by the assumption that $\operatorname{supp} \psi(s_1)$ varies continuously with s_1 [see Green (1973)]. This independence axiom, however, greatly simplifies matters without altering the substance of the results. The last part of (4) says that the agent is really uncertain about future spot prices and excludes linear dependencies in expected prices. It is a mild postulate.

Define now D as the set of prices $s_1 = (p_1, q_1)$ in S_1 which are positive ($s_1 \gg 0$) and such that the vector $q_1 / |q_1|$ of relative prices of forward purchases belongs to Π. Clearly, D is open in Δ^{ℓ} and convex. More importantly, D is precisely the set of prices for which the agent's behaviour is well defined. The fact that prices have to be positive is not surprising since preferences were assumed to be monotonic. The fact that $q_1 / |q_1|$ must belong to Π is more informative. As the proof of the next proposition will show this is a consequence of the possibility of an *arbitrage* on forward markets. Indeed, when $q_1/|q_1|$ does not belong to Π, the agent believes that unlimited profitable arbitrage is possible, and thus $\zeta(s_1)$ is empty. When $q_1/|q_1|$ belongs to Π, any attempt to make a profit on forward markets involves the possibility of a loss with a positive subjective probability. It is then intuitively clear that the assumption stating that the agent does not deliberately plan to be bankrupt imposes in such a case a bound on his forward transactions, and thus that $\zeta(s_1)$ is well defined.

Proposition 2

Under Assumption 6, and if $e_1 \gg 0$, (i) $\zeta(s_1)$ is non-empty if and only if $s_1 \in D$. The correspondence ζ from D to R^ℓ is compact and convex-valued, upper hemi-continuous and satisfies Walras' Law, $s_1 \cdot \zeta(s_1) = 0$. (ii) Consider any sequence s_1^k in D tending to some s in ∂D, and any sequence $z^k \in \zeta(s_1^k)$. Then $\bar{s}_1 \cdot z^k$ diverges to $+\infty$ for every $\bar{s}_1$ in D.

Proof

The case $\ell_2 = 1$ is trivial since arbitrage is then impossible, so I shall assume $\ell_2 \geqq 2$. If some component of s_1 is zero, $\zeta(s_1)$ is empty since preferences are monotonic. Suppose now that $s_1 \gg 0$ and that $q_1/|q_1|$ does not belong to Π. Since Π is convex and open relative to int Δ^{ℓ_2}, it follows from a well-known separation theorem that there exists $\bar{b}$ in R^{ℓ_2}, $\bar{b} \neq 0$, which satisfies $q_1 \cdot \bar{b} \leqq 0$ and is such that $p_2 \cdot \bar{b} \geqq 0$, for all p_2 in Π with strict inequality for some p_2 in this set. Accordingly, any action $a_1 = (x_1, b_1)$ in $\beta_1(s_1)$ must be dominated by the action $(x_1, b_1 + \bar{b}) \in \beta_1(s_1)$, and $\zeta(s_1)$ must be empty. Assume next that $s_1 \gg 0$ and that $q_1/|q_1|$ belongs to Π. Then for every $\bar{b} \neq 0$ such that $q_1 \cdot \bar{b} \leqq 0$, there exists p_2 in Π such that $p_2 \cdot \bar{b} < 0$. This implies that $\beta_1(s_1)$ is bounded, and therefore that $\zeta(s_1)$ is non-empty. Indeed, assume that, on the contrary, there exists an unbounded sequence $a_1^k = (x_1^k, b_1^k)$ in $\beta_1(s_1)$, such that $|a_1^k|$, the norm of a_1^k, diverges to $+\infty$. Then any accumulation point $\bar{a} = (\bar{x}, \bar{b})$ of the sequence $(a_1^k/|a_1^k|)$ satisfies $p_1 \cdot \bar{x} + q_1 \bar{b} \leqq 0$, which implies $\bar{b} \neq 0$ and $q_1 \cdot \bar{b} \leqq 0$, and $p_2 \cdot \bar{b} \geqq 0$, for all p_2 in Π, which contradicts the fact that $q_1/|q_1|$ belongs to Π.

This establishes that $\zeta(s_1)$ is non-empty if and only if s_1 belongs to D. It is then easy to check that (i) holds, by applying the argument proving Proposition 1 of Section 2.3. [To get continuity, replace S_1 by D and S_2 by int $\Delta^{\ell_2} \times R_+^{\ell_2}$. To get convexity, remark that β_2 satisfies (ii) of Assumption 5 whenever a_1' and a_1'' belong to $\beta_1(s_1)$. Walras' Law follows from the monotonicity of preferences.]

I wish now to show (ii). Suppose that, on the contrary, there are sequences s_1^k in D tending to $s \in \partial D$, $z \in \zeta(s_1^k)$, and $\bar{s}_1 \in D$ such that $\bar{s}_1 \cdot z^k$ is bounded above. This implies that z^k is bounded. Indeed, if there were a subsequence (same notation) such that $|z_1^k|$ diverges to $+\infty$, any accumulation point $(\bar{x}, \bar{b})$ of the sequence

$(z_1^k/|z_1^k|)$ would satisfy $\bar{b}\neq 0$, $\bar{q}_1\cdot\bar{b}\leqq 0$ and $p_2\cdot\bar{b}\geqq 0$, for all p_2 in Π, which would contradict the fact that s_1 belongs to D. If the sequence z^k is bounded, one can assume, without loss of generality, that it converges to z. By continuity, $z\in\zeta(s)$. But this would contradict the fact that s does not belong to D. This completes the proof of the proposition. Q.E.D.

Consider now an economy composed of m agents, $i=1,\ldots,m$, each of whom satisfies the assumptions of Proposition 2. For each s_1 in S_1, aggregate excess demand is defined by $\zeta(s_1)=\sum_{i=1}^m \zeta^i(s_1)$. Then an equilibrium is a price system s_1 such that $0\in\zeta(s_1)$. For each agent, one can define the sets Π^i and D^i as above. It is then clear that given the other assumptions, a necessary and sufficient condition for the existence of an equilibrium is the occurrence of some agreement among agents about future spot prices. In other words, it is necessary and sufficient that the intersection of all Π^i be non-empty. Indeed, if it is empty, the intersection D of all D^i is empty too, and no equilibrium can exist, since aggregate excess demand is empty at all prices. If the intersection of all Π^i is non-empty, aggregate excess demand $\zeta(s_1)$ is non-empty if and only if s_1 belongs to D, and it is clear from Proposition 2 that it satisfies all assumptions of the market equilibrium lemma of Section 3.1. Therefore an equilibrium exists in that case. This leads to the following theorem:

Theorem 1

Consider an economy composed of m agents, $i=1,\ldots,m$ each of whom satisfies Assumption 6 and $e_1^i\gg 0$. Then a necessary and sufficient condition for the existence of an equilibrium is that the intersection of all Π^i be non-empty.

The methods developed in the analysis of this simple model are important because they appear to be applicable to a wide variety of models, in particular to the study of money and financial markets [Hammond (1977), Hart (1974)]. I will give an illustration in the next section.

While the qualitative conclusion reached in this simple model is rather neat and appealing, there are a few issues which need to be studied if we wish to understand better the functionings of forward and financial markets.

The first problem is to extend this class of models to the case where the agents make plans several periods ahead and exchange forward contracts maturing at different future dates. A systematic study of this point would be useful, since it has obvious connections with the theory of the term structure of interest rates, as Green himself pointed out in a series of examples [Green (1971); see also Younès (1972a)].

Serious problems are posed by the issue of bankruptcy. One of the problems arises when one considers agents who are endowed with previously contracted debts and claims when the market of period 1 opens. Some agents may be bankrupt at some configurations of prices, and some default rules must be explicitly built in the model in that case [Green (1974), Grandmont (1970)]. The existing studies appear to use purely mechanical rules in order to determine whether an agent is bankrupt and the extent of bankruptcy. As Ledyard (1974) pointed out in his comments on Green's model, it would be desirable to have a theory which would determine the extent of bankrupty as a decision of the agents themselves.

Another interesting issue arises when one allows an agent to make plans which involve the prospect of going bankrupt at a later date with some probability. At least two difficulties arise here: the agent's behaviour will exhibit fundamental non-convexities, and the agent may be willing to take unlimited short positions. To see these points, it may be worth looking at an example, adapted from Grandmont (1970). Consider the model described in this section, in the simple case where $\ell_1 = \ell_2 = 1$. If one allows an agent to contemplate the prospect of going bankrupt at date 2, the constraint sets become $\beta_1(s_1) = \{a_1 = (x_1, b_1) | s_1 \cdot a_1 \leq p_1 \cdot e_1\}$ and $\beta_2(a_1, s_2) = \{a_2 = x_2 | x_2 \leq \max(0, e_2 + b_1)\}$.

It is now easy to see that there is a basic non-convexity in the problem, which arises from the fact that β_2 is no longer "linear" [(ii) of Assumption 5] with respect to a_1, even when a_1 is restricted to vary only in $\beta_1(s_1)$. For instance, consider $u(x_1, x_2) - f(w(x_1, x_2))$, where $f(w) = 1 - \exp(-w)$, and $w(x_1, x_2) = x_1 + \delta x_2$. Then the indifference curves associate with the expected utility $u^*(a_1, s_1, s_2)$ of Section 2.2 will be given by $x_1 + \delta(e_1 + b_1) = k$, if $e_2 + b_1 \geq 0$, $x_1 = k$, otherwise, where k is an arbitrary constant. The expected utility $v(a_1, s_1)$ will thus display non-convexities. The example makes clear too that the agent will choose $b_1 = -\infty$ for any value of s_1.

The problem of unlimited short positions in futures markets arises when the only penalty incurred by a bankrupt agent is to be forced to choose a zero consumption bundle in the period of default, and when the agent cannot be satiated. One is thus led to introduce extraeconomic penalties which depend upon the *extent* of bankruptcy and/or to assume that consumption sets are compact. The problem of non-convexities will remain, however, and can be dealt with by using methods as in Starr (1969). Although there are some studies along these lines [Green (1974), Grandmont (1970)], much more research is needed on these topics.

Finally, the problem of bankruptcy raises questions about the mere notion of *competitive* futures markets [Gale (1976b)]. In the above model it was assumed that any level of forward contracts could be achieved at the given prices q_1. But, since the risk of default obviously depends on the size of forward commitments made by a particular agent (or a class of agents), it is not clear why the prices of forward contracts should be independent of their size. This type of argument appears to show that the imperfect competition must be taken into account when analysing forward markets. The issue is fundamental and must be investigated thoroughly.

3.3 Temporary competitive equilibrium and money

I noted above that in temporary equilibrium models agents can exchange forward contracts involving the delivery or receipt of money, as when they decide to save in the form of cash balances, bank deposits, financial assets and/or to borrow. The approach accordingly provides a sound basis for the study of monetary phenomena. An important problem [Hahn (1965)] is, in particular, to find conditions ensuring the existence of a competitive equilibrium with a positive money price although money has no intrinsic value [Grandmont (1974)]. More interestingly, the method leaves room for the introduction of specific agents who perform the functions of banks or of financial intermediaries and can influence the stock of money by granting loans or by open market operations [Grandmont and Laroque (1975, 1976a)]. It is then possible to study the influence on the economy of various monetary policies and to discuss on a precise basis such monetary issues as the validity of the quantity theory of money or the existence of a "liquidity trap". In order to show the power of the method when applied to monetary theory, I shall conduct the argument within a specific model studied by Grandmont and Laroque (1976a).

I consider an exchange economy where trade takes place sequentially over time. At each date consumers can exchange n perishable consumption goods on spot markets and can save part of their income by holding perpetuities (a promise to pay to the bearer a unit of money in each period) and/or fiat money. There is another agent (the bank) who pays interest on perpetuities and who can influence the outstanding money stock by buying or selling perpetuities. The problem is to study the existence and properties of a competitive equilibrium on the $\ell=n+2$ markets at a given date, say $t=1$.

Consider a representative consumer at date 1, and assume that he is planning only one period ahead. This restriction is inessential, for the following analysis is valid for an arbitrary finite planning horizon. It is important to note that this assumption does not imply that the agent believes that the world will come to an end at the end of his horizon. As a matter of fact, the model could be reformulated at little cost in the framework of an overlapping generations model of the Samuelson type. In that case, the economy would go on for ever, but each agent would make plans only for his lifetime.

The representative consumer knows at date 1 his endowment $e_1 \in R^n_+$ of consumption goods and his stock of assets $b_0=(b_{01}, b_{02}) \in R^2_+$, which results from his past decisions, where b_{01} is his stock of perpetuities and b_{02} is his stock of money, including the interest payments on the perpetuities b_{01}.

It is easy to check that the agent's decision problem fits exactly into the framework described in Section 2. The signal received by the agent at date 1 is $s_1=(p_1, q_1)$ and represents the prices of consumption goods p_1 and those of the assets q_1. Accordingly, $S_1=\Delta^\ell$. In period 2 the agent will perceive a signal

$s_2=(p_2,q_2,e_2)$, where (p_2,q_2) is the price system and e_2 his endowment of consumption goods at that date. Thus, $S_2=\Delta^{\ell}\times R^n_+$, and the agent's expectation function is given by a map ψ as in Section 2.

An action $a=(x,b)$ in every period describes the agent's consumption x at that date and the stock of assets he wishes to hold until the next period. The consequences of any couple of actions (a_1,a_2) are described by the corresponding consumption stream (x_1,x_2). Therefore $A_1=A_2=R^{\ell}_+$ and $C=R^n_+\times R^n_+$. The agent's preferences are defined on $\mathfrak{M}(C)$ as in Section 2.1.

The constraints faced by the agent in each period are budget constraints:

$$\beta_1(s_1)=\{a_1\in A_1 \mid p_1\cdot x_1+q_1\cdot b_1\leqq p_1\cdot e_1+q_1\cdot b_0\},$$

$$\beta_2(a_1,s_2)=\{a_2\in A_2 \mid p_2\cdot x_2+q_2\cdot b_2\leqq p_2\cdot e_2+q_2\cdot \tilde{b}_1\}$$

where $\tilde{b}=(b_{11},b_{12}+b_{11})$ represents the agent's initial stock of assets at date 2, after the interest payments on perpetuities.

The agent's decision problem is thus the same as in Section 2, and one can define for each $s_1\in S_1$ a set $\alpha(s_1)$ of optimal actions, and a set of excess demands

$$\zeta(s_1)=\{z \mid z=(x_1-e_1,b_1-b_0),(x_1,b_1)\in\alpha(s_1)\}.$$

Assume now that there are m consumers, $i=1,\dots,m$. If, for each of them, one defines a set of excess demands $\zeta^i(s_1)$, aggregate excess demand is by definition $\zeta(s_1)=\Sigma_{i=1}^m\zeta^i(s_1)$.

The operations of the other agent, called the "bank", are restricted to the markets for money and for perpetuities. The bank's action $a=(x,b)\in R^{\ell}$ in the period under consideration represents his net supply of all commodities and must satisfy, for every s_1, the accounting identity $q_1\cdot b=0$. The feasible set of the bank's actions is thus

$$\beta(s_1)=\{a=(x,b)\in R^{\ell} \mid x=0,q_1\cdot b=0\}.$$

The bank's short-run monetary policy in period 1 is taken as exogenous and is described by a relation η which associates a (possibly empty) subset $\eta(s_1)$ of $\beta(s_1)$ to every s_1.

An *equilibrium corresponding to a monetary policy* η is defined as a price system $s_1\in S_1$ such that $0\in\zeta(s_1)-\eta(s_1)$.

I state now the assumptions satisfied by a representative consumer:

Assumption 7

(1) The preferences on $\mathfrak{M}(C)$ satisfy the expected utility hypothesis (Assumption 1 of 2.1). In addition, any von Neumann–Morgenstern utility u is concave and monotone.

(2) The expectation function ψ satisfies the continuity assumption (Assumption 2) of 2.1.

(3) For every s_1 in S_1, $\psi(s_1)$ assigns probability one to the set of s_2 such that $p_2 \gg 0$.
(4) For every s_1 in S_1, $\psi(s_1)$ assigns a positive probability to the set of s_2 such that the price money q_{22} is positive.

The only really new part of these conditions is (4), which says that every consumer expects the price of money to be positive in the future with some probability, even when the price of money is currently zero. We shall see that it is the key condition to get an equilibrium with a positive price for money. It can be justified on the ground that expectations depend also on observations made before date 1, when past prices of money were positive.

One can study different monetary policies in this context. For instance, one could assume that the bank wishes to sell a given quantity $\bar{b}$ of perpetuities on the market, and to let the price system adjust freely. In that case, $\eta(s_1)$ is defined as $\{a=(x,b)\in\beta(s_1)|b_1=\bar{b}_1\}$. Or the bank could decide to devote a given quantity of money $\bar{b}_2$ to its purchases of perpetuities. This would fix its net supply of money, $b_2=\bar{b}_2$. Equilibrium can be easily studied under such assumptions. In what follows I shall concentrate on the case where the bank wishes to peg at some level $r \geqq 0$ the interest rate on perpetuities, i.e. the inverse of the money price of perpetuities. In that case, the bank's net supply will be completely elastic if $rq_{11}=q_{12}$, and undefined otherwise:

$$\eta_r(s_1)=\beta(s_1) \text{ if } rq_{11}=q_{12},$$
$$= \text{the empty set, otherwise.}$$

The existence of a competitive equilibrium with a positive price of money is a fundamental problem in any model involving money and banking institutions since the non-existence of such equilibria would mean the collapse of the banking system under consideration due to the presence of great inflationary pressures. Another important issue is the power of the bank to control the interest rate (or the money supply). The following theorem provides an answer to these questions. A key assumption for the validity of the result is (4) of Assumption 7.

Theorem 2

Consider an economy with m consumers, $i=1,\ldots,m$, each of whom satisfies Assumption 7, and such that $\Sigma_{i=1}^{m}(e_1^i, b_0^i) \gg 0$. Then, (i) if $r=0$, there is no equilibrium corresponding to η_r, and (ii) if $r>0$, there exists an equilibrium corresponding to η_r, and every such equilibrium is such that $s_1 \gg 0$.

Proof

I sketch only the proof, which is easy. First, it is straight-forward to check that, under the assumptions of the theorem, $\zeta(s_1)$ is non-empty if and only if $s_1 \gg 0$ [with the convention that $\zeta(s_1)$ is empty whenever $\zeta^i(s_1)$ is empty for some i]. Accordingly, no equilibrium can exist when $r=0$, since this would imply a zero price for money. When $r>0$, one can reduce the problem to the search for an equilibrium in $n+1$ markets (those for goods and money) by assuming that the market for perpetuities is always in equilibrium. In that case, for any $s_1 \gg 0$, with $rq_{11}=q_{12}$, and any (x_1, b_1) in $\zeta(s_1)$, the bank's net supply of money is then $-(b_{11}/r)$, in which case the vector of aggregate excess demands on the goods and the money markets is $(x_1, b_{12}+(b_{11}/r))$. Therefore to each such s_1 one can associate a set of aggregate excess demands on these markets by: $\{(x_1, m_1)\in R^{n+1} | m_1=b_{12}+(b_{11}/r), (x_1, b_{11}, b_{12})\in\zeta(s_1)\}$.

Now, for any vector (p_1, q_{12}) in the set $D=\{(p_1, q_{12})\in \text{int } R^{n+1}_{+} | \Sigma_h p_{1h}+q_{12}(1+(1/r))=1\}$, one can associate a vector s_1 in S_1 which satisfies $s_1 \gg 0$, $rq_{11}=q_{12}$ by defining $q_{11}=(q_{12}/r)$, and conversely. For each such vector $\sigma=(p_1, q_{12})\in D$ consider the corresponding s_1 and the associated set of aggregate excess demands (x_1, m_1) for goods and money. This defines a correspondence ζ^* from D into R^{n+1}, and an equilibrium is determined by a vector σ such that $0\in\zeta^*(\sigma)$. The existence of an equilibrium is then proved by applying the Market Equilibrium Lemma of Section 3.1 to the correspondence ζ^* and the set D, which is affine to the interior of the unit simplex of R^{n+1}_{+}. This completes the proof of the theorem. Q.E.D.

The foregoing result claims the viability of the banking system under consideration provided that, in particular, point (4) of Assumption 7 is verified. This axiom states that every consumer believes that the price of money will be positive with some probability, even when the current price of money is zero. It implies some degree of "inelasticity" of price expectations with respect to current prices, which does not seem unreasonable if one thinks that expectations depend upon past prices as well. Such a condition on expectations appears in many models of this type. This does not exclude, of course, the possibility of designing specific models where there is an equilibrium with a positive price of money, although (4) of Assumption 7 does not hold [see Hool (1976, 1979) for such a model involving constraints on transactions].

The following typical example, which is adapted from Grandmont (1974), may help the reader to understand the need for a condition of this type. It pictures a particular case of the economy considered in this section where price expectations are highly sensitive to current prices, contrary to what was postulated in (4) of Assumption 7. As a result, the inflationary pressures are so great that the only equilibrium positions of the economy involve a zero price of money (i.e. monetary prices are infinite), meaning the complete breakdown of the banking system.

Consider a representative consumer, satisfying all parts of Assumption 7 except (4) who believes that the future price system (p_2, q_2) and his endowment e_2 at date 2 will surely be equal to the current price system (p_1, q_1) and his current endowment e_1. Assume that, in addition, his utility function $u(x_1, x_2)$ is separable and can be written in the form $w(x_1)+\delta w(x_2)$, where $0<\delta\leq 1$ (impatience).

Then, for any $s_1 \gg 0$, any optimal action $a_1=(x_1, b_1)$ must satisfy $b_{12}=0$ (the demand for money is zero because money as an asset is dominated by perpetuities which yield a positive and sure rate of interest) and $b_{11} \leqq (b_{01}/2)$ (this relation results from the agent's preference for present consumption). If one considers now a collection of m agents, who satisfy these properties, their aggregate behaviour will result in an excess supply of perpetuities and of money at all prices $s_1 \gg 0$. It is clear that the bank cannot bring into equilibrium simultaneously these two markets when $s_1 \gg 0$, since its operations must fulfill the constraints embodied in the definition of $\beta(s_1)$. Therefore, the only possible equilibrium positions of this economy involve a zero price of money and a zero price of perpetuities (the existence of such an equilibrium is easy to prove).

3.3.1. *The quantity theory of money*

The short-run quantity theory of money, as proposed by Patinkin (1965), is essentially a statement about the homogeneity properties of the set of temporary equilibrium positions in a monetary economy at a given date, with respect to initial endowments of money and financial assets. Patinkin obtained his results by postulating specific properties of the agents' utility functions with respect to money holdings and prices. Since these utility functions are similar to the expected utility functions v that were introduced in Section 2.2, the present framework will allow me to discuss Patinkin's theory on a precise basis [for details, see Grandmont (1974)]. It will be argued that Patinkin's theory rests on shakey foundations, since it is essentially based upon the assumption that the agents' expectations about money prices are unit elastic with respect to current money prices. First, such a high degree of elasticity appears to be restrictive since price expectations depend upon past prices as well. Second, it appears to involve some logical problems, for it contradicts (4) of Assumption 7, in which case the existence of an equilibrium with a positive price for money may not guaranteed.

In order to pursue the discussion, it is convenient to work with money prices: for every s_1 such that the price of money q_{12} is positive, the corresponding monetary price system s_1^* is defined as (s_1/q_{12}). Let us first look at the agents' expected utility functions introduced in Section 2.2,

$$v(a_1, s_1^*)=v(x_1, b_1, p_1^*, q_1^*)$$

which play the same role in the present model as Patinkin's utility functions. Patinkin assumed that only "real" assets entered the agents' utility functions. A more general formulation, leading to Patinkin's results, would be to assume the homogeneity of v with respect to asset holdings and current money prices of goods, i.e.

$$v(x_1, \lambda b_1, \lambda p_1^*, q_1^*)=\lambda^{\alpha} v(x_1, b_1, p_1^*, q_1^*), \quad \text{all } \lambda>0.$$

It is easily seen that such a property is true if expectations about money prices of goods are *unit elastic* with respect to current money prices of goods, while expectations about the interest rate are *inelastic* with respect to the same variables, but that it fails in general to hold otherwise [for details see Grandmont (1974) which can be easily adapted].

Patinkin's statement of the quantity theory amounts to saying in the present model that if $s_1^* = (p_1^*, q_1^*)$ is a monetary equilibrium price system of the economy, then $(\lambda p_1^*, q_1^*)$ is a monetary equilibrium price system of the new economy obtained by changing b_0^i in λb_0^i, for all i, and thus for all $\lambda > 0$. Again, it is easily seen that the statement is true under the above assumptions about the elasticity of expected prices and interest rate with respect to current monetary prices of goods, but it is not in general true when these conditions are violated.

To conclude, Patinkin's theory appears to be valid only under quite restrictive assumptions on expectations. Moreover, these assumptions contradict (4) of Assumption 7, in which case the set of equilibria with a positive price of money, which is the object studied by the quantity theory, may be empty. The range of application of this theory appears therefore to be narrow. Of course, the foregoing argument does not exclude the possibility that the quantity theory is *locally* valid (for some λ's near 1), nor does it exclude the possibility of constructing models possessing equilibria with a positive price of money where the quantity theory holds [e.g. see Hool (1976a)]. Finally, if the quantity theory is interpreted as a homogeneity property of *stationary* states of a money economy, then it is possible to show that it is valid under quite general conditions [see Grandmont (1974), Grandmont and Younès (1972), and Grandmont and Laroque (1973, 1975)].

3.3.2 The "liquidity trap"

The liquidity trap is usually defined as a property of the agents' demand for money which diverges supposedly to infinity when the interest rate on perpetuities tends to zero or to a positive but very low level. Can this phenomenon occur in the present model?

First, we note that it can occur only when the interest rate goes to zero, since $\zeta(s_1)$ is non-empty and compact for every $s_1 \gg 0$. Therefore, consider a sequence $s_1^k \gg 0$ which tends to $s_1 \in S_1$, such that $r^k = (q_{12}^k / q_{11}^k)$ tends to zero. This implies that the price of money q_{12}^k converges itself to zero, in which case, for every sequence $z^k \in \zeta(s_1^k)$, the norm of z^k diverges to $+\infty$. But this does not mean that the aggregate demand for money itself necessarily increases without bound. This will be the case indeed if $p_1 \gg 0$, $q_{11} > 0$ and $q_{12} = 0$. But otherwise, one can construct non-pathological examples where, for instance, the aggregate demand for money is zero along the sequence s_1^k [Grandmont and Laroque (1976a)].

Thus, under the usual definition, a liquidity trap exists in some cases, while in others it does not. There is, however, another interpretation of the concept of a liquidity trap [Patinkin (1965)], which allows asserting neatly its existence in the present model. For every interest rate on perpetuities $r>0$, Theorem 2 ensures the existence of a temporary equilibrium. Let $\mu(r)$ be the corresponding set of equilibrium aggregate money stocks held by the consumers. Then,

Proposition 3

Let r^k be a sequence of positive interest rates tending to zero. Under the assumptions of Theorem 2, then $\lim M^k = +\infty$ for every sequence $M^k \in \mu(r^k)$.

Proof

Assume that, on the contrary, there is a sequence $M^k \in \mu(r^k)$ which is bounded. Consider an associated sequence of equilibrium price systems s_1^k, and the associated equilibrium actions (a_1^{ik}, a^k) of the consumers and of the bank. Each trader's consumption x_1^{ik} is bounded, since total consumption must equal $\Sigma_i e_1^i$. The bank's supply of money b_2^k is bounded because M^k is bounded. Its supply of perpetuities b_1^k, which is equal to $(-r^k b_2^k)$ tends to zero. Therefore, each consumer's holding of assets b_1^{ik} is bounded too. But it can easily be shown that there must be an i such that the sequence $|a^{ik}|$ is unbounded. This leads to a contradiction. Q.E.D.

The economic mechanism underlying this result is easy to understand if one looks at the bank's budget constraint, which can be written $(b_1^k/r^k)+b_2^k=0$. Indeed, the fact that M^k tends to infinity is equivalent to $\lim b_2^k=+\infty$. This implies that $b_1^k<0$ for k large enough. That is to say, when r^k tends to zero, the consumers eventually become net sellers of part of their stock of perpetuities, and the bank must create money in amounts which increase without limit in order to bring the perpetuities market into equilibrium.

In this model a liquidity trap in the sense of Proposition 3 can occur only when r tends to zero, since, according to Theorem 2, equilibrium money stocks are finite for every positive r. Following a suggestion of Younès (1972a), it is possible to modify the model so that a liquidity trap exists at a positive interest rate. For that, it suffices to allow the consumers to borrow by issuing perpetuities. If a trader believes at date 1 that the interest rate in period 2 will surely be above some value $\bar{r}_2$, independent of the current price system, then he will issue an infinite amount of perpetuities and hold the proceeds in the form of money balances if the interest rate r_1 in period 1 falls below $\bar{r}_1=\bar{r}_2/(1+\bar{r}_2)$. In that case, the bank would be unable to drive r_1 below $\bar{r}_1$, and a liquidity trap would occur when r_1 tends to $\bar{r}_1$ from above. Such a model can easily be studied by the techniques developed in this paper.

3.3.3. Conclusion

The foregoing model illustrates the power of the temporary equilibrium method for the study of monetary issues. The analysis was focused on the property of money to be a store of value and on the fact, which is essential in modern economic systems, that money is issued (produced) by the banking system as a counterpart of its purchases of assets. The study was conducted within the framework of an extremely rudimentary banking system involving only one banking institution. While many lessons were learnt by studying this simple model, it would be desirable to extend the analysis to the case of several banking institutions. This would permit studying on a precise basis an issue like, for instance, the control of the money supply by a central bank in the presence of commercial banks.

But there is an important function of money which I have not mentioned up to now, i.e. its function as a medium of exchange. The issue is fundamental for only its study can permit us to understand why agents are willing to hold idle cash balances when positive and sure interest earning assets are available. The question has been analysed within the framework of sequence economies with perfect foresight [Hahn (1971, 1973), Honkapohja (1977), Kurz (1974), and Starrett (1973)]. A few studies have been devoted to the problem within the framework of temporary equilibrium theory. Some of them use the method proposed by Clower (1967), which consists in imposing directly a constraint on an agent's transactions, stating that the value of an agent's purchases during a period cannot exceed his initial cash balance plus some fixed proportion of the value of his sales during the same period. This constraint is intended to reflect the assumptions that money is used as an intermediary in every exchange, and that different commodities are exchanged at different localisations or "trading posts" [Grandmont and Younès (1972, 1973), Hool (1974a, 1974b, 1975, 1976, 1979)]. Although the method is rudimentary, it appears to have some power since it allows us to study the interaction between the different properties of money, and, in particular, to make precise a few issues such as Friedman's theory of optimal cash balances and the validity of the classical dichotomy [Grandmont and Younès (1972, 1973)]. One interesting point of Hool's research [see, in particular, Hool (1979)] is the fact that he shows the possibility of constructing a model where price expectations are unit elastic with respect to current prices and which possesses equilibria with a positive price of money. All these remarks show the importance of studying in a finer way in the future the role of money in the process of exchange by introducing explicitly transaction costs and, more generally, transaction technologies. Although a few studies exist on this topic [Müller and Schweizer (1978), Paquin (1976, 1977)], more work is needed in this area.

3.4 Temporary competitive equilibrium and production

The works that I have reviewed so far were all concerned with the study of pure exchange economies involving consumers trading in an uncertain world in various institutional setups. Although there is still much work to be done in this area in order to reach a general and concise formulation, the studies that I mentioned in the two preceding sections display a striking unity, and one can hope that a convincing synthesis is not too far ahead. The reason for this relative success is the fact, I believe, that there seems to be an agreement among researchers about the nature of the basic economic unit in exchange economies (the consumer) and the kind of motives (objective function) which one can attribute to it. Although there have been a number of attempts to integrate firms in temporary equilibrium models [Morishima (1964), whose work has been generalised recently by Diewert (1978), Arrow and Hahn (1971), Stigum (1969a, 1969b, 1972), Sondermann (1974) and Grandmont and Laroque (1976b)], or in the theory of sequences of markets operating under certainty by using the perfect foresight approach [Radner (1972, 1974) and Gale (1975)], the result is far less impressive. This is because there is much less agreement about the proper modelling of a firm operating in sequence economies with incomplete futures markets. I shall therefore, without being formal, confine myself in this section to a very brief review of the main characteristics of the works which have been done on this topic and to some remarks concerning the main problems that we face.

Let me begin with a few well-known remarks. In the Arrow–Debreu model, a firm is viewed as an economic unit whose task is to combine inputs of some commodities to produce other commodities, and which distributes its profits to its shareholders. Under the assumptions of the model, stock markets would be inactive if they were opened and, given a price system, all stockholders of a firm would unanimously choose a production plan which maximized profit if they were asked to do so. This fact allows representation of a firm in the Arrow–Debreu model as an abstract entity, distinct from its stockholders, whose goal it is to maximise profits on its production set. A related point is that, since the consumers can discount their wealth back to the initial date, the pattern of dividend payments over time is a matter of indifference to the shareholders. Consequently, one can think that all profits are distributed at the initial date.

The situation is much more complicated in temporary equilibrium theory. A firm is still, of course, a unit whose task is to combine inputs at some date in order to produce other commodities at the same date and/or at a later date. The description of the technological possibilities of a firm therefore does not differ much from the one given in the Arrow–Debreu model, except that the time structure of the production set may be described more specifically. But what is new is that:

(i) The traders may have different expectations about the profitability of different firms and thus stock markets may be active [Stigum (1969a, 1969b), Arrow and Hahn (1971), Sondermann (1974), Radner (1972) and Gale (1975)].

(ii) The firms face a budget constraint in each period. In order to finance their production plan they can issue bonds redeemable in money and/or issue new equity.
(iii) The pattern of dividend payments over time is no longer a matter of indifference to the shareholders.

Therefore, in addition to a production plan, a firm typically has to choose a financial plan and a stream of dividend payments. The strategy adopted by the researchers in the field was to transpose the formulation of the Arrow–Debreu model and to view a firm as an abstract entity distinct from its stockholders, with expectations and an objective function of its own, the latter being a more or less complicated function of dividend payments (Stigum, Gale), of profits (Arrow–Hahn, Radner, Sondermann), and even of the firm's asset–debt structure at the end of its planning horizon (Stigum). This approach is obviously unsatisfactory, and much more work is needed on this topic if we wish to reach a better understanding of the functionings of the firm as an organization.

3.5. Bibliographical note

In addition to the works that were described above, it is worthwhile to mention a few interesting studies which focus attention not only on the existence and properties of a temporary competitive equilibrium at a given date, but also on the evolution over time of sequences of such equilibria. A first category of works analyses the existence and efficiency properties of stationary trajectories in such economies in relation with the role of money in the exchange process [Grandmont and Younès (1972, 1973), Paquin (1977)] or with monetary policy [Grandmont and Laroque (1975)]. Another interesting problem is that of the convergence of the sequence of temporary equilibria towards a stationary state in relation to the properties of the agents' expectation functions. This issue has been studied within the framework of a simple model with Samuelsonian overlapping generations and with outside money by Fuchs and Laroque (1976) and by Fuchs (1976, 1977a, 1977b, 1979a, 1979b).

4. Temporary equilibrium with quantity rationing

The method described in Section 3 takes as a point of departure that prices move fast enough in each period so as to match supply and demand on all markets. We obtain in this way a sequence of temporary competitive equilibria. This method enabled us to make progress towards a greater realism by comparison with the traditional general competitive equilibrium theory. It allowed us in particular to include in our models an important "disequilibrium" feature, since at each date the agents' plans for the future are not transmitted to the market and therefore may be incompatible.

This method is, however, powerless if we wish to take into account a phenomenon which is important in practice, i.e. the unemployment of resources (excess supply of labour, excess productive capacities, etc.), and can be observed in real economies no matter what their type of organization. Following the impulse given by Clower (1965), Leijonhufvud (1968), and Patinkin (1965), a number of recent works have studied "non-tâtonnement" models [Hahn and Negishi (1962)] which consider situations where exchanges may take place at prices which do not equilibrate supply and demand in the classical sense. These models yield, accordingly, a sound microeconomic basis for the analysis of phenomena such as unemployment which were studied traditionally mainly by macroeconomists. In these models, in addition to the price system, quantitative constraints on trades as perceived by the agents play a central role.

Important progress has been achieved by the study of an extreme case, where prices are temporarily fixed in each period. This corresponds to the assumption which is often made in Keynesian models that in the short run quantities move much faster than prices. The structure of these models is, in general, the following one. At the outset of each period, prices are quoted and cannot change during the period. At these prices, supplies and demands expressed ex ante by the agents may be incompatible. In such a case, the ex post equilibrium at that date is achieved by quantity rationing. Once such an equilibrium is established, transactions take place, and prices may be revised at the beginning of the next period by the agents who control them in view of the information generated by the exchanges of the current period.

As noted above, these models turn out to be powerful tools for analysing a few issues which belonged traditionally to the realm of macroeconomic theory. In particular, it is possible to obtain a temporary equilibrium involving an excess supply both on the markets for the firms' output and on the labour market (unemployment of a Keynesian type). But these models are also able to generate other situations where, for instance, there is an excess supply for labour and an excess demand for the firms' output (stagflation), or where there is an excess demand on both markets (inflation) [see, in particular, Barro and Grossman (1971, 1974, 1976), Benassy (1973, 1975a), Malinvaud (1977), Negishi (1978), and Younès (1970b)]. This equilibrium concept therefore appears to be meaningful since it seems to enable us to unify macroeconomic theories which looked fundamentally different beforehand. It appears to be useful too for the study of bottlenecks which may arise in economies where prices are administered by the Central Planning Bureau [Portes (1976)].

In this section I first present the equilibrium concept underlying this type of model by adapting the works of Benassy (1975a) and of Drèze (1975) (Section 4.1). I then study the application of this method to the study of the microeconomic foundations of macroeconomic theory by adapting a model analysed by Benassy (1978) (Section 4.2).

4.1. Market equilibrium with quantity rationing

Consider an economy at a given date. There are ℓ commodities to be exchanged, $h=1,\dots,\ell$, commodity ℓ being money. Prices which were quoted at the outset of the period are described by a vector p in the interior of the simplex Δ^ℓ of R^ℓ_+.

The set of feasible net trades in the period for a given representative agent is described by a subset Z of R^ℓ. The agent's final transaction, z, must thus belong to Z and satisfy the budget constraint $p\cdot z=0$.

In addition to the price system this agent perceives for each commodity h other than money, quantitative constraints $\underline{z}_h \leqq 0$ and $\bar{z}_h \geqq 0$ which set lower and upper bounds on his trade of commodity h. These constraints generate two vectors $\underline{z}\in R^{\ell-1}_-$ and $\bar{z}\in R^{\ell-1}_+$. I will denote by s the vector $(p,\underline{z},\bar{z})$, an element of $S=\operatorname{int}\Delta^\ell\times R^{\ell-1}_-\times R^{\ell-1}_+$. Given this vector, the agent's final transaction must then belong to the set

$$\beta(s)=\{z\in Z\,|\,p\cdot z=0,\ \underline{z}_h\leqq z_h\leqq \bar{z}_h,\ h\neq\ell\}.$$

An important feature of the model is that no quantitative constraint is perceived in the case of money. The idea underlying this axiom stems from the fact that money is a medium of exchange. Indeed, a particular interpretation of the model is that there are $\ell-1$ separated trading posts or markets, one for each commodity other than money, where the agent exchanges commodity h against money at the ruling prices. Although these markets are separated, the agent exchanges simultaneously on all of them. For each $h\neq\ell$, let $t(h)$ be the elementary transaction describing an exchange of one unit of commodity h against (p_h/p_ℓ) units of money: $t(h)$ is a vector of R^ℓ given by $t_h(h)=1$, $t_\ell(h)=-(p_h/p_\ell)$ and $t_k(h)=0$ for $k\neq h,\ell$. Then every trade z such that $p\cdot z=0$ can be written $z=\sum_{h=1}^{h=\ell-1}z_h t(h)$, and conversely. According to this interpretation one can view z_h as the intensity of the transaction of commodity h against money, and the constraints $\underline{z}_h$ and $\bar{z}_h$ can be interpreted as constraints on this intensity. Money then plays the role of a medium of exchange like bank deposits, and one can imagine that payments on each trading post are made by cheques or by using a credit card. The agent's final money holding must then satisfy some constraints (like non-negativity for instance) which are embodied in the definition of the feasible set of net trades Z.

The agent's preferences are defined on net trades made in the period and are represented by a function which depends upon z and the signals perceived by the agent during the period (and, of course, implicitly on the signals perceived in the previous periods). This function must be interpreted as an expected utility index derived from a multiperiod decision problem, as in Section 2.2. In general, the agent may have some information about the other agents' offers to trade and final transactions. For simplicity of exposition, I assume that only the quantitative constraints $\underline{z}$ and $\bar{z}$ perceived by the agent have an influence on his expectations. Accordingly, whatever the agent's information is during the period, his preferences will depend only on the signal $s=(p,\underline{z},\bar{z})$ previously introduced. They will be represented in the sequel by an expected utility index $v\colon Z\times S\to R$.

Given a signal $s=(p,\underline{z},\bar{z})$ in S, the set of *constrained* excess demands expressed by the agent $\zeta(s)$ is then defined as the set of net trades which maximise the function $v(z,s)$ with respect to z subject to the constraint $z\in\beta(s)$.

It is now possible to make precise when a perceived constraint is binding. Heuristically, the constraint $z_h\leqq\bar{z}_h$ is binding if the agent can be made better off by relaxing it. More precisely, let $\bar{\beta}_h(s)$ be the set obtained from $\beta(s)$ by dropping this constraint:

$$\bar{\beta}_h(s)=\{z\in Z\mid p\cdot z=0,\ \underline{z}_k\leqq z_k\leqq\bar{z}_k,\ k\neq h,\ \ell \text{ and } \underline{z}_h\leqq z_h\}.$$

Then the constraint $\bar{z}_h$ is binding if there exists z' in $\bar{\beta}_h(s)$ such that $v(z',s)>v(z,s)$ for all z in $\beta(s)$. A similar definition can be stated for a constraint $\underline{z}_h$.

4.1.1. *Market equilibrium with quantity rationing*

Consider an economy at a given date involving m agents, $i=1,\dots,m$, like the one described above. The idea underlying the equilibrium concept which I introduce now is rather natural. Indeed, it assumes essentially that only agents on one "side" of each market can perceive binding constraints. More precisely, given the price system $p\in\operatorname{int}\Delta^{\ell}$,

A K-equilibrium is a collection of net trades (z^i) *and a collection of signals* $(s^i=(p,\underline{z}^i,\bar{z}^i))$ *such that*:

(1) $z^i\in\zeta^i(s^i)$, *for all* i, *and* $\sum_i z^i=0$.
(2) *If for some* $h\neq\ell$, *there is an* i *such that the constraint* $\bar{z}^i_h$ *is binding, then for every* j *the constraint* $\underline{z}^j_h$ *is not binding.*

One must notice that this concept involves an equilibrium notion (at fixed prices) since it assumes that in the case of an excess demand on some market no seller is left with an unsold commodity. This assumption is valid only if there is a significant (and costless) exchange of information and market adjustment among the traders. Another reason which shows that we are dealing with an equilibrium concept is the fact that the situation on some markets depends crucially upon the constraints perceived on others (spillover effects). For instance, the demand for consumption goods will be influenced by the existence and the extent of unemployment on the labour market. These cursory remarks clearly show that the allocations described in the above definition are the result of an equilibrating process acting on quantities.

The above equilibrium concept does not make precise how shortages are distributed among the agents. This feature will lead, in general, to a continuum of K-allocations, as the following simple example shows. Assume $\ell=2$, $m=3$, and that at the ruling prices the first agent wishes to sell 20 units of commodity 1, while the two other agents would each like to buy 20 units of this commodity. In such a case there is a continuum of K-equilibria. Each of them is characterized by the fact that the first agent sells 20 units of commodity 1 to the two others, while agents 2 and 3 buy respectively $z_1^2 \geqq 0$ and $z_1^3 \geqq 0$ units of that commodity, with $z_1^2 + z_1^3 = 20$. In order to obtain a more specific theory, therefore, one must describe how shortages are distributed among agents. This leads to the concept of a *rationing scheme*, which is presented next, and to the K-equilibria which are associated with it.

4.1.2. Rationing schemes

A rationing scheme associates to trade offers made by the agents, the quantitative constraints $(\underline{z}^i, \bar{z}^i)$ which limit their final transactions. Let $\tilde{z}^i \in R^{\ell-1}$ be the trade offer made by agent i for all commodities h other than money. The fact that trade offers are only made for commodities $h \neq \ell$ corresponds to the implicit role of money as a medium of exchange that I mentioned above. I will describe below a possible way to determine these $\tilde{z}^i$, but it may be useful to note at this stage that these trade offers may violate the quantitative constraints perceived by the agents. Let $\tilde{z} = (\tilde{z}^i) \in R^{(\ell-1)m}$ be the collection of these trade offers. A rationing scheme will associate to $\tilde{z}$ the constraints $\bar{z}_h^i = \bar{F}_h^i(\tilde{z}) \geqq 0$ and $\underline{z}_h^i = \underline{F}_h^i(\tilde{z}) \leqq 0$ for each agent i and each commodity $h \neq \ell$. The basic idea underlying this concept is the implicit assumption that the ith agent's final transaction z_h^i for commodity $h \neq \ell$ will be equal to the minimum of his trade offer $\tilde{z}_h^i$ and of the perceived constraint $\bar{F}_h^i(\tilde{z})$ if he is a demander of that commodity ($\tilde{z}_h^i \geqq 0$), and to the maximum of $\tilde{z}_h^i$ and of $\underline{F}_h^i(\tilde{z})$ otherwise. Since a rationing scheme is designed mainly to bring about the consistency of the agents' ex post transactions, the sum over the agents of the resulting z_h^i should be zero. The rationing scheme should also obey the basic principle underlying the notion of a K-equilibrium, namely that only one side of each market is rationed. In other words, if for some $h \neq \ell$ there is an agent i such that $\tilde{z}_h^i > \bar{F}_h^i(\tilde{z})$, then one should have $\tilde{z}_h^j \geqq \underline{F}_h^j(\tilde{z})$, for all j. To sum up,

A rationing scheme is defined by two functions $\bar{F}$: $R^{(\ell-1)m} \to R_{+}^{(\ell-1)m}$ *and* $\underline{F}$: $R^{(\ell-1)m} \to R_{-}^{(\ell-1)m}$ *which satisfy for every* $\tilde{z}$ *and every* $h \neq \ell$:

(a) $\sum_{\tilde{z}_h^i \geqq 0} \min(\tilde{z}_h^i, \bar{F}_h^i(\tilde{z})) + \sum_{\tilde{z}_h^i \leqq 0} \max(\tilde{z}_h^i, \underline{F}_h^i(\tilde{z})) = 0$
and

(b) *if* $\tilde{z}_h^i > \bar{F}_h^i(\tilde{z})$ *for some* i, *then* $\tilde{z}_h^j \geqq \underline{F}_h^i(\tilde{z})$ *for all* j.

This concept can be illustrated by a few simple examples. Consider the case where an excess demand appears on the market $h \neq \ell$, i.e. where $\tilde{z}_h = \sum_i \tilde{z}_h^i > 0$, and let $I = \{i_1, \ldots, i_n\}$ be the set of demanders ($\tilde{z}_h^i \geqq 0$). The rationing scheme is said to be uniform if $\bar{F}_h^i(\tilde{z}) = \bar{F}_h^j(\tilde{z})$, for all i, j in I. If the rationing scheme states that agent i_1 must be served first, agent i_2 second, and so on, one says that this is a *queue*. In such a case, if the sellers' total supply on market h is $x = \sum_i \max(0, -\tilde{z}_h^i)$, one has $\bar{F}_h^{i_1}(\tilde{z}) = x$, $\bar{F}_h^{i_2}(\tilde{z}) = x - \min(\tilde{z}_h^{i_1}, \bar{F}_h^{i_1}(\tilde{z}))$, and so on. Finally, a *proportional* rationing scheme postulates that $\bar{F}_h^i(\tilde{z}) = \lambda x$, for all i in I, where $0 \leqq \lambda < 1$ is obtained by dividing the total supply x by total demand $\tilde{z}_h + x$.

4.1.3. *K-Equilibria corresponding to a rationing scheme*

Intuitively, one wishes to define a K-equilibrium corresponding to a given rationing scheme $(\underline{F}, \bar{F})$ as a K-equilibrium (z^i, s^i) and a collection of trade offers $\tilde{z} = (\tilde{z}^i)$ such that the perceived constraints satisfy $\bar{z}_h^i = \bar{F}_h^i(\tilde{z})$, $\underline{z}_h^i = \underline{F}_h^i(\tilde{z})$, and the final transactions $z_h^i = \min(\tilde{z}_h^i, \bar{F}_h^i(\tilde{z}))$ whenever $\tilde{z}_h^i \geqq 0$, $z_h^i = \max(\tilde{z}_h^i, \underline{F}_h^i(\tilde{z}))$ whenever $\tilde{z}_h^i \leqq 0$. In order to achieve this goal, one has to make precise the way the agents formulate their trade offers $\tilde{z}^i$.

I shall describe a model of this type which exploits the idea of *effective demand* that underlies Keynesian theorising as it was proposed by Clower (1965), and generalized by Barro and Grossman (1971) and by Benassy (1973). Clower made the simple and very important remark that in Keynesian theory effective demand for goods takes into account the level of unemployment, i.e. the constraints perceived by the workers on the labour market. If one extends this idea to the present context one is led naturally to consider, for a given signal perceived by agent i, $s^i = (p, \underline{z}^i, \bar{z}^i)$, the set $\beta_h^i(s^i)$ obtained from $\beta^i(s^i)$ by deleting the constraints corresponding to commodity $h \neq \ell$:

$$\beta_h^i(s^i) = \{z \in Z^i \mid p \cdot z = 0, \underline{z}_k^i \leqq z_k \leqq \bar{z}_k^i, k \neq h, \ell\}.$$

Then, the ith agent's effective demand on market h is defined as the hth component of the net trades which maximise $v^i(z, s^i)$ with respect to z, subject to $z \in \beta_h^i(s^i)$. These effective demands form a subset $\zeta_h^i(s^i)$ of the real line. When this operation is repeated for all markets $h \neq \ell$, one obtains a set of effective demand vectors $\zeta^i(s^i)$, a subset of $R^{\ell-1}$ which is defined as the product of the $\zeta_h^i(s^i)$ for $h = 1, \ldots, \ell-1$.

A K-equilibrium defined by (z^i) *and* $s^i=(p,\underline{z}^i,\bar{z}^i)$ *corresponds to the rationing scheme* $(\underline{F},\bar{F})$, *if there exists a collection* $\tilde{z}=(\tilde{z}^i)$ *of trade offers with* $\tilde{z}^i\in\tilde{\zeta}^i(s^i)$ *for all i, such that for all i and* $h\neq\ell$:

(i) $\bar{z}^i_h=\bar{F}^i_h(\tilde{z})$ *and* $\underline{z}^i_h=\underline{F}^i_h(\tilde{z})$ *and*
(ii) $z^i_h=\min(\tilde{z}^i_h,\bar{F}^i_h(\tilde{z}))$ *when* $\tilde{z}^i_h\geqq 0$, *and* $z^i_h=\max(\tilde{z}^i_h,\underline{F}^i_h(\tilde{z}))$ *when* $\tilde{z}^i_h\leqq 0$.

The following result gives sufficient conditions for the existence of such an equilibrium.

Lemma 2

Assume that for all i, (i) Z^i is bounded below, closed, convex and $0\in Z^i$ and (ii) the function v^i is jointly continuous with respect to (z^i,s^i), and quasi-concave with respect to z^i. Then there exists a K-equilibrium corresponding to the rationing scheme $(\underline{F},\bar{F})$ provided that the functions $\underline{F}$ and $\bar{F}$ are continuous.

Proof

Assume first that the v^i are strictly quasi-concave in z^i. The correspondences ζ^i and $\tilde{\zeta}^i$ can then be viewed as functions, and it is easy to check their continuity. Consider an element $\sigma=((z^i),(\underline{z}^i,\bar{z}^i),(\tilde{z}^i))$, where, for all i, $z^i\in Z^i$ and $p\cdot z^i=0$, $\underline{z}^i\in R^{\ell-1}_-$, $\bar{z}^i\in R^{\ell-1}_+$, $\tilde{z}^i\in R^{\ell-1}$. To this vector one can associate a point of the same space in the following way. To $\tilde{z}=(\tilde{z}^i)$ one associates the new constraints $\bar{z}^i_h=\bar{F}^i_h(\tilde{z})$ and $\underline{z}^i_h=\underline{F}^i_h(\tilde{z})$, and then, from these new constraints, a new vector of constrained excess demands z^i and of effective excess demands $\tilde{z}^i$. The function which is so defined is continuous and takes values in a compact convex subset of an Euclidean space of appropriate dimension. The restriction on this function to this subset has a fixed point. It is easy to check that this fixed point determines a K-equilibrium corresponding to the scheme $(\underline{F},\bar{F})$.

In order to cover the case where the v^i are quasi-concave, one remarks that every function $v(z,s)$ is quasi-concave in z and continuous in (z,s) and can be approximated by a sequence of functions $v^k(z,s)$ which are continuous in (z,s), strictly quasi-concave in (z,s), and such that $\lim v^k(z^k,s^k)=v(z,s)$ whenever $\lim z^k=z$ and $\lim s^k=s$. If one chooses an approximation of this type for each v^i, it suffices to apply the lemma to each element of the sequence k. The sequence of K-equilibria which is obtained in this way is bounded, and by continuity any limit point of this sequence is a K-equilibrium corresponding to the scheme $(\underline{F},\bar{F})$. Q.E.D.

The above model describes how the plans of the different agents can be made compatible ex post by means of a rationing scheme when there are temporary price rigidities. This theory is a consistent useful framework. It displays, however, a few weaknesses. The main one is related to how the agents are assumed to formulate their trade offers. In the above model these "effective demands" are formed market by market without taking into account their consequences, i.e. the final transactions which will be the result of them. As soon as one takes into account this link, the somewhat artificial character of these effective demands appears immediately. I will limit myself to the illustration of this point by a few remarks.

Consider a K-equilibrium associated to a given scheme $(\underline{F}, \overline{F})$, and assume that there is an excess demand on market h, i.e. $\sum_i \tilde{z}^i_h > 0$. Assume, in addition, that $\overline{F}^i_h(\tilde{z})$ is non-decreasing with respect to $\tilde{z}^i_h$ when this quantity is non-negative. There are two cases. If there is a rationed agent, i.e. an i such that $\overline{F}^i_h(\tilde{z}) < \tilde{z}^i_h$, and $\tilde{z}'$ obtained from $\tilde{z}$ by increasing $\tilde{z}^i_h$ by a positive quantity, such that $\overline{F}^i_h(\tilde{z}') > \overline{F}^i_h(\tilde{z})$, one can say that the rationing scheme is *manipulable* at $\tilde{z}$ by the agent i. Otherwise, the rationing scheme is said to be *non-manipulable*. A rationing scheme which is uniform or obeys a queuing process is non-manipulable. On the other hand, a proportional rationing scheme is manipulable. It is intuitively clear that in the case of a non-manipulable rationing scheme, under reasonable assumptions on the scheme $(\underline{F}, \overline{F})$, the agents who are rationed on some market can increase their trade offers on that market without changing the final allocation. In such a case the particular definition of effective demand which was employed in the model is rather arbitrary. On the other hand, in the case of a manipulable rationing scheme there must be an agent who can improve his situation by increasing his effective demand. In that case the use of the concept of effective demand made in the model is somewhat irrational from the agent's viewpoint.

One may think that one can improve upon the foregoing formulation by designing a model where each agent would know the rationing scheme and would send to the market trade offers which, given the other agents' trade offers, would result in a vector of final transactions most preferred by him. This would yield a concept of a Nash equilibrium in the space of trade offers. This approach, however, involves difficulties (in particular, non-convexities) some of which are fundamental. The first point is that the no trade equilibrium is then always an equilibrium of such a system. The problem is then to find conditions ensuring the existence of a non-trivial equilibrium. But one is faced with the same kind of problems as before. Indeed, if the rationing scheme is non-manipulable, under reasonable assumptions a Nash equilibrium will involve final transactions identical to those yielded by the model in this section. But in that case, trade offers are arbitrary. On the other hand, if the rationing scheme is manipulable, a Nash equilibrium where trade actually takes place may not exist in general, e.g. in the case where rationing is proportional and where $\ell=2$ [Benassy (1977), Böhm and Lévine (1976), and Heller and Starr (1976)]. It might be possible to find non-trivial Nash equilibria in the case of a manipulable rationing scheme if it is influenced by random factors, but the issue is not completely understood as yet [Green (1977), Svensson (1977)].

To sum up this discussion, the concept of equilibrium with quantity rationing which was presented in this section looks reliable if the rationing scheme is non-manipulable, at least if one is interested only in final transactions arising from the model. However, in such a case trade offers sent to the market appear arbitrary to some extent. On the other hand, the model is not very robust in the case of a manipulable rationing scheme, which seems in agreement with what one could have expected beforehand.

4.2. *Equilibrium with quantity rationing and macroeconomic theory*

The basic concept of equilibrium described in Section 4.1 is a powerful tool for studying markets in "disequilibrium". Accordingly, a significant number of recent works have been devoted to the construction of models which are similar to those that are employed in macroeconomic theory. The goal of these studies is to achieve a better understanding of the origins of unemployment and/or of inflation by analysing models where the agents' microeconomic characteristics (tastes, technology, and expectations), their behaviour and the institutional framework within which they interact, is clearly specified.

This research has taught us many important lessons. However, since it is still in progress, I believe it is better not to attempt to make a synthesis of the recent works. Therefore, I will limit the scope of this section to the study of a simple "macroeconomic" example which illustrates forcefully what appears to be the fundamental novelty of this approach compared to traditional macroeconomic theories, i.e. the possibility that one can describe with the help of the same conceptual model situations which are as different as Keynesian unemployment, stagflation or general inflation. This example has been extensively analysed in the literature (Barro and Grossman, Benassy, Malinvaud, and Younès). The specification which I shall use is very close to that of Benassy (1978).

There are three aggregated commodities: a (consumption) good, labour, and money, the prices of which are respectively p, w, and 1. There are three aggregated agents too: a firm, a consumer, and the government.

The firm produces within the period under consideration the consumption good with labour according to the production function $Y=F(L)$ which is strictly concave, increasing, continuously differentiable and such that $F(0)=0$, $F'(0)=+\infty$. The firm seeks to maximise its short-run profit $\Pi=pY-wL$, which is distributed to the consumer during the period.

The consumer consumes the quantity C of the good, offers L units of labour and decides to hold M units of money for the next period. The maximum amount of labour that the consumer can physically offer is L_0, and his initial money stock is M_0. Accordingly, any action (C, L, M) of the consumer during the period must fulfil the constraints $pC+M=M_0+wL+\Pi$ and $L \leqq L_0$, where Π is the firm's short-run profit.

The consumer's preferences on actions are represented by a utility function $v(C, L, M)$ which has to be considered as an expected utility index derived from a multiperiod choice problem as in Section 2. I will adopt the following specification: $v(C, L, M)=\alpha_1 \log C+\alpha_2 \log M$, $\alpha_i>0$, $\alpha_1+\alpha_2=1$, which postulates that labour does not entail any disutility.

Finally, the government's public consumption G is taken as exogenous, and it is financed by an issue of ΔM units of money, according to the equation $pG=\Delta M$.

Under the above assumptions, provided that $G<\alpha_2 F(L_0)$, which I shall assume from now on, there is a unique (temporary) competitive equilibrium defined by (p^*, w^*) such that

$$p^*=\frac{\alpha_1 M_0}{\alpha_2 F(L_0)-G}, \qquad w^*=p^* F'(L_0).$$

Output Y^* is then equal to $F(L_0)$, employment L^* to L_0 and private consumption C^* to $F(L_0)-G$. One remarks immediately than an increase in public consumption does not affect output or employment or the real wage, but implies an equal decrease in private consumption and an increase in the money price and wage.

Assume now that (p, w) is fixed at the outset of the period at a value which differs from (p^*, w^*). A disequilibrium is bound to appear in the economy, and the ex post compatibility of the agents' transactions will be ensured by quantity rationing. One can then apply in a simple manner the model which was presented in Section 4.1. In order to fix the ideas, I shall assume that when there is an excess demand on the good's market, the government is served first.

In this simple case three regimes can appear. *Keynesian unemployment*, where there is an excess supply for the consumption good and for labour. *Classical unemployment*, or *stagflation*, where there is excess supply on the market for labour, and an excess demand on the good's market. Finally, *repressed inflation*, where there is an excess demand on both markets. The fourth case which could theoretically appear (an excess demand for labour and an excess supply for the good) cannot arise in this model, for it would imply that the firm would perceive binding constraints on these two markets, which is impossible since the firm's activity is bound to satisfy $Y=F(L)$. It would reappear if the firm were allowed to invest or to store part of its output.

4.2.1. *Keynesian unemployment*

This regime is characterised by an excess supply on both markets. Therefore, the firm perceives a binding constraint $Y \leqq \bar{Y}$ on his output, where $\bar{Y}$ is equal to aggregate effective demand $C+G$. On the other hand, the consumer perceives a binding constraint on his labour supply $L \leqq \bar{L}$, where $\bar{L}$ is equal to the firm's labour demand and thus to $F^{-1}(\bar{Y})$.

Accordingly, output Y and employment L will be equal to $\bar{Y}$ and $\bar{L}$, respectively, and they are linked by $Y=F(L)$. Aggregate effective demand is then given by $\alpha_1(M_0/p+Y)+G$ and must equal output, which implies

$$\text{(a)}\quad Y=\frac{\alpha_1}{\alpha_2}\frac{M_0}{p}+\frac{G}{\alpha_2} \quad \text{and} \quad L=F^{-1}(Y).$$

One must, of course, check that the resulting constraints $Y \leqq Y$ and $L \leqq \bar{L}$ are indeed binding, which implies some restrictions on (p,w). For the consumer, one must verify that $\bar{L} \leqq L_0$, or $F(\bar{L}) \leqq F(L_0)$, which can be written $(\alpha_1/\alpha_2)(M_0/p)+G/\alpha_2 \leqq F(L_0)$. This is equivalent to $M_0/p \leqq M_0/p^*$. As for the firm, one must check that $\bar{Y}$ is less than the output level which would result from the free maximisation of the firm's profit at the quoted p and w, and which is implicitly given by $w/p=F'(L)$ and $Y=F(L)$. If Φ is the function which associates to any level of output the marginal productivity of the labour necessary to produce it $(\Phi(y)=F'(F^{-1}(y)))$, this is equivalent to

$$\frac{w}{p} \leqq \Phi\left(\frac{\alpha_1}{\alpha_2}\frac{M_0}{p}+\frac{G}{\alpha_2}\right).$$

One easily verifies that $\Phi'<0$.

To sum up, if one considers a level of initial real money stock M_0/p and a real wage w/p such that

$$\frac{M_0}{p} \leqq \frac{M_0}{p^*} \quad \text{and} \quad \frac{w}{p} \leqq \Phi\left(\frac{\alpha_1}{\alpha_2}\frac{M_0}{p}+\frac{G}{\alpha_2}\right),$$

one can associate with them a Keynesian unemployment equilibrium where output and employment are given by (a). This region of the plane $(M_0/p, w/p)$ is limited by two curves which go through the point $(M_0/p^*, w^*/p^*)$ corresponding to the competitive equilibrium.

The formulae (a) allow an immediate study of the impact of a few economic policies. First, one verifies the traditional theory of the Keynesian multiplier, since an increase ΔG in public consumption implies an increase in output of $\Delta G/\alpha_2 > \Delta G$. Secondly, a decrease in p stimulates output and employment. On the other hand, a change in the money wage has no effect. This stems from the assumption made in this simple model that profit is instantaneously distributed to the consumer. Accordingly, a change in w influences the wage bill–profit ratio but has no impact on total income, hence on aggregate effective demand. In a more complex model, where for instance profit would be retained by the firm in the form of money for distribution in further periods, a change in w would influence the output level and employment [see, for example, Malinvaud (1977)], who finds in a similar example that an increase in w stimulates economic activity by increasing aggregate demand].

4.2.2. Classical unemployment

This regime is characterised by an excess demand on the good's market and by an excess supply for labour. The consumer thus perceives two binding constraints: $C \leqq \bar{C}$ and $L \leqq \bar{L}$. By contrast, the firm does not perceive any constraint. Therefore, output Y and employment L are given by

$$\text{(b)} \quad \frac{w}{p} = F'(L) \quad \text{and} \quad Y = F(L).$$

Private consumption C, which is equal to $\bar{C}$, is then given by $Y - G$. Since C must be non-negative, this imposes the condition $w/p \leqq \Phi(G)$, which I shall assume. Finally, $L = \bar{L}$.

It remains to verify that the resulting constraints, $C \leqq \bar{C}$ and $L \leqq \bar{L}$, are indeed binding. In the case of labour one must check that $\bar{L} \leqq L_0$, which is the case when $w/p \geqq F'(L_0)$, or equivalently $w/p \geqq w^*/p^*$. The constraint $C \leqq \bar{C}$ is binding if aggregate effective demand $\alpha_1(M_0/p + Y) + G$ exceeds actual output Y, which can be written

$$\frac{w}{p} \geqq \Phi\left(\frac{\alpha_1}{\alpha_2}\frac{M_0}{p} + \frac{G}{\alpha_2}\right).$$

To sum up, to any point of the plane $(M_0/p, w/p)$ in the region limited by the inequalities

$$\frac{w}{p} \geqq \Phi\left(\frac{\alpha_1}{\alpha_2}\frac{M_0}{p} + \frac{G}{\alpha_2}\right) \quad \text{and} \quad \frac{w}{p} \geqq \frac{w^*}{p^*},$$

one can associate a classical unemployment equilibrium where output and employment are given by (b). These formulae make clear the economic policy advocated by classical economists in order to restore full employment, since only a decrease in the real wage can increase the output level. On the contrary, an increase in public spending has no effect on output and employment, decreasing private consumption by an equal amount. This kind of ıeasoning was often made by classical economists, too.

4.2.3. *Repressed inflation*

In this regime there is an excess demand on both markets. Thus, the consumer perceives a binding constraint $C \leqq C$ and the firm finds his labour demand constrained by $L \leqq \bar{L}$. Since $\bar{L}$ describes the consumer's effective labour supply, employment L $(=\bar{L})$ must be equal to L_0, output Y to $F(L_0)$ and private consumption C $(=\bar{C})$ to $F(L_0)-G$.

One must here again express the fact that the resulting perceived constraints are indeed binding. The constraint $L \leqq \bar{L}$ will be binding if $w/p \leqq F'(L_0)$ or, equivalently, $w/p \leqq w^*/p^*$. On the other hand, the constraint $C \leqq \bar{C}$ is binding if aggregate effective demand $\alpha_1(M_0/p+F(L_0))+G$ exceeds actual output $F(L_0)$, which is expressed by the inequality $M_0/p \geqq M_0/p^*$. Accordingly, at any point of the plane $(M_0/p, w/p)$ such that $w/p \leqq w^*/p^*$ and $M_0/p \geqq M_0/p^*$, there is a regime of repressed inflation, where employment is equal to L_0 and output to $F(L_0)$. In this region an increase in public spending has no effect on output but depresses private consumption by the same amount.

To sum up, if one forgets the constraints $w/p \leqq \Phi(G)$, which appeared in the study of classical unemployment, one finds that the plane $(M_0/p, w/p)$ is divided in three regions, and that in each region one and only one regime will prevail. The properties of each regime are very different. In particular, economic policies aimed at increasing output and employment in the case of Keynesian unemployment and of classical unemployment are of a very different nature, if not contradictory.

The study of this simple example illustrates the power of the concept of equilibrium with quantity rationing since it permits the unification within the same conceptual framework of theories of employment (Keynesian, classical) which appeared as fundamentally distinct beforehand.

However, although this approach seems quite promising at the conceptual level, it would be wise not to generalise hastily the conclusions which are obtained from the study of specific analytical examples similar to the one analysed in this section. Such is the case, in particular, for the impact of economic policies upon employment, such as an increase in public spending or a change in the nominal wage. For instance, if one can show that in some cases an increase in the nominal wage stimulated economic activity in a regime of Keynesian unemployment [Malinvaud (1977)], other equally plausible specifications of the same type of model can lead to the reverse result [K. Hildenbrand and W. Hildenbrand (1978)], and thus destroy one of the appealing contrasts between Keynesian and classical unemployment.

These remarks show the need for pursuing theoretical as well as empirical research in this area before this type of model becomes operational. A possible line of research would be to take into account more explicitly the intertemporal character of the agents' economic activities and therefore the essential role of expectations for the determination of employment [for a first step in this direction, see Grandmont and Laroque (1976b)]. Another improvement of the model would be to include a more realistic monetary sector. A first step could be to take into account a simple banking system, as in Section 3.3.

The crucial weakness of this type of model appears, however, to be the absence of a satisfactory theory of price and wage formation. A number of studies in this area postulate that prices move from one period to another in a mechanistic way in response to observed excess effective demand. This viewpoint appears unsatisfactory on several grounds. First, it cannot lead to a true theory of price formation, since it reintroduces a *deus ex machina* which is responsible for their determination and which is similar to the Walrasian auctioneer. Secondly, the stationary states of the resulting dynamic system cannot display unemployment. If one wishes to explain the occurrence of a persistent disequilibrium with such a model one is compelled to make use of arguments assuming more or less the instability of the system, arguments which appear at present belong to the domain of speculation.

An alternative approach, which seems promising but also more difficult, is to combine the fixprice method with an assumption of imperfect competition [Benassy (1976a), Grandmont and Laroque (1976b), Hahn (1976a, 1978), Iwai (1974a, 1974b), and Negishi (1978)]. One can then hope to obtain a satisfactory explanation of the existence of a lasting unemployment, or of such relations as the so-called Phillips curve. This approach raises many difficult problems which are still open (the treatment of imperfect competition being one of the weak points of economic theory), and it requires a major research effort.

4.3. Bibliographical note

The first formal studies of the existence of a general equilibrium with quantity rationing are due to Benassy (1973, 1975a), Drèze (1975), Glustoff (1968) and Younès (1970a, 1970b) [see also Braverman (1972)]. The presentation made in Section 4.1 is adapted from Drèze, who considered the more general case where money prices can vary within some a priori given bounds, and from Benassy. Younès analyses the efficiency properties of this type of equilibrium and links them to the role of money in the exchange process, an issue which was emphasised by Clower and Leijonhufvud. This problem was also studied by Benassy (1975a), Malinvaud and Younès (1974, 1975). An analysis of this type of equilibrium by means of cooperative game theory can be found in Grandmont, Laroque and Younès (1978). A study of this equilibrium concept in terms of a Nash equilibrium was made by Böhm and Lévine (1976) and Heller and Starr (1976). Stochastic rationing is considered by Green (1977) and Svensson (1977).

There are a number of applications of this method to aggregated models, as in Section 4.2. The first ones appear to be due to Barro and Grossman (1971) and to Younès (1970b). For a systematic presentation of this topic, the reader is referred to Barro and Grossman (1976), Haga (1976), Malinvaud (1977), and Muellbauer and Portes (1978).

A few attempts aiming at including imperfect competition in this type of model in a general equilibrium framework have been made by Benassy (1976a), Grandmont and Laroque (1976b), and Hahn (1977a, 1978).

References

Arrow, K. and J. R. Green (1973), "Notes on Expectations Equilibria in Bayesian Settings", Working Paper, Department of Economics, Stanford University, 1973.

Arrow, K., and F. H. Hahn (1971), General competitive analysis. San Francisco: Holden Day.

Balch, M., D. McFadden, and S. W. Yu (eds.) (1974), Essays on economic behavior under uncertainty. Contributions to Economic Analysis. Amsterdam: North-Holland.

Barro, R. J. and H. I. Grossman (1971), "A general disequilibrium model of income and employment", American Economic Review, 61:82–93.

Barro, R. J. and H. I. Grossman (1974), "Suppressed inflation and the supply multiplier", Review of Economic Studies, 41:87–104.

Barro, R. J. and H. I. Grossman (1976), Money, employment and inflation. Cambridge: Cambridge University Press.

Benassy, J. P. (1973), Disequilibrium theory. Unpublished Ph.D. Dissertation, University of California, Berkeley.

Benassy, J. P. (1975a), "Neo-Keynesian disequilibrium in a monetary economy", Review of Economic Studies, 42:503–523.

Benassy, J. P. (1975b), "Disequilibrium exchange in barter and monetary economies", Economic Inquiry, 13:131–156.

Benassy, J. P. (1976a), "The disequilibrium approach to monopolistic price setting and general monopolistic equilibrium", Review of Economic Studies, 43:69–81.

Benassy, J. P. (1976b), "Regulation of the wage profits conflict and the unemployment inflation dilemma in a dynamic disequilibrium model", Economie Appliquée, 29:409–444.

Benassy, J. P. (1977), "On quantity signals and the foundations of effective demand theory", The Scandinavian Journal of Economics, 1977:147–168.

Benassy, J. P. (1978), "A neokeynesian model of price and quantity determination in disequilibrium", in: G. Schwödiauer, ed., Equilibrium and Disequilibrium in Economy Theory. Boston: D. Reidel Pub. Co.

Billingsley, P. (1968), Convergence of probability measures. New York: Wiley.

Bliss, C. P. (1974), "Capital theory in the short run", Department of Economics, University of Essex.

Bohm, V. (1978), "Disequilibrium dynamics in a simple macroeconomic model", Journal of Economic Theory, 17:179–199.

Bohm, V. and J. P. Levine (1976), "Temporary equilibria with quantity rationing", CORE Discussion Paper 7614, Catholic University of Louvain, Belgium.

Braverman, E. M. (1972), "A production model with disequilibrium prices", Ekonomika i Matematicheskia Metody, 2:41–64.

Chetty, V. K. and D. Dasgupta (1978): "Temporary competitive equilibrium in a monetary economy with uncertain technology and many planning periods", Journal of Mathematical Economics, 5:23–42.

Christiansen, D. S. (1974), Some aspects of the theory of short run equilibrium. Unpublished Ph.D. Dissertation, Department of Economics, Stanford University.

Christiansen, D. S. (1975), "Temporary equilibrium: A stochastic dynamic programming approach", Discussion Paper, Department of Economics, University of Rochester.
Christiansen, D. S. and M. Majumdar (1977), "On shifting temporary equilibrium", Journal of Economic Theory, 16:1–9.
Clower, R. W. (1965), "The Keynesian counterrevolution: A theoretical appraisal", in: F. H. Hahn and Brechling, eds., The theory of interest rates. London: Macmillan.
Clower, R. W. (1967), "A reconsideration of the microfoundations of monetary theory", Western Economic Journal, 6:1–9.
Debreu, G. (1956), "Market-equilibrium", Proceedings of the National Academy of Sciences of the U.S.A., 42:876–878.
Debreu, G. (1959), Theory of value. New York.
Dehez, P. and J. Gabszewicz (1977), "On disequilibrium savings and public consumption", CORE Discussion Paper, Catholic University of Louvain.
Delbaen, F. (1974), "Continuity of expected utility", in: J. Drèze, ed., Allocation under uncertainty, equilibrium and optimality. London: Macmillan.
Diewert, W. E. (1978), "Walras' theory of capital formation and the existence of a temporary equilibrium", in: G. Schwödiauer, ed., Equilibrium and disequilibrium in economic theory. Boston: D. Reidel Pub. Co.
Dixit, A. (1976), "Public finance in a temporary Keynesian equilibrium model", Journal of Economic Theory, 12:242–258.
Dixit, A. (1978), "The balance in a model of temporary equilibrium with rationing", Review of Economic Studies, 45:393–404
Drandakis, E. M. (1966), "On the competitive equilibrium in a monetary economy", International Economic Review, 7:304–328.
Drèze, J. (1974a), "Investment under private ownership: Optimality, equilibrium and stability", in: J. Drèze, ed., Allocation under uncertainty, equilibrium and optimality, Proceedings of an I.E.A. Workshop on Economic Theory, Bergen, Norway. London: Macmillan.
Drèze, J. (1974b), Allocation under uncertainty, equilibrium and optimality, Proceedings of an I.E.A. Workshop in Economic Theory, Bergen, Norway, 1971. London: Macmillan.
Drèze, J. (1975), "Existence of an equilibrium under price rigidity and quantity rationing", International Economic Review, 16:301–320.
Fitzroy, F. R. (1973), "A framework for temporary equilibrium", Alfred Weber Institute, University of Heidelberg.
Friedman, M. (1969), The optimum quantity of money and other essays. Chicago: Aldine.
Fuchs, G. (1976), "Asymptotic stability of stationary temporary equilibria and changes in expectations", Journal of Economic Theory, 12:201–216.
Fuchs, G. (1977a), "Dynamic role and evolution of expectations", in: Systèmes dynamiques et modèles economiques, Proceedings of a C.N.R.S. International Meeting, Vol. 259, 183.
Fuchs, G. (1977b), "Formation of expectations. A model in temporary general equilibrium theory", Journal of Mathematical Economics, 4:167–188.
Fuchs, G. (1979a), "Dynamics of expectations in temporary general equilibrium theory", Journal of Mathematical Economics, 6:229–252.
Fuchs, G. (1979b), "Are error learning behaviours stabilizing?", Journal of Economic Theory, 3:300–317.
Fuchs, G., and G. Laroque (1976), "Dynamics of temporary equilibria and expectations", Econometrica, 44:1157–1178.
Futia, C. A. (1977), "Excess supply equilibria", Journal of Economic Theory, 14:200–220.
Gale, Douglas (1975), "Rational expectations and the rate of return", Christ's College, Cambridge, England.
Gale, Douglas (1976a), "New concepts of equilibrium for economics with bankruptcy", Working Paper, Churchill College, Cambridge.
Gale, Douglas (1976b), "Bankruptcy, rationing and sequences of incomplete markets: Some survey notes", Working Paper, Churchill, Cambridge.
Gale, Douglas (1976c), "Economies with trading uncertainty", CRMS Working Paper, University of California, Berkeley.

Gepts, S. J. (1977), "An aggregate model of rationing", CORE Discussion Paper 7704, Catholic University of Louvain.

Glustoff, E. (1968), "On the existence of a Keynesian equilibrium", Review of Economic Studies, 35:327–334.

Grandmont, J. M. (1970), On the temporary competitive equilibrium. Unpublished Ph.D Dissertation, CRMS Working Paper, University of California, Berkeley.

Grandmont, J. M. (1972), "Continuity properties of a von Neumann–Morgenstern utility", Journal of Economic Theory, 4:45–57.

Grandmont, J. M. (1974), "On the short run equilibrium in a monetary economy", in: J. Drèze, ed., Allocation under uncertainty, equilibrium, and optimality. London: Macmillan.

Grandmont, J. M. (1977), "Temporary general equilibrium theory", Econometrica, 45:535–572.

Grandmont, J. M. and W. Hildenbrand (1976), "Stochastic processes of temporary equilibria", Journal of Mathematical Economics, 1:247–277.

Grandmont, J. M. and A. P. Kirman (1973), "Foreign exchange markets: A temporary equilibrium approach", CORE Discussion Paper, 7308.

Grandmont, J. M. and G. Laroque (1973), "Money in the pure consumption loan model", Journal of Economic Theory, 6:382–395.

Grandmont, J. M. and G. Laroque (1975), "On money and banking", Review of Economic Studies, 42:207–236.

Grandmont, J. M. and G. Laroque (1976a), "On the liquidity trap", Econometrica, 44:129–135.

Grandmont, J. M. and G. Laroque (1976b), "On Keynesian temporary equilibria", Review of Economic Studies, 43:53–67.

Grandmont, J. M., G. Laroque, and Y. Younès (1978), "Equilibrium with quantity rationing and recontracting", Journal of Economic Theory, 19:84–102.

Grandmont, J. M. and Y. Younès (1972), "On the role of money and the existence of a monetary equilibrium", Review of Economic Studies, 39:355–372.

Grandmont, J. M. and Y. Younès (1973), "On the efficiency of a monetary equilibrium", Review of Economic Studies, 40:149–165.

Green, J. R. (1971), "A simple general equilibrium model of the term structure of interest rates", Harvard Discussion Paper 183, Harvard University.

Green, J. R. (1973), "Temporary general equilibrium in a sequential trading model with spot and future transactions", Econometrica, 41:1103–1123.

Green, J. R. (1974), "Preexisting contracts and temporary general equilibrium", in: M. Balch, D. MacFadden, and S. W. Yu, eds., Essays on Economic Behavior under Uncertainty, Contributions to Economic Analysis. Amsterdam: North-Holland.

Green, J. R. (1977), "On the theory of effective demand", Working Paper, Harvard University.

Greenberg, J. and H. Muller (1977), "Equilibria under price rigidities and externalities", CORE Discussion Paper 7720, Catholic University of Louvain.

Grossman, H. I. (1971), "Money, interest and prices in market disequilibrium", Journal of Political Economy, 79:943–961.

Grossman, H. I. (1972), "A choice-theoretic model of an income investment accelerator", American Economic Review, 62:630–641.

Haga, H. (1976), A disequilibrium-equilibrium model with money and bonds, Lecture Notes in Economics and Mathematical Systems. Berlin: Springer-Verlag.

Hahn, F. H. (1965), "On some problems of proving the existence of an equilibrium in a monetary economy", in: F. H. Hahn and Brechling, eds., The theory of interest rates. London: Macmillan.

Hahn, F. H. (1971), "Equilibrium with transaction costs", Econometrica, 39:417–439.

Hahn, F. H. (1973), "On transaction costs, inessential sequence economies and money", Review of Economic Studies, 40:449–461.

Hahn, F. H. (1973), On the notion of equilibrium in economics. Cambridge: Cambridge University Press.

Hahn, F. H. (1977a), "Exercises in conjectural equilibria", The Scandinavian Journal of Economics, 1977:210–226.

Hahn, F. H. (1977b), "Unsatisfactory equilibria", mimeo. Churchill College, Cambridge.

Hahn, F. H. (1978), "On non-Walrasian equilibria", Review of Economic Studies, 45:1–17.

Hahn, F. H. and Brechling (eds.) (1965), The theory of interest rates. London: Macmillan.
Hahn, F. H., and T. Negishi (1962), "A theorem on non tatonnement stability", Econometrica, 30:463–469.
Hammond, P. (1977), "General asset markets and the existence of temporary Walrasian equilibrium", Working Paper, University of Essex.
Hart, O. D. (1974), "On the existence of equilibrium in a securities model", Journal of Economic Theory, 9:293–311.
Heller, W. P. and R. M. Starr (1976), "Unemployment equilibrium with rational expectations", IMSSS Working Paper 66, Stanford University.
Helpman, E. and J. J. Laffont (1975), "On moral hazard in general equilibrium theory", Journal of Economic Theory, 10:8–23.
Hicks, J. (1946), Value and capital, 2nd edn. Oxford: Clarendon Press.
Hicks, J. (1965), Capital and growth. Oxford: Oxford University Press.
Hildenbrand, W. (1971), "Random preferences and equilibrium analysis", Journal of Economic Theory, 3:414–429.
Hildenbrand, K. and W. Hildenbrand, (1978), "On Keynesian equilibria with unemployment and quantity rationing", Journal of Economic Theory, 18:255–277.
Honkapohja, S. (1977), "Competitive equilibria, Pareto efficiency and the core in a monetary sequence economy with transaction costs", Department of Economics, University of Helsinki.
Honkapohja, S. (1979), "On the dynamics of disequilibria in a macro model of flexible wages and prices", in New Trends in Dynamic System Theory and Economics, M. Aoki and A. Marzollo, (eds.), Academic Press.
Hool, B. (1974a), Money, financial assets and general equilibrium in a sequential market economy. Unpublished Ph.D. Dissertation, Department of Economics, University of California, Berkeley.
Hool, B. (1974b), "Temporary Walrasian equilibrium in a monetary economy", Social Systems Research Institute, University of Wisconsin, Madison.
Hool, B. (1975), "Temporal general equilibrium with money and financial assets", SSRI, University of Wisconsin, Madison.
Hool, B. (1976), "Money, expectations and the existence of a temporary equilibrium", Review of Economic Studies, 43:439–445.
Hool, B. (1979), "Liquidity, speculation and the demand for money", Journal of Economic Theory, 21:73–87.
Hool, B. (1980), "Monetary and fiscal policies in equilibria with rationing", International Economic Review, 21:301–317.
Iwai, K. (1974a), "Towards Keynesian microdynamics of price, wage, sales and employment", Cowles Discussion Paper, Yale University.
Iwai, K. (1974b), "The firm in uncertain markets and its price, wage and employment adjustments", Review of Economic Studies, 41:257–276.
Jordan, J. S. (1976), "Temporary competitive equilibrium and the existence of self-fulfilling expectations", Journal of Economic Theory, 12:455–471.
Jordan, J. S. (1977), "The continuity of optimal dynamic decision rules", Econometrica, 45:1365–1376.
Keynes, J. M. (1936), The general theory of money, interest and employment.
Kurz, M. (1974), "Equilibrium in a finite sequence of markets with transaction cost", Econometrica, 42:1–20.
Laffont, J. J. (1975), "Optimism and experts versus adverse selection in a competitive economy", Journal of Economic Theory, 10:284–308.
Laroque, G. (1976), "The fixed price equilibria: Some results in local comparative statics", IMSSS Technical Report 223, Stanford University.
Laroque, G. and H. Polemarchakis (1978), "On the structure of the set of fixed price equilibria", Journal of Mathematical Economics, 5:53–70.
Ledyard, J. O. (1974), "On sequences of temporary equilibria", in: M. Balch, D. McFadden and S. W. Yu, eds., Essays on economic behavior under uncertainty, Contributions to Economic Analysis. Amsterdam: North-Holland.
Leijonhufvud, A. (1968), On Keynesian economics and the economics of Keynes. Oxford: Oxford University Press. 1968.

Malinvaud, E. (1977), The theory of unemployment reconsidered. Oxford: Basil Blackwell.
Malinvaud, E. and Y. Younès: "Une nouvelle formulation générale pour l'etude des fondements microéconomiques de la macroéconomie", INSEE and CEPREMAP, Paris.
Malinvaud, E. and Y. Younès (1975), "A New formulation for the microeconomic foundations of macroeconomics", Paper presented at an IEA Conference on the Microeconomic Foundations of Macroeconomics, S'Agaro, Spain, April 1975.
Morishima, M. (1964), Equilibrium, stability and growth. Oxford: Clarendon Press.
Muellbauer, J. and R. Portes (1978), "Macroeconomic models with quantity rationing", Economic Journal, 88:788–821.
Muller, H. and U. Schweizer (1978), "Temporary equilibrium in a money economy", Journal of Economic Theory, 19:267–286.
Negishi, T. (1978), "Existence of an under employment equilibrium", in: G. Schwödiauer, ed., Equilibrium and disequilibrium economic theory. Boston: D. Reidel Pub. Co.
Paquin, L. T. (1976), "A monetary theory of value based on transaction costs", Working paper 255, Institute for Economic Research, Queen's University, Kingston, Ontario, Canada.
Paquin, L. T. (1977), "A monetary theory of value", Working Paper 271, Institute for Economic Research, Queen's University, Kingston, Ontario, Canada.
Parthasarathy, K. (1967), Probability measures on metric spaces. New York: Academic Press.
Patinkin, D. (1965), Money, interest and prices, 2nd edn. New York: Harper and Row.
Portes, R. (1976), "Macroeconomic equilibrium and disequilibrium in centrally planned economies", Discussion Paper 45, Department of Economics, Birbeck College, London.
Radner, R. (1967), "Equilibre des marchés à terme et au comptant en cas d'incertitude", in: Cahiers du Séminaire d'Econométrie. Paris: CNRS.
Radner, R. (1968), "Competitive equilibrium under uncertainty", Econometrica, 36:31–58.
Radner, R. (1970), "Problems in the theory of markets under uncertainty", American Economic Review, 60:454–460.
Radner, R. (1972), "Existence of equilibrium plans, prices and price expectations in a sequence of markets", Econometrica, 40:289–303.
Radner, R. (1974), "Market equilibrium under uncertainty: Concepts and problems", in: M. D. Intriligator and D. A. Hendrick, eds., Frontiers of quantitative analysis, Vol. II. Amsterdam: North-Holland.
Samuelson, P. A. (1958), "An exact consumption loan model of interest with or without the social contrivance of money", Journal of Political Economy, 66:467–482.
Schwödiauer, G. (ed.) (1978), Equilibrium and disequilibrium in economic theory. Boston: D. Reidel Pub. Co.
Sondermann, D. (1974), "Temporary competitive equilibrium under uncertainty", in: J. Drèze, ed., Allocation under uncertainty, equilibrium, and optimality. London: Macmillan.
Starr, R. M. (1969), "Quasi equilibrium in markets with non convex preferences", Econometrica, 37:25–38.
Starrett, D. (1973), "Inefficiency and the demand for money in a sequence economy", Review of Economic Studies, 40:437–448.
Stigum, G. (1969a), "Entrepreneurial choice over time under conditions of uncertainty", International Economic Review, 10:426–442.
Stigum, B. (1969b), "Competitive equilibria under uncertainty", Quarterly Journal of Economics, 83:533–561.
Stigum, B. (1972), "Resources allocation under uncertainty", International Economic Review, 13:431–459.
Stigum, B. (1974), "Competitive resource allocation over time under uncertainty", in: M. Balch, D. McFadden and S. W. Yu, eds., Essays on economic behavior under uncertainty, Contributions to Economic Analysis. Amsterdam: North-Holland.
Svensson, L. E. O. (1976), "Sequences of temporary equilibria, stationary point expectations, and Pareto efficiency", Journal of Economic Theory, 13:169–183.
Svensson, L. E. O. (1976), "On competitive markets and intertemporal allocations", Institute for International Economic Studies, Stockholm.
Svensson, L. E. O. (1977), "Effective demand on stochastic rationing", Seminar Paper 76, Institute for

International Economic Studies, Stockholm.
Varian, H. R. (1976), "On persistent disequilibrium", Journal of Economic Theory, 10:218–227.
Varian, H. R. (1977), "Non-Walrasian equilibria", Econometrica, 45:573–590.
Younès, Y. (1970a), "Sur une notion d'equilibre utilisable dans le cas où les agents economiques ne sont pas assurés de la compatibilité de leurs plans", CEPREMAP, Paris (revised, 1973).
Younès, Y. (1970b), "Sur les notions d'equilibre et de déséquilibre utilisées dans les modèles décrivant l'evolution d'une economie capitaliste", CEPREMAP, Paris.
Younès, Y. (1972a), "Intérêt et monnaie externe dans une economie d'echanges au comptant en equilibre Walrasien de court terme", CEPREMAP, Paris (English version, 1973).
Younès, Y. (1972b), "Monnaie et motif de précaution dans une economie d'echanges oú les ressources des agents sont aléatoires", CEPREMAP, Paris.
Younès, Y. (1975), "On the role of money in the process of exchange and the existence of a non-Walrasian equilibrium", Review of Economic Studies, 42:489–501.

Chapter 20

EQUILIBRIUM UNDER UNCERTAINTY*

ROY RADNER

Bell Laboratories

1. Introduction

1.1. Equilibrium and uncertainty

In chapter X of *Value and Capital*, Hicks distinguished two senses of equilibrium: (1) temporary equilibrium, in which at a given date supply equals demand, and (2) equilibrium over time, which he defined by the "condition that prices realized on (each date) are the same as those which were previously expected to rule at that date". Hicks also considered the condition of stationarity (i.e., constancy of prices and quantities through time) as a candidate for an equilibrium concept, but he rejected it as not being so fundamental as the condition of equilibrium over time. Thus he wrote: "A stationary state is in full equilibrium, not merely when demands equal supplies at the currently established prices, but also when the same prices continue to rule at all dates – when prices are constant over time." He considered the test of equilibrium over time to be "more important than mere arithmetical sameness or difference", noting that it "does imply constant prices in a stationary economy, but does not necessarily imply constant prices in an economy subject to change".

Unfortunately, Hicks did not face squarely the problems and issues associated with uncertainty, preferring instead to utilize the approximation of certainty equivalents and point expectations. As he put it: "This is not an absolutely satisfactory way of dealing with risk – I feel myself that there ought to be an Economics of Risk on beyond the Dynamic Economics we shall work out here – but it does suffice to show that the investigations we are about to make are not devoid of applicability."

*This paper is based on research supported in part by the National Science Foundation, and represents a substantial revision and augmentation of chapter 2 in Intriligator and Kendrick (1974). I am grateful for the comments of Mr. Nirvikar Singh, who read a preliminary draft. The views expressed here are those of the author, and should not be interpreted as the views of Bell Laboratories.

Handbook of Mathematical Economics, vol. II, edited by K.J. Arrow and M.D. Intriligator

The recent spurt of interest in sequences of markets under uncertainty suggests that a number of economic theorists are prepared to take up the subject again where Hicks left off. The Arrow-Debreu model (see below) gives us a solid foundation of equilibrium analysis upon which to build. Modern decision theory and probability theory enable us to deal more effectively with the issues raised by uncertainty, and with the conceptual problems of describing an economy that is both "stationary" and "subject to change". In this chapter I shall review several concepts of equilibrium that have appeared in the recent literature on markets and uncertainty. My goal is not so much to survey this literature as to present the various concepts in a unified framework. I shall also indicate a number of unsolved problems. Although I have not attempted to summarize the existing literature, the reader will find bibliographic notes at the end of the chapter.

I have limited the scope of my exposition in several ways. I have primarily (but not entirely) restricted my consideration to models of general equilibrium. I have not dealt with the literature on (1) individual behavior in the face of uncertainty, (2) the "economics of information", or (3) the dynamics of temporary or momentary equilibrium under *certainty*. The main focus will be on various theories of equilibrium under uncertainty in a sequence of markets, in the absence of complete forward markets for contingent delivery.

1.2. Plan of the chapter

The remainder of Section 1 will be devoted to (1) an informal review of the Arrow–Debreu model of competitive equilibrium under uncertainty, and its extension to the case in which different economic agents have different information, (2) a brief summary of the limitations of that model, and (3) an informal introduction to concepts of equilibrium in a sequence of markets. After the basic theoretical framework for studying sequences of markets under uncertainty is introduced in Section 2, topics (1) and (2) are reviewed more formally in Section 3, together with the notion of temporary equilibrium.

Many of the conceptual and theoretical problems of equilibrium in a sequence of markets arise even in the case of pure exchange. Accordingly, for that special case Section 4 presents a systematic account of equilibrium of plans and expectations in a sequence of markets, and an introduction to the theory of rational expectations equilibrium. The additional problems that arise when one considers production are more briefly sketched in Section 5.

I think it would be generally agreed that, even for the case of certainty, the theory of adjustment and stability of competitive markets is in a less satisfactory state than are the theories of existence and optimality of general equilibrium. The situation is no better for the case of uncertainty, where it is natural to pose the questions in terms of the convergence of a stochastic process to a stationary process. Nevertheless, the three examples described in Section 6 provide some interesting insights into the nature of stability under uncertainty, as well as encouraging signs of progress in this difficult area.

Bibliographical notes are gathered in Section 7, which supplements the references that are scattered throughout the body of the chapter. To supplement this incomplete bibliography, and to complement the limited treatment of uncertainty in this chapter, the reader should consult Chapters 6, 13, 19, 23, and 27 of this Handbook, and also the excellent recent review by Hirshleifer and Riley (1979).

1.3. *Informal review of the Arrow–Debreu model of a complete market for present and future contingent delivery*

One of the notable intellectual achievements of economic theory during the past 20 years has been the rigorous elaboration of the Walras–Pareto theory of value, i.e. the theory of the existence and optimality of competitive equilibrium. Although many economists and mathematicians contributed to this development, the resulting edifice owes so much to the pioneering and influential work of Arrow and Debreu that in this chapter I shall refer to it as the "Arrow–Debreu theory". [For a comprehensive treatment, together with references to previous work, see Debreu (1959), and Arrow and Hahn (1971).]

The Arrow–Debreu theory was not originally put forward for the case of uncertainty, but an ingenious device introduced by Arrow (1953), and further elaborated by Debreu (1953), enabled the theory to be reinterpreted to cover the case of uncertainty about the availability of resources and about consumption and production possibilities. [See Debreu (1959, ch. 7) for a unified treatment of time and uncertainty.]

The basic idea is that commodities are to be distinguished, not only by their physical characteristics and by the location and dates of their availability and/or use, but also by the *environmental event* in which they are made available and/or used. For example, ice cream made available (at a particular location on a particular date) if the weather is hot may be considered to be a different commodity from the same kind of ice cream made available (at the same location and date) if the weather is cold. We are thus led to consider a list of "commodities" that is greatly expanded by comparison with the corresponding case of certainty about the environment. The standard arguments of the theory of competitive equilibrium, applied to an economy with this expanded list of "commodities", then require that we envisage a "price" for each "commodity" in the list, or, more precisely, a set of price ratios specifying the rate of exchange between each pair of "commodities".

Just what institutions could, or do, effect such exchanges is a matter of interpretation that is, strictly speaking, outside the model. I shall present one straightforward interpretation, and then comment briefly on an alternative interpretation.

First, however, it will be useful to give a more precise account of the concepts of *environment* and *event* that I shall be employing. (A mathematical model is presented in Section 2.) The description of the "physical world" is decomposed into three sets of variables: (1) decision variables, which are controlled (chosen) by economic agents; (2) environmental variables, which are not controlled by any economic agent; and (3) all other variables, which are completely determined (possibly jointly) by decisions and environmental variables. A *state* of the environment is a complete specification (history) of the environmental variables from the beginning to the end of the economic system in question. An *event* is a set of states; for example, the event "the weather is hot in New York on 1, July 1970" is the set of all possible histories of the environment in which the temperature in New York during the day of 1, July 1970 reaches a high of at least (say) 75°F. Granting that we cannot know the future with certainty, at any given date there will be a family of *elementary observable* (knowable) events, which can be represented by a partition of the set of all possible states (histories) into a family of mutually exclusive subsets. It is natural to assume that the partitions corresponding to successive dates are successively finer, which represents the accumulation of information about the environment.

We shall imagine that a "market" is organized before the beginning of the physical history of the economic system. An elementary contract in this market will consist of the purchase (or sale) of some specified number of units of a specified commodity to be delivered at a specified location and date, if and only if a specified elementary event occurs. Payment for this purchase is to be made now (at the beginning), in "units of account", at a specified price quoted for the commodity-location-date-event combination. Delivery of the commodity in more than one elementary event is obtained by combining a suitable set of elementary contracts. For example, if delivery of one quart of ice cream (at a specified location and date) in hot weather costs $1.50 (now), and delivery of one quart in non-hot weather costs $1.10, then *sure* delivery of one quart (i.e., whatever be the weather) costs $1.50+$1.10=$2.60.

There are two groups of economic agents in the economy: *producers* and *consumers*. A producer chooses a *production plan*, which determines his input and/or output of each commodity at each date in each elementary event. (I shall henceforth suppress explicit reference to location, it being understood that the location is specified in the term "commodity".) For a given set of prices, the *present value* of a production plan is the sum of the values of the inputs minus the sum of the values of the outputs. Each producer is characterized by a set of production plans that are (physically) feasible for him – his *production possibility set*.

A consumer chooses a *consumption plan*, which specifies his consumption of each commodity at each date in each elementary event. Each consumer is characterized by (1) a set of consumption plans that are (physically, psychologically, etc.) feasible for him, his *consumption possibility set*; (2) *preferences* among the alternative plans that are feasible for him; (3) his *endowment of physical resources*, i.e. a specification of the quantity of each commodity, e.g., labor, at each date in each event with which he is exogenously endowed; and (4) his *shares* in producers' profits, i.e. a specification for each producer of the fraction of the present value of that producer's production plan that will be credited to the consumer's account. (For any one producer, the sum of the consumers' shares is unity.) For given prices and given production plans of all the producers, the *present net worth* of a consumer is the total value of his resources plus the total value of his shares of the present values of producer's production plans.

An *equilibrium* of the economy is a set of prices, a set of production plans (one for each producer), and a set of consumption plans (one for each consumer), such that (1) each producer's plan has maximum present value in his production possibility set; (2) each consumer's plan maximizes his preferences within his consumption possibility set, subject to the additional (budget) contraint that the present cost of his consumption plan does not exceed his present net worth; and (3) for each commodity at each date in each elementary event, the total demand equals the total supply, i.e. the total planned consumption equals the sum of the total resource endowments and the total planned net output (where inputs are counted as negative outputs).

Notice that (1) producers and consumers are "price takers"; (2) for given prices there is no uncertainty about the present value of a production plan or of given resource endowments, nor about the present cost of a consumption plan; (3) therefore, for given prices and given producers' plans, there is no uncertainty about a given consumer's present net worth; and (4) since a consumption plan may specify that, for a given commodity at a given date, the quantity consumed is to vary according to the event that actually occurs, a consumer's preferences among plans will reflect not only his "tastes", but also his subjective beliefs about the likelihoods of different events and his attitude toward risk [see, for example, Savage (1954)].

If follows that beliefs and attitudes toward risk play no role in the assumed behavior of producers. On the other hand, beliefs and attitudes toward risk do play a role in the assumed behavior of consumers, although for given prices and production plans each consumer knows his (single) budget constraint with certainty.

I shall call the model just described an "Arrow–Debreu" economy. One can demonstrate, under "standard conditions": (1) the existence of an equilibrium, (2) the Pareto-optimality of an equilibrium, and (3) that, roughly speaking, every Pareto-optimal choice of production and consumption plans is an equilibrium relative to some price system for some distribution of resource endowments and shares [see, for example, Debreu (1959 chs. 5 and 6) and Debreu (1962)].

In the above interpretation of the Arrow–Debreu economy, all accounts are settled before the history of the economy begins, and there is no incentive to revise plans, reopen the "market", or trade in shares. There is an alternative interpretation, which will be of interest in connection with the rest of this chapter, but which corresponds to exactly the same formal model. In this second interpretation there is a single commodity at each date – let us call it "gold" – that is taken as a numeraire at that date. A "price system" has two parts: (1) for each date and each elementary event at that date there is a price, to be paid in gold at date 1, for one unit of gold to be delivered at the specified date and event; (2) for each commodity, date, and event at that date, there is a price, to be paid in gold at that date and event. The first part of the price system can be interpreted as "insurance premiums", and the second part as "spot prices" at the given date and event. The "insurance" interpretation is to be made with some reservation, however, since there is no real object being insured, and no limit to the amount of insurance that an individual may take out against the occurrence of a given event. For this reason, the first part of the price system might be better interpreted as reflecting a combination of *betting odds* and *interest rates*.

Although the second part of the price system might be interpreted as "spot prices", it would be a mistake to think of the *determination of the equilibrium values* of these prices as being deferred in real time to the dates to which they refer. The definition of equilibrium requires that the agents have access to the complete system of prices when choosing their plans. In effect, this requires that at the beginning of time all agents have available a (common) forecast of the equilibrium spot prices that will prevail at every future date and event.

1.4. Extension of the Arrow–Debreu model to the case in which different agents have different information

In an Arrow–Debreu economy at any one date each agent will have incomplete information about the state of the environment, but all the agents will have the *same* information. This last assumption is not tenable if we are to take good account of the effects of uncertainty in an economy. I shall now sketch how, by a simple reinterpretation of the concepts of production possibility set and consumption possibility set, we can extend the theory of the Arrow–Debreu economy to allow for differences in information among the economic agents.

For each date the information that will be available to a given agent at that date may be characterized by a partition of the set of states of the environment. To be consistent with our previous terminology, we should assume that each such *information partition* must be at least as coarse as the partition that describes the elementary events at that date, i.e. each set in the information partition must contain a set in the elementary event partition for the same date.

For example, each set in the event partition at a given date might specify the high temperature at that date, whereas each set in a given agent's information partition might specify only whether this temperature was higher than 75°F or not. Or the event partition at a given date might specify the temperature at each date during the past month, whereas the information partition might specify only the mean temperature over the past month.

An agent's information restricts his set of feasible plans in the following manner. Suppose that at a given date the agent knows only that the state of the environment lies in a specified set A (one of the sets in his information partition at that date), and suppose (as would be typical) that the set A contains several of the elementary events that are in principle observable at that date. Then any action that the agent takes at that date must necessarily be the same for all elementary events in the set A. In particular, if the agent is a consumer, then his consumption of any specified commodity at that date must be the same in all elementary events contained in the information set A; if the agent is a producer, then his input or output of any specified commodity must be the same for all events in A. (I am assuming that consumers know what they consume, and producers what they produce, at any given date.)

Let us call the sequence of information partitions for a given agent his *information structure*, and let us say that this structure is *fixed* if it is given independently of the actions of himself or any other agent. Furthermore, in the case of a fixed information structure, let us say that a given plan (consumption or production) is *compatible* with that structure if it satisfies the conditions described in the previous paragraph at each date.

Suppose that the consumption and production possibility sets of the Arrow–Debreu economy are interpreted as characterizing, for each agent, those plans that would be feasible if he had "full information" (i.e. if his information partition at each date coincided with the elementary event partition at that date). The set of feasible plans for any agent with a fixed information structure can then be obtained by restricting him to those plans in the "full information" possibility set that are also compatible with his given information structure.

From this point on, all the machinery of the Arrow–Debreu economy (with some minor technical modifications) can be brought to bear on the present model. In particular, we obtain a theory of existence and optimality of competitive equilibrium *relative to a fixed structure of information for the economic agents*. I shall call this the "extended Arrow–Debreu economy".[1]

[1]The use of this terminology is not intended to imply that either Arrow or Debreu approve of this way of extending their model to incorporate informational differences among agents.

1.5. Choice of information

There is no difficulty in principle in incorporating the choice of information structure into the *model* of the extended Arrow–Debreu economy. I doubt, however, that it is reasonable to assume that the technological conditions for the acquisition and use of information generally satisfy the hypotheses of the standard theorems on the existence and optimality of competitive equilibrium.

The acquisition and use of information about the environment typically require the expenditure of goods and services, i.e. of commodities. If one production plan requires more information for its implementation than another (i.e. requires a finer information partition at one or more dates), then the list of (commodity) inputs should reflect the increased inputs for information. In this manner a set of feasible production plans can reflect the possibility of choice among alternative information structures.

Unfortunately, the acquisition of information often involves a "set-up cost", i.e the resources needed to obtain the information may be independent of the scale of the production process in which the information is used. This set-up cost will introduce a non-convexity in the production possibility set, and thus one of the standard conditions in the theory of the Arrow–Debreu economy will not be satisfied [see Radner (1968, section 9)]. (See Section 5.2 for a brief discussion of other sources of non-convexity associated with the choice of information.)

There is another interesting class of cases in which an agent's information structure is not fixed, namely cases in which the agent's information at one date may depend upon production or consumption decisions taken at previous dates, but all actions can be scaled down to any desired size. Unfortunately, space limitations prevent a discussion of this class in the present chapter.

1.6. Critique of the extended Arrow–Debreu economy

If the Arrow–Debreu model is given a literal interpretation, then it clearly requires that the economic agents possess capabilities of imagination and calculation that exceed reality by many orders of magnitude. Related to this is the observation that the theory requires in principle a complete system of insurance and future markets, which appears to be too complex, detailed, and refined to have any practical significance. A further obstacle to the achievements of a complete insurance market is the phenomenon of "moral hazard" (see Section 3.5).

A second line of criticism is that the theory does not take account of at least three important institutional features of modern capitalist economies: (1) money, (2) the stock market, and (3) active markets at every date.

These two lines of criticsm have an important connection, which suggests how the Arrow–Debreu theory might be improved. If, as in the Arrow–Debreu model, each production plan has a sure unambiguous present value at the beginning of time, then consumers have no interest in trading in shares, and there is no point in a stock market. If all accounts can be settled at the beginning of time, then there is no need for money during the subsequent life of the economy; in any case, the standard "motives" for holding money do not apply.

On the other hand, once we recognize explicitly that there is a *sequence* of markets, one for each date, and not one of them complete (in the Arrow–Debreu sense), then certain phenomena and institutions, not accounted for in the Arrow–Debreu model, become reasonable. First, there is uncertainty about the prices that will hold in future markets, as well as uncertainty about the environment.

Second, producers do not have a clear-cut natural way of comparing net revenues at different dates and states. Stockholders have an incentive to establish a stock exchange, since it enables them to change the way their future revenues depend on the states of the environment. As an alternative to selling his shares in a particular enterprise, a stockholder may try to influence the management of the enterprise in order to make the production plan conform better to his own subjective probabilities and attitude toward risk.

Third, consumers will typically not be able to discount all of their "wealth" at the beginning of time, because (1) their shares of producers' future (uncertain) net revenues cannot be so discounted, and (2) they cannot discount all of their future resource endowments. Consumers will be subject to a *sequence* of budget constraints, one for each date (rather than to a single budget constraint relating present cost of his consumption plan to present net worth, as in the Arrow–Debreu economy).

Fourth, economic agents may have an incentive to speculate on the prices in future markets, by storing goods, hedging, etc. Instead of storing goods, an agent may be interested in saving part of one date's income, *in units of account*, for use on a subsequent date, if there is an institution that makes this possible. There will thus be a demand for "money" in the form of demand deposits.

Fifth, agents will be interested in forecasting the prices in markets at future dates. These prices will be functions of both the state of the environment and the decisions of (in principle, all) economic agents up to the date in question. In fact, agents may be able to learn something about the information and decisions of other agents by observing market prices.

1.7. Sequences of markets

Consider now a sequence of markets at successive dates. Suppose that no market at any one date is complete in the Arrow–Debreu sense, i.e. at every date and for every commodity there will be some *future* dates and some events at those future dates for which *it will not be possible to make current contracts for future delivery contingent upon those events*. In such a model, several types of "equilibrium" concept suggest themselves. First, we may think of a sequence of "momentary" equilibria in which the current market is cleared at each date. The prices at which the current market is cleared at any one date will depend upon (among other things) the expectations that the agents hold concerning prices in future markets (to be distinguished from futures prices on the current market!). We can represent a given agent's expectations in a precise manner as a function (schedule) that indicates what the prices will be at a given future date in each elementary event at that date. This includes, in particular, the representation of future prices as random variables, if we admit that the uncertainty of the agent about future events can be scaled in terms of subjective probabilities [see Savage 1954)].

In the evolution of a sequence of momentary equilibria, each agent's expectations will be successively revised in the light of new information about the environment and about current prices. Therefore, the evolution of the economy will depend upon the rules or processes of expectation formation and revision used by the agents. In particular, there might be interesting conditions under which such a sequence of momentary equilibria would converge, in some sense, to a (stochastic) steady state. This steady state, e.g. stationary probability distribution of prices, would constitute a second concept of equilibrium.

A third concept of equilibrium emerges if we investigate the possibility of consistency among the expectations and plans of the various agents. I shall say that the agents have *common* expectations if they associate the same (future) prices to the same events. (Note that this does not necessarily imply that they agree on the joint probability distribution of future prices, since different agents might well assign different subjective probabilities to the same event.) I shall say that the plans of the agents are *consistent* if, for each commodity, each date, and each event at that date, the planned supply of that commodity at that date in that event equals the planned demand, and if a corresponding condition holds for the stock markets. An *equilibrium of plans, prices, and price expectations* is a set of prices on the current market, a set of common expectations for the future, and a consistent set of individual plans, one for each agent, such that, given the current prices and price-expectations, each individual agent's plan is optimal for him, subject to an appropriate sequence of budget constraints. If agents are using equilibrium prices to make inferences about the environment, then this equilibrium takes the special form of a so-called *rational expectations equilibrium*.

As I have already indicated in Section 1.2, the remainder of this chapter will be devoted to more formal descriptions of these models and equilibrium concepts, together with a discussion of the associated questions of existence, optimality, and stability.

2. Events, commodities, commitments, and prices

Consider an economy extending through a finite or infinite set of dates, t, in an environment with a set S of alternative states. The possible states of the environment are represented as sequences, $s=(s_t)$, where s_t is interpreted as *the state of the environment at date t*, and s as the *complete history of the environment*. For each t, the set of alternative states s_t will be denoted by S_t.

Economic magnitudes (consumption, prices, etc.) that can be observed at date t can depend upon the past and present states of the environment, but not upon the future. Thus, let the *elementary events at date t* be the partial sequences, $e_t=(\ldots, s_{t-1}, s_t)$, of states of the environment up through date t; an *event at date t* is a set of elementary events at date t.

I shall assume here that, for each date t, the set S_t of alternative states of the environment at t is finite. If the history of the environment extends only finitely far into the past, then for any date there are only finitely many events at that date. On the other hand, if the history of the environment extends infinitely far into the past, then for every date there are infinitely many (indeed, uncountably many) elementary events at that date. In this case, it is usual not to consider every set of elementary events to be "observable" (measurable), but only a sigma-field, say $\boldsymbol{F}_t$, of such sets;[2] here $\boldsymbol{F}_t$ will be taken to be the sigma-field generated by all *cylinders*, i.e. Cartesian products $\times_{u \leq t} A_u$, such that $A_u = S_u$ for all but finitely many dates u. (Note that a *cylinder* is an event that depends on the states of the environment at only finitely many dates.) In either case, whether the past is finite or infinite, $\boldsymbol{F}_t$ will denote the set of (observable, i.e. measurable) events at date t.

For each date there is a finite set of "commodities", numbered $1, \ldots, H$. Depending on the particular context, these commodities may or may not include various kinds of money and other financial assets. A *simple commitment* at date t in (elementary) event e_t specifies the number of units of commodity h that a trader will deliver "to the market" at date $u(\geq t)$ in some event F in $\boldsymbol{F}_u$. A negative delivery is to be interpreted as a receipt. For $u=t$ we have a "spot" commitment, whereas for $u>t$ we have a "forward" commitment. The set of commodities at each date could, in principle, depend on the date and on the event that occurred at that date. To keep an already complicated notation from getting completely out of hand, I shall suppose that the set of commodities is the same at every date–event pair.

At every date and in every elementary event at that date there will be a market in commitments, with future delivery possibly contingent upon the occurrence of

[2] For basic facts about measure theory, see, for example, Halmos (1950) or Chapter 5 of this Handbook. Recall that a *sigma-field* of sets is a non-empty family of sets closed under the formation of complements and countable unions. (Some authors, including Halmos, call this a sigma-algebra.)

some future events. Thus, the set of markets will be indexed by the set M of possible date–event pairs. Formally, let M be the set of all pairs (t, e) such that t is a date and e is an elementary event at date t. There is a natural partial ordering of M; for $m=(t,e)$ and $n=(u,f)$ in M, say that $m \leqslant n$ if $t \leqslant u$ and f is a refinement of e, that is to say, the partial history e is an initial subsequence of the partial history f. Thus, $m \leqslant n$ if the date–event pair n could possibly follow the date–event pair m (or is the same as m).

One observes that in actual markets not all possible commitments are traded (see Section 1). In the present model I shall take the set of tradeable commitments at any particular date–event pair to be given exogenously. For each date–event pair $m=(t,e)$ in M, each date $u \geqslant t$, and each commodity h, let $\boldsymbol{M}^h_{mu}$ be a family of (not necessarily elementary) events at date u, which is given exogenously, and has the following properties:

(a) $\boldsymbol{M}^h_{mu}$ is either empty or is a subfield of $\boldsymbol{F}_u$;

(b) $\boldsymbol{M}^h_{mt} = \boldsymbol{F}_t$ if $m=(t,e)$, i.e. all spot commitments are allowed; and

(c) $\boldsymbol{M}^h_{mu}$ is finite if $m=(t,e)$ and $t<u$. (2.1)

Part (c) of (2.1) is a technical assumption which can be relaxed at the expense of using more powerful mathematical methods. I shall call the sets $\boldsymbol{M}^h_{mu}$ the *market fields*.

Assume further that, if at a date–event pair m one may make a commitment for delivery at date v contingent upon some event F, then at a later date–event pair $n \geqslant m$ one may do the same. Formally:

$$\text{if } m=(t,e),\ n=(u,f),\ m \leqslant n,\ v \geqslant u, \quad \text{and} \quad F \text{ is in } \boldsymbol{M}^h_{mv},$$
$$\text{then } F \text{ is in } \boldsymbol{M}^h_{nv}. \qquad (2.2)$$

One might suppose that it would not be allowed to make a commitment at a date–event pair $m=(t,e)$ for delivery at some subsequent date u contingent upon the occurrence of an event F at date u if, given the event e a date t the event F at u were impossible. However, such a constraint might not be realistic if the parties to the contract had incomplete and different information at date t about which elementary event had occurred at that date (a situation that I shall discuss in Sections 3.4 and 3.5). Therefore, I shall not rule out such commitments a priori. Caveat emptor!

Finally, I shall assume (unless notice is given to the contrary) that there is an upper bound, L, on allowable commitments, where L is a given positive number. Such an upper bound is natural; for example, a commitment to deliver a quantity vastly greater than the total supply of the commodity would not be credible to moderately well-informed traders.

To sum up, a simple commitment at a date–event pair $m=(t,e)$ to deliver some quantity of commodity h at date u, contingent upon the occurrence of event F at date u, is *allowable* if

m is in M and $u \geq t$,

F belongs to $\boldsymbol{M}^h_{mu}$, and

the quantity committed does not exceed L. (2.3)

The concept of a *measurable function* provides a compact way of representing allowable commitments. To motivate the use of this concept in the present context, let M be an event at date u, and consider a simple commitment to deliver r units of commodity h at date u if and only if event F occurs. Such a commitment could be regarded as a function defined on all elementary events at date u, which takes the value r on the set F and the value 0 elsewhere. A more general form of commitment would be any real-valued function z defined on the set of elementary events at date u, with the interpretation that the quantity $z(e)$ is to be delivered if elementary event e occurs. If $\boldsymbol{M}$ is a finite subfield of events at date u, then there is a finite partition of the set of elementary events at date u such that every event in M is a union of elements of that partition. Any sum of simple commitments that are allowable with respect to $\boldsymbol{M}$ would be a function defined on elementary events with the property that it is constant on elements of the partition that generates $\boldsymbol{M}$. Equivalently, it would have the property that, for any real number r, the set of elementary events in which the amount delivered is r is a set in $\boldsymbol{M}$. This motivates the following definition: if $\boldsymbol{F}$ is a sigma-field of events at date u, and z is a real-valued function on the set of elementary events at u, then z is measurable with respect to $\boldsymbol{F}$ (or just $\boldsymbol{F}$-measurable) if, for every real number r, the set of elementary events e such that $z(e) \leq r$, is in $\boldsymbol{F}$. (One can show that the set of $\boldsymbol{F}$-measurable functions is closed under addition and under multiplication by real numbers.)[3]

We can now generalize our definition of allowable commitment: an *allowable commitment* at a date–event pair $m=(t,e)$ for delivery of a commodity h at date $u \geq t$ is a real-valued function on the set of elementary events at date u that is measurable with respect to $\boldsymbol{M}^h_{mu}$ and bounded by L. (2.4)

If two traders exchanged commitments, one would normally say that they concluded a contract, and the terms of the contract would determine relative

[3]In most of this chapter all of the market fields will be finite, and therefore it may seem a bit heavy-handed to introduce the mathematical concept of measurable function at this point. However, this conceptual machinery will lead the reader in a natural way into the cases in which the market fields are not finite, where more advanced probabilistic methods are needed.

prices of the commodities involved. For the purposes of this paper, it will be convenient to imagine that the traders make commitments "to the market" at prices in *units of account*, with payment made at the time the commitment is made.

The concept of a measurable function also provides a compact way of representing prices. Consider a particular date–event pair m, a subsequent date u, a commodity h, and a corresponding allowable commitment z. If p is a real-valued function on the set of elementary events at u that is measurable with respect to $\boldsymbol{M}^h_{mu}$, and we interpret $p(e)$ as the price (at m) of commodity h delivered at date u if and only if elementary event e occurs, then the *value of the commitment* z is the inner product

$$pz = \sum_e p(e)z(e), \tag{2.5}$$

provided that there are only finitely many elementary events at date u [otherwise, the sum in (2.5) would have to be replaced by some suitable integral].

For every date–event pair $m=(t,e)$ in M, let Z_m denote the vector space carrying all commitments allowable at m, i.e. the vector space of all arrays z^h_{mu} such that

$$\begin{aligned} &u \geqslant t; \qquad h=1,\ldots,H, \\ &z^h_{mu} \text{ is measurable with respect to } \boldsymbol{M}^h_{mu}. \end{aligned} \tag{2.6}$$

The contracts concluded by a given trader on the market at m will result in an array of net commitments, which can be represented by a vector, say z_m, in Z_m. Correspondingly, the prices ruling in the market at m can also be represented by a vector, say p_m, in Z_m, which will be called the *system of prices at m*. The *net revenue* of the trader at m is the inner product, $p_m z_m$, of these two vectors.

The above model, which is adapted from Radner (1972), will serve as the basic framework for the rest of the paper.

3. Market equilibrium

3.1. Temporary market equilibrium

Let us now place ourselves at a particular date–event pair, m, and consider how prices and quantities might be determined as an equilibrium on the corresponding market at m. Imagine that a particular price system at m, say p_m, is "announced" for market m. The commitments made by traders in the past, their currently available resources, their experiences with past prices, their expectations regarding future prices and resource availability, and their preferences for present and future consumption, will together result in a *total* excess market supply, say z_m. Roughly speaking, we shall say that the market is in temporary equilibrium at m if the excess market supply is zero.[4]

In general treatments of equilibrium theory, one typically admits that satisfaction of preferences subject to budget constraints does not necessarily determine, for each price system, a unique excess supply vector for each trader. Therefore, the supply behavior of the traders will be described by a *total excess supply correspondence*, which maps price vectors into sets of excess total supply vectors.

Since prices are quoted in units of account, it is natural to assume that two proportional price vectors would give rise to the same sets of excess supply vectors. The holding of assets can be represented by contracts in which current delivery is exchanged for a promise of future delivery; in this case, the assets must be included in the list of commodities. I shall, furthermore, make the traditional assumption that prices must be non-negative. Accordingly, let P_m denote the unit simplex of Z_m, i.e. the set of all non-negative vectors in Z_m the sum of whose coordinates is unity. A price vector at m is restricted to be a vector in P_m. The (total) excess supply correspondence, σ_m, is a mapping from P_m to the subsets of Z_m.

Since prices have been restricted to be non-negative, it is natural to admit that, in equilibrium, excess supply of a commodity could be positive, but only if the corresponding price were zero.

We are now in a position to define temporary market equilibrium at a date–event pair m, with supply behavior summarized by a total excess supply correspondence, σ_m. It is understood that σ_m reflects behavior given the traders' information about past prices and events. A *temporary market equilibrium at m* is a pair (z_m^*, p_m^*), such that

(a) z_m^* is in Z_m, and p_m^* is in P_m;
(b) z_m^* is in $\sigma_m(p_m^*)$;
(c) $z_m^* \geqslant 0$; and
(d) for any coordinate of z_m^* that is strictly positive, the corresponding coordinate of p_m^* is zero. (3.1)

Before discussing in detail alternative theories of individual supply behavior, it is useful to anticipate properties that the total excess supply correspondence might be expected to have, as well as properties that would be sufficient to guarantee the existence of a temporary equilibrium.

First, we have *Walras' Law*:

$$pz=0, \quad \text{for all } p \text{ in } P_m \text{ and } z \text{ in } \sigma(p). \tag{3.2}$$

[4] Note that z_m is the algebraic sum of the individual excess supply vectors.

This can be thought of as following from individual budget constraints, or as a convention that expresses the idea that all contracts at m are undertaken with respect to the same price system.

Second, one can place an a priori bound on the set of possible equilibria. The requirement that every coordinate of every trader's excess supply vector be allowable implies an upper bound on the total excess supply, equal to IL, where I is the number of traders and L is the bound in (2.4). This upper bound, and condition (c) of (2.1), together imply a uniform bound on all possible equilibria. Therefore, there is a bounded subset of Z_m to which one could restrict the total excess supply correspondence without changing the set of the bound L; this point will be discussed later.

The following proposition gives sufficient conditions for the existence of a temporary market equilibrium.

Theorem 3.1

Let σ_m be the total excess supply correspondence for a given market in M. There exists a temporary market equilibrium at m if σ_m has the following properties:

(a) There is a closed bounded subset, $\hat{Z}_m$, of Z_m such that σ_m is an upper hemicontinuous correspondence from P_m to $\hat{Z}_m$.
(b) For every p in P_m, $\sigma_m(p)$ is non-empty and convex.
(c) (Walras' Law) For every p in P_m and z in $\sigma_m(p)$, $pz=0$.

The proof of this proposition is essentially the same as that in Debreu (1959, prop. (1), section 5.6, p. 82). Indeed, for any particular market m, there is nothing about the formal model expressed in Theorem 3.1 that is peculiar to a *temporary* equilibrium, as distinct from an "Arrow–Debreu equilibrium". Thus, we see that the basic Arrow–Debreu framework is formally adequate for a discussion of the existence of temporary market equilibrium. It remains to discuss those properties of individual behavior that would imply the hypotheses of Theorem 3.1 concerning the total excess supply correspondence.

3.2 The Arrow–Debreu model

The Arrow–Debreu model of a complete market for present and future contingent delivery of commodities may be considered as a special case of our model of market equilibrium at a single date. Suppose that the dates run from 0 to $T-1$, and let us place ourselves at the initial date, $t=0$. Assume, furthermore, that there is no uncertainty about the environment at the initial date. (This assumption will be relaxed in the next section.) It follows from the assumptions in Section 2 that every market field M^h_{mu} is finite, and hence that every vector space Z_m is finite-dimensional. To describe the condition that there is no uncertainty at date 0, the set S_0 is assumed to have only one element; thus there is only one date–event pair at date 0. This date–event pair will also be denoted by the symbol 0.

The assumption that "markets are complete" is expressed formally by

$$M^h_{0u} = F_u, \quad \text{for all } u \geq 0. \tag{3.3}$$

This assumption has important implications. We shall see that, as a consequence of this assumption, the Arrow–Debreu model of markets with uncertainty becomes formally equivalent to a model with certainty. A complete argument requires a model of equilibrium in a sequence of markets (see Section 4), but an informal argument can be presented here.

The "complete markets" assumption (3.3) above, and (2.2) of Section 2, together imply that markets are complete at *every* date–event pair, i.e. that

$$M^h_{mu} = F_u, \quad \text{for all } m \text{ and all dates } u \text{ subsequent to } m. \tag{3.4}$$

Let (z^*_0, p^*_0) be a temporary equilibrium at 0, as defined in Section 2, and consider a subsequent date–event pair m. Every commitment that is allowable at m is also allowable at 0. Let p^*_m be the restriction of the price system p^*_0 to the vector space Z_m, suitably normalized. Given the price system p^*_m, would an agent want to undertake some new commitments at m that were not already planned at date 0? The answer is no. First, any commitments allowable at m were also allowable at 0. Second, the agent's budget constraint at m would require that the sum total of any new commitments at m have net value zero at the price system p^*_m; but in that case they would also have net value zero at the price system p^*_0. Hence, any new combination of commitments that is allowable and financially feasible at m was already allowable and financially feasible at date 0 (under the price system p^*_0). It follows that, with the complete-markets assumption, there will be no incentive to reopen the markets after date 0, and a temporary equilibrium at date 0 suffices to determine agents' deliveries and receipts for the entire life of the economy. Furthermore, the market at date zero is formally equivalent to a market under certainty. This equivalence is established by labelling commodities delivered at different date–event pairs as different "commodities".

In Section 1 I have already sketched a critique of the realism of the Arrow–Debreu model of complete markets. I turn now to a more detailed discussion of information and expectations in models of sequences of incomplete markets.

3.3. *Information and expectations*

Recall that we placed ourselves at a particular date–event pair, and that the excess supply correspondence at that date–event pair reflects the traders' information about past prices and about the history of the environment up through that date. If a given trader's excess supply correspondence is generated by preference satisfaction, then the relevant preferences will be conditional upon the information available. If, furthermore, the trader's preferences can be scaled in terms of utility and subjective probability, and conform to the Expected Utility Hypothesis, then the relevant probabilities are the conditional probabilities given the available information.[5] These conditional probabilities express the trader's expectations regarding the future. Although a general theoretical treatment of our problem does not necessarily require us to assume that traders' preferences conform to the Expected Utility Hypothesis, it will be helpful in the following heuristic discussion to keep in mind this particular interpretation of expectations.

A trader's expectations concern both future environmental events and future prices. Regarding expectations about future environmental events, there is no conceptual problem. According to the Expected Utility Hypothesis, each trader is characterized by a subjective probability measure on the set of complete histories of the environment. Since, by definition, the evolution of the environment is exogenous, a trader's conditional subjective probability of a future event, given the information to date, is well defined.

It is not so obvious how to proceed with regard to traders' expectations about future prices. I shall contrast two possible approaches. In the first, which I shall call the *perfect foresight* approach, let us assume that the behavior of traders is such as to determine, for each complete history, s, of the environment, a unique corresponding sequence of price systems, say $\phi_t^*(e_t)$. If the "laws" governing the economic system are known to all, then every trader can calculate the sequence of functions ϕ_t^*. In this case, at any date–event pair a trader's expectations regarding future prices are well defined in terms of the functions ϕ_t^* and his conditional subjective probability measures on histories of the environment, given his current information. Traders need not agree on the probabilities of future environmental events, and therefore they need not agree on the probability distribution of future prices, *but they must agree on which future prices are associated with which events*. I shall call this last type of agreement the condition of *common price expectation functions*.

Thus, the perfect foresight approach implies that, in equilibrium, traders have common price expectation functions. These price expectation functions indicate, for each date–event pair, what the equilibrium price system would be in the

[5] See, for example Marschak and Radner (1972, ch. 2, section 5), and Radner (1968).

corresponding market at that date–event pair. Pursuing this line of thought, it follows that, in equilibrium, the traders would have strategies (plans) such that, if these strategies were carried out, the markets would be cleared at each date–event pair [in the sense of (3.1)]. The question of the existence of such an equilibrium of plans and price expectations is explored in Sections 4 and 5.

Furthermore, the present model, as described in Section 2, can be refined so as to represent the possibility that different traders have *different* information about the environment. For example, this can be done by associating to each trader at each date t a subfield of $\boldsymbol{F}_t$, representing the events that the individual trader can observe at date t. (See Section 3.4 below.) In particular, one could express in this way the hypothesis that a trader cannot observe the individual preferences and resource endowments of other traders. Indeed, one can also introduce into the description of the state of the environment variables that, for each trader, represent his alternative hypotheses about the "true laws" of the economic system. In this way the condition of common price expectation functions can lose much of its apparent restrictiveness.

The situation in which traders enter the market with different non-price information presents an opportunity for agents to learn about the environment from prices, since current market prices reflect, in a possibly complicated manner, the non-price information signals received by the various agents. To take an extreme example, the "inside information" of a trader in a securities market may lead him to bid up the price to a level higher than it otherwise would have been. In this case, an astute market observer might be able to infer that an insider had obtained some favorable information, just by careful observation of the price movements. More generally, *an economic agent who has a good understanding of the market is in a position to use market prices to make inferences about the (non-price) information received by other agents.*

These inferences are derived, explicitly or implicitly, from an individual's "model" of the relationship between the non-price information received by market participants and the market prices. On the other hand, the true relationship is determined by the individual agents' behavior, and hence by their individual models. Furthermore, economic agents have the opportunity to revise their individual models in the light of observations and published data. Hence, there is a feedback from the true relationship to the individual models. An equilibrium of this system, in which the individual models are identical with the true model, is called a *rational expectations equilibrium*.

This concept of equilibrium is more subtle, of course, then the ordinary concept of the equilibrium of supply and demand. In a rational expectations equilibrium, not only are prices determined so as to equate supply and demand, but individual economic agents correctly perceive the true relationship between the non-price information received by the market participants and the resulting equilibrium market prices. This constrasts with the ordinary concept of equilibrium in which the agents respond to prices but do not attempt to infer other agents' non-price information from the actual market prices.

Research on rational expectations equilibrium is quite recent, and the subject has not been fully explored. Nevertheless, several important insights have already been obtained; these insights will be sketched in Sections 4.2–4.4 and 6.4.

Although it is capable of describing a richer set of institutions and behavior than is the Arrow–Debreu model, the perfect foresight approach is contrary to the spirit of much of competitive market theory in that it postulates that individual traders must be able to forecast, in some sense, the equilibrium prices that will prevail in the future under all alternative states of the environment. Even if one grants the extenuating circumstances mentioned in previous paragraphs, this approach still seems to require of the traders a capacity for imagination and computation far beyond what is realistic. An equilibrium of plans and price expectations might be appropriate as a conceptualization of the ideal goal of indicative planning, or of a long-run steady state toward which the economy might tend in a stationary stochastic environment.

These last considerations lead us in a different direction, which I shall call the *bounded rationality* approach. This approach is much less well defined, but expresses itself in terms of various retreats from the hypothesis of "fully rational" behavior by traders, e.g. by assuming that the traders' planning horizons are severely limited, or that their expectation formation follows some simple rules-of-thumb. An example is described in Section 3.6.

3.4. *The extended Arrow–Debreu model*

Recall that, in the Arrow–Debreu model, it was assumed that the information available to all traders at any particular date t was the same for all traders. This information was represented by a sigma-field, say $\boldsymbol{F}_t$, of events observable at t. To generalize the Arrow–Debreu model to the case in which different traders have possibly different information, it is natural to represent trader i's information at date t by a sigma-field $\boldsymbol{F}_{it}$, depending on both i and t. As we shall now see, this requires only a reinterpretation of the Arrow–Debreu model, provided that the information of each trader is determined *exogenously*.

In order to discuss conditions for the existence and optimality of equilibrium, I need to introduce at this point a formal model of the preferences and consumption plans of the traders. Fix attention, for the moment, on a particular trader, say i. For any date t and any elementary event e_t at that date, let $x_{it}(e_t)$ denote the H-dimensional vector of quantities of commodities $1,\dots,H$ consumed by trader i in event e_t at event t. Thus, the function x_{it} maps elementary events at date t into R^H, the Euclidean space of dimension equal to the number of commodities. The array $x_i = (x_{it})$ will be called the *consumption plan* of trader i.

Let X_i denote the set of feasible consumption plans for trader i, or trader i's *consumption set*. This consumption set will reflect the various constraints – e.g. biological, psychological, or sociological – that limit trader i's consumption. In this respect, X_i will correspond precisely to the usual concept of consumption set in the Arrow–Debreu model. In addition, X_i will reflect the information available to trader i, as follows. In accordance with standard practice in the theory of statistical decision, we can represent the information that trader i will have at date t by a sigma-field of sets of elementary events at that date; let $\boldsymbol{F}_{it}$ denote this sigma-field.[6] To express the idea that the trader can plan to make his actual consumption depend on information that he will receive in the future, and to reflect the assumption that he can observe his own consumption when it occurs, I require that x_{it} must be measurable with respect to $\boldsymbol{F}_{it}$. This requirement will be called the condition of *informational feasibility*.

It is useful to note that, for a given sigma-field $\boldsymbol{F}_{it}$, informational feasibility (at date t) constitutes a set of linear constraints on the function x_{it}. This is most easily seen in the case in which the set of elementary events at date t is finite. In that case, the function x_{it} can be though of as a vector the dimension of which is equal to the product of the number of commodities (H) and the number of elementary events. If the set F is an element of the partition that generates $\boldsymbol{F}_{it}$, then trader i's information does not distinguish between two elementary events in F, and so trader i's consumption must be the same in those two events. The collection of all such equality conditions (given $\boldsymbol{F}_{it}$) is precisely equivalent to the condition that x_{it} be measurable with respect to $\boldsymbol{F}_{it}$. This implies that the set of informationally feasible functions x_{it} is a linear subspace of the set of functions from elementary events to R^H. It is this last property that generalizes to the case in which the set of elementary events at date t is not finite. It follows that trader i's consumption set, X_i, lies in the linear subspace of consumption plans x_i such that, for each t, x_{it} is measurable with respect to $\boldsymbol{F}_{it}$.

Each trader i is assumed to have a preference pre-ordering on his consumption set. Following standard conventions, I shall denote this preference pre-ordering by $\precsim_i$. A special case of interest is the one in which a trader's preference pre-ordering is induced by the expected utility of his consumption plan. Let ψ_i be a probability measure on the set S of complete histories, s, of the environment (see Section 2). This probability measure may reflect the personal beliefs (at date 0) of trader i, concerning the relative likelihoods of different (future) histories, or it may reflect trader i's own information at date 0, or both. For a given history, s, of the environment, let $x_i(s)$ denote the realized history of trader i's consumption; this is a vector in a space of dimension HT, where T is the number of dates.

[6]See, for example, Savage (1954, chs. 6 and 7) and Marschak and Radner (1972, chs. 2 and 3). If the set of elementary events at t is finite, then F_t is generated by a finite partition. More detailed information corresponds to a finer sigma-field (or equivalently, in the finite case, to a finer partition).

Trader i's *utility function* is a real-valued function, u_i, on R^{HT}, and the *expected utility* (to trader i) of the consumption plan x_i is the expected value (with respect to the probability measure ψ_i) of $u_i[x_i(s)]$. Of two production plans, a trader prefers the one with the higher expected utility, and is indifferent between them if their expected utilities are equal. In what follows, it will not be required that each trader's preferences be consistent with an expected utility calculation, but it will be useful to have this special case in mind when we come to various assumptions about the traders' preferences among consumption plans.

Finally, for every trader i, let $w_{it}(e_t)$ denote trader i's *endowment of commodities at date t in elementary event* e_t. Just as in the case of consumption, it is natural to suppose that the function w_{it} is measurable with respect to $\boldsymbol{F}_{it}$, since it is natural to suppose that a trader can observe his own realized endowment when the corresponding date arrives. Let w_i denote the sequence (w_{it}), which will be called i's *endowment*.

For each date t, let $\boldsymbol{F}_t$ be the smallest sigma-field of sets of elementary events at t that contains $\boldsymbol{F}_{it}$ for each i, let Z_t be the set of $\boldsymbol{F}_t$-measurable functions mapping elementary events at t into R^H, and let Z be the Cartesian product of the sets Z_t. The sigma-field $\boldsymbol{F}_t$ represents the information that is available jointly to the traders, that is to say, all the information that would be available if the traders pooled their information at date t. By the definition of the sigma-fields $\boldsymbol{F}_t$, every trader's endowment is a point in Z, and every trader's consumption set is a subset of Z.

From this point on (in Section 3.4) I shall explicitly restrict the discussion to the case in which there are finitely many dates and finitely many elementary events at each date. In this case, the set Z, considered as a vector space,[7] is finite-dimensional.

In the extended Arrow–Debreu model, as in the model of Section 3.2, I shall assume that all markets are complete. As noted in Section 3.2, in this case one can, without loss of generality, restrict attention to the one market at date 0. Accordingly, one may take a price system (see Section 2) to be an element of Z. Corresponding to each consumption plan x_i of trader i is a trade plan, z_i, defined by $z_i = w_i - x_i$.

As this definition shows, the fact that in the extended Arrow–Debreu model one is dealing with essentially one market implies that there is a one-to-one correspondence between consumption plans and trade plans, so that a separate consideration of trade plans is superfluous. As will be seen in Section 4, the

[7]To spell this out, let L_t be the set of all functions that map elementary events at date t into vectors in the commodity space R^H, and let L be the Cartesian product of the sets L_t. A point x in L has coordinates $x_t^h(e_t)$, where h denotes a commodity, t denotes a date, and e_t denotes an elementary event at date t. From the discussion of information feasibility above it is clear that Z is a linear subspace of L.

separate consideration of trade plans will become necessary once we drop the assumption that markets are complete.

Following the terminology of Debreu (1959), I shall call an I-tuple (x_i) of consumption plans (one for each trader) *attainable* if, first, each trader's consumption plan is in his consumption set, and second, total consumption does not exceed total endowments,[8] i.e.

$$\sum_i x_i \leqslant \sum_i w_i. \tag{3.5}$$

Since there is only one market, each trader i faces a single budget constraint,

$$px_i \leqslant pw_i, \tag{3.6}$$

where p is the price system.

The extended Arrow–Debreu model can now be seen to be *formally* equivalent to the Arrow–Debreu model with certainty. The theory of equilibrium in such a model is well known, but in order to interpret some of the standard hypotheses of that theory in the context of uncertainty, I shall recall here one set of conditions sufficient for the existence of an equilibrium, and also make some remarks concerning the optimality properties of an equilibrium.

To define trader i's (gross) demand correspondence, for each price system p let $\delta_i(p)$ denote the (possibly empty) set of most preferred consumption vectors in X_i that also satisfy the budget constraint (3.6). An *equilibrium* is a pair $\{p,(x_i^*)\}$, where p is a non-zero, *non-negative* price system, (x_i) is an attainable I-tuple of consumption plans, and

$$x_i^* \text{ is in } \delta_i(p), \qquad i=1,\dots,I, \tag{3.7}$$

$$p\sum_i (w_i - x_i^*) \geqslant 0. \tag{3.8}$$

Note that conditions (3.5) and (3.8), together with the non-negativity of p, imply that, for every commodity h, every date t, and every elementary event e_t at date t, either the excess supply

$$\sum_i \left[w_{it}^h(e_t) - x_{it}^{*h}(e_t)\right]$$

is zero, or the corresponding price, $p_t^h(e_t)$, is zero.

[8] The inequality in (3.5), as distinct from a strict equality, implies some kind of "free disposal". Note that, at any date, the total amount to be disposed may not be measurable with respect to the information of any single trader, so that the disposal activity might have to be performed by some agent who has finer information than any single trader has. The reader who finds this feature objectionable is referred to Radner (1968) for a version of this model without the free disposal assumption.

Theorem 3.2

An equilibrium exists if, for every trader i, the following conditions are satisfied:

(a) X_i is closed and convex, and there is a vector, say $\bar{x}_i$, such that $x_i \geqslant \bar{x}_i$ for all x_i in X_i;
(b) for every feasible consumption plan there is another, also feasible, that trader i strictly prefers to the first;
(c) trader i's preference pre-ordering is continuous and convex;[9]
(d) w_i is in X_i;
(e) there is an $\tilde{x}_i$ in X_i such that $\tilde{x}_i \leqslant w_i$;

furthermore,[10]

(f) $\Sigma_i (w_i - \tilde{x}_i) \gg 0$.

Theorem 3.2 is a special case of Theorem 4.1 below; it can also be proved directly from Theorem 3.1, by standard methods.

The convexity of preferences can be interpreted as an absence of risk-loving (i.e. as either risk aversion or risk-neutrality). Recall that a preference pre-ordering $\lesssim$ on a subset X of Z is *convex* if, for all points x and y in X, if x is strictly preferred to y then any convex combination of x and y that is different from y is also preferred to y. Suppose that x and x' are consumption plans, that F is a set of histories of the environment, that F' is the complement of F, and that

$$x(s) = \begin{cases} r, & \text{for } s \text{ in } F, \\ r', & \text{for } s \text{ in } F', \end{cases}$$

$$x'(s) = \begin{cases} r', & \text{for } s \text{ in } F, \\ r, & \text{for } s \text{ in } F'. \end{cases}$$

Suppose, furthermore, that a trader were indifferent between x and x'; we might interpret this by saying that the trader believed that the events F and F' were equally likely.[11] (Note that x' is obtained from x by reversing the consequences of the two events, F and F'.) Now consider the consumption plan x'' defined by

$$x'' \equiv (\tfrac{1}{2})x + (\tfrac{1}{2})x'.$$

[9]A pre-ordering $\lesssim$ on a subset of X of Z is *continuous* if, for every X, the set of x' in X such that $x' \lesssim x$, and the set of x' in X such that $x \lesssim x'$, are both (relatively) closed in X. For a definition of convex pre-ordering, see text below.

[10]Here and elsewhere if x is a vector, $x \geqq 0$ means $x > 0$ and $x \neq 0$, and $x \gg 0$ means every coordinate of x is strictly positive. Similarly, if x and y are vectors, $x \geqq y$ means $x - y \geqq 0$, etc.

[11]For this interpretation to be valid, it would be necessary that the trader not be indifferent between r and r'. This discussion of the convexity assumption is based upon Debreu (1959, p. 101).

Notice that

$$x''(s)=\frac{r+r'}{2}, \quad \text{for all } s,$$

so that x'' yields the consumption vector $(r+r')/2$ with certainty. Convexity and continuity of the trader's preferences would imply that x'' would be at least as good as both x and x'. Thus, the trader would, in this instance, be either averse to risk or neutral towards risk.

Since the extended Arrow-Debreu model is formally equivalent to the Arrow-Debreu model with certainty, one can also apply the standard theorems on the relationship between equilibrium and Pareto optimality.[12]

Recall that an attainable I-tuple of consumption plans is Pareto optimal if there is no other attainable I-tuple of consumption plans that makes no trader worse off and makes at least one trader strictly be better off. Attainability is defined relative to the consumption sets $X_1,\ldots, X_I$, which reflect the information available to the several traders. Thus, Pareto optimality is here defined *relative to the available information.*

Let us consider, for a moment, the effect of increasing the information available to a trader. Such an increase will typically be reflected in a refinement[13] of one or more of his sigma-fields $\boldsymbol{F}_{it}$. This in turn will be reflected in an enlargement of his set, X_i, of feasible consumption plans, and hence in an enlargement of the set of attainable I-tuples of consumption plans. However, it cannot be concluded that a new equilibrium corresponding to the increased information will be Pareto superior to the old one.[14]

Thus far, the extended Arrow-Debreu model has been elaborated here in the context of a fixed information structure for each trader. There would be no difficulty, in principle, of further extending the model to describe a situation in which traders had a choice among alternative information structures, with different information structures requiring different expenditures of resources. However, the technology of the acquisition and use of information is such that the corresponding consumption set will typically fail to be convex.[15]

[12]See, for example, Debreu (1959, chs. 6 and 7).

[13]For conditions that characterize an increase in information, see Marschak and Radner (1972, ch. 2).

[14]This point is discussed by Hirshleifer (1971, 1975).

[15]See Radner (1968).

3.5. *Private information, moral hazard, self-selection, and the incentive to reopen markets*

It is unusual for a trader to be able to conclude a contract for delivery contingent up on an event that only he can observe, i.e. on private information. It is, in fact, an implication of the extended Arrow–Debreu model that such contracts will be essentially excluded in equilibrium. This restriction, and the assumption that only purely exogenous events may be used as contingencies for delivery, together severely restrict the set of contracts that will appear in equilibrium. In real markets, various types of contract, not formally admitted in the extended Arrow–Debreu model, have the effect of partially circumventing these restrictions.

For a formal discussion of private information, it turns out to be more convenient to define information that is *not* private. However, to motivate the definitions that follow, I shall first consider a special case. Suppose that all of the information in the economy is represented by a long list of variables, called information signals, and that any one trader's information consists of some subset of information signals. A signal is "private" if it is observed by *only one* trader. Among those signals that are not private, we might call "public" those that are observed by *every* trader. Of course, not every "non-private" signal need be public.

To extend these ideas to the representation of information in terms of sigma-fields, recall that, for a given date t, $\boldsymbol{F}_{it}$ denotes the sigma-field that represents trader i's information at t. For any collection of sigma-fields, let their *pooled information* be the smallest sigma-field that contains every member of the collection, and let their *common information* be the largest sigma-field that is contained in every member of the collection. Let $\boldsymbol{F}'_{it}$ denote the pooled information of all traders other than i, and let $\boldsymbol{F}''_{it}$ denote the information common to $\boldsymbol{F}_{it}$ and $\boldsymbol{F}'_{it}$. I shall call $\boldsymbol{F}''_{it}$ trader i's *non-private information*.

Consider an equilibrium of the extended Arrow–Debreu model such that excess supply is actually equal to zero (and not merely non-negative). In this case, it is easy to verify that for every date each trader's net commitment must be measurable with respect to his non-private information. This statement must be suitably modified in the case in which some contracts are in excess supply in equilibrium (I omit the details); a paraphrase of the precise modification is that, for each trader, any non-zero net commitment that has a positive price must be for delivery (or receipt) contingent upon his non-private information. The information common to *all* traders might be called *public*. A fortiori, any commitment for delivery contingent upon public information is, in principle, available in equilibrium to every trader.

In practice, since traders make contracts with other traders, and not with an abstract "market", delivery will be contingent upon information that is common to the two traders in question, which is a further restriction. Indeed, if the contract is to be enforced at law, the common information must usually be such as can be verified ex post by a third party according to the rules of evidence, which constitutes an even narrower restriction.

Finally, recall that, in the extended Arrow–Debreu model, information is assumed to be about exogenous events, i.e. events whose occurrence or non-occurrence is not affected by the traders' decisions. The cumulative effect of all of these restrictions is that the set of markets needed to implement an *equilibrium* of the extended Arrow–Debreu model might be considerably smaller than the set of markets that is a priori allowable. This mitigates, to some extent, the criticism that the Arrow–Debreu model requires that activity of "too many" markets.

On the other hand, in real markets one observes various types of contract not admitted in the Arrow–Debreu model that have the effect of partially circumventing the above restrictions. First, many contracts specify performance contingent upon endogenous events, and even events whose occurrence may depend on the decisions of the contracting parties. Contracts of this last type are said to be subject to *moral hazard*, and often contain provisions that limit the quantities that may be traded, or incorporate other forms of "non-linear" pricing. (Particular examples that have been studied from a theoretical point of view include insurance, sharecropping, professional services, and the market for skilled workers. A number of these examples fall under the rubric of the "principal-agent model". See the bibliographic notes for references.)

Second, even without the presence of moral hazard, differences in private information constitute one form of differences in preferences, and sellers may have an incentive to use non-linear pricing to appeal differentially to potential buyers with different private information. However, the buyers may reveal private information by their choice of contract (self-selection) or by their performance on previous contracts (favorable or adverse selection). This consideration leads to interesting theoretical questions regarding both the existence (or non-existence) and the optimality (or non-optimality) of equilibrium.

Finally, at any date at which traders obtain private information they may have an incentive to make new trades that were not implementable by contracts for contingent delivery in the equilibrium of the extended Arrow–Debreu model. This suggests that, to the extent that private information is important, the extended Arrow–Debreu model is not an adequate model of market equilibrium with uncertainty, and that a more appropriate model would incorporate a sequence of markets, which is the focus of the remainder of this chapter.

3.6. *An example of the bounded rationality approach: The Grandmont model of temporary equilibrium*[16]

To illustrate what I have called the *bounded rationality approach* I shall now present a model of temporary equilibrium due to Grandmont (1970, 1974).

[16]For a systematic treatment of the theory of temporary general equilibrium, see Chapter 19 of this Handbook.

Basically, in this example the assumption of bounded rationality takes the form that each trader looks only one period ahead in the formulation of plans. Furthermore, uncertainty about future prices is not explicitly related to uncertainty about the environment.

Consider dates 1 and 2 (or, more generally, t and $t+1$), and suppose that at each date there are N physical commodities (goods and services) and one form of "money". In the notation of section 2, $H=N+1$; let money be commodity number H. Before I give a precise formulation of the rest of the model, I shall briefly describe the situation that the model is intended to formalize. The traders have non-negative endowments of physical commodities and money at each date. Only spot trades of physical commodities are allowed, but at date 1 traders may make commitments to deliver (or receive) money at date 2. However, commitments to deliver money at date 2 may not be made contingent upon any event at date 2. Traders do not consume money at either date, but may carry money forward from date 1 to date 2. Such money balances must be non-negative. At date 1, traders are uncertain about the prices that will prevail at date 2, although the (exogenous) sources of this uncertainty are not specified explicitly. However, a trader's subjective probability distribution of prices at date 2 is influenced by the prices that actually prevail at date 1. Thus, prices at date 1 influence a trader's demand at date 1, not only through the budget constraint at date 1, but also through the trader's expectations about prices at date 2, which in turn influence the "indirect utility" of money balances carried forward from date 1 to date 2.

The primary goal of the model is to explain, through an equilibrium existence theorem, the positive price of money relative to physical commodities. After a precise statement of this result, I shall interpret the model in terms of our more general framework.

Because uncertainty does not enter the situation in a way that is neatly covered by the model of Section 2, I shall use a notation here that corresponds only approximately to that of Section 2. For each trader i, commodity h, and date $t(=1,2)$, let x_{it}^h and w_{it}^h be the corresponding consumption and resource endowment, respectively, of trader i. The resource endowments are given non-negative numbers, and consumption of each commodity at each date is constrained to be non-negative. Furthermore, consumers do not actually consume money. Let x_{it} denote the vector with coordinates $x_{it}^1,\dots, x_{it}^N$, and let w_{it} denote the vector with coordinates $w_{it}^1,\dots,w_{it}^H$. Thus, each w_{it} is a given point in R_+^H, the non-negative orthant of H-dimensional Euclidean space, and each x_{it} is constrained to be a point in R_+^N, the non-negative orthant of N-dimensional Euclidean space.

As we shall see, each trader i will typically be uncertain about his consumption at date 2. His preferences will be assumed to conform to the Expected Utility Hypothesis, i.e. there exists a real-valued function, u_i, defined on R_+^{2N}, such that trader i's preferences on non-negative consumption pairs (x_{i1}, x_{i2}) with x_{i2}

possibly random, is represented by the expected utility, $Eu(x_{i1}, x_{i2})$, where the mathematical expectation is calculated according to the trader's subjective probability measure.

At date 1, only spot trades are allowed for physical commodities ($h=1,\ldots,N$); for money, traders may also make commitments for *sure* delivery at date 2. In the notation of Section 2, M_{12}^h is empty if $h=1,\ldots,N$, and M_{12}^H contains only the whole space S_2 and the empty set. Let b_i denote the amount of money that trader i will *receive* at date 2 as a consequence of trades made at date 1; in other words, b_i is the negative of the amount of money that i has committed himself at date 1 to deliver at date 2. We can also interpret b_i to be the size of the money balance that i carries forward from date 1 to date 2. In this model, Grandmont constrains money balances carried forward to be non-negative. Furthermore, in order to express the idea that money balances provide no interest, Grandmont constrains the price at date 1 of money delivered at date 2 to equal the spot price of money at date 1. Let p_1 and p_2 denote the vectors of spot prices at dates 1 and 2, respectively.

At date 1, trader i chooses a consumption vector, x_{i1}, and a money balance, b_i, subject to the budget constraint

$$p_1(x_{i1}, b_i)=p_1 w_{i1}. \tag{3.9}$$

In making this choice, trader i anticipates that at date 2 he will choose a consumption vector, x_{i2}, to maximize the utility $u_i(x_{i1}, x_{i2})$ subject to the budget constraint

$$p_2(x_{i2},0)=p_2 w_{i2}+p_2^H b_i. \tag{3.10}$$

Notice that at date 1 the trader does not anticipate that at date 2 he may want to carry a positive money balance forward to date 3. For each price vector p_1, let $\gamma_i(p_1)$ be the *set* of such optimal choices at date 1.

If i chooses the pair (x_{i1}, b_i), then his excess supply of goods and money is $w_{i1}-(x_{i1}, b_i)=z_{i1}$. Corresponding to the set $\gamma(p_1)$ of optimal choices is a set $\sigma_i(p_1)$ of excess supply vectors z_{i1}. The *total excess supply correspondence*, σ, is defined by

$$\sigma(p_1)=\sum_i \sigma_i(p_1), \tag{3.11}$$

where in (3.11) the indicated sum is a vector sum of sets.

Let P denote the unit simplex in R^H, i.e. the set of non-negative vectors in R^H whose coordinates sum to unity.

A temporary *equilibrium* at date 1 is a price vector p_1^* together with a set of choices (x_{i1}^*, b_i^*), one for each trader i, such that

(a) p_1^* is in P,
(b) for each i, (x_{i1}^*, b_i^*) is in $\gamma_i(p_1)$,
(c) $\Sigma_i[w_{i1} - (x_{i1}^*, b_i^*)] \geq 0$, and
(d) $p_1^* \Sigma_i[w_{i1} - (x^*_{i1}, b^*_i)] = 0$. (3.12)

I now describe a set of assumptions sufficient to guarantee the existence of a temporary equilibrium. Actually, Grandmont is interested in an equilibrium in which the price of money is strictly positive, and he gives conditions that imply the existence of an equilibrium in which *all* prices are strictly positive.

In Grandmont's model, the state of the environment is not explicitly introduced as a variable. However, at date 1 each trader is uncertain about the price vector, p_2, that will prevail at date 2.This uncertainty is expressed as a (subjective) probability measure on the set of possible price vectors, p_2. Many aspects of the history of the economy will influence this probability distribution; in particular, it will be influenced by the vector p_1 of prices at date 1. Thus, let $\Phi_1(\cdot; p_1)$ be the probability measure that expresses i's uncertainty at date 1 about prices at date 2, given that the vector of prices at date 1 is p_1. For every p_1 in P, $\Phi_i(\cdot; p_1)$ is a probability measure on P.

Since p_2 appears in the budget constraint at date 2, the trader is also uncertain (at date 1) about the consumption that he will choose at date 2, given his choice of (x_{i1}, b_i). Let trader i's consumption at date 2 be $\xi_{i2}(p_2; x_{i1}, b_i)$. At date 1, given the price vector p_1, trader i is assumed to choose (x_{i1}, b_i) non-negative to maximize the expected utility $\mathrm{E}u_i[x_{i1}, \xi_{i2}(p_2; x_{i1}, b_i)]$, subject to the constraint (3.9).

A trader i will be called *regular* if the following conditions are satisfied:

(a) His resource endowments in physical commodities are strictly positive, i.e. $w_{it}^h > 0$, for $h = 1, \ldots, N$, and $t = 1, 2$.
(b) His utility function u_i is continuous, concave, and strictly increasing in each argument; furthermore, his preferences for consumption in the two periods are independent.
(c) For every strictly positive price vector p_1, the probability measure $\Phi_i(\cdot; p_1)$ assigns probability 1 to the set of strictly positive price vectors p_2; thus for any p_1 in P_{++}, the interior of p, $\Phi_i(P_{++}; p_1) = 1$.
(d) Considered as a mapping from P_{++} to the set of probability measures on P_{++}, Φ_i is weakly continuous. (3.13)

I now introduce a condition that expresses the idea that trader i's expectations about future prices are not "too sensitive" to current prices.

In view of assumption (3.13), Φ_i will henceforth be considered a mapping from P_{++} to the set of probability measures on P_{++}. Such a mapping will be called *uniformly tight* if for every number $\varepsilon>0$ there exists a compact subset C_ε of P_{++} such that, for all p_1 in P_{++}, $\Phi_i(C_\varepsilon; p_1)>1-\varepsilon$. Notice that Φ_i is uniformly tight if, for example, it does not in fact depend on p_1, or more generally, if there exists a compact subset of C of P_{++} such that, for all p_1 in P_{++}, $\Phi_i(C; p_1)=1$. Also, if trader i has "point expectations", i.e. if for every p_1, $\Phi_i(\cdot\,; p_1)$ assigns probability 1 to a single price vector p_2, then uniform tightness of Φ_i is inconsistent with a "unitary elasticity of expectations".

Theorem 3.3

If all traders are regular, and, for at least one trader i,

(a) Φ_i is uniformly tight, and
(b) $w_{i1}^{II}>0$ (i has a positive endowment of money at date 1),

then there exists a temporary equilibrium at date 1 with a strictly positive price vector p_1.

Notice that each trader is assumed to expect the price of money at date 2 to be positive, even though he does not anticipate demanding money at date 2. Of course, if all traders anticipated that no trader would demand money at date 2 (at any positive price), then it would not be rational for any trader to expect the price of money to be positive at date 2; hence at date 1 the price of money to be delivered at date 2 would be zero, and hence the spot price of money at date 1 would be zero (recall that money is not consumed). This apparent contradiction can be resolved in two ways. First, we may imagine that each trader does in fact realize that he (and others) will want to carry money forward from date 2 to date 3, but ignores this in his own decision problem in order to simplify it. Second, we may imagine that each trader is in the market for only two periods (more generally, for a finite number of periods), and that he anticipates that at date 2 new traders will enter the market and will demand money. The second interpretation, in terms of "overlapping generations", does not quite fit the model, since the model assumes that the traders at date 1 and date 2 are the same. Nevertheless, both interpretations are in the spirit of the bounded rationality approach, in the sense that expectations about future prices are not explicitly related to forecasts of future environmental events, or of individual or market behavior.

The following example, due to Grandmont, shows that a condition like that of uniform tightness of expectations is needed. In this example, each consumer has "point expectations", with a "unitary elasticity of expectations"; formally, for every trader i, and every p in P_{++},

$$\Phi_i(\{p\}; p) = 1. \tag{3.14}$$

Let there be only one physical commodity ($N=1$), and let i's endowment of this commodity at each date be $g_i(>0)$. Furthermore, denote i's endowment of money at date 1 by $\bar{b}_i > 0$, and suppose that his endowment of money at date 2 is 0. Thus

$$w_{i1} = (g_i, \bar{b}_i); \qquad w_{i2} = (g_i, 0). \tag{3.15}$$

Denote the price of the physical commodity at date t, by q_t, and the price of money by r_t. Thus

$$p_t = (q_t, r_t). \tag{3.16}$$

Given that $p_1 = (q, r)$, each trader expects, with subjective certainty, that p_2 will also be equal to (q, r). If x_{it} denotes i's consumption of the physical commodity at date t, and b_i the money balance that he carries forward from date 1 to date 2, then his budget constraints, given the prices q and r at both dates, are

$$\begin{aligned} qx_{i1} + rb_i &\leq qg_i + r\bar{b}_1, \\ qx_{i2} \qquad &\leq qg_i + rb_i. \end{aligned} \tag{3.17}$$

Finally, suppose that his utility function is of the "Cobb–Douglas" type,

$$u_i(x_{i1}, x_{i2}) = x_{i1}^{\alpha_i} x_{i2}^{1-\alpha} \tag{3.18}$$

where α_i is a given number, $0 < \alpha_i < 1$.

Since each trader's expectations are held with certainty, given q and r, he wishes to choose consumptions x_{i1} and x_{i2}, and a money balance b_i, all non-negative, to maximize (3.18) subject to the budget constraints (3.17). This maximum will occur with consumption positive in both periods and with equality in the budget constraints. If $\alpha_i \geq \frac{1}{2}$, then trader i will try to have at least as much consumption at date 1 as at date 2; if $r > 0$, then this would imply $b_i \leq (\bar{b}_i/2)$. Therefore, if $\alpha_i \geq \frac{1}{2}$, for every i, and $r > 0$, then the excess supply of money at date 1 would be at least $\left(\sum_i \bar{b}_i/2\right)$, so that no equilibrium at date 1 could be achieved at a positive price for money.

The reader is referred to Grandmont (1970) for a proof of Theorem 3.3. The main difficulty in applying Theorem 3.3 is the demonstration that the excess supply correspondence is upper hemicontinuous, since the excess supply for an individual trader at date 1 depends on p_1 both through the budget constraint at date 1 and through the trader's expectations about p_2, which in turn depend on p_1. Assumption (3.13d), that $\Phi_i(\cdot\,; p_1)$ is weakly continuous[17] in p_1, plays a crucial role here. To get strict positivity of prices, Grandmont uses both the familiar assumption of strict monotonicity of preferences and the assumption of uniform tightness.

Since expectations about future prices are not explicitly related to the working of the future market, in this model there is no natural way to analyse the phenomenon of "rational expectations". Thus the mappings Φ_i are exogenous. Models with endogenous expectations are described in Sections 4 and 6.4.

4. Equilibrium of plans and expectations

4.1. *Equilibrium in a sequence of markets with common information*

I now turn to the *perfect foresight approach*, which was briefly described in Section 3.3. The present subsection focuses on the theoretical problems inherent in studying equilibria of a *sequence* of markets, no one of which is complete in the Arrow–Debreu sense (Section 3.2). In order to focus on some of these problems, I shall assume here that, at every date, all traders have the same information. It may seem strange to make such an assumption, since I have argued in Section 3.5 that informational differences among traders is one reason (but not the only one) that trading takes place in a sequence of markets, rather than in one grand initial market. However, informational differences among traders can be incorporated into the model of this subsection in much the same way that the extended Arrow–Debreu model incorporates such differences into the original Arrow–Debreu model. To do so explicitly at this point would, I think, obscure rather than clarify the issues raised by the sequential nature of the markets. One issue that is peculiarly related to differential information, namely the possibility that traders might draw inferences from equilibrium prices about other traders' information, will be postponed until Section 4.2.

In the perfect foresight approach one represents a trader's preferences about present and future prices by a *price expectation function*, which gives the price system at each date–event pair. Each trader is assumed to make a plan that specifies his consumption and trade commitments for each date–event pair. The plans of the traders are *consistent* if the total planned excess supply of every trade commitment is zero at every date–event pair. Roughly speaking, an equilibrium of plans and price expectations is a common price expectation function and a

[17]See Billingsley (1968) and Parthasarathy (1967) for material on weak convergence of probability measures.

consistent set of plans, one for each trader, such that, given the common price expectation function, each trader's plan is optimal for him, subject to an appropriate sequence of budget constraints.

Suppose that the economy extends through a finite set of dates, $t=1,\dots,T$. A *consumption plan* for a trader is a vector $x=(x_m)$, where m ranges over the set M of all elementary date–event pairs and where, for each m, x_m is a vector in R^H, interpreted as the vector of quantities of commodities consumed by the trader at date–event pair m. A *trade plan* is a vector $z=(z_m)$, where, for each m in M, z_m is a vector of commitments on the market at m (see Section 2). A *plan* for a trader is a pair (x, z), where x is a consumption plan and z is a trade plan.

Let $Z=\times_m Z_m$ (see Section 2), and $X=R^{|M|H}$. A consumption plan is a point in X, a trade plan is a point in Z, and a price expectation function is a point in Z. Since the set of dates is finite, the assumptions of Section 2 imply that the set M of elementary date–event pairs is finite, and that Z is finite-dimensional.

The assumptions that will be made below will imply that, at each date–event pair m, demand in the market at m is homogeneous of degree zero in the prices at m; furthermore, in equilibrium prices will be non-negative. Hence it will be convenient at the outset to restrict the vector of prices at m to the set P_m of non-negative vectors in Z_m whose coordinates sum to unity. Let P denote the Cartesian product of the sets P_m, as m runs over the set of all date–event pairs.

A trader i is characterized by:

(1) a consumption set X_i, a subset of X;
(2) a preference pre-ordering, $\precsim_i$, on X; and
(3) a vector $w_i=(w_{im})$ of resource endowments, where for each elementary date–event pair m in M, w_{im} is a vector in R^H.

The concept of an allowable commitment was introduced in Section 2, Condition (2.6); a trade plan will be called *allowable* if every coordinate is an allowable commitment. Furthermore, it is natural to require that a trader not make commitments that he will not be able to honor. Recall that for two date–event pairs $m=(t, e)$ and $n=(u, f)$ in M, $m\leqq n$ if $t\leqq u$ and e is an initial subsequence of f; for a trade plan z write

$$\bar{z}_n^h = \sum_{(t,e)\leqq n} z_{tu}^h(f); \qquad \bar{z}_n = (\bar{z}_n^h). \tag{4.1}$$

The number $\bar{z}_n^h$ is the total net delivery of commodity h at date–event pair n committed in contracts on or before n.

A consumption–trade plan (x_i, z_i) is *feasible* for i, given the price system p, if:

(a) x_i is in X_i;
(b) z_i is allowable [see (2.4) and (4.1)];

(c) $\bar{z}_{in} \leqq w_{in} - x_{in}$, for every n in M; and
(d) $p_m z_{im} = 0$, for every m in M. (4.2)

The set of plans feasible for i, given p, will be denoted by $\Gamma_i(p)$.

Condition (4.2c) requires that the trader not plan to deliver at any date–event pair more than he would have available from his resources after subtracting his consumption. The inequality in (4.2c) expresses "free disposal". Condition (4.2d) is a sequence of budget constraints, one for each date–event pair.

The behavior of a trader is summarized in his *behavior correspondence*, denoted by γ_i, which is defined by

> for each price system p in p, $\gamma_i(p)$ is the set of plans (x, z) in $\Gamma_i(p)$ that are optimal with respect to his preferences. (4.3)

Note that $\gamma_i(p)$ may be empty for some p.

I shall assume that, for each trader i,

> X_i is closed and convex, and there is a vector $\bar{x}_i$ such that $x \geqq \bar{x}_i$ for all x in X_i; (4.4)
>
> for every x in X_i and every m in M, there is an x' in X_i, differing from x only at m, such that $x'_i >_i x_i$; (4.5)
>
> there is an $\tilde{x}_i$ in X_i such that $\tilde{x}_i \ll w_i$; (4.6)
>
> and his pre-ordering is continuous and convex. (4.7)

The assumption (4.5) of "non-satiation" at each date–event pair implies that the trader has a positive subjective probability for each event (provided his preference pre-ordering permits of a scaling in terms of subjective probability and utility). Recall that assumption that $\lesssim_i$ is convex implies that the trader does not exhibit a "preference for risk" (see Section 3.4).

An *equilibrium of plans and price expectations* is an array $[(x_i, z_i), p]$ of plans (one for each trader) and a price expectation function such that:[18]

(a) p is in P;
(b) (x_i, z_i) is in $\gamma_i(p)$, for each i; and
(c) $\Sigma_i z_{im} \geqq 0$, and $p_m \Sigma_i z_{im} = 0$, for every m in M. (4.8)

Theorem 4.1

If Assumptions (4.4)–(4.7) are satisfied, then the pure exchange economy has an equilibrium of plans and price expectations.

[18]When there is no risk of ambiguity, a reference to "i" is to be understood as a reference to a trader, i.e. to a number from 1 to I.

Theorem 4.1 cannot be derived as a direct application of Theorem 3.1, because there is a separate budget constraint for each date–event pair (4.2d). Nevertheless, a similar method can be used, which I shall now sketch.[19] For every trader i, define his *excess supply correspondence* by

$$\sigma_i(p)=\{z|(x,z)\in\gamma_i(p), \text{ for some } x\}, \tag{4.9}$$

and define the *total excess supply correspondence* by

$$\sigma(p)=\sum_i \sigma_i(p). \tag{4.10}$$

For every m there is a compact subset, say Z_m, of Z_m such that one can restrict the total supply correspondence to have values in $\hat{Z}\equiv \times_m \hat{Z}_m$ without changing the set of equilibria. [This is a technique familiar from Debreu (1959, ch. 5)]. Let $\hat{\sigma}$ denote the corresponding restriction of σ. For every z in $\hat{Z}$, define $\mu(z)$ to be the set of $p=(p_m)$ in P that minimize $\sum_m p_m z_m$ in P. Notice that $p^*=(p_m^*)$ is in $\mu(z)$ if and only if, for every m, p_m^* minimizes $p_m z_m$ in P_m. If we now apply the Kakutani fixed-point theorem to the correspondence $\sigma\times\mu$, as in the proof of Theorem 3.1, we get, corresponding to (4.8),

$$p_m z_m^* \geqq p_m^* z_m^*, \quad \text{for every } m \text{ and every } p_m \text{ in } p_m. \tag{4.11}$$

From (4.2d), summed over all traders i, we get Walras' Law *at each* m:

$$p_m^* z_m^* = 0, \quad \text{all } m \text{ in } M. \tag{4.12}$$

We can now proceed, at each m, as in the proof of Theorem 3.1.

In the standard Arrow–Debreu model an equilibrium is an optimum, under very general conditions. However, Hart (1975) has given an example of an equilibrium of plans and price expectations that is not optimal, even though the example satisfies the standard conditions on preferences, etc.

4.2. *Rational expectations equilibrium in a market with differential information*

Imagine now that we are to extend the sequence-of-markets model to allow for informational differences among the traders, much as the Arrow–Debreu model was extended in Section 3.4. The same apparatus for representing differential information will be applicable, and most of the same remarks in Sections 3.4 and 3.5 concerning the nature of equilibria will also be applicable.

[19]For a complete proof, see Radner (1972).

However, one issue that was not discussed there may be of particular importance when there is a sequence of markets. In the sequence economy, equilibrium prices at different dates will typically be statistically dependent, so that some traders would be able to improve their forecasts of future prices by taking account of this dependence, i.e. by conditioning their forecasts on present and past equilibrium prices as well as on their own (exogenous) non-price information signals. In particular, the value today of an asset that can be bought today and resold at some future date depends on the anticipated probability distribution of its future price, which in turn may well be correlated with its current equilibrium price. This would be the case if the asset had some qualities that were imperfectly known to some traders today, but would become public information in the future.

As a matter of fact, this phenomenon is not restricted to assets that are traded at more than one date. For example, the entire supply of some vintage wine may be bought up by private persons (i.e. not dealers) quite soon after it is bottled, and well before it is to be drunk. The buyers will then have made their purchases before knowing with certainty the eventual quality of the wine, but some buyers will have better information than others. Under these circumstances, the buyers with poor non-price information may improve their decisions by "judging quality by price." Effectively, such buyers will be attempting to infer from the (equilibrium) price something about the information available to the better informed buyers. Such inferences, however, will have a feedback on the equilibrium price, so that the relationship between the traders' information and the equilibrium price will be affected.

These considerations have led to the study of so-called "rational expectations equilibria", about which I made some introductory remarks in Section 3.3 (which I shall not repeat here). Although the most important applications of this concept are to markets in securities and other assets that are traded at many dates, it is possible to elucidate some of the more basic difficulties concerning existence of equilibria in a simpler model without explicit reference to a sequence of markets. This is what I shall do in the present subsection, where I present a model of asset trading in which, generically, there exists a rational expectations equilibrium that reveals to all traders the initial information possessed by all of the traders taken together. In this model there are finitely many states of initial information, and the concept of rational expectations used requires that each trader know the relationship between initial information and equilibrium prices.

The fact that there are only finitely many states of initial information plays a crucial role in the argument that establishes the generic existence of equilibrium. More generally, it would seem that what is important is that the dimension of the space of initial information should be less than the dimension of the price space [see Allen (1977, 1979)], although this topic has not been fully explored. In Section 4.3 I provide a robust example of non-existence of rational expectations equilibrium, and in Section 4.4 I briefly describe Futia's analysis of a sequence-of-markets model with linear excess demand functions.

The proof of generic existence with finitely many states of initial information demonstrates, in addition, a remarkable information efficiency property of equilibrium. Generically, *a rational expectations equilibrium reveals to all traders the information possessed by all of the traders taken together*. Seen in a broader context, this property of equilibrium might cast doubt on the incentives for a trader to obtain information about the environment prior to entering the market, provided he could count on other traders obtaining the same information, which would then be revealed by market prices. But if each trader reasoned in this way, then no trader would obtain prior information, and so there would be no prior information for market prices to reveal! However, to examine this question more carefully one needs a model that reflects the dynamics of market adjustment and price formation [see, for example, Jordan (1979b)].

The proof of these results on the generic existence and informational efficiency of rational expectations equilibrium is based on an auxiliary proposition that has some independent interest. Roughly speaking, this auxiliary proposition concerns the comparison of ordinary exchange equilibria under uncertainty in which traders have (subjective) probability beliefs about the payoff-relevant state of the environment. The auxiliary proposition gives conditions under which, generically, two exchange economies that differ only in the traders' probability distributions will have different equilibrium prices.

I shall first sketch the model and the argument leading to the main results. This will be followed by a precise description of the model. [For details of the proof see Radner (1979).]

4.2.1. *Pure exchange under uncertainty*

Consider first a conventional model of a pure exchange economy with uncertainty. The decision problem for trader i is to choose a vector of assets, which will be called his *portfolio*. The eventual utility to i of his portfolio is uncertain at the time he purchases it. This is expressed by saying that his utility depends on his portfolio and on the exogenous environment. Each trader has a subjective probability distribution on the set of alternative environments, and we suppose that his criterion for choosing among alternative portfolios is expected utility. Given his initial endowment of assets, and given a vector of asset prices, he will demand a portfolio that maximizes his expected utility subject to the budget constraint that the cost of his portfolio not exceed the value of his initial endowment. For simplicity, I shall suppose that only non-negative portfolios are allowed (no short sales). His *excess supply* is the vector of differences between his initial endowments and the assets he demands. Suppose that his utility function is sufficiently regular so that, for any vector of prices, his excess supply is unique. Let p denote the vector of asset prices, let π denote the array of subjective probability distributions of the environment (one for each trader), and let $Z(p, \pi)$ denote the corresponding *total excess supply*, i.e. the (vector) sum of the individual traders' excess supplies. Given a probability array π, an *equilibrium* is a price vector for which the total excess supply is zero, i.e. a solution p of the equation system

$$Z(p, \pi)=0. \tag{4.13}$$

With sufficient regularity conditions on the traders' utility functions, at least one equilibrium will exist for every probability array.

The auxiliary proposition deals with the question: Under what conditions must two different probability arrays lead to different corresponding equilibria? Call a pair (π_1, π_2) of probability arrays *confounding* if there is a solution (p_1, p_2) of the equation system

$$Z(p_1, \pi_1)=0; \qquad Z(p_2, \pi_2)=0; \qquad p_1=p_2. \tag{4.14}$$

In other words, for a confounding pair of probability arrays, there exists a single price vector that is an equilibrium for each of them. On the other hand, if a pair of probability arrays is *not* confounding, then *any* corresponding pair of equilibrium price vectors will be distinct.

It is easy to give examples of "textbook" utility functions for which confounding pairs of probability arrays exist [see Radner (1979)]. However, since the equation system (4.14) has more equations than unknowns, it would seem plausible that, if the equations were in "general position" then no solution would exist. To make this idea precise, assume that the set E of alternative states of the environment is finite; then a probability distribution on E can be represented as a point in a vector space of dimension $\#E-1$, where $\#E$ denotes the number of states in E (recall that the probabilities must sum to unity). A probability array can be represented by a point in a space of dimension $I(\#E-1)$, where I is the number of traders, and the set Π_0 of pairs of probability arrays lies in a space of dimension $2I(\#E-1)$. The equation system (4.14) is thus parameterized by points in Π_0.

Since asset prices are relative, it will be understood that they are to be normalized, say by taking the sum to be unity. Hence, if there are n assets, there are $(n-1)$ independent prices, so (4.14) has $2(n-1)$ unknowns (for every parameter point in Π_0). On the other hand, there are nominally $3n$ equations in (4.14). However, if the utility functions are such that every budget is exhausted, then the value of excess demand will always be zero, even out of equilibrium (Walras' Law). Furthermore, as noted above, there are only $(n-1)$ independent prices, so the condition $p_1=p_2$ represents only $(n-1)$ independent equations. Hence there are at most $3(n-1)$ independent equations in (4.14). This is still larger than the number of unknowns, however, so that one would not typically expect (4.14) to have a solution.

Call a subset of Π_0 *negligible* if its closure has Lebesgue measure zero. The auxiliary proposition gives conditions under which *the set of confounding pairs of probability arrays is negligible*. In terms of the equation system (4.14), this conclusion can be interpreted as follows. Let C_0 be the set of confounding pairs in Π_0, and let $\bar{C}_0$ be the closure of C_0 in Π_0. If a parameter point is in C_0 (i.e. confounding), then every neighborhood of it has a parameter point for which (4.14) has no solution. On the other hand, if a parameter point is not in $\bar{C}_0$ (and is therefore not confounding), then there is some neighborhood of it such that for *all* points in the neighborhood the system (4.14) has no solution. (In addition, there may be points of $\bar{C}_0$ that are not confounding, i.e. not in C_0.)

In another terminology, if a property holds except on a negligible set, one says that it holds *generically*. In this terminology, the conclusion of the auxiliary proposition is that, *generically*, *different probability arrays give rise to different equilibrium prices*.

4.2.2. *Full communication equilibria and revealing prices*

At the next stage in the analysis, I introduce the concepts of full communication equilibrium and revealing prices. Consider a pure exchange situation similar to the one just described, except that before the market activity takes place some exogenous information about the environment is made available to *all* the traders. To express the idea that this information may be incomplete or noisy, let f denote the information signal, and suppose that every trader has a subjective joint probability distribution of the signal f and the environment e. [Strictly speaking, since the exogenous information signal f is also part of the traders' environment, one should now call e *the payoff-relevant part of the environment* (see Marschak (1963)), but for simplicity I shall continue to call e the *environment*.] Given the information signal f, each trader's preferences among portfolios will be determined by his *conditional* expected utility, using his conditional probability distribution of e given f. Denote this conditional distribution by (the vector) π_{fi}, and for each f denote the array (π_{fi}) of distributions, one for each trader, by π_f. Corresponding to each signal f, an equilibrium price vector, p_f, is a solution of the equation system

$$Z(p_f, \pi_f)=0. \tag{4.15}$$

A full communication equilibrium (FCE) is a mapping $\hat{\phi}$, from information signals to price vectors such that, for each signal f, $\hat{\phi}(f)$ is an equilibrium for f, i.e. a solution p_f of (4.15). (The reason for the terminology "full communication" will become apparent later.) I shall say that a FCE is *revealing* if it is one-to-one, i.e. it maps different signals into distinct price vectors. (Recall that all price vectors are normalized.) Thus, if market prices are determined by a revealing FCE, then one can infer the underlying signal from observing the market prices.

Assume that the set F of alternative information signals is finite, and let π denote the (finite) array of probability arrays π_f. The set Π of arrays π has dimension $(\#F)I(\#E-1)$. Also, if F is finite, the function $\hat{\phi}$ is actually a finite-dimensional vector, with $(\#F)n$ coordinates (where n is the number of assets); because of the normalization of prices, $\hat{\phi}$ in fact lies in a set of dimension $(\#F)(n-1)$. Thus, a point π in Π is a vector of parameters for the system of equations

$$Z[\hat{\phi}(f), \pi_f]=0, \quad \text{for every } f \text{ in } S; \tag{4.16}$$

this is a system of finitely many equations in finitely many unknowns.

Observe that, for any parameter point π, every FCE is revealing if and only if, for every pair (f, f') of distinct signals, the pair $(\pi_f, \pi_{f'})$ is not confounding. It is easy to check that if a set $C_0(f, f')$ of pairs $(\pi_f, \pi_{f'})$ is negligible in the corresponding space of dimension $2I(\#E-1)$, then the set $C(f, f')$ of points π in Π for which $(\pi_f, \pi_{f'})$ is in $C_0(f, f')$ is also negligible. Let C be the set of points π in Π such that, for some distinct f and f', the pair $(\pi_f, \pi_{f'})$ is confounding. Since the set F is finite, and the union of finitely many negligible sets is negligible, it follows that, if for every distinct f and f' the set of confounding pairs is negligible, then the set C is also negligible. Hence the conclusion of the first auxiliary proposition (above) implies that, *generically, in* Π, *a full communication equilibrium is revealing.*

4.2.3. *Differential information and rational expectations equilibrium*

At the final stage in the analysis, I consider the situation in which different traders come to the market with possibly different exogenous information signals. Let f_i denote the exogenous information signal available to trader i, and let $f=(f_1,\ldots, f_I)$ denote the total exogenous information available in the market. Thus, each trader may have only a part of the total information available. Each trader i has a subjective joint probability distribution of f and e.

As a preliminary thought experiment, imagine that, given his information signal f_i, each trader chose among portfolios according to his conditional expected utility, using his conditional probability distribution of e given f_i. This would generate an excess supply function for each trader i, given f_i, and thus would generate a total excess supply function for all traders, given f. An "exogenous information equilibrium" price vector, given f, would be one for which this total excess demand would be zero. For each f let $\phi(f)$ be a corresponding exogenous equilibrium price vector, and suppose that every trader behaves in such a way that $\phi(f)$ is in fact the market price if f is realized and each trader i observes f_i.

Now imagine that, after a number of independent realizations of this situation, a particular trader, say number 1, becomes "sophisticated" and realizes that there is a regular relationship between the total information signal and the market price; this relationship is, of course, described by the mapping ϕ. Trader 1 would then be able to infer something about the total information signals from his observations of the market price, $\phi(f)$ (unless, of course, the market price were the same for all f). This would change his excess demand function, since the market price would not only enter his budget constraint, but would also provide a supplementary signal – in addition to his exogenous signal f_i – on which to condition his expected utility. But if his excess demand were not an insignificant part of the total, this "sophisticated" behavior would change the total excess demand function, too, which would change the relation ϕ between total exogenous information and market price! In fact, if *all* traders became sophisticated in this manner, then the original exogenous equilibrium price vectors $\phi(f)$ would typically no longer clear the market given the total exogenous information signal f. What would be required would be a "forecast function" ϕ that would be self-fulfilling.

These preliminary considerations motivate the following formal definitions. A *forecast function* ϕ is a mapping that associates with each total exogenous information signal f a price vector $\phi(f)$. Given a forecast function ϕ, suppose that each trader i chooses among portfolios according to his conditional expected utility given the (augmented) information $[f_i, \phi(f)]$.

This behavior will generate, for each total signal and each trader, an excess supply; call the resulting total excess supply the *sophisticated excess supply*. It should be emphasized that this sophisticated excess supply depends on the forecast function ϕ and on the particular total signal f; denote it by $\sigma(f,\phi)$. A rational expectations equilibrium (REE) is a forecast function such that the corresponding sophisticated excess supply is zero for every total information signal, i.e. a function ϕ such that

$$\sigma(f,\phi)=0, \quad \text{for all } f \text{ in } F. \tag{4.17}$$

Note that sophisticated supply behavior requires the knowledge of the entire function ϕ to determine the supply corresponding to any single f, and that the equation system (4.17) is *a system of simultaneous equations in the values of the function* ϕ.

We can now see the connection between full communication and rational expectations equilibria. Consider a FCE $\hat{\phi}$ based on the total information signals. In our present terminology, the FCE $\hat{\phi}$ is a particular forecast function. Recall that $\hat{\phi}$ is *revealing* if it is one-to-one. Suppose that the FCE were revealing; then for every trader and every total information signal f, the conditional probability distribution of e given $\hat{\phi}(f)$ would be the same as the conditional probability distribution of e given f. Hence the sophisticated supply of trader i given f_i and $\hat{\phi}(f)$ would equal his ordinary supply given f. By the equilibrium property of $\hat{\phi}$, the ordinary total excess supply, given that every trader knows f and that the price vector is $\hat{\phi}(f)$, is zero. It follows that the FCE $\hat{\phi}$ is a forecast function that satisfies (4.17). Thus, we see that *a revealing full communication equilibrium is a rational expectations equilibrium.*

One can now state the main result about REEs as a corollary of the preceding observation and the auxiliary proposition. Under the assumption of the auxiliary proposition, *generically, there exists a rational expectations equilibrium that is revealing.*

4.2.4. A precise model

Consider a pure exchange economy with I traders. The exogenously determined characteristics of the economy, called here the *environment*, will be represented as a point e in a finite set E. Before making any trade, each agent i has available exogenously some information, f_i, a point in a finite set F_i. I shall call f_i the *signal* received by trader i. For trader i, the variables $e, f_1, \ldots, f_I$ have a joint probability distribution, say Q_i. In principle, Q_i could be different for different traders, reflecting different subjective beliefs. Let F be the Cartesian product of the signal sets F_i; then Q_i is a probability measure on the Cartesian product $E \times F$. A point f in F represents all the exogenous information available jointly to all the traders.

The decision problem for trader i is to choose a vector, say x_i, of "assets." Assume that there are H assets, and that x_i must be non-negative. Trader i's initial endowment is w_i. If p denotes the asset price vector, then i's budget constraint is

$$px_i \leq pw_i. \tag{4.18}$$

The eventual value (utility) to i of x_i depends upon the environment e; denote it by $u_i(x_i, e)$. Trader i's criterion for choosing among alternative portfolios is his conditional expected utility, given the information available to him. This information includes both the (exogenous) signal f_i *and the market price.*

For any particular behavior on the part of the traders, the equilibrium market price will be a function of f, the joint signal. Let $\boldsymbol{P}$ denote the set of all non-negative price vectors, normalized so that the sum of all prices is unity. Call a function from F to $\boldsymbol{P}$ a *forecast function*. Given a forecast function ϕ, a signal f_i, and a price vector p, trader i's *demand* is the vector x_i that maximizes his conditional expected utility, given f_i and $\phi(f)=p$,

$$\mathrm{E}_{Q_i}\{U_i(x_i,e)|f_i,\phi(f)=p\}, \tag{4.19}$$

subject to the budget constraint (4.18). (I shall make assumptions that guarantee that each trader's demand is unique, given his signal and the price vector.) If trader i's demand is x_i, the total market *excess* supply is

$$\sum_i (w_i - x_i).$$

Given the forecast function ϕ, excess supply will depend on the joint signal f, directly through the individual traders' signals f_i, and indirectly through the price vector $p=\phi(f)$. Therefore, I shall denote the total excess demand by $\sigma(f,\phi)$. A rational expectations equilibrium (REE) is a forecast function ϕ^* such that

$$\sigma(f,\phi^*)=0, \quad \text{for every } f \text{ in } F. \tag{4.20}$$

I shall now specify the model in somewhat more detail, in order to represent more precisely a situation of trading in assets that are to be sold (or valued) at a later date. Suppose that the vector x_i has the form (c_i, y_i) where

$c_i = i$'s expenditure on current consumption, and
$y_i = i$'s portfolio, with K different assets (in the strict sense of the word).

Thus, c_i is a non-negative real number, and y_i is a non-negative K-dimensional vector. The value of one unit of asset k next period will be v^k; however, at the time of the current market traders are uncertain about the vector $v=(v^k)$ of future asset values. The vector v can take on one of finitely many values v_e, e in E; e is the *payoff-relevant environment*. If trader i chooses (c_i, y_i), and environment e obtains, then the future value of his portfolio will be the inner product $v_e' y$; assume that the utility of this outcome to him is

$$u_i(x_i,e)=U_{0i}(c_i)+U_i(v_e' y_i). \tag{4.21}$$

We may interpret $U_i(Y)$ as the indirect utility of starting next period with wealth Y.

One can postulate conditions that will guarantee that at equilibrium all prices are positive. In this case it would be legitimate to normalize prices so that the price of current consumption is 1; let q denote the vector of asset prices.

Recall that w_i denotes trader i's vector of initial endowments. We may interpret the first coordinate of w_i as i's endowment of "cash". If the coordinates of w_i are numbered $0,\ldots, K$, and $W_i \equiv (w_i^1,\ldots, w_i^K)$ (i's asset endowment), then trader i's budget constraint is

$$c_i + q'y_i \leqq w_1^0 + q'W_i. \tag{4.22}$$

Having normalized the price of consumption to be unity, only the K asset prices are variable. It is therefore convenient to change our terminology slightly, and to call a forecast function from F to R_+^K. Given ϕ, q, and f_i, a trader i's demand is a (c, y) that maximizes his conditional expected utility

$$\mathrm{E}_{Q_i}\{U_{i0}(c) + U_i(v_e'y) | f_i, f \text{ in } \phi^{-1}(a)\}, \tag{4.23}$$

subject to the budget constraint (4.22). Let $\xi_{if_i}(q, \phi)$ denote i's demand for assets; the *total* excess demand for assets is

$$\sigma_f(q, \phi) \equiv \sum_i [pW_i - \xi_{if_i}(q, \phi)]. \tag{4.24}$$

A rational expectations equilibrium (REE) is a forecast function ϕ^* such that

$$\sigma_f[\phi^*(f), \phi^*] = 0, \quad \text{every } f \text{ in } F. \tag{4.25}$$

Note that (4.25) is a set of simultaneous equations in all of the values of the forecast function ϕ^*.

One is now in a position to carry out the program of proof outlined above. Under suitable regularity conditions [see Radner (1979)], one can prove;

1. The set of confounding pairs (π_1, π_2) is negligible in Π_0.
2. Except for a negligible subset of Π, for every probability array in Π every corresponding full communication equilibrium is revealing.
3. Except for a negligible subset of Π, for every probability array in Π there is a corresponding rational expectations equilibrium that is revealing.

The above propositions do not, in themselves, exclude the existence of non-revealing rational expectations equilibria. However, under suitable conditions one can show that such equilibria occur in a negligible set.

4.3. *A robust example of non-existence of rational expectations equilibrium*

If the set of environments and information signals is not finite, and the dimension of the price space does not exceed the dimension of the signal space, then the existence of rational expectations equilibrium need not be generic, as the following example shows.[20]

Suppose that there are two traders ($i=1,2$) and two "assets". For the purposes of this subsection only, denote trader i's "portfolio" by $x_i=(y_i, z_i)$, and assume that his preferences are of the "Cobb–Douglas" form, representable by

$$u_i(x_i, e)=\alpha_i(e)\log y_i+[1-\alpha_i(e)]\log z_i, \qquad 0<\alpha_i(e)<1, \tag{4.26}$$

where e in E denotes the payoff-relevant state of the environment [cf. (3.18), but take the logarithm]. Suppose that the environment e is represented by a real number between 0 and 1, so that E is the unit interval. In the utility functions (4.26), assume that

$$\alpha_1(e)=\frac{1+e}{3}; \qquad \alpha_2(e)=\frac{2-e^2}{3}. \tag{4.27}$$

The information structure is such that trader 1 learns e but trader 2 does not. In other words, trader i's information signal is e, whereas trader 2's signal is a constant. Let trader 2 have a uniform prior probability distribution of e on the unit interval. Finally, let the initial endowment of each trader be $(1,1)$.

Again, normalize prices so that the price vector is $p=(q,1-q)$. We may take prices to be non-negative, so that q is between 0 and 1. Thus, a forecast function may be taken to be a function[21] from E (the unit interval) to the set of possible prices of the first "asset" (also the unit interval).

Since the endowment of each trader is $(1,1)$, the wealth of each trader is $q+(1-q)=1$, independent of prices. Let ϕ be a forecast function. If the price of asset 1 is $q=\phi(e)$, then trader 1's demand is easily verified to be

$$\begin{aligned} y_1&=\frac{\alpha_1(e)}{q}=\frac{1+e}{3q}, \\ z_1&=\frac{1-\alpha_1(e)}{1-q}=\frac{2-e}{3(1-q)}. \end{aligned} \tag{4.28}$$

Trader 2, not knowing e directly, will maximize his conditional expected utility given $\phi(e)=q$. Let

$$\begin{aligned} \bar{\alpha}_2&=\mathrm{E}[\alpha_2(e)|\phi(e)=q], \\ \bar{e}^2&\equiv\mathrm{E}[e^2|\phi(e)=q]; \end{aligned} \tag{4.29}$$

[20] This example is based on Jordan and Radner (1979).
[21] More precisely, a measurable function.

then trader 2's demand, given $\phi(e)=q$, is

$$\begin{aligned} y_2 &= \frac{\bar{\alpha}_2}{q} = \frac{2-\bar{e}^2}{3q}, \\ z_2 &= \frac{1-\bar{\alpha}_2}{1-q} = \frac{1+\bar{e}^2}{3(1-q)}. \end{aligned} \tag{4.30}$$

Excess supply equals zero if

$$y_1+y_2=2, \qquad z_1+z_2=2. \tag{4.31}$$

Putting together (4.28)–(4.31), and solving for q, one gets

$$q=(\tfrac{1}{6})(3+e-\bar{e}^2).$$

Thus, the condition for a forecast function ϕ to be a rational expectations equilibrium is

$$\phi(e)=(\tfrac{1}{6})\big(3+e-\mathrm{E}\big[e^2|\phi(e)\big]\big). \tag{4.32}$$

I shall now show that the equilibrium condition (4.32) for a REE has no solution. We can rewrite (4.32) as

$$e=6\phi(e)+\mathrm{E}\big[e^2|\phi(e)\big]-3. \tag{4.33}$$

Therefore, if $\phi(e)=\phi(e')$, then $e=e'$, or equivalently, if $e\neq e'$, then $\phi(e)\neq\phi(e')$. In other words if ϕ is a REE then it is one-to-one. But if ϕ is one-to-one, then $\mathrm{E}[e^2|\phi(e)]=e^2$, and therefore, by (4.32),

$$\phi(e)=(\tfrac{1}{6})(3+e-e^2), \tag{4.34}$$

which is *not* one-to-one; indeed, from (4.34), $\phi(e)=\phi(\frac{1}{2}-e)$. Hence there is no solution ϕ to (4.32), and hence no REE.

Now consider a whole family of economies, obtained from the foregoing example by changing the utility functions of the traders.

To be precise, let U be the set of all utility functions u with arguments y, z, and e (where y and z are positive real numbers, and e is in the unit interval) such that:

(i) u is twice continuously differentiable;
(ii) for every state e, $u(y, z, e)$ is strictly increasing and strictly concave in y and z; and

(iii) for every state e, the indifference curves do not touch the y and z axes.

Consider the set A of economies obtained by specifying a utility function u_1 for trader 1 and a utility function u_2 for trader 2, with u_1 and u_2 in U. (Formally, A is the Cartesian product $U \times U$.) We may think of A as a very large, indeed infinite-dimensional, "parameter space". In particular, the foregoing example is a particular economy in A; call it A^0. One can show that, if one puts a suitable topology on U (namely, the C^2 Whitney topology), and the corresponding product topology on A, then *there is an open set* A^0 *in* A *containing* A^0, *such that for every economy in* A^0 *there exists no REE*. Roughly speaking, one can perturb the utility function in the example A^0 a small amount in any "direction" and still not have an economy for which a REE exists.

One can, in fact, extend the example to allow the traders' initial endowments and beliefs to vary as well, but I omit the details.

4.4. *Rational expectations equilibria in a stationary linear model*

A thorough analysis of rational expectations equilibria in a sequence of markets has been given for a special case in which there is a market for a single commodity, the excess supply equations are linear (in a sense to be explained), and the underlying payoff-relevant events and the information signals are stationary Gaussian processes [Futia (1979)]. I shall give here a very brief account of the model and some of Futia's results.

Consider a doubly infinite sequence of dates, t, with a market for a single commodity at each date, and let p_t denote the price of the commodity at date t. Each trader has a (random) non-negative endowment of the commodity at each date; let e_t denote the total endowment at t. Each trader i also observes at each date some information signal, f_{it}, a random vector.

Before describing the excess supply function, I shall describe the underlying probability space and the family of stochastic processes that will be considered. Let ε_{tj} be a family of independent Gaussian (normal) random variables, each with zero mean and unit variance, where $j=1,\ldots, J$, and $-\infty<t<\infty$, and let ε_t denote the J-dimensional vector with coordinates ε_{tj}. Let H be the family of all random variables of the form

$$\sum_t \alpha_t \cdot \varepsilon_t, \tag{4.35}$$

where (α_t) is a doubly infinite sequence of J-dimensional vectors whose lengths are square-summable. Note that such a random variable is Gaussian, with zero mean and variance equal to

$$\sum_t \alpha_t \cdot \alpha_t .$$

For a random variable h in H, given by (4.35), define the *shift of* h, denoted by Th, by

$$Th = \sum_t \alpha_t \cdot \varepsilon_{t-1}. \tag{4.36}$$

A sequence (h_t) of random variables in H will be called *stationary* if there is a random variable h in H such that $h_t = T^t h$, for all t. A sequence of random vectors with coordinates in H will be called *stationary* if it is stationary coordinate-by-coordinate. We may take the underlying probability space to be the space S of sequence (ε_t). It is to be understood that all random variables considered in this subsection are in H.

I shall assume that the sequences (e_t) and (f_{it}) are stationary. For each trader i let $\boldsymbol{F}_{it}$ denote the sigma field generated by the random vector $\{f_{im}: m \leqq t\}$, and let $\boldsymbol{F}_t$ denote the sigma-field obtained by pooling the sigma-fields $\boldsymbol{F}_{it}$, $i = 1, \ldots, I$. Since it is natural to suppose that each trader can observe his own endowment, it is equally natural to assume that, for every t, e_t is measurable with respect to $\boldsymbol{F}_t$. Finally, let $\boldsymbol{P}_t$ denote the sigma-field generated by the current and past prices, namely $\{p_m: m \leqslant t\}$.

I shall consider prices that form a stationary sequence of random variables (in H). Trader i's conditional expectation of p_{t+1} given his information at date t (including his past and current information signals, and the past and current prices) is

$$\tilde{p}_{it} = \mathrm{E}\{p_{t+1} | \boldsymbol{F}_{it}, \boldsymbol{P}_t\}. \tag{4.37}$$

Assume that the total excess supply at date t is

$$\alpha p_t - \sum_i \tilde{p}_{it} + e_t, \tag{4.38}$$

where α is a parameter that is strictly greater than I. A (stationary) rational expectations equilibrium is a stationary price sequence such that excess demand at every date is (almost surely) zero.

Just as in Section 4.2, one can define a full communication equilibrium (FCE) in which, at every date, each trader has the pooled information $\boldsymbol{F}_t$. One can show that, *in this model, there exists a unique FCE, say* $\hat{p} = (\hat{p}_t)$. Call the FCE *informative* if, for every i and t,

$$\mathrm{E}\{\hat{p}_{t+1} | \boldsymbol{F}_{it}, \boldsymbol{P}_t\} = \mathrm{E}\{\hat{p}_{t+1} | \boldsymbol{F}_t\}. \tag{4.39}$$

If $p^* = (p_t^*)$ is a rational expectations equilibrium, call p^* *symmetric* if, for every i, j and t, traders i and j would (almost surely) make the same conditional forecast of p_{t+1}^*; in other words, $\hat{p}_{it}^* = \hat{p}_{jt}^*$.

Theorem 4.2

There exists a symmetric rational expectations equilibrium if and only if the corresponding full communication equilibrium is informative; in this case the two equilibria are equal.

Since the condition (4.39) is well-defined for any price sequence, it also makes sense to ask whether a REE is informative. However, by definition, an informative REE is symmetric; hence, by Theorem 4.2, a REE is informative if and only if it is the FCE. I should emphasize that an informative equilibrium need not be revealing (in the sense of Section 4.2). Nevertheless, an informative equilibrium enables each trader to make as good a conditional prediction of the next period's price as if he has the pooled information of all traders.

The space H of random variables can be identified with the space of square-summable sequences (a_t), and hence is a Hilbert space. Furthermore, every stationary sequence of random variables in H is generated by shifting a single random variable in H. Finally, it can be shown that every closed subspace of H can be identified with the sigma-field generated by some collection of random variables in H. These considerations lead to a natural topology on the set of models I have just described. With this topology one has:

Theorem 4.3

The set of models with no REE, and the set of models with only asymmetric REE's, each have a non-empty interior.

Theorem 4.3 shows that, even in the special case of a linear model with Gaussian random variables, the situation in which the environment–information space is infinite is quite different from the one in which it is finite. In particular, it is no longer the case that the existence of REEs is generic.

Finally, I should point out that the same methods of analysis can be applied to the case in which price information is used only with a lag.

5. Production

5.1. *Introduction*

In Section 1 I sketched how production is treated in the Arrow–Debreu model. A production plan for a producer specifies his net output at every date–event pair. Since the markets for contingent delivery are complete, for any given price system the present value of a production plan is determined, so the ordering of production plans by present value is well-defined. Although these present values are shared among the consumers, there is no difference of opinion about the values of these shares (given a price system), and so there is no scope for a non-trivial market in shares. These comments are also applicable to the extended Arrow–Debreu model.

Markets for shares can be modelled in a more significant way once one begins to take account of a sequence of markets. Some of the basic ideas are introduced in Section 5.3 in the context of a two-period model in which the production plan of each firm is fixed. In this constrained framework a limited optimality property of the stock market can be explained. A general sequence-of-markets model with production and stock markets is described in Section 5.4, which is really an extension of Section 4.1. Here the existence and optimality of equilibrium is problematic, and some of these issues are discussed further in Section 5.5.

The concept of valuation equilibrium, which is closely related to that of competitive equilibrium equilibrium (see Section 5.3), arises naturally in the theory of optimal economic growth. Here the prices that figure in the equilibrium are the "shadow prices" or Kuhn–Tucker–Lagrange multipliers. Some relevant references for the theory of growth under uncertainty are given in the bibliographic notes. See also Chapter 26 of this Handbook.

5.2. *Production in the Arrow–Debreu model*

At an abstract level, production can be introduced into the Arrow–Debreu model very simply. For producer j, let y_{jt} denote the random vector of net outputs at date t, and let y_j denote the sequence (y_{jt}). The random vector y_{jt} must be measurable with respect to the sigma-field $\boldsymbol{F}_t$ that represents the information available to all agents at date t. The sequence y_j will be called the *production plan* of producer j; he has available to him some set, Y_j, of feasible production plans, called his *production set*.

Recall that a price system in the Arrow–Debreu model is a sequence, $p=(p_t)$, of vectors such that p_t is measurable with respect to $\boldsymbol{F}_t$. The *present value* of the production plan y with respect to the price system p is the sum of inner products, as in (2.5),

$$py=\sum_t p_t y_t=\sum_{(t,e)} p_t(e)y_t(e).$$

Each consumer i is endowed both with a vector w_i of commodities and a vector $f_i=(f_i^j)$ of shares, where the coordinates of f_i are fractions, and the sum of the shares in any one firm is unity. Given the price system p, and the production plans y_j, one for each firm j, consumer i's wealth in units of account is

$$pw_i+\sum_j f_j py_j.$$

Each consumer chooses a most preferred consumption plan in his consumption set (see Section 3) subject to the constraint that the cost of his consumption plan not exceed his wealth in units of account.

In this manner the model of an economy with uncertainty has been reduced formally to a model with certainty. The extension of this to take account of differences among the traders in the information available to them (the extended Arrow–Debreu model) is straightforward. I have already informally discussed the limitations of the Arrow–Debreu model in Section 1, and shall not repeat those comments here. However, this is a good point to briefly explain why a model of a production set that expresses a *choice* among information structures is likely to be non-convex.

Actually, there are several sources of possible non-convexity. First, the acquisition of information may require an expenditure of resources that is independent of the scale of production. In other words, the expenditure for information may be a fixed cost which typically destroys convexity of the production set.

Second, suppose that a producer has available a family of alternative information structures, to each of which corresponds a production set. The producer's total production set is thus the union of the individual sets. Even if the individual sets were convex, their union need not, in general. be convex, and would be "unlikely" to be so if there were only finitely many alternative information structures.

If the individual production sets vary continuously with the information structure, there are special circumstances under which their union is convex. Nevertheless, some recent theoretical explorations suggest that non-convexity may be common when there is a choice of information structure [see Radner and Stiglitz, (1976) and R. Wilson (1975)], even when the other aspects of the situation are quite regular.

5.3. Stock markets and optimal allocation in a two-period model with fixed production plans

In the absence of complete markets for contingent future delivery of commodities, a market for the shares of future output or profits of production units provides to consumers a limited opportunity for trading in contingent deliveries. I shall introduce the theoretical analysis of stock markets under uncertainty with a very simple two-date model that brings out their allocative role for consumers. In this model, the plans of the production units are given, the outputs of the production units at the second date are uncertain, and at date 1 consumers trade in commodities for current consumption and in shares of physical outputs of the firms at date 2. I shall show that such a market can sustain a *constrained* Pareto-optimal allocation of commodities available for consumption at both dates; the nature of the constraint is that the proportion of each production unit's output at date 2 allocated to any one consumer may not vary from one state of the environment to another. This "optimality theorem" has not been extended to general models of sequences of markets with production, but nevertheless it suggests an interesting insight into the usefulness of stock markets as an institution for allocation under uncertainty.

Consider now a model with 2 dates, H commodities at each date, m consumers, and n production units. The plan for each production unit is given. After setting aside the inputs into production, there is a total quantity w^h of commodity h available for allocation to the consumers; let w denote the vector (w^h). There is uncertainty about the outputs of the several production units of date 2. This is formalized by assuming that there is a finite set S of alternative states of the environment, and that the vector of outputs of units j at date 2 depends on the state s that actually obtains; let $y_j(s)$ denote this vector. The total quantity of commodity h available for consumption at date in 2 in state s is

$$y^h(s)=\sum_j y_j^h(s). \tag{5.1}$$

Normally, one would consider all possible allocations of the vector $y(s)$ at date 2 in state s. However, we shall consider a constrained set of allocations of the form

$$x_{i2}(s)=\sum_j f_i^j y_j(s), \tag{5.2}$$

where f_i^j is the fraction of the output of unit j allocated to consumer i, and this fraction is the same for all states s.

Let $|S|$ denote the number of states in S. A *consumption plan* for consumer i is a pair (x_{i1}, x_{i2}), where x_{i1}, a vector in R^H, represents i's consumption at date 1, and $x_{i2}=(x_{i2}(s))$, a vector in $R^{H|S|}$, represents i's consumption at date 2 in each state s in S. Let X_i denote i's set of feasible consumption plans; X_i is a subset of $R^{H(1+|S|)}$. We shall assume that i has a preference pre-ordering $\lesssim_i$, on X_i.

Let F denote the set of vectors $f=(f^j)$ in R^H such that, for each j, $0\leqq f^j\leqq 1$. An *allocation* to i is a pair (x_{i1}, f_i) such that x_{i1} is in R^H and f_i is in F. An allocation determines i's consumption plan according to (5.2). Let A_i denote the set of allocations to i that correspond to consumption plans in X_i; the preference pre-ordering on X_i includes a corresponding preference pre-ordering on A_i, which will also be denoted by $\lesssim_i$.

An *allocation* is an m-tuple of allocations to the several consumers. An allocation $((x_{i1}, f_i))$ is *attainable* if

(a) for every i, (x_{i1}, f_i) is in A_i;
(b) for every j, $\Sigma_i f_i^j = 1$; and
(c) $\Sigma_i x_{i1} = w$.

I shall denote by A the set of attainable allocations. An attainable allocation a^* is *Pareto optimal* if there is *no other* attainable a' such that no consumer prefers a' to a^* and at least one consumer strictly prefers a' to a^*.

We shall be interested in characterizing Pareto-optimal allocations as equilibria relative to some price system. To make this precise, I use the concept of valuation equilibrium. Let p denote the vector of commodity prices at date 1; and let v denote the vector of share prices, i.e. v^j is the price of 100% of the output at date 2 of production unit j. A *valuation equilibrium* is an $(m+2)$-tuple $(((x_{i1}^*, f_i^*)), p, v)$, where

(a) $((x_{i1}^*, f_i^*))$ is an attainable allocation; and
(b) for every i if (x_{i1}, f_i) is in A_i, and if $(x_{i1}, f_i) \gtrsim_i (x_{i1}^*, f_i^*)$

then

$$px_{i1} + vf_i \geqq px_{i1}^* + vf_i^*. \tag{5.3}$$

Condition (5.3) says that, at the given commodity and share prices, any allocation to i that he wants as much as (x_{i1}^*, f_i^*) also costs as much.[22]

With the model formulated in this way, one can apply directly the standard theory of the relationship between Pareto-optimal allocations and valuation equilibria [see, for example Debreu (1959, ch. 6)]. An allocation $a_i \equiv (x_{i1}, f_i)$ in A_i will be called a *satiation allocation for* i if there is no other allocation in A_i that is strictly preferred by i to a_i. Let $a = (a_i)$ denote an attainable allocation. A valuation equilibrium (a, p, v) will be said to be *above minimum cost* if, for every i, the cost of a_i (given p and v) is strictly greater than the minimum cost of allocations in A_i.

Theorem 5.1

Assume that, for every i, the consumption set X_i is convex and the preference preordering $\lesssim_i$ is continuous and convex. If (a, p, v) is a valuation equilibrium that is above minimum cost, and such that no a_i is a satiation allocation for i, then a is Pareto optimal. Conversely, if a is a Pareto-optimal allocation such that,

[22] The term "valuation equilibrium" was introduced by Debreu (1954). His definition was, however, slightly different; essentially, my condition (5.3b) was replaced by the condition that an allocation that costs no more than (x_{il}^*, f_i^*) is not preferred to it.

for some i, a_i is not a satiation allocation for i, then there exists a price system $(p, \underset{\cdot}{v})$ such that (a, p, v) is a valuation equilibrium.

Theorem 5.1 provides no direct information about the efficacy of the stock market in guiding the choice of production plans, nor is this simple two-date model adequate for an analysis of the problems of a sequence of markets. In the next section we consider the question of the existence of equilibrium in a sequence of markets.

5.4. *Equilibrium of plans and price expectations with production and stock markets*

I shall now sketch a model of equilibrium of plans and price expectations along the lines of Section 4.1, but incorporating production decisions and stock markets.[23] It will be seen that, in this model, the existence of equilibrium is not assured, even with "neoclassical" assumptions about producers and consumers.

Consider again the model of Section 2, but here the traders are producers and consumers. A producer is characterized by a set of technologically feasible (intertemporal) production plans; a production plan specifies the net output of every commodity at every date–event pair. Each producer chooses a production plan and a trade plan; these two must be consistent in the sense that, at every date–event pair, the total of past and present contracts for current delivery must equal current net output. Given a price system $p=(p_m)$, a particular trade plan $z_j=(z_{jm})$ for producer j determines a *net revenue vector* $r_j=(r_{jm})$, where

$$r_{jm}=p_m z_{jm}. \tag{5.4}$$

A key assumption about producer behavior is that each producer has a preference pre-ordering on the set of all possible revenue vectors. This preference pre-ordering is taken as a datum (see comments below), and expresses the producer's beliefs concerning the relative likelihoods of events, as well as his attitude toward risk. Given a commodity price system, each producer is assumed to choose a production–trade plan that maximizes his preferences among corresponding revenue vectors.

At each date–event pair m, each producer's net revenue r_{jm} is distributed (in units of account) among the "shareholders" (the consumers) according to shares held at the immediately preceding date–event pair. These shares are traded on a stock market at each date–event pair, and shareholders retain their shares from one date to the next if they do not sell them. If "retained earnings" are interpreted as inputs into current production, then r_{jm} can be interpreted as producer j's total dividend at date–event pair m. Notice that in this model,

[23]For a detailed development, see Radner (1972).

as distinct from that of Section 5.3, consumers receive their dividends in units of account rather than in physical commodities.

Each consumer chooses a consumption plan, a trade plan, and a portfolio plan. The consumption and trade plans were described in Section 4.1. At any date–event pair $m=(t, e)$, the *portfolio* of consumer i is a vector $f_{im}=(f^j_{im})$, where f^j_{im} is consumer i's share of producer j's net revenue at date $(t+1)$, i.e. in *every* elementary event at date $(t+1)$. A portfolio at any date–event pair is thus represented by a vector in F, the set of vectors (f^j) in R^n such that, for *every* j, $0 \leqq f^j \leqq 1$. A *portfolio plan* for consumer i is a vector $f_i=(f_{im})$ such that, for every m, f_{im} is in F. Let v^j_m denote the price of 100% of the shares of producer j at date–event pair m; $v_m=(v^j_m)$ is the vector of share prices at m, and $v=(v_m)$ is the *share price system*. Given v_m, the market value of i's portfolio f_{im} at m is the inner product $v_m f_{im}$.

Each consumer faces a budget constraint at each date–event pair $m=(t, e)$. On the credit side at m are three items: (1) the market value of his portfolio carried forward from the *previous* date $(t-1)$, calculated at the *current* share prices v_m; (2) his share of dividends at m, as determined by that portfolio; and (3) the market value of the *positive* components of z_{im}, calculated at the commodity prices p_m. On the debit side at m are two items: (1) the cost of his *new* portfolio at m; and (2) the market value of the *negative* components of z_{im}. At every m his debits must not exceed his credits. A consumer is characterized by a consumption set, a preference pre-ordering, and resource endowments, just as in Section 4.1, and in addition he has initial endowments of shares at the first date. Given a commodity price system and a share price system, each consumer is assumed to choose a consumption–trade–portfolio plan to maximize his preferences (concerning consumption) subject to the budget constraints at all date–event pairs.

An *equilibrium of plans and price expectations* is an array of plans, one for each consumer and producer, and a system (p, v) of commodity and share prices such that, given p and v, each agent's plan is optimal, and such that, at each date–event pair, the excess supply of every commodity and every share is zero. Recall that $p=(p_m)$ and $v=(v_m)$ can be interpreted as *common price expectation functions* (see Section 3.3).

Closely related to the concept of equilibrium is that of *pseudo-equilibrium*; a pseudo-equilibrium is defined just as an equilibrium, except that the condition of market clearing is replaced by the following condition: at each date–event pair the value of total excess supply is smaller at the pseudo-equilibrium prices than at any other prices. (At each date–event pair the current commodity and share prices are jointly normalized, e.g. by taking their sum to be unity.) It can be shown that one can naturally interpret the value of excess supply as the difference between planned savings and planned investment.

Theorem 5.2

If

(a) every consumer satisfies conditions (4.4)–(4.7) of Section 4.1, and furthermore initially holds a positive endowment of every share;
(b) the total initial supply of shares is entirely held by consumers;
(c) every producer's production possibility set is closed and convex, and satisfies the condition of free disposal,[24]
(d) the total production possibility set satisfies the condition of irreversibility of production,[25]
(e) every producer's preferences on net revenue can be represented by a continuous, strictly concave utility function;

then there exists a pseudo-equilibrium. Furthermore, a pseudo-equilibrium in which all share prices are positive ($v \gg 0$), and the share market is cleared at the initial date, is an equilibrium.

The reader is referred to Radner (1972) for a proof of Theorem 5.2. The main difficulty is the proof that the budget sets of the consumers are continuous correspondences; once this is done the proof can proceed along lines similar to that of Theorem 3.1.

The gap between pseudo-equilibrium and equilibrium is discussed in the next section.

5.5. *Comments on the assumptions and on alternative formulations*

In the present model the "shareholders" have *un*limited liability, and therefore have a status more like that of partners than of shareholders, as these terms are usually understood. One way to formulate limited liability for shareholders is to impose the constraint on producers that their net revenues be non-negative at each date–event pair. The reason this formulation has not been adopted is that I have not yet found reasonable conditions under which the producers' correspondences could be shown to be upper hemicontinuous. This is analogous to the problem that arises when, for a given price system, the consumer's budget constraints force him to be on the boundary of his consumption set. In the case of the consumer, this situation is avoided by a condition such as (4.7) [for weaker conditions, see Debreu (1959, notes to ch. 5, pp. 88–89) and Debreu (1962)]. However, for the case of the producer, it is not considered unusual in the

[24, 25] For definitions of these terms see Debreu (1959, ch. 3) or Radner (1972, p. 295).

standard theory of the firm that, especially in equilibrium, the maximum profit achievable at the given price system could be zero (e.g. in the case of constant returns to scale).

It would be interesting to have conditions on the producers and consumers that would directly guarantee the existence of an equilibrium, not just a pseudo-equilibrium. In other words, under what conditions would the share markets be cleared at every date–event pair? Notice that if there is an excess supply of shares of a given producer j at a date–event pair (t, e), then at date $(t+1)$ only part of the producer's revenue will be "distributed". One would expect this situation to arise only if his revenue is to be negative in at least one event at date $t+1$; thus at such a date–event pair the producer would have a deficit covered neither by "loans" (i.e. not offset by forward contracts) nor by shareholders' contributions. In other words, the producer would be "bankrupt" at that point.

One approach might be to eliminate from a pseudo-equilibrium all producers for whom the excess supply of shares is not zero at some date–event pair, and then search for an equilibrium with the smaller set of producers, etc. successively reducing the set of producers until an equilibrium is found. This procedure has the trivial consequence that an equilibrium always exists, since it exists for the case of pure exchange[26] (the set of producers is empty!) This may not be the most satisfactory resolution of the problem, but it does point up the desirability of having some formulation of the possibility of "exit" for producers who are not doing well.

Although the present model does not allow for "exit of producers (except with the modification described in the preceding paragraph), it does allow for "entrance" in the following limited sense. A producer may have zero production up to some date, but plans to produce thereafter; this is not inconsistent with a positive demand for shares at preceding dates.

The creation of new "equity" in an enterprise is also allowed for in a limited sense. A producer may plan for a large investment at a given date–event pair, with a negative revenue. If the total supply of shares at the preceding date–event pair is nevertheless taken up by the market, this investment may be said to have been "financed" by shareholders.

The assumptions of Section 5.4 describe a model of producer behavior that is not influenced by the shareholders or (directly) by the prices of shares. A common alternative hypothesis is that a producer tries to maximize the current market value of his enterprise. There seem to me to be at least two difficulties with this hypothesis. First, there are different market values at different date–event pairs, so it is not clear how these can be maximized simultaneously. Second, the market value of an enterprise at any date–event pair is a price, which is supposed to be determined, along with other prices, by an equilibrium of supply and

[26] Provided consumers have free disposal; see Section 4.1.

demand. The "market-value-maximizing" hypothesis would seem to require the producer to predict, in some sense, the effect of a change in his plan on a price *equilibrium*; in this case, the producers would no longer be price-takers, and one would need some sort of theory of general equilibrium for monopolistic competition.

I have already alluded, at the end of Section 4.1, to the possible non-optimality of equilibrium. Indeed, the question of what concept of optimality is most appropriate here has not been fully explored.

6. Adjustment, stability, and stationary disequilibrium

6.1. Introduction

Thus far I have restricted my attention to cases in which at every date the market is at least in temporary equilibrium. If the time interval between dates is short relative to the speed of adjustment of prices, then before prices can adjust to a temporary equilibrium the economy will already be at the next date, and the environment will have changed. One will then observe a process of repeated incomplete adjustment, together with stochastic changes in the environment, and the economy will always be in disequilibrium in the sense that markets will never (or rarely) clear. Nevertheless, even in this case of repeated disequilibrium one would want to distinguish situations in which prices and quantities fluctuated in some "steady" manner around long-run averages, from situations in which prices or quantities, or both, fluctuated with greater and greater variance, or increased without bound. To describe the first situation it is natural to use the concept of a *stationary stochastic process*, which is the generalization to the case of uncertainty of the concept of a deterministic equilibrium. However, it is important to emphasize that the stationarity of a stochastic process does not rule out fluctuations of varying period and amplitude.

One will also be interested in conditions under which a stochastic process of adjustment is "stable" in the sense that it converges to a stationary process. I shall discuss here three examples of such stability. In the first (Section 6.2), each tâtonnement iteration is followed by a change in the environment, and the successive environments are independent and identically distributed. One can interpret the example as a model of a sequence of spot markets in which no stocks are carried over from one date to the next. The price at each date depends on the price and the excess supply at the previous date, and the excess supply at each date depends on the current price and state of the environment. The resulting sequence of prices is a Markov process. This example has the property that in "equilibrium", i.e. when the process is stationary, expected (or long-run average) excess supply need not be zero.

In the second example (Section 6.3), inventories of the commodities are possible, and excess supplies at any date are added to inventories. With assumptions that are reminiscent of "diagonal dominance" conditions, one can demonstrate a strong form of stability in the Markovian case, and a weaker form of stability in the non-Markovian case. Because of the carryover of excess supplies, expected excess supply is zero in equilibrium.

The third example (Section 6.4) concerns adjustment towards a rational expectations equilibrium in a special linear-Gaussian model like that of Section 4.4.

6.2. *Stochastic adjustment without inventories*

In this subsection I shall describe an example of stochastic adjustment without inventories that has been explored by Green and Majumdar.[27] Consider a sequence of markets in which, for each date t, p_t is the vector of prices, z_t is the vector of excess supplies, and s_t is the exogenous state of the environment. The successive states of the environment are assumed to be independent and identically distributed. The successive prices and excess supplies are determined by

$$z_t = \zeta(p_t, s_t); \qquad p_{t+1} = \alpha(p_t, z_t), \tag{6.1}$$

where ζ is a given *excess supply function*, and α is a given *price adjustment function*. The two equations in (6.1) can be combined into a single relation that determines the next price vector as a function of the current price vector and state of the environment:

$$p_{t+1} = \beta(p_t, s_t) = \alpha[p_t, \zeta(p_t, s_t). \tag{6.2}$$

Since the successive states of the environment are independent and identically distributed, the successive price vectors form a Markov process. The (marginal) probability distribution, say π_t, of p_t will in general depend on t and on the initial distribution, π_0. Let P_0 denote the set of normalized price vectors at any one date. A probability measure π on P_0 is called *invariant* if $\pi_0 = \pi$ implies that $\pi_t = \pi$, for all t. We may call an invariant probability measure an *equilibrium price distribution*. This terminology is consistent with the terminology of Section 6.1 in the sense that, if π_0 is an equilibrium price distribution, then the sequence of price vectors forms a stationary process.

It is important to emphasize, however, that the price process may be stationary, *and yet the long-run average of the excess supply need not be zero*!

I now present a set of assumptions on the excess supply and price adjustment functions that will guarantee the existence of an equilibrium distribution. Recall

[27] The specific conditions for stability described here are taken from Majumdar (1972).

that every S_t is a copy of S_0, and that I have assumed for simplicity that S_0 is finite.

(a) $\zeta(\cdot, s)$ is continuous on P_0 for every s in S_0.
(b) If $p_n \to p$ in P_0, and p is not strictly positive, then the length of $\zeta(p_n, s)$ increases without limit as n increases, for all s in S_0.
(c) For every commodity h there is a $\delta_h > 0$ such that for any price vector p, $p^h \leqq \delta_h$ implies that $\zeta^h(p, s) > 0$, for all s in S_0.
(d) α is continuous on $P_0 \times Z_0$.
(e) There exists an $\varepsilon > 0$, less than every δ_h in (c), such that for every commodity h, every p in P_0, and every z in Z_0,

$$|\alpha^h(p, z) - p^h| < \varepsilon.$$

(f) $-[\alpha^h(p, x) - p^h]$ is sign-preserving for z^h.

Theorem 6.1

Under assumptions (a)–(f), there exists an equilibrium price distribution.

Theorem 6.2

If one adds the assumption that, for every s in S_0, $\beta(\cdot, s)$ is a contraction mapping on P_0, then for any initial price distribution π_0, the sequence π_t will converge to an equilibrium price distribution that is independent of π_0, i.e. the price process is stable.

The reader is referred to Green and Majumdar (1975) for proofs of these theorems.

6.3. *Stochastic adjustment with inventories*[28]

In a large number of markets, inventories and back-orders serve as buffers against random variations in supply and demand. I shall use the term *stock* to denote the level of an inventory (positive stock) or back-orders (negative stock). The levels and movements of stocks provide important signals about past and current excess demands, and these signals in turn influence the movements of prices. On the other hand, prices are influenced by signals other than stocks, and themselves serve as signals that influence supplies and demands, and thus also the stocks.

The assumptions made here about price and stock adjustment are similar in spirit to the Diagonal Dominance assumption used by Lionel McKenzie to study

[28]This subsection is based on Hildenbrand and Radner (1979).

the stability of a tâtonnement adjustment process in the deterministic case. [See McKenzie, (1960); see also Arrow, Block, and Hurwicz (1959).] These assumptions are not directly comparable with Diagonal Dominance, however, because of the difference in the model.[29] The assumptions about adjustment can be paraphrased as follows: for every commodity, when its stock is sufficiently high (low) its own money price will fall (rise) *on the average*, and when its price is sufficiently high (low) its stock will rise (fall) on the average.

I shall also make use of two other assumptions: (B) the (one-period) increments of prices and stocks are uniformly bounded, and (M) the stochastic process of prices and stocks is Markovian, with stationary transition probabilities and a discrete state space. With these assumptions one can show that the Markov process converges to a stationary process, which as I have noted is the stochastic analog of an "equilibrium". The assumptions do not preclude a multiplicity of equilibria, but they do insure that there are only finitely many of them. In general, from any starting state there will be a probability distribution on the set of ultimate equilibria, i.e. on the set of alternative stationary processes to which the process may converge.

In this model, excess supply is not mentioned explicitly. If excess supply (positive or negative) in each period is added to the current stock, and there is no attrition of stocks, then the stationarity of stocks implies that long-run average excess supply will be zero. This will also be the case if attrition is suitably symmetric.

Even without the Markovian assumption (M) one can prove a stability-like result: there is a bounded set D such that, from every initial state, every trajectory (almost surely) enters D, and the expected time to the first entry is finite.

Consider a sequence of markets; at each date t the state of the market is characterized by a price vector, p_t, and a stock vector, v_t. Prices are constrained to be non-negative. Stocks may be positive or negative; negative stocks are interpreted as accepted but unfilled orders. For the purpose of this subsection only, the *state of the economy* at date t is denoted by $s_t=(p_t, v_t)$. (Note that this departs from the notation elsewhere in this chapter in that s_t is not purely exogenous.)

I shall make four assumptions about the sequence (s_t), which are listed below. The first (S) is that (s_t) is a stochastic process. In the second part of the theorem I strengthen this assumption by assuming that the stochastic process (s_t) is *Markovian* with stationary transition probabilities and discrete state space. The second assumption, (B), is technical, but essential; it requires the increments in the variables p_t^h, v_t^h to be uniformly bounded. The third and fourth assumptions [(PA) and (SA)] concern price and stock adjustment, respectively. These last two

[29] In addition to being deterministic, the model used by McKenzie (and others in that literature) differs from ours in that the state variables are prices and *excess demands*, rather than prices and stocks. For a general discussion of such tâtonnement models, see, for example, Arrow and Hahn (1971, chs. 11 and 12).

assumptions can be paraphrased as follows: for every commodity, when *its* stock is sufficiently high (low), its own price will fall (rise) *on the average*, and when its price is sufficiently high (low) its stock will rise (fall) *on the average*.

I now give the precise statements of these four assumptions:

(S) *The process* (p_t, v_t) *is a stochastic process defined* on some probability space $(\Omega, \boldsymbol{F}, \Phi)$. Let $\boldsymbol{F}_t$ denote the sub-sigma-field of $\boldsymbol{F}$ generated by partial histories of the process up through date t, i.e. $\boldsymbol{F}_t$ is the smallest sub-sigma-field of $\boldsymbol{F}$ such that the random variables $s, \ldots, s_t$ are measurable. All statements about random variables defined on $(\Omega, \boldsymbol{F}, \Phi)$ are to be understood as holding Φ-almost surely.

(B) *The coordinates of* p_t *and* v_t *have uniformly bounded increments*. This bound will be denoted by γ, i.e. for every h and t, $|p_{t+1}^h - p_t^h| \leqslant \gamma$ and $|v_{t+1}^h - v_t^h| \leqslant \gamma$.

(PA) *For every commodity h there exist numbers* $v^{*h} > 0, \alpha^{*h} > 0, v_*^h \leqslant 0$ and $\alpha_*^h > 0$ such that:

(1) $\mathrm{E}(p_{t+1}^h | \boldsymbol{F}_t) \leqslant \max\{0, p_t^h - \alpha^{*h}\}$ on $\{v_t^h \geqslant v^{*h}\}$;
(2) $\mathrm{E}(p_{t+1}^h | \boldsymbol{F}_t) \geqslant p_t^h - \alpha_*^h$ on $\{v_t^h \leqslant v_*^h\}$.

(SA) *For every commodity h there are positive numbers* $p^{*h}, \beta^{*h}, p_*^h, \beta_*^h$ with $p_*^h < p^{*h}$ such that:

(1) $\mathrm{E}(v_{t+1}^h | \boldsymbol{F}_t) \geqslant v_t^h + \beta^{*h}$ on $\{p_t^h \geqslant p^{*h}\}$;
(2) $\mathrm{E}(v_{t+1}^h | \boldsymbol{F}_t) \leqslant \max(0, v_t^h - \beta_*^h)$ on $\{p_t^h \leqslant p_*^h\}$.

Theorem 6.3.

Assumption (S), (B), (PA), and (SA) imply that there exists a bounded set $D \subset R^H \times R^H$ of states of the stochastic process (p_t, v_t) such that Φ-almost every trajectory intersects D and the expected time for the first intersection is finite, i.e.

$$\mathrm{E}(T) < \infty, \quad \text{where } T(\omega) = \inf\{t | (p_t(\omega), v_t(\omega)) \in D\}.$$

Furthermore, if the stochastic process (p_t, v_t) is Markovian with stationary transition probabilities and discrete state space, then there is a partition of the state space into finitely many positive recurrent classes and a set of transient states that contains no closed set.

The first part of the theorem can be strengthened. One can show that the set D is *positive recurrent* in the sense that every trajectory spends a positive fraction of time D, i.e. there exists a number $c>0$ such that

$$\liminf_t \frac{1}{t} \sum_{k=0}^{t-1} 1_D(s_k) \geqslant c, \qquad \Phi\text{-a.s.},$$

where the symbol 1_D denotes the indicator function for the set D. [For a proof of this stronger result, see Föllmer (1977).]

6.4. *Adjustment towards rational expectations equilibrium*

The theory of rational expectations equilibrium (Section 4.2) postulates behavior that makes heavy demands on the rationality of the economic agents, insofar as it requires them not only to make the usual expected utility calculations in the standard model of a market with uncertainty, but also to have a correct model of the joint distribution of equilibrium prices, their own initial information, and the eventual values of the assets in which they are trading – their "market models". In a theory of adjustment towards a rational expectations equilibrium (REE), what are the appropriate assumptions about the agents' rationality during the "learning" process? As agents revise their market models, the true market model changes in a way that, in principle, depends on the revision rules of all of the agents. Thus, a theory of thorough-going rationality would seem to point to a treatment of the learning and adjustment process as a sequential game with incomplete and imperfect information.[30]

In my opinion such an approach would be unrealistic and contrary to the spirit of a process of adjustment and learning. A more realistic alternative would envisage some form of bounded rationality during the adjustment process, which, if stable, would converge to a fully rational equilibrium. I shall sketch here two closely related processes of this type, in the context of a linear model similar to that of Section 4.4, but even simpler.[31]

Consider again I traders in successive markets for a single commodity; the markets will be dated $t=1,2,\ldots$, ad. inf. In contrast with the model of Section 4.4, the eventual "value" of a unity of commodity bought at date t will not be its price in a future spot market, but rather some exogenous random variable r_t (as in the case of wine laid down for future consumption). Let f_{it} denote trader i's exogenous information signal at t, and p_t the price of the commodity at t; as in Section 4.4, $\boldsymbol{F}_{it}$ will denote the sigma-field generated by the current and past

[30] See Harsanyi (1967, 1968) for the now-standard analysis of equilibrium in games with incomplete information.

[31] Existence and comparative statics of REE in linear models of this type have been studied by Grossman and others; see Section 7 for bibliographic notes. The results sketched here for the present model are based on unpublished notes by Radner, and Bray (1980).

information signals, $\{f_{im}: m \leqq t\}$, and $\boldsymbol{P}_t$ will denote the sigma-field generated by the current and past prices, $\{p_m: m \leqq t\}$. Define $\tilde{r}_{it}$ to be i's forecast of r_t, given $\boldsymbol{F}_{it}$ and $\boldsymbol{P}_t$, and assume that i's excess demand at t is the linear function

$$\delta \tilde{r}_{it} - \delta' p_t, \quad \text{where } \delta, \delta' > 0. \tag{6.3}$$

(For simplicity of exposition I have assumed that the parameters δ and δ' in the demand function are the same for all traders. See Section 7.6.)

Imagine that, in addition to any possibly negative excess demand by the traders who are explicitly described above, there is an exogenous supply, e_t, which is a random variable. The total excess supply at date t is therefore

$$\alpha p_t - \delta \sum_i \tilde{r}_{it} + e_t, \quad \text{where } \alpha \equiv I\delta'. \tag{6.4}$$

Assume further that, for each i and t,

$$f_{it} = r_t + y_{it}, \tag{6.5}$$

that the random variables $e_t, r_t, y_{1t}, \ldots, y_{It}$ are all Gaussian and mutually independent (among themselves and through time), and that the successive $(I+2)$-tuples $(e_t, r_t, y_{1t}, \ldots, y_{It})$ are identically distributed. (One may interpret f_{it} as a "measure" of r_t based on i's non-price information, and y_{it} as the measurement error.) To simplify the exposition, and without essential loss of generality (with the qualification noted below), I assume that all of the above random variables have mean zero; define

$$\varepsilon = \mathrm{E}e_t^2; \qquad \rho - \mathrm{E}r_t^2; \qquad \eta_i = \mathbf{E}y_{it}^2. \tag{6.6}$$

Because of the intertemporal independence assumption, in a REE the forecast $\tilde{r}_{it}$ would depend only on f_{it} and p_t, and because of the Gaussian assumption, would be linear, say

$$\tilde{r}_{it} = a_i^* + b_i^* p_{it} + c_{it}^* f_{it}. \tag{6.7}$$

This last, together with the condition that excess supply, (6.4), be zero, implies that

$$\left(\alpha - \delta \sum_i b_i^*\right) p_t - \delta \sum_i c_i^* f_{it} + e_t - \delta \sum_i a_i^* = 0. \tag{6.8}$$

Eq. (6.8) determines the "equilibrium" aspect of the REE; the "rational expectations" aspect is expressed by [32]

$$\mathbf{E}\{r_t | f_{it}, p_t\} = a_i^* + b_i^* p_t + c_i^* f_{it}, \qquad i=1,\ldots,I. \tag{6.9}$$

It is easy to check that (6.8) and (6.9) imply that

$$\mathbf{E}p_t = 0; \qquad a_i^* = 0, \qquad i=1,\ldots,I. \tag{6.10}$$

Eqs. (6.7)–(6.10) can now be summarized by

$$p_t = \frac{\delta \sum_i c_i^* f_{it} - e_t}{\alpha - \delta \sum_i b_i^*}, \tag{6.11}$$

$$\mathbf{E}\{r_t | f_{it}, p_t\} = b_i^* p_t + c_i^* f_{it}, \tag{6.12}$$

any solution of which is defined to be a REE. In spite of the apparent simplicity of the model, a complete analysis of the solutions of (6.11) and (6.12) does not (to my knowledge) yet exist. However, one can show that a solution exists (see Section 7.6). Incidentally, this provides an example of a REE that is not revealing (see Section 4.2).

Equations (6.11) and (6.12) suggest a first adjustment process, which is unrealistic but will serve to clarify some of the difficulties. Suppose that for some number of dates agent i uses a forecast function,

$$\tilde{r}_{it} = b_i p_t + c_i f_{it}, \qquad i=1,\ldots,I, \tag{6.13}$$

such that the coefficients (b_i, c_i) do not necessarily constitute a REE, i.e. are not a solution of (6.11) and (6.12). At each date at which the forecast functions (6.13) are used, the market clearing price will be

$$p_t = \frac{\delta \sum_i c_i f_{it} - e_t}{\alpha - \delta \sum_i b_i}; \tag{6.14}$$

call this a *temporary expectations equilibrium* (*TEE*). Suppose now that the agents use their respective forecast functions (6.13) for a large number of dates, so that agent i effectively learns the joint (Gaussian) distribution of (f_{it}, p_t, r_t). (Strictly speaking, an infinite number of dates would be required!) One can show that, for

[32] The definition of REE used here is slightly different from the one used in Section 4.2. The difference is discussed at the end of this section.

this distribution, the conditional expectation of r_t given p_t and f_{it} is

$$E\{r_t | p_t, f_{jt}\} = b_j' p_t + c_j' f_{jt}, \tag{6.15}$$

where

$$c_j' = \left[\frac{1}{\Delta_j}\right] \rho\delta^2 \left[\frac{\sum_i c_i^2 \eta_i}{I^2} + \frac{\mu}{\delta^2} - \frac{\bar{c} c_j \eta_j}{I}\right],$$

$$b_j' = \left[\frac{1}{\Delta_j}\right] (\delta' - \delta\bar{b}) \delta\rho\eta_j \left[\bar{c} - \frac{c_j}{I}\right],$$

$$\Delta_j = \delta^2 \left[\bar{c}\rho\eta_j \left[\bar{c} - \frac{2c_j}{I}\right] + (\rho + \eta_j) \left[\frac{\sum_i c_i^2 \eta_i}{I^2} + \frac{\mu}{\delta}\right] - \frac{c_j^2 \eta_j^2}{I^2}\right],$$

$$\bar{b} = \frac{1}{I} \sum_i b_i; \qquad \bar{c} = \frac{1}{I} \sum_i b_i; \qquad \mu = \frac{\varepsilon}{I^2}. \tag{6.16}$$

Note that each c_j' is a ratio of quadratic functions of all the coefficients c_i, and each b_j' is a function of both sets of coefficients. If we define the adjustment process by the difference equation

$$c_j(n+1) = c_j'(n); \qquad b_j(n+1) = b_j'(n),$$
$$j = 1, \ldots, I, \qquad n = 1, 2, \ldots, \text{ad. inf.}, \tag{6.17}$$

then any limit of this sequence is a REE. One can show that the system (6.17) may be locally stable or unstable, depending on the parameters. Roughly speaking, the system may be unstable if, *in the REE*, the coefficients b_i^* of current price are "too large" compared to the coefficients c_i^* of current non-price information. This can occur, for example, if the variance ε of the exogenous supply is sufficiently small.

The difference equation (6.17) approximately represents a situation in which agents revise their forecast coefficients infrequently, but only on the basis of experience since the last revision. The following difference equation represents a situation in which, at each revision, agents average over *all* of their previous experience:

$$c_j(n+1) = \left[\frac{1}{n+1}\right] (c_j'(n) + c_j(n) + \cdots + c_j(1)), \tag{6.18}$$

and similarly for $b_j(n+1)$. One can show that, at least for certain special cases, (6.18) is stable when (6.17) is not.

A more satisfactory theory of adjustment would take account of the fact that agents must revise their forecast functions after only a *finite* number of periods, so that their successive coefficients constitute a *stochastic process*, rather than a deterministic one as in (6.17) or (6.18). For example, consider the process in which, after each date, each agent i uses ordinary least squares to estimate new forecast (regression) coefficients from all of the previous observations (f_{it}, p_t, r_t). For the purposes of this section, I shall call this the OLS process.

Unfortunately, the OLS process leads to highly non-linear stochastic difference equations, whose asymptotic behavior has not yet been analyzed at even the low level of generality of the present model. However, Bray (1980) has succeeded in demonstrating the stability of the OLS process in a simpler example.[33] Suppose that there are two sets of traders, the "informed", denoted by $\boldsymbol{N}$, and the "uninformed", denoted by $\boldsymbol{U}$. All informed tracers have the *same* non-price information.

$$f_{it} = r_t + y_t \equiv f_t, \qquad i \text{ in } \boldsymbol{N}, \tag{6.19}$$

whereas the uninformed traders have *no* non-price information,

$$f_{it} = \text{constant}, \qquad i \text{ in } \boldsymbol{U}. \tag{6.20}$$

Assume that the random variables r_t, y_t, and e_t are independent and Gaussian, and that the successive triples (r_t, y_t, e_t) are identically distributed, with mean zero and respective variances ρ, η, and ε. Assume, finally, that *the structure of information is common knowledge.*[34]

An implication of the last assumption is that the informed traders can infer that, in the TEE, the price p_t can depend only on the current and past non-price signals f_t (which are known to all informed traders), and hence provides no information about r_t not already revealed by f_t. Hence an informed trader can do no better than to use the forecast function

$$\tilde{r}_{it} = \mathrm{E}\{r_t | f_t\} = \left(\frac{\rho}{\rho + \eta}\right) f_t, \qquad i \text{ in } \boldsymbol{N}. \tag{6.21}$$

The uninformed traders' forecast functions must at most depend on prices, and in an REE will depend on current price. Thus, suppose that the uninformed traders' forecast functions are

$$\tilde{r}_{it} = b_{it} p_t, \qquad i \text{ in } \boldsymbol{U}. \tag{6.22}$$

[33] In order to unify the exposition in this section, I have used a version of Bray's model that is slightly simpler than the one in her paper.

[34] For a precise definition of "common knowledge", see Aumann (1976).

It is straightforward to verify that the equation for a TEE corresponding to (6.14) is

$$p_t = \frac{\delta N\left(\frac{\rho}{\rho+\eta}\right) f_t - e_t}{\alpha - \delta \sum_{i \in U} b_{it}}, \tag{6.23}$$

where $N = \#\boldsymbol{N}$ is the number of informed traders. Given the TEE (6.23), the true regression of r_t on p_t is

$$E(r_t | p_t) = b_t' p_t, \tag{6.24}$$

where

$$\begin{aligned} b_t' &= k\left[\frac{\delta'}{(1-\nu)\delta} - \bar{b}_t\right], \\ \bar{b}_t &= \frac{\sum_{i \in U} b_{it}}{I - N}, \\ \nu &= N/I, \\ k &= \frac{\nu(1-\nu)\delta^2\rho^2}{\nu^2\delta^2\rho^2 + \mu(\rho+\eta)}. \end{aligned} \tag{6.25}$$

[These equations correspond to (6.15) and (6.16).] Note that ν is the *fraction* of informed traders.

From (6.25) we see that in a REE the coefficients b_{it} will be the same for all uninformed traders, say b^*, where b^* is the (unique) solution of

$$b^* = k\left[\frac{\delta'}{(1-\nu)\delta} - b^*\right],$$

or

$$b^* = \frac{k\delta'}{(1+k)(1-\nu)\delta}. \tag{6.26}$$

In this example, the OLS adjustment process is defined as follows. The initial coefficients (b_{i1}) for the uninformed traders are given a priori. At each date $t+1 \geqq 1$ the price is determined by the TEE equation (6.23), and at each date $t > 1$

the coefficient $b_{i,t+1}$ is the ordinary least squares estimate of b in the equation $r=bp$, fitted to the data points $(p_1, r_1),\ldots,(p_t, r_t)$, i.e.

$$b_{i,t+1}=\frac{\sum_{n=1}^{t} r_n p_n}{\sum_{n=1}^{t} p_n^2}, \qquad t>1. \tag{6.27}$$

Note that for $t>1$, all uninformed traders will have the same coefficients b_{it}.

Theorem 6.4

For the OLS adjustment process defined by (6.23) and (6.27),

$$\lim_{t\to\infty} b_{it}=b^*, \quad \text{a.s.}$$

[For a proof of Theorem 6.4, see Bray (1980).] Bray also considers the case in which f_t and e_t are negatively correlated. In this case the parameter corresponding to k in (6.25) may be negative. A REE exists only if $k\neq-1$, and the convergence of the OLS process has been proved only for $k>-1$. It is conjectured, but not yet proved, that the OLS process is unstable for $k<-1$.

It is of interest to compare the asymptotic properties of the OLS process in the present example ($k>0$) with those of the deterministic process corresponding to (6.17). Here it is easy to see from (6.25) that the latter is stable if and only if $k<1$. On the other hand, the deterministic process corresponding to (6.18) is stable for *all* $k>0$. In some sense, the OLS process is a stochastic analogue of this second deterministic process.

The assumption that the structure of information is common knowledge plays an important role in achieving the simplicity of the Bray model. Without this knowledge the informed traders would not be justified in forecasting r_t from f_t alone, i.e. in ignoring current price, at least when the market is not in a REE. The effect of abandoning this assumption would be to make the OLS process multidimensional rather than one-dimensional.

The assumption that the means of r_t and e_t are zero plays a similar, but more innocuous, role. Suppose that we allow these means to be different from zero. Without loss of generality, one may take $\mathrm{E}f_{it}=\mathrm{E}r_t$, since one may replace f_{it} by $\mathrm{E}\{r_t|f_{it}\}$. Suppose that at date t trader i uses the forecast function

$$\tilde{r}_{it}=a_{it}+b_{it}p_t+c_{it}f_{it}. \tag{6.28}$$

If $\tilde{r}_{it}$ had the same (unconditional) mean as r_t (unconditionally unbiased forecast), then, from (6.28),

$$\mathrm{E}r_t=a_{it}+b_{it}\mathrm{E}p_t+c_{it}\mathrm{E}r_t. \tag{6.29}$$

It is straightforward to show that, for a TEE [use (6.14)], (6.29) implies

$$\mathrm{E}p_t = \left(\frac{1}{\alpha}\right)(\delta I \mathrm{E}r_t - \mathrm{E}e_t), \tag{6.30}$$

so that *the expected price does not depend on the particular coefficients in the traders' forecast functions*, provided that they are unconditionally unbiased, i.e. that (6.29) is satisfied. Thus, given b_{it} and c_{it}, a trader who knows (6.30) can choose a_{it} so that (6.29) is satisfied, provided all other traders are doing the same. This requirement is of course much less demanding informationally than the REE condition. Once (6.29) is satisfied, all random variables (including p_t) can be measured from their means, and there is no loss of generality in assuming those means to be zero.

I shall close this section with some remarks on the subtle but important difference between the definition of REE used here and the one used in Section 4.2. Suppressing the date index, let $\tilde{f}$ denote the I-tuple $(f_1, \ldots, f_I)$, the total non-price information available to the market. Recall that a REE in the sense of Section 4.2 – which I shall call here REE1 – is a function ϕ that maps the total market information $\tilde{f}$ into a price p, such that, if

$$\tilde{r}_i = \mathrm{E}\{r \mid f_i, \phi(\tilde{f})\}, \tag{6.31}$$

and price is determined by

$$p = \phi(\tilde{f}), \tag{6.32}$$

then total excess supply, (6.4), will be zero, almost surely, i.e.

$$\alpha p - \delta \sum_i \tilde{r}_i + e = 0, \quad \text{a.s.} \tag{6.33}$$

On the other hand, in the present section a REE is an I-tuple of coefficient pairs (b_i, c_i) satisfying (6.11) and (6.12); call this a REE2. To every REE2, $\{(b_i^*, c_i^*)\}$, corresponds a REE1, with $p = \phi(\tilde{f})$ given by (6.11).

For an example of a REE1 that is not a REE2, consider the case in which there is no uncertainity in the exogenous supply, e, and all traders have equally inaccurate non-price signals; in the present model this translates into $\varepsilon = 0$ and, for all i, $\eta_i = \eta$. (Note that $\varepsilon = 0$ implies that $\mu = 0$, and that $e = m = 0$, a.s.) Since $e = 0$, the market-clearing condition (6.33) is equivalent to

$$p = \left(\frac{\delta}{\alpha}\right)\sum_i \tilde{r}_i = \left(\frac{\delta}{\delta'}\right)\left(\frac{1}{I}\sum_i \tilde{r}_i\right). \tag{6.34}$$

Notice that, with our current assumptions,

$$\mathrm{E}\{r \mid f_i\} = \left(\frac{\rho}{\rho+\eta}\right) f_i, \qquad i=1,\dots,I. \tag{6.35}$$

This suggests consideration of a mapping of the form

$$\phi^0(\tilde{f}) = \left(\frac{K}{I}\right) \sum_i f_i. \tag{6.36}$$

One can verify directly that

$$\tilde{r}_i^0 \equiv \mathrm{E}\{r \mid f_i, \phi^0(\hat{f})\} = \left(\frac{1}{K}\right)\left(\frac{\rho}{\rho+\eta/I}\right) \phi^0(\tilde{f}) \tag{6.37}$$

[note that $\phi(\tilde{f})$ is a sufficient statistic for r]. If trader i used the forecast function (6.37), then the market-clearing price would be, from (6.34),

$$\left(\frac{\delta}{\delta'}\right)\left(\frac{1}{I}\right) \sum_i r_i^0 = \left(\frac{1}{K}\right)\left(\frac{\delta}{\delta'}\right)\left(\frac{\rho}{\rho+\eta/I}\right) \phi^0(\tilde{f}). \tag{6.38}$$

For a REE1, the right-hand side of (6.38) should equal $\phi^0(\tilde{f})$, and for this one takes

$$K = \left(\frac{\delta}{\delta'}\right)\left(\frac{\rho}{\rho+\eta/I}\right). \tag{6.39}$$

However, this does not constitute a REE2, as defined by (6.11) and (6.12). The forecast function (6.37) corresponds to taking

$$b_i = \left(\frac{1}{K}\right)\left(\frac{\rho}{\rho+\eta/I}\right) = \frac{\delta'}{\delta}, \qquad c_i = 0, \tag{6.40}$$

but with these coefficients, the right-hand side (6.11) is not well-defined, since the numerator and denominator are both zero.

One way of interpreting the distinction between REE1 and REE2 is that in a REE2 the non-price information of the traders must be reflected in the price *through the traders' demand functions*, whereas no such restriction is imposed in the definition of REE1.[35] With each $c_i = 0$, as in (6.40), each trader's demand is

[35] See Section 7.6.

independent of his own non-price information (given the price), and therefore the equilibrium price cannot reflect any non-price information. Indeed, with the mapping ϕ^0 of (6.36) and (6.39), each trader's demand function is identically zero, as an inspection of (6.3) shows!

7. Bibliographic notes

The following bibliographic notes supplement the references that have already been cited at various points in the chapter. I have not, however, attempted a complete survey of the relevant literature. As I mentioned in Section 1, to supplement this incomplete bibliography, and to complete the limited treatment of uncertainty in this chapter, the reader should consult Chapters 6, 13, 19, 23, and 27 of this Handbook, and also the recent review article by Hirshleifer and Riley (1979).

7.1. Notes to Section 1

I have already referred to the important contribution by Hicks to the study of temporary equilibrium [Hicks (1939)]. These ideas were followed up by Patinkin (1956). The dynamics of temporary equilibrium under certainty were studied by Hahn (1966), Shell and Stiglitz (1967), and Uzawa (1968), who brought out serious problems of instability in models with heterogeneous capital goods.

Recent attempts to develop a microeconomic foundation for monetary theory seem to be more related to the concept of temporary equilibrium than to the Arrow–Debreu model with complete markets, beginning in particular with Hahn (1965) and Clower (1967). A good idea of recent research on this topic can be obtained from various contributions to the conference volume edited by Kareken and Wallace (1980), as well as from the volume's bibliography.

The extension of the Arrow–Debreu model to the case in which different agents have different information was described in Radner (1967, 1968), which also contained critiques of this type of model. A systematic treatment of equilibrium in a sequence of markets (with uncertainty), but without informational differences among the agents, was given in Radner (1972). Before that, Muth (1961) had investigated a linear-Gaussian model of a sequence of markets for a single good, introducing the notion of "rational expectations". Lucas and Prescott (1971) extended Muth's idea to a more general sequential model of investment, output, and prices in a one-good industry.

The sequential nature of markets has also been attributed to the presence of transactions costs, even without uncertainty. Hahn (1971) demonstrated the existence of equilibrium in a sequence of markets with transactions costs. In Hahn's model, the resource costs of a transaction are proportional to the size of the transaction, and the presence of such costs induces traders to postpone commitments closer to the date of actual delivery, thus creating the need for a sequence of markets. In a similar spirit, the effects of transactions costs have been analyzed by Foley (1970) and by Kurz (1974).

The investigation of market equilibrium in which the agents made inferences about the environment from equilibrium prices was introduced in Radner (1967), in the context of a two-period model. This idea was taken up independently by Lucas (1972) and Green (1973), and further investigated by Grossman and others. (Additional bibliographic notes on this topic will be found in Section 7.4 below.)

For a provocative essay on the theoretical and conceptual problems that arise in the study of equilibrium in economics I refer the reader to Hahn's Inaugural Lecture (1973).

An interest in uncertainty leads naturally to a concern with the economics of information, a field in which Jacob Marschak has pioneered [see Marschak (1964, 1968, 1971); Marschak and Radner (1972)]. Indeed, Marschak's interest in information can be traced to his studies in monetary theory (1938, 1949). Important surveys of the economics of information, from different points of view, will be found in Hirshleifer (1973), Rothschild (1973), and Hirshleifer and Riley (1979).

Systematic accounts of the role of information in theories of decision and organization may be found in the volume edited by McGuire and Radner (1972).

7.2. Notes to Section 2

The model of events, commodities, commitments, and prices in Section 2 is adapted from Radner (1972). In the special case of the Arrow–Debreu model, with finitely many dates and events, the corresponding framework was introduced in Arrow (1953) and Debreu (1953). In the case in which there are infinitely many events and/or dates, the commodity space is infinite dimensional. For such spaces, the representation of a price system as a linear functional was introduced by Debreu (1954).

7.3. Notes to Section 3

Sections 3.1 and 3.6 are adapted from Radner (1974a). A renewed interest in the rigorous analysis of temporary equilibrium with uncertainty was signalled by Stigum (1969) and Grandmont (1970); for an account of subsequent work see Chapter 19 of this Handbook.

The representation of information in terms of partitions or, more generally, sigma-fields is well known in the theory of statistical decision. The best unified treatment of this topic with the theories of subjective probability and the Expected Utility Hypothesis is still that of Savage (1954). A briefer and more elementary treatment, which also deals with groups of decision-makers with different information, can be found in Marschak and Radner (1972).

The economics of moral hazard was discussed by Arrow (1963, 1965, 1968) and Pauly (1968). The first three references have been reprinted in Arrow (1971), which is also recommended for stimulating theoretical treatments of other aspects of risk in economics. The implications of moral hazard have been more recently examined under the rubric of the "principal–agent problem"; see Ross (1973), Hurwicz and Shapiro (1978), Holmström (1979), Shavell (1979), Radner (1980), and the references cited in those papers. For a more general organizational setting of the problem, see Groves (1973). An early forerunner of the principal–agent literature was Simon (1951). The problem also has a formal analogue in parts of the theory of optimal income taxation; see Mirrlees (1971) and his Chapter 24 of this Handbook. The phenomena of *adverse selection* and *signalling* are related to, but conceptually distinguishable from, that of moral hazard; for research on these topics the reader is referred to Akerlof (1970), Rothschild and Stiglitz (1976), Wilson (1977, 1980), Spence (1974), and Riley (1975), together with the references cited there and in the survey by Hirshleifer and Riley (1979).

7.4. Notes on Section 4

Section 4.1 is based on Radner (1972). The papers by Muth (1961) and Lucas and Prescott (1971) have been noted above in Section 7.1. Most of the literature on equilibrium in a sequence of markets deals with the case in which durable assets are traded, and thus formally goes beyond the simple model of pure exchange in Section 4.1. Important results on an exchange economy with assets can be found in Bewley (1980a, 1980b), and in references cited there. (See also the notes to Section 5.) A general theorem on the existence of *stationary* equilibrium in a sequence of markets is given in Radner (1974, section 4).

Section 4.2 is based on Radner (1979). Allen (1980) has significantly generalized this result to demonstrate, under certain regularity conditions, the generic existence of revealing rational expectations equilibria if the dimension of the price space exceeds the dimension of the parameter space that characterizes the set of economies. Earlier papers by Radner (1967), Lucas (1972), and Green (1973) have been noted in Section 7.1. In a series of papers beginning with his Ph.D. dissertation (1975), Grossman, and Grossman and Stiglitz, explored various aspects of rational expectations equilibria (REE), including (1) circumstances under which equilibrium prices would be sufficient statistics even though not strictly revealing, and (2) the reduction in the incentives to acquire private information associated with the revelation of that information through the REE prices. The references to these and related papers can be found in the paper by Grossman (1981).

Radner (1967) noted that, if traders conditioned their expected utility on market prices, then their demands could be discontinuous as a function of prices. He did not, however, give a specific example of non-existence of REE; this was done by Kreps (1977). Rothschild, in a private communication, suggested that in the Kreps example the non-existence of REE was due to a "degeneracy" in the system of equilibrium equations; this observation led to the generic existence theorem of Radner (1979). "Robust" examples of non-existence of REE were given by Green (1977) and by Jordan and Radner (1979). Section 4.3 is based on the latter paper.

Section 4.4 summarizes some results in Futia (1981). The methods of that paper have been applied to an equilibrium theory of stochastic business cycles in Futia (1979).

Work on endogenous expectations, REE, and the "efficient markets hypothesis" that is related to the material of Section 4 has been done by Jordan (1976, 1977, 1979a, 1979b, 1979c, 1980) and by Border and Jordan (1979). Among other things, the work by Jordan examines the conditions for existence of REE *throughout* a wide class of economic environments (not just generically in the class), and also considers more general classes of endogenous data structures than that generated by market equilibrium.

For reviews of literature on rational expectations in macroeconomic theory, see Shiller (1978) and Kantor (1979).

In this chapter the term "rational expectations equilibrium" has been reserved for the type of market equilibrium described in Section 4.2, in which traders make inferences about the environment from market prices. The reader should be warned, however, that the same term has been applied to the equilibrium in a sequence of markets described in Section 4.1, as well as to other kinds of equilibria involving "expectations".

Additional references about REE are given in Section 7.6.

7.5. Notes to Section 5

Non-convexities associated with the choice of information have been described by Radner (1968), R. Wilson (1975), and Radner and Stiglitz (1976). Non-optimality of equilibrium in economies with incomplete markets has been examined by Hart (1975) and Diamond (1980). Hart has also shown that, without a bound on the positions that traders can take, equilibrium may not exist.

Theorem 5.1 was stated without proof in Radner (1967). Recall that, in that theorem, the plans of the production units were given. In general it does not appear that one can rely on the stock market to sustain, in a decentralized fashion, the choice of both production and consumption decisions that are Pareto optimal even in some constrained sense. Of course, the answer to this question depends on one's definitions of "decentralized" and "constrained Pareto optimum". In particular, there is no consensus on the definition of "decentralized behavior of production units" in this context. (See, for example, Section 5.5 above for a discussion of the criterion of maximizing stock market value.) The special case of a two-date model in which each production unit chooses only the scale of its activity (i.e. each unit has essentially only one composite input and one composite output) has received considerable attention, in connection with the hypothesis of maximizing the value of the firm on the stock market. In this context, Diamond (1967) showed that, if the uncertainty in each firm's production is confined to a multiplicative factor in its production function, then a constrained Pareto optimum can be sustained by an equilibrium in the commodity and stock markets. On the other hand, Stiglitz (1972) and Jensen and Long (1972) gave examples of the non-optimality of equilibrium. Drèze (1974) pointed out that the determination of a constrained Pareto optimum in this type of model is in general not a convex programming problem, and that there can be allocations satisfying the first-order conditions for an optimum that are not, in fact, optimal. This is, of course, related to the examples of non-optimal market equilibria referred to above. He also showed that the choice of production plans is formally related to the choice of the output of a public good, and studied the convergence and incentive-compatibility properties of a decentralized tâtonnement adjustment process for achieving a local optimum. [This process is related to the process studied by Drèze and De la Vallée Poussin (1970) in the public goods case.]

The relationship between stock market equilibrium and constrained Pareto optimality has been further studied by Gevers (1972), Grossman and Hart (1979), and others (see the last paper for further references). This work underscores the importance of the choice of hypothesis concerning the objective function of the firm. Grossman and Hart (1979) assume that "each firm's objective function (is) a weighted sum of shareholders' private valuations of the firm's future production stream, where each shareholder uses his own marginal rate of substitution between present and future income to discount future profits, and where the weights are (the) consumers' initial shareholdings." With this specification of firms' objectives, they characterize a sense in which equilibria are optimal. (In addition, they can demonstrate the existence of an equilibrium, avoiding the difficulties in Theorem 5.2 of Section 5.4 above.) In the case of shareholder unanimity about the choice of production plans, this objective reduces to the maximization of the market value of the firm's shares. For other work on the theory of stock markets, including conditions under which there will be shareholder unanimity, see Leland (1974), Merton and Subrahmanyam (1974), Ekern and Wilson (1974), and Radner (1974b).

Bewley (1980c) has studied a model of equilibrium in a sequence of markets that is mid-way between the Arrow–Debreu model and the model of Section 5.4; there are no forward markets for contingent claims, but each trader has only a single budget constraint which requires that his long-run average expenditure per period not exceed his long-run average income. In Bewley's model, a *stationary equilibrium* is a stationary stochastic process of temporary equilibria. Bewley shows that, roughly speaking, if consumers maximize long-run average utility per period, and firms maximize long-run average profit per period, then a stationary equilibrium exists and is Pareto optimal.

Rothenberg and Smith (1971) provide an interesting analysis of comparative statics for a two-factor, two-good model with uncertainty either about the total labor supply or about a random parameter in one of the production functions. In general, the effects of an increase in uncertainty depend on the third derivatives of the production functions; nevertheless, Rothenberg and Smith are able to show how, in their model, alternative assumptions about the technology influence the effects of uncertainty on expected profits in each of the two sectors, on the expected total value of output, and on the expected returns to the two factors. They also show that the effects of uncertainty in a *partial equilibrium* context may be reversed when the comparative statics of a *full equilibrium* are considered.

The concept of valuation equilibrium arises naturally in the consideration of optimal growth under uncertainty; here the (stochastic) prices that figure in the equilibrium are the "shadow prices" or Kuhn–Tucker–Lagrange multipliers. Prices can also be used to characterize capital accumulation that satisfies the weaker condition of intertemporal efficiency. Contributions to the theory of optimal capital accumulation under uncertainty have been made by Brock and Mirman (1973), Radner (1973), Dana (1974), Jeanjean (1974), Zilcha (1976a, 1976b), Brock and Majumdar (1978), and Majumdar and Radner (1980). For the case of efficient growth see Majumdar (1972). Other papers and references will be found in Los and Los (1974), especially the papers by Dynkin, Evstigneev, and Tacsar.

7.6. Notes to Section 6

Section 6.2 is based on Green and Majumdar (1975), which was preceded by Green (1971) and Majumdar (1972).

In a related development, Grandmont and Hildenbrand (1974), studied a stochastic process of temporary equilibria corresponding to an "overlapping generations" model in which the agents have random endowments. With the aid of an assumption that the temporary equilibrium in one period depends continuously on the state of the economy in the previous period, they were able to demonstrate the ergodicity of the process. One can interpret such a process as being in temporary equilibrium at every date with respect to current excess demand, but in permanent (but stable) *dis*equilibrium with respect to fulfillment of expectations. This analysis has been extended by Hellwig (1979).

Section 6.3 is based on Hildenbrand and Radner (1979).

The linear model of Section 6.4 is suggested by Grossman (1977), and is almost identical to the one used by Hellwig (1980). The linear individual demand functions (6.3) can be justified for traders with constant absolute risk aversion, but then one would, in general, have different coefficients for different traders. Thus, the model of Section 6.4 is not strictly consistent with individual utility maximization unless all the traders have the same constant absolute risk aversion and the same variance of noise in their private (non-price) information. In Hellwig (1980), the traders may differ in these parameters. He demonstrates the existence of at least one REE, and analyzes how REEs depend on the parameters of the traders and on the number of traders. Related results for the linear model are to be found in Grossman (1978) and in Grossman and Stiglitz (1980). They use the special information structure with "informed and uninformed traders" that is used in Theorem 6.4.

The stability analyses of the adjustment processes (6.17) and (6.18) have not appeared before; for approaches in a similar spirit, see Brock (1972), DeCanio (1979), and Bray (1980).

Theorem 6.4 is due to Bray (1980). Numerical results of a similar nature have been obtained with Monte-Carlo methods by Cyert and Degroot (1974). Blume and Easley (1980) prove a corresponding theorem for the case in which there are finitely many alternative states of information and the environment (as in Section 4.2), and in which each trader has, a priori, finitely many hypotheses about the underlying economic model, including an REE model that is common to all of the traders' sets of hypotheses.

The criticism of REE that is noted at the end of Section 6.4 was, to my knowledge, first pointed out by Beja (1976), and is also mentioned by Hellwig (1980).

Note added in proof

I would like to call attention to the following recent (as yet unpublished) contributions to the theory of rational expectations equilibrium (REE), in addition to those already mentioned in Sections 4 and 7. For the case in which the dimension of the space of pooled (nonprice) information exceeds the dimension of the price space, Jordan (1981) has provided regularity conditions under which, for a suitable topology on the set of economies, there is a residual set of economies that have REEs. Furthermore, although these REEs will typically not be revealing, they can be taken to be approximately revealing, for any desired degree of approximation. (Recall that a subset of a topological space is *residual* if it is a countable intersection of open and dense subsets.) Two other papers give conditions for the existence of *approximate* equilibria, to any desired degrees of approximation, without restrictions on the dimensions of the information and message spaces. Roughly speaking, in Allen (1981) excess demand is approximately zero, but the expectations are strictly correct, whereas in Anderson and Sonnenschein (1981), excess demand is always zero, but the expectations are only approximately correct. Anderson and Sonnenschein prefer to interpret their result as establishing the existence of exact REE for a different idealization of the equilibrium concept.

References

Akerlof, G. A. (1970), "The market for 'lemons': Qualitative uncertainty and the market mechanism", Quarterly Journal of Economics, 84:488–500.
Arrow, K. J. (1953), "Le rôle de valeurs boursières pour la répartition la meilleure des risques", Econométrie, Paris, Centre National de la Recherche Scientifique, 41–48. English translation "The role of securities in the optimal allocation of risk-bearing", Review of Economic Studies, 31:91–96.
Arrow, K. J. (1963), "Uncertainty and the welfare economics of medical care", American Economic Review, 53:941–973.
Arrow, K. J. (1965), Aspects of the theory of risk-bearing. Helsinki: Yrjo Johansson Lectures Series.
Arrow, K. J. (1971), Essays in the theory of risk-bearing. Chicago: Markham.
Arrow, K. J. and F. H. Hahn (1971). General competitive analysis. San Francisco: Holden-Day.
Aumann, R. J. (1976), "Agreeing to disagree", Annals of Statistics, 4:1236–1239.
Beja, A. (1976), "The limited information efficiency of market processes", Working Paper No. 43. Research Program in Finance, University of California, Berkeley (unpublished).
Berge, C. (1966), Espaces topologiques. Paris: Dunod.
Bewley, T. F. (1980a), "The optimum quantity of money", in: J. Kareken and N. Wallace, eds., Models of monetary economics. Federal Reserve Bank of Minneapolis.
Bewley, T. F. (1980b), "The martingale property of asset prices", Dept. of Economics, Northwestern University (unpublished).
Bewley, T. F. (1981), "Stationary equilibrium", Journal of Economic Theory, 24:265–295.
Billingsley, P. (1968), Convergence of probability measures. New York: Wiley.
Blume, L. E. and D. Easley (1980), "Learning to be rational", CREST Working Paper No. C18 (revised), Dept, of Economics, University of Michigan (unpublished).
Border, K. C. and J. S. Jordan (1979), "Expectations equilibrium with expectations based on past data", Journal of Economic Theory, 22:395–406.
Bray, M. (1980), "Learning, estimation, and the stability of rational expectations", Graduate School of Business, Stanford University (unpublished).
Brock, W. A. (1972), "On models of expectations that arise from maximizing behavior of economic agents over time", Journal of Economic Theory, 5:348–376.
Brock, W. A. and L. J. Mirman (1973), "Optimal growth and uncertainty: the Ramsey–Weiszäcker case", International Economic Review, October.

Brock, W. A. and M. Majumdar (1978),"Global asymptotic stability results for multisector models of optimal growth under uncertainty when future utilities are discounted", Journal of Economic Theory, 18:225–243.
Clower, R. (1967), "A reconsideration of the microfoundations of monetary theory", Western Economic Journal, 6:1–8.
Cyert, R. M. and M. H. DeGroot (1974), "Rational expectations and Bayesian analysis", Journal of Political Economics, 82:521–536.
Dana, R.-A. (1974), "Evaluation of development programs in a stationary stochastic economy with bounded primary resources", in: J. Los, and M. W. Los, eds., Mathematical methods in economics. Amsterdam: North-Holland.
Debreu, G. (1953), "Une économie de l'incertain", mimeo. Paris: Electricité de France.
Debreu, G. (1954), "Valuation equilibrium and Pareto optimum", Proceedings of the National Academy of Science, U.S.A., 40:588–592.
Debreu, G. (1959), Theory of Value. New York: Wiley Reprinted by Yale University Press, New Haven (1971).
Debreu, G. (1962), "New concepts and techniques for equilibrium analysis", International Economic Review, 3:257–273.
DeCanio, S. J. (1979), "Rational expectations and learning from experience", Quarterly Journal of Economics, 92:47–57.
Diamond, P. A. (1967), "The role of a stock market in a general equilibrium model with technological uncertainty", American Economic Review, 57:759–765.
Diamond, P. A. (1980), "Efficiency with uncertain supply", Review of Economic Studies, 47:645–652.
Drèze, J. (1974), "Investment under private ownership: optimality, equilibrium and stability", in: J. Drèze, ed., Allocation under uncertainty. Macmillan.
Drèze, J. and De la Vallée Poussin, (1971), "A tâtonnement process for public goods", Review of Economic Studies, 38:133–150.
Ekern, S. and R. Wilson (1974), "On the theory of the firm in an economy with incomplete markets", Bell Journal of Economics, 5:171–180.
Foley, D. K. (1970), "Economic equilibrium with costly marketing", Journal of Economic Theory, 2:276–291.
Futia, C. (1979), "Stochastic business cycles", Bell Laboratories, Murray Hill, NJ (unpublished).
Futia, C. (1981), "Rational expectations in stationary linear models", Econometrica, 49:171–192.
Gevers, L. (1974), "Competitive equilibrium of the stock exchange and Pareto-efficiency", in: J. Drèze, ed., Allocation under uncertainty. Macmillan.
Grandmont, J.-M., 1970, "On the temporary competitive equilibrium". Working Paper No. 305, Center for Research in Management Science, University of California, Berkeley, August (unpublished).
Grandmont, J.-M. (1974), "Shortrun equilibrium analysis in a monetary economy", in: J. Drèze, ed., Allocation under uncertainty. Macmillan.
Grandmont, J.-M. and W. Hildenbrand (1974), "Stochastic processes of temporary equilibria", Journal of Mathematical Economics, 1:247–278.
Green, J. R. (1971), "The nature and existence of stochastic equilibria", Discussion Paper No. 193, Harvard Institute of Economic Research, Harvard University (unpublished).
Green, J. R. (1973), "Information, efficiency, and equilibrium", Discussion Paper 284, Harvard Institute of Economic Research, Harvard University (unpublished).
Green, J. R. (1977), "The nonexistence of informational equilibria", Review of Economic Studies, 44:451–463.
Green, J. R. and M. Majumdar (1975), "The nature of stochastic equilibria", Econometrica, 43:647–660.
Grossman, S. J. (1975), The existence of futures markets, noisy rational expectations, and informational externalities. Ph.D. Dissertation, Dept. of Economics, University of Chicago.
Grossman, S. J. (1977), "The existence of futures markets, noisy rational expectations, and informational externalities", Review of Economic Studies, 44:431–449.
Grossman, S. J. (1978), "Further results on the informational efficiency of competitive stock markets", Journal of Economic Theory, 18:81–101.

Grossman, S. J. (1981), "An introduction to the theory of rational expectations under asymmetric information", Review of Economic Studies.
Grossman, S. J. and O. D. Hart (1979), "A theory of competitive equilibrium in stock market economies", Econometrica, 47:293–330.
Grossman, S. J. and J. E. Stiglitz (1980), "On the impossibility of informationally efficient markets", American Economic Review, 70:393–408.
Groves, T. (1973), "Incentives in teams", Econometrica, 41:617–631.
Guesnerie, R. and J.-Y. Jaffray (1974), "Optimality of equilibrium of plans, prices, price expectations", in: J. Drèze, ed., Allocation under uncertainty. Macmillan.
Hahn, F. H. (1965), "On some problems of proving the existence of an equilibrium in a monetary economy", in: Hahn and Brechling, eds., The Theory of Interest Rates. London: Macmillan.
Hahn, F. H. (1966), "Equilibrium dynamics with heterogeneous capital goods", Quarterly Journal of Economics, 30:633–646.
Hahn, F. H. (1971), "Equilibrium with transactions costs", Econometrica, 39:417–440.
Hahn, F. H. (1973), On the notion of equilibrium in economics. Cambridge: Cambridge University Press.
Hart, O. D. (1975), "On the optimality of equilibrium when the market structure is incomplete", Journal of Economic Theory, 11:418–443.
Hellwig, M. F. (1979), "Stochastic processes of temporary equilibria: a note", Discussion Paper No. 46, Dept. of Economics, University of Bonn (unpublished).
Hellwig, M. F. (1980), "On the aggregation of information in capital markets", Journal of Economic Theory, 22:477–498.
Hicks, J. R. (1939), Value and Capital, 2nd edn. Oxford: Clarendon Press.
Hildenbrand, W. and R. Radner (1979), "Stochastic stability of market adjustment in disequilibrium", in: J. R. Green and J. A. Scheinkman, eds., General equilibrium, growth, and trade. New York: Academic Press.
Hirshleifer, J. (1973), "Where are we in the theory of information?", American Economic Review, Papers and Proceedings, 63:31–40.
Hirshleifer, J. and J. G. Riley (1979), "The analytics of uncertainty and information – an expository survey", Journal of Economic Literature, 17:1375–1421.
Holmström, B. (1979), "Moral hazard and observability", Bell Journal of Economics, 10:74–91.
Hurwicz, L. and L. Shapiro, 1978, "Incentive structures maximizing residual gain under incomplete information", Bell Journal of Economics, 9:180–191.
Jeanjean, P. (1974), "Optimal development programs under uncertainty: The undiscounted case", Journal of Economic Theory, 7:66–92.
Jensen, M. and J. Long (1972), "Corporate investment under uncertainty and Pareto optimality of stock markets", Bell Journal of Economic and Management Science, 3:151–174.
Jordan, J. S. (1976), "Temporary competitive equilibrium and the existence of self-fulfilling expectations", Journal of Economic Theory, 12:455–471.
Jordan, J. S. (1977), "Expectations equilibria and informational efficiency for stochastic environments", Journal of Economic Theory, 16:354–372.
Jordan, J. S. (1979a), "Admissible market data structures: a complete characterization", Dept. of Economics, University of Minnesota.
Jordan, J. S. (1979b), "A dynamic model of expectations equilibrium", Dept. of Economics, University of Minnesota.
Jordan, J. S. (1979c), "On the efficient markets hypothesis", Dept. of Economics, University of Minnesota (to appear in Econometrica).
Jordan, J. S. (1980), "On the predictability of economic events", Econometrica, 48:955–972.
Jordan, J. S. and R. Radner (1979), "The nonexistence of rational expectations equilibria: a robust example", Dept. of Economics, University of Minnesota (unpublished).
Kantor, B. (1979), "Rational expectations and economic thought", Journal of Economic Literature, 17:1422–1441.
Kareken, J. and N. Wallace (eds.) (1980), Models of monetary economics. Federal Reserve Bank of Minneapolis.
Kelley, J. L. and Namioka (1963), Linear topological spaces. Princeton: Van Nostrand.

Kreps, D. M. (1977), "A note on 'fulfilled expectations' equilibria", Journal of Economic Theory, 14:32–43.
Kurz, M. (1974), "Equilibrium with transaction cost and money in a single market exchange economy", Journal of Economic Theory, 7:418–452.
Leland, H. (1974), "Production theory and the stock market", Bell Journal of Economics, 5:125–144.
Los, J. and M. W. Los (eds.) (1974), Mathematical methods in economics. Amsterdam: North-Holland.
Lucas, R. E. (1972), "Expectations and the neutrality of money", Journal of Economic Theory, 4:103–124.
Majumdar, M. K. (1972), "Equilibrium and adjustment in random economies", Technical Report No. 58, Institute of Mathematical Studies in the Social Sciences, Stanford University (unpublished).
Majumdar, M. (1972), "Some general theorems on efficiency prices with an infinite-dimensional commodity space", Journal of Economic Theory, 5:1–13.
Majumdar, M. and R. Radner (1980), "Stationary optimal policies with discounting in a stochastic activity analysis model", Dept. of Economics, Cornell University, Ithaca NY (unpublished)
Marschak J. (1949) "Role of liquidity under complete and incomplete information", American Economic Review, 39:182–195.
Marschak, J., 1964, "Problems in information economics", in: Bonini, Jaedicke, and Wagner, eds., Management controls: New directions in basic research. New York: McGraw-Hill.
Marschak, J. (1968), "Economics of inquiring, communicating, and deciding", American Economic Review, 58:1–18.
Marschak, J. (1971), "Economics of information systems", in: M. Intriligator, ed., Frontiers of Quantitative Economics, Vol I, Amsterdam: North-Holland Publishing Company.
Marschak, J. and H. Markower (1938), "Assets, prices, and monetary theory", Economica, 5:261–288. Reprinted in G. Stigler and K. Boulding (eds.), Readings in Price Theory. Chicago: Irwin, 1952.
Marschak, J. and R. Radner, (1972), Economic theory of teams. New Haven: Yale University Press.
McGuire, C. B. and R. Radner (1972), Decision and organization. Amsterdam: North-Holland Publishing Company.
Merton, R. and M. Subrahmanyaw, (1974), "Optimality of capital markets", Bell Journal of Economics, 5:145–170.
Mirrlees, J. A. (1971), "An exploration in the theory of optimum income taxation", Review of Economic Studies, 38:175–208.
Muth, J. (1961), "Rational expectations and the theory of price movements", Econometrica, 29:315–335.
Neveu, J. (1965), Mathematical foundations of the calculus of probability. San Francisco: Holden-Day.
Parthasarathy, K. R., 1967, Probability measures on metric spaces. New York: Academic Press.
Patinkin, D. (1965), Money, Interest, and Prices, 2nd edn. New York: Harper and Row.
Pauly, M. V. (1968), "The economics of moral hazard", American Economic Review, 58:531–537.
Radner, R. (1967), "Equilibre des marchés à terme et au comptant en cas d'incertitude", Cahiers d'Econométrie, C.N.R.S., Paris, 4:35–52.
Radner, R. (1968), "Competitive equilbrium under uncertainty", Econometrica, 36:31–58.
Radner, R. (1970), "Problems in the theory of markets under uncertainty", American Economic Review, 60:454–460.
Radner, R. (1972), "Existence of equilibrium of plans, prices, and price expectations in a sequence of markets", Econometrica, 40:289–303.
Radner, R. (1973), "Optimal stationary consumption with stochastic production and resources", Journal of Economic Theory, 6:68–90.
Radner, R. (1974a), "Market equilibrium and uncertainty: concepts and problems", in: M. D. Intriligator and D. A. Kendrick, eds., Frontiers of quantitative economics, Vol. II. Amsterdam: North-Holland, ch. 2.
Radner, R. (1974b), "A note on unanimity of stockholders' preferences among alternative production plans", Bell Journal of Economics, 5:181–186.
Radner, R. (1979), "Rational expectations equilibrium: generic existence and the information revealed by prices", Econometrica, 47:655–678.
Radner, R. (1980), "Monitoring cooperative agreements in a repeated principal-agent relationship", Economic Research Center, Bell Laboratories, Murray Hill, N.J.

Radner, R. and J. E. Stiglitz (1976), "A nonconcavity in the value of information", Dept. of Economics, University of California, Berkely (to appear in M. Boyer and R. E. Kihlstrom (eds.), Bayesian models in economic theory. Amsterdam: North-Holland.)
Riley, J. G. (1975), "Competitive signalling", Journal of Economic Theory, 10:174–186.
Ross, S. (1973), "The economic theory of agency", American Economic Review, 63:134–139.
Rothenberg, T. J. and K. R. Smith (1971), "The effect of uncertainty on resource allocation in a general equilibrium model", Quarterly Journal of Economics, 85:440–459.
Rothschild, M. (1973), "Models of market organization with imperfect information: a survey", Journal of Political Economy, 81:1283–1308.
Rothschild, M. and J. E. Stiglitz (1971), "Equilibrium in competitive insurance markets: An essay on the economics of imperfect information", Quarterly Journal of Economics, 90:629–649.
Savage, L. J. (1954), Foundations of statistics. New York: Wiley.
Shavell, S. (1979), "Risk sharing and incentives in the principal and agent relationship", Bell Journal of Economics, 10:55–73.
Shell, K. and J. E. Stiglitz (1967), "The allocation of investment in a dynamic economy", Quarterly Journal of Economics, 81:592–609.
Shiller, R. J. (1978), "Rational expectations and the dynamic structure of macroeconomic models", Journal of Monetary Economics, 4:1–44.
Simon, H. A. (1951), "A formal theory of the employment relation", Econometrica, 19:293–305.
Spence, M. (1974), "Competitive and optimal responses to signals: an analysis of efficiency and distribution", Journal of Economic Theory, 7:296–332.
Sitgum, B. P. (1969), "Competitive equilibrium under uncertainty", Quarterly Journal of Economics, 83:533–561.
Uzawa, H. (1968), "Market allocation and optimum growth", Australian Papers, 7:17–27.
Wilson, C. A. (1977), "A model of insurance markets with incomplete information", Journal of Economic Theory, 16:167–207.
Wilson, C. A. (1980), "The nature of equilibrium in markets with adverse selection", Bell Journal of Economics, 11:108–130.
Wilson, R. (1975), "Informational economies of scale", Bell Journal of Economics, 6:184–195.
Zilcha, I. (1976a), "Characterization by prices of optimal programs under uncertainty", Journal of Mathematical Economics, 3:173–184.
Zilcha, I. (1976b), "On competitive prices in a multi-sector economy with stochastic production and resources", Review of Economic Studies, 43:431–438.

Supplementary references added in proof

Allen, B. (1981), "Approximate equilibria in microeconomic rational expectations models", Department of Economics, University of Pennsylvania (unpublished).
Anderson, R. M. and H. Sonnenschein (1981), "On the existence of rational expectations equilibrium", Department of Economics, Princeton University (unpublished).
Jordan, J. S. (1981), "On the generic existence of rational expectations equilibrium", Department of Economics, University of Minnesota (unpublished).

Chapter 21

THE COMPUTATION OF EQUILIBRIUM PRICES: AN EXPOSITION*

HERBERT E. SCARF

Yale University

1. The general equilibrium model

A demonstration of the existence of equilibrium prices for a general Walrasian model of competitive behavior necessarily makes use of some variant of Brouwer's fixed point theorem as an essential step in the argument. The strategy for calculating equilibrium prices is to render that step constructive by a numerical approximation of the fixed point implied by Brouwer's theorem, or one of its alternatives. We begin this chapter with a brief review of the competitive model and the role of fixed point theorems. [See Debreu (1959).]

The typical consumer will be assumed to have a set of preferences for *commodity bundles* $x=(x_1,\ldots,x_n)$ in the non-negative orthant of n-dimensional space. These preferences can be described either by an abstract *preference relationship*, $\succsim$, satisfying a series of plausible axioms, or by a specific *utility indicator* $u(x)$.

It is customary to assume that, prior to production and trade, the typical consumer will own a non-negative vector of commodities w, his *endowment*, whose evaluation by means of market prices forms the basis for that consumer's income. More specifically, if the non-negative vector of prices $\pi=(\pi_1,\ldots,\pi_n)$ is expected to prevail then the consumer can obtain an income

$$I=\sum_1^n \pi_i\cdot w_i=\pi\cdot w$$

by the sale of his endowment bundle w. This income is then used by the consumer for the purchase of an alternative bundle of commodities in such a way as to provide him with the highest possible utility level.

Given a vector of prices for all of the commodities in the economy the consumer is faced with a maximization problem: to find a vector of commodities

*This research was supported by a grant from the National Science Foundation.

Handbook of Mathematical Economics, vol. II, edited by K.J. Arrow and M.D. Intriligator

x which maximizes his utility function (or is maximal according to his preference relationship) subject to the *budget constraint*

$$\pi \cdot x \leqslant \pi \cdot w.$$

The problem faced by the consumer is clearly unchanged if the price vector π is replaced by $\lambda\pi$, with λ an arbitrary positive number. This provides us with a degree of freedom in normalizing prices in any fashion which happens to be convenient. For example, prices can be normalized so as to lie on the *unit sphere* $\sum_1^n \pi_i^2 = 1$ (meaningless from an economic point of view, but mathematically quite useful) or on the *unit simplex* $\sum_1^n \pi_i = 1$. For definiteness we shall adopt the latter normalization throughout most of this chapter.

Under quite acceptable assumptions, such as continuity and strict convexity of preferences, the solution to the consumer's maximization problem will be a single-valued *demand function* $x(\pi)$ which is continuous on the unit simplex and which satisfies the identity

$$\sum_1^n \pi_i x_i(\pi) \equiv \sum_1^n \pi_i w_i,$$

stating that the consumer's entire income will be spent on the purchase of commodities. (This is, of course, not meant to rule out the possibility that the consumer may choose to save some of his current income for future purchases – an aspect of consumer behavior which can be captured by declaring that certain commodities become available in the future.)

For the purpose of computing equilibrium prices the demand functions $x(\pi)$ are frequently more natural to work with than the underlying utility function or preference relationship. A general equilibrium model must be fully specified in order to obtain a numerical solution. On the consumer side this is typically done by providing a numerical or algebraic description of the functions $x(\pi)$. A few examples of some of the more familiar utility functions and their associated demand functions may be appropriate at this point.

The Cobb–Douglas utility function

(1) $u(x) = x_1^{a_1} \dots x_n^{a_n}$, with $a_i \geqslant 0$, $\sum a_i = 1$.

A consumer with this utility function will spend the same fraction a_i of his income on the ith good, independently of relative prices. Since his income is given by $\pi \cdot w$, this implies that his demand for the ith good is given by

$$x_i(\pi) = \frac{a_i \pi \cdot w}{\pi_i}.$$

(There is an insignificant technical problem that these demand functions and others which are subsequently presented exhibit a discontinuity on the boundary of the unit simplex.)

The fixed-proportions utility function

(2) $u(x)=\min[x_1/a_1,\ldots,x_n/a_n]$, with $a_i>0$.

A consumer with preferences exhibiting such strict complementarity wishes to purchase a commodity bundle proportional to the vector $(a_1,\ldots,a_n)$. His demands are therefore given by

$$x_i(\pi)=a_i\cdot\pi\cdot w/\pi\cdot a.$$

The C.E.S. utility function

(3) $u(x_1,\ldots,x_n)=\left(\sum_1^n a_i^{1/b}x_i^{(b-1)/b}\right)^{b/(b-1)}$, with $a_i>0$ and $b\leq 1$.

This utility function includes the two previous examples for special values of the parameter b. The demand functions $x_i(\pi)$ are given by

$$x_i(\pi)=\frac{a_i\pi\cdot w}{\pi_i^b\sum_j\pi_j^{1-b}a_j}.$$

A model of equilibrium will generally involve a number of consuming units – individuals, aggregations of individuals, or countries involved in foreign trade – each of whom has a stock of assets w, prior to production and trade, and each of whose demands are functions of the prevailing prices π. If the jth consumer's assets are represented by w^j and his demand functions by $x^j(\pi)$, then the *market demand functions* are defined by

$$x(\pi)=\sum_j x^j(\pi).$$

The market demand functions specify the total consumer demand, for the goods and services in the economy, as a function of all prices.

Market demand functions cannot be arbitrarily specified if they are derived from individual demand functions by the process of aggregation described above. For example, if the individual demand functions are continuous the market demand functions will be continuous as well. Moreover, the market demand functions will satisfy the identity

$$\pi\cdot x(\pi)\equiv\pi\cdot w$$

(if $w=\sum w^j$) known as *Walras' Law* – resulting from the assumption that the income available for consumer purchases is accounted for fully by the sale of privately owned assets.

The *market excess demand functions*, representing the difference between market demands and the supply of commodities prior to production, are defined by

$$\xi(\pi)=x(\pi)-w.$$

Our analysis may be summarized as follows:

1.1. [*Properties*]

The market excess demand functions will satisfy the following conditions:

1. They are defined and continuous everywhere in the positive orthant other than the origin.
2. They are homogeneous of degree zero, i.e. $\xi(\lambda\pi)=\xi(\pi)$, for any positive λ.
3. They must satisfy the Walras Law:

$$\pi\cdot\xi(\pi)\equiv 0.$$

In this form, the Walras Law has an interesting geometrical interpretation when prices are normalized to lie on the unit sphere (Figure 1.1). If the excess demand $\xi(\pi)$ is drawn as a vector originating at π, the geometric interpretation of the Walras Law is that this vector is tangent to the unit sphere at π. The excess demands determine, therefore, a continuous vector field on that part of the unit sphere lying in the non-negative orthant.

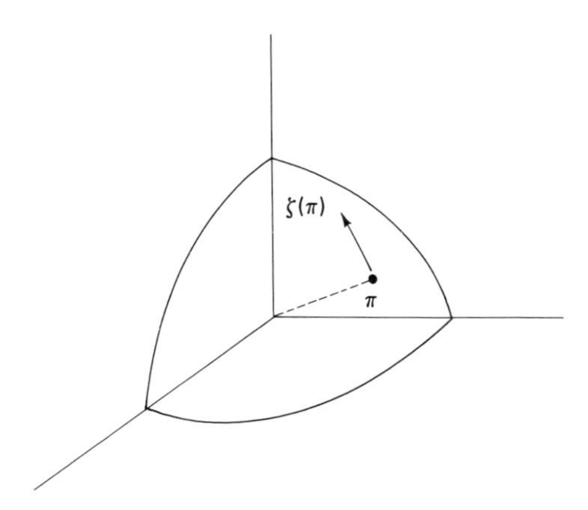

Figure 1.1.

A description of the consumer side of the economy, for the purposes of the general equilibrium model, is adequately provided by the market excess demand functions. To complete the model a specification of the productive techniques available to the economy must also be given. We shall return to this point after examining the pure trade model in which production is absent and consumers simply exchange the commodities which they initially own. For such a model the total supply is given by the vector w and the market demand by the functions $x(\pi)$. An equilibrium price vector π^* is one which equilibrates supply and demand for all commodities. We have the following formal definition which allows for the possibility that some of the commodities have a zero price and are in excess supply.

1.2. [*Definition*]

A non-zero price vector π^* is an *equilibrium price vector* for the pure trade model if $\xi_i(\pi^*)\leq 0$, for all i, and is equal to zero when $\pi_i^* > 0$.

If all of the coordinates of π^* are positive, then π^* is a zero of the vector field ξ on the unit sphere. The existence of a zero for an arbitrary vector field on the sphere is a subtle topological question. It is therefore a matter of some significance, in assessing the difficulty of demonstrating the existence of equilibrium prices (and in their calculation), to ask how general a vector field can be obtained by the process of aggregating individual demand functions. This question was first studied by Sonnenschein (1973), and was given a definitive answer in subsequent papers by Mantel (1974), Debreu (1974), and McFadden, Mantel, Mas-Colell and Richter (1975). There are some technical difficulties on the boundary of the positive orthant, but given an arbitrary continuous vector field on the sphere and an arbitrary open subset of the positive orthant, there will be a collection of n consumers with preferences and initial holdings whose individual demand functions sum to the preassigned vector field on the open subset. In other words, market excess demand functions are essentially arbitrary aside from the conditions described in 1.1.

Walras suggested that an equilibrium price vector could be found by a "*tâtonnement*" *process*: the continued revision of non-equilibrium prices on the basis of the discrepancy between supply and demand. If π is a price vector for which $\xi_i(\pi)\neq 0$, then a formalization of our intuitive understanding of the price mechanism would suggest that π_i be increased if $\xi_i(\pi)>0$ and decreased if $\xi_i(\pi)<0$. The continuous analogue of this process is a system of differential equations of the following sort:

$$d\pi_i/dt=\xi_i(\pi),$$

with some modification if the price vector happens to lie on the boundary of the positive orthant. The solution is a curve which simply follows the vector field on the sphere. Given the arbitrariness of this vector field, however, there is no difficulty in designing a model of exchange in which the Walrasian adjustment mechanism leads to the most capricious behavior with no tendency whatsoever towards equilibrium. For example, we can prescribe an arbitrary differentiable curve on the positive part of the sphere and construct a model of exchange such that the adjustment process follows that curve. The possibilities are quite staggering in large dimensions and surely imply that the price adjustment mechanism can be used neither to provide a proof of the existence of competitive equilibria nor an effective computational procedure for the general case.

Production can be introduced in the general equilibrium model in a variety of ways. Perhaps the simplest is to describe the productive techniques available to the economy by an *activity analysis model* based on a matrix of the following type

$$A=\begin{bmatrix} -1 & 0 & \dots & 0 & \dots & a_{1j} & \dots & a_{1k} \\ 0 & -1 & \dots & 0 & \dots & a_{2j} & \dots & a_{2k} \\ \vdots & \vdots & & \vdots & & \vdots & & \vdots \\ 0 & 0 & \dots & -1 & \dots & a_{nj} & \dots & a_{nk} \end{bmatrix}.$$

Each column of A represents a known technical mode of production, which can be employed at an arbitrary non-negative level. Inputs into production are represented by negative entries in a given column and outputs by positive entries. (The first n activities are *disposal activities*.) The result of using techniques 1 through k at levels $x_1,\dots,x_k$ is a *net production plan*, Ax, where $x=(x_1,\dots,x_k)$. The negative entries in Ax represent demands for factors of production which must be contributed by consumers from their stock of initial assets w. The positive entries represent increases in the stock of other assets, which can then be distributed among the consumers. An equilibrium price vector π^* is one which equilibrates the net consumer demand $\xi(\pi^*)$ with the net supply Ax. It also induces the correct selection of productive techniques on the grounds of decentralized profit maximization. We have the following formal definition of equilibrium:

1.3. [*Definition*]

A non-zero price vector π^* and a non-negative vector of activity levels x^* represent *equilibrium prices and activity levels* if

1. $\xi(\pi^*)=Ax^*$.
2. $\pi^*A\leqslant 0$.
3. $\pi^*Ax^*=0$.

The second of these conditions says that all potential activities make a non-positive profit when evaluated by means of equilibrium prices π^*, and the third condition that those activities which are actually used at equilibrium make a profit of zero. The demonstration that equilibrium prices and activity levels exist – under suitable assumptions on the matrix A – is more subtle than that of the case of pure trade, and typically makes use of Kakutani's fixed point theorem rather than Brouwer's theorem. The arguments apply equally well to a description of production given by a closed convex cone [see, for example, Debreu (1959)] rather than an activity analysis model.

2. Brouwer's fixed point theorem

Let S be a closed, bounded, convex set in n-dimensional Euclidean space, which is mapped into itself by the continuous mapping $x \to f(x)$. *Brouwer's theorem* asserts the existence of a fixed point, i.e. a point for which $\hat{x}=f(\hat{x})$. In order to illustrate the theorem we consider a few simple examples.

(1) Let S be the closed unit interval $[0,1]$. A continuous mapping of this interval into itself is defined by an ordinary continuous function $f(x)$ whose values also lie in the unit interval (Figure 2.1). A fixed point of this mapping is a point where the graph $(x, f(x))$ intersects the 45° line from the origin, or a point where $g(x)=f(x)-x$ is equal to zero. But $g(0)\geqslant 0$, and $g(1)\leqslant 0$, so that the function must be equal to zero at some point in the unit interval. Brouwer's theorem is a generalization of this well-known property of continuous functions to higher dimensions.

(2) Let S be the unit simplex $\{x=(x_1,\ldots,x_n)|x_i\geqslant 0, \sum x_j=1\}$, and let A be a square matrix with non-negative entries, whose column sums are equal to unity.

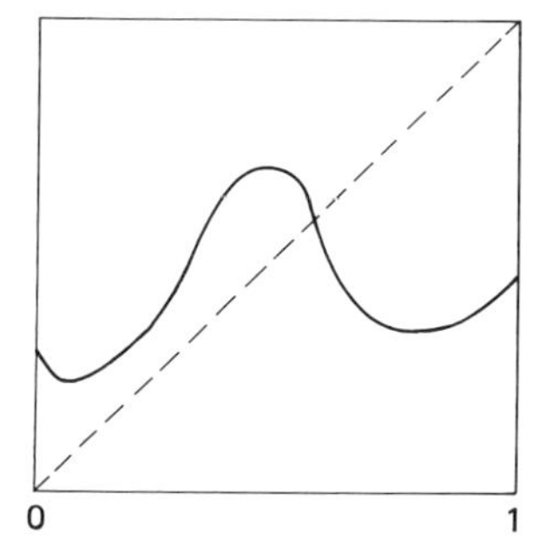

Figure 2.1.

The mapping

$$f(x)=Ax$$

is continuous and maps the simplex into itself. The conditions for Brouwer's theorem are satisfied and there will be a fixed point $\hat{x}=A\hat{x}$. In this particular case Brouwer's theorem implies that a non-negative square matrix with column sums of unity must have a characteristic root of 1 with an associated non-negative characteristic vector, a result which can be obtained by simpler arguments.

(3) In this example we show how Brouwer's theorem can be used to demonstrate the existence of equilibrium prices for a model of exchange. We consider prices normalized so as to lie on the unit simplex S. For each π in S, $\xi(\pi)$ will be the market excess demand associated with this price. ξ is assumed to be continuous on the simplex and to satisfy the identity $\pi\cdot\xi(\pi)\equiv 0$.

Define a mapping as follows

$$f_i(\pi)=\frac{\pi_i+\max[0,\xi_i(\pi)]}{1+\sum_j \max[0,\xi_j(\pi)]}.$$

The fact that f is continuous follows from the continuity of the market excess demand functions. It should also be clear that $f(\pi)$ is on the unit simplex if π is. Brouwer's theorem can therefore be applied, and we deduce the existence of a price vector $\hat{\pi}=f(\hat{\pi})$. We shall demonstrate that $\hat{\pi}$ is an equilibrium price for the model of exchange.

If c is defined to be

$$c=1+\sum_j \max[0,\xi_j(\hat{\pi})]\geqslant 1,$$

then

$$c\hat{\pi}_i=\hat{\pi}_i+\max[0,\xi_i(\hat{\pi})]$$

or

$$(c-1)\hat{\pi}_i=\max[0,\xi_i(\hat{\pi})].$$

But c must be equal to 1, for if it were strictly larger this last relationship would imply that

$$\xi_i(\hat{\pi})>0, \quad \text{for each } i \text{ with } \hat{\pi}_i>0,$$

violating the Walras Law $\sum \hat{\pi}_i \xi_i(\hat{\pi})=0$. If $c=1$, however, it follows that $\sum_j \max[0, \xi_j(\hat{\pi})]=0$, or $\xi_j(\hat{\pi})\leq 0$, for all j.

We see that the existence theorem for exchange economies is an immediate consequence of Brouwer's theorem. It is instructive to note that the converse is also correct. The following argument [Uzawa (1962)] shows that Brouwer's theorem follows from the existence of an equilibrium price vector for an arbitrary exchange economy.

Let $x \to f(x)$ be a continuous mapping of the unit simplex into itself. Define the functions

$$\xi_i(x)=f_i(x)-\lambda(x)x_i,$$

where

$$\lambda(x)=\sum_j x_j f_j(x) \Big/ \sum_j x_j^2.$$

Then ξ_i are defined and continuous everywhere on the unit simplex and, by construction, satisfy the Walras Law $\sum x_i \xi_i(x)=0$. From the existence theorem for competitive equilibria, we conclude that there is an $\hat{x}$ with $\xi_i(\hat{x})\leq 0$, and equal to zero if $\hat{x}_i>0$. But then

$$f_i(\hat{x})\leq \lambda(\hat{x})\hat{x}_i,$$

with equality if $\hat{x}_i>0$. If $\hat{x}_i=0$ we must also have equality, since otherwise $f_i(\hat{x})<0$, and f would not map the unit simplex into itself. It follows that

$$f_i(\hat{x})=\lambda(\hat{x})\hat{x}_i.$$

But since $\sum f_i(\hat{x})=\sum \hat{x}_i=1$, it follows that $\lambda(\hat{x})=1$ and that $\hat{x}$ is indeed a fixed point of the mapping.

Since Brouwer's theorem and the theorem asserting the existence of equilibrium prices are equivalent to each other, any effective numerical procedure for computing equilibrium prices must at the same time be an algorithm for computing fixed points of a continuous mapping.

3. An algorithm for computing fixed points

In this section we present our first algorithm for approximating a fixed point of a continuous mapping of the unit simplex into itself. There is no loss in generality in restricting our attention to the unit simplex. If the mapping $f(x)$ is defined for

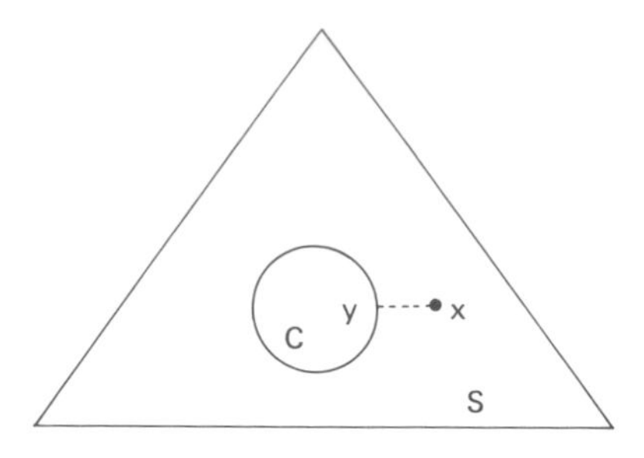

Figure 3.1.

a more general closed, bounded, convex set C, we embed C in a larger simplex S (as in Figure 3.1), and define a mapping $g(x)$ of S into itself in the following way:

(1) if $x \in C$, then $g(x) = f(x)$, and
(2) if $x \notin C$, then $g(x) = f(y)$,

where y is that unique point in C which is closest, in the sense of Euclidean distance, to x.

Since y varies continuously with x it is clear that the new mapping satisfies the conditions of Brouwer's theorem and has a fixed point $\hat{x} = g(\hat{x})$. But since $g(\hat{x}) \in C$ it follows that $\hat{x} \in C$, and is a fixed point of the original mapping f.

A *simplex* is defined to be the convex hull of its n vertices $v^1, \ldots, v^n$, i.e. the set of points of the form

$$x = \sum \alpha_j v^j, \quad \text{with } \alpha_j \geqslant 0 \text{ and } \sum \alpha_j = 1.$$

The vertices are assumed to be linearly independent in the sense that each vector in the simplex has one and only one representation in the above form. The simplex, which has dimension $n-1$ in our notation, has n faces of dimension

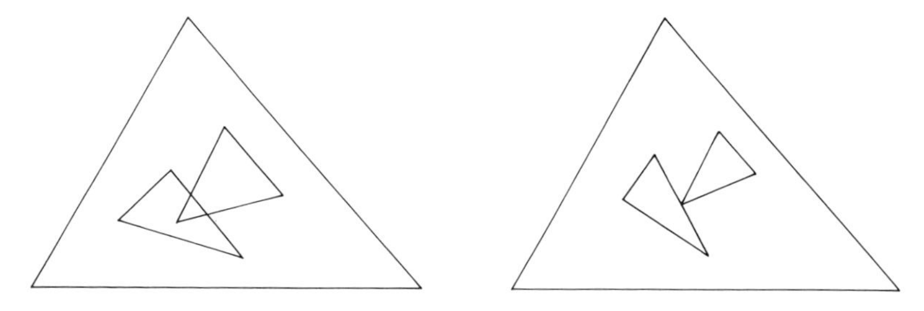

Figure 3.2.

$n-2$. The face opposite the vertex v^j consists of those vectors whose representation as a convex combination of the vertices has $\alpha_j = 0$.

More generally the simplex has a number of subfaces of lower dimension. For any subset T of $(1,2,\dots,n)$ there is a face of the simplex consisting of those vectors with $\alpha_j = 0$, for j in T.

We now consider a finite collection of simplices $S^1,\dots,S^k$ contained in the large simplex S. If no restriction is placed on this collection than two of its members may have interior points in common, or their intersection may not consist of an entire face of one or the other of them (see Figure 3.2). For the purpose of demonstrating Brouwer's theorem we shall require the collection of simplices to form what is known as a "simplicial subdivision", by ruling out these forms of intersection.

3.1. [Definition]

A collection of simplices $S^1,\dots,S^k$ is called a *simplicial subdivision* of S if

(1) S is contained in the union of the simplices $S^1,\dots,S^k$, and
(2) the intersection of any two simplices is either empty or a full face of both of them.

Figure 3.3 illustrates a typical simplicial subdivision of the simplex.

The method we shall adopt to demonstrate Brouwer's theorem and to provide an effective computational procedure for approximating fixed points makes use of a combinatorial lemma involving simplicial subdivisions of the simplex. There are alternative computational methods, notably those of Kellog, Li and Yorke, (1977) and Smale (1976), which avoid simplicial subdivisions completely and use instead the methods of differential topology. It is not yet clear which of these techniques,

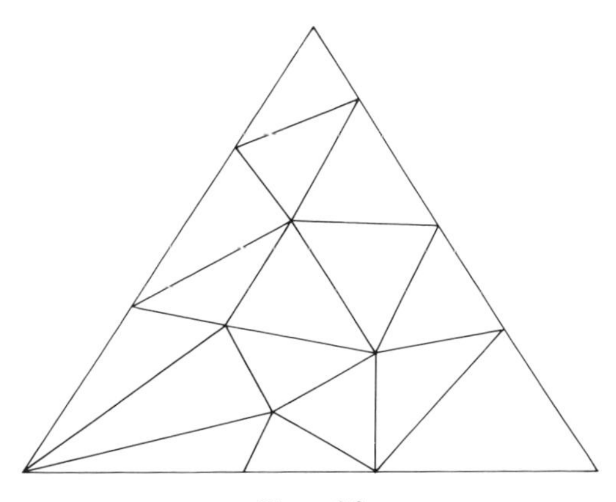

Figure 3.3.

which have greater similarities than might appear on the surface, is superior from a computational point of view.

To appreciate the combinatorial approach to Brouwer's theorem let us examine the simplest case, namely a continuous mapping of the unit interval into itself. A simplicial subdivision of the unit interval is given by a monotonic sequence of points, say $v^3 < v^4 < \cdots < v^k$ (I have used the notation v^1 for the lower end point of the unit interval and v^2 for the upper). In Figure 3.4 I have placed below each vertex v^j an integer label which is either 1 or 2. The rule for associating the integral label is to label v^j with 1 if $f(v^j) \geqslant v^j$ and 2 if $f(v^j) < v^j$. The combinatorial lemma, which is trivial for this one-dimensional problem, is simply that there must exist one small interval, in this case (v^3, v^4), whose two end points are differently labeled. If the grid were very fine, this combinatorial lemma would permit us to select two points, very close together, for which $f(x) - x$ has opposite signs. Any point in the interval will serve as an approximate fixed point of the mapping, with the degree of approximation [the closeness of x to $f(x)$] controlled by the fineness of the grid, and the modulus of continuity of f. Of course, a proof of the existence of a true rather than an approximate fixed point involves the continued refinement of the grid and the selection of a convergent subsequence of approximants.

Our first generalization of this elementary combinatorial observation will involve the special class of simplicial subdivisions specified in the following definition:

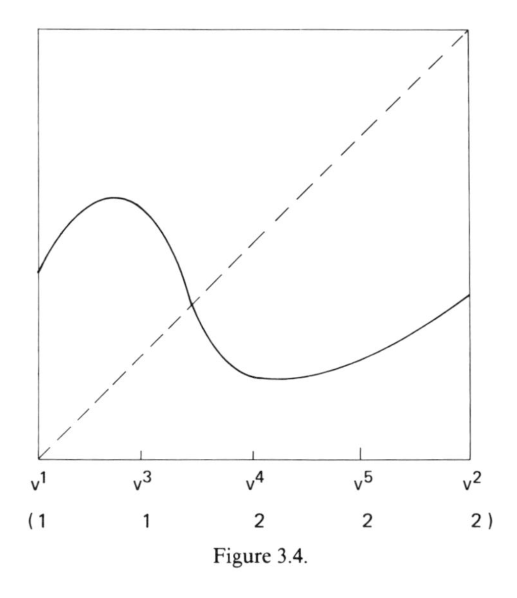

Figure 3.4.

3.2. [*Definition*]

A simplicial subdivision of the simplex $S=\{x|x=(x_1,\ldots,x_n)|x_i\geq 0, \sum x_i=1\}$ will be said to be *restricted* if no vertices of the subdivision, other than the n unit vectors, lie on the boundary of S.

An example of a restricted simplicial subdivision is given in Figure 3.5.

Let us consider a restricted subdivision of S, with $v^1,\ldots,v^n$ representing the first n unit vectors, and with remaining vertices $v^{n+1},\ldots,v^k$. We shall assume that each vertex of the subdivision has associated with it an integer label $\ell(v^j)$ selected from the set $(1,2,\ldots,n)$. When, below, we apply our argument to Brouwer's theorem, the label associated with a given vertex will depend on the continuous mapping $f(x)$. For the present, however, the labels will be completely arbitrary aside from the condition that $\ell(v^j)=j$, for $j=1,2,\ldots,n$. In other words, the only vertices whose labels are prescribed in advance are those on the boundary of S.

The remarkable combinatorial lemma which lies behind Brouwer's theorem is the following:

3.3. [*Lemma*]

Let the simplicial subdivision be restricted, and let the integer labels $\ell(v^j)$ be arbitrary members of the set $(1,2,\ldots,n)$ aside from the condition that $\ell(v^j)=j$, for $j=1,2,\ldots,n$. Then there exists at least one simplex in the simplicial subdivision all of whose labels are distinct.

Avoiding for the moment the question of why one would want to find a simplex with distinct labels, we shall demonstrate the existence of such a simplex

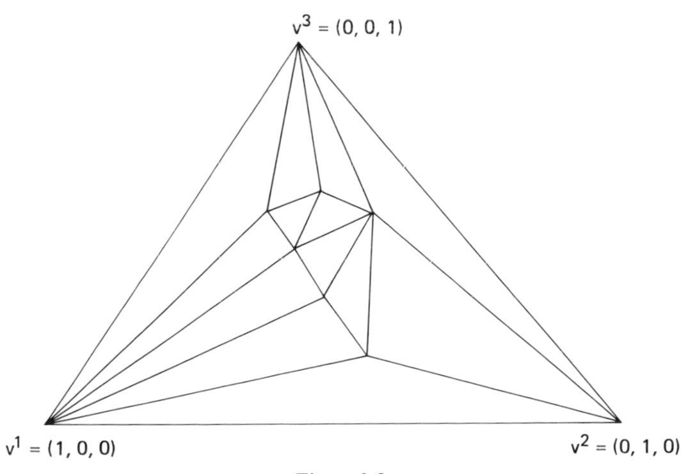

Figure 3.5.

by an explicit computational procedure based on an argument which first appeared in Lemke and Howson (1964) and Lemke, (1965).

We begin by considering that unique simplex in the subdivision whose vertices are the $n-1$ unit vectors $v^2, v^3, \ldots, v^n$ and a single other vertex, say v^j. The vertices $v^2, v^3, \ldots, v^n$ will have the labels $2, 3, \ldots, n$. If the vertex v^j is labeled 1 the algorithm terminates immediately with a simplex whose labels are distinct, i.e. a *completely labeled simplex*. Otherwise $\ell(v^j)$ will be one of the integers $2, 3, \ldots, n$. We proceed by eliminating the vertex whose label agrees with $\ell(v^j)$, arriving at a new simplex in the subdivision. Again we determine the label associated with the new vertex which has just been introduced. If this label is 1, we terminate; otherwise the process continues by eliminating that old vertex in the sudivision whose label agrees with that of the vertex just introduced.

At each stage of the algorithm we will be faced with a simplex in the subdivision whose n vertices bear the $n-1$ labels $2, 3, \ldots, n$. A pair of the vertices, one of which has just been introduced, will have a common label. We continue by removing that member of this pair which has not just been introduced.

Figure 3.6 illustrates a typical example of a path of simplices followed by the algorithm.

There are a finite number of simplices in the subdivision. We shall demonstrate that the algorithm never returns to a simplex which it previously encounters. Assuming this to be correct, the algorithm must terminate after a finite number of iterations. But termination can only occur if we have reached a simplex all of whose labels are different, or if we arrive at a simplex $n-1$ of whose vertices lie on the boundary of S, with the remaining vertex about to be removed. Such a

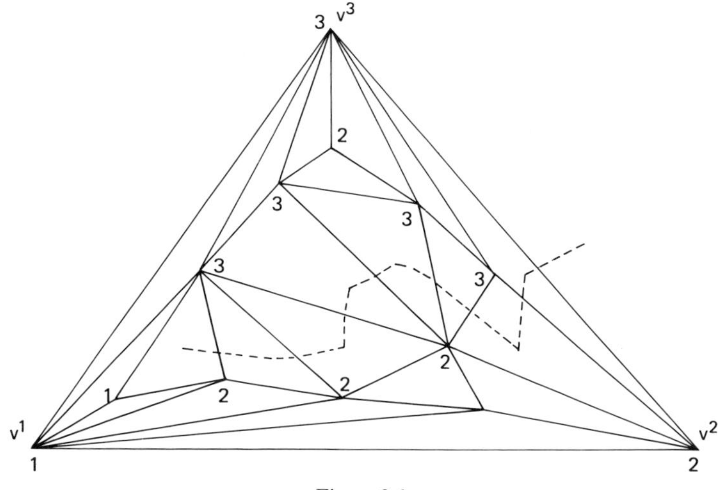

Figure 3.6.

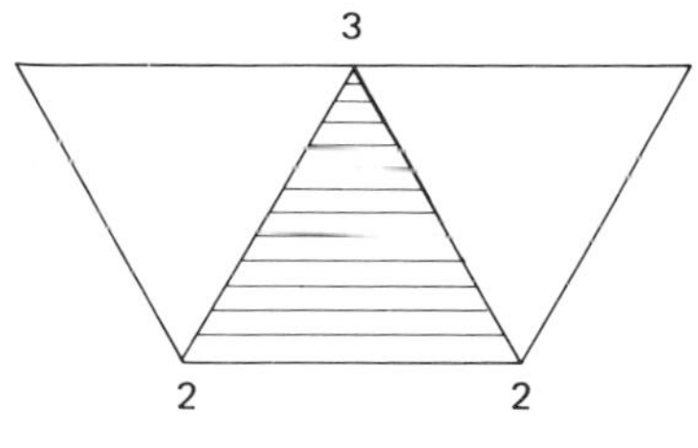

Figure 3.7.

boundary simplex would, however, have to contain the vertices $v^2,\ldots,v^n$, since if the vertex v^1 appears in a simplex encountered in the algorithm the label 1 would have been obtained, and the algorithm would have terminated.

We see, therefore, that a proof that the algorithm terminates in a finite number of iterations, with a simplex whose labels are distinct, consists in showing that the algorithm never returns to the same simplex. Consider the *first* simplex which is revisited (as in Figure 3.7). If it is not the initial simplex it can be arrived at in one of two possible ways: through one or another of the adjacent simplices with $n-1$ distinct labels. But both of these adjacent simplices would have been encountered during the first visit and our simplex is therefore not the first simplex to be revisited. A similar argument demonstrates that the initial simplex (with vertices $v^2, v^3,\ldots, v^n$ and one other vertex) is not the first simplex to be revisited. This concludes the proof of Lemma 3.3.

The customary proofs of Brouwer's theorem make use of a combinatorial lemma known as Sperner's lemma [see, for example, Berge (1959)] which is a generalization of Lemma 3.3 to arbitrary simplicial subdivisions of the simplex. Consider now a simplicial subdivision many of whose vertices lie on the boundary of S. The vertices v^j will again be labeled with integer labels $\ell(v^j)$ selected from the set $(1,2,\ldots,n)$. The vertices, which are interior to S, will have arbitrary labels (as in Figure 3.8). In order to guarantee the existence of a simplex all of whose labels are distinct the vertices on the boundary faces of S will have labels which are restricted according to the following theorem:

3.4. [*Sperner's lemma*]

Let v be an arbitrary vertex in the subdivision and assume that $\ell(v)$ is one of the indices i for which $v_i>0$. Then there exists a simplex in the subdivision with distinct labels.

The assumption of Sperner's lemma is that a vertex on a face of S generated by a subset of the unit vectors must have a label corresponding to one of these unit vectors. The assumption is illustrated in Figure 3.8, in which the shaded simplex is completely labeled.

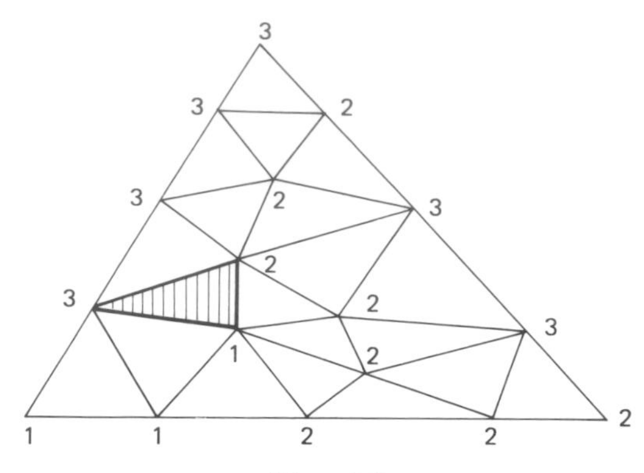

Figure 3.8.

Sperner's lemma may be demonstrated from Lemma 3.3 by the following simple argument. We embed the unit simplex in a larger simplex S' with vertices $s^1, \ldots, s^n$. We then extend the simplicial subdivision of S to a *restricted* simplicial subdivision of S' by introducing a number of simplices obtained in the following way: take an arbitrary subset T of $(1, 2, \ldots, n)$ with $t < n$ members. Consider a collection of $n-t$ vertices on the face of S defined by $x_i = 0$, for $i \in T$, and which lie in a single simplex of the subdivision of S. These $n-t$ vertices in S are augmented as in Figure 3.9 by the t vertices s^i for $i \in T$ in order to define a simplex in the larger subdivision.

Lemma 3.3 will be applied to the subdivision of S'. We must associate a distinct label with each of the new vertices $s^1, \ldots, s^n$, and we wish to do this in such a fashion that the completely labeled simplex obtained by applying Lemma 3.3 will contain none of the new vertices.

Let us define $\ell(s^i)$ to be equal to $i+1$ modulo n. In other words, $\ell(s^1) = 2$, $\ell(s^2) = 3, \ldots, \ell(s^n) = 1$. A completely labeled simplex may then contain the vertices s^i, for i in T, and $n-t$ other vertices on the face $x_i = 0$, for $i \in T$. Since these remaining vertices will, by the assumption of Sperner's lemma, have labels different from the members of T, it follows that the collection of vertices s^i in the completely labeled simplex must bear all of the labels in T. This is in contradiction to $\ell(s^i) = i+1$ modulo n unless T is the empty set and the completely labeled simplex is a member of the original simplicial subdivision of S. This completes the proof.

Having demonstrated Sperner's lemma by means of an explicit computational procedure we turn now to Brouwer's theorem. Let us consider a continuous mapping $x \to f(x)$ of the unit simplex into itself. The mapping is specified by n functions $f_i(x)$, for $i = 1, \ldots, n$, which are defined on the unit simplex, are continuous, and satisfy

$$f_i(x) \geqslant 0, \quad \text{for } i = 1, \ldots, n,$$

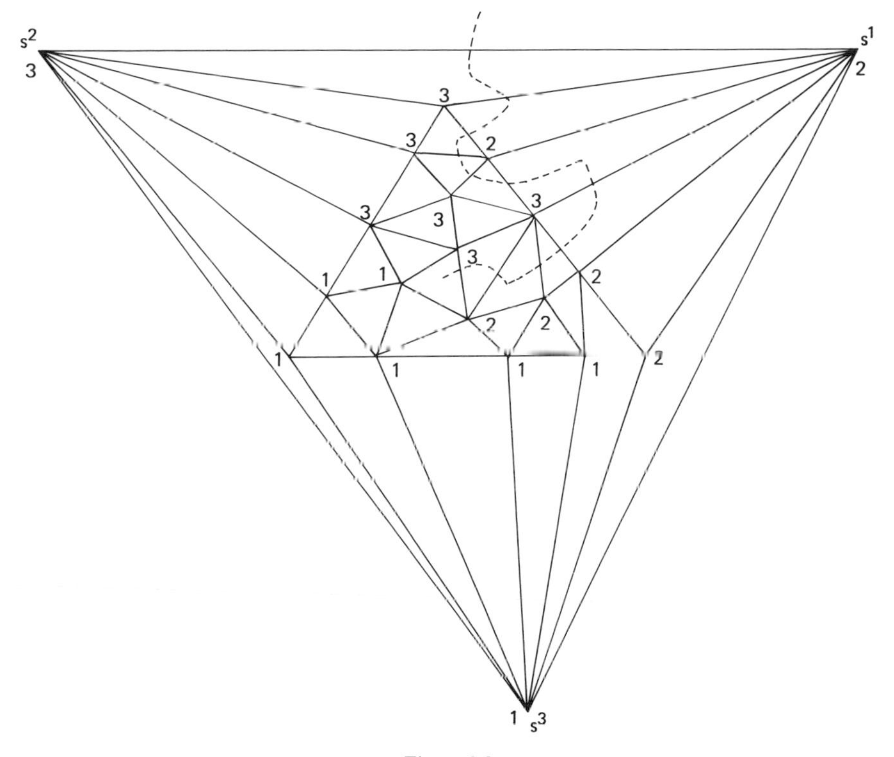

Figure 3.9.

and

$$\sum f_i(x) \equiv 1.$$

A simplicial subdivision of the simplex is given, with vertices $v^1, \dots, v^n, v^{n+1}, \dots, v^k$. We assume that none of these vertices are fixed points of the mapping f, since otherwise Brouwer's theorem would be trivially correct. It follows that for each vertex v we must have

$$v_i > f_i(v)$$

for at least one index i. We define the label associated with the vertex v to be one of these indices. If the vertex lies on the boundary of the simplex, with say $v_j = 0$, then the label associated with v must be different from j. The assumptions of Sperner's lemma are satisfied, and the algorithm on which our proof of Sperner's lemma is based may be used to determine a completely labeled simplex.

If the simplicial subdivision has a very fine mesh, in the sense that the distance between any pair of points in the same simplex is quite small, then any point in the completely labeled simplex will be close to its image and serve as an approximate fixed point of the mapping. [An upper bound for $\| f(x)-x \|$ is given in Scarf (1973) as a function of the mesh size and the modulus of continuity of the mapping.] In order to demonstrate Brouwer's theorem completely we must consider a sequence of subdivisions whose mesh tends to zero. Each such subdivision will yield a completely labeled simplex and, as a consequence of the compactness of the unit simplex, there is a convergent subsequence of completely labeled simplices all of whose vertices tend to a single point x^*. (This is, of course, the non-constructive step in demonstrating Brouwer's theorem, rather than providing an approximate fixed point.) The point x^* is the limit of points bearing all of the labels $1,2,\ldots,n$. Since f is continuous it follows that

$$x_i^* \geqslant f_i(x^*), \quad \text{for all } i.$$

Since $\sum x_i^* = \sum f_i(x^*)$, this can only happen if $x_i^* = f_i(x^*)$, for $i=1,\ldots,n$. This concludes the proof of Brouwer's theorem.

Let us now show that the existence of equilibrium prices for a model of exchange follows directly from Sperner's lemma without using Brouwer's theorem as an intermediary tool. We consider the continuous excess demand functions $\xi_1(\pi),\ldots,\xi_n(\pi)$ defined on the unit price simplex

$$S=\{\pi=(\pi_1,\ldots,\pi_n) \mid \pi_i \geqslant 0, \sum \pi_i = 1\},$$

and satisfying the Walras Law

$$\sum \pi_i \xi_i(\pi) \equiv 0.$$

S is subjected to a simplicial subdivision of fine mesh, with vertices $\pi^1,\ldots,\pi^n,\pi^{n+1},\ldots,\pi^k$. We shall label a vertex π of this subdivision with an integer i for which $\xi_i \leqslant 0$.

In order to satisfy the hypothesis of Sperner's lemma we must show that for any price vector π there is at least one coordinate with $\pi_i > 0$ and $\xi_i(\pi) \leqslant 0$. This will provide us with a labeling consistent with Lemma 3.4. But this is certainly correct since otherwise every coordinate i with $\pi_i > 0$ would have $\xi_i(\pi) > 0$, violating the Walras Law.

We may therefore conclude that there is a simplex with distinct labels. If the subdivisions are refined we may select a subsequence of these completely labeled simplices whose vertices tend, in the limit, to a price vector π^*. Since the excess demand functions are continuous $\xi_i(\pi^*) \leqslant 0$, for all i. π^* is indeed an equilibrium price vector.

We shall conclude this section by demonstrating an alternative combinatorial lemma, which may be viewed as a dual form of Sperner's lemma. As before we

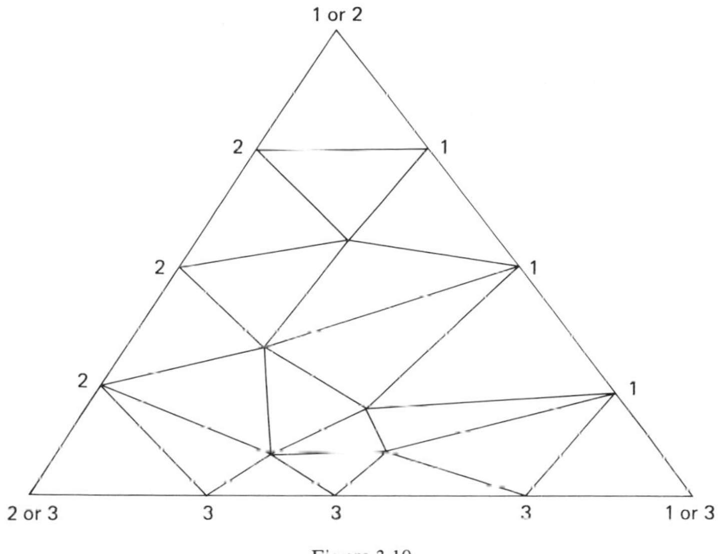

Figure 3.10.

consider a simplicial subdivision of the simplex S many of whose vertices lie on the boundary. The vertices v^j which are interior to the simplex will be given an arbitrary integer label $\ell(v^j)$ selected from the set $(1,2,\dots,n)$. As distinct from Sperner's lemma, however, a vertex on the boundary will be given an integer label associated with one of its coordinates which is equal to zero (see Figure 3.10).

In order to guarantee the existence of a completely labeled simplex, the subdivision must be sufficiently fine. For example, if there is only one simplex in the subdivision, the simplex S itself, then the vertices can clearly be labeled, consistently with the above rule, and with some label omitted.

3.5. [*Lemma*]

Assume that no simplex in the subdivision has a non-empty intersection with every face $x_i=0$. Let $\ell(v)$ be an arbitrary member of the set $(1,2,\dots,n)$ if v is a vertex which is interior to S. If v is on the boundary then $\ell(v)$ is equal to one of the indices i for which $v_i=0$. Then there exists at least one simplex all of whose labels are distinct.

As in the proof of Sperner's lemma, we embed S in a larger simplex S' with vertices $s^1,\dots,s^n$. The simplicial subdivision is extended to a restricted subdivision as before, i.e. we consider $n-t$ vertices on the face defined by $x_i=0$, for i in some index set T (with $t=|T|$), and which lie in a single simplex of the original subdivision. We then augment these vertices by the t vertices s^i, for $i\in T$.

The vertices $s^1,\dots,s^n$ will be labeled according to the rule $\ell(s^i)=i$, differing from the rule used in the proof of Sperner's lemma. Lemma 3.3 is then applied to this restricted subdivision so as to produce a completely labeled simplex. In order to demonstrate Lemma 3.5 we must show that none of the vertices $s^1, \ldots, s^n$ is contained in the completely labeled simplex.

Assume, to the contrary, that the simplex contains s^i, for $i \in T$, and $n-t$ vertices of a simplex in the original subdivision lying on the face $x_i = 0$, for $i \in T$. Since s^i bears the label i, the $n-t$ vertices must bear all of the labels in T^c. But these vertices all lie on the boundary of S, and their reason for bearing a given label must be that the corresponding coordinate is equal to 0. It follows that the convex hull of these $n-t$ vertices has a non-empty intersection with each face $x_i = 0$, for $i = 1, 2, \ldots, n$. This contradicts the assumption of Lemma 3.5.

Lemma 3.5 may be used as a substitute for Sperner's lemma in proving Brouwer's theorem or in demonstrating the existence of equilibrium prices in a model of exchange. In the latter problem a price vector π which is a vertex of the subdivision interior to S is given a label i for which $\xi_i(\pi)$ is maximal. The price vectors on the boundary receive a label corresponding to a zero coordinate. A completely labeled simplex is then found for each of a sequence of subdivisions whose mesh tends to zero. Consider a subsequence of completely labeled simplices all of whose vertices tend to the common price vector $\pi^* = (\pi_1^*, \ldots, \pi_n^*)$. For any coordinate i with $\pi_i^* > 0$ we must have $\xi_i(\pi^*) \geqslant \xi_j(\pi^*)$ for all j. It follows that $\xi_i(\pi^*)$ are all equal to some common value c for those i with $\pi_i^* > 0$, and that $\xi_i(\pi^*) \leqslant c$ if $\pi_i^* = 0$. But from the Walras Law,

$$0 = \sum \pi_i^* \xi_i(\pi^*) = c,$$

and π^* is indeed an equilibrium price vector.

4. The non-retraction theorem

Brouwer's theorem is frequently demonstrated by an appeal to a well-known topological theorem, the *non-retraction theorem*, which asserts that there is no continuous mapping of the unit simplex which carries all points in the simplex into the boundary, and which maps each boundary point into itself [Hirsch (1963)]. It will be instructive for us to see the relationship between these two theorems and to interpret our combinatorial Lemma 3.3 as an example of the non-retraction theorem.

To see that Brouwer's theorem follows from the non-retraction theorem let $x \to f(x)$ be a continuous mapping of the unit simplex into itself. If the mapping has no fixed points we can define a continuous retraction to the boundary of S in the following way (see Figure 4.1)

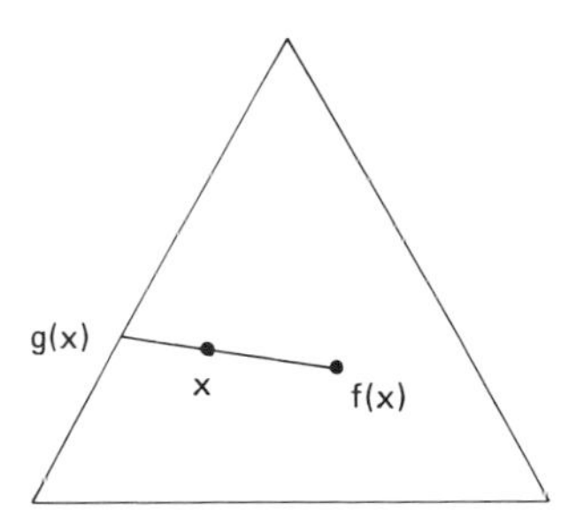

Figure 4.1.

For each x draw the straight line originating at the image $f(x)$ and passing through x. The straight line will intersect the boundary of the simplex at a point which we call $g(x)$. Clearly, $g(x)$ is a continuous function of x, and for any x on the boundary of the simplex $g(x)\equiv x$; g is therefore a continuous retraction of x onto the boundary. Since this cannot be, the original mapping must have a fixed point.

Conversely, Brouwer's theorem may be used to demonstrate the non-existence of a continuous retraction g. We simply compose g with a continuous mapping of the boundary of the simplex onto itself which has no fixed points. [For example, the mapping $(x_1, x_2, \dots, x_n) \to (x_n, x_1, \dots, x_{n-1})$.] The composite mapping will have no fixed points and therefore contradicts Brouwer's theorem.

Now let us return to the situation described in Lemma 3.3. The unit simplex is given a *restricted* simplicial subdivision with vertices $v^1, \dots, v^n, v^{n+1}, \dots, v^k$. All of these vertices, aside from the first n, are assumed to be strictly interior to the simplex. In addition, each vertex is given an integer label $\ell(v)$, which is selected without restriction from the set $(1, 2, \dots, n)$, aside from the requirement that $\ell(v^j) = j$, for $j = 1, \dots, n$.

These integer labels may be used to define a continuous piecewise linear mapping $g(x)$ of the unit simplex into itself. For each vertex v in the subdivision we define $g(v) = v^{\ell(v)}$; in other words, to be that one of the first n vertices whose label agrees with that of v. We then extend the mapping linearly throughout each simplex in the subdivision. If a simplex in the subdivision has vertices $v^{j_1}, v^{j_2}, \dots, v^{j_n}$, then a point in the simplex has the form

$$x = \alpha_1 v^{j_1} + \alpha_2 v^{j_2} + \dots + \alpha_n v^{j_n},$$

with $\alpha_i \geqslant 0$, $\sum \alpha_i = 1$. We then define

$$g(x) = \alpha_1 g(v^{j_1}) + \alpha_2 g(v^{j_2}) + \dots + \alpha_n g(v^{j_n}).$$

The mapping g is clearly continuous, and, as a consequence of the assumption that the subdivision is restricted, $g(x) \equiv x$, for every boundary point of the simplex. This follows from the observation that a boundary point of S is a convex combination of a subset of the first n vectors, each of which is mapped into itself by g.

If Lemma 3.3 were false, no simplex in the subdivision would have a full set of labels. But the n vertices of a simplex for which, say, the label 1 is missing would all be mapped, under g, into that face of the boundary of S given by $x_1 = 0$. Every point on such a simplex would also be mapped, by our construction, into the same face. We see therefore that the absence of a simplex with a complete set of labels permits us to construct a continuous piecewise linear retraction of the simplex into itself.

The constructive argument given for Lemma 3.3 may also be interpreted in terms of $g(x)$, as first suggested by Hirsch (1963). Let c be a vector on the boundary of S, whose first coordinate is zero, and whose remaining coordinates are strictly positive. We shall demonstrate that there exists a continuous path $x(t)$ which is linear in each simplex of the subdivision of the simplex through which it passes and which satisfies $g(x(t)) \equiv c$. Moreover, the path will have one end point at c, the other end point in a completely labeled simplex, and trace out precisely that sequence of simplices which appear in the proof of Lemma 3.3.

The simplices arising in Lemma 3.3 have the property that their n vertices bear the $n-1$ labels $2, 3, \ldots, n$. One of these labels will appear on two distinct vertices. Two faces of the simplex can be constructed by dropping either one of these two vertices; the $n-1$ vertices on each of these faces will have a full set of labels $2, 3, \ldots, n$. It follows that there is an interior point on each of these two faces for which $g(x) = c$. Since the mapping is linear, the straight line segment connecting these two points will also satisfy $g(x) = c$ (See Figure 4.2).

This continuous piecewise linear path starts at the boundary point c and passes through a sequence of simplices bearing the $n-1$ labels $2, 3, \ldots, n$. It terminates upon reaching a completely labeled simplex since such a simplex would have no

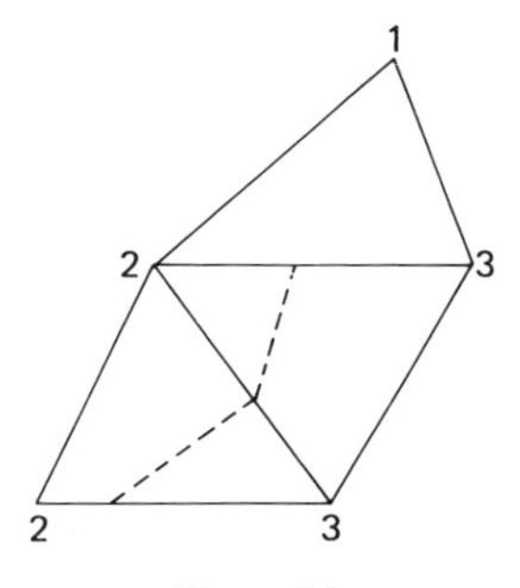

Figure 4.2.

other face on which $g(x)=c$ has a solution. This does in fact provide an alternative proof for Lemma 3.3 since if there were no completely labeled simplex the path could never terminate.

The relationship between fixed point theorems and the solution of systems of equations will be explored below.

5. A specific subdivision

The algorithm described in Section 3 approximates a fixed point of a continuous mapping by moving systematically through a sequence of simplices in the simplicial subdivision. At each iteration we are given a particular simplex with vertices, say, $v^{j_1},\ldots,v^{j_n}$. Depending on the particular assignment of integer labels, a specific vertex in the simplex is removed. We move to that unique simplex in the subdivision which has $n-1$ vertices in common with the remaining vertices of the original simplex. If the method is to be implemented effectively on a computer the simplicial subdivision must be such that this replacement operation is easy to carry out. Moreover, the vertices of the simplicial subdivision must be scattered with some regularity throughout the unit simplex so that the degree of approximation is independent of the region of the simplex in which the approximate fixed point happens to lie. The particular subdivision, which was brought to the attention of researchers in this field by Hansen (1968) and Kuhn (1968) solves these two problems admirably.

We begin our discussion by describing this particular simplicial subdivision for n-dimensional Euclidean space. The n unit vectors will be denoted by $e^1, e^2,\ldots, e^n$. With this notation a simplex will have $n+1$ vertices $v^0, v^1,\ldots, v^n$, which will consist of the points in R^n with integral coordinates.

5.1. [Definition]

A typical *simplex* in the subdivision will be defined by an integral point b (called the *base point*) and a permutation $(\varphi_1,\ldots,\varphi_n)$ of the integers $1,2,\ldots,n$. The vertices are then given by

$$
\begin{aligned}
v^0 &= b,\\
v^1 &= v^0 + e^{\varphi_1},\\
v^2 &= v^1 + e^{\varphi_2},\\
&\vdots\\
v^n &= v^{n-1} + e^{\varphi_n}.
\end{aligned}
$$

Figure 5.1 illustrates this simplicial subdivision for the plane. The shaded simplex has the base point $b=(1,0)$ and the permutation $(\varphi_1,\varphi_2)=(2,1)$. Its

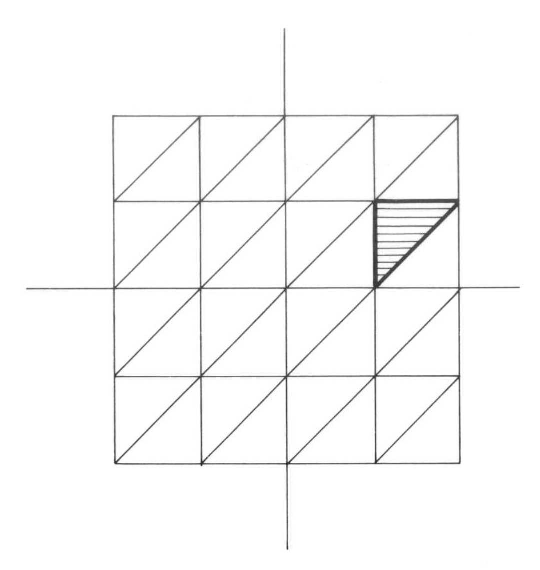

Figure 5.1.

vertices consist therefore of

$$v^0=b \qquad =(1,0),$$

$$v^1=v^0+e^2=(1,1),$$

$$v^2=v^1+e^1 \ =(2,1).$$

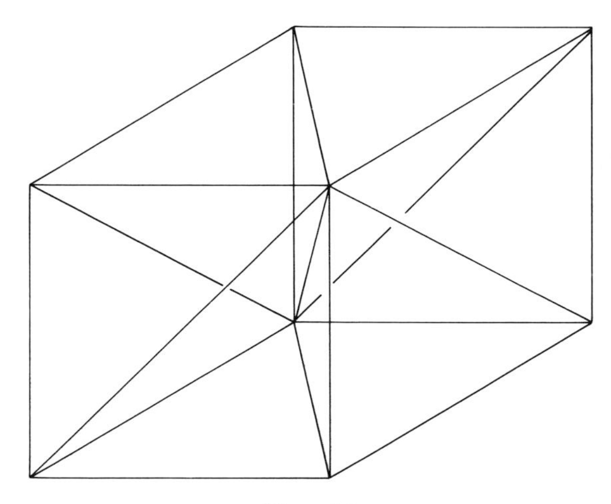

Figure 5.2.

There are two distinct types of simplices in the plane: one for each of the two permutations of (1,2). In general there will be $n!$ distinct simplices with every simplex in the subdivision equivalent under translation to one of them. Figure 5.2 illustrates the six distinct simplices for $n=3$.

Let us consider a simplex with vertices

$$\begin{aligned} v^0 &= b, \\ v^1 &= v^0 + e^{\varphi_1}, \\ &\vdots \\ v^j &= v^{j-1} + e^{\varphi_j}, \\ v^{j+1} &= v^j + e^{\varphi_{j+1}}, \\ &\vdots \\ v^n &= v^{n-1} + e^{\varphi_n}, \end{aligned}$$

and inquire about the replacement for a given vertex v^j; for the moment we assume that $0<j<n$. The new simplex will contain the vertices $v^0, v^1, \ldots, v^{j-1}$. In order to bring this about it is sufficient to have the same base point b and to have the new permutation $\hat{\varphi}_1, \hat{\varphi}_2, \ldots, \hat{\varphi}_n$ agree with the old as far as its first $j-1$ members are concerned, i.e. $\hat{\varphi}_1 = \varphi_1, \ldots, \hat{\varphi}_{j-1} = \varphi_{j-1}$.

In order not to have v^j in the new simplex we must have $\hat{\varphi}_j \neq \varphi_j$. But $v^{j+1} = v^{j-1} + e^{\varphi_j} + e^{\varphi_{j+1}}$. Since v^{j+1} must remain in the new simplex it follows that $(\hat{\varphi}_j, \hat{\varphi}_{j+1}) = (\varphi_{j+1}, \varphi_j)$. In other words, the adjacent simplex with the same vertices, other than v^j, is found by taking the same base point b and the permutation in which φ_j and φ_{j+1} are transposed. The replacement $\hat{v}^j$ is then given by

$$\begin{aligned} \hat{v}^j &= v^{j-1} + e^{\varphi_{j+1}} \\ &= v^{j-1} + v^{j+1} - v^j. \end{aligned}$$

This is a remarkably simple description of the replacement which can be programmed for the computer with great ease.

If v^0 is being replaced, then we consider the new base point $\hat{b} = v^1$, and the permutation

$$(\hat{\varphi}_1, \hat{\varphi}_2, \ldots, \hat{\varphi}_n) = (\varphi_2, \ldots, \varphi_n, \varphi_1)$$

resulting in the vertices

$$\begin{aligned} \hat{v}^0 &= \hat{b} = v^1, \\ \hat{v}^1 &= \hat{v}^0 + e^{\hat{\varphi}_1} = v^1 + e^{\varphi_2} = v^2, \\ &\vdots \\ \hat{v}^{n-1} &= \hat{v}^{n-2} + e^{\hat{\varphi}_{n-1}} = v^{n-1} + e^{\varphi_n} = v^n \end{aligned}$$

and

$$\hat{v}^n = \hat{v}^{n-1} + e^{\hat{\varphi}_n} = v^n + e^{\varphi_1}.$$

We see that in this case the replacement for v^0 is given by

$$v^n + e^{\varphi_1} = v^n + v^1 - v^0.$$

A similar argument shows that the replacement for v^n is given by $v^0 + v^{n-1} - v^n$. These rules may be summarized by the following theorem:

5.2. [*Theorem*]

Let the vertices of a simplex in the simplicial subdivision be given by $v^0, v^1, \ldots, v^n$, according to Definition 5.1. The replacement for an arbitrary vertex v^j is given by

$$v^{j-1} + v^{j+1} - v^j,$$

with the superscripts interpreted modulo n.

In practice, we store the vertices of the current simplex as columns in a matrix. When a given column is removed it is replaced by the sum of its two adjacent columns minus itself, with the interpretation that columns 0 and n are adjacent. The following examples illustrate this remark.

$$\begin{bmatrix} 0 & 1 & 1 & 1 \\ 0 & 0 & 1 & 1 \\ 0 & 0 & 0 & 1 \end{bmatrix}$$

$$\downarrow$$

$$\begin{bmatrix} 0 & 1 & 1 & 1 \\ 0 & 0 & 0 & 1 \\ 0 & 0 & 1 & 1 \end{bmatrix}$$

$$\downarrow$$

$$\begin{bmatrix} 2 & 1 & 1 & 1 \\ 1 & 0 & 0 & 1 \\ 1 & 0 & 1 & 1 \end{bmatrix}$$

$$\downarrow$$

$$\begin{bmatrix} 2 & 2 & 1 & 1 \\ 1 & 1 & 0 & 1 \\ 1 & 2 & 1 & 1 \end{bmatrix}.$$

From the point of view of numerical computation, nothing more than this explicit rule for the replacement operation is really required. To be complete, however, it may be useful to present the details of an argument which verifies that we have indeed constructed a simplicial subdivision of R^n. We begin by showing that an arbitrary vector $x=(x_1,\dots,x_n)$ is contained in at least one simplex of the subdivision.

5.3. [*Theorem*]

A necessary and sufficient condition that x be contained in the simplex defined by the base 0 and the permutation $\varphi_1,\dots,\varphi_n$ is that

$$1\geqslant x_{\varphi_1}\geqslant x_{\varphi_2}\geqslant\dots\geqslant x_{\varphi_n}\geqslant 0.$$

If x is a convex combination of the vertices $v^0, v^1,\dots, v^n$ defined by Definition 5.1, with $b=0$, then

$$\begin{aligned} x &= \sum_{j=0}^{n} \alpha_j \left(\sum_{\ell=1}^{j} e^{\varphi_\ell} \right) \\ &= \sum_{\ell=1}^{n} e^{\varphi_\ell} \left(\sum_{j=\ell}^{n} \alpha_j \right), \end{aligned}$$

with $\alpha_j\geqslant 0$, $\sum_0^n \alpha_j=1$. It follows that

$$x_{\varphi_\ell}=\sum_{j-\ell}^{n}\alpha_j,$$

and therefore that

$$x_{\varphi_\ell}-x_{\varphi_{\ell+1}}=\alpha_\ell\geqslant 0, \quad \text{for } \ell=1,\dots,n-1.$$

Also, $x_{\varphi_n}=\alpha_n\geqslant 0$ and, in addition, $1\geqslant\sum_{\ell=1}^n\alpha_\ell=x_{\varphi_1}$. We must therefore have

$$1\geqslant x_{\varphi_1}\geqslant x_{\varphi_2}\geqslant\dots\geqslant x_{\varphi_n}\geqslant 0.$$

On the other hand, if this inequality is satisfied then defining

$$\begin{aligned} &\alpha_\ell=x_{\varphi_\ell}-x_{\varphi_{\ell+1}}, \quad \text{for } \ell=1,\dots,n-1, \\ &\alpha_n=x_{\varphi_n}, \end{aligned}$$

and

$$\alpha_0=1-x_{\varphi_1}$$

will provide us with a set of weights which represents x as a convex combination of the vertices $v^0, \ldots, v^n$. This demonstrates Theorem 5.3.

We see that an arbitrary vector in the unit cube will be contained in at least one simplex. By adopting a different base point a simplex can be found which contains an arbitrary vector in R^n.

Each of the $n!$ permutations $\varphi_1, \ldots, \varphi_n$ defines a simplex in the subdivision lying inside of the unit cube, to consist of those vectors with

$$1 \geqslant x_{\varphi_1} \geqslant x_{\varphi_2} \geqslant \ldots \geqslant x_{\varphi_n} \geqslant 0.$$

A vector is interior to a particular simplex if all of the weights α_j are strictly positive, i.e. if all of the above inequalities are strict. It follows therefore that no two simplices in the subdivision have interior points in common.

In order to demonstrate that we have a simplicial subdivision we must show that the intersection of any two simplices is a full face of both of them. A face of one of our simplices is obtained by setting a particular subset of the α's equal to zero and letting the remaining α's range over all non-negative values summing to 1. This is equivalent to insisting that a particular subset of the $n+1$ inequalities

$$1 \geqslant x_{\varphi_1} \geqslant x_{\varphi_2} \geqslant \ldots \geqslant x_{\varphi_n} \geqslant 0$$

be equalities.

The intersection of two simplices in the unit cube, one defined by the permutation $\varphi_1, \ldots, \varphi_n$ and the other by $\hat{\varphi}_1, \ldots, \hat{\varphi}_n$, will consist of those vectors in the unit cube which simultaneously satisfy the above inequalities and

$$1 \geqslant x_{\hat{\varphi}_1} \geqslant x_{\hat{\varphi}_2} \geqslant \ldots \geqslant x_{\hat{\varphi}_n} \geqslant 0.$$

For example, if $n=4$, $\varphi=(4,3,2,1)$ and $\hat{\varphi}=(3,2,4,1)$, then the intersection consists of those vectors in the cube which simultaneously satisfy

$$1 \geqslant x_4 \geqslant x_3 \geqslant x_2 \geqslant x_1 \geqslant 0$$

and

$$1 \geqslant x_3 \geqslant x_2 \geqslant x_4 \geqslant x_1 \geqslant 0.$$

Together these imply

$$1 \geqslant x_4 = x_3 = x_2 \geqslant x_1 \geqslant 0$$

defining a full subface of each simplex. This argument, when posed in general terms, demonstrates the following theorem:

5.4. [*Theorem*]

The collection of simplices defined by Definition 5.1 is a simplicial subdivision of R^n.

In order to implement our constructive approach to fixed point theorems, we require a simplicial subdivision of the unit simplex

$$S=\left\{x=(x_0,\ldots,x_n)|x_i\geqslant 0,\sum_0^n x_i=1\right\},$$

and with two vectors in the same simplex close to each other. We shall construct such a simplicial subdivision with vertices $(k_0/D,k_1/D,\ldots,k_n/D)$, where D is a large positive integer and $k_0,\ldots,k_n$ are non-negative integers summing to D. To do this we simply modify the definition given in 5.1 by letting the base point b be an integral vector in R^{n+1} whose coordinates sum to D, and by redefining $e^1,\ldots,e^n$ to be the n vectors in R^{n+1} given by

$$e^1=(1,-1,0,\ldots,0),$$

$$e^2=(0,1,-1,\ldots,0),$$

$$\vdots$$

$$e^n=(0,0,0,\ldots,1,-1).$$

The simplices defined by

$$k^0=b,$$

$$k^1=k^0+e^{\varphi_1},$$

$$\vdots$$

$$k^n=k^{n-1}+e^{\varphi_n},$$

will have all of their vertices on the plane $\Sigma_0^n k_j=D$ and will form a simplicial subdivision of this plane since they are obtained from our previous subdivision of R^n by a linear transformation.

We notice that the ith coordinate of any two vertices of the same simplex are either equal or differ by 1. This implies that none of the coordinate hyperplanes $x_i=0$ intersects any of the simplices in an interior point; we can therefore select those simplices which are unambiguously in the non-negative orthant. After division by D this provides us with a simplicial subdivision of the unit simplex with the property that two vectors in the same simplex x and x' satisfy $|x_i - x_i'| \leq 1/D$ for all i.

It should be noticed that the replacement operation described in Theorem 5.2 holds in precisely the same form for this simplicial subdivision. A sequence of replacements is illustrated in Figure 5.3.

In this figure $D=4$; the numerators of the vertices in the sequence of simplices are as follows:

$$\begin{bmatrix} 4 & 3 & 3 \\ 0 & 0 & 1 \\ 0 & 1 & 0 \end{bmatrix}$$

$$\downarrow$$

$$\begin{bmatrix} 2 & 3 & 3 \\ 1 & 0 & 1 \\ 1 & 1 & 0 \end{bmatrix}$$

$$\downarrow$$

$$\begin{bmatrix} 2 & 2 & 3 \\ 1 & 2 & 1 \\ 1 & 0 & 0 \end{bmatrix}$$

$$\downarrow$$

$$\begin{bmatrix} 2 & 2 & 1 \\ 1 & 2 & 2 \\ 1 & 0 & 1 \end{bmatrix}$$

$$\downarrow$$

$$\begin{bmatrix} 2 & 1 & 1 \\ 1 & 1 & 2 \\ 1 & 2 & 1 \end{bmatrix}$$

$$\downarrow$$

$$\begin{bmatrix} 0 & 1 & 1 \\ 2 & 1 & 2 \\ 2 & 2 & 1 \end{bmatrix}.$$

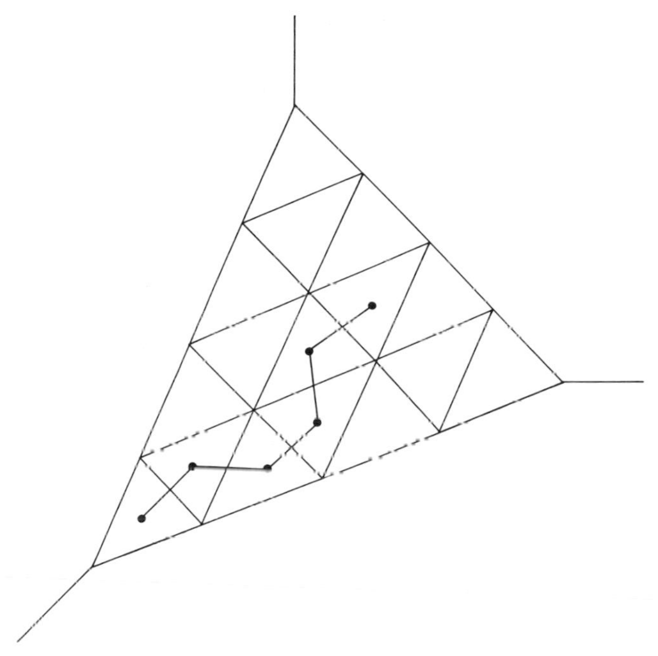

Figure 5.3.

6. A numerical example

In this section I shall give an example of a general equilibrium model which I shall solve by the use of Lemmas 3.4 and 3.5 and the particular simplicial subdivision of the preceding section. The previous discussion of Sperner's lemma involved embedding the unit simplex in a larger simplex and extending the simplicial subdivision to a restricted subdivision of the larger simplex (see Figure 3.9). Since the new subdivision will not be of the type described in the previous section, the very elementary replacement operation cannot be employed without some modification. Before presenting the numerical example I shall describe Kuhn's method [Kuhn (1968)] for initiating the algorithm, which avoids this difficulty and, in addition, permits one to start the calculation at any boundary point of the simplex.

Let the unit simplex $S=\{x=(x_1,\ldots,x_n)|x_i\geqslant 0, \sum_{i=1}^n x_i=1\}$ be given a simplicial subdivision of the type described in the previous section. We select a large positive integer D and consider as vertices in the subdivision the points $(k_1/D,\ldots,k_n/D)$ with $k_1,\ldots,k_n$ non-negative integers summing to D. The numerators of the vertices of any particular simplex in the subdivision can be written as the columns in an $n\times n$ matrix

$$K=\begin{bmatrix} k_1^1 & \dots & k_1^n \\ \vdots & & \vdots \\ k_n^1 & & k_n^n \end{bmatrix}.$$

The arguments of the previous section imply that the columns of K can be ordered in such a way that $k^j = k^{j-1} + e^{\varphi_j}$, where e^i is a vector whose $(i-1)$st coordinate is $+1$, whose ith coordinate is -1 and whose remaining coordinates are 0. Moreover $\varphi_2,\dots,\varphi_n$ is a permutation of the set $(2,\dots,n)$. In other words, each column is identical with its predecessor (interpreted modulo n) except for the entries in two successive rows, the first of which is a single unit higher and the second a single unit lower than the entries in the preceding column.

The vertices k^j/D are given an integer label $\ell(k^j)$ taken from the set $(1,2,\dots,n)$ and subject to the requirement that a vector receive a label corresponding to one of its positive coordinates. Let us attempt to initiate our search for a completely labeled simplex at a point $(0, k_2^*,\dots,k_n^*)$ on that face of the simplex whose first coordinate is 0.

We begin by extending the simplex to include those vertices whose first coordinate is given by $k_1=-1$ (see Figure 6.1). These additional vertices will be given an integer label, from the set $(2,\dots,n)$, in such a way that there is a *unique* simplex in the extended subdivision which consists of a single vertex with $k_1=0$, $(n-1)$ vertices with $k_1=-1$, and with the latter set of vertices bearing all of the labels $2,\dots,n$.

6.1. [*Labeling rule*]

A vertex given by $(-1, k_2,\dots,k_n)$, with $k_i \geqslant 0$, for $i=2,\dots,n$ and $\sum_2^n k_j = D+1$, will be labeled with the first integer for which $k_i > k_i^*$.

The columns of the matrix

$$\begin{bmatrix} -1 & \dots & -1 & -1 & 0 \\ k_2^* & & k_2^* & k_2^*+1 & k_2^* \\ k_3^* & & k_3^*+1 & k_3^* & k_3^* \\ \vdots & & \vdots & \vdots & \vdots \\ k_n^*+1 & & k_n^* & k_n^* & k_n^* \end{bmatrix}$$

represent a simplex in this extended subdivision. The first $n-1$ vectors bear the labels $n, n-1,\dots,2$; the final vector bears a label different from 1. The algorithm can be initiated at this simplex and can only run into difficulty if it encounters

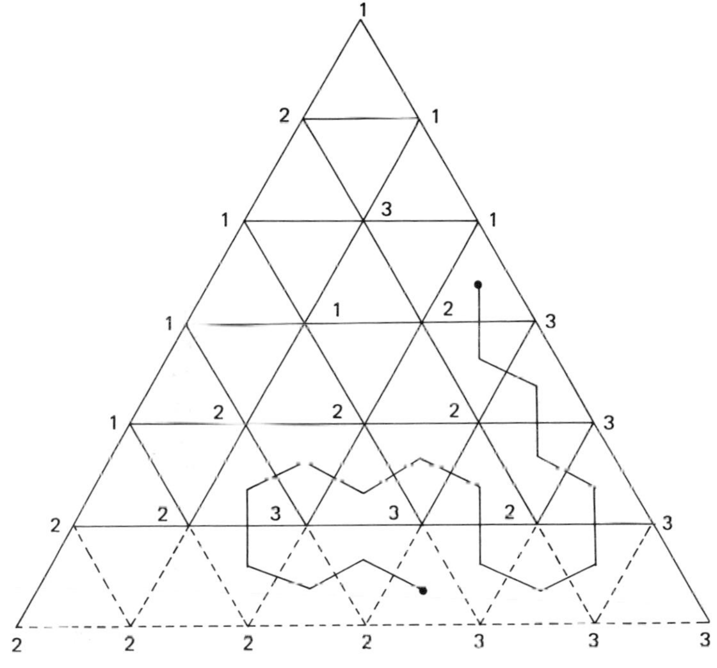

Figure 6.1.

another simplex

$$\begin{bmatrix} -1 & \cdots & -1 & 0 \\ k_2^1 & & k_2^{n-1} & k_2^n \\ \vdots & & \vdots & \vdots \\ k_n^1 & & k_n^{n-1} & k_n^n \end{bmatrix}$$

whose first $n-1$ columns bear all of the labels $n,\dots,2$, say in that particular order. For this to be true, however, we must have

$$k_{n-j+1}^j > k_{n-j+1}^*, \quad \text{for } j=1,\dots,n-1.$$

Since no two entries in the same row of this matrix differ by more than unity,

$$k_i^j \geqslant k_i^*, \quad \text{for } j=1,\dots,n-1 \text{ and } i=2,\dots,n.$$

But $\sum_{i=2}^n k_i^j = \sum_{i=2}^n k_i^* + 1$, and it follows that the second matrix is identical to the one initiating the algorithm.

Table 6.1.

Consumer	Initial stock of commodities									
1	0.6	0.2	0.2	20.0	0.1	2.0	9.0	5.0	5.0	15.0
2	0.2	11.0	12.0	13.0	14.0	15.0	16.0	5.0	5.0	9.0
3	0.4	9.0	8.0	7.0	6.0	5.0	4.0	5.0	7.0	12.0
4	1.0	5.0	5.0	5.0	5.0	5.0	5.0	8.0	3.0	17.0
5	8.0	1.0	22.0	10.0	0.3	0.9	5.1	0.1	6.2	11.0

Let us now consider an example of an exchange economy involving 5 types of consumers and 10 commodities. The ith consumer will own, prior to trade, a specific vector w^i, as given by Table 6.1.

The ith consumer's demand functions will be derived from a C.E.S. utility function

$$u_i(x) = \sum_{j=1}^{10} a_{ij}^{1/b_i} x_j^{1-1/b_i}.$$

Table 6.2 describes the parameters a_{ij} for each consumer. The parameters b are given by: $1, 2.0$; $2, 1.3$; $3, 3.0$; $4, 0.2$; $5, 0.6$.

The problem was solved by selecting $D=200$, and initiating the algorithm at the center of the face $x_1=0$. After 6063 iterations the following matrix, whose columns are the numerators of the vertices of a completely labeled simplex, was found:

$$\begin{bmatrix} 39 & 39 & 39 & 39 & 40 & 39 & 39 & 39 & 39 & 39 \\ 22 & 22 & 22 & 22 & 22 & 23 & 22 & 22 & 22 & 22 \\ 19 & 19 & 19 & 19 & 19 & 19 & 20 & 19 & 19 & 19 \\ 8 & 8 & 8 & 8 & 8 & 8 & 8 & 9 & 8 & 8 \\ 23 & 23 & 23 & 23 & 23 & 23 & 23 & 23 & 24 & 23 \\ 15 & 15 & 15 & 15 & 15 & 15 & 15 & 15 & 15 & 16 \\ 24 & 23 & 23 & 23 & 23 & 23 & 23 & 23 & 23 & 23 \\ 21 & 22 & 21 & 21 & 21 & 21 & 21 & 21 & 21 & 21 \\ 20 & 20 & 21 & 20 & 20 & 20 & 20 & 20 & 20 & 20 \\ 9 & 9 & 9 & 10 & 9 & 9 & 9 & 9 & 9 & 9 \end{bmatrix}$$

Each of these columns, after division by 200, may be used as an approximate equilibrium price vector, yielding a vector of excess demands which is fairly close to zero. A better approximation is obtained by averaging these prices, using a variation of Newton's method, to yield

$$\pi = (0.187, 0.109, 0.099, 0.043, 0.117, 0.077, 0.117, 0.102, 0.099, 0.049).$$

At this price vector the excess demands are equal to zero for two decimal places.

Table 6.2.

Consumer	Utility parameters									
1	1.0	1.0	3.0	0.1	0.1	1.2	2.0	1.0	1.0	0.07
2	1.0	1.0	1.0	1.0	1.0	1.0	1.0	1.0	1.0	1.0
3	9.9	0.1	5.0	0.2	6.0	0.2	8.0	1.0	1.0	0.2
4	1.0	2.0	3.0	4.0	5.0	6.0	7.0	8.0	9.0	10.0
5	1.0	13.0	11.0	9.0	4.0	0.9	8.0	1.0	2.0	10.0

The algorithm underlying Lemma 3.5 can also be used to solve this problem. In this latter algorithm a vertex of the subdivision k^j/D receives a label corresponding to one of its zero coordinates if it is on the boundary of the unit simplex and otherwise a label corresponding to that commodity with the largest excess demand. To be specific, we shall agree on the following convention:

6.2. [*Labeling rule for the second algorithm*]

The label associated with a vertex on the boundary of the simplex will be the *first* coordinate which is equal to zero.

Our earlier arguments for this version of the algorithm involved embedding the unit simplex in a larger simplex to which the simplicial subdivision is extended. The particular form of our simplicial subdivision, however, permits us to work directly with the unit simplex and to show that the algorithm never leads us outside of the simplex.

Let us define $d=D-n+2$. The columns of the following matrix represents a simplex in the subdivision:

$$\begin{bmatrix} d & d+1 & d+1 & \dots & d+1 & d \\ 1 & 0 & 1 & & 1 & 1 \\ 1 & 1 & 0 & & 1 & 1 \\ \vdots & \vdots & \vdots & & \vdots & \vdots \\ 1 & 1 & 1 & & 0 & 0 \\ 0 & 0 & 0 & & 0 & 1 \end{bmatrix}$$

$$\text{Labels} \quad n \quad 2 \quad 3 \quad \dots \quad n-1 \quad n-1$$

According to our labeling rule, the label 1 does not appear, all of the remaining labels do appear, and the label $n-1$ is associated with the last two columns. If the nth column is removed we are carried outside of the unit simplex; we begin the algorithm, therefore, by removing column $n-1$.

The algorithm can only run into difficulty if we encounter another matrix which contains a row $(0,0,0,\dots,0,1)$ and if the last column is to be removed because it shares a label with one of the other columns. If, however, this row is the ith row in the matrix, and $i<n$, then each of the first $n-1$ columns will

receive a label $\leqslant i$. The label n can only appear in the last column, which, consequently, will not be removed.

We can therefore assume that the row $(0,0,0,\dots,0,1)$ is the nth row in the matrix. If the last column is to be removed, the first $n-1$ columns of the matrix will bear the labels $2,3,\dots,n$, and since each one of these columns is on the boundary of the simplex, the corresponding label will be borne because it is the first coordinate equal to zero in that column. It follows that each row of the matrix, other than the first, must have a zero entry, and therefore that these rows are composed entirely of zeros and ones.

The column bearing the label n must then be given by

$$\begin{bmatrix} d \\ 1 \\ 1 \\ 1 \\ \vdots \\ 1 \\ 0 \end{bmatrix},$$

i.e. the first column in our initial matrix.

The column bearing the label 2 must have at least two zero entries in it, one in the second row, and one in the last. Since the entries in rows $2,3,\dots,n$ are zero or one, if there were a third zero entry it would follow that the entry in row 1 would be $\geqslant d+2$. Since this is impossible we see that the column bearing the label 2 must be the second column in our original matrix.

If this argument is continued, we see that the first $n-1$ columns of the matrix leading us outside of the simplex must be identical with those of our initial matrix. But this face can be completed to a simplex in the subdivision only by the addition of the last column of the original matrix. It follows that an exit from the unit simplex can only occur by returning to the initial simplex; this is impossible from the general arguments of Section 3.

When this version of the algorithm was applied to the previous example, again with $D=200$, termination was achieved in 2996 iterations, with the completely labeled simplex whose numerators are displayed below:

$$\begin{bmatrix}
36 & 36 & 36 & 35 & 36 & 36 & 36 & 36 & 36 & 36 \\
22 & 22 & 22 & 22 & 21 & 22 & 22 & 22 & 22 & 22 \\
20 & 20 & 20 & 20 & 20 & 19 & 20 & 20 & 20 & 20 \\
9 & 9 & 9 & 9 & 9 & 9 & 8 & 9 & 9 & 9 \\
23 & 23 & 23 & 23 & 23 & 23 & 23 & 22 & 23 & 23 \\
16 & 16 & 16 & 16 & 16 & 16 & 16 & 16 & 15 & 16 \\
24 & 24 & 24 & 24 & 24 & 24 & 24 & 24 & 24 & 23 \\
20 & 21 & 21 & 21 & 21 & 21 & 21 & 21 & 21 & 21 \\
20 & 19 & 20 & 20 & 20 & 20 & 20 & 20 & 20 & 20 \\
10 & 10 & 9 & 10 & 10 & 10 & 10 & 10 & 10 & 10
\end{bmatrix}.$$

The more rapid termination of this version of the algorithm is entirely attributable to the fact that the vertex (1,0,...,0) is a better approximation to the true equilibrium price than the vector initiating the previous algorithm and not to any inherent superiority of this version.

Both algorithms require a fairly substantial amount of computation in order to yield modest accuracy. In the following sections I shall describe a number of variations of the basic algorithm which provide a dramatic improvement in computational speed, in addition to extending the class of problems to which these methods can be applied.

7. Vector labels

The algorithms with which we have been concerned associate with each vertex of a simplicial subdivision an integer label selected from the set $(1,2,\ldots,n)$. When applied, for example, to determining equilibrium prices for a model of exchange the excess demands $\xi_1(\pi),\ldots,\xi_n(\pi)$ are evaluated, and the associated label corresponds, say, to that commodity with the largest excess demand.

It is clearly inefficient from a computational point of view to replace the vector of excess demands, whose evaluation may in itself be quite time consuming, by a single integer. The vector labeling procedures to be discussed in this section permit us to utilize all of the information in the excess demand vector with the expectation of greater accuracy in less computing time. In addition they are applicable to a large class of problems, including, for example, the approximation of a fixed point implied by Kakutani's theorem, to which the earlier methods cannot readily be applied.

Let us consider, as in Figure 7.1, a *restricted* simplicial subdivision of the unit simplex, with vertices $v^1,\ldots,v^n,v^{n+1},\ldots,v^k$. Each vertex v^j will have associated with it a label $\ell(v^j)$, which is a vector in n-dimensional space. In the applications of the vector labeling algorithms these labels will depend on the particular problem being solved. For the moment, however, the labels are arbitrary, aside from the stipulation that the unit vectors defining the simplex receive the labels $\ell(v^j)=v^j$, for $j=1,2,\ldots,n$. We shall also be given, in advance, a specific positive vector b in R^n.

Our previous algorithms attempted to determine a simplex in the subdivision, with vertices $v^{j_1},v^{j_2},\ldots,v^{j_n}$ and whose labels $\ell(v^{j_1}),\ell(v^{j_2}),\ldots,\ell(v^{j_n})$ were distinct. By a solution to the vector labeling problem we mean the determination of a simplex with vertices $v^{j_1},\ldots,v^{j_n}$, for which the system of linear equations

$$y_{j_1}\ell(v^{j_1})+\cdots+y_{j_n}\ell(v^{j_n})=b,$$

has a non-negative solution $y_{j_1},\ldots,y_{j_n}$. The following theorem provides sufficient conditions on the labels $\ell(v^j)$ for the problem to have a solution.

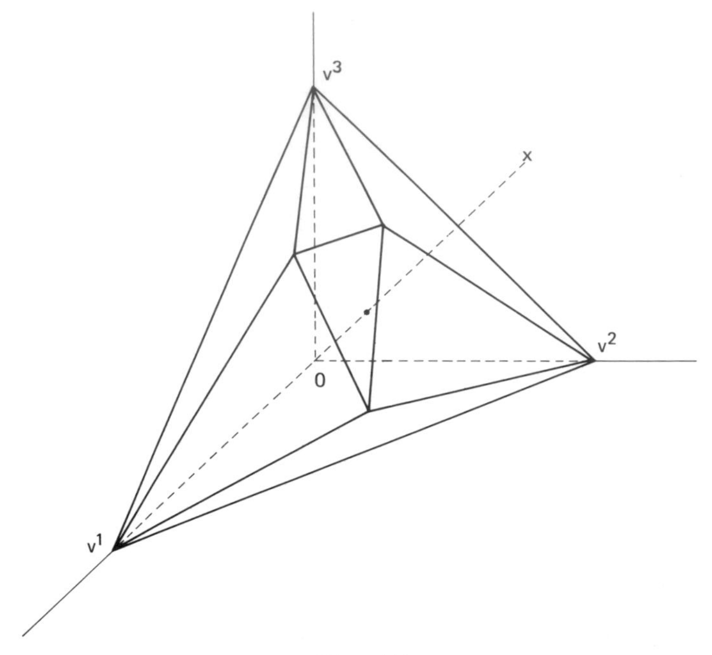

Figure 7.1.

7.1. [*Theorem*]

Consider a restricted subdivision of the unit simplex with vertices $v^1,\ldots,v^n,v^{n+1},\ldots,v^k$. Each vertex v^j has associated with it a label $\ell(v^j)\in R^n$; the first n vectors have as labels $\ell(v^j)=v^j$. Assume that

$$\sum_{j=1}^{k} y_j \ell(v^j) \leqslant 0,$$

and $y_j \geqslant 0$, for $j=1,\ldots,k$, implies $y_j=0$, for all j. Then for any positive vector $b\in R^n$, there is a simplex in the subdivision with vertices $v^{j_1},\ldots,v^{j_n}$, such that

$$\sum_i y_{j_i} \ell(v^{j_i}) = b$$

has a non-negative solution.

Theorem 7.1 has an interesting geometric interpretation which illuminates the method we shall provide for determining a correct simplex. Let us define a mapping $g(x_1,\ldots,x_n)$ from the non-negative orthant of R^n into R^n in the following way:

7.2. [*Definition*]

$g(x)$ is defined to be the unique mapping satisfying

(1) $g(v^j)=\ell(v^j)$ for each vertex in the subdivision,
(2) $g(x)$ is linear in each simplex in the subdivision, and
(3) $g(x)$ is homogeneous of degree 1, i.e. $g(\lambda x)=\lambda g(x)$ for $\lambda \geqslant 0$.

To determine $g(x)$ for any non-negative x (different from 0) we find that λ for which λx is on the unit simplex, and therefore contained in a particular simplex with vertices $v^{j_1},\ldots,v^{j_n}$. We then write

$$\lambda x = y_{j_1} v^{j_1} + \cdots + y_{j_n} v^{j_n},$$

with $y \geqslant 0$ and summing to 1, and define

$$g(\lambda x) = y_{j_1} \ell(v^{j_1}) + \cdots + y_{j_n} \ell(v^{j_n}).$$

With this interpretation we see that Theorem 7.1 is equivalent to asserting the existence of a vector x^* with $g(x^*)=b$. The validity of this assertion is a consequence of the following two properties which are immediate from the definition of $g(x)$:

7.3. [*Properties*]

(1) $g(x)=x$, for x on the boundary of the positive orthant, and
(2) $g(x) \leqslant 0$, implies $x=0$.

The mapping $g(x)$ is piecewise linear. The non-negative orthant of R^n is partitioned into a finite set of polyhedral cones – each one of which has n extreme rays through the origin – and such that g is linear in each cone.

Let $P=R^n_+ \times [0,1]$, the product of the non-negative orthant and the closed unit interval. The set P is naturally subdivided into a finite number of "wedges" (products of cones and the unit interval) as Figure 7.2 indicates for $n=2$.

We shall extend g to a mapping G of the set P into n space in such a way that G is linear in each wedge.

7.4. [*Definition*]

Let c be a vector in R^n, strictly larger than b in each coordinate. For $(x_1,\ldots,x_n,t) \in P$ we define

$$G(x,t) = g(x) + tc.$$

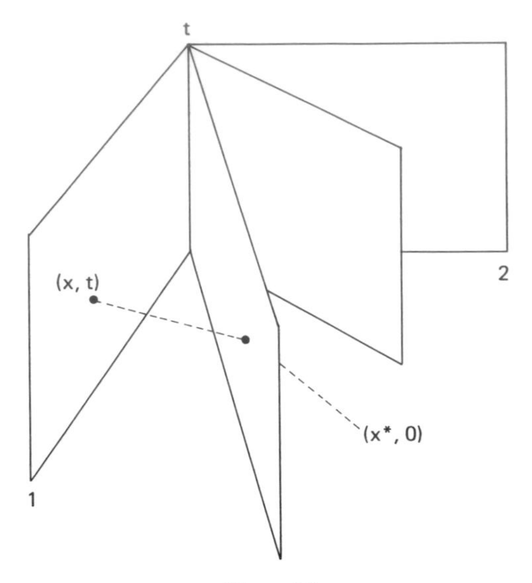

Figure 7.2.

With this definition a solution to the vector labeling problem is a vector x^* for which $G(x^*,0)=b$. The mapping G will permit us to trace out the one-dimensional family of solutions to $G(x,t)=b$ and to show that there is at least one member of the family with $t=0$.

The boundary of the set P consists of a number of hyperplanes: the two hyperplanes given by $t=0$ and $t=1$, and the n hyperplanes defined by $x_i=0$. Let us ask whether there are any solutions to $G(x,t)=b$ on the boundary of P – other than $t=0$.

A solution on the upper boundary would require $g(x)+c=b$ and since $c>b$ this implies $g(x)<0$, contrary to the second property in 7.3. To examine the solutions on that part of the boundary defined by $x_i=0$, we make use of the first property in 7.3, i.e. that $g(x)=x$ on the boundary of the non-negative orthant. For such points $G(x,t)=b$ is equivalent to $x+tc=b$. These equations have a solution obtained by selecting that typically unique index, say i^*, for which (b_i/c_i) is minimal, setting $t=b_{i^*}/c_{i^*}$ and $x=b-ct$ (Figure 7.3).

We see that there is a unique solution to the equations $G(x,t)=b$ on the boundary of P, other than that part of the boundary on which $t=0$. The algorithm for finding x^* consists simply of following the piecewise linear path defined by $G(x,t)=b$ from this particular boundary point until it intersects a point on the boundary with $t=0$.

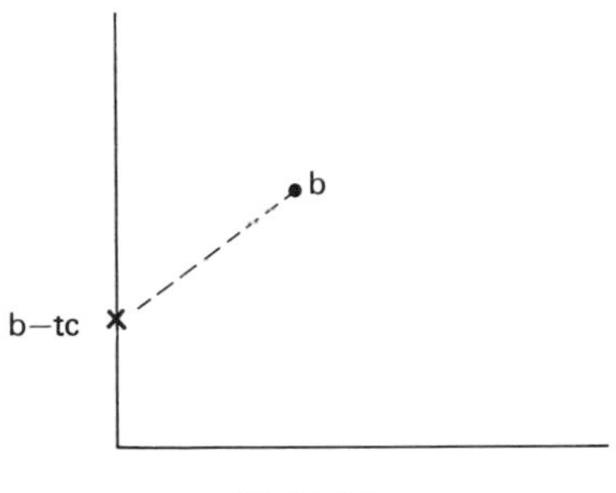

Figure 7.3.

Each wedge is defined by a simplex in the subdivision with vertices $v^{j_1},\ldots,v^{j_n}$. If

$$x=y_{j_1}v^{j_1}+\cdots+y_{j_n}v^{j_n},$$

then

$$g(x)-y_{j_1}\ell(v^{j_1})+\cdots+y_{j_n}\ell(v^{j_n}).$$

In this wedge the equations $G(x,t)=b$ therefore take the form

$$y_{j_1}\ell(v^{j_1})+\cdots+y_{j_n}\ell(v^{j_n})+tc=b,$$

with $y_{j_i}\geqslant 0$ and $0\leqslant t\leqslant 1$. The path enters the wedge on a particular boundary face of the wedge for which one of the y variables is equal to zero and exits when another variable is equal to zero. The precise determination of these two faces can be facilitated by adopting the terminology of linear programming. Let us define a matrix L with n rows and with $k+1$ columns, where k is the number of vertices in the simplicial subdivision. The jth column of L (for $j\leqslant k$) is the vector $\ell(v^j)$. Column $k+1$ is defined to be c.

Assume, to be specific, that the path enters the wedge at a point where $y_{j_1}=0$. This corresponds to the statement that columns $(j_2,\ldots,j_n,k+1)$ form a feasible basis for the system of linear equations $Ly=b$. The path will exit at a point where $y_{j_1}>0$ and some other variable is set equal to zero; but this is precisely the same as bringing the column $\ell(v^{j_1})$ into the feasible basis and removing one of the columns in the basis by an ordinary linear programming pivot step. If the column $\ell(v^{j_i})$ is removed the adjacent wedge is found by replacing the vertex v^{j_i} in the simplex defined by $v^{j_1},\ldots,v^{j_n}$. If the $(k+1)$st column is removed the algorithm terminates since at this point we obtain a solution with $t=0$.

To summarize, at each stage of the algorithm we are given a specific simplex with vertices $v^{j_1},\ldots,v^{j_n}$. The columns of L associated with $n-1$ of these vertices and the vector c form a feasible basis for $Ly=b$. We proceed by bringing into the basis the column associated with the missing vertex, replacing the vertex whose column has been eliminated from the feasible basis and continuing. The initial configuration is the simplex $v^1,\ldots,v^n$ with v^{i^*} absent and say v' included, since these $n-1$ vertices form a feasible basis along with column c. The algorithm terminates when column c is removed from the feasible basis by the pivot step.

To be completely tidy about our argument for Theorem 7.1, two observations are in order. First of all the pivot steps which are required can always be carried out. For if not the path would contain an infinite half-line in a single wedge. The variable t would necessarily be constant along this line so that $g(x)$ would be constant along a half-line extending to infinity, violating the second property of 7.3. Secondly, any degeneracy in the system $Ly=b$ must be avoided by one of the devices of linear programming. Degeneracy means that the exit from a given wedge may lead to more than one adjacent wedge with a bifurcation (as in Figure 7.4) of the piecewise linear path.

It should also be remarked that the parameter t need not be monotonic along the path and that the set of solutions to $G(x,t)=b$ may contain piecewise linear components other than the one constructed by our algorithm. These matters are discussed in detail in Eaves and Scarf (1976).

In order to utilize the vector labeling algorithm it must be adapted to the regular simplicial subdivision introduced in Section 5. We do this by associating with a vertex v, on the boundary of the unit simplex, the vector label $\ell(v)$ equal to the ith unit vector in R^n if v_i is the *first* coordinate equal to zero. The algorithm begins, as in Section 6, with the simplex, the numerators of whose vertices are given by

$$\begin{array}{rccccc} & \left[\begin{array}{ccccc} d & d+1 & \cdots & d+1 & d \\ 1 & 0 & & 1 & 1 \\ 1 & 1 & & 1 & 1 \\ \vdots & \vdots & & \vdots & \vdots \\ 1 & 1 & & 0 & 0 \\ 0 & 0 & & 0 & 1 \end{array}\right] \\ \text{Labels} & \begin{array}{ccccc} e^n & e^2 & \cdots & e^{n-1} & e^{n-1}, \end{array} \end{array}$$

and whose associated vector labels are all unit vectors. The vector labels associated with the first $(n-1)$ vertices form a feasible basis, along with the vector c, for the system $Ly=b$, if c is selected so that $b_1/c_1 \leqslant b_i/c_i$. The first step is to pivot into the basis the vector associated with the last column. We remove the vertex associated with column $n-1$, find the new simplex, and continue. The

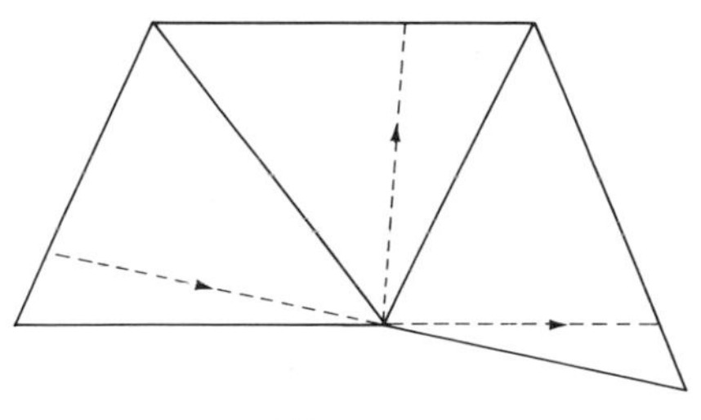

Figure 7.4.

arguments of Section 6 may be used to show that we never depart from the unit simplex and terminate after a finite number of iterations with a simplex whose associated vector labels form a feasible basis for $Ly=b$.

8. The general equilibrium model with production

The methods of the previous section provide an exceptionally flexible technique for the numerical solution of a wide variety of problems. They can be used to approximate equilibrium prices in a general Walrasian model – with or without production, to approximate fixed points of an upper semi-continuous point to set mapping, to solve non-linear programming problems, and to determine a vector in the core of an n-person game. Limitations of space make it impossible to describe anything other than the application of vector labeling methods to the determination of equilibrium prices, and even in this example we shall be forced to restrict our attention to one of the many approaches which have been suggested. The reader who is curious about other applications can consult Scarf (1973) or other items in the bibliography at the end of the paper.

Consider a general equilibrium model specified by a strictly positive vector w of assets prior to production, a set of market demand functions $x(\pi)$ (assumed continuous, homogeneous of degree 1, and to satisfy the Walras' Law $\pi\cdot x(\pi)\equiv \pi\cdot w$), and an activity analysis matrix

$$A=\begin{bmatrix} -1 & \cdots & 0 & \cdots & a_{1j} & \cdots \\ 0 & & 0 & & a_{2j} & \cdots \\ \vdots & & \vdots & & \vdots & \\ 0 & & -1 & & a_{nj} & \cdots \end{bmatrix}.$$

We make the conventional assumption that $Ay\geqslant 0$, $y\geqslant 0$, implies $y=0$. A price vector π^* and a non-negative vector of activity levels y^* will be a solution to the general equilibrium problem if

(1) $x(\pi^*)=w+Ay^*$, and
(2) $\pi^* A \leqslant 0$.

The following assignment of vector labels to the vertices of a simplicial subdivision of the price simplex will provide an approximate solution:

8.1 [*Labeling rule*]

Let π be a vertex of the simplicial subdivision. If π is on the boundary of the simplex then $\ell(\pi)$ is the ith unit vector where i is the first coordinate of π equal to zero. If π is interior to the simplex we find that column a^j in the activity analysis matrix for which $\pi \cdot a^j$ is maximal. If $\pi \cdot a^j > 0$, then $\ell(\pi) = -a^j$. If $\pi \cdot a^j \leqslant 0$, then $\ell(\pi) = x(\pi)$. Moreover, the vector b is defined to be w.

In order to apply Theorem 7.1 we must verify that the vector labels $\ell(\pi^j)$ satisfy the condition that $\sum y_j \ell(\pi^j) \leqslant 0$, $y_j \geqslant 0$ implies $y_j = 0$, for all j. This can easily be shown to be a consequence of the assumption made about the activity analysis matrix plus the fact that for each price vector π, $x(\pi) \geqslant 0$, and is positive for at least one coordinate.

When the algorithm is applied we determine a simplex with vertices $\pi^{j_1}, \ldots, \pi^{j_n}$. Several of these vertices will have associated vector labels which are the market demand functions evaluated at the corresponding price; the remaining vertices have labels which are the negatives of certain columns in the activity analysis matrix. With an obvious change in notation the equations $Ly=b$ can be written as

$$\sum y_j x(\pi^j) - \sum y'_\ell a^\ell \leqslant w.$$

The Labeling Rule 8.1 permits us to make the following statement. In the above inequality $y_j > 0$ implies that at prices π^j *all* activities make a profit $\leqslant 0$, and $y'_\ell > 0$ if for some vertex of the final simplex a^ℓ is the activity which makes the largest profit, and it is positive. Moreover the ith row will be an equality unless some vertex of the final simplex has its ith coordinate equal to zero.

It may be shown that the activity levels y'_ℓ and a suitable average of the prices in the final simplex are an approximate equilibrium in the sense that the two defining properties are approximately satisfied. This is a rather tedious argument, however, which can be circumvented by imagining that the problem has been solved for an infinite sequence of grids whose mesh tends to zero. We then select a subsequence of solutions, whose vertices tend to a price vector π^* and whose activity levels tend to a vector y^*. The above inequalities become

$$y^* x(\pi^*) - \sum y^*_\ell a^\ell \leqslant w.$$

The labeling rule now provides us with the following information:

8.2 [Consequences of the labeling rule]

(1) If $y^*>0$, then $\pi^*A\leqslant 0$.
(2) If $y_\ell^*>0$, then a^ℓ is a column which maximizes profit at prices π^*, and $\pi^*a^\ell\geqslant 0$.
(3) If $\pi_i^*>0$, then the ith row in the above inequality is, in fact, an equality.

In order to show that π^* and $\{y_\ell^*\}$ form a competitive equilibrium we begin by demonstrating that $y^*>0$.

If $y^*=0$, we see that $-\sum y_\ell^* a^\ell\leqslant w$ and as a consequence of (3):

$$-\sum y_\ell^*\pi^*\cdot a^\ell=\pi^*\cdot w.$$

But this is impossible since the right-hand side is positive and the left non-positive. We see that $y^*>0$ and therefore $\pi^*A\leqslant 0$. In conjunction with (2) we can now argue that $\pi^*\cdot a^\ell=0$ for any activity with $y_\ell^*>0$. It follows that $\sum y_\ell^*\pi^*\cdot a^\ell=0$, and as a consequence

$$y^*\pi^*\cdot x(\pi^*)=\pi^*\cdot w.$$

Making use of Walras' Law, y^* is therefore equal to 1, and our inequalities become

$$x(\pi^*)\leqslant w+\sum y_\ell^* a^\ell,$$

with equality if the ith price is positive. This concludes the argument that π^*, y^* are an equilibrium solution.

In order to illustrate this algorithm we consider an example, discussed in detail in Scarf (1967a), of a general equilibrium model with production. There are six commodities which can be described as follows:

(1) Capital available at the end of the period.
(2) Capital available at the beginning of the period.
(3) Skilled labor.
(4) Unskilled labor.
(5) Non-durable consumer goods.
(6) Durable consumer goods.

The activity analysis model of production is given by the following matrix, in which the six columns representing disposal activities have been omitted.

Commodity	Activities 7	8	9	10	11	12	13	14
1	4	4	1.6	1.6	1.6	0.9	7	8
2	−5.3	−5	−2	−2	−2	−1	−4	−5
3	−2	−1	−2	−4	−1	0	−3	−2
4	−1	−6	−3	−1	−8	0	−1	−8
5	0	0	6	8	7	0	0	0
6	4	3.5	0	0	0	0	0	0

In addition to the specification of production possibilities, we assume that there are five consumers who, at the beginning of the period own positive quantities of goods 2, 3, 4, 6 according to Table 8.1.

Each consumer will be assumed to have a C.E.S. utility function, of the type used in Section 6 and with the set of parameters shown in Table 8.2.

The parameters b are given by:

Consumer	b
1	1.2
2	1.6
3	0.8
4	0.5
5	0.6

To solve this problem a denominator $D=200$ was selected, and the Labeling Rule 8.1 was used. After approximately 2200 iterations the simplex whose

Table 8.1.

Consumer	Commodity 2	3	4	6
1	3	5	0.1	1
2	0.1	0.1	7	2
3	2	6	0.1	1.5
4	1	0.1	8	1
5	6	0.1	0.5	2

Table 8.2.

Consumer	Utility parameters					
1	4	0	0.2	0	2	3.6
2	0.4	0	0	0.6	4	1
3	2	0	0.5	0	2	1.5
4	5	0	0	0.2	5	4.5
5	3	0	0	0.2	4	2

numerators are given by the columns of the following matrix was obtained:

$$\begin{bmatrix} 44 & 44 & 44 & 44 & 44 & 45 \\ 47 & 48 & 48 & 48 & 48 & 47 \\ 35 & 34 & 35 & 35 & 35 & 35 \\ 12 & 12 & 11 & 12 & 12 & 12 \\ 22 & 22 & 22 & 21 & 22 & 22 \\ 40 & 40 & 40 & 40 & 39 & 39 \end{bmatrix}.$$

Five of these six columns have associated vector labels which are the negatives of activities 7, 9, 10, 11 and 13. The corresponding weights y_j' can be used as approximations to the equilibrium activity levels. In order to obtain an approximate equilibrium price vector the columns were averaged, to obtain:

Equilibrium prices

0.220833 0.238333 0.174167 0.059167 0.109167 0.198333

These prices were used to calculate the table of profitabilities (Table 8.3) and to evaluate the consumer market demand functions (Table 8.4).

As can be seen, the approximation is not exceptionally accurate; in particular, the discrepancy between supply and demand is fairly large for certain commodities. An improvement in accuracy can be obtained by selecting a much finer grid, but this is a very expensive procedure since it requires us to initiate the algorithm

Table 8.3.

Activity	Level	Profit
7	0.468253	0.006000
8	0.0	−0.143333
9	3.127146	0.005833
10	0.188899	−0.005833
11	0.165613	−0.006667
12	0.0	−0.039583
13	0.365860	0.010833
14	0.0	−0.246667

Table 8.4.

Commodity	Demand	Supply
1	11.188389	10.004686
2	0.0	1.191501
3	1.099712	2.090414
4	2.731755	3.970645
5	23.079518	21.433357
6	9.705318	9.373014

at a vertex of the simplex and to discard the information we have already obtained as to the approximate location of the equilibrium price vector. Alternatively there are numerical methods – essentially adaptations of Newton's method to systems of non-linear inequalities – which can be used to improve the current approximation. But neither of these approaches are remotely as good as the algorithms of Merrill and Eaves which are described in the next section.

9. The algorithms of Merrill and Eaves

The computational methods we have discussed in the previous sections have two major drawbacks. First of all they require the algorithm to be initiated at a vertex of the unit simplex – or, as in Kuhn's version of integer labeling, at a boundary point of the simplex. If an answer is obtained with a fixed grid, whose accuracy is inadequate for the problem at hand, the algorithm must be restarted with a finer grid and the results of the previous calculations discarded completely. The algorithms introduced by Merrill (1972) and Eaves (1972) permit the computation to be initiated at an arbitrary point on the simplex and allow a continual refinement of the grid. They yield a vast improvement in computational speed over the earlier algorithms which require a fixed simplicial decomposition and are used in virtually all practical applications of fixed point methods.

Merrill's method, which is essentially identical to the "sandwich" method subsequently introduced by Kuhn and MacKinnon (1975), is particularly simple to describe. In order to introduce the basic idea of Merrill's method, let us consider the case in which the unit simplex is one-dimensional and where the basic problem is to be solved by integer labeling techniques. The interval $\{(x_1, x_2)|x_i \geqslant 0, x_1 + x_2 = 1\}$ is subdivided with some preassigned grid size D. The vertices in the subdivision will be of the form $(k_1/D, k_2/D)$ with k_1 and k_2 non-negative integers summing to D. The numerators of the vertices of a typical simplex in the subdivision will be given by

$$\begin{bmatrix} k_1 & k_1+1 \\ k_2 & k_2-1 \end{bmatrix}.$$

The vertex defined by $(0, D)$ will be given the label 1 and $(D,0)$ the label 2. The customary integer labeling methods starts at one of these extreme vertices and moves into the simplex until we first encounter a vertex with the other label.

Let us imagine that we have an initial guess as to where the true answer lies, given by the vector $(k_1^*/D, k_2^*/D)$, with k_1^*, k_2^* non-negative integers adding to D. Merrill's algorithm incorporates this additional information by drawing the simplex $k_1 + k_2 = D-1$, and subdividing the resulting two dimensional figure (see Figure 9.1).

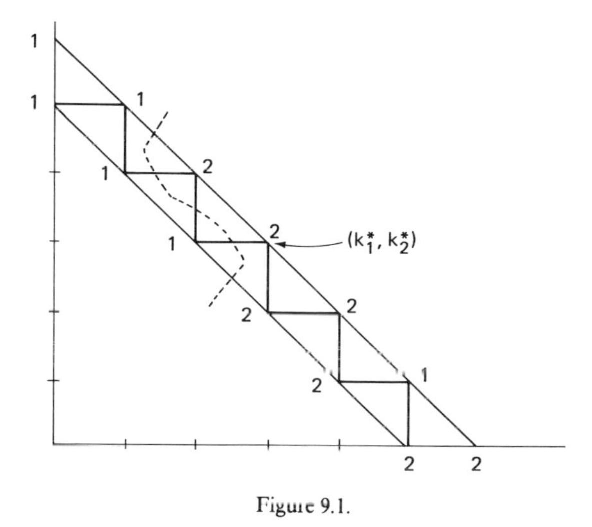

Figure 9.1.

In Figure 9.1, $D=6$, and the integer labels on the upper simplex are assumed to be derived from the problem at hand. The integer labels on the lower simplex will be based on the initial guess (k_1^*, k_2^*) which in this particular example we assume to be (3,3). On the lower level the pair (k_1, k_2) will be given the label i, if i is the first coordinate for which $k_i < k_i^*$.

Merrill's algorithm moves through a sequence of two-dimensional simplices, each defined by three vertices. We begin at the simplex

$$\begin{bmatrix} k_1^* & k_1^*-1 & k_1^* \\ k_2^* & k_2^* & k_2^*-1 \end{bmatrix},$$

two of whose vertices are on the artificial level and one vertex on the original simplex. The initial simplex has been constructed in such a way that the two vertices on the artificial level bear the labels 1 and 2, one of which is shared with the vector (k_1^*, k_2^*). We remove that vertex on the artificial level with the doubled label and continue until we are forced to exit through the original face of the "sandwich". At that point we have found a completely labeled simplex whose labels derive from the original problem. If the accuracy which has been achieved is not adequate we repeat the calculation, with say a doubled grid size, and with the new guess given by a choice of one of the vertices in the completely labeled simplex. Since the grid size grows exponentially we can expect a high degree of precision in a relatively small number of iterations.

The layer which is subdivided is part of a two-dimensional simplex, obtained by introducing an additional coordinate $k_0 = D - k_1 - k_2$. With this notation the initial simplex may be represented by

$$\begin{bmatrix} 0 & 1 & 1 \\ k_1^* & k_1^*-1 & k_1^* \\ k_2^* & k_2^* & k_2^*-1 \end{bmatrix}.$$

This permits us to make use of the regular simplicial subdivision based on the vertices (k_0, k_1, k_2) with $\Sigma_0^2 k_i = D$, and, in particular, to exploit the simplicity of the replacement step. Of course, vectors on the artificial level will have $k_0 = 1$, and those on the original level $k_0 = 0$.

Let us generalize these observations by considering a problem of size n, and by making use of vector labeling. We are concerned with the set of non-negative integers $(k_1, \ldots, k_n)$ summing to D. Each such vector has associated with it a vector label $\ell(k)$ in R^n, derived from the problem at hand, and satisfying the condition that $\ell(k)$ is the ith unit vector in R^n if k_i is the first coordinate equal to zero. We wish to determine a simplex in the subdivision such that a non-negative linear combination of the associated vector labels equals a preassigned positive vector b. In addition, we are given an initial guess $(k_1^*, \ldots, k_n^*)$ as to the solution.

The set of vertices is enlarged by considering an additional coordinate k_0 and the regular simplicial decomposition based on the set of vertices $(k_0, k_1, \ldots, k_n)$ with $k_i \geqslant 0$, $\Sigma_0^n k_i = D$. Only those vertices with $k_0 = 0, 1$ are relevant for the computation and these are associated with vector labels in R^n (not R^{n+1}) according to the following rule:

9.1. [*Labeling rule for Merrill's algorithm*]

A vector $k = (k_0, k_1, \ldots, k_n)$ will, if $k_0 = 0$, receive the label $\ell(k_1, \ldots, k_n)$. If $k_0 = 1$ the vector label is the ith unit vector in R^n if i is the first coordinate for which $k_i < k_i^*$.

The initial simplex is given by the columns of the following $(n+1) \times (n+1)$ matrix:

$$\begin{bmatrix} 0 & 1 & \cdots & 1 \\ k_1^* & k_1^*-1 & & k_1^* \\ \vdots & \vdots & & \vdots \\ k_n^* & k_n^* & & k_n^*-1 \end{bmatrix}.$$

If the rows and columns of this matrix are indexed from $0, \ldots, n$, we see that columns $1, \ldots, n$ have as vector labels the n unit vectors in R^n. These vector labels form a feasible basis for the system $Ly = b$. The vector label associated with the zeroth column derives from the original problem.

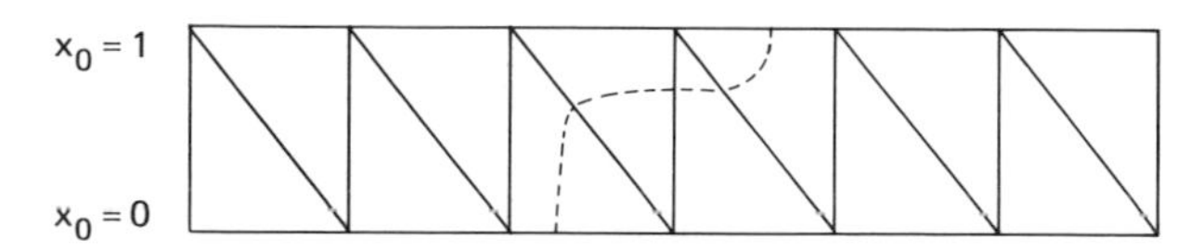

Figure 9.2.

The first step of the algorithm is to introduce this latter vector into the feasible basis, by a pivot step. One of the other vectors is eliminated and its replacement found by determining an adjacent simplex. The vector label associated with the new vertex is brought into the feasible basis, and we continue.

The algorithm terminates when the simplex contains n vectors whose coordinate is 0 (i.e. vectors whose labels come from the original problem) and a single vector whose zeroth coordinate is 1 and whose associated column has just been removed from the feasible basis by a pivot step. The labels associated with these n vectors therefore form a feasible basis for $Ly = b$, and we have obtained a solution to the original problem. If the accuracy is not adequate the algorithm is repeated with a higher value of D.

The argument for the second algorithm of Section 6 can be used to show that we never depart from the "sandwich" other than through the top face. Demonstrating that the algorithm cannot cycle is also a simple application of the types of analysis used in previous sections.

Merrill's method may be viewed in terms of homotopy theory. We take the product of the unit simplex and the closed unit interval [0, 1] and subject it to a

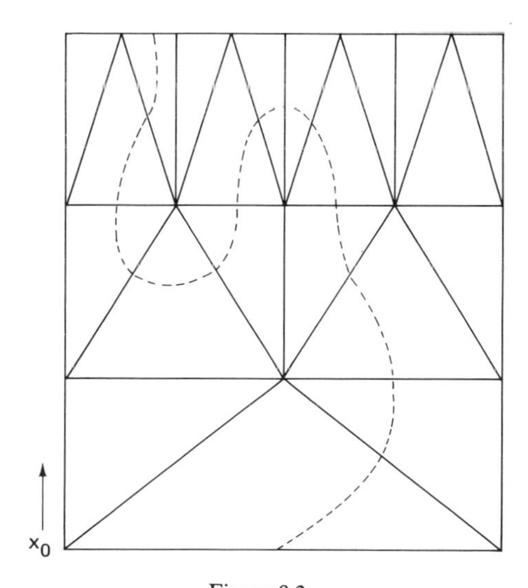

Figure 9.3.

simplicial subdivision all of whose vertices lie on one of the two bounding planes. We construct a piecewise linear mapping of this product into the unit simplex which approximates the true mapping on the face $x_0=0$. On the face $x_0=1$ the mapping is chosen so as to have a unique fixed point. Merrill's method essentially traces the fixed point from one face to another, as in Figure 9.2.

Eaves' method, on the other hand, involves a simplicial decomposition of the product of the unit simplex and the half line $[0,\infty)$ with the grid becoming finer as the coordinate x_0 increases (Figure 9.3). Eaves' method has the distinct advantage that the vector labels obtained at the previous level are retained. This means that the Jacobian of the mapping is utilized rather than a single estimate of the fixed point. Eaves' method does, however, require the construction of a simplicial decomposition of the product space with exponentially decreasing grid size and this is rather complicated to work with.

In applying Merrill's algorithm to the general equilibrium model of the last section we began with a small value of D and an initial estimate of $k^*=(9,9,8,8,8)$. The grid size was then tripled a total of nine successive times. In roughly 2000 iterations – essentially the same as that required by a fixed grid size of 200 – the following simplex was obtained:

$$\begin{bmatrix} 216827 & 216827 & 216827 & 216827 & 216827 & 216828 \\ 247077 & 247077 & 247077 & 247078 & 247078 & 247077 \\ 158471 & 158472 & 158472 & 158471 & 158471 & 158471 \\ 54073 & 54072 & 54072 & 54072 & 54073 & 54073 \\ 104398 & 104398 & 104399 & 104399 & 104398 & 104398 \\ 203304 & 203304 & 203303 & 203303 & 203303 & 203303 \end{bmatrix}.$$

Any of these columns can be taken as an estimate of the equilibrium price vector, after division by the sum of the entries in the column,

Equilibrium prices					
0.220319	0.251057	0.161024	0.054943	0.106080	0.206578

Table 8.6.

Activity	Level	Profit
7	0.463533	−0.000003
8	0.0	−0.141668
9	3.939607	−0.000001
10	0.006050	−0.000002
11	0.0	−0.007615
12	0.0	−0.052769
13	0.438389	−0.000006
14	0.0	−0.254323

Table 8.7.

Commodity	Demand	Supply
1	11.234238	11.235904
2	0.0	−0.001595
3	1.155703	1.154355
4	2.974338	2.973208
5	23.682987	23.686038
6	9.354235	9.354134

These prices may then be used to calculate the profits shown in Table 8.5 in which the activity levels are derived from the weights in the final feasible basis.

Table 8.7 illustrates the relationship between supply and demand based on these prices and activity levels. As can be seen the degree of approximation has been substantially improved – with no increase in computational cost – over the answer based on a fixed grid size.

References

Allgower, E. L. (1974), "Application of a fixed point search algorithm", in: S. Karamardin, ed., Fixed Points: Algorithms and Applications, New York: Academic Press.

Allgower, E. L. and M. M. Jeppson (1973), "The approximation of solutions of nonlinear elliptic boundary value problems with several solutions", Springer Lecture Notes, Vol. 333, 1–20.

Berge, C. (1959), Espace topologiques, functions multivoques. Paris: Dunod.

Cohen, D. I. A. (1967), "On the Sperner lemma", Journal of Combinatorial Theory, 2:585–587.

Cottle, R. W. and G. B. Dantzig (1970), "A generalization of the linear complementarity problem", Journal of Combinatorial Theory, 8:79–90.

Dantzig, G. B. and A. S. Manne (1974), "A complementarity algorithm for an optimal capital path with invariant proportions", Journal of Economic Theory, 9:312–323.

Debreu, G. (1959), Theory of value. New York. John Wiley & Sons, Inc.

Debreu, G. (1974a), "Excess demand functions", Journal of Mathematical Economics, 1:15–21.

Debreu, G. (1974b), "Four aspects of the mathematical theory of economic equilibrium", Proceedings of International Congress of Mathematicians, Vancouver, 65–77.

Dohmen, J. and J. Schoeber (1975), Approximated fixed points, EIT 55, Department of Econometrics, Tilburg University, Holland.

Eaves, B. C. (1970), "An odd theorem", Proceedings of the American Mathematical Society, 26:509–513.

Eaves, B. C. (1971a), "Computing Kakutani fixed points", SIAM Journal of Applied Mathematics, 21:236–244.

Eaves, B. C. (1971b), "On the basic theory of complementarity", Mathematical Programming, 1:68–75.

Eaves, B. C. (1972), "Homotopies for computation of fixed points", Mathematical Programming, 3:1–22.

Eaves, B. C. and R. Saigal (1972), "Homotopies for computation of fixed points on unbounded regions", Mathematical Programming, 3:225–237.

Eaves, B. C. and H. Scarf (1976), "The solution of systems of piecewise linear equations", Mathematics of Operations Research, 1:1–27.

Fisher, M. L. and F. J. Gould (1974), "A simplicial algorithm for the nonlinear complementarity problem", Mathematical Programming, 6:281–300.

Fisher, M. L., F. J. Gould and J. W. Tolle (1974), "A new simplicial approximation algorithm with restarts: Relations between convergence and labelling", in: Proceedings of a Conference on Computing Fixed Points with Applications, Department of Mathematical Sciences, Clemson University, Clemson, S.C.

Freidenfelds, J. (1974), "A set intersection theory and applications", Mathematical Programming, 7:199–211.

Garcia, C. B. (1975), "Continuation methods for simplicial mappings", Graduate School of Business, University of Chicago.

Garcia, C. B. (1976), "A hybrid algorithm for the computation of fixed points", Management Science, 22:606–613.

Garcia, C. B. and W. I. Zangwill (1976), "Determining all solutions to certain systems of nonlinear equations", Graduate School of Business, University of Chicago.

Ginsburgh, V. and J. Waelbroeck (1975), "A general equilibrium model of world trade, Part I, Full format computation of economic equilibria", Cowles Discussion Paper No. 412, Yale University.

Ginsburgh, V. and J. Waelbroeck (1974), "Computational experience with a large general equilibrium model", CORE Discussion Paper No. 7420, Heverlee, Belgium.

Hansen, T. (1968), On the approximation of a competitive equilibrium. Ph.D. Thesis, Yale University.

Hansen, T. (1974), "On the approximation of Nash equilibrium points in an N-person noncooperative game", SIAM Journal of Applied Mathematics, 26:622–637.

Hansen, T. and T. C. Koopmans (1972), "On the definition and computation of a capital stock invariant under optimization", Journal of Economic Theory, 5:487–523.

Hansen, T. and H. Scarf (1969), "On the applications of a recent combinatorial algorithm", Cowles Foundation Discussion Paper No. 272, Yale University.

Hirsch, M. W. (1963), "A proof of the nonretractibility of a cell onto its boundary", Proceedings of the American Mathematical Society, 14:364–365.

Jeppson, M. M. (1972), "A search for the fixed points of a continuous mapping", in: R. H. Day and S. M. Robinson, eds., Mathematical topics in economic theory and computation. Philadelphia: Siam.

Kannai, Y. (1970), "On closed coverings of simplexes", SIAM Journal of Applied Mathematics, 19:459–461.

Karamardian, S. (ed.) (1977), Computing fixed points with applications. New York: Academic Press.

Karamardian, S. (1969), "The nonlinear complementarity problem with applications, Parts I and II", Journal of Optimization Theory and Applications, 4:87–98, 167–181.

Kellogg, R. B., T. Y. Li and J. Yorke (1977), "A method of continuation for calculating a Brouwer fixed point", in: S. Karamadian, ed., Computing fixed points with applications. New York: Academic Press.

Kojima, A. (1975), A unification of the existence theorems of the nonlinear complementarity problem", Mathematical Programming, 9:257–277.

Kuhn, H. W. (1960), "Some combinatorial lemmas in topology", IBM Journal of Research and Development, 4:518–524.

Kuhn, H. W. (1968), "Simplicial approximation of fixed points", Proceedings of the National Academy of Sciences of the USA, 61:1238–1242.

Kuhn, H. W. and J. G. MacKinnon (1975), "The sandwich method for finding fixed points", Journal of Optimization Theory and Applications, 17:189–204.

Lemke, C. E. (1965), Bimatrix equilibrium points and mathematical programming", Management Sciences, 11:681–689.

Lemke, C. E. and J. T. Howson, Jr. (1964), "Equilibrium points of bimatrix games", SIAM Journal of Applied Mathematics, 12:413–423.

McFadden, D., A. Mas-Colell, R. Mantel, and M. K. Richter (1975), "A characterization of community excess demand functions", Journal of Economic Theory, 9:361–374.

MacKinnon, J. G. (1975), "Solving economic general equilibrium models by the sandwich method", Proceedings of a Conference on Computing Fixed Points with Applications, Department of Mathematical Sciences, Clemson University, Clemson, S.C.

Mantel, R. R. (1968), "Toward a constructive proof of the existence of equilibrium in a competitive economy", Yale Economic Essays, 8:155–200.

Mantel, R. R. (1974), "On the characterization of aggregate excess demand", Journal of Economic Theory, 9:348–353.

Merrill, O. H. (1972), Applications and extensions of an algorithm that computes fixed points of certain upper semi-continuous point to set mappings. Ph.D. Dissertation, Department of Industrial Engineering, University of Michigan.

Saigal, R. (1976), "On paths generated by fixed point algorithms", Mathematics of Operations Research, 1:359–380.

Saigal, R. and C. B. Simon (1973), "Generic properties of the complementarity problem", Mathematical Programming, 4:324–335.

Scarf, H. (1967a), "On the computation of equilibrium prices", in: Ten economic studies in the tradition of Irving Fisher. New York: John Wiley & Sons, Inc.

Scarf, H. (1967b), "The approximation of fixed points of a continuous mapping", SIAM Journal of Applied Mathematics, 15:1328–1343.

Scarf, H. (1967c), "The core of an N person game", Econometrica, 35:50–69.

Scarf H. (with the collaboration of T. Hansen) (1973), Computation of economic equilibria. New Haven: Yale University Press.

Shapley, L. S. (1974), "A note on the Lemke–Howson algorithm", Mathematical Programming Study, 1:175–189.

Shoven, J. B. and J. Whalley (1972), "A general equilibrium calculation of the effects of differential taxation of income from capital in the U.S.", Journal of Public Economics, 1:281–321.

Smale, S. (1976), "A convergent process of price adjustment and global Newton methods", Journal of Mathematical Economics, 3:107–120.

Sonnenschein, H. (1973), "Do Walras' identity and continuity characterize the class of community excess demand functions?", Journal of Economic Theory, 6:345–354.

Todd, M. J. (1976a), The computation of fixed points and applications, Lecture Notes in Economics and Mathematical Systems No. 124 (Mathematical Economics). New York: Springer-Verlag.

Todd, M. J. (1976b), "Orientation in complementary pivot algorithms", Mathematics of Operations Research, 1:54–66.

Todd, M. J. (1976c), "On triangulations for computing fixed points", Mathematical Programming, 10:322–346.

Uzawa, H. (1962), "Walras' existence theorem and Brouwer's fixed point theorem", Economic Studies Quarterly, 13:1.

Wilmuth, R. J. (1973), The computations of fixed points. Ph.D. Thesis, Department of Operations Research, Stanford University.

Wilson, R. (1971), "Computing equilibria of N-person games", SIAM Journal of Applied Mathematics, 21:80–87.

LIST OF THEOREMS

INDEX